EXPONENTIAL AND LOGARITHMIC FUNCTIONS

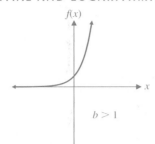

Exponential function
$f(x) = b^x$

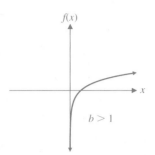

Exponential function
$f(x) = b^x$

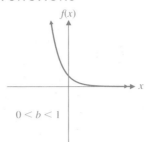

Logarithmic function
$f(x) = \log_b x$

REPRESENTATIVE POLYNOMIAL FUNCTIONS (DEGREE > 2)

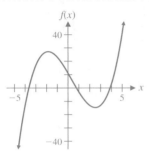

Third-degree polynomial
$f(x) = x^3 - x^2 - 14x + 11$

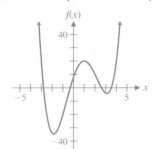

Fourth-degree polynomial
$f(x) = x^4 - 3x^3 - 9x^2 + 23x + 8$

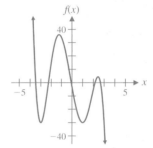

Fifth-degree polynomial
$f(x) = -x^5 - x^4 + 14x^3 + 6x^2 - 45x - 3$

REPRESENTATIVE RATIONAL FUNCTIONS

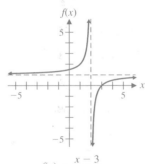

$f(x) = \dfrac{x - 3}{x - 2}$

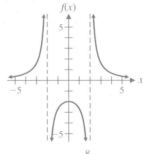

$f(x) = \dfrac{8}{x^2 - 4}$

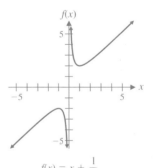

$f(x) = x + \dfrac{1}{x}$

GRAPH TRANSFORMATIONS

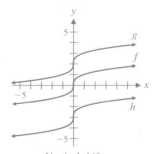

Vertical shift
$g(x) = f(x) + 2$
$h(x) = f(x) - 3$

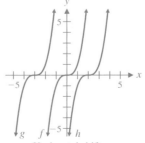

Horizontal shift
$g(x) = f(x + 3)$
$h(x) = f(x - 2)$

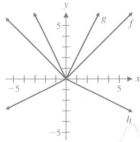

Stretch, shrink and reflection
$g(x) = 2f(x)$
$h(x) = -0.5f(x)$

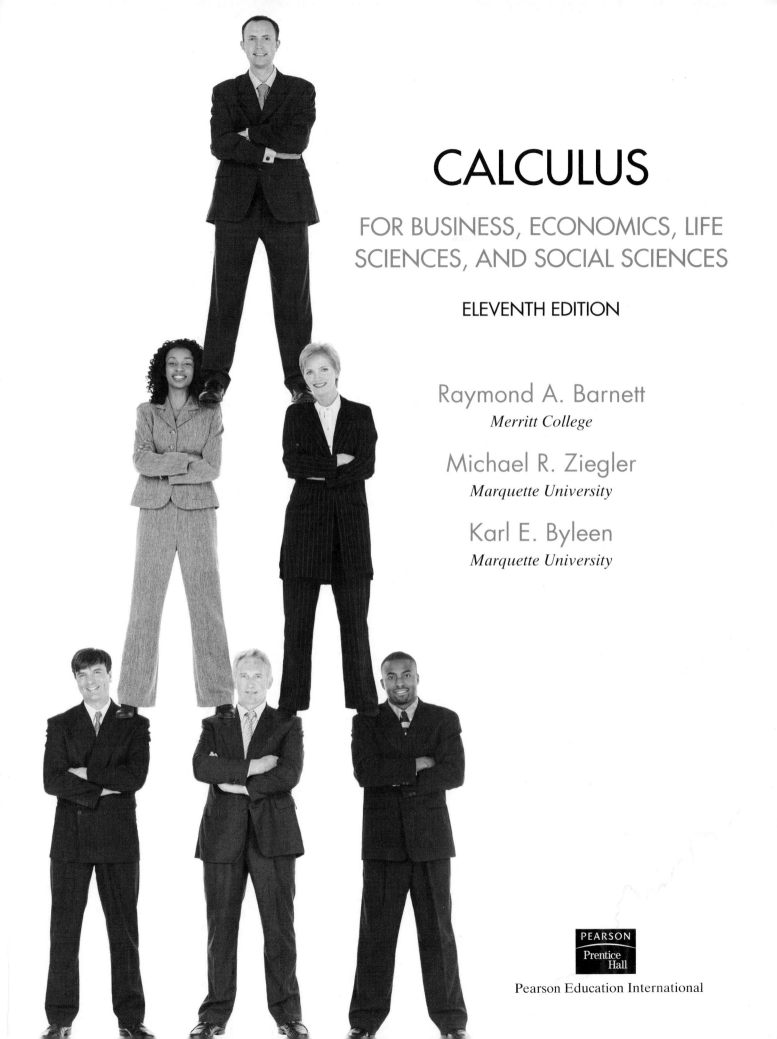

CALCULUS

FOR BUSINESS, ECONOMICS, LIFE
SCIENCES, AND SOCIAL SCIENCES

ELEVENTH EDITION

Raymond A. Barnett
Merritt College

Michael R. Ziegler
Marquette University

Karl E. Byleen
Marquette University

PEARSON
Prentice
Hall

Pearson Education International

Acquisitions Editor: Chuck Synovec
Vice President and Editorial Director: Mathematics: Christine Hoag
Project Manager: Michael Bell
Editorial Assistant/Print Supplements Editor: Joanne Wendelken
Production Editor: Raegan Keida Heerema
Senior Managing Editor: Linda Mihatov Behrens
Assistant Managing Editor: Bayani Mendoza de Leon
Manufacturing Buyer: Maura Zaldivar
Associate Director of Operations: Alexis Heydt-Long
Marketing Manager: Wayne Parkins
Marketing Assistant: Jennifer de Leeuwerk
Art Director: John Christiana
Interior and Cover Designer: 511 Design
Art Editor: Thomas Benfatti
Creative Director: Juan López
Manager, Cover, Visual Research & Permissions: Karen Sanatar
Art Studio: Scientific Illustrators

COVER IMAGE CREDITS

Stock Market Monitor	Digital Vision / Getty Images, Inc
Money	Digital Vision / Getty Images, Inc
Woman and Microscope	Digital Vision / Getty Images, Inc
Simatron Tiger	Digital Vision / Getty Images, Inc

ISBN-10: 0-13-206974-1
ISBN-13: 978-0-13-206974-8

Pearson Education LTD., *London*
Pearson Education Australia PTY, Limited, *Sydney*
Pearson Education Singapore, Pte. Ltd
Pearson Education North Asia Ltd, *Hong Kong*
Pearson Education Canada, Ltd., *Toronto*
Pearson Educación de Mexico, S.A. de C.V.
Pearson Education - Japan, *Toyko*
Pearson Education Malaysia, Pte. Ltd
Pearson Education, Upper Saddle River, New Jersey

CONTENTS

PREFACE   xi

PART 1   A LIBRARY OF ELEMENTARY FUNCTIONS

1   Linear Equations and Graphs   2

1-1 Linear Equations and Inequalities   3
1-2 Graphs and Lines   13
1-3 Linear Regression   29
Chapter 1 Review   42
Review Exercise   43

2   Functions and Graphs   46

2-1 Functions   47
2-2 Elementary Functions: Graphs and Transformations   63
2-3 Quadratic Functions   76
2-4 Exponential Functions   93
2-5 Logarithmic Functions   105
Chapter 2 Review   119
Review Exercise   121

PART 2   CALCULUS

3   Limits and the Derivative   128

3-1 Introduction to Limits   129
3-2 Continuity   144
3-3 Infinite Limits and Limits at Infinity   155
3-4 The Derivative   169
3-5 Basic Differentiation Properties   183
3-6 Differentials   192
3-7 Marginal Analysis in Business and Economics   199
Chapter 3 Review   209
Review Exercise   210

4   Additional Derivative Topics   216

4-1 The Constant *e* and Continuous Compound Interest   217
4-2 Derivatives of Exponential and Logarithmic Functions   224
4-3 Derivatives of Products and Quotients   232
4-4 The Chain Rule   241
4-5 Implicit Differentiation   251
4-6 Related Rates   257
4-7 Elasticity of Demand   263
Chapter 4 Review   270
Review Exercise   272

5 Graphing and Optimization   274

5-1 First Derivative and Graphs   275
5-2 Second Derivative and Graphs   293
5-3 L'Hôpital's Rule   311
5-4 Curve-Sketching Techniques   321
5-5 Absolute Maxima and Minima   334
5-6 Optimization   342
    Chapter 5 Review   355
    Review Exercise   356

6 Integration   359

6-1 Antiderivatives and Indefinite Integrals   360
6-2 Integration by Substitution   372
6-3 Differential Equations; Growth and Decay   384
6-4 The Definite Integral   394
6-5 The Fundamental Theorem of Calculus   404
    Chapter 6 Review   417
    Review Exercise   419

7 Additional Integration Topics   423

7-1 Area between Curves   424
7-2 Applications in Business and Economics   434
7-3 Integration by Parts   446
7-4 Integration Using Tables   452
    Chapter 7 Review   459
    Review Exercise   460

8 Multivariable Calculus   463

8-1 Functions of Several Variables   463
8-2 Partial Derivatives   473
8-3 Maxima and Minima   482
8-4 Maxima and Minima Using Lagrange Multipliers   490
8-5 Method of Least Squares   499
8-6 Double Integrals over Rectangular Regions   509
8-7 Double Integrals over More General Regions   519
    Chapter 8 Review   528
    Review Exercise   530

9 Trigonometric Functions   533

9-1 Trigonometric Functions Review   534
9-2 Derivatives of Trigonometric Functions   542
9-3 Integration of Trigonometric Functions   547
    Chapter 9 Review   552
    Review Exercise   553

Basic Algebra Review 555

Self-Test on Basic Algebra 555

A-1 Algebra and Real Numbers 556
A-2 Operations on Polynomials 562
A-3 Factoring Polynomials 569
A-4 Operations on Rational Expressions 576
A-5 Integer Exponents and Scientific Notation 582
A-6 Rational Exponents and Radicals 588
A-7 Quadratic Equations 595

Special Topics 604

B-1 Sequences, Series, and Summation Notation 604
B-2 Arithmetic and Geometric Sequences 610
B-3 Binomial Theorem 617

Tables 621

Table I Basic Geometric Formulas 621
Table II Integration Formulas 623

Answers A-1

Index I-1

Applications Index AI-1

PART ONE A LIBRARY OF ELEMENTARY FUNCTIONS*

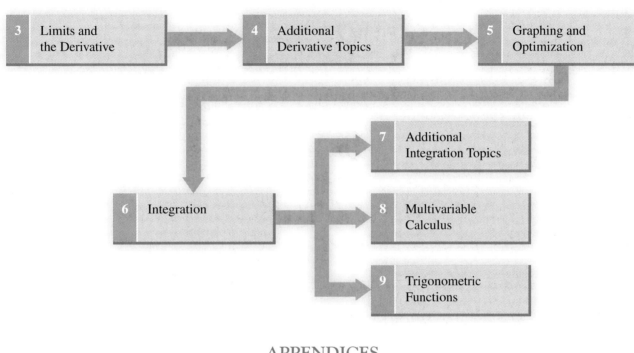

1 Linear Equations and Graphs → **2** Functions and Graphs

PART TWO CALCULUS

3 Limits and the Derivative → **4** Additional Derivative Topics → **5** Graphing and Optimization

6 Integration

7 Additional Integration Topics

8 Multivariable Calculus

9 Trigonometric Functions

APPENDICES

A Self-Test Basic Algebra Review

B Special Topics

---

* Selected topics from Part One may be referred to as needed in Part Two or reviewed systematically before starting Part Two.

The eleventh edition of *Calculus for Business, Economics, Life Sciences, and Social Sciences* is designed for a one-term course in calculus and for students who have had 1–2 years of high school algebra or the equivalent. The choice and independence of topics make the text readily adaptable to a variety of courses (see the Chapter Dependency Chart on page ix). It is one of three books in the authors' college mathematics series.

Improvements in this edition evolved out of the generous response from a large number of users of the last and previous editions as well as survey results from instructors, mathematics departments, course outlines, and college catalogs. Fundamental to a book's growth and effectiveness is classroom use and feedback. Now in its eleventh edition, *Calculus for Business, Economics, Life Sciences, and Social Sciences* has had the benefit of having a substantial amount of both.

EXAMPLES AND MATCHED PROBLEMS

Over 300 completely worked examples are used to introduce concepts and to demonstrate problem-solving techniques. Many examples have multiple parts, significantly increasing the total number of worked examples. The examples are **annotated** and the problem-solving steps are clearly identified. **Think Boxes** (dashed boxes) are used to enclose steps that are usually performed mentally (see sections 1-1 and 2-1).

Each example is followed by a similar **matched problem** for the student to work while reading the material. This actively involves the student in the learning process. The answers to these matched problems are included at the end of each section for easy reference.

EXERCISE SETS

The book contains **over 4,300 carefully selected and graded exercises.** Many problems have multiple parts, significantly increasing the total number of exercises. Each exercise set is designed so that every student will experience success. Exercise sets are divided into A (routine, easy mechanics), B (more difficult mechanics), and C (difficult mechanics and some theory) levels.

APPLICATIONS

A major objective of this book is to give the student substantial experience in modeling and solving real-world problems. Enough applications are included to convince even the most skeptical student that mathematics is really useful (see the Applications Index inside the back cover).

Almost **every exercise set contains application problems,** including applications from business and economics, life sciences, and social sciences. An instructor with students from all three disciplines can let them choose applications from their own field of interest; if most students are from one of the three areas, then special emphasis can be placed there. Most of the applications are simplified versions of actual real-world problems taken from professional journals and books. No specialized experience is required to solve any of the applications.

EXPLORE AND DISCUSS

Every section contains **Explore & Discuss problems** interspersed at appropriate places to encourage the student to think about a relationship or process before a result is stated, or to investigate additional consequences of a development in the text. Verbalization of mathematical concepts, results, and processes is encouraged in these Explore & Discuss problems, as well as in some matched problems, and in some problems in almost every exercise set. This serves to foster critical thinking skills. The Explore & Discuss material also can be used as in-class or out-of-class group activities. Problems in the exercise sets that require verbalization are indicated by color problem numbers.

TECHNOLOGY

Although access to a graphing calculator is not assumed, it is likely that many students will want to make use of a graphing calculator or mathematical software. To assist these students, optional graphing calculator activities are included in appropriate places in the book. These include brief discussions in the text, examples or portions of examples solved on a graphing calculator, and exercises for the student to solve. Linear regression is introduced in Section 1-3, and regression techniques on a graphing calculator are used at appropriate points to illustrate mathematical modeling with real data. All the optional graphing calculator material is clearly identified by and can be omitted without loss of continuity, if desired.

All **graphs** are computer-generated to ensure mathematical accuracy. Graphing calculator screens displayed in the text are actual output from a graphing calculator.

EMPHASIS AND STYLE

The text is written for student comprehension. Great care has been taken to write a book that is mathematically correct and accessible to students. Emphasis is on computational skills, ideas, and problem solving rather than mathematical theory. Most derivations and proofs are omitted except where their inclusion adds significant insight into a particular concept. General concepts and results are usually presented only after particular cases have been discussed.

ADDITIONAL PEDAGOGICAL FEATURES

Boxes are used to highlight important **definitions, results,** and **step-by-step processes** (see Sections 1-1 and 2-3). **Caution statements** appear throughout the text where student errors often occur (see Sections 3-1 and 3-7). An **insight feature,** appearing in nearly every section, makes explicit connections with previous knowledge. **Functional use of color** improves the clarity of many illustrations, graphs, and developments, and guides students through certain critical steps (see Sections 2-1 and 3-1). **Boldface type** is used to introduce new terms and highlight important comments. **Chapter review sections** include a comprehensive summary of important terms, symbols, and concepts, keyed to completely worked examples, followed by a comprehensive set of review exercises. Answers to most review exercises are included in the back of the book; each answer contains a reference to the section in which that type of problem is discussed. Answers to all other odd-numbered problems are also in the back of the book. Answers to application problems in linear programming include both the mathematical model and the numeric answer.

CONTENT

The text begins with the development of a library of elementary functions in Chapters 1 and 2, including their properties and uses. We encourage students to investigate mathematical ideas and processes graphically and numerically, as well as algebraically. This development lays a firm foundation for studying mathematics both in this book and in future endeavors. Depending on the syllabus for the course and the background of the students, some or all of this material can be covered at the beginning of a course, or selected portions can be referred to as needed later in the course.

The material in Part Two (Calculus) consists of differential calculus (Chapters 3–5), integral calculus (Chapters 6–7), multivariable calculus (Chapter 8), and a brief discussion of differentiation and integration of trigonometric functions (Chapter 9). In general, Chapters 3–6 must be covered in sequence; however, certain sections can be omitted or given brief treatments, as pointed out in the discussion that follows (see the Chapter Dependency Chart on page ix).

Chapter 3 introduces the derivative. The first three sections cover limits (including infinite limits and limits at infinity), continuity, and the limit properties that are essential to understanding the definition of the derivative in Section 3-4. The remaining sections of the chapter cover basic rules of differentiation, differentials, and applications of derivatives in business and economics. The interplay between graph-ical, numerical, and algebraic concepts is emphasized here and throughout the text.

In Chapter 4 the derivatives of exponential and logarithmic functions are obtained before the product rule, quotient rule, and chain rule are introduced. Implicit differ-entiation is introduced in Section 4-5 and applied to related rates problems in Sec-tion 4-6. Elasticity of demand is introduced in Section 4-7. The topics in these last three sections of Chapter 4 are not referred to elsewhere in the text and can be omitted.

Chapter 5 focuses on graphing and optimization. The first two sections cover first-derivative and second-derivative graph properties. L'Hôpital's rule is discussed in Section 5-3. A graphing strategy is presented and illustrated in Section 5-4. Optimization is covered in Sections 5-5 and 5-6, including examples and problems involving end-point solutions.

Chapter 6 introduces integration. The first two sections cover antidifferentiation techniques essential to the remainder of the text. Section 6-3 discusses some appli-cations involving differential equations that can be omitted. The definite integral is defined in terms of Riemann sums in Section 6-4 and the fundamental theorem of calculus is discussed in Section 6-5. As before, the interplay between graphical, nu-meric, and algebraic properties is emphasized. These two sections also are required for the remaining chapters in the text.

Chapter 7 covers additional integration topics and is organized to provide maximum flexibility for the instructor. The first section extends the area concepts introduced in Chapter 6 to the area between two curves and related applications. Section 7-2 cov-ers three more applications of integration, and Sections 7-3 and 7-4 deal with additional techniques of integration. Any or all of the topics in Chapter 7 can be omitted.

The first five sections of Chapter 8 deal with differential multivariable calculus and can be covered any time after Section 5-6 has been completed. Sections 8-6 and 8-7 require the integration concepts discussed in Chapter 6.

Chapter 9 provides brief coverage of trigonometric functions that can be incor-porated into the course, if desired. Section 9-1 provides a review of basic trigono-metric concepts. Section 9-2 can be covered any time after Section 5-3 has been completed. Section 9-3 requires the material in Chapter 6.

Appendix A contains a self-test and a concise review of basic algebra that also may be covered as part of the course or referred to as needed. As mentioned above, Appendix B contains additional topics that can be covered in conjunciton with cer-tain sections in the text, if desired.

ERROR CHECK

Because of the careful checking and proofing by a number of mathematics instruc-tors (acting independently), the authors and publisher believe this book to be sub-stantially error-free. For any errors remaining, the authors would be grateful if they were sent to: Karl E. Byleen, 9322 W. Garden Court, Hales Corners, WI 53130; or by e-mail, to: kbyleen@wi.rr.com

ACKNOWLEDGMENTS

In addition to the authors many others are involved in the successful publication of a book.

We wish to thank the following reviewers of the 11th edition:

Maria Terrell, *Cornell University*
Jeffrey Lynn Johnson
Jon Prewett, *University of Wyoming*
Fred M. Wright, *Iowa State University*
Darryl Egley, *North Harris College*
Gregory Goeckel, *Presbyterian College*
Steven Klassen, *Missouri Western State University*

James Martin, *Christopher Newport University*

Cynthia Schultz, *Illinois Valley Community College*

Virginia Hanning, *San Jacinto College*

Garland Guyton, *Montgomery College*

Bruce Hedman, *University of Connecticut*

Lauren Fern, *University of Montana*

Wesley W. Maiers, *Valparaiso University*

Gary R. Penner, *Richland College*

We also wish to thank our colleagues who have provided input on previous editions:

Chris Boldt, Bob Bradshaw, Bruce Chaffee, Robert Chaney, Dianne Clark, Charles E. Cleaver, Barbara Cohen, Richard L. Conlon, Catherine Cron, Lou D' Alotto, Madhu Deshpande, Kenneth A. Dodaro, Michael W. Ecker, Jerry R. Ehman, Lucina Gallagher, Martha M. Harvey, Sue Henderson, Lloyd R. Hicks, Louis F. Hoelzle, Paul Hutchins, K. Wayne James, Robert H. Johnston, Robert Krystock, Inessa Levi, James T. Loats, Frank Lopez, Roy H. Luke, Wayne Miller, Mel Mitchell, Linda M. Neal, Ronald Persky, Kenneth A. Peters, Jr., Dix Petty, Tom Plavchak, Bob Prielipp, Thomas Riedel, Stephen Rodi, Arthur Rosenthal, Sheldon Rothman, Elaine Russell, John Ryan, Daniel E. Scanlon, George R. Schriro, Arnold L. Schroeder, Hari Shanker, Joan Smith, J. Sriskandarajah, Steven Terry, Beverly Vredevelt, Delores A. Williams, Caroline Woods, Charles W. Zimmerman, Pat Zrolka, and Cathleen A. Zucco-Tevelot.

We also express our thanks to:

Hossein Hamedani for providing a careful and thorough check of all the mathematical calculations in the book.

Garret Etgen, Hossein Hamedani, David Schneider, and Mary Ann Teel for developing the supplemental manuals that are so important to the success of a text.

Jeanne Wallace for accurately and efficiently producing most of the manuals that supplement the text.

Brian Morris and his staff at Scientific Illustrators for their effective illustrations and accurate graphs.

All the people at Prentice Hall who contributed their efforts to the production of this book, especially Sally Yagan, Chuck Synovec, acquisitions editor; and Raegan Heerema, production editor.

Producing this new edition with the help of all these extremely competent people has been a most satisfying experience.

R. A. Barnett
M. R. Ziegler
K. E. Byleen

PART ONE

A Library Of Elementary Functions

Linear Equations and Graphs

CHAPTER

1

1-1 Linear Equations and Inequalities

1-2 Graphs and Lines

1-3 Linear Regression

Chapter 1 Review

Review Exercise

INTRODUCTION

We begin by discussing some algebraic methods for solving equations and inequalities. Next, we introduce coordinate systems that allow us to explore the relationship between algebra and geometry. Finally, we use this algebraic–geometric relationship to find equations that can be used to describe real-world data sets. We also consider many applied problems that can be solved using the concepts discussed in this chapter.

Section 1-1   LINEAR EQUATIONS AND INEQUALITIES

- Linear Equations
- Linear Inequalities
- Applications

The equation

$$3 - 2(x + 3) = \frac{x}{3} - 5$$

and the inequality

$$\frac{x}{2} + 2(3x - 1) \geq 5$$

are both first degree in one variable. In general, a **first-degree, or linear, equation** in one variable is any equation that can be written in the form

$$\textbf{Standard form:} \quad ax + b = 0 \qquad a \neq 0 \tag{1}$$

If the equality symbol, =, in (1) is replaced by $<, >, \leq,$ or $\geq$, the resulting expression is called a **first-degree, or linear, inequality.**

A **solution** of an equation (or inequality) involving a single variable is a number that when substituted for the variable makes the equation (or inequality) true. The set of all solutions is called the **solution set.** When we say that we **solve an equation** (or inequality), we mean that we find its solution set.

Knowing what is meant by the solution set is one thing; finding it is another. We start by recalling the idea of equivalent equations and equivalent inequalities. If we perform an operation on an equation (or inequality) that produces another equation (or inequality) with the same solution set, then the two equations (or inequalities) are said to be **equivalent.** The basic idea in solving equations and inequalities is to perform operations on these forms that produce simpler equivalent forms, and to continue the process until we obtain an equation or inequality with an obvious solution.

■ Linear Equations

Linear equations are generally solved using the following equality properties:

THEOREM 1   EQUALITY PROPERTIES   An equivalent equation will result if

1. The same quantity is added to or subtracted from each side of a given equation.
2. Each side of a given equation is multiplied by or divided by the same nonzero quantity.

Several examples should remind you of the process of solving equations.

EXAMPLE 1   **Solving a Linear Equation**   Solve and check:

$$8x - 3(x - 4) = 3(x - 4) + 6$$

SOLUTION

$$
\begin{aligned}
8x - 3(x - 4) &= 3(x - 4) + 6 && \text{Remove parentheses.} \\
8x - 3x + 12 &= 3x - 12 + 6 && \text{Combine like terms.} \\
5x + 12 &= 3x - 6 && \text{Subtract } 3x \text{ from both sides.} \\
2x &= -18 && \text{Divide both sides by 2.} \\
x &= -9
\end{aligned}
$$

CHECK

$$8x - 3(x - 4) = 3(x - 4) + 6$$
$$8(-9) - 3[(-9) - 4] \overset{?}{=} 3[(-9) - 4] + 6$$
$$-72 - 3(-13) \overset{?}{=} 3(-13) + 6$$
$$-33 \overset{?}{=} -33$$

■

MATCHED PROBLEM 1 | Solve and check: $3x - 2(2x - 5) = 2(x + 3) - 8$

Explore & Discuss 1

According to equality property 2, multiplying both sides of an equation by a nonzero number always produces an equivalent equation. What is the smallest positive number that you could use to multiply both sides of the following equation to produce an equivalent equation without fractions?

$$\frac{x + 1}{3} - \frac{x}{4} = \frac{1}{2}$$

EXAMPLE 2   **Solving a Linear Equation** Solve and check: $\dfrac{x + 2}{2} - \dfrac{x}{3} = 5$

SOLUTION   What operations can we perform on

$$\frac{x + 2}{2} - \frac{x}{3} = 5$$

to eliminate the denominators? If we can find a number that is exactly divisible by each denominator, we can use the multiplication property of equality to clear the denominators. The LCD (least common denominator) of the fractions, 6, is exactly what we are looking for! Actually, any common denominator will do, but the LCD results in a simpler equivalent equation. Thus, we multiply both sides of the equation by 6:

$$6\left(\frac{x + 2}{2} - \frac{x}{3}\right) = 6 \cdot 5 \quad *$$
$$6 \cdot \frac{\overset{3}{(x + 2)}}{2} - 6 \cdot \frac{x}{\underset{1}{3}}\overset{2}{} = 30$$

$$3(x + 2) - 2x = 30 \qquad \text{Remove parentheses.}$$
$$3x + 6 - 2x = 30 \qquad \text{Combine like terms.}$$
$$x + 6 = 30 \qquad \text{Subtract 6 from both sides.}$$
$$x = 24$$

CHECK

$$\frac{x + 2}{2} - \frac{x}{3} = 5$$
$$\frac{24 + 2}{2} - \frac{24}{3} \overset{?}{=} 5$$
$$13 - 8 \overset{?}{=} 5$$
$$5 \overset{?}{=} 5$$

■

MATCHED PROBLEM 2 | Solve and check: $\dfrac{x + 1}{3} - \dfrac{x}{4} = \dfrac{1}{2}$

In many applications of algebra, formulas or equations must be changed to alternative equivalent forms. The following examples are typical.

---

* Dashed boxes are used throughout the book to denote steps that are usually performed mentally.

EXAMPLE 3	**Solving a Formula for a Particular Variable** Solve the amount formula for simple interest, $A = P + Prt$, for

(A) r in terms of the other variables

(B) P in terms of the other variables

SOLUTION   (A) $A = P + Prt$ *Reverse equation.*

$\qquad\qquad\qquad P + Prt = A$ *Now isolate r on the left side.*

$\qquad\qquad\qquad\qquad Prt = A - P$ *Divide both members by Pt.*

$$r = \frac{A - P}{Pt}$$

(B) $A = P + Prt$ *Reverse equation.*

$\qquad\qquad\qquad P + Prt = A$ *Factor out P (note the use of the distributive property).*

$\qquad\qquad\qquad P(1 + rt) = A$ *Divide by (1 + rt).*

$$P = \frac{A}{1 + rt}$$

MATCHED PROBLEM 3	Solve $M = Nt + Nr$ for

(A) t (B) N

■ Linear Inequalities

Before we start solving linear inequalities, let us recall what we mean by $<$ (less than) and $>$ (greater than). If a and b are real numbers, we write

$$a < b \quad \text{\textit{a is less than b.}}$$

if there exists a positive number p such that $a + p = b$. Certainly, we would expect that if a positive number was added to any real number, the sum would be larger than the original. That is essentially what the definition states. If $a < b$, we may also write

$$b > a \quad \text{\textit{b is greater than a.}}$$

EXAMPLE 4	**Inequalities**

(A) $3 < 5$ *Since 3 + 2 = 5*

(B) $-6 < -2$ *Since −6 + 4 = −2*

(C) $0 > -10$ *Since −10 < 0*

MATCHED PROBLEM 4	Replace each question mark with either $<$ or $>$.

(A) $2\,?\,8$ (B) $-20\,?\,0$ (C) $-3\,?\,-30$

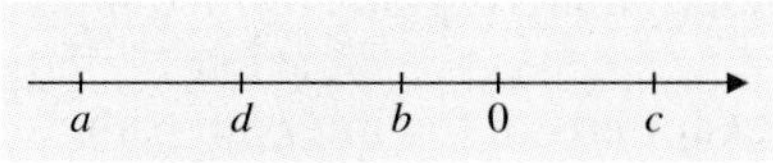

FIGURE 1 $a < b, c > d$

The inequality symbols have a very clear geometric interpretation on the real number line. If $a < b$, then a is to the left of b on the number line; if $c > d$, then c is to the right of d (Fig. 1). Check this geometric property with the inequalities in Example 4.

Explore & Discuss **2**   Replace ? with $<$ or $>$ in each of the following:

(A) $-1\,?\,3$ and $2(-1)\,?\,2(3)$

(B) $-1\,?\,3$ and $-2(-1)\,?\,-2(3)$

(C) $12 \, ? -8$ and $\dfrac{12}{4} \, ? \dfrac{-8}{4}$

(D) $12 \, ? -8$ and $\dfrac{12}{-4} \, ? \dfrac{-8}{-4}$

Based on these examples, describe verbally the effect of multiplying both sides of an inequality by a number.

Now let us turn to the problem of solving linear inequalities in one variable. Recall that a **solution** of an inequality involving one variable is a number that, when substituted for the variable, makes the inequality true. The set of all solutions is called the **solution set.** When we say that we **solve an inequality,** we mean that we find its solution set. The procedures used to solve linear inequalities in one variable are almost the same as those used to solve linear equations in one variable but with one important exception, as noted in property 3 below.

THEOREM 2   INEQUALITY PROPERTIES

An equivalent inequality will result and the **sense or direction will remain the same** if each side of the original inequality

   1. Has the same real number added to or subtracted from it.
   2. Is multiplied or divided by the same *positive* number.

An equivalent inequality will result and the **sense or direction will reverse** if each side of the original inequality:

   3. Is multiplied or divided by the same *negative* number.

Note: Multiplication by 0 and division by 0 are not permitted.

Thus, we can perform essentially the same operations on inequalities that we perform on equations, with the exception that **the sense of the inequality reverses if we multiply or divide both sides by a negative number.** Otherwise, the sense of the inequality does not change. For example, if we start with the true statement

$$-3 > -7$$

and multiply both sides by 2, we obtain

$$-6 > -14$$

and the sense of the inequality stays the same. But if we multiply both sides of $-3 > -7$ by -2, the left side becomes 6 and the right side becomes 14, so we must write

$$6 < 14$$

to have a true statement. Thus, the sense of the inequality reverses.

If $a < b$, the **double inequality** $a < x < b$ means that $a < x$ **and** $x < b$; that is, x is between a and b. **Interval notation** is also used to describe sets defined by inequalities, as shown in Table 1.

The numbers a and b in Table 1 are called the **endpoints** of the interval. An interval is **closed** if it contains all its endpoints and **open** if it does not contain any of its endpoints. Thus, the intervals $[a, b]$, $(-\infty, a]$, and $[b, \infty)$ are closed and the intervals (a, b), $(-\infty, a)$, and (b, ∞) are open. Note that the symbol ∞ (read infinity) is not a number. When we write $[b, \infty)$ we are simply referring to the interval that starts at b and continues indefinitely to the right. We never refer to ∞ as an endpoint and we never write $[b, \infty]$. The interval $(-\infty, \infty)$ is the entire real number line.

Note that an endpoint of a line graph in Table 1 has a square bracket through it if the endpoint is included in the interval and a parenthesis through it if it is not.

TABLE 1

Interval Notation	Inequality Notation	Line Graph
$[a, b]$	$a \leq x \leq b$	
(a, b)	$a \leq x < b$	
(a, b)	$a < x \leq b$	
(a, b)	$a < x < b$	
$(-\infty, a)$	$x \leq a$	
$(-\infty, a)$	$x < a$	
$(b, \infty)*$	$x \geq b$	
(b, ∞)	$x > b$	

INSIGHT

The notation $(2, 7)$ has two common mathematical interpretations: the ordered pair with first coordinate 2 and second coordinate 7, and the open interval consisting of all real numbers between 2 and 7. The choice of interpretations is usually determined by the context in which the notation is used. The notation $(2, -7)$ also can be interpreted as an ordered pair, but not as an interval. In interval notation, it is always assumed that the left endpoint is less than the right endpoint. Thus, $(-7, 2)$ is correct interval notation, but $(2, -7)$ is not.

EXAMPLE 5 **Interval and Inequality Notation, and Line Graphs**

(A) Write $[-2, 3)$ as a double inequality and graph.

(B) Write $x \geq -5$ in interval notation and graph.

SOLUTION (A) $[-2, 3)$ is equivalent to $-2 \leq x < 3$.

(B) $x \geq -5$ is equivalent to $[-5, \infty)$.

MATCHED PROBLEM 5 (A) Write $(-7, 4]$ as a double inequality and graph.

(B) Write $x < 3$ in interval notation and graph.

Explore & Discuss **3** The solution to Example 5B shows the graph of the inequality $x \geq -5$. What is the graph of $x < -5$? What is the corresponding interval? Describe the relationship between these sets.

EXAMPLE 6	**Solving a Linear Inequality** Solve and graph:

$$2(2x + 3) < 6(x - 2) + 10$$

SOLUTION

$2(2x + 3) < 6(x - 2) + 10$	Remove parentheses.
$4x + 6 < 6x - 12 + 10$	Combine like terms.
$4x + 6 < 6x - 2$	Subtract $6x$ from both sides.
$-2x + 6 < -2$	Subtract 6 from both sides.
$-2x < -8$	Divide both sides by -2 and reverse
$x > 4 \quad \text{or} \quad (4, \infty)$	the sense of the inequality.

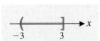

Notice that in the graph of $x > 4$, we use a parenthesis through 4, since the point 4 is not included in the graph.

MATCHED PROBLEM 6	Solve and graph: $3(x - 1) \leq 5(x + 2) - 5$

EXAMPLE 7	**Solving a Double Inequality** Solve and graph: $-3 < 2x + 3 \leq 9$

SOLUTION We are looking for all numbers x such that $2x + 3$ is between -3 and 9, including 9 but not -3. We proceed as before except that we try to isolate x in the middle:

$$-3 < 2x + 3 \leq 9$$

$$-3 - 3 < 2x + 3 - 3 \leq 9 - 3$$

$$-6 < 2x \leq 6$$

$$\frac{-6}{2} < \frac{2x}{2} \leq \frac{6}{2}$$

$$-3 < x \leq 3 \quad \text{or} \quad (-3, 3]$$

MATCHED PROBLEM 7	Solve and graph: $-8 \leq 3x - 5 < 7$

Note that a linear equation usually has exactly one solution, while a linear inequality usually has infinitely many solutions.

APPLICATIONS

To realize the full potential of algebra, we must be able to translate real-world problems into mathematical forms. In short, we must be able to do word problems.

There are many different types of applications—so many, in fact, that no single approach applies to all. However, the following suggestions may help you get started:

Procedure for Solving Word Problems

1. Read the problem carefully and introduce a variable to represent an unknown quantity in the problem. Often the question asked in a problem will indicate the best way to introduce this variable.

2. Identify other quantities in the problem (known or unknown) and, whenever possible, express unknown quantities in terms of the variable you introduced in step 1.

3. Write a verbal statement using the conditions stated in the problem and then write an equivalent mathematical statement (equation or inequality).

4. Solve the equation or inequality and answer the questions posed in the problem.

5. Check the solution(s) in the original problem.

EXAMPLE 8

Purchase Price   John purchases a computer from an online store for $851.26, including a $57 shipping charge and 5.2% state sales tax. What is the purchase price of the computer?

SOLUTION

Step 1. After reading the problem, we decide to let x represent the purchase price of the computer.

Step 2. Identify quantities in the problem.

$$\text{Shipping charges: } \$57$$
$$\text{Sales tax: } 0.052x$$
$$\text{Total cost: } \$851.26$$

Step 3. Write a verbal statement and an equation.

$$\text{Price} + \text{Shipping Charges} + \text{Sales Tax} = \text{Total Order Cost}$$
$$x \quad + \quad 57 \quad + \quad 0.052x \quad = 851.26$$

Step 4. Solve the equation and answer the question.

$$x + 57 + 0.052x = 851.26 \quad \textit{Collect like terms.}$$
$$1.052x + 57 = 851.26 \quad \textit{Subtract 57 from both sides.}$$
$$1.052x = 794.26 \quad \textit{Divide both sides by 1.052.}$$
$$x = 755$$

The price of the computer is $755.

Step 5. Check the answer in the original problem.

$$\text{Price} = \$755.00$$
$$\text{Shipping charges} = \quad \$57.00$$
$$\underline{\text{Tax } 0.052 \cdot 755 = \quad \$39.26}$$
$$\text{Total} = \$851.26$$

MATCHED PROBLEM 8

Mary paid 8.5% sales tax and a $190 title and license fee when she bought a new car for a total of $28,400. What is the purchase price of the car?

The next example involves the important concept of **break-even analysis,** which is encountered in several places in this text. Any manufacturing company has **costs, C,** and **revenues, R.** The company will have a **loss** if $R < C$, will **break even** if $R = C$, and will have a **profit** if $R > C$. Costs involve **fixed costs,** such as plant overhead, product design, setup, and promotion; and **variable costs,** which are dependent on the number of items produced at a certain cost per item.

EXAMPLE 9

Break-Even Analysis   A recording company produces compact disks (CDs). One-time fixed costs for a particular CD are $24,000, which includes costs such as recording, album design, and promotion. Variable costs amount to $6.20 per CD and include the manufacturing, distribution, and royalty costs for each disk actually manufactured and sold to a retailer. The CD is sold to retail outlets at $8.70 each. How many CDs must be manufactured and sold for the company to break even?

SOLUTION   **Step 1.** Let x = number of CDs manufactured and sold.

Step 2.
$$C = \text{cost of producing } x \text{ CDs}$$
$$R = \text{revenue (return) on sales of } x \text{ CDs}$$
$$\text{Fixed costs} = \$24{,}000$$
$$\text{Variable costs} = \$6.20x$$
$$C = \text{Fixed costs} + \text{variable costs}$$
$$= \$24{,}000 + \$6.20x$$
$$R = \$8.70x$$

Step 3. The company breaks even if $R = C$; that is, if
$$\$8.70x = \$24{,}000 + \$6.20x$$

Step 4.
$$8.7x = 24{,}000 + 6.2x \qquad \text{Subtract 6.2x from both sides.}$$
$$2.5x = 24{,}000 \qquad\qquad \text{Divide both sides by 2.5.}$$
$$x = 9{,}600$$

The company must make and sell 9,600 CDs to break even.

Step 5. Check:

Costs	Revenue
$24{,}000 + 6.2(9{,}600)$	$8.7(9{,}600)$
$= \$83{,}520$	$= \$83{,}520$

MATCHED PROBLEM 9   How many CDs would a recording company have to make and sell to break even if the fixed costs are $18,000, variable costs are $5.20 per CD, and the CDs are sold to retailers for $7.60 each?

EXAMPLE 10   **Consumer Price Index**   The Consumer Price Index (CPI) is a measure of the average change in prices over time from a designated reference period, which equals 100. The index is based on prices of basic consumer goods and services, and is published at regular intervals by the Bureau of Labor Statistics. Table 2 lists the CPI for several years from 1960 to 2000. What net annual salary in 2000 would have the same purchasing power as a net annual salary of $13,000 in 1960? Compute the answer to the nearest dollar.

SOLUTION   **Step 1.** Let x = the purchasing power of an annual salary in 2000.

Step 2. Annual salary in 1960 = $13,000

$$\text{CPI in 1960} = 29.6$$
$$\text{CPI in 2000} = 172.2$$

TABLE 2 CPI (1982–1984 = 100)

Year	Index
1960	29.6
1970	38.8
1980	82.4
1990	130.7
2000	172.2

Step 3. The ratio of a salary in 2000 to a salary in 1960 is the same as the ratio of the CPI in 2000 to the CPI in 1960.

$$\frac{x}{13{,}000} = \frac{172.2}{29.6} \qquad \text{Multiply both sides by 13,000.}$$

Step 4.
$$x = 13{,}000 \cdot \frac{172.2}{29.6}$$
$$= \$75{,}628 \text{ per year}$$

Step 5.

Salary Ratio	CPI Ratio
$\dfrac{75{,}628}{13{,}000} = 5.81754$	$\dfrac{172.2}{29.6} = 5.81757$

[*Note:* The slight difference in these ratios is due to rounding the 2000 salary to the nearest dollar.]

| MATCHED PROBLEM 10 | What net annual salary in 1970 would have had the same purchasing power as a net annual salary of $100,000 in 2000? Compute the answer to the nearest dollar. |

Answers to Matched Problems   **1.** $x = 4$ **2.** $x = 2$ **3.** (A) $t = \dfrac{M - Nr}{N}$ (B) $N = \dfrac{M}{t + r}$

4. (A) $<$ (B) $<$ (C) $>$

5. (A) $-7 < x \le 4$;   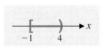   (B) $(-\infty, 3)$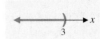

6. $x \ge -4$ or $[-4, \infty)$

7. $-1 \le x < 4$ or $[-1, 4)$ **8.** $26,000

9. 7,500 CDs   **10.** $22,532

Exercise 1-1

A   *Solve Problems 1–6.*

1. $2m + 9 = 5m - 6$ **2.** $3y - 4 = 6y - 19$

3. $2x + 3 < -4$ **4.** $5x + 2 > 1$

5. $-3x \ge -12$ **6.** $-4x \le 8$

Solve Problems 7–10 and graph.

7. $-4x - 7 > 5$ **8.** $-2x + 8 < 4$

9. $2 \le x + 3 \le 5$ **10.** $-4 < 2y - 3 < 9$

Solve Problems 11–26.

11. $\dfrac{x}{3} - \dfrac{1}{2} = \dfrac{1}{3}$ **12.** $\dfrac{m}{5} - 2 = \dfrac{3}{5}$

13. $\dfrac{x}{3} > \dfrac{-5}{4}$ **14.** $\dfrac{y}{-2} \le -1$

15. $\dfrac{y}{3} = 4 - \dfrac{y}{6}$ **16.** $\dfrac{x}{4} = 9 - \dfrac{x}{2}$

B

17. $10x + 25(x - 3) = 275$

18. $-3(4 - x) = 5 - (x + 1)$

19. $3 - y \le 4(y - 3)$ **20.** $x - 2 \ge 2(x - 5)$

21. $\dfrac{x}{5} - \dfrac{x}{6} = \dfrac{6}{5}$ **22.** $\dfrac{y}{4} - \dfrac{y}{3} = \dfrac{1}{2}$

23. $\dfrac{m}{5} - 3 < \dfrac{3}{5} - \dfrac{m}{2}$ **24.** $\dfrac{u}{2} - \dfrac{2}{3} < \dfrac{u}{3} + 2$

25. $0.1(x - 7) + 0.05x = 0.8$

26. $0.03(2x + 1) - 0.05x = 12$

Solve Problems 27–30 and graph.

27. $2 \le 3x - 7 < 14$

28. $-4 \le 5x + 6 < 21$

29. $-4 \le \frac{9}{5}C + 32 \le 68$

30. $-1 \le \frac{2}{3}t + 5 \le 11$

C   *Solve Problems 31–38 for the indicated variable.*

31. $3x - 4y = 12$; for y

32. $y = -\frac{2}{3}x + 8$; for x

33. $Ax + By = C$; for y $(B \ne 0)$

34. $y = mx + b$; for m

35. $F = \frac{9}{5}C + 32$; for C

36. $C = \frac{5}{9}(F - 32)$; for F

37. $A = \frac{2}{3}(Bm - Bn)$; for B

38. $X = \frac{1}{5}(3CD - C)$; for C

Solve Problems 39 and 40 and graph.

39. $-3 \le 4 - 7x < 18$

40. $-10 \le 8 - 3u \le -6$

41. What can be said about the signs of the numbers a and b in each case?

(A) $ab > 0$ (B) $ab < 0$

(C) $\dfrac{a}{b} > 0$ (D) $\dfrac{a}{b} < 0$

42. What can be said about the signs of the numbers $a, b,$ and c in each case?

(A) $abc > 0$ (B) $\dfrac{ab}{c} < 0$

(C) $\dfrac{a}{bc} > 0$ (D) $\dfrac{a^2}{bc} < 0$

43. Replace each question mark with $<$ or $>$, as appropriate:

(A) If $a - b = 2$, then a ? b.

(B) If $c - d = -1$, then c ? d.

44. For what c and d is $c + d < c - d$?

45. If both a and b are positive numbers and b/a is greater than 1, then is $a - b$ positive or negative?

46. If both a and b are negative numbers and b/a is greater than 1, then is $a - b$ positive or negative?

In Problems 47–52, discuss the validity of each statement. If the statement is true, explain why. If not, give a counterexample.

47. If the intersection of two open intervals is nonempty, then their intersection is an open interval.

48. If the intersection of two closed intervals is nonempty, then their intersection is a closed interval.

49. The union of any two open intervals is an open interval.

50. The union of any two closed intervals is a closed interval.

51. If the intersection of two open intervals is nonempty, then their union is an open interval.

52. If the intersection of two closed intervals is nonempty, then their union is a closed interval.

Applications

53. *Puzzle.* A jazz concert brought in $165,000 on the sale of 8,000 tickets. If the tickets sold for $15 and $25 each, how many of each type of ticket were sold?

54. *Puzzle.* An all-day parking meter takes only dimes and quarters. If it contains 100 coins with a total value of $14.50, how many of each type of coin are in the meter?

55. *IRA.* You have $500,000 in an IRA (Individual Retirement Account) at the time you retire. You have the option of investing this money in two funds: Fund A pays 5.2% annually and Fund B pays 7.7% annually. How should you divide your money between Fund A and Fund B to produce an annual interest income of $34,000?

56. *IRA.* Refer to Problem 55. How should you divide your money between Fund A and Fund B to produce an annual interest income of $30,000?

57. *Inflation.* If the price change of cars parallels the change in the CPI (see Table 2 in Example 10), what would a car sell for (to the nearest dollar) in 2000 if a comparable model sold for $5,000 in 1970?

58. *Inflation.* If the price change in houses parallels the CPI (see Table 2 in Example 10), what would a house valued at $200,000 in 2000 be valued at (to the nearest dollar) in 1960?

59. *Retail and wholesale prices.* Retail prices in a department store are obtained by marking up the wholesale price by 40%. That is, retail price is obtained by adding 40% of the wholesale price to the wholesale price.

(A) What is the retail price of a suit if the wholesale price is $300?

(B) What is the wholesale price of a pair of jeans if the retail price is $77?

60. *Retail and sale prices.* Sale prices in a department store are obtained by marking down the retail price by 15%. That is, sale price is obtained by subtracting 15% of the retail price from the retail price.

(A) What is the sale price of a hat that has a retail price of $60?

(B) What is the retail price of a dress that has a sales price of $68?

61. *Equipment rental.* A golf course charges $52 for a round of golf using a set of their clubs and $44 if you have your own clubs. If you buy a set of clubs for $270, how many rounds must you play with your clubs to recover the cost of the clubs?

62. *Equipment rental.* The local supermarket rents carpet cleaners for $20 a day. These cleaners use shampoo in a special cartridge that sells for $16 and is available only from the supermarket. A home carpet cleaner can be purchased for $300. Shampoo for the home cleaner is readily available for $9 a bottle. Past experience has shown that it takes two shampoo cartridges to clean the 10-foot-by-12-foot carpet in your living room with the rented cleaner. Cleaning the same area with the home cleaner will consume three bottles of shampoo. If you buy the home cleaner, how many times must you clean the living room carpet to make buying cheaper than renting?

63. *Sales commissions.* One employee of a computer store is paid a base salary of $2,000 a month plus an 8% commission on all sales over $7,000 during the month. How much must the employee sell in one month to earn a total of $4,000 for the month?

64. *Sales commissions.* A second employee of the computer store in Problem 63 is paid a base salary of $3,000 a month plus a 5% commission on all sales during the month.

(A) How much must this employee sell in one month to earn a total of $4,000 for the month?

(B) Determine the sales level at which both employees receive the same monthly income.

(C) If employees can select either of these payment methods, how would you advise an employee to make this selection?

65. *Break-even analysis.* A publisher for a promising new novel figures fixed costs (overhead, advances, promotion, copy editing, typesetting, and so on) at $55,000, and variable costs (printing, paper, binding, shipping) at $1.60 for each book produced. If the book is sold to distributors for $11 each, how many must be produced and sold for the publisher to break even?

66. *Break-even analysis.* The publisher of a new book called *Muscle-Powered Sports* figures fixed costs at $92,000 and variable costs at $2.10 for each book produced. If the book is sold to distributors for $15 each, how many must be sold for the publisher to break even?

67. *Break-even analysis.* The publisher in Problem 65 finds that rising prices for paper increase the variable costs to $2.10 per book.

(A) Discuss possible strategies the company might use to deal with this increase in costs.

(B) If the company continues to sell the books for $11, how many books must they sell now to make a profit?

(C) If the company wants to start making a profit at the same production level as before the cost increase, how much should they sell the book for now?

68. *Break-even analysis.* The publisher in Problem 66 finds that rising prices for paper increase the variable costs to $2.70 per book.

(A) Discuss possible strategies the company might use to deal with this increase in costs.

(B) If the company continues to sell the books for $15, how many books must they sell now to make a profit?

(C) If the company wants to start making a profit at the same production level as before the cost increase, how much should they sell the book for now?

69. *Wildlife management.* A naturalist for a fish and game department estimated the total number of rainbow trout in a certain lake using the popular capture–mark–recapture technique. He netted, marked, and released 200 rainbow trout. A week later, allowing for thorough mixing, he again netted 200 trout and found 8 marked ones among them. Assuming that the proportion of marked fish in the second sample was the same as the proportion of all marked fish in the total population, estimate the number of rainbow trout in the lake.

70. *Ecology.* If the temperature for a 24 hour period at an Antarctic station ranged between $-49°F$ and $14°F$ (that is, $-49 \le F \le 14$), what was the range in degrees Celsius? [*Note: $F = \frac{9}{5}C + 32$.*]

71. *Psychology.* The IQ (intelligence quotient) is found by dividing the mental age (MA), as indicated on standard tests, by the chronological age (CA) and multiplying by 100. For example, if a child has a mental age of 12 and a chronological age of 8, the calculated IQ is 150. If a 9-year-old girl has an IQ of 140, compute her mental age.

72. *Psychology.* Refer to Problem 71. If the IQ of a group of 12-year-old children varies between 80 and 140, what is the range of their mental ages?

73. *Anthropology.* In their study of genetic groupings, anthropologists use a ratio called the **cephalic index.** This is the ratio of the breadth B of the head to its length L (looking down from above) expressed as a percentage. A study of the Gurung community of Nepal published in the *Kathmandu University Medical Journal* in 2005 found that the average head length of males was 18 cm and their head breadths varied between 12 and 18 cm. Find the range of the cephalic index for males. Round endpoints to one decimal place.

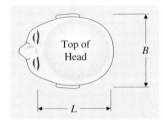

Figure for 73–74

74. *Anthropology.* Refer to Problem 73. The same study found that the average head length of females was 17.4 cm and their head breadths varied between 15 and 20 cm. Find the range of the cephalic index for females. Round endpoints to one decimal place.

Section 1-2 GRAPHS AND LINES

- Cartesian Coordinate System
- Graphs of $Ax + By = C$
- Slope of a Line
- Equations of Lines: Special Forms
- Applications

In this section we will consider one of the most basic geometric figures—a line. When we use the term *line* in this book, we mean *straight line*. We will learn how to recognize and graph a line, and how to use information concerning a line to find its equation. Examining the graph of any equation often results in additional insight into the nature of the equation's solutions.

■ Cartesian Coordinate System

Recall that to form a **Cartesian** or **rectangular coordinate system,** we select two real number lines, one horizontal and one vertical, and let them cross through their origins as indicated in Figure 1. Up and to the right are the usual choices for the positive directions. These two number lines are called the **horizontal axis** and the **vertical axis,** or, together, the **coordinate axes.** The horizontal axis is usually referred to as the **x axis** and the vertical axis as the **y axis,** and each is labeled accordingly. Other labels may be used in certain situations. The coordinate axes divide the plane into four parts called **quadrants,** which are numbered counterclockwise from I to IV (see Fig. 1).

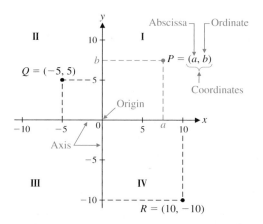

FIGURE 1 The Cartesian (rectangular) coordinate system

Now we want to assign *coordinates* to each point in the plane. Given an arbitrary point P in the plane, pass horizontal and vertical lines through the point (Fig. 1). The vertical line will intersect the horizontal axis at a point with coordinate a, and the horizontal line will intersect the vertical axis at a point with coordinate b. These two numbers written as the ordered pair* (a, b) form the **coordinates** of the point P. The first coordinate, a, is called the **abscissa** of P; the second coordinate, b, is called the **ordinate** of P. The abscissa of Q in Figure 1 is -5, and the ordinate of Q is 5. The coordinates of a point can also be referenced in terms of the axis labels. The **x coordinate** of R in Figure 1 is 10, and the **y coordinate** of R is -10. The point with coordinates $(0, 0)$ is called the **origin.**

The procedure we have just described assigns to each point P in the plane a unique pair of real numbers (a, b). Conversely, if we are given an ordered pair of real numbers (a, b), then, reversing this procedure, we can determine a unique point P in the plane. Thus,

There is a one-to-one correspondence between the points in a plane and the elements in the set of all ordered pairs of real numbers.

This is often referred to as the **fundamental theorem of analytic geometry.**

■ Graphs of Ax + By = C

In Section 1-1, we called an equation of the form $ax + b = 0$ $(a \neq 0)$ a linear equation in one variable. Now we want to consider linear equations in two variables:

---

* An **ordered pair** of real numbers is a pair of numbers in which the order is specified. We now use (a, b) as the coordinates of a point in a plane. In Section 1-1, we used (a, b) to represent an interval on a real number line. These concepts are not the same. You must always interpret the symbol (a, b) in terms of the context in which it is used.

DEFINITION **Linear Equations in Two Variables** A **linear equation in two variables** is an equation that can be written in the **standard form**

$$Ax + By = C$$

where A, B, and C are constants (A and B not both 0) and x and y are variables.

A **solution** of an equation in two variables is an ordered pair of real numbers that satisfy the equation. For example, (4, 3) is a solution of $3x - 2y = 6$. The **solution set** of an equation in two variables is the set of all solutions of the equation. The **graph** of an equation is the graph of its solution set.

Explore & Discuss **1**

(A) As noted earlier, (4, 3) is a solution of the equation

$$3x - 2y = 6$$

Find three more solutions of this equation. Plot these solutions in a Cartesian coordinate systems. What familiar geometric shape could be used to describe the solution set of this equation?

(B) Repeat part (A) for the equation $x = 2$.

(C) Repeat part (A) for the equation $y = -3$.

In Explore–Discuss 1, you should have recognized that the graph of each equation is a (straight) line. Theorem 1 confirms this fact.

THEOREM 1 **GRAPH OF A LINEAR EQUATION IN TWO VARIABLES** The graph of any equation of the form

$$Ax + By = C \quad \text{(A and B not both 0)} \qquad (1)$$

is a line, and any line in a Cartesian coordinate system is the graph of an equation of this form.

If $A \neq 0$ and $B \neq 0$, then equation (1) can be written as

$$y = -\frac{A}{B}x + \frac{C}{B} = mx + b, m = 0$$

a form we will use often. If $A = 0$ and $B \neq 0$, then equation (1) can be written as

$$y = \frac{C}{B}$$

and its graph is a **horizontal line**. If $A \neq 0$ and $B = 0$, then equation (1) can be written as

$$x = -\frac{A}{B}$$

and its graph is a **vertical line**. To graph equation (1), or any of its special cases, plot any two points in the solution set and use a straightedge to draw the line through these two points. The points where the line crosses the axes are often the easiest to find. The y intercept* is the y coordinate of the point where the graph crosses the y axis and the x intercept is the x coordinate of the point where the graph crosses the x axis. To find the y intercept, let $x = 0$ and solve for y. To find the x intercept, let $y = 0$ and solve for x. It is a good idea to find a third point as a check point.

---

* If the x intercept is a and the y intercept is b, then the graph of the line passes through the points $(a, 0)$ and $(0, b)$. It is common practice to refer to both the numbers a and b and the points $(a, 0)$ and $(0, b)$ as the x and y intercepts of the line.

EXAMPLE 1 **Using Intercepts to Graph a Line** Graph: $3x - 4y = 12$

SOLUTION

x	y	
0	−3	y intercept
4	0	x intercept
8	3	Check point

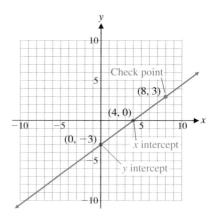

MATCHED PROBLEM 1 Graph: $4x - 3y = 12$

 The icon in the margin is used throughout this book to identify optional graphing calculator activities that are intended to give you additional insight into the concepts under discussion. You may have to consult the manual for your calculator* for the details necessary to carry out these activities.

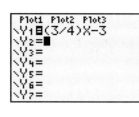 EXAMPLE 2 **Using a Graphing Calculator** Graph $3x - 4y = 12$ on a graphing calculator and find the intercepts.

SOLUTION First, we solve $3x - 4y = 12$ for y.

$$3x - 4y = 12 \qquad \text{Add } -3x \text{ to both sides.}$$
$$-4y = -3x + 12 \qquad \text{Divide both sides by } -4.$$
$$y = \frac{-3x + 12}{-4} \qquad \text{Simplify.}$$
$$y = \frac{3}{4}x - 3 \tag{2}$$

Now we enter the right side of equation (2) in a calculator (Fig. 2A), enter values for the window variables (Fig. 2B), and graph the line (Fig. 2C).

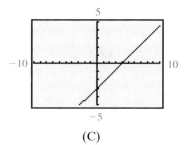

(A)   (B)   (C)

FIGURE 2 Graphing a line on a graphing calculator

Next we use two calculator routines to find the intercepts: TRACE (Fig. 3A) and zero (Fig. 3B).

---

* We used a Texas Instruments graphing calculator from the TI-83/84 family to produce the graphing calculator screens in the book. Manuals for most graphing calculators are readily available on the Internet.

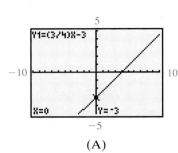   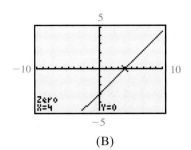

(A)   (B)

FIGURE 3 Using TRACE and zero on a graphing calculator ▬

MATCHED PROBLEM 2 Graph $4x - 3y = 12$ on a graphing calculator and find the intercepts.

EXAMPLE 3   **Horizontal and Vertical Lines**

(A) Graph $x = -4$ and $y = 6$ simultaneously in the same rectangular coordinate system.

(B) Write the equations of the vertical and horizontal lines that pass through the point $(7, -5)$.

SOLUTION (A)

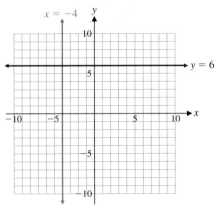

(B) Horizontal line through $(7, -5)$: $y = -5$

Vertical line through $(7, -5)$: $x = 7$ ▬

MATCHED PROBLEM 3   (A) Graph $x = 5$ and $y = -3$ simultaneously in the same rectangular coordinate system.

(B) Write the equations of the vertical and horizontal lines that pass through the point $(-8, 2)$.

■ Slope of a Line

If we take two points $P_1(x_1, y_1)$ and $P_2(x_2, y_2)$, on a line, then the ratio of the change in y to the change in x as the point moves from point P_1 to point P_2 is called the **slope** of the line. In a sense, slope provides a measure of the "steepness" of a line relative to the x axis. The change in x is often called the **run** and the change in y the **rise.**

DEFINITION   **Slope of a Line** If a line passes through two distinct points $P_1(x_1, y_1)$ and $P_2(x_2, y_2)$, then its slope is given by the formula

$$m = \frac{y_2 - y_1}{x_2 - x_1} \qquad x_1 \neq x_2$$

$$= \frac{\text{vertical change (rise)}}{\text{horizontal change (run)}}$$

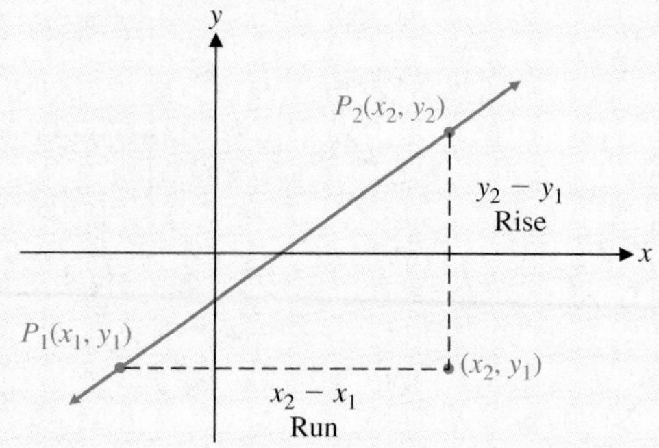

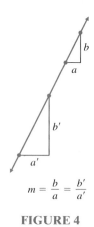

For a horizontal line, y does not change; hence, its slope is 0. For a vertical line, x does not change; hence, $x_1 = x_2$ and its slope is not defined. In general, the slope of a line may be positive, negative, 0, or not defined. Each case is illustrated geometrically in Table 1.

TABLE 1	Geometric Interpretation of Slope			
Line	Rising as x moves from left to right	Falling as x moves from left to right	Horizontal	Vertical
Slope	Positive	Negative	0	Not defined
Example	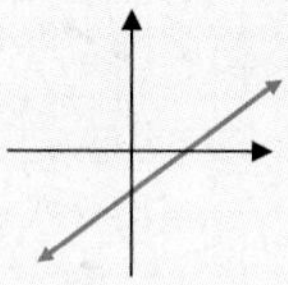	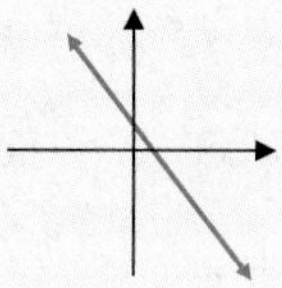	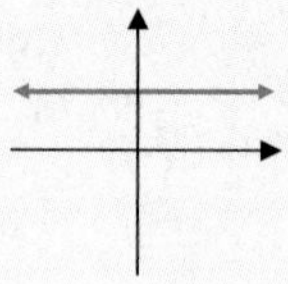	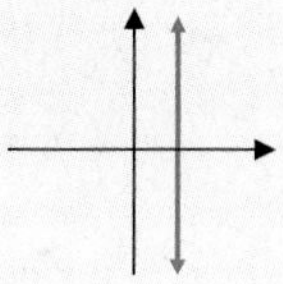

INSIGHT

One property of real numbers discussed in Appendix A, Section A-1, is

$$\frac{-a}{-b} = -\frac{-a}{b} = -\frac{a}{-b} = \frac{a}{b}, \quad b \neq 0$$

This property implies that it does not matter which point we label as P_1 and which we label as P_2 in the slope formula. For example, if $A = (4, 3)$ and $B = (1, 2)$, then

$$B = P_2 = (1, 2) \qquad A = P_2 = (4, 3)$$
$$A = P_1 = (4, 3) \qquad B = P_1 = (1, 2)$$
$$m = \frac{2 - 3}{1 - 4} = \frac{-1}{-3} = \frac{1}{3} = \frac{3 - 2}{4 - 1}$$

A property of similar triangles (see Table II in Appendix C) ensures that the slope of a line is the same for any pair of distinct points on the line (Fig. 4).

$$m = \frac{b}{a} = \frac{b'}{a'}$$

FIGURE 4

EXAMPLE 4 **Finding Slopes** Sketch a line through each pair of points, and find the slope of each line.

(A) $(-3, -2), (3, 4)$ (B) $(-1, 3), (2, -3)$

(C) $(-2, -3), (3, -3)$ (D) $(-2, 4), (-2, -2)$

SOLUTION   (A)

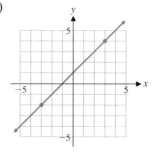

$$m = \frac{4 - (-2)}{3 - (-3)} = \frac{6}{6} = 1$$

(B)

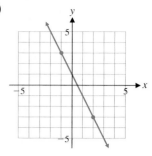

$$m = \frac{-3 - 3}{2 - (-1)} = \frac{-6}{3} = -2$$

(C)

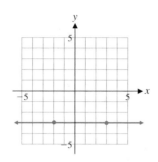

$$m = \frac{-3 - (-3)}{3 - (-2)} = \frac{0}{5} = 0$$

(D)

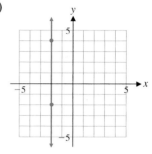

$$m = \frac{-2 - 4}{-2 - (-2)} = \frac{-6}{0}$$

Slope is not defined.   ∎

MATCHED PROBLEM 4   Find the slope of the line through each pair of points.

(A) $(-2, 4), (3, 4)$ (B) $(-2, 4), (0, -4)$

(C) $(-1, 5), (-1, -2)$ (D) $(-1, -2), (2, 1)$

■ Equations of Lines: Special Forms

Let us start by investigating why $y = mx + b$ is called the *slope-intercept form* for a line.

Explore & Discuss **2**

(A) Graph $y = x + b$ for $b = -5, -3, 0, 3,$ and 5 simultaneously in the same coordinate system. Verbally describe the geometric significance of b.

(B) Graph $y = mx - 1$ for $m = -2, -1, 0, 1,$ and 2 simultaneously in the same coordinate system. Verbally describe the geometric significance of m.

(C) Using a graphing utility, explore the graph of $y = mx + b$ for different values of m and b.

As you can see from Explore–Discuss 2, constants m and b in $y = mx + b$ have special geometric significance, which we now explicitly state.

If we let $x = 0$, then $y = b$, and we observe that the graph of $y = mx + b$ crosses the y axis at $(0, b)$. The constant b is the y *intercept*. For example, the y intercept of the graph of $y = -4x - 1$ is -1.

To determine the geometric significance of m, we proceed as follows: If $y = mx + b$, then by setting $x = 0$ and $x = 1$, we conclude that $(0, b)$ and $(1, m + b)$ lie on its graph (Fig. 5). Hence, the slope of this line is given by:

$$\text{Slope} = \frac{y_2 - y_1}{x_2 - x_1} = \frac{(m + b) - b}{1 - 0} = m$$

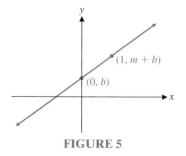

FIGURE 5

Thus, m is the slope of the line given by $y = mx + b$.

DEFINITION | **Slope-Intercept Form**

The equation

$$y = mx + b \quad m = \text{Slope}, b = y \text{ intercept} \qquad (3)$$

is called the **slope-intercept form** of an equation of a line.

EXAMPLE 5 | **Using the Slope-Intercept Form**

(A) Find the slope and y intercept, and graph $y = -\frac{2}{3}x - 3$.

(B) Write the equation of the line with slope $\frac{2}{3}$ and y intercept -2.

SOLUTION | (A) Slope $= m = -\frac{2}{3}$ | (B) $m = \frac{2}{3}$ and $b = -2$;

y intercept $= b = -3$ | thus, $y = \frac{2}{3}x - 2$

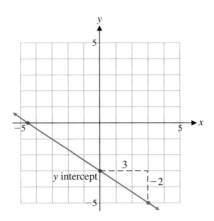

MATCHED PROBLEM 5 | Write the equation of the line with slope $\frac{1}{2}$ and y intercept -1. Graph.

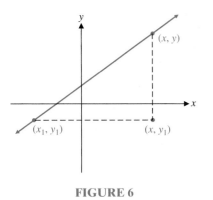

FIGURE 6

Suppose that a line has slope m and passes through a fixed point (x_1, y_1). If the point (x, y) is any other point on the line (Fig. 6), then

$$\frac{y - y_1}{x - x_1} = m$$

That is,

$$y - y_1 = m(x - x_1) \qquad (4)$$

We now observe that (x_1, y_1) also satisfies equation (4) and conclude that equation (4) is an equation of a line with slope m that passes through (x_1, y_1).

DEFINITION | **Point-Slope Form**

An equation of a line with slope m that passes through (x_1, y_1) is

$$y - y_1 = m(x - x_1) \qquad (4)$$

which is called the **point-slope form** of an equation of a line.

The point-slope form is extremely useful, since it enables us to find an equation for a line if we know its slope and the coordinates of a point on the line or if we know the coordinates of two points on the line.

| EXAMPLE 6 | **Using the Point-Slope Form** |

(A) Find an equation for the line that has slope $\frac{1}{2}$ and passes through $(-4, 3)$. Write the final answer in the form $Ax + By = C$.

(B) Find an equation for the line that passes through the two points $(-3, 2)$ and $(-4, 5)$. Write the resulting equation in the form $y = mx + b$.

SOLUTION   (A) Use $y - y_1 = m(x - x_1)$. Let $m = \frac{1}{2}$ and $(x_1, y_1) = (-4, 3)$. Then

$$y - 3 = \tfrac{1}{2}[x - (-4)]$$
$$y - 3 = \tfrac{1}{2}(x + 4) \qquad \text{Multiply by 2.}$$
$$2y - 6 = x + 4$$
$$-x + 2y = 10 \quad \text{or} \quad x - 2y = -10$$

(B) First, find the slope of the line by using the slope formula:

$$m = \frac{y_2 - y_1}{x_2 - x_1} = \frac{5 - 2}{-4 - (-3)} = \frac{3}{-1} = -3$$

Now use $y - y_1 = m(x - x_1)$ with $m = -3$ and $(x_1, y_1) = (-3, 2)$:

$$y - 2 = -3[x - (-3)]$$
$$y - 2 = -3(x + 3)$$
$$y - 2 = -3x - 9$$
$$y = -3x - 7$$

| MATCHED PROBLEM 6 | (A) Find an equation for the line that has slope $\frac{2}{3}$ and passes through $(6, -2)$. Write the resulting equation in the form $Ax + By = C, A > 0$. |

(B) Find an equation for the line that passes through $(2, -3)$ and $(4, 3)$. Write the resulting equation in the form $y = mx + b$.

The various forms of the equation of a line that we have discussed are summarized in Table 2 for convenient reference.

TABLE 2   Equations of a Line

Standard form	$Ax + By = C$	A and B not both 0
Slope-intercept form	$y = mx + b$	Slope: m; y intercept: b
Point-slope form	$y - y_1 = m(x - x_1)$	Slope: m; point: (x_1, y_1)
Horizontal line	$y = b$	Slope: 0
Vertical line	$x = a$	Slope: undefined

APPLICATIONS

We will now see how equations of lines occur in certain applications.

| EXAMPLE 7 | **Cost Equation**   The management of a company that manufactures roller skates has fixed costs (costs at 0 output) of $300 per day and total costs of $4,300 per day at an output of 100 pairs of skates per day. Assume that cost C is linearly related to output x. |

(A) Find the slope of the line joining the points associated with outputs of 0 and 100; that is, the line passing through $(0, 300)$ and $(100, 4,300)$.

(B) Find an equation of the line relating output to cost. Write the final answer in the form $C = mx + b$.

(C) Graph the cost equation from part (B) for $0 \le x \le 200$.

SOLUTION (A) $m = \dfrac{y_2 - y_1}{x_2 - x_1}$

$$= \dfrac{4{,}300 - 300}{100 - 0}$$

$$= \dfrac{4{,}000}{100} = 40$$

(B) We must find an equation of the line that passes through $(0, 300)$ with slope 40. We use the slope-intercept form:

$$C = mx + b$$
$$C = 40x + 300$$

(C)

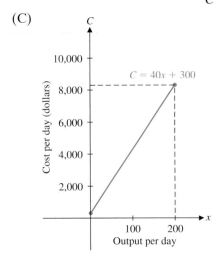

In Example 7, the *fixed cost* of $300 per day covers plant cost, insurance, and so on. This cost is incurred whether or not there is any production. The *variable cost* is $40x$, which depends on the day's output. Since increasing production from x to $x + 1$ will increase the cost by $40 (from $40x + 300$ to $40x + 340$), the slope 40 can be interpreted as the **rate of change** of the cost function with respect to production x.

MATCHED PROBLEM 7 Answer parts (A) and (B) in Example 7 for fixed costs of $250 per day and total costs of $3,450 per day at an output of 80 pairs of skates per day.

EXAMPLE 8 **Supply and Demand** Table 3 lists the supply and demand for wheat in the United States during two recent years. Assume that the relationship between supply and price is linear and the relationship between demand and price is also linear.

(A) Find a linear supply equation of the form $p = mx + b$, where p is the price per bushel in dollars and x is the corresponding supply in billions of bushels.

(B) Find a linear demand equation of the form $p = mx + b$, where p is the price per bushel in dollars and x is the corresponding demand in billions of bushels.

(C) Graph the supply and demand equations in the same coordinate system and find their point of intersection.

TABLE 3 U.S. Wheat Supply and Demand

Year	Supply (mil bu)	Demand (mil bu)	Price ($/bu)
1997	2,480	2,300	3.38
1999	2,300	2,390	2.48

Source: The ProExporter Network®

SOLUTION (A) To find a supply equation in the form $p = mx + b$, we must first find two points of the form (x, p) that supply line passes through. From Table 3, $(2,480, 3.38)$ and $(2,300, 2.48)$ are two such points. First, find the slope of the line:

$$m = \frac{3.38 - 2.48}{2,480 - 2,300} = \frac{0.9}{180} = 0.005$$

Now, use the point-slope form to find the equation of the line:

$$p - p_1 = m(x - x_1) \qquad (x_1, p_1) = (2,300, 2.48)$$
$$p - 2.48 = 0.005(x - 2,300)$$
$$p - 2.48 = 0.005x - 11.5$$
$$p = 0.005x - 9.02 \qquad \textit{Price-supply equation}$$

Check:

We check the price–supply equation by verifying that each of the given points satisfies the equation:

$(2,480, 3.38)$ $(2,300, 2.48)$

$3.38 \overset{?}{=} 0.005(2,480) - 9.02$ $2.48 \overset{?}{=} 0.005(2,300) - 9.02$

$3.38 \overset{\checkmark}{=} 3.38$ $2.48 \overset{\checkmark}{=} 2.48$

(B) From Table 3, $(2,300, 3.38)$ and $(2,390, 2.48)$ are two points on the linear demand equation.

$$m = \frac{3.38 - 2.48}{2,300 - 2,390} = \frac{0.9}{-90} = -0.01$$

$$p - p_1 = m(x - x_1) \qquad (x_1, p_1) = (2,390, 2.48)$$
$$p - 2.48 = -0.01(x - 2,390)$$
$$p - 2.48 = -0.01x + 23.9$$
$$p = -0.01x + 26.38 \qquad \textit{Price–demand equation}$$

Check:

$(2,300, 3.38)$ $(2,390, 2.48)$

$3.38 \overset{?}{=} -0.01(2,300) + 26.38$ $2.48 \overset{?}{=} -0.01(2,390) + 26.38$

$3.38 \overset{\checkmark}{=} 3.38$ $2.48 \overset{\checkmark}{=} 2.48$

(C) From part (A), we plot the points $(2,480, 3.38)$ and $(2,300, 2.48)$ and then draw a line through them. We do the same with the points $(2,300, 3.38)$ and $(2,390, 2.48)$ from part (B) (Fig. 7). (Note that we restricted the axes to intervals that contain these data points—a common practice in applications.)

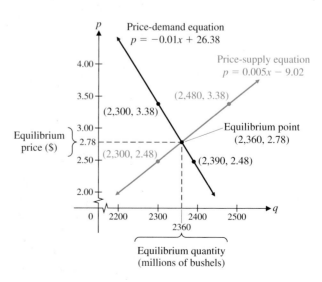

FIGURE 7 Graphs of supply and demand equations

To find the intersection point of these lines, we first find x by solving

Price–Supply	Price–Demand

$$0.005x - 9.02 = -0.01x + 26.38$$
$$0.015x = 35.4$$
$$x = 2{,}360 \quad \text{million bushels}$$

Then we use the price–supply equation to find p when $x = 2{,}360$:

$$p = 0.005x - 9.02$$
$$p = 0.005(2{,}360) - 9.02 = \$2.78$$

As a check, we use the price–demand equation to find p when $x = 2{,}360$:

$$p = -0.01x + 26.38$$
$$= -0.01(2{,}360) + 26.38 = \$2.78$$

The lines intersect at $(2{,}360, 2.87)$. See Figure 7.

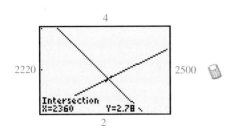

FIGURE 8
Finding an intersection point

In Figure 8, we use INTERSECT on a graphing calculator to find the intersection point.

In a free competitive market, the price of a product is determined by the relationship between supply and demand. The price tends to stabilize at the point of intersection of the demand and supply equations. This point is called the **equilibrium point**, the corresponding price is called the **equilibrium price**, and the common value of supply and demand is called the **equilibrium quantity**. All of these are illustrated in Figure 7. ▬

MATCHED PROBLEM 8   Use the data in Table 4 to find

(A) A linear supply equation of the form $p = mx + b$.
(B) A linear demand equation of the form $p = mx + b$.
(C) The equilibrium point.

TABLE 4 U.S. Wheat Supply and Demand

Year	Supply (mil bu)	Demand (mil bu)	Price ($/bu)
1991	1980	2,420	3.00
1995	2,180	2,380	4.60

Source: The ProExporter Network®

Answers to Matched Problems   **1.**

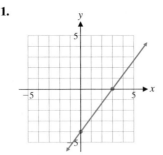

2. y intercept $= -4$, x intercept $= 3$

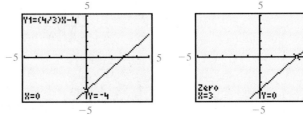

3. (A)

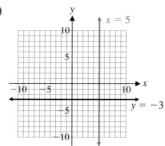

(B) Horizontal line: $y = 2$; vertical line: $= -8$

4. (A) 0   (B) -4 (C) Not defined   (D) 1

5. $y = \frac{1}{2}x - 1$

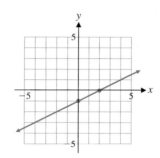

6. (A) $2x - 3y = 18$

  (B) $y = 3x - 9$

7. (A) $m = 40$

  (B) $C = 40x + 250$

8. (A) $p = 0.008x - 12.84$

  (B) $p = -0.04x + 99.8$

  (C) $(2,347, 5.93)$

Exercise 1-2

A *Problems 1–4 refer to graphs (A)–(D).*

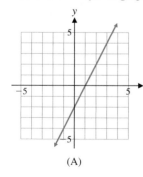

(A)

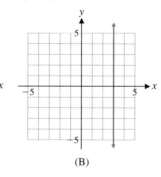

(B)

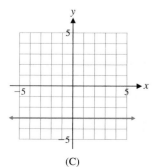

(C)

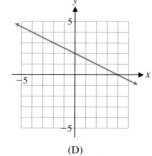

(D)

1. Identify the graph(s) of lines with a negative slope.

2. Identify the graph(s) of lines with a positive slope.

3. Identify the graph(s) of any lines with slope zero.

4. Identify the graph(s) of any lines with undefined slope. What can you say about their slopes?

In Problems 5–8, sketch a graph of each equation in a rectangular coordinate system.

5. $y = 2x - 3$ **6.** $y = \frac{x}{2} + 1$

7. $2x + 3y = 12$ **8.** $8x - 37 = 24$

In Problems 9–12, find the slope and y intercept of the graph of each equation.

9. $y = 3x + 1$ **10.** $y = \frac{x}{5} - 2$

11. $y = -\frac{3}{7}x - 6$ **12.** $y = 0.7x + 5$

In Problems 13–16, write an equation of the line with the indicated slope and y intercept.

13. Slope $= -2$ **14.** Slope $= \frac{3}{4}$
  y intercept $= 3$ y intercept $= -5$

15. Slope $= \frac{4}{3}$ **16.** Slope $= -5$
  y intercept $= -4$ y intercept $= 9$

In Problems 17–20, use the graph of each line to find the x intercept, y intercept, and slope. Write the slope-intercept form of the equation of the line.

17.

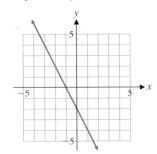

`18.`

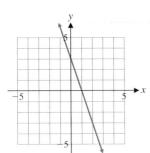

19.

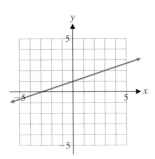

20.

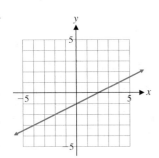

B *Sketch a graph of each equation or pair of equations in Problems 21–26 in a rectangular coordinate system.*

21. $y = -\frac{2}{3}x - 2$ **22.** $y = -\frac{3}{2}x + 1$

23. $3x - 2y = 10$ **24.** $5x - 6y = 15$

25. $x = 3; y = -2$ **26.** $x = -3; y = 2$

In Problems 27–30, find the slope of the graph of each equation.

27. $4x + y = 3$ **28.** $5x - y = -2$

29. $3x + 5y = 15$ **30.** $2x - 3y = 18$

31. Given $Ax + By = 12$, graph each of the following three cases in the same coordinate system.

(A) $A = 2$ and $B = 0$

(B) $A = 0$ and $B = 3$

(C) $A = 3$ and $B = 4$

32. Given $Ax + By = 24$, graph each of the following three cases in the same coordinate system.

(A) $A = 6$ and $B = 0$

(B) $A = 0$ and $B = 8$

(C) $A = 2$ and $B = 3$

33. Graph $y = 25x + 200$, $x \geq 0$.

34. Graph $y = 40x + 160$, $x \geq 0$.

35. (A) Graph $y = 1.2x - 4.2$ in a rectangular coordinate system.

(B) Find the x and y intercepts algebraically to one decimal place.

(C) Graph $y = 1.2x - 4.2$ in a graphing utility.

(D) Find the x and y intercepts to one decimal place using TRACE and the zero command on your graphing utility.

36. (A) Graph $y = -0.8x + 5.2$ in a rectangular coordinate system.

(B) Find the x and y intercepts algebraically to one decimal place.

(C) Graph $y = -0.8x + 5.2$ in a graphing utility.

(D) Find the x and y intercepts to one decimal place using TRACE and the zero command on your graphing utility.

(E) Using the results of parts (A) and (B) or (C) and (D), find the solution set for the linear inequality

$$-0.8x + 5.2 < 0$$

In Problems 37–40, write the equations of the vertical and horizontal lines through each point.

37. $(4, -3)$ **38.** $(-5, 6)$

39. $(-1.5, -3.5)$ **40.** $(2.6, 3.8)$

In Problems 41–46, write the equation of the line through each indicated point with the indicated slope. Write the final answer in the form $y = mx + b$.

41. $m = -4; (2, -3)$ **42.** $m = -6; (-4, 1)$

43. $m = \frac{3}{2}; (-4, -5)$ **44.** $m = \frac{4}{3}; (-6, 2)$

45. $m = 0; (-1.5, 4.6)$ **46.** $m = 0; (3.1, -2.7)$

In Problems 47–54,

(A) *Find the slope of the line that passes through the given points.*

(B) *Find the standard form of the equation of the line.*

47. $(2, 5)$ and $(5, 7)$ **48.** $(1, 2)$ and $(3, 5)$

49. $(-2, -1)$ and $(2, -6)$ **50.** $(2, 3)$ and $(-3, 7)$

51. $(5, 3)$ and $(5, -3)$ **52.** $(1, 4)$ and $(0, 4)$

53. $(-2, 5)$ and $(3, 5)$ **54.** $(2, 0)$ and $(2, -3)$

55. Discuss the relationship among the graphs of the lines with equation $y = mx + 2$, where m is any real number.

56. Discuss the relationship among the graphs of the lines with equation $y = -0.5x + b$, where b is any real number.

Applications

57. *Simple interest.* If $P (the principal) is invested at an interest rate of r, then the amount A that is due after t years is given by

$$A = Prt + P$$

If $100 is invested at 6% ($r = 0.06$), then

$$A = 6t + 100, t \geq 0$$

(A) What will $100 amount to after 5 years? After 20 years?

(B) Sketch a graph of $A = 6t + 100$ for $0 \leq t \leq 20$.

(C) Find the slope of the graph and interpret verbally.

58. *Simple interest.* Use the simple interest formula from Problem 57. If $1,000 is invested at 7.5% ($r = 0.075$), then $A = 75t + 1,000, t \geq 0$.

(A) What will $1,000 amount to after 5 years? After 20 years?

(B) Sketch a graph of $A = 75t + 1,000$ for $0 \leq t \leq 20$.

(C) Find the slope of the graph and interpret verbally.

59. *Cost analysis.* A doughnut shop has a fixed cost of $124 per day and a variable cost of $0.12 per doughnut. Find the total daily cost of producing x doughnuts. How many doughnuts can be produced for a total daily cost of $250?

60. *Cost analysis.* A small company manufactures picnic tables. The weekly fixed cost is $1,200 and the variable cost is $45 per table. Find the total daily cost of producing x picnic tables. How many picnic tables can be produced for a total weekly cost of $4,800?

61. *Cost analysis.* A plant can manufacture 80 golf clubs per day for a total daily cost of $7,647 and 100 golf clubs per day for a total daily cost of $9,147.

(A) Assuming that daily cost and production are linearly related, find the total daily cost of producing x golf clubs.

(B) Graph the total daily cost for $0 \leq x \leq 200$.

(C) Interpret the slope and y intercept of this cost equation.

62. *Cost analysis.* A plant can manufacture 50 tennis rackets per day for a total daily cost of $3,855 and 60 tennis rackets per day for a total daily cost of $4,245.

(A) Assuming that daily cost and production are linearly related, find the total daily cost of producing x tennis rackets.

(B) Graph the total daily cost for $0 \leq x \leq 100$.

(C) Interpret the slope and y intercept of this cost equation.

63. *Business—Markup policy.* A drug store sells a drug costing $85 for $112 and a drug costing $175 for $238.

(A) If the markup policy of the drug store is assumed to be linear, write an equation that expresses retail price R in terms of cost C (wholesale price).

(B) What does a store pay (to the nearest dollar) for a drug that retails for $185?

64. *Business—Markup policy.* A clothing store sells a shirt costing $20 for $33 and a jacket costing $60 for $93.

(A) If the markup policy of the store is assumed to be linear, write an equation that expresses retail price R in terms of cost C (wholesale price).

(B) What does a store pay for a suit that retails for $240?

65. *Business—Depreciation.* A farmer buys a new tractor for $157,000 and assumes that it will have a trade-in value of $82,000 after 10 years. The farmer uses a constant rate of depreciation (commonly called **straight-line depreciation**—one of several methods permitted by the IRS) to determine the annual value of the tractor.

(A) Find a linear model for the depreciated value V of the tractor t years after it was purchased.

(B) What is the depreciated value of the tractor after 6 years?

(C) When will the depreciated value fall below $70,000?

(D) Graph V for $0 \leq t \leq 20$ and illustrate the answers from parts (A) and (B) on the graph.

66. *Business—Depreciation.* A charter fishing company buys a new boat for $224,000 and assumes that it will have a trade in value of $115,200 after 16 years.

(A) Find a linear model for the depreciated value V of the boat t years after it was purchased.

(B) What is the depreciated value of the tractor after 10 years?

(C) When will the depreciated value fall below $100,000?

(D) Graph V for $0 \leq t \leq 30$ and illustrate the answers from (A) and (B) on the graph.

67. *Boiling point.* The temperature at which water starts to boil is called its **boiling point** and is linearly related to the altitude. Water boils at 212°F at sea level and at 193.6°F at an altitude of 10,000. (*Source:* `biggreenegg.com`)

(A) Find a relationship of the form $T = mx + b$ where T is degrees Fahrenheit and x is altitude in thousands of feet.

(B) Find the boiling point at an altitude at of 3,500.

(C) Find the altitude if the boiling point is 200°F.

(D) Graph T and illustrate the answers to (B) and (C) on the graph.

68. *Boiling point.* The temperature at which water starts to boil is also linearly related to barometric pressure. Water boils at 212°F at a pressure of 29.9 inHg (inches of mercury) and at 191°F at a pressure of 28.4 inHg. (*Source:* `biggreenegg.com`)

(A) Find a relationship of the form $T = mx + b$, where T is degrees Fahrenheit and x is pressure in inches of mercury.

(B) Find the boiling point at a pressure of 31 inHg.

(C) Find the pressure if the boiling point is 199°F.

(D) Graph T and illustrate the answers to (B) and (C) on the graph.

69. *Flight conditions.* In stable air, the air temperature drops about 3.6°F for each 1,000-foot rise in altitude. (*Source:* Federal Aviation Administration)

(A) If the temperature at sea level is 70°F, write a linear equation that expresses temperature T in terms of altitude A in thousands of feet.

(B) At what altitude is the temperature 34°F?

70. *Flight navigation.* The airspeed indicator on some aircraft is affected by the changes in atmospheric pressure at different altitudes. A pilot can estimate the true airspeed by observing the indicated airspeed and adding to it about 1.6% for every 1,000 feet of altitude. (*Source:* Megginson Technologies Ltd.)

(A) If a pilot maintains a constant reading of 200 miles per hour on the airspeed indicator as the aircraft climbs from sea level to an altitude of 10,000 feet, write a linear equation that expresses true airspeed T (in miles per hour) in terms of altitude A (in thousands of feet).

(B) What would be the true airspeed of the aircraft at 6,500 feet?

71. *Demographics.* The average number of persons per household in the United States has been shrinking steadily for as long as statistics have been kept and is approximately linear with respect to time. In 1980, there were about 2.8 persons per household and in 2005, about 2.6. (*Source:* U.S. Census Bureau)

(A) If N represents the average number of persons per household and t represents the number of years since 1980, write a linear equation that expresses N in terms of t.

(B) Use this equation to estimate household size in the year 2030.

72. *Demographics.* The **median** household income divides the households into two groups, the half whose income is less than or equal to the median and the half whose income is greater than the median. The median household income in the United States grew from about $34,000 in 1980 to about $45,000 in 2005. (*Source*: U.S. Census Bureau)

(A) If I represents the median household income and t represents the number of years since 1980, write a linear equation that expresses I in terms of t.

(B) Use this equation to estimate median household income in the year 2030.

73. *Cigarette smoking.* The percentage of female cigarette smokers in the United States declined from 21% in 2000 to 18.5% in 2004. (*Source*: Centers for Disease Control)

(A) Find a linear equation relating percentage of female smokers (f) to years since 2000 (t).

(B) Find the year in which the percentage of female smokers falls below 15%.

(C) Graph the equation for $0 \le t \le 15$ and illustrate the answer to part (B) on the graph.

74. *Cigarette smoking.* The percentage of male cigarette smokers in the United States declined from 25.7% in 2000 to 23.4% in 2004. (*Source:* Centers for Disease Control)

(A) Find a linear equation relating percentage of male smokers (m) to years since 2000 (t).

(B) Find the year in which the percentage of male smokers falls below 20%.

(C) Graph the equation for $0 \le t \le 15$ and illustrate the answer to part (B) on the graph.

75. *Supply and demand.* Use the barley market data in Table 5 to find

(A) A linear supply equation of the form $p = mx + b$.

(B) A linear demand equation of the form $p = mx + b$.

(C) The equilibrium point.

(D) Graph the supply equation, demand equation, and equilibrium point in the same coordinate system.

TABLE 5	U.S. Barley Supply and Demand		
Year	Supply (mil bu)	Demand (mil bu)	Price ($/bu)
1990	7,500	7,900	2.28
1991	7,900	7,800	2.37

Source: U.S. Department of Agriculture

76. *Supply and demand.* Use the corn market data in Table 6 to find

(A) A linear supply equation of the form $p = mx + b$.

(B) A linear demand equation of the form $p = mx + b$.

(C) The equilibrium point.

(D) Graph the supply equation, demand equation, and equilibrium point in the same coordinate system.

TABLE 6	U.S. Corn Supply and Demand		
Year	Supply (mil bu)	Demand (mil bu)	Price ($/bu)
1998	9,800	9,300	1.94
1999	9,400	9,500	1.82

Source: The ProExporter Network®

77. *Physics.* Hooke's law states that the relationship between the stretch s of a spring and the weight w causing the stretch is linear (a principle upon which all spring scales are constructed). For a particular spring, a 5-pound weight causes a stretch of 2 inches, while with no weight the stretch of the spring is 0.

(A) Find a linear equation that expresses s in terms of w.

(B) What is the stretch for a weight of 20 pounds?

(C) What weight will cause a stretch of 3.6 inches?

78. *Physics.* The distance d between a fixed spring and the floor is a linear function of the weight w attached to the bottom of the spring. The bottom of the spring is 18 inches from the floor when the weight is 3 pounds and 10 inches from the floor when the weight is 5 pounds.

(A) Find a linear equation that expresses d in terms of w.

(B) Find the distance from the bottom of the spring to the floor if no weight is attached.

(C) Find the smallest weight that will make the bottom of the spring touch the floor. (Ignore the height of the weight.)

79. *Energy consumption.* Table 7 lists U.S. oil imports as a percentage* of total energy consumption for selected years.

TABLE 7	
Year	Oil Imports (%)
1960	9
1970	12
1980	21
1990	24
2000	29

Source: Energy Information Administration

---

* We use percentage because energy is measured in quadrillion BTUs (British Thermal Units) and one quadrillion is 1,000,000,000,000,000. Percentages are easier to comprehend.

Let x represent years since 1960 and y represent the corresponding percentage of oil imports.

(A) Find the equation of the line through (0, 9) and (40, 29), the first and last data points in Table 7.

(B) Find the equation of the line through (0, 9) and (10, 12), the first and second data points in Table 7.

(C) Graph the lines from parts (A) and (B) and the data points (x, y) from Table 7 in the same coordinate system.

(D) Use each equation to predict oil imports as a percentage of total consumption in 2020.

(E) Which of the two lines seems to better represent the data in Table 7? Discuss.

80. *Energy production.* Table 8 lists U.S. crude oil production as a percentage of total U.S. energy production for selected years. Let x represent years since 1960 and y represent the corresponding percentage of oil production.

(A) Find the equation of the line through (0, 35) and (40, 17), the first and last data points in Table 8.

(B) Find the equation of the line through (0, 35) and (10, 32), the first and second data points in Table 8.

(C) Graph the lines from parts (A) and (B) and the data points (x, y) from Table 8 in the same coordinate system.

(D) Use each equation to predict oil imports as a percentage of total consumption in 2020.

(E) Which of the two lines seems to better represent the data in Table 8? Discuss.

TABLE 8	
Year	**Oil Production (%)**
1960	35
1970	32
1980	27
1990	22
2000	17

Source: Energy Information Administration

Section 1-3 LINEAR REGRESSION

- Slope as a Rate of Change
- Linear Regression

Mathematical modeling is the process of using mathematics to solve real-world problems. This process can be broken down into three steps (Fig. 1):

Step 1. *Construct* the **mathematical model,** a problem whose solution will provide information about the real-world problem.

Step 2. *Solve* the mathematical model.

Step 3. *Interpret* the solution to the mathematical model in terms of the original real-world problem.

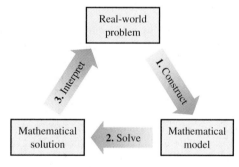

FIGURE 1

In more complex problems, this cycle may have to be repeated several times to obtain the required information about the real-world problem. In this section we will discuss one of the simplest mathematical models, a linear equation. With the aid of a graphing calculator or computer, we also will learn how to analyze a linear model based on real-word data.

Slope as a Rate of Change

If x and y are related by the equation $y = mx + b$, where m and b are constants with $m \neq 0$, then x and y are **linearly related**. If (x_1, y_1) and (x_2, y_2) are two distinct points on this line, then the slope of the line is

$$m = \frac{y_2 - y_1}{x_2 - x_1} = \frac{\text{Change in } y}{\text{Change in } x} \tag{1}$$

In applications, ratio (1) is called the **rate of change** of y with respect to x. Since the slope of a line is unique, **the rate of change of two linearly related variables is constant.** Here are some examples of familiar rates of change: miles per hour, revolutions per minute, price per pound, passengers per plane, and so on. If the relationship between x and y is not linear, ratio (1) is called the **average rate of change** of y with respect to x.

EXAMPLE 1

Estimating Body Surface Area Appropriate doses of medicine for both animals and humans are often based on body surface area (BSA). Since weight is much easier to determine than BSA, veterinarians use the weight of an animal to estimate BSA. The following linear equation expresses BSA for canines in terms of weight:*

$$a = 16.12w + 375.6$$

where a is BSA in square inches and w is weight in pounds.

(A) Interpret the slope of the BSA equation.
(B) What is the effect of a one pound increase in weight?

SOLUTION

(A) The rate of change BSA with respect to weight is 16.12 square inches per pound.
(B) Since slope is the ratio of rise to run, increasing w by 1 pound (run) increases a by 16.12 square inches (rise). ▬

MATCHED PROBLEM 1

The equation $a = 28.55w + 118.7$ expresses BSA for felines in terms of weight, where a is BSA in square inches and w is weight in pounds.

(A) Interpret the slope of the BSA equation.
(B) What is the effect of a one pound increase in weight?

Explore & Discuss 1

As illustrated in Example 1A, the slope m of a line with equation $y = mx + b$ has two intereptations:

1. m is the rate of change of y with respect to x.
2. Increasing x by one unit will change y by m units.

How are these two intereptations related?

Parachutes are used to deliver cargo to areas that cannot be reached by other means of conveyance. The **rate of descent** of the cargo is the rate of change of altitude with respect to time. The absolute value of the rate of descent is called the **speed** of the cargo. At low altitudes, the altitude of the cargo and the time in the air are linearly related. The appropriate rate of descent varies widely with the item. Bulk food (rice, flour, beans, etc.) and clothing can tolerate nearly any rate of descent under 40 ft/sec. Machinery and electronics (pumps, generators, radios, etc.) should generally be dropped at 15 ft/sec or less. Butler Tactical Parachute Systems, Roanoke, Virginia, manufactures a variety of canopies for dropping cargo. The following example uses information taken from one of the company's brochures.

EXAMPLE 2

Finding the Rate of Descent A 100-pound cargo of delicate electronic equipment is dropped from an altitude of 2,880 feet and lands 200 seconds later.

(A) Find a linear model relating altitude a (in feet) and time in the air t (in seconds).
(B) How fast is the cargo moving when it lands?

---

*Based on data from Veterinary Oncology Consultants, PTY LTD.

SOLUTION (A) If $a = mt + b$ is the linear equation relating altitude a and time in air t, then the graph of this equation must pass through the following points:

$$(t_1, a_1) = (0, 2{,}880) \quad \text{Cargo is dropped from plane.}$$
$$(t_2, a_2) = (200, 0) \quad \text{Cargo lands.}$$

The slope of this line is

$$m = \frac{a_2 - a_1}{t_2 - t_1} = \frac{0 - 2{,}880}{200 - 0} = -14.4$$

and the equation of this line is

$$a - 0 = -14.4(t - 200)$$
$$a = -14.4t + 2{,}880$$

(B) The rate of descent is the slope $m = -14.4$, so the speed of the cargo at landing is $|-14.4| = 14.4$ ft/sec. ▄

MATCHED PROBLEM 2   A 400-pound load of grain is dropped from an altitude of 2,880 feet and lands 80 seconds later.

(A) Find a linear model relating altitude a (in feet) and time in the air t (in seconds).

(B) How fast is the cargo moving when it lands?

■ Linear Regression

In real-world applications we often encounter numerical data in the form of a table. The very powerful mathematical tool *regression analysis*, can be used to analyze numerical data. In general, **regression analysis** is a process for finding a function that provides a useful model for a set of data points. Graphs of equations are often called **curves** and regression analysis is also referred to as **curve fitting**. In the next example, we use a linear model obtained by using **linear regression** on a graphing calculator.

EXAMPLE 3   **Diamond Prices** Prices for round-shaped diamonds taken from an online trader are given in Table 1.

TABLE 1 Round–Shaped Diamond Prices	
Weight (carats)	**Price**
0.5	$2,790
0.6	$3,191
0.7	$3,694
0.8	$4,154
0.9	$5,018
1.0	$5,898

Source: www.tradeshop.com

(A) A linear model for the data in Table 1 is given by

$$p = 6{,}140c - 480 \tag{2}$$

where p is the price of a diamond weighing c carats. (We will discuss the source of models like this later in this section.) Plot the points in Table 1 on a Cartesian coordinate system, producing a *scatter plot*, and graph the model on the same axes.

(B) Interpret the slope of the model in (2).

(C) Use the model to estimate the cost of a 0.85-carat diamond and the cost of a 1.2-carat diamond. Round answers to the nearest dollar.

(D) Use the model to estimate the weight of a diamond (to two decimal places) that sells for $4,000.

SOLUTION (A) A **scatter plot** is simply a graph of the points in Table 1 (Fig. 2A). To add the graph of the model to the scatter plot, we find any two points that satisfy equation (2) [we choose $(0.4, 1,976)$ and $(1.1, 6,274)$]. Plotting these points and drawing a line through them gives us Figure 2B.

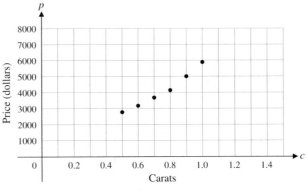

(A) Scatter plot

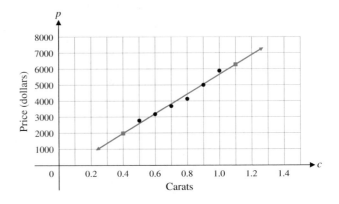

(B) Linear model

FIGURE 2

(B) The rate of change of the price of a diamond with respect to its weight is 6,140. Increasing the weight by one carat will increase the price by about $6,140.

(C) The graph of the model (Fig. 2B) does not pass through any of the points in the scatter plot, but it comes close to all of them. [Verify this by evaluating equation (2) at $c = 0.5, 0.6, \ldots, 1.$] So we can use equation (2) to approximate points not in Table 1.

$$c = 0.85 \qquad\qquad\qquad c = 1.2$$
$$p \approx 6,140(0.85) - 480 \qquad p \approx 6,140(1.2) - 480$$
$$= \$4,739 \qquad\qquad\qquad = \$6,888$$

A 0.85-carat diamond will cost about $4,739 and a 1.2-carat diamond will cost about $6,888.

(D) To find the weight of a $4,000 diamond, we solve the following equation for c:

$$6{,}140c - 480 = 4{,}000 \qquad \textit{Add 480 to both sides.}$$
$$6{,}140c = 4{,}480 \qquad \textit{Divide both sides by 6,140.}$$
$$c = \frac{4{,}480}{6{,}140} \approx 0.73 \qquad \textit{Rounded to two decimal places.}$$

A $4,000 diamond will weigh about 0.73 carats.

MATCHED PROBLEM 3   Prices for emerald-shaped diamonds taken from an online trader are given in Table 2. Repeat Example 3 for this data with the linear model

$$p = 5{,}600c - 1{,}100$$

where p is the price of an emerald-shaped diamond weighing c carats.

TABLE 2	Emerald-Shaped Diamond Prices
Weight (carats)	**Price**
0.5	$1,677
0.6	$2,353
0.7	$2,718
0.8	$3,218
0.9	$3,982
1.0	$4,510

Source: **www.tradeshop.com**

The model we used in Example 3 was obtained by using a technique called **linear regression** and the model is called the **regression line**. This technique produces a line that is the **best fit** for a given data set. We will not discuss the theory behind this technique, nor the meaning of *best fit*. Although you can find a linear regression line by hand, we prefer to leave the calculations to a graphing calculator or a computer. Don't be concerned if you don't have either of these electronic devices. We will supply the regression model in most of the applications we discuss, as we did in Example 3.

Explore & Discuss 2

We used linear regression to produce the model in Example 3. If you have a graphing calculator that supports linear regression, then you can find this model. The linear regression process varies greatly from one calculator to another. Consult the user's manual* for the details of linear regression. The screens in Figure 3 are related to the construction of the model in Example 3 on a Texas Instruments TI-84 Plus.

(A) Produce similar screens on your graphing calculator.

(B) Do the same for Matched Problem 3.

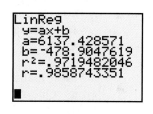 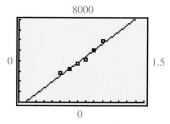

(A) Entering the data   (B) Finding the model   (C) Graphing the data and the model

FIGURE 3 Linear regression on a graphing calculator

---

* Manuals for most graphing calculators are readily available on the Internet.

In Example 3, we used the regression model to approximate points that were not given in Table 1 but would fit between points in the table. This process is called **interpolation**. In the next example, we use a regression model to approximate points outside the given data set. This process is called **extrapolation**, and the approximations are often referred to as **predictions**.

| EXAMPLE 4 | **Carbon Monoxide Emissions** Table 3 contains information about carbon monoxide emissions in recent years. The transportation emissions include emissions from all trains, planes, automobiles, and any other types of transportation. The linear regression model for the transportation emissions (after rounding) is |

$$C = 152 - 4t$$

where C is the carbon monoxide emissions (in millions of short tons*) and t is time in years with $t = 0$ corresponding to 1980.

(A) Interpret the slope of the regression line as a rate of change.

(B) Use the regression model to predict the emissions from the transportation sector in 2010.

TABLE 3 U.S. Carbon Monoxide Emissions (millions of short tons)

	1980	1985	1990	1995	2000
Transportation	148	138	114	88	72
Nontransportation	38	39	40	39	42

Source: U.S. Bureau of Transportation Statistics

SOLUTION
(A) The slope $m = -4$ is the rate of change of emissions with respect to time. Since the slope is negative and the emissions are given in millions of tons, the emissions are decreasing at a rate of $4(1,000,000) = 4,000,000$ tons per year.

(B) If $t = 30$, then

$$C = 152 - 4(30) = 32 \qquad \text{or} \qquad 32,000,000 \text{ tons}$$

So approximately 32,000,00 tons of carbon monoxide will be emitted in 2010.

| MATCHED PROBLEM 4 | Repeat Example 4 using the linear regression model

$$C = 0.16t + 38$$

where C is the carbon monoxide emissions that are not caused by transportation (in millions of short tons) and time in years with $t = 0$ corresponding to 1980.

Explore & Discuss **3**

Use the Internet or a library to expand the data in Table 3 to include years after 2004. Compare the actual values for the years you find to the model's predicted values. Discuss the accuracy of predictions given by the regression model as time goes by.

Forest managers must estimate things like growth, volume, yield, and forest potential. One common measure is the **diameter of a tree at breast height** (Dbh), which is defined as the diameter of the tree at a point 4.5 feet above the ground on the uphill side of the tree. Example 5 is concerned with using Dbh to estimate the height of balsam fir trees.

---

* A short ton (or a U.S. ton) is 2,000 pounds, a long ton (or a U.K. ton) is 2,240 pounds, and a metric tonne is 1,000 kilograms or 2204.6 pounds.

EXAMPLE 5   **Forestry**   A linear regression model for the height of balsam fir trees is

$$h = 3.8d + 18.73$$

where d is Dbh in inches and h is the height in feet.

(A) Interpret the slope of this model.

(B) What is the effect of a 1-inch increase in Dbh?

(C) Estimate the height of a balsam fir with a Dbh of 8 inches. Round your answer to the nearest foot.

(D) Estimate the Dbh of a balsam fir that is 30 feet tall. Round your answer to the nearest inch.

SOLUTION   (A) The rate of change of height with respect to breast height diameter is 3.8 feet per inch.

(B) Height increases by 3.8 feet.

(C) We must find h when $d = 8$:

$$h = 3.8d + 18.73 \qquad \text{Substitute } d = 8.$$
$$h = 3.8(8) + 18.73 \qquad \text{Evaluate.}$$
$$h = 49.13 \approx 49 \text{ ft}$$

(D) We must find d when $h = 30$:

$$h = 3.8d + 18.73 \qquad \text{Substitute } h = 30.$$
$$30 = 3.8d + 18.73 \qquad \text{Subtract 18.73 from both sides.}$$
$$11.27 = 3.8d \qquad \text{Divide both sides by 3.8.}$$
$$d = \frac{11.27}{3.8} \approx 3 \text{ in}$$

The data used to produce the regression model in Example 5 are from a stand of trees in the Jack Haggerty Forest at Lakeland University in Canada (Table 4). We used the popular spreadsheet Excel to produce a scatter plot of the data in Table 4 and to find the regression model (Fig. 4).

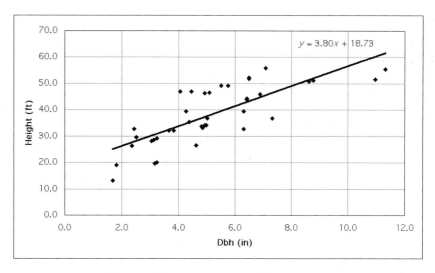

FIGURE 4 Linear regression with a spreadsheet

TABLE 4	Balsam Fir						
Dbh (in)	**Height (ft)**	**Dbh (in)**	**Height (ft)**	**Dbh (in)**	**Height (ft)**	**Dbh (in)**	**Height (ft)**
6.5	51.8	6.4	44.0	3.1	19.7	4.6	26.6
8.6	50.9	4.4	46.9	7.1	55.8	4.8	33.1
5.7	49.2	6.5	52.2	6.3	32.8	3.1	28.5
4.9	46.3	4.1	46.9	2.4	26.2	3.2	29.2
6.4	44.3	8.8	51.2	2.5	29.5	5.0	34.1
4.1	46.9	5.0	36.7	6.9	45.9	3.0	28.2
1.7	13.1	4.9	34.1	2.4	32.8	4.8	33.8
1.8	19.0	3.8	32.2	4.3	39.4	4.4	35.4
3.2	20.0	5.5	49.2	7.3	36.7	11.3	55.4
5.1	46.6	6.3	39.4	10.9	51.5	3.7	32.2

MATCHED PROBLEM 5

Figure 5 shows the scatter plot for a stand of white spruce trees in the Jack Haggerty Forest at Lakeland University in Canada. A regression model produced by a spreadsheet (Fig. 5), after rounding, is

$$h = 1.8d + 34$$

where d is Dbh in inches and h is the height in feet.

(A) Interpret the slope of this model.
(B) What is the effect of a 1-inch increase in Dbh?
(C) Estimate the height of a white spruce with a Dbh of 10 inches. Round your answer to the nearest foot.
(D) Estimate the Dbh of a white spruce that is 65 feet tall. Round your answer to the nearest inch.

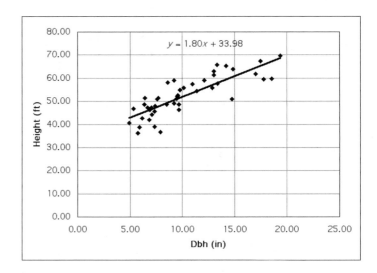

FIGURE 5 Linear regression for white spruce trees

Answers to Matched Problems

1. (A) The rate of change BSA with respect to weight is 28.55 square inches per pound.
(B) Increasing w by 1 pound increases a by 28.55 square inches.

2. (A) $a = -36t + 2,880$ (B) 36 ft/sec

3. (A)

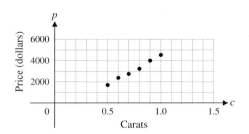

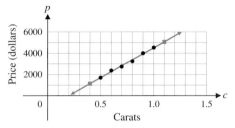

(B) The rate of change of the price of a diamond with respect to its weight is $5,600. Increasing the weight by one carat will increase the price by about $5,600.

(C) $3,660; $5,620   (D) 0.91 carats

4. (A) The slope $m = 0.16$ is the rate of change of emissions with respect to time. Since the slope is positive and the emissions are given in millions of tons, the emissions are increasing at a rate of $0.16(1,000,000) = 160,000$ tons per year.

(B) Approximately 42,800,00 tons of carbon monoxide will be emitted in 2010.

5. (A) The slope is 1.8 so the rate of change of height with respect to breast height diameter is 1.8 feet per inch.

(B) Height increases by 1.8 feet.   (C) 52 ft   (D) 17 in

Exercise 1-3

Applications

1. *Ideal weight.* In 1983 Dr. J. D. Robinson published the following estimate of the ideal body weight of a woman:

49 kg + 1.7 kg for each inch over 5 feet

(A) Find a linear model for Robinson's estimate of the ideal weight of a women using w for ideal body weight (in kilograms) and h for height over 5 feet (in inches).

(B) Interpret the slope of the model.

(C) If a woman is 5'4" tall, what does the model predict her weight to be?

(D) If a woman weighs 60 kilograms, what does the model predict her height to be?

2. *Ideal weight.* Dr. J. D. Robinson also published the following estimate of the ideal body weight of a man:

52 kg + 1.9 kg for each inch over 5 feet

(A) Find a linear model for Robinson's estimate of the ideal weight of a man using w for ideal body weight (in kilograms) and h for height over 5 feet (in inches).

(B) Interpret the slope of the model.

(C) If a man is 5'8" tall, what does the model predict his weight to be?

(D) If a man weighs 70 kilograms, what does the model predict his height to be?

3. *Underwater pressure.* At sea level, the weight of the atmosphere exerts a pressure of 14.7 pounds per square inch, commonly referred to as 1 **atmosphere of pressure.** As an object descends in water, pressure P and depth d are linearly related. In salt water, the pressure at a depth of 33 feet is 2 atmospheres or 29.4 pounds per square inch.

(A) Find a linear model that relates pressure P (in pounds per square inch) to depth d (in feet).

(B) Interpret the slope of the model.

(C) Find the pressure at a depth of 50 feet.

(D) Find the depth at which the pressure is 4 atmospheres.

4. *Underwater pressure.* Refer to Problem 3. In fresh water, the pressure at a depth of 34 feet is 2 atmospheres or 29.4 pounds per square inch.

(A) Find a linear model that relates pressure P (in pounds per square inch) to depth d (in feet).

(B) Interpret the slope of the model.

(C) Find the pressure at a depth of 50 feet.

(D) Find the depth at which the pressure is 4 atmospheres.

5. *Rate of descent—Parachutes.* At low altitudes the altitude of a parachutist and time in the air are linearly related. A jump at 2,880 ft using the U.S. Army's T-10 parachute system lasts 120 seconds.

(A) Find a linear model relating altitude a (in feet) and time in the air t (in seconds).

(B) Find the rate of descent for a T-10 system.

(C) Find the speed of the parachutist at landing.

6. *Rate of descent—Parachutes.* The U.S Army is considering a new parachute, the Advanced Tactical Parachute System (ATPS). A jump at 2,880 ft using the ATPS system lasts 180 seconds.

   (A) Find a linear model relating altitude *a* (in feet) and time in the air *t* (in seconds).

   (B) Find the rate of descent for an ATPS system parachute.

   (C) Find the speed of the parachutist at landing.

7. *Speed of sound.* The speed of sound through air is linearly related to the temperature of the air. If sound travels at 331 m/sec at 0°C and at 343 m/sec at 20°C, construct a linear model relating the speed of sound (*s*) and the air temperature (*t*). Interpret the slope of this model. *Source:* Engineering Toolbox

8. *Speed of sound.* The speed of sound through sea water is linearly related to the temperature of the water. If sound travels at 1,403 m/sec at 0°C and at 1,481 m/sec at 20°C, construct a linear model relating the speed of sound (*s*) and the air temperature (*t*). Interpret the slope of this model. *Source:* Engineering Toolbox

9. *Energy production.* Table 5 lists U.S. fossil fuel production as a percentage of total energy production for selected years. A linear regression model for this data is

$$y = -0.36x + 94.6$$

where *x* represents years since 1960 and *y* represents the corresponding percentage of oil imports.

TABLE 5 U.S. Fossil Fuel Production

Year	Production (%)
1960	93
1970	93
1980	88
1990	83
2000	80

Source: Energy Information Administration

(A) Draw a scatter plot of the data and a graph of the model on the same axes.

(B) Interpret the slope of the model.

(C) Use the model to predict fossil fuel production in 2010.

(D) Use the model to estimate the year in which fossil fuel production will fall below 70% of total energy production.

10. *Energy consumption.* Table 6 lists U.S. fossil fuel consumption as a percentage of total energy consumption for selected years. A linear regression model for this data is

$$y = -0.22x + 94$$

where *x* represents years since 1960 and *y* represents the corresponding percentage of fossil fuel consumption.

(A) Draw a scatter plot of the data and a graph of the model on the same axes.

(B) Interpret the slope of the model.

(C) Use the model to predict fossil fuel consumption in 2010.

TABLE 6 U.S. Fossil Fuel Consumption

Year	Production (%)
1960	93
1970	94
1980	89
1990	86
2000	86

Source: Energy Information Administration

(D) Use the model to estimate the year in which fossil fuel consumption will fall below 80% of total energy consumption.

11. *Cigarette smoking.* The data in Table 7 shows that the percentage of female cigarette smokers in the United Statess declined from 21% in 2000 to 18.5% in 2004.

    (A) Applying linear regression to the data for females in Table 7 produces the model

    $$f = -0.65t + 21.18$$

    where *f* is percentage of female smokers and *t* is time in years. Draw a scatter plot of the female smoker data and a graph of the regression model on the same axes for $0 \leq t \leq 15$.

    (B) Estimate the year in which the percentage of female smokers falls below 15% and illustrate the answer on your graph from part (A).

TABLE 7 Percentage of Smoking Prevalence among U.S. Adults

Year	Males (%)	Females (%)
2000	25.7	21.0
2001	25.2	20.7
2002	25.2	20.0
2003	24.1	19.2
2004	23.4	18.5

Source: Centers for Disease Control

12. *Cigarette smoking.* The data in Table 7 shows that the percentage of male cigarette smokers in the United State declined from 25.7% in 2000 to 23.4% in 2004.

    (A) Applying linear regression to the data for males in Table 7 produces the model

    $$m = -0.57t + 25.86$$

    where *m* is percentage of male smokers and *t* is time in years. Draw a scatter plot of the male smoker data and a graph of the regression model for $0 \leq t \leq 15$.

    (B) Estimate the year in which the percentage of male smokers falls below 20% and illustrate the answer on your graph from part (A).

13. *Real estate.* Table 8 contains recent average and median purchase prices for a house in Texas. A linear regression model for the average purchase price is

    $$y = 5.3x + 145$$

    where *x* is years since 2000 and *y* is average purchase price (in thousands of dollars).

TABLE 8	Texas Real Estate Prices	
Year	Average Price (in thousands)	Median Price (in thousands)
2000	$146	$112
2001	$150	$120
2002	$156	$125
2003	$160	$128
2004	$164	$130
2005	$174	$136

Source: Real Estate Center at Texas A&M University

(A) Draw a scatter plot of the average price data and a graph of the model on the same axes.

(B) Predict the average price in 2010.

(C) Interpret the slope of the model.

14. *Real estate.* A linear regression model for the median purchase price in Table 8 is

$$y = 4.4x + 114$$

where x is years since 2000 and y is median purchase price (in thousands of dollars).

(A) Plot the median price data and the model on the same axes.

(B) Predict the median price in 2010.

(C) Interpret the slope of the model.

15. *Licensed drivers.* Table 9 contains the state population and the number of licensed drivers in the state (both in millions) for the states with population under 1 million in 2004. The regression model for this data is

$$y = 0.63x + 0.08$$

where x is the state population and y is the number of licensed drivers in the state.

TABLE 9	Licensed Drivers in 2004	
State	Population	Licensed Drivers
Alaska	0.66	0.48
Delaware	0.83	0.53
Montana	0.93	0.71
North Dakota	0.63	0.46
South Dakota	0.77	0.56
Vermont	0.62	0.55
Wyoming	0.51	0.38

Source: Bureau of Transportation Statistics

(A) Draw a scatter plot of the data and a graph of the model on the same axes.

(B) If the population of Idaho in 2004 was about 1.4 million, use the model to estimate the number of licensed drivers in Idaho in 2004 to the nearest thousand.

(C) If the number of licensed drivers in Rhode Island in 2004 was about 0.74 million, use the model to estimate the population of Rhode Island in 2004 to the nearest thousand.

16. *Licensed drivers.* Table 10 contains the state population and the number of licensed drivers in the state (both in millions) for the states with population over 10 million in 2004. The regression model for this data is

$$y = 0.62x + 0.78$$

where x is the state population and y is the number of licensed drivers in the state.

TABLE 10	Licensed Drivers in 2004	
State	Population	Licensed Drivers
California	36	23
Florida	17	13
Illinois	13	8
Michigan	10	7
New York	19	11
Ohio	11	8
Pennsylvania	12	8
Texas	22	15

Source: Bureau of Transportation Statistics

(A) Draw a scatter plot of the data and a graph of the model on the same axes.

(B) If the population of Minnesota in 2004 was about 5 million, use the model to estimate the number of licensed drivers in Minnesota in 2004 to the nearest thousand.

(C) If the number of licensed drivers in Wisconsin in 2004 was about 3.9 million, use the model to estimate the population of Wisconsin in 2004 to the nearest thousand.

17. *Revenue.* A linear regression model for the revenue data in Table 11 is

$$R = 27.6t + 205$$

where R is total annual revenue and t is time since 1/31/02 in years.

TABLE 11	Wal-Mart Stores, Inc.				
Billions of U.S. Dollars	12 Months Ending 1/31/02	12 Months Ending 1/31/03	12 Months Ending 1/31/04	12 Months Ending 1/31/05	12 Months Ending 1/31/06
Total Revenue	206	232	259	288	316
Gross Profit	45	51	58	65	72

Source: Reuters

(A) Draw a scatter plot of the data and a graph of the model on the same axes.

(B) Predict Wal-Mart's annual revenue for the period ending 1/31/10.

18. *Profit.* A linear regression model for the gross profit data in Table 11 is

$$R = 6.8t + 44.6$$

where R is gross profit and t is time since 1/31/02 in years.

(A) Draw a scatter plot of the data and a graph of the model on the same axes.

(B) Predict Wal-Mart's annual gross profit for the period ending 1/31/10.

19. *Freezing temperature.* Ethylene glycol and propylene glycol are liquids used in antifreeze and deicing solutions. Ethylene glycol is listed as a hazardous chemical by the Environmental Protection Agency, while propylene glycol is generally regarded as safe. Table 12 lists the freezing temperature for various concentrations (as a percentage of total weight) of each chemical in a solution used to deice airplanes. A linear regression model for the ethylene glycol data in Table 12 is

$$E = -0.55T + 31$$

where E is the percentage of ethylene glycol in the deicing solution and T is the temperature at which the solution freezes.

(A) Draw a scatter plot of the data and a graph of the model on the same axes.

(B) Use the model to estimate the freezing temperature to the nearest degree of a solution that is 30% ethylene glycol.

(C) Use the model to estimate the percentage of ethylene glycol in a solution that freezes at 15°F.

TABLE 12

Freezing Temperature (°F)	Ethylene Glycol (%Wt.)	Propylene Glycol (%Wt.)
−50	56	58
−40	53	55
−30	49	52
−20	45	48
−10	40	43
0	33	36
10	25	29
20	16	19

Source: T. Labuza, University of Minesota

20. *Freezing temperature.* A linear regression model for the propylene glycol data in Table 12 is

$$P = -0.54T + 34$$

where P is the percentage of propylene glycol in the deicing solution and T is the temperature at which the solution freezes.

(A) Draw a scatter plot of the data and a graph of the model on the same axes.

(B) Use the model to estimate the freezing temperature to the nearest degree of a solution that is 30% propylene glycol.

(C) Use the model to estimate the percentage of propylene glycol in a solution that freezes at 15°F.

21. *Forestry.* The figure contains a scatter plot of 100 data points for black spruce trees and the linear regression model for this data.

(A) Interpret the slope of the model.

(B) What is the effect of a 1-inch increase in Dbh?

(C) Estimate the height of a black spruce with a Dbh of 15 inches. Round your answer to the nearest foot.

(D) Estimate the Dbh of a black spruce that is 25 feet tall. Round your answer to the nearest inch.

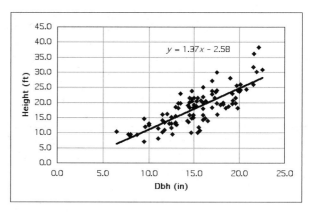

Figure for 21; black spruce
Source: Lakeland University

22. *Forestry.* The figure contains a scatter plot of 100 data points for black walnut trees and the linear regression model for this data.

(A) Interpret the slope of the model.

(B) What is the effect of a 1-inch increase in Dbh?

(C) Estimate the height of a black walnut with a Dbh of 12 inches. Round your answer to the nearest foot.

(D) Estimate the Dbh of a black walnut that is 25 feet tall. Round your answer to the nearest inch.

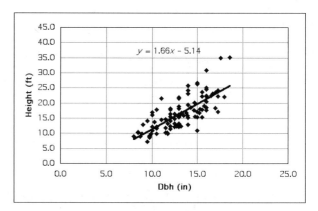

Figure for 22; black walnut
Source: Kagen Research

23. *Cable television.* Table 13 shows the increase in both price and revenue for cable television in the United States. The figure shows a scatter plot and a linear regression model for the average monthly price data in Table 13.

(A) Interpret the slope of the model.

(B) Use the model to predict the average monthly price (to the nearest dollar) in 2015.

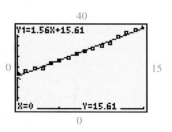

Figure for 23

TABLE 13 Cable Television Price and Revenue

Year	Average Monthly Price (dollars)	Annual Total Revenue (billions of dollars)
1990	16.78	17.58
1991	18.10	19.43
1992	19.08	21.08
1993	19.39	22.84
1994	21.62	23.13
1995	23.07	25.42
1996	24.41	27.71
1997	26.48	30.49
1998	27.51	33.50
1999	28.92	36.92
2000	30.08	40.86
2001	31.58	43.52
2002	34.52	49.43
2003	36.59	51.30
2004	38.23	57.60
2005	39.96	63.09

24. *Cable television.* The figure shows a scatter plot and a linear regression model for the annual revenue data in Table 13.

(A) Interpret the slope of the model.

(B) Use the model to predict the annual revenue (to the nearest billion dollars) in 2015.

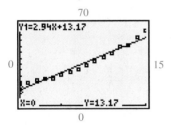

Figure for 24

25. *College enrollment.* Table 14 lists the fall enrollment in degree-granting institutions by gender and the figure contains a scatter plot and a regression line for each data set.

(A) Interpret the slope of each model.

(B) Predict (to the nearest million) both the male enrollment and the female enrollment in 2010.

(C) Estimate the first year that female enrollment exceeds male enrollment.

TABLE 14 Fall Enrollment (millions of students)

Year	Male	Female
1970	5.04	3.54
1975	6.15	5.04
1980	5.87	6.22
1985	5.82	6.43
1990	6.28	7.53
1995	6.34	7.92
2000	6.72	8.59

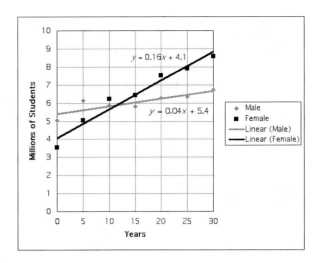

Figure for 25; fall enrollment by gender

26. *Graduate school enrollment.* Table 15 lists the fall graduate student enrollment in degree-granting institutions by gender and the figure contains a scatter plot and a regression line for each data set.

(A) Interpret the slope of each model.

(B) Predict (to one decimal place) both the male enrollment and the female enrollment in 2010.

(C) Estimate the first year that female enrollment exceeded male enrollment.

TABLE 15 Fall Graduate School Enrollment (millions of students)

Year	Male	Female
1970	0.63	0.40
1975	0.70	0.56
1980	0.68	0.56
1985	0.68	0.70
1990	0.74	0.85
1995	0.77	0.96
2000	0.78	1.07

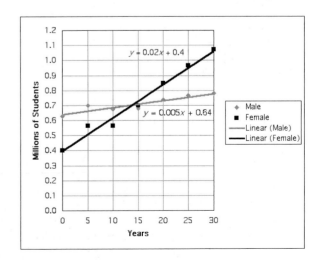

Figure for 26; fall enrollment by gender

Problems 27–30 require a graphing calculator or a computer that can calculate the linear regression line for a given data set.

27. *Olympic Games.* Find a linear regression model for the men's 100-meter freestyle data given in Table 16, where x is years since 1968 and y is winning time (in seconds). Do the same for the women's 100-meter freestyle data. (Round regression coefficients to three decimal places.) Do these models indicate that the women will eventually catch up with the men? If so, when? Do you think this will actually occur?

TABLE 16   Winning Times in Olympic Swimming Events

	100-Meter Freestyle		200-Meter Backstroke	
	Men	**Women**	**Men**	**Women**
1968	52.20	60.0	2:09.60	2:24.80
1972	51.22	58.59	2:02.82	2:19.19
1976	49.99	55.65	1:59.19	2:13.43
1980	50.40	54.79	2:01.93	2:11.77
1984	49.80	55.92	2:00.23	2:12.38
1988	48.63	54.93	1:59.37	2:09.29
1992	49.02	54.65	1:58.47	2:07.06
1996	48.74	54.50	1.58.54	2:07.83
2000	48.30	53.83	1:56.76	2:08.16
2004	48.17	53.84	1:54.76	2.09.16

Source: www.infoplease.com

28. *Olympic Games.* Find a linear regression model for the men's 200-meter backstroke data given in Table 16, where x is years since 1968 and y is winning time (in seconds). Do the same for the women's 200-meter backstroke data. (Round regression coefficients to four decimal places.) Do these models indicate that the women will eventually catch up with the men? If so, when? Do you think this will actually occur?

29. *Supply and demand.* Table 17 contains price–supply data and price–demand data for corn. Find a linear regression model for the price–supply data where x is supply (in billions of bushels) and y is price (in dollars). Do the same for the price-demand data. (Round regression coefficients to two decimal places.) Find the equilibrium price for corn.

TABLE 17   Supply and Demand for U.S. Corn

Price ($/bu)	Supply (billion bu)	Price ($/bu)	Demand (billion bu)
2.15	6.29	2.07	9.78
2.29	7.27	2.15	9.35
2.36	7.53	2.22	8.47
2.48	7.93	2.34	8.12
2.47	8.12	2.39	7.76
2.55	8.24	2.47	6.98

Source: www.usda.gov/nass/pubs/histdata.htm

30. *Supply and demand.* Table 18 contains price–supply data and price–demand data for soybeans. Find a linear regression model for the price–supply data where x is supply (in billions of bushels) and y is price (in dollars). Do the same for the price–demand data. (Round regression coefficients to two decimal places.) Find the equilibrium price for soybeans.

TABLE 18   Supply and Demand for U.S. Soybeans

Price ($/bu)	Supply (billion bu)	Price ($/bu)	Demand (billion bu)
5.15	1.55	4.93	2.60
5.79	1.86	5.48	2.40
5.88	1.94	5.71	2.18
6.07	2.08	6.07	2.05
6.15	2.15	6.40	1.95
6.25	2.27	6.66	1.85

Source: www.usda.gov/nass/pubs/histdata.htm

CHAPTER 1 REVIEW

Important Terms, Symbols, and Concepts

1-1 Linear Equations and Inequalities   **Examples**

- A **first-degree, or linear, equation** in one variable is any equation that can be written in the form

$$\text{Standard form:} \quad ax + b = 0 \quad a \neq 0$$

If the equality sign in the standard form is replaced by $<$, $>$, $\leq$, or $\geq$, the resulting expression is called a **first-degree, or linear, inequality.**

- A **solution** of an equation (or inequality) involving a single variable is a number that when substituted for the variable makes the equation (inequality) true. The set of all solutions is called the **solution set.**

- If we perform an operation on an equation (or inequality) that produces another equation   Ex. 1, p. 3
  (or inequality) with the same solution set, then the two equations (or inequalities) are **equivalent.**   Ex. 2, p. 4
  Equations are solved by adding or subtracting the same quantity to both sides, or by multiplying   Ex. 3, p. 5
  both sides by the same *nonzero* quantity until an equation with an obvious solution is obtained.

- Inequalities are solved in the same manner as equations with one important exception.   Ex. 5, p. 7
  If both sides of an inequality are multiplied by the same *negative* number   Ex. 6, p. 8

or divided by the same *negative* number, then the direction or sense of the inequality will reverse ($<$ becomes $>$, $\geq$ becomes $\leq$, and so on).

Ex. 7, p. 8

- A suggested strategy (p. 8-9) can be used to solve many word problems.

Ex. 8, p. 9

- A company breaks even if revenues R = costs C, makes a profit if $R > C$, and incurs a loss if $R < C$.

Ex. 10, p. 9

1-2 Graphs and Lines

- A **Cartesian or rectangular coordinate system** is formed by the intersection of a horizontal real number line, usually called the **x axis**, and a vertical real number line, usually called the **y axis**, at their origins. The axes determine a plane and divide this plane into four quadrants. Each point in the plane corresponds to its **coordinates** —an ordered pair (a, b) determined by passing horizontal and vertical lines through the point. The **abscissa** or **x coordinate** a is the coordinate of the intersection of the vertical line and the x axis, and the **ordinate** or **y coordinate** b is the coordinate of the intersection of the horizontal line and the y axis. The point with coordinates $(0, 0)$ is called the **origin.**

Fig 1, p. 14

- The **standard form** for a linear equation in two variables is $Ax + By = C$, with A and B not both zero. The graph of this equation is a line, and every line in a Cartesian coordinate system is the graph of a linear equation.

Ex. 1, p. 16
Ex. 2, p. 16

- The graph of the equation $x = a$ is a **vertical line** and the graph of $y = b$ is a **horizontal line.**

Ex. 3, p. 17

- If (x_1, y_1) and (x_2, y_2) are two distinct points on a line, then $m = (y_2 - y_1)/(x_2 - x_1)$ is the **slope** of the line.

Ex. 4, p. 18

- The equation $y = mx + b$ is the **slope-intercept form** of the equation of the line with slope m and y intercept b.

Ex. 5, p. 20

- The **point-slope form** of the equation of the line with slope m that passes through (x_1, y_1) is $y - y_1 = m(x - x_1)$.

Ex. 6, p. 20

- In a competitive market, the intersection of the supply equation and the demand equation is called the **equilibrium point**, the corresponding price is called the **equilibrium price**, and the common value of supply and demand is called the **equilibrium quantity.**

Ex. 8, p. 22

1-3 Linear Regression

- If the variables x and y are related by the equation $y = mx + b$, then x and y are **linearly related** and the slope m is the **rate of change** of y with respect to x.

Ex. 1, p. 30
Ex. 2, p. 30

- **Regression analysis** is used to fit a curve to a data set, usually with the aid of a graphing calculator or a computer. A graph of the points in a data set is called a **scatter plot**. A regression model can be used to **interpolate** between points in a data set or to **extrapolate** or predict points outside the data set.

Ex. 3, p. 31
Ex. 4, p. 34
Ex. 5, p. 35

REVIEW EXERCISE

Work through all the problems in this chapter review and check answers in the back of the book. Answers to all review problems are there, and following each answer is a number in italics indicating the section in which that type of problem is discussed. Where weaknesses show up, review appropriate sections in the text.

A

1. Solve $2x + 3 = 7x - 11$.

2. Solve $\dfrac{x}{12} - \dfrac{x - 3}{3} = \dfrac{1}{2}$.

3. Solve $2x + 5y = 9$ for y.

4. Solve $3x - 4y = 7$ for x.

Solve Problems 5–7 and graph on a real number line.

5. $4y - 3 < 10$

6. $-1 < -2x + 5 \leq 3$

7. $1 - \dfrac{x - 3}{3} \leq \dfrac{1}{2}$

8. Sketch a graph of $3x + 2y = 9$.

9. Write an equation of a line with x intercept 6 and y intercept 4. Write the final answer in the form $Ax + By = C$.

10. Sketch a graph of $2x - 3y = 18$. What are the intercepts and slope of the line?

11. Write an equation in the form $y = mx + b$ for a line with slope $-\dfrac{2}{3}$ and y intercept 6.

12. Write the equations of the vertical line and the horizontal line that pass through $(-6, 5)$.

13. Write the equation of a line through each indicated point with the indicated slope. Write the final answer in the form $y = mx + b$.

(A) $m = -\dfrac{2}{3}; (-3, 2)$ (B) $m = 0; (3, 3)$

14. Write the equation of the line through the two indicated points. Write the final answer in the form $Ax + By = C$.

(A) $(-3, 5), (1, -1)$ (B) $(-1, 5), (4, 5)$

(C) $(-2, 7), (-2, -2)$

B *Solve Problems 15–19.*

15. $3x + 25 = 5x$

16. $\dfrac{u}{5} = \dfrac{u}{6} + \dfrac{6}{5}$

17. $\dfrac{5x}{3} - \dfrac{4 + x}{2} = \dfrac{x - 2}{4} + 1$

18. $0.05x + 0.25(30 - x) = 3.3$

19. $0.2(x - 3) + 0.05x = 0.4$

Solve Problems 20–24 and graph on a real number line.

20. $2(x + 4) > 5x - 4$

21. $3(2 - x) - 2 \leq 2x - 1$

22. $\dfrac{x + 3}{8} - \dfrac{4 + x}{2} > 5 - \dfrac{2 - x}{3}$

23. $-5 \leq 3 - 2x < 1$

24. $-1.5 \leq 2 - 4x \leq 0.5$

25. Given $Ax + By = 30$, graph each of the following cases on the same coordinate axes.

(A) $A = 5$ and $B = 0$

(B) $A = 0$ and $B = 6$

(C) $A = 6$ and $B = 5$

26. Describe the graphs of $x = -3$ and $y = 2$. Graph both simultaneously in the same coordinate system.

27. Describe the lines defined by the following equations:

(A) $3x + 4y = 0$ (B) $3x + 4 = 0$

(C) $4y = 0$ (D) $3x + 4y - 36 = 0$

C *Solve Problems 28 and 29 for the indicated variable.*

28. $A = \dfrac{1}{2}(a + b)h$; for $a (h \neq 0)$

29. $S = \dfrac{P}{1 - dt}$; for $d(dt \neq 1)$

30. For what values of a and b is the inequality $a + b < b - a$ true?

31. If a and b are negative numbers and $a > b$, then is a/b greater than 1 or less than 1?

32. Graph $y = mx + b$ and $y = -\dfrac{1}{m} + b$ simultaneously in the same coordinate system for b fixed and several different values of $m, m \neq 0$. Describe the apparent relationship between the graphs of the two equations.

APPLICATIONS

33. *An investor has $300,000 to invest.* If part is invested at 5% and the rest at 9%, how much should be invested at 5% to yield 8% on the total amount?

34. *Break-even analysis.* A producer of educational CDs is producing an instructional CD. The producer estimates that it will cost $90,000 to record the CD and $5.10 per unit to copy and distribute the CD. If the wholesale price of the CD is $14.70, how many CDs must be sold for the producer to break even?

35. *Sports medicine.* A simple rule of thumb for determining your maximum safe heart rate (in beats per minute) is to subtract your age from 220. While exercising, you should maintain a heart rate between 60% and 85% of your maximum safe rate.

(A) Find a linear model for the minimum heart rate m that a person of age x years should maintain while exercising.

(B) Find a linear model for the maximum heart rate M that a person of age x years should maintain while exercising.

(C) What range of heartbeats should you maintain while exercising if you are 20 years old?

(D) What range of heartbeats should you maintain while exercising if you are 50 years old?

36. *Linear depreciation.* A bulldozer was purchased by a construction company for $224,000 and is assumed to have a depreciated value of $100,000 after 8 years. If the value is depreciated linearly from $224,000 to $100,000,

(A) Find the linear equation that relates value V (in dollars) to time t (in years).

(B) What would be the depreciated value of the system after 12 years?

37. *Business—Pricing.* A sporting goods store sells tennis rackets that cost $130 for $208 and court shoes that cost $50 for $80.

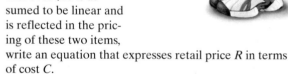

(A) If the markup policy of the store for items that cost over $10 is assumed to be linear and is reflected in the pricing of these two items, write an equation that expresses retail price R in terms of cost C.

(B) What would be the retail price of a pair of in-line skates that cost $120?

(C) What would be the cost of a pair of cross-country skis that had a retail price of $176?

(D) What is the slope of the graph of the equation found in part (A)? Interpret the slope relative to the problem.

38. *Income.* A salesperson receives a base salary of $200 per week and a commission of 10% on all sales over $3,000 during the week. Find the weekly earnings for weekly sales of $2,000 and for weekly sales of $5,000.

39. *Price–demand.* The weekly demand for mouthwash in a chain of drug stores is 1,160 bottles at a price of $3.79 each. If the price is lowered to $3.59, the weekly demand increases to 1,320 bottles. Assuming that the relationship between the weekly demand x and the price per bottle p is linear, express p in terms of x. How many bottles would the stores sell each week if the price were lowered to $3.29?

40. *Freezing temperature.* Methanol, also known as wood alcohol, can be used as a fuel for suitably equipped vehicles. It was also widely used as an antifreeze in cars until the late 1930s and is still used today in special applications, such as pipelines. Table 1 lists the freezing temperature for various concentrations (as a percentage of total weight) of methanol in water. A linear regression model for the data in Table 1 is

$$T = 40 - 2M$$

where M is the percentage of methanol in the solution and T is the temperature at which the solution freezes.

TABLE 1	
Methanol (%Wt)	**Freezing temperature (°F)**
0	32
10	20
20	0
30	−15
40	−40
50	−65
60	−95

Source: Ashland Inc.

(A) Draw a scatter plot of the data and a graph of the model on the same axes.

(B) Use the model to estimate the freezing temperature to the nearest degree of a solution that is 35% methanol.

(C) Use the model to estimate the percentage of methanol in a solution that freezes at −50°F.

41. *Internet usage.* Table 2 lists the time spent on the Internet by college students in recent years. Linear regression models for the data in Table 2 are

Men: $h = 1.3t + 5.8$

Women: $h = 1.5t + 3.7$

where t is years since 1997.

TABLE 2	Hours per Week on the Interent	
Year	**Men**	**Women**
1997	6.5	4.4
1998	6.2	5.1
1999	8.5	6
2000	9.7	8.4
2001	11.9	9.2
2002	11.9	11.9
2003	13.8	12.4
2004	15.4	14.8

Source: Studentmonitor.com

(A) Interpret the slope of each model.

(B) Draw a scatter plot of both data sets and a graph of both models in the same coordinate systems.

(C) In which year will time spent by women surpass time spent by men?

42. *Consumer Price Index.* The United States Consumer Price Index (CPI) in recent years is given in Table 3. A scatter plot of the data and the linear regression line are shown in the figure.

(A) Interpret the slope of the model.

(B) Predict the CPI in 2010.

TABLE 3	Consumer Price Index (1982–1984 = 100)
Year	**CPI**
1990	130.7
1991	136.2
1992	140.3
1993	144.5
1994	148.2
1995	152.4
1996	156.9
1997	160.5
1998	163.0
1999	166.6
2000	172.2
2001	177.1
2002	179.9
2003	184.0
2004	188.9
2005	195.3

Source: U.S. Bureau of Labor Statistics

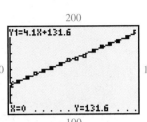

Figure for 42

43. *Forestry.* The figure contains a scatter plot of 20 data points for white pine trees and the linear regression model for this data.

(A) Interpret the slope of the model.

(B) What is the effect of a 1-inch increase in Dbh?

(C) Estimate the height of a white pine tree with a Dbh of 25 inches. Round your answer to the nearest foot.

(D) Estimate the Dbh of a white pine tree that is 15 feet tall. Round your answer to the nearest inch.

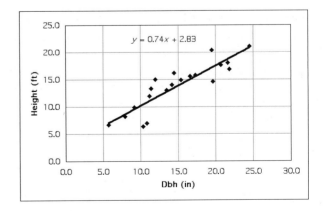

Figure for 43

Functions and Graphs

CHAPTER 2

2-1　Functions

2-2　Elementary Functions: Graphs and Transformations

2-3　Quadratic Functions

2-4　Exponential Functions

2-5　Logarithmic Functions

Chapter 2 Review

Review Exercise

INTRODUCTION

The function concept is one of the most important ideas in mathematics. The study of mathematics beyond the elementary level requires a firm understanding of a basic list of elementary functions, their properties, and their graphs. See the inside front cover of this book for a list of the functions that form our library of elementary functions. Most functions in the list will be introduced to you by the end of Chapter 2 and should become a part of your mathematical toolbox for use in this and most future courses or activities that involve mathematics. A few more elementary functions may be added to these in other courses, but the functions listed inside the front cover are more than sufficient for all the applications in this text.

Section 2-1   FUNCTIONS

- ■ Equations in Two Variables
- ■ Definition of a Function
- ■ Functions Specified by Equations
- ■ Function Notation
- ■ Applications

We introduce the general notion of a *function* as a correspondence between two sets. Then we restrict attention to functions for which the two sets are both sets of real numbers. The most useful are those functions that are specified by equations in two variables. We discuss the terminology and notation associated with functions, graphs of functions, and applications to economics.

■ Equations in Two Variables

In Chapter 1 we found that the graph of an equation of the form $Ax + By = C$, where A and B are not both zero, is a line. Because a line is determined by any two of its points, such an equation is easy to graph: Just plot *any* two points in its solution set and sketch the unique line through them.

More complicated equations in two variables, for example, $y = 9 - x^2$ or $x^2 = y^4$, are more difficult to graph. To **sketch the graph** of an equation, we plot enough points from its solution set in a rectangular coordinate system so that the total graph is apparent and then connect these points with a smooth curve. This process is called **point-by-point plotting.**

EXAMPLE 1   **Point-by-Point Plotting**   Sketch the graph of each equation.

(A) $y = 9 - x^2$ (B) $x^2 = y^4$

SOLUTIONS   (A) Make up a table of solutions—that is, ordered pairs of real numbers that satisfy the given equation. For easy mental calculation, choose integer values for x.

x	−4	−3	−2	−1	0	1	2	3	4
y	−7	0	5	8	9	8	5	0	−7

After plotting these solutions, if there are any portions of the graph that are unclear, plot additional points until the shape of the graph is apparent. Then join all the plotted points with a smooth curve as shown in Figure 1. Arrowheads are

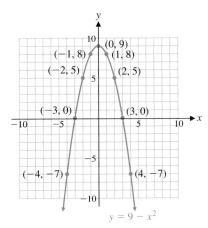

FIGURE 1 $y = 9 - x^2$

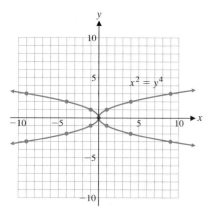

FIGURE 2 $x^2 = y^4$

used to indicate that the graph continues beyond the portion shown here with no significant changes in shape.

(B) Again we make a table of solutions—here it may be easier to choose integer values for y and calculate values for x. Note, for example, that if $y = 2$, then $x = \pm 4$; that is, the ordered pairs $(4, 2)$ and $(-4, 2)$ are both in the solution set.

x	± 9	± 4	± 1	0	± 1	± 4	± 9
y	-3	-2	-1	0	1	2	3

We plot these points and join them with a smooth curve (Fig. 2). ▬

MATCHED PROBLEM 1

Sketch the graph of each equation.

(A) $y = x^2 - 4$ (B) $y^2 = \dfrac{100}{x^2 + 1}$

Explore & Discuss **1**

To graph the equation $y = -x^3 + 3x$, we use point-by-point plotting to obtain

x	y
-1	-2
0	0
1	2

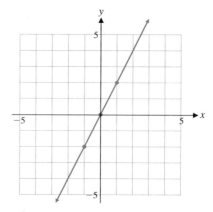

(A) Do you think this is the correct graph of the equation? If so, why? If not, why?

(B) Add points on the graph for $x = -2, -1.5, -0.5, 0.5, 1.5,$ and 2.

(C) Now, what do you think the graph looks like? Sketch your version of the graph, adding more points as necessary.

(D) Graph this equation on a graphing calculator and compare it with your graph from part (C).

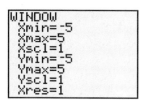

(A)

```
Plot1 Plot2 Plot3
\Y1■3X-X^3
\Y2=
\Y3=
\Y4=
\Y5=
\Y6=
\Y7=
```

(A)

```
WINDOW
Xmin=-5
Xmax=5
Xscl=1
Ymin=-5
Ymax=5
Yscl=1
Xres=1
```

(B)

FIGURE 3

The icon in the margin is used throughout this book to identify optional graphing calculator activities that are intended to give you additional insight into the concepts under discussion. You may have to consult the manual for your graphing calculator for the details necessary to carry out these activities. For example, to graph the equation in Explore–Discuss 1 on most graphing calculators, you first have to enter the equation (Fig. 3A) and the window variables (Fig. 3B).

As Explore–Discuss 1 illustrates, the shape of a graph may not be apparent from your first choice of points on the graph. Using point-by-point plotting, it may be difficult to find points in the solution set of the equation, and it may be difficult to determine when you have found enough points to understand the shape of the graph. We will supplement the technique of point-by-point plotting with a detailed analysis of several basic equations, giving you the ability to sketch graphs with accuracy and confidence.

■ Definition of a Function

Central to the concept of function is correspondence. You have already had experiences with correspondences in daily living. For example,

To each person there corresponds an annual income.

To each item in a supermarket there corresponds a price.

To each student there corresponds a grade-point average.

To each day there corresponds a maximum temperature.

For the manufacture of x items there corresponds a cost.

For the sale of x items there corresponds a revenue.

To each square there corresponds an area.

To each number there corresponds its cube.

One of the most important aspects of any science is the establishment of correspondences among various types of phenomena. Once a correspondence is known, predictions can be made. A cost analyst would like to predict costs for various levels of output in a manufacturing process; a medical researcher would like to know the correspondence between heart disease and obesity; a psychologist would like to predict the level of performance after a subject has repeated a task a given number of times; and so on.

What do all the examples above have in common? Each describes the matching of elements from one set with the elements in a second set. Consider the tables of the cube, square, and square root given in Tables 1–3.

Tables 1 and 2 specify functions, but Table 3 does not. Why not? The definition of the term *function* will explain.

TABLE 1			TABLE 2			TABLE 3	
Domain		**Range**	**Domain**		**Range**	**Domain**	**Range**
Number		*Cube*	*Number*		*Square*	*Number*	*Square root*
−2	⟶	−8	−2		4	0 ⟶ 0	
−1	⟶	−1	−1		1	1 ⟶ 1, −1	
0	⟶	0	0		0		
1	⟶	1	1			4 ⟶ 2, −2	
2	⟶	8	2			9 ⟶ 3, −3	

DEFINITION   **Function**

A **function** is a correspondence between two sets of elements such that to each element in the first set there corresponds one and only one element in the second set.

The first set is called the **domain,** and the set of corresponding elements in the second set is called the **range.**

Tables 1 and 2 specify functions, since to each domain value there corresponds exactly one range value (for example, the cube of −2 is −8 and no other number). On the other hand, Table 3 does not specify a function, since to at least one domain value there corresponds more than one range value (for example, to the domain value 9 there corresponds −3 and 3, both square roots of 9).

Explore & Discuss 2   Consider the set of students enrolled in a college and the set of faculty members of that college. Suppose we define a correspondence between the two sets by saying that a student corresponds to a faculty member if the student is currently enrolled in a course taught by that faculty member. Is this correspondence a function? Discuss.

Functions Specified by Equations

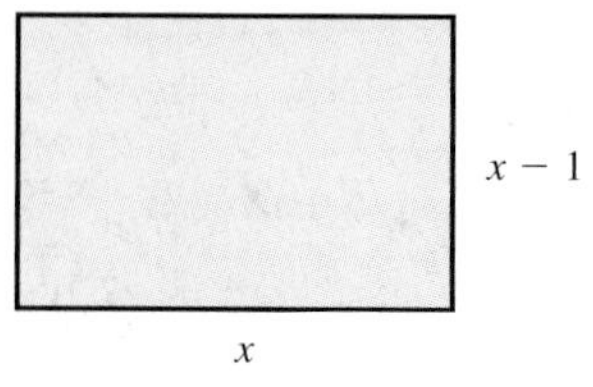

Most of the functions in this book will have domains and ranges that are (infinite) sets of real numbers. The **graph** of such a function is the set of all points (x, y) in the Cartesian plane such that x is an element of the domain and y is the corresponding element in the range. The correspondence between domain and range elements is often specified by an equation in two variables. Consider, for example, the equation for the area of a rectangle with width 1 inch less than its length (Fig. 4). If x is the length, then the area y is given by

$$y = x(x - 1) \qquad x \geq 1$$

FIGURE 4

For each **input** x (length), we obtain an **output** y (area). For example,

If $x = 5$,   then $y = 5\,(5 - 1) = 5 \cdot 4 = 20.$

If $x = 1$,   then $y = 1\,(1 - 1) = 1 \cdot 0 = 0.$

If $x = \sqrt{5}$,   then $y = \sqrt{5}\,(\sqrt{5} - 1) = 5 - \sqrt{5}$

$$\approx 2.7639.$$

The input values are domain values, and the output values are range values. The equation assigns each domain value x a range value y. The variable x is called an *independent variable* (since values can be "independently" assigned to x from the domain), and y is called a *dependent variable* (since the value of y "depends" on the value assigned to x). In general, any variable used as a placeholder for domain values is called an **independent variable;** any variable that is used as a placeholder for range values is called a **dependent variable.**

When does an equation specify a function?

DEFINITION   **Functions Specified by Equations**

If in an equation in two variables, we get exactly one output (value for the dependent variable) for each input (value for the independent variable), then the equation specifies a function. The graph of such a function is just the graph of the specifying equation.

If we get more than one output for a given input, the equation does not specify a function.

EXAMPLE 2   **Functions and Equations**   Determine which of the following equations specify functions with independent variable x.

(A) $4y - 3x = 8$, x a real number   (B) $y^2 - x^2 = 9$, x a real number

SOLUTION   (A) Solving for the dependent variable y, we have

$$4y - 3x = 8$$
$$4y = 8 + 3x \tag{1}$$
$$y = 2 + \frac{3}{4}x$$

Since each input value x corresponds to exactly one output value ($y = 2 + \frac{3}{4}x$), we see that equation (1) specifies a function.

(B) Solving for the dependent variable y, we have

$$y^2 - x^2 = 9$$
$$y^2 = 9 + x^2 \qquad (2)$$
$$y = \pm\sqrt{9 + x^2}$$

Since $9 + x^2$ is always a positive real number for any real number x and since each positive real number has two square roots,* to each input value x there corresponds two output values ($y = -\sqrt{9 + x^2}$ and $y = \sqrt{9 + x^2}$). For example, if $x = 4$, then equation (2) is satisfied for $y = 5$ and for $y = -5$. Thus, equation (2) does not specify a function. ■

MATCHED PROBLEM 2 Determine which of the following equations specify functions with independent variable x.

(A) $y^2 - x^4 = 9$, x a real number   (B) $3y - 2x = 3$, x a real number

Since the graph of an equation is the graph of all the ordered pairs that satisfy the equation, it is very easy to determine whether an equation specifies a function by examining its graph. The graphs of the two equations we considered in Example 2 are shown in Figure 5.

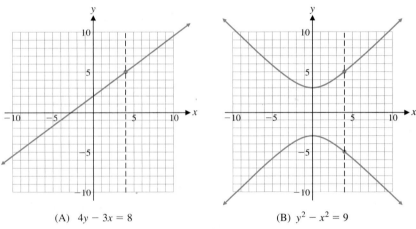

(A) $4y - 3x = 8$ (B) $y^2 - x^2 = 9$

FIGURE 5

In Figure 5A notice that any vertical line will intersect the graph of the equation $4y - 3x = 8$ in exactly one point. This shows that to each x value there corresponds exactly one y value and confirms our conclusion that this equation specifies a function. On the other hand, Figure 5B shows that there exist vertical lines that intersect the graph of $y^2 - x^2 = 9$ in two points. This indicates that there exist x values to which there correspond two different y values and verifies our conclusion that this equation does not specify a function. These observations are generalized in Theorem 1.

THEOREM 1   VERTICAL-LINE TEST FOR A FUNCTION

An equation specifies a function if each vertical line in the coordinate system passes through at most one point on the graph of the equation.

If any vertical line passes through two or more points on the graph of an equation, then the equation does not specify a function.

---

* Recall that each positive real number N has two square roots: $\sqrt{N}$, the principal square root, and $-\sqrt{N}$, the negative of the principal square root (see Appendix A, Section A-6).

The function graphed in Figure 5A is an example of a *linear function*. The vertical-line test implies that equations of the form $y = mx + b$, where $m \neq 0$, specify functions; they are called **linear functions.** Similarly, equations of the form $y = b$ specify functions; they are called **constant functions,** and their graphs are horizontal lines. The vertical-line test implies that equations of the form $x = a$ do not specify functions; note that the graph of $x = a$ is itself a vertical line.

Explore & Discuss 3

The definition of a function specifies that to each element in the domain there corresponds one and only one element in the range.

(A) Give an example of a function such that to each element of the range there correspond exactly two elements of the domain.

(B) Give an example of a function such that to each element of the range there corresponds exactly one element of the domain.

In Example 2, the domains were explicitly stated along with the given equations. In many cases, this will not be done. Unless stated to the contrary, we shall adhere to the following convention regarding domains and ranges for functions specified by equations:

If a function is specified by an equation and the domain is not indicated, then we assume that the domain is the set of all real number replacements of the independent variable (inputs) that produce real values for the dependent variable (outputs). The range is the set of all outputs corresponding to input values.

In many applied problems the domain is determined by practical considerations within the problem (see Example 7).

EXAMPLE 3

Finding a Domain Find the domain of the function specified by the equation $y = \sqrt{4 - x}$, assuming that x is the independent variable.

SOLUTION For y to be real, $4 - x$ must be greater than or equal to 0; that is,

$$4 - x \geq 0$$
$$-x \geq -4$$
$$x \leq 4 \qquad \text{Sense of inequality reverses when both sides are divided by } -1.$$

Thus,

Domain: $x \leq 4$ (inequality notation) or $(-\infty, 4]$ (interval notation) ▬

MATCHED PROBLEM 3

Find the domain of the function specified by the equation $y = \sqrt{x - 2}$, assuming x is the independent variable.

▪ Function Notation

We have just seen that a function involves two sets, a domain and a range, and a correspondence that assigns to each element in the domain exactly one element in the range. We use different letters to denote names for numbers; in essentially the same way, we will now use different letters to denote names for functions. For example, f and g may be used to name the functions specified by the equations $y = 2x + 1$ and $y = x^2 + 2x - 3$:

$$f\colon \quad y = 2x + 1$$
$$g\colon \quad y = x^2 + 2x - 3$$

(3)

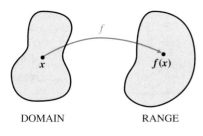

DOMAIN   RANGE

FIGURE 6

If x represents an element in the domain of a function f, then we frequently use the symbol

$$f(x)$$

in place of y to designate the number in the range of the function f to which x is paired (Fig. 6). This symbol does *not* represent the product of f and x. The symbol $f(x)$ is read as "f of x," "f at x," or "the value of f at x." Whenever we write $y = f(x)$, we assume that the variable x is an independent variable and that both y and $f(x)$ are dependent variables.

Using function notation, we can now write functions f and g in (3) in the form

$$f(x) = 2x + 1 \qquad \text{and} \qquad g(x) = x^2 + 2x - 3$$

Let us find $f(3)$ and $g(-5)$. To find $f(3)$, we replace x with 3 wherever x occurs in $f(x) = 2x + 1$ and evaluate the right side:

$$f(x) = 2x + 1$$
$$f(3) = 2 \cdot 3 + 1$$
$$= 6 + 1 = 7 \qquad \textit{For input 3, the output is 7.}$$

Thus,

$$f(3) = 7 \quad \textit{The function f assigns the range value 7 to the domain value 3.}$$

To find $g(-5)$, we replace each x by -5 in $g(x) = x^2 + 2x - 3$ and evaluate the right side:

$$g(x) = x^2 + 2x - 3$$
$$g(-5) = (-5)^2 + 2(-5) - 3$$
$$= 25 - 10 - 3 = 12 \quad \textit{For input} -5, \textit{the output is 12.}$$

Thus,

$$g(-5) = 12 \quad \textit{The function g assigns the range value 12 to the domain value} -5.$$

It is very important to understand and remember the definition of $f(x)$:

For any element x in the domain of the function f, the symbol $f(x)$ represents the element in the range of f corresponding to x in the domain of f. If x is an input value, then $f(x)$ is the corresponding output value. If x is an element that is not in the domain of f, then f is not defined at x and $f(x)$ does not exist.

EXAMPLE 4

Function Evaluation If

$$f(x) = \frac{12}{x - 2} \qquad g(x) = 1 - x^2 \qquad h(x) = \sqrt{x - 1}$$

then

(A) $f(6) = \dfrac{12}{6 - 2}^* = \dfrac{12}{4} = 3$

(B) $g(-2) = 1 - (-2)^2 = 1 - 4 = -3$

(C) $h(-2) = \sqrt{-2 - 1} = \sqrt{-3}$

But $\sqrt{-3}$ is not a real number. Since we have agreed to restrict the domain of a function to values of x that produce real values for the function, -2 is not in the domain of h and $h(-2)$ does not exist.

(D) $f(0) + g(1) - h(10) = \dfrac{12}{0 - 2} + (1 - 1^2) - \sqrt{10 - 1}$

$$= \frac{12}{-2} + 0 - \sqrt{9}$$

$$= -6 - 3 = -9$$

---

* Dashed boxes are used throughout the book to represent steps that are usually performed mentally.

| MATCHED PROBLEM 4 | Use the functions in Example 3 to find

(A) $f(-2)$ (B) $g(-1)$ (C) $h(-8)$ (D) $\dfrac{f(3)}{h(5)}$

| EXAMPLE 5 | **Finding Domains** Find the domains of functions $f, g,$ and h:

$$f(x) = \frac{12}{x - 2} \qquad g(x) = 1 - x^2 \qquad h(x) = \sqrt{x - 1}$$

SOLUTION *Domain of f:* $12/(x - 2)$ represents a real number for all replacements of x by real numbers except for $x = 2$ (division by 0 is not defined). Thus, $f(2)$ does not exist, and the domain of f is the set of all real numbers except 2. We often indicate this by writing

$$f(x) = \frac{12}{x - 2} \qquad x \neq 2$$

Domain of g: The domain is R, the set of all real numbers, since $1 - x^2$ represents a real number for all replacements of x by real numbers.

Domain of h: The domain is the set of all real numbers x such that $\sqrt{x - 1}$ is a real number—that is, such that

$$x - 1 \geq 0$$
$$x \geq 1 \quad \text{or} \quad [1, \infty)$$

| MATCHED PROBLEM 5 | Find the domains of functions $F, G,$ and H:

$$F(x) = x^2 - 3x + 1 \qquad G(x) = \frac{5}{x + 3} \qquad H(x) = \sqrt{2 - x}$$

In addition to evaluating functions at specific numbers, it is important to be able to evaluate functions at expressions that involve one or more variables. For example, the **difference quotient**

$$\frac{f(x + h) - f(x)}{h} \qquad \text{x and x + h in the domain of f, h ≠ 0}$$

is studied extensively in calculus.

> **INSIGHT**
>
> In algebra, you learned to use parentheses for grouping variables. For example,
>
> $$2(x + h) = 2x + 2h$$
>
> Now we are using parentheses in the function symbol $f(x)$. For example, if $f(x) = x^2$, then
>
> $$f(x + h) = (x + h)^2 = x^2 + 2xh + h^2$$
>
> Note that $f(x) + f(h) = x^2 + h^2 \neq f(x + h)$. That is, the function name f does not distribute across the grouped variables $(x + h)$ as the "2" does in $2(x + h)$ (see Appendix A, Section A-2).

Explore & Discuss **4** Let x and h be real numbers.

(A) If $f(x) = 4x + 3$, which of the following is true?
   1. $f(x + h) = 4x + 3 + h$
   2. $f(x + h) = 4x + 4h + 3$
   3. $f(x + h) = 4x + 4h + 6$

(B) If $g(x) = x^2$, which of the following is true?

    1. $g(x + h) = x^2 + h$
    2. $g(x + h) = x^2 + h^2$
    3. $g(x + h) = x^2 + 2hx + h^2$

(C) If $M(x) = x^2 + 4x + 3$, describe the operations that must be performed to evaluate $M(x + h)$.

EXAMPLE 6 **Using Function Notation** For $f(x) = x^2 - 2x + 7$, find

(A) $f(a)$ (B) $f(a + h)$ (C) $f(a + h) - f(a)$ (D) $\dfrac{f(a + h) - f(a)}{h}, \quad h \neq 0$

SOLUTION (A) $f(a) = a^2 - 2a + 7$

(B) $f(a + h) = (a + h)^2 - 2(a + h) + 7 = a^2 + 2ah + h^2 - 2a - 2h + 7$

(C) $f(a + h) - f(a) = (a^2 + 2ah + h^2 - 2a - 2h + 7) - (a^2 - 2a + 7)$

$\qquad\qquad\qquad\qquad = 2ah + h^2 - 2h$

(D) $\dfrac{f(a + h) - f(a)}{h} = \dfrac{2ah + h^2 - 2h}{h} = \boxed{\dfrac{h(2a + h - 2)}{h}}$ *Because* $h \neq 0, \dfrac{h}{h} = 1.$

$\qquad\qquad\qquad = 2a + h - 2$ ▪

MATCHED PROBLEM 6 Repeat Example 6 for $f(x) = x^2 - 4x + 9$.

APPLICATIONS

We now turn to the important concepts of **break-even** and **profit–loss** analysis, which we will return to a number of times in this book. Any manufacturing company has **costs,** C, and **revenues,** R. The company will have a **loss** if $R < C$, will **break even** if $R = C$, and will have a **profit** if $R > C$. Costs include **fixed costs** such as plant overhead, product design, setup, and promotion; and **variable costs,** which are dependent on the number of items produced at a certain cost per item. In addition, **price–demand** functions, usually established by financial departments using historical data or sampling techniques, play an important part in profit–loss analysis. We will let x, the number of units manufactured and sold, represent the independent variable. Cost functions, revenue functions, profit functions, and price–demand functions are often stated in the following forms, where $a, b, m,$ and n are constants determined from the context of a particular problem:

Cost Function

$$C = (\text{fixed costs}) + (\text{variable costs})$$
$$= a + bx$$

Price–Demand Function

$$p = m - nx \quad \text{x is the number of items that can be sold at \$$p$ per item.}$$

Revenue Function

$$R = (\text{number of items sold}) \times (\text{price per item})$$
$$= xp = x(m - nx)$$

Profit Function

$$P = R - C$$
$$= x(m - nx) - (a + bx)$$

Example 7 and Matched Problem 7 explore the relationships among the algebraic definition of a function, the numerical values of the function, and the graphical representation of the function. The interplay among algebraic, numeric, and graphic viewpoints is an important aspect of our treatment of functions and their use. In Example 7, we also see how a function can be used to describe data from the real world, a process that is often referred to as *mathematical modeling*. The material in this example will be returned to in subsequent sections so that we can analyze it in greater detail and from different points of view.

EXAMPLE 7

Price–Demand and Revenue Modeling A manufacturer of a popular digital camera wholesales the camera to retail outlets throughout the United States. Using statistical methods, the financial department in the company produced the price–demand data in Table 4, where p is the wholesale price per camera at which x million cameras are sold. Notice that as the price goes down, the number sold goes up.

TABLE 4 Price–Demand	
x (Millions)	*p* ($)
2	87
5	68
8	53
12	37

TABLE 5 Revenue	
x (Millions)	*R(x)* (Millions $)
1	90
3	
6	
9	
12	
15	

Using special analytical techniques (regression analysis), an analyst arrived at the following price–demand function that models the Table 4 data:

$$p(x) = 94.8 - 5x \qquad 1 \le x \le 15 \qquad (5)$$

(A) Plot the data in Table 4. Then sketch a graph of the price–demand function in the same coordinate system.

(B) What is the company's revenue function for this camera, and what is the domain of this function?

(C) Complete Table 5, computing revenues to the nearest million dollars.

(D) Plot the data in Table 5. Then sketch a graph of the revenue function using these points.

(E) Plot the revenue function on a graphing calculator.

SOLUTION (A)

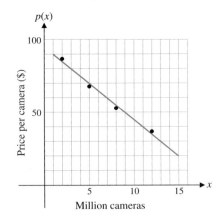

FIGURE 7 Price–demand

In Figure 7, notice that the model approximates the actual data in Table 4, and it is assumed that it gives realistic and useful results for all other values of x between 1 million and 15 million.

(B) $R(x) = xp(x) = x(94.8 - 5x)$ million dollars

Domain: $1 \leq x \leq 15$

[Same domain as the price–demand function, equation (5).]

(C)

TABLE 5 Revenue

x (Millions)	R(x) (Million $)
1	90
3	239
6	389
9	448
12	418
15	297

(D)

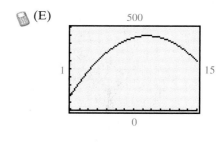

(E)

MATCHED PROBLEM 7

The financial department in Example 6, using statistical techniques, produced the data in Table 6, where $C(x)$ is the cost in millions of dollars for manufacturing and selling x million cameras.

TABLE 6 Cost Data

x (Millions)	C(x) (Million $)
1	175
5	260
8	305
12	395

Using special analytical techniques (regression analysis), an analyst produced the following cost function to model the data:

$$C(x) = 156 + 19.7x \qquad 1 \leq x \leq 15 \qquad (6)$$

(A) Plot the data in Table 6. Then sketch a graph of equation (6) in the same coordinate system.

(B) What is the company's profit function for this camera, and what is its domain?

(C) Complete Table 7, computing profits to the nearest million dollars.

TABLE 7 Profit

x (Millions)	P(x) (Million $)
1	−86
3	
6	
9	
12	
15	

(D) Plot the points from part (C). Then sketch a graph of the profit function through these points.

(E) Plot the profit function on a graphing calculator.

Answers to Matched Problems **1.** (A)

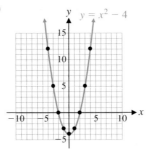

(B)

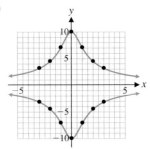

2. (A) Does not specify a function

(B) Specifies a function

3. $x \geq 2$ (inequality notation) or $[2, \infty)$ (interval notation)

4. (A) -3

(B) 0

(C) Does not exist

(D) 6

5. Domain of F: R; domain of G: all real numbers except -3; domain of H: $x \leq 2$ (inequality notation) or $(-\infty, 2]$ (interval notation)

6. (A) $a^2 - 4a + 9$

(B) $a^2 + 2ah + h^2 - 4a - 4h + 9$

(C) $2ah + h^2 - 4h$

(D) $2a + h - 4$

7. (A)

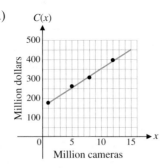

(B) $P(x) = R(x) - C(x) = x(94.8 - 5x) - (156 + 19.7x)$; domain: $1 \leq x \leq 15$

(C)

TABLE 7 Profit	
x (Millions)	**$P(x)$ (Million $)**
1	−86
3	24
6	115
9	115
12	25
15	−155

(D) $P(x)$

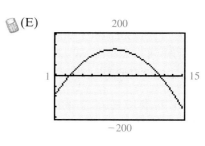

(E)

Exercise 2-1

A *In Problems 1–8, use point-by-point plotting to sketch the graph of each equation.*

1. $y = x + 1$

2. $x = y + 1$

3. $x = y^2$

4. $y = x^2$

5. $y = x^3$

6. $x = y^3$

7. $xy = -6$

8. $xy = 12$

Indicate whether each table in Problems 9–14 specifies a function.

9.

Domain	Range
3	0
5	1
7	2

10.

Domain	Range
−1	5
−2	7
−3	9

11.

Domain	Range
3	5
	6
4	7
5	8

12.

Domain	Range
8	0
9	1
	2
10	3

13.

Domain	Range
3	
6	5
9	
12	6

14.

Domain	Range
−2	
−1	
0	6
1	

Indicate whether each graph in Problems 15–20 specifies a function.

15.

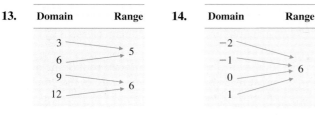

16.

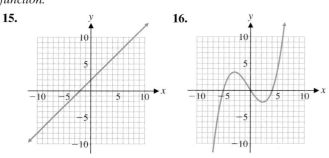

17.

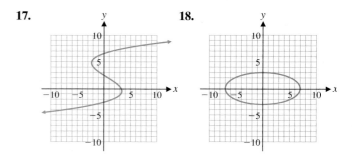

18.

19.

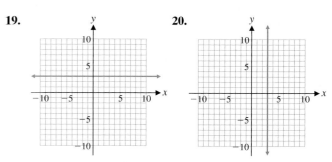

20.

In Problems 21–30, each equation specifies a function. Determine whether the function is linear, constant, or neither.

21. $y = 3 - 7x$

22. $y = \pi$

23. $y = -6$

24. $y = 5x - \frac{1}{2}(4 - x)$

25. $y = 8x - 1 + \dfrac{1}{x}$

26. $y = 3x + \frac{1}{2}(5 - 6x)$

27. $y = \dfrac{1 + x}{2} + \dfrac{1 - x}{3}$

28. $y = \dfrac{1}{2x + 3}$

29. $y = x^2 + (1 - x)(1 + x) + 1$

30. $y = x^2 - 9$

In Problems 31 and 32 which of the indicated correspondences define functions? Explain.

31. Let P be the set of residents of Pennsylvania and let R and S be the set of members of the U.S. House of Representatives and the set of members of the U.S. Senate, respectively, elected by the residents of Pennsylvania.

(A) A resident corresponds to the congressperson representing the resident's congressional district.

(B) A resident corresponds to the senator representing the resident's state.

32. Let P be the set of patients in a hospital, let D be the set of doctors on the hospital staff, and N be the set of nurses on the hospital staff.

(A) A patient corresponds to the doctor if that doctor admitted the patient to the hospital.

(B) A patient corresponds to the nurse if that nurse cares for the patient.

In Problems 33–40, use point-by-point plotting to sketch the graph of each function.

33. $f(x) = 1 - x$ **34.** $f(x) = \dfrac{x}{2} - 3$

35. $f(x) = x^2 - 1$ **36.** $f(x) = 3 - x^2$

37. $f(x) = 4 - x^3$ **38.** $f(x) = x^3 - 2$

39. $f(x) = \dfrac{8}{x}$ **40.** $f(x) = \dfrac{-6}{x}$

In Problems 41 and 42, the three points in the table are on the graph of the indicated function f. Do these three points provide sufficient information for you to sketch the graph of $y = f(x)$? Add more points to the table until you are satisfied that your sketch is a good representation of the graph of $y = f(x)$ on the interval $[-5, 5]$.

41.

x	-1	0	1
$f(x)$	-1	0	1

$f(x) = \dfrac{2x}{x^2 + 1}$

42.

x	0	1	2
$f(x)$	0	1	2

$f(x) = \dfrac{3x^2}{x^2 + 2}$

43. Let $f(x) = 100x - 5x^2$ and $g(x) = 150 + 20x$, $0 \le x \le 20$.

(A) Evaluate $f(x), g(x),$ and $f(x) - g(x)$ for $x = 0, 5, 10, 15, 20$.

(B) Graph $y = f(x), y = g(x),$ and $y = f(x) - g(x)$ on the interval $[0, 20]$.

44. Repeat Problem 43 for $f(x) = 200x - 10x^2$ and $g(x) = 200 + 50x$.

In Problems 45–52, use the following graph of a function f to determine x or y to the nearest integer, as indicated. Some problems may have more than one answer.

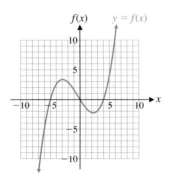

45. $y = f(-5)$ **46.** $y = f(4)$

47. $y = f(5)$ **48.** $y = f(-2)$

49. $0 = f(x)$ **50.** $3 = f(x), x < 0$

51. $-4 = f(x)$ **52.** $4 = f(x)$

If $f(x) = 2x - 3$ and $g(x) = x^2 + 2x$, find each of the expressions in Problems 53–64.

53. $f(2)$ **54.** $f(1)$

55. $f(-1)$ **56.** $g(1)$

57. $g(-3)$ **58.** $g(-2)$

59. $f(1) + g(2)$ **60.** $f(3) - g(3)$

61. $g(3) \cdot f(0)$ **62.** $g(0) \cdot f(-2)$

63. $\dfrac{g(-2)}{f(-2)}$ **64.** $\dfrac{g(-3)}{f(2)}$

B *In Problems 65–70, find the domain of each function.*

65. $F(x) = 2x^3 - x^2 + 3$ **66.** $H(x) = 7 - 2x^2 - x^4$

67. $f(x) = \dfrac{x - 2}{x + 4}$ **68.** $g(x) = \dfrac{x + 1}{x - 2}$

69. $g(x) = \sqrt{7 - x}$ **70.** $F(x) = \dfrac{1}{\sqrt{5 + x}}$

71. Two people are discussing the function

$$f(x) = \frac{x^2 - 4}{x^2 - 9}$$

and one says to the other, "$f(2)$ exists but $f(3)$ does not." Explain what they are talking about.

72. Referring to the function in Problem 71, do $f(-2)$ and $f(-3)$ exist? Explain.

The verbal statement "function f multiplies the square of the domain element by 3 and then subtracts 7 from the result" and the algebraic statement "$f(x) = 3x^2 - 7$" define the same function. In Problems 73–76, translate each verbal definition of a function into an algebraic definition.

73. Function g subtracts 5 from twice the cube of the domain element.

74. Function f multiplies the domain element by -3 and adds 4 to the result.

75. Function G multiplies the square root of the domain element by 2 and subtracts the square of the domain element from the result.

76. Function F multiplies the cube of the domain element by -8 and adds 3 times the square root of 3 to the result.

In Problems 77–80, translate each algebraic definition of the function into a verbal definition.

77. $f(x) = 2x - 3$ **78.** $g(x) = -2x + 7$

79. $F(x) = 3x^3 - 2\sqrt{x}$ **80.** $G(x) = 4\sqrt{x} - x^2$

Determine which of the equations in Problems 81–90 specify functions with independent variable x. For those that do, find the domain. For those that do not, find a value of x to which there corresponds more than one value of y.

81. $4x - 5y = 20$　　**82.** $3y - 7x = 15$

83. $x^2 - y = 1$　　**84.** $x - y^2 = 1$

85. $x + y^2 = 10$　　**86.** $x^2 + y = 10$

87. $xy - 4y = 1$　　**88.** $xy + y - x = 5$

89. $x^2 + y^2 = 25$　　**90.** $x^2 - y^2 = 16$

91. If $F(t) = 4t + 7$, find
$$\frac{F(3 + h) - F(3)}{h}$$

92. If $G(r) = 3 - 5r$, find
$$\frac{G(2 + h) - G(2)}{h}$$

93. If $Q(x) = x^2 - 5x + 1$, find
$$\frac{Q(2 + h) - Q(2)}{h}$$

94. If $P(x) = 2x^2 - 3x - 7$, find
$$\frac{P(3 + h) - P(3)}{h}$$

If $f(x) = x^2 - 1$, find and simplify each expression in Problems 95–106.

95. $f(5)$　　　　**96.** $f(-3)$

97. $f(2 + 5)$　　**98.** $f(3 - 6)$

99. $f(2) + f(5)$　　**100.** $f(3) - f(6)$

101. $f(f(1))$　　**102.** $f(f(-2))$

103. $f(2x)$　　　**104.** $f(-3x)$

105. $f(x + 1)$　　**106.** $f(1 - x)$

C　*In Problems 107–112, find and simplify each of the following.*

(A) $f(x + h)$

(B) $f(x + h) - f(x)$

(C) $\dfrac{f(x + h) - f(x)}{h}$

107. $f(x) = 4x - 3$　　**108.** $f(x) = -3x + 9$

109. $f(x) = 4x^2 - 7x + 6$　　**110.** $f(x) = 3x^2 + 5x - 8$

111. $f(x) = x(20 - x)$　　**112.** $f(x) = x(x + 40)$

Problems 113–116, refer to the area A and perimeter P of a rectangle with length l and width w (see the figure).

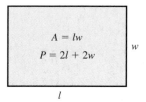

$$A = lw$$
$$P = 2l + 2w$$

113. The area of a rectangle is 25 square inches. Express the perimeter $P(w)$ as a function of the width w, and state the domain of this function.

114. The area of a rectangle is 81 square inches. Express the perimeter $P(l)$ as a function of the length l, and state the domain of this function.

115. The perimeter of a rectangle is 100 meters. Express the area $A(l)$ as a function of the length l, and state the domain of this function.

116. The perimeter of a rectangle is 160 meters. Express the area $A(w)$ as a function of the width w, and state the domain of this function.

Applications

117. *Price–demand.* A company manufactures memory chips for microcomputers. Its marketing research department, using statistical techniques, collected the data shown in Table 8, where p is the wholesale price per chip at which x million chips can be sold. Using special analytical techniques (regression analysis), an analyst produced the following price–demand function to model the data:

$$p(x) = 75 - 3x \qquad 1 \le x \le 20$$

TABLE 8　Price–Demand

x (Millions)	p ($)
1	72
4	63
9	48
14	33
20	15

Plot the data points in Table 8, and sketch a graph of the price–demand function in the same coordinate system. What would be the estimated price per chip for a demand of 7 million chips? For a demand of 11 million chips?

118. *Price–demand.* A company manufactures "notebook" computers. Its marketing research department, using statistical techniques, collected the data shown in Table 9, where p is the wholesale price per computer at which x thousand computers can be sold. Using special analytical techniques (regression analysis), an analyst produced the following price–demand function to model the data:

$$p(x) = 2{,}000 - 60x \qquad 1 \le x \le 25$$

TABLE 9　Price–Demand

x (Thousands)	p ($)
1	1,940
8	1,520
16	1,040
21	740
25	500

Plot the data points in Table 9, and sketch a graph of the price–demand function in the same coordinate system. What would be the estimated price per computer for a demand of 11 thousand computers? For a demand of 18 thousand computers?

119. *Revenue.*

(A) Using the price–demand function

$$p(x) = 75 - 3x \quad 1 \le x \le 20$$

from Problem 117, write the company's revenue function and indicate its domain.

(B) Complete Table 10, computing revenues to the nearest million dollars.

TABLE 10 Revenue	
x (Millions)	*R(x)* (Million $)
1	72
4	
8	
12	
16	
20	

(C) Plot the points from part (B) and sketch a graph of the revenue function through these points. Choose millions for the units on the horizontal and vertical axes.

120. *Revenue.*

(A) Using the price–demand function

$$p(x) = 2,000 - 60x \quad 1 \le x \le 25$$

from Problem 118, write the company's revenue function and indicate its domain.

(B) Complete Table 11, computing revenues to the nearest thousand dollars.

TABLE 11 Revenue	
x (Thousands)	*R(x)* (Thousand $)
1	1,940
5	
10	
15	
20	
25	

(C) Plot the points from part (B) and sketch a graph of the revenue function through these points. Choose thousands for the units on the horizontal and vertical axes.

121. *Profit.* The financial department for the company in Problems 117 and 119 established the following cost function for producing and selling *x* million memory chips:

$$C(x) = 125 + 16x \text{ million dollars}$$

(A) Write a profit function for producing and selling *x* million memory chips, and indicate its domain.

(B) Complete Table 12, computing profits to the nearest million dollars.

TABLE 12 Profit	
x (Millions)	*P(x)* (Million $)
1	−69
4	
8	
12	
16	
20	

(C) Plot the points in part (B) and sketch a graph of the profit function through these points.

122. *Profit.* The financial department for the company in Problems 118 and 120 established the following cost function for producing and selling *x* thousand "notebook" computers:

$$C(x) = 4,000 + 500x \text{ thousand dollars}$$

(A) Write a profit function for producing and selling *x* thousand "notebook" computers, and indicate the domain of this function.

(B) Complete Table 13, computing profits to the nearest thousand dollars.

TABLE 13 Profit	
x (Thousands)	*P(x)* (Thousand $)
1	−2,560
5	
10	
15	
20	
25	

(C) Plot the points in part (B) and sketch a graph of the profit function through these points.

123. *Packaging.* A candy box is to be made out of a piece of cardboard that measures 8 by 12 inches. Equal-sized squares *x* inches on a side will be cut out of each corner, and then the ends and sides will be folded up to form a rectangular box.

(A) Express the volume of the box *V(x)* in terms of *x*.

(B) What is the domain of the function *V* (determined by the physical restrictions)?

(C) Complete Table 14.

TABLE 14 Volume	
x	*V(x)*
1	
2	
3	

(D) Plot the points in part (C) and sketch a graph of the volume function through these points.

124. *Packaging.* Refer to Problem 123.

(A) Table 15 shows the volume of the box for some values of *x* between 1 and 2. Use these values to estimate to one

TABLE 15 Volume	
x	*V(x)*
1.1	62.524
1.2	64.512
1.3	65.988
1.4	66.976
1.5	67.5
1.6	67.584
1.7	67.252

decimal place the value of x between 1 and 2 that would produce a box with a volume of 65 cubic inches.

(B) Describe how you could refine this table to estimate x to two decimal places.

(C) Carry out the refinement you described in part (B) and approximate x to two decimal places.

125. *Packaging.* Refer to Problems 123 and 124.

(A) Examine the graph of $V(x)$ from Problem 123D and discuss the possible locations of other values of x that would produce a box with a volume of 65 cubic inches. Construct a table like Table 15 to estimate any such value to one decimal place.

(B) Refine the table you constructed in part (A) to provide an approximation to two decimal places.

126. *Packaging.* A parcel delivery service will only deliver packages with length plus girth (distance around) not exceeding 108 inches. A rectangular shipping box with square ends x inches on a side is to be used.

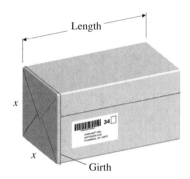

(A) If the full 108 inches is to be used, express the volume of the box $V(x)$ in terms of x.

(B) What is the domain of the function V (determined by the physical restrictions)?

(C) Complete Table 16.

TABLE 16 Volume

x	$V(x)$
5	
10	
15	
20	
25	

(D) Plot the points in part (C) and sketch a graph of the volume function through these points.

127. *Muscle contraction.* In a study of the speed of muscle contraction in frogs under various loads, noted British biophysicist and Nobel Prize winner A. W. Hill determined that the weight w (in grams) placed on the muscle and the speed of contraction v (in centimeters per second) are approximately related by an equation of the form

$$(w + a)(v + b) = c$$

where $a, b,$ and c are constants. Suppose that for a certain muscle, $a = 15, b = 1,$ and $c = 90$. Express v as a function of w. Find the speed of contraction if a weight of 16 grams is placed on the muscle.

128. *Politics.* The percentage s of seats in the House of Representatives won by Democrats and the percentage v of votes cast for Democrats (when expressed as decimal fractions) are related by the equation

$$5v - 2s = 1.4 \qquad 0 < s < 1, \quad 0.28 < v < 0.68$$

(A) Express v as a function of s, and find the percentage of votes required for the Democrats to win 51% of the seats.

(B) Express s as a function of v, and find the percentage of seats won if Democrats receive 51% of the votes.

Section 2-2 ELEMENTARY FUNCTIONS: GRAPHS AND TRANSFORMATIONS

- A Beginning Library of Elementary Functions
- Vertical and Horizontal Shifts
- Reflections, Stretches, and Shrinks
- Piecewise-Defined Functions

The functions

$$g(x) = x^2 - 4 \qquad h(x) = (x - 4)^2 \qquad k(x) = -4x^2$$

all can be expressed in terms of the function $f(x) = x^2$ as follows:

$$g(x) = f(x) - 4 \qquad h(x) = f(x - 4) \qquad k(x) = -4f(x)$$

In this section we will see that the graphs of functions $g, h,$ and k are closely related to the graph of function f. Insight gained by understanding these relationships will help us analyze and interpret the graphs of many different functions.

■ **A Beginning Library of Elementary Functions**

As you progress through this book, and most any other mathematics course beyond this one, you will repeatedly encounter a relatively small list of elementary functions. We will identify these functions, study their basic properties, and include them in a library of elementary functions (see the inside front cover). This library will become an important addition to your mathematical toolbox and can be used in any course or activity where mathematics is applied.

We begin by placing six basic functions in our library.

DEFINITION | **Basic Elementary Functions**

$$f(x) = x \qquad \text{Identity function}$$
$$h(x) = x^2 \qquad \text{Square function}$$
$$m(x) = x^3 \qquad \text{Cube function}$$
$$n(x) = \sqrt{x} \qquad \text{Square root function}$$
$$p(x) = \sqrt[3]{x} \qquad \text{Cube root function}$$
$$g(x) = |x| \qquad \text{Absolute value function}$$

These elementary functions can be evaluated by hand for certain values of x and with a calculator for all values of x for which they are defined.

EXAMPLE 1 | **Evaluating Basic Elementary Functions** Evaluate each basic elementary function at

(A) $x = 64$

(B) $x = -12.75$

Round any approximate values to four decimal places.

SOLUTION (A) $f(64) = 64$

$h(64) = 64^2 = 4{,}096$ *Use a calculator.*

$m(64) = 64^3 = 262{,}144$ *Use a calculator.*

$n(64) = \sqrt{64} = 8$

$p(64) = \sqrt[3]{64} = 4$

$g(64) = |64| = 64$

(B) $f(-12.75) = -12.75$

$h(-12.75) = (-12.75)^2 = 162.5625$ *Use a calculator.*

$m(-12.75) = (-12.75)^3 \approx -2{,}072.6719$ *Use a calculator.*

$n(-12.75) = \sqrt{-12.75}$ *Not a real number.*

$p(-12.75) = \sqrt[3]{-12.75} \approx -2.3362$ *Use a calculator.*

$g(-12.75) = |-12.75| = 12.75$

MATCHED PROBLEM 1 Evaluate each basic elementary function at

(A) $x = 729$ (B) $x = -5.25$

Round any approximate values to four decimal places.

REMARK   Most computers and graphing calculators use ABS(x) to represent the absolute value function. The following representation can also be useful:

$$|x| = \sqrt{x^2}$$

Figure 1 shows the graph, range, and domain of each of the basic elementary functions.

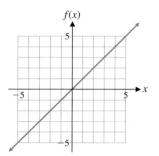

(A) **Identity function**
$f(x) = x$
Domain: R
Range: R

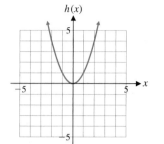

(B) **Square function**
$h(x) = x^2$
Domain: R
Range: $[0, \infty)$

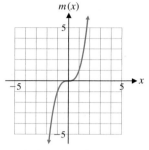

(C) **Cube function**
$m(x) = x^3$
Domain: R
Range: R

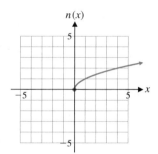

(D) **Square root function**
$n(x) = \sqrt{x}$
Domain: $[0, \infty)$
Range: $[0, \infty)$

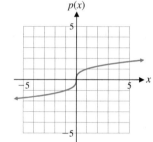

(E) **Cube root function**
$p(x) = \sqrt[3]{x}$
Domain: R
Range: R

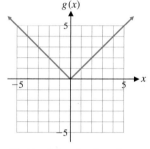

(F) **Absolute value function**
$g(x) = |x|$
Domain: R
Range: $[0, \infty)$

FIGURE 1 Some basic functions and their graphs
Note: Letters used to designate these functions may vary from context to context; R is the set of all real numbers.

INSIGHT

Absolute Value In beginning algebra, absolute value is often interpreted as distance from the origin on a real number line (see Appendix A, Section A-1).

If $x < 0$, then $-x$ is the *positive* distance from the origin to x and if $x > 0$, then x is the positive distance from the origin to x. Thus,

$$|x| = \begin{cases} -x & \text{if } x < 0 \\ x & \text{if } x \geq 0 \end{cases}$$

■ Vertical and Horizontal Shifts

If a new function is formed by performing an operation on a given function, then the graph of the new function is called a **transformation** of the graph of the original function. For example, graphs of both $y = f(x) + k$ and $y = f(x + h)$ are transformations of the graph of $y = f(x)$.

Explore & Discuss **1**   Let $f(x) = x^2$.

(A) Graph $y = f(x) + k$ for $k = -4, 0$, and 2 simultaneously in the same coordinate system. Describe the relationship between the graph of $y = f(x)$ and the graph of $y = f(x) + k$ for k any real number.

(B) Graph $y = f(x + h)$ for $h = -4, 0$, and 2 simultaneously in the same coordinate system. Describe the relationship between the graph of $y = f(x)$ and the graph of $y = f(x + h)$ for h any real number.

EXAMPLE 2   **Vertical and Horizontal Shifts**

(A) How are the graphs of $y = |x| + 4$ and $y = |x| - 5$ related to the graph of $y = |x|$? Confirm your answer by graphing all three functions simultaneously in the same coordinate system.

(B) How are the graphs of $y = |x + 4|$ and $y = |x - 5|$ related to the graph of $y = |x|$? Confirm your answer by graphing all three functions simultaneously in the same coordinate system.

SOLUTION   (A) The graph of $y = |x| + 4$ is the same as the graph of $y = |x|$ shifted upward 4 units, and the graph of $y = |x| - 5$ is the same as the graph of $y = |x|$ shifted downward 5 units. Figure 2 confirms these conclusions. [It appears that the graph of $y = f(x) + k$ is the graph of $y = f(x)$ shifted up if k is positive and down if k is negative.]

(B) The graph of $y = |x + 4|$ is the same as the graph of $y = |x|$ shifted to the left 4 units, and the graph of $y = |x - 5|$ is the same as the graph of $y = |x|$ shifted to the right 5 units. Figure 3 confirms these conclusions. [It appears that the graph of $y = f(x + h)$ is the graph of $y = f(x)$ shifted right if h is negative and left if h is positive—the opposite of what you might expect.]

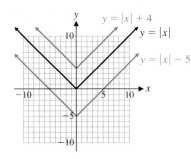

FIGURE 2 Vertical shifts

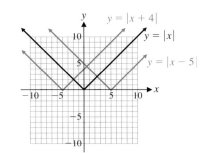

FIGURE 3 Horizontal shifts

MATCHED PROBLEM 2   (A) How are the graphs of $y = \sqrt{x} + 5$ and $y = \sqrt{x} - 4$ related to the graph of $y = \sqrt{x}$? Confirm your answer by graphing all three functions simultaneously in the same coordinate system.

(B) How are the graphs of $y = \sqrt{x + 5}$ and $y = \sqrt{x - 4}$ related to the graph of $y = \sqrt{x}$? Confirm your answer by graphing all three functions simultaneously in the same coordinate system.

Comparing the graphs of $y = f(x) + k$ with the graph of $y = f(x)$, we see that the graph of $y = f(x) + k$ can be obtained from the graph of $y = f(x)$ by **vertically translating** (shifting) the graph of the latter upward k units if k is positive and downward $|k|$ units if k is negative. Comparing the graphs of $y = f(x + h)$ with the graph of $y = f(x)$, we see that the graph of $y = f(x + h)$ can be obtained from the graph of $y = f(x)$ by **horizontally translating** (shifting) the graph of the latter h units to the left if h is positive and $|h|$ units to the right if h is negative.

| EXAMPLE 3 | **Vertical and Horizontal Translations (Shifts)** The graphs in Figure 4 are either horizontal or vertical shifts of the graph of $f(x) = x^2$. Write appropriate equations for functions $H, G, M,$ and N in terms of f. |

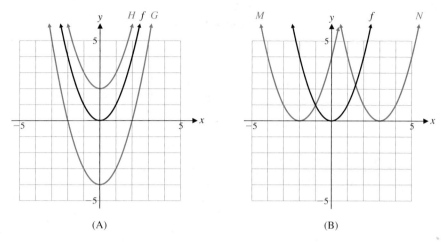

(A)   (B)

FIGURE 4 Vertical and horizontal shifts

SOLUTION   Functions H and G are vertical shifts given by

$$H(x) = x^2 + 2 \qquad G(x) = x^2 - 4$$

Functions M and N are horizontal shifts given by

$$M(x) = (x + 2)^2 \qquad N(x) = (x - 3)^2$$

| MATCHED PROBLEM 3 | The graphs in Figure 5 are either horizontal or vertical shifts of the graph of $f(x) = \sqrt[3]{x}$. Write appropriate equations for functions $H, G, M,$ and N in terms of f. |

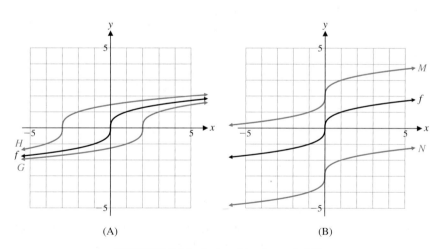

(A)   (B)

FIGURE 5 Vertical and horizontal shifts

■ Reflections, Stretches, and Shrinks

We now investigate how the graph of $y = Af(x)$ is related to the graph of $y = f(x)$ for different real numbers A.

Explore & Discuss **2**

(A) Graph $y = Ax^2$ for $A = 1, 4$, and $\frac{1}{4}$ simultaneously in the same coordinate system.

(B) Graph $y = Ax^2$ for $A = -1, -4$, and $-\frac{1}{4}$ simultaneously in the same coordinate system.

(C) Describe the relationship between the graph of $h(x) = x^2$ and the graph of $G(x) = Ax^2$ for A any real number.

Comparing $y = Af(x)$ to $y = f(x)$, we see that the graph of $y = Af(x)$ can be obtained from the graph of $y = f(x)$ by multiplying each ordinate value of the latter by A. The result is a **vertical stretch** of the graph of $y = f(x)$ if $A > 1$, a **vertical shrink** of the graph of $y = f(x)$ if $0 < A < 1$, and a **reflection in the x axis** if $A = -1$. If A is a negative number other than -1, then the result is a combination of a reflection in the x axis and either a vertical stretch or a vertical shrink.

EXAMPLE 4	**Reflections, Stretches, and Shrinks**

(A) How are the graphs of $y = 2|x|$ and $y = 0.5|x|$ related to the graph of $y = |x|$? Confirm your answer by graphing all three functions simultaneously in the same coordinate system.

(B) How is the graph of $y = -2|x|$ related to the graph of $y = |x|$? Confirm your answer by graphing both functions simultaneously in the same coordinate system.

SOLUTION

(A) The graph of $y = 2|x|$ is a vertical stretch of the graph of $y = |x|$ by a factor of 2, and the graph of $y = 0.5|x|$ is a vertical shrink of the graph of $y = |x|$ by a factor of 0.5. Figure 6 confirms this conclusion.

(B) The graph of $y = -2|x|$ is a reflection in the x axis and a vertical stretch of the graph of $y = |x|$. Figure 7 confirms this conclusion.

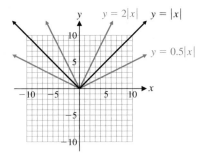

FIGURE 6 Vertical stretch and shrink

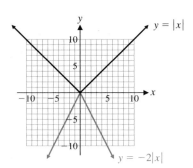

FIGURE 7 Reflection and vertical stretch

MATCHED PROBLEM 4

(A) How are the graphs of $y = 2x$ and $y = 0.5x$ related to the graph of $y = x$? Confirm your answer by graphing all three functions simultaneously in the same coordinate system.

(B) How is the graph of $y = -0.5x$ related to the graph of $y = x$? Confirm your answer by graphing both functions in the same coordinate system.

The various transformations considered above are summarized in the following box for easy reference:

SUMMARY   **GRAPH TRANSFORMATIONS**

Vertical Translation:

$$y = f(x) + k \quad \begin{cases} k > 0 & \text{Shift graph of } y = f(x) \text{ up } k \text{ units.} \\ k < 0 & \text{Shift graph of } y = f(x) \text{ down } |k| \text{ units.} \end{cases}$$

Horizontal Translation:

$$y = f(x + h) \quad \begin{cases} h > 0 & \text{Shift graph of } y = f(x) \text{ left } h \text{ units.} \\ h < 0 & \text{Shift graph of } y = f(x) \text{ right } |h| \text{ units.} \end{cases}$$

Reflection:

$$y = -f(x) \quad \text{Reflect the graph of } y = f(x) \text{ in the } x \text{ axis.}$$

Vertical Stretch and Shrink:

$$y = Af(x) \quad \begin{cases} A > 1 & \text{Stretch graph of } y = f(x) \text{ vertically} \\ & \text{by multiplying each ordinate value by } A. \\ 0 < A < 1 & \text{Shrink graph of } y = f(x) \text{ vertically} \\ & \text{by multiplying each ordinate value by } A. \end{cases}$$

Explore & Discuss **3**   Use a graphing calculator to explore the graph of $y = A(x + h)^2 + k$ for various values of the constants A, h, and k. Discuss how the graph of $y = A(x + h)^2 + k$ is related to the graph of $y = x^2$.

| **EXAMPLE 5** | **Combining Graph Transformations** Discuss the relationship between the graphs of $y = -|x - 3| + 1$ and $y = |x|$. Confirm your answer by graphing both functions simultaneously in the same coordinate system. |

SOLUTION   The graph of $y = -|x - 3| + 1$ is a reflection in the x axis, a horizontal translation of 3 units to the right, and a vertical translation of 1 unit upward of the graph of $y = |x|$. Figure 8 confirms this description.

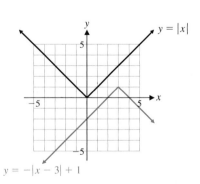

FIGURE 8 Combined transformations

| MATCHED PROBLEM 5 | The graph of $y = G(x)$ in Figure 9 on the next page involves a reflection and a translation of the graph of $y = x^3$. Describe how the graph of function G is related to the graph of $y = x^3$ and find an equation of the function G. |

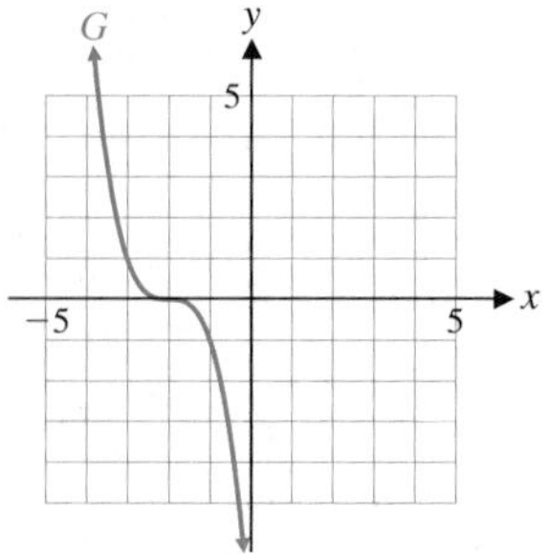

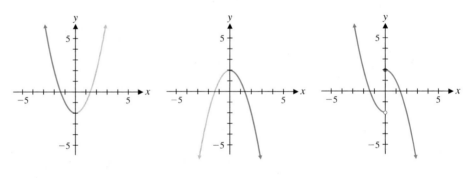

FIGURE 9 Combined transformations

Piecewise-Defined Functions

Earlier we noted that the absolute value of a real number x can be defined as

$$|x| = \begin{cases} -x & \text{if } x < 0 \\ x & \text{if } x \geq 0 \end{cases}$$

Notice that this function is defined by different rules for different parts of its domain. Functions whose definitions involve more than one rule are called **piecewise-defined functions.** Graphing one of these functions involves graphing each rule over the appropriate portion of the domain (Fig. 10). In Figure 10C, notice that an open dot is used to show that the point $(0, -2)$ is not part of the graph and a solid dot is used to show that $(0, 2)$ is part of the graph.

As the next example illustrates, piecewise-defined functions occur naturally in many applications.

(A) $y = x^2 - 2$ (B) $y = 2 - x^2$ (C) $y = \begin{cases} x^2 - 2 & \text{if } x < 0 \\ 2 - x^2 & \text{if } x \geq 0 \end{cases}$

FIGURE 10 Graphing a piecewise-defined function

EXAMPLE 6

Natural Gas Rates Easton Utilities uses the rates shown in Table 1 to compute the monthly cost of natural gas for each customer. Write a piecewise definition for the cost of consuming x CCF (cubic hundred feet) of natural gas and graph the function.

TABLE 1 Charges per Month
$0.7866 per CCF for the first 5 CCF
$0.4601 per CCF for the next 35 CCF
$0.2508 per CCF for all over 40 CCF

SOLUTION If $C(x)$ is the cost, in dollars, of using x CCF of natural gas in one month, then the first line of Table 1 implies that

$$C(x) = 0.7866x \quad \text{if } 0 \leq x \leq 5$$

Note that $C(5) = 3.933$ is the cost of 5 CCF. If $5 < x \leq 40$, then $x - 5$ represents the amount of gas that cost $0.4601 per CCF, $0.4601(x - 5)$ represents the cost of this gas, and the total cost is

$$C(x) = 3.933 + 0.4601(x - 5)$$

If $x > 40$, then

$$C(x) = 20.0365 + 0.2508(x - 40)$$

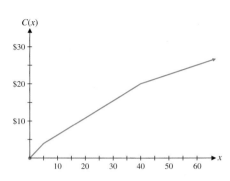

$C(x)$

$30

$20

$10

10 20 30 40 50 60 x

FIGURE 11 Cost of purchasing x CCF of natural gas

where $20.0365 = C(40)$, the cost of the first 40 CCF. Combining all these equations, we have the following piecewise definition for $C(x)$:

$$C(x) = \begin{cases} 0.7866x & \text{if } 0 \leq x \leq 5 \\ 3.933 + 0.46501\,(x - 5) & \text{if } 5 < x \leq 40 \\ 20.0365 + 0.2508\,(x - 40) & \text{if } 40 < x \end{cases}$$

To graph C, first note that each rule in the definition of C represents a transformation of the identity function $f(x) = x$. Graphing each transformation over the indicated interval produces the graph of C shown in Figure 11. ∎

MATCHED PROBLEM 6

Natural Gas Rates Trussville Utilities uses the rates shown in Table 2 to compute the monthly cost of natural gas for residential customers. Write a piecewise definition for the cost of consuming x CCF of natural gas and graph the function.

TABLE 2   Charges per Month
$0.7675 per CCF for the first 50 CCF
$0.6400 per CCF for the next 150 CCF
$0.6130 per CCF for all over 200 CCF

Answers to Matched Problems

1. (A) $f(729) = 729$, $h(729) = 531{,}441$, $m(729) = 387{,}420{,}489$,
    $n(729) = 27$, $p(729) = 9$, $g(729) = 729$
   (B) $f(-5.25) = -5.25$, $h(-5.25) = 27.5625$,
    $m(-5.25) = -144.7031$, $n(-5.25)$ is not a real number,
    $p(-5.25) = -1.7380$, $g(-5.25) = 5.25$

2. (A) The graph of $y = \sqrt{x} + 5$ is the same as the graph of $y = \sqrt{x}$ shifted upward 5 units, and the graph of $y = \sqrt{x} - 4$ is the same as the graph of $y = \sqrt{x}$ shifted downward 4 units. The figure confirms these conclusions.

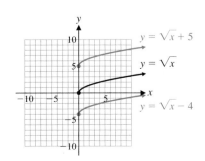

(B) The graph of $y = \sqrt{x + 5}$ is the same as the graph of $y = \sqrt{x}$ shifted to the left 5 units, and the graph of $y = \sqrt{x - 4}$ is the same as the graph of $y = \sqrt{x}$ shifted to the right 4 units. The figure confirms these conclusions.

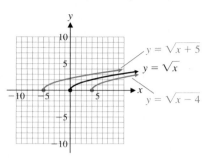

3. $H(x) = \sqrt[3]{x + 3}, G(x) = \sqrt[3]{x - 2}, M(x) = \sqrt[3]{x} + 2, N(x) = \sqrt[3]{x} - 3$

4. (A) The graph of $y = 2x$ is a vertical stretch of the graph of $y = x$, and the graph of $y = 0.5x$ is a vertical shrink of the graph of $y = x$. The figure confirms these conclusions.

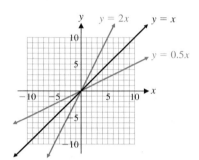

(B) The graph of $y = -0.5x$ is a vertical shrink and a reflection in the x axis of the graph of $y = x$. The figure confirms this conclusion.

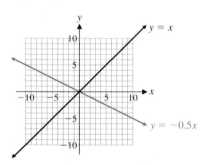

5. The graph of function G is a reflection in the x axis and a horizontal translation of 2 units to the left of the graph of $y = x^3$. An equation for G is $G(x) = -(x + 2)^3$.

6. $C(x) = \begin{cases} 0.7675x & \text{if } 0 \le x \le 50 \\ 38.375 + 0.64\,(x - 50) & \text{if } 50 < x \le 200 \\ 134.375 + 0.613\,(x - 200) & \text{if } 200 < x \end{cases}$

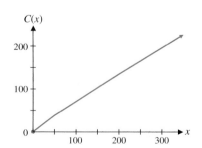

Exercise 2-2

A *Without looking back in the text, indicate the domain and range of each of the functions in Problems 1–8.*

1. $f(x) = 2x$

2. $g(x) = -0.3x$

3. $h(x) = -0.6\sqrt{x}$

4. $k(x) = 4\sqrt{x}$

5. $m(x) = 3|x|$

6. $n(x) = -0.1x^2$

7. $r(x) = -x^3$

8. $s(x) = 5\sqrt[3]{x}$

Graph each of the functions in Problems 9–20 using the graphs of functions f and g below.

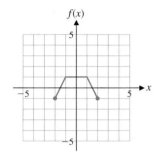

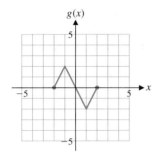

9. $y = f(x) + 2$

10. $y = g(x) - 1$

11. $y = f(x + 2)$

12. $y = g(x - 1)$

13. $y = g(x - 3)$

14. $y = f(x + 3)$

15. $y = g(x) - 3$

16. $y = f(x) + 3$

17. $y = -f(x)$

18. $y = -g(x)$

19. $y = 0.5g(x)$

20. $y = 2f(x)$

B *In Problems 21–28, indicate verbally how the graph of each function is related to the graph of one of the six basic functions in Figure 1 on page 65. Sketch a graph of each function.*

21. $g(x) = -|x + 3|$

22. $h(x) = -|x - 5|$

23. $f(x) = (x - 4)^2 - 3$

24. $m(x) = (x + 3)^2 + 4$

25. $f(x) = 7 - \sqrt{x}$

26. $g(x) = -6 + \sqrt[3]{x}$

27. $h(x) = -3|x|$

28. $m(x) = -0.4x^2$

Each graph in Problems 29–36 is the result of applying a sequence of transformations to the graph of one of the six basic functions in Figure 1 on page 65. Identify the basic function and describe the transformation verbally. Write an equation for the given graph.

29.

30.

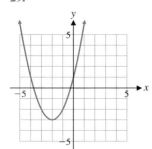

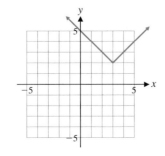

31.

32.

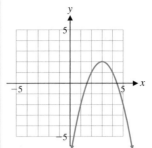

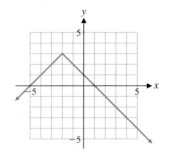

33.

34.

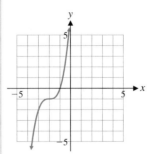

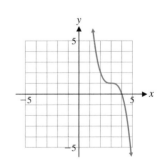

35.

36.

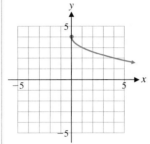

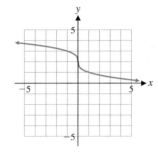

In Problems 37–42, the graph of the function g is formed by applying the indicated sequence of transformations to the given function f. Find an equation for the function g and graph g using $-5 \le x \le 5$ and $-5 \le y \le 5$.

37. The graph of $f(x) = \sqrt{x}$ is shifted 2 units to the right and 3 units down.

38. The graph of $f(x) = \sqrt[3]{x}$ is shifted 3 units to the left and 2 units up.

39. The graph of $f(x) = |x|$ is reflected in the x axis and shifted to the left 3 units.

40. The graph of $f(x) = |x|$ is reflected in the x axis and shifted to the right 1 unit.

41. The graph of $f(x) = x^3$ is reflected in the x axis and shifted 2 units to the right and down 1 unit.

42. The graph of $f(x) = x^2$ is reflected in the x axis and shifted to the left 2 units and up 4 units.

Graph each function in Problems 43–48.

43. $f(x) = \begin{cases} 2 - 2x & \text{if } x < 2 \\ x - 2 & \text{if } x \geq 2 \end{cases}$

44. $g(x) = \begin{cases} x + 1 & \text{if } x < -1 \\ 2 + 2x & \text{if } x \geq -1 \end{cases}$

45. $h(x) = \begin{cases} 5 + 0.5x & \text{if } 0 \leq x \leq 10 \\ -10 + 2x & \text{if } x > 10 \end{cases}$

46. $h(x) = \begin{cases} 10 + 2x & \text{if } 0 \leq x \leq 20 \\ 40 + 0.5x & \text{if } x > 20 \end{cases}$

47. $h(x) = \begin{cases} 2x & \text{if } 0 \leq x \leq 20 \\ x + 20 & \text{if } 20 < x \leq 40 \\ 0.5x + 45 & \text{if } x > 40 \end{cases}$

48. $h(x) = \begin{cases} 4x + 20 & \text{if } 0 \leq x \leq 20 \\ 2x + 60 & \text{if } 20 < x \leq 100 \\ -x + 360 & \text{if } x > 100 \end{cases}$

C *Each of the graphs in Problems 49–54 involves a reflection in the x axis and/or a vertical stretch or shrink of one of the basic functions in Figure 1 on page 65. Identify the basic function, and describe the transformation verbally. Write an equation for the given graph.*

49.

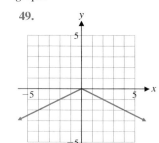

50.

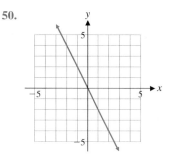

51.

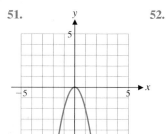

52.

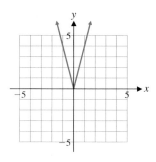

53.

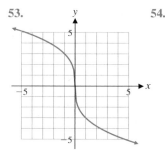

54.

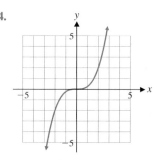

Changing the order in a sequence of transformations may change the final result. Investigate each pair of transformations in Problems 55–60 to determine if reversing their order can produce a different result. Support your conclusions with specific examples and/or mathematical arguments.

55. Vertical shift; horizontal shift

56. Vertical shift; reflection in y axis

57. Vertical shift; reflection in x axis

58. Vertical shift; vertical stretch

59. Horizontal shift; reflection in y axis

60. Horizontal shift; vertical shrink

Applications

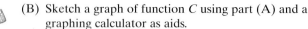

61. *Price–demand.* A retail chain sells CD players. The retail price $p(x)$ (in dollars) and the weekly demand x for a particular model are related by

$$p(x) = 115 - 4\sqrt{x} \qquad 9 \leq x \leq 289$$

(A) Describe how the graph of function p can be obtained from the graph of one of the basic functions in Figure 1 on page 65.

(B) Sketch a graph of function p using part (A) as an aid.

62. *Price–supply.* The manufacturers of the CD players in Problem 61 are willing to supply x players at a price of $p(x)$ as given by the equation

$$p(x) = 4\sqrt{x} \qquad 9 \leq x \leq 289$$

(A) Describe how the graph of function p can be obtained from the graph of one of the basic functions in Figure 1 on page 65.

(B) Sketch a graph of function p using part (A) as an aid.

63. *Hospital costs.* Using statistical methods, the financial department of a hospital arrived at the cost equation

$$C(x) = 0.00048(x - 500)^3 + 60{,}000 \qquad 100 \leq x \leq 1{,}000$$

where $C(x)$ is the cost in dollars for handling x cases per month.

(A) Describe how the graph of function C can be obtained from the graph of one of the basic functions in Figure 1 on page 65.

(B) Sketch a graph of function C using part (A) and a graphing calculator as aids.

64. *Price–demand.* A company manufactures and sells in-line skates. Its financial department has established the price–demand function

$$p(x) = 190 - 0.013(x - 10)^2 \qquad 10 \leq x \leq 100$$

where $p(x)$ is the price at which x thousand pairs of skates can be sold.

(A) Describe how the graph of function p can be obtained from the graph of one of the basic functions in Figure 1 on page 65.

(B) Sketch a graph of function p using part (A) and a graphing calculator as aids.

65. *Electricity rates.* Table 3 shows the electricity rates charged by Monroe Utilities in the summer months. The base is a fixed monthly charge, independent of the kWh (kilowatt-hours) used during the month.

(A) Write a piecewise definition of the monthly charge $S(x)$ for a customer who uses x kWh in a summer month.

(B) Graph $S(x)$.

TABLE 3   Summer (July–October)
Base charge, $8.50
First 700 kWh or less at 0.0650/kWh
Over 700 kWh at 0.0900/kWh

66. *Electricity rates.* Table 4 shows the electricity rates charged by Monroe Utilities in the winter months.

(A) Write a piecewise definition of the monthly charge $W(x)$ for a customer who uses x kWh in a winter month.

TABLE 4   Winter (November–June)
Base charge, $8.50
First 700 kWh or less at 0.0650/kWh
Over 700 kWh at 0.0530/kWh

(B) Graph $W(x)$.

67. *State income tax.* Table 5 shows a recent state income tax schedule for married couples filing a joint return in the state of Kansas.

(A) Write a piecewise definition for the tax due $T(x)$ on an income of x dollars.

(B) Graph $T(x)$.

(C) Find the tax due on a taxable income of $40,000. Of $70,000.

TABLE 5 Kansas State Income Tax		
SCHEDULE I—MARRIED FILING JOINT		
If taxable income is		
Over	But Not Over	Tax Due Is
$0	$30,000	3.50% of taxable income
$30,000	$60,000	$1,050 plus 6.25% of excess over $30,000
$60,000		$2,925 plus 6.45% of excess over $60,000

68. *State income tax.* Table 6 shows a recent state income tax schedule for individuals filing a return in the state of Kansas.

TABLE 6 Kansas State Income Tax		
SCHEDULE II—SINGLE, HEAD OF HOUSEHOLD, OR MARRIED FILING SEPARATE		
If taxable income is		
Over	But Not Over	Tax Due Is
$0	$15,000	3.50% of taxable income
$15,000	$30,000	$525 plus 6.25% of excess over $15,000
$30,000		$1,462.50 plus 6.45% of excess over $30,000

(A) Write a piecewise definition for the tax due $T(x)$ on an income of x dollars.

(B) Graph $T(x)$.

(C) Find the tax due on a taxable income of $20,000. Of $35,000.

(D) Would it be better for a married couple in Kansas with two equal incomes to file jointly or separately? Discuss.

69. *Physiology.* A good approximation of the normal weight of a person 60 inches or taller but not taller than 80 inches is given by $w(x) = 5.5x - 220$, where x is height in inches and $w(x)$ is weight in pounds.

(A) Describe how the graph of function w can be obtained from the graph of one of the basic functions in Figure 1, page 65.

(B) Sketch a graph of function w using part (A) as an aid.

70. *Physiology.* The average weight of a particular species of snake is given by $w(x) = 463x^3$, $0.2 \le x \le 0.8$, where x is length in meters and $w(x)$ is weight in grams.

(A) Describe how the graph of function w can be obtained from the graph of one of the basic functions in Figure 1, page 65.

(B) Sketch a graph of function w using part (A) as an aid.

71. *Safety research.* Under ideal conditions, if a person driving a vehicle slams on the brakes and skids to a stop, the speed of the vehicle $v(x)$ (in miles per hour) is given approximately by $v(x) = C\sqrt{x}$, where x is the length of skid marks (in feet) and C is a constant that depends on the road conditions and the weight of the vehicle. For a particular vehicle, $v(x) = 7.08\sqrt{x}$ and $4 \le x \le 144$.

(A) Describe how the graph of function v can be obtained from the graph of one of the basic functions in Figure 1, page 65.

(B) Sketch a graph of function v using part (A) as an aid.

72. *Learning.* A production analyst has found that on the average it takes a new person $T(x)$ minutes to perform a particular assembly operation after x performances of the operation, where $T(x) = 10 - \sqrt[3]{x}$, $0 \le x \le 125$.

(A) Describe how the graph of function T can be obtained from the graph of one of the basic functions in Figure 1, page 65.

(B) Sketch a graph of function T using part (A) as an aid.

Section 2-3 QUADRATIC FUNCTIONS

- Quadratic Functions, Equations, and Inequalities
- Properties of Quadratic Functions and Their Graphs
- Applications
- More General Functions: Polynomial and Rational Functions

If the degree of a linear function is increased by one, we obtain a *second-degree function,* usually called a *quadratic function,* another basic function that we will need in our library of elementary functions. We will investigate relationships between quadratic functions and the solutions to quadratic equations and inequalities. Other important properties of quadratic functions will also be investigated, including maximum and minimum properties. We will then be in a position to solve important practical problems such as finding production levels that will produce maximum revenue or maximum profit.

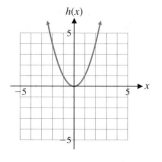

FIGURE 1 Square function $h(x) = x^2$

Quadratic Functions, Equations, and Inequalities

The graph of the square function $h(x) = x^2$ is shown in Figure 1. Notice that the graph is symmetric with respect to the y axis and that $(0, 0)$ is the lowest point on the graph. Let's explore the effect of applying a sequence of basic transformations to the graph of h.

Explore & Discuss **1**

Indicate how the graph of each function is related to the graph of the function $h(x) = x^2$. Find the highest or lowest point, whichever exists, on each graph.

(A) $f(x) = (x - 3)^2 - 7 = x^2 - 6x + 2$
(B) $g(x) = 0.5(x + 2)^2 + 3 = 0.5x^2 + 2x + 5$
(C) $m(x) = -(x - 4)^2 + 8 = -x^2 + 8x - 8$
(D) $n(x) = -3(x + 1)^2 - 1 = -3x^2 - 6x - 4$

Graphing the functions in Explore–Discuss 1 produces figures similar in shape to the graph of the square function in Figure 1. These figures are called *parabolas.* The functions that produced these parabolas are examples of the important class of *quadratic functions,* which we now define.

DEFINITION **Quadratic Functions** If $a, b,$ and c are real numbers with $a \ne 0$, then the function

$$f(x) = ax^2 + bx + c \qquad \text{Standard form}$$

is a **quadratic function** and its graph is a **parabola.**

> ### INSIGHT
>
> If x is any real number, then $ax^2 + bx + c$ is also a real number. According to the agreement on domain and range in Section 2-1, the domain of a quadratic function is R, the set of real numbers.

We will discuss methods for determining the range of a quadratic function later in this section. Typical graphs of quadratic functions are illustrated in Figure 2.

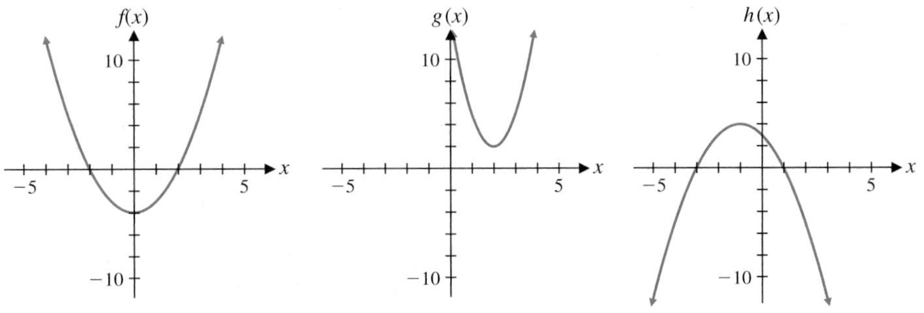

(A) $f(x) = x^2 - 4$ (B) $g(x) = 3x^2 - 12x + 14$ (C) $h(x) = 3 - 2x - x^2$

FIGURE 2 Graphs of quadratic functions

INSIGHT

The x intercepts of a linear function can be found by solving the linear equation $y = mx + b = 0$ for $x, m \neq 0$ (see Section 1-2). Similarly, the x intercepts of a quadratic function can be found by solving the quadratic equation $y = ax^2 + bx + c = 0$ for $x, a \neq 0$. Several methods for solving quadratic equations are discussed in Appendix A, Section A-7. The most popular of these is the **quadratic formula.**

If $ax^2 + bx + c = 0$, $a \neq 0$, then

$$x = \frac{-b \pm \sqrt{b^2 - 4ac}}{2a}, \text{ provided } b^2 - 4ac \geq 0$$

EXAMPLE 1   **Intercepts, Equations, and Inequalities**

(A) Sketch a graph of $f(x) = -x^2 + 5x + 3$ in a rectangular coordinate system.

(B) Find x and y intercepts algebraically to four decimal places.

(C) Graph $f(x) = -x^2 + 5x + 3$ in a standard viewing window.

(D) Find the x and y intercepts to four decimal places using TRACE and zero on your graphing calculator.

(E) Solve the quadratic inequality $-x^2 + 5x + 3 \geq 0$ graphically to four decimal places using the results of parts (A) and (B) or (C) and (D).

(F) Solve the equation $-x^2 + 5x + 3 = 4$ graphically to four decimal places using INTERSECT on your graphing calculator.

SOLUTION   (A) Hand-sketching a graph of f:

x	y
-1	-3
0	3
1	7
2	9
3	9
4	7
5	3
6	-3

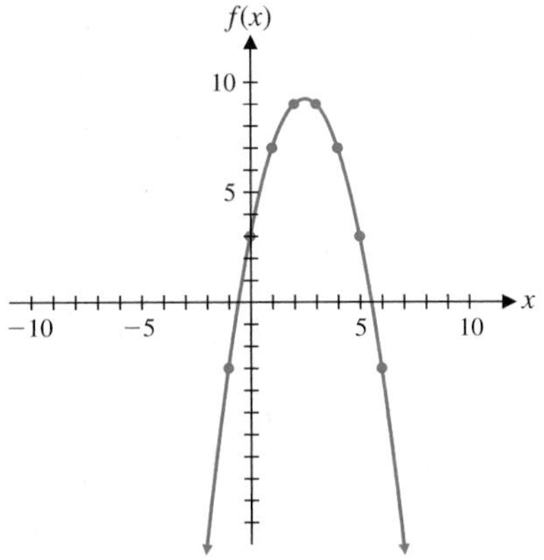

(B) Finding intercepts algebraically:

$$y \text{ intercept:} \quad f(0) = -(0)^2 + 5(0) + 3 = 3$$
$$x \text{ intercepts:} \quad f(x) = 0$$
$$-x^2 + 5x + 3 = 0 \qquad \textit{Quadratic equation}$$

$$x = \frac{-b \pm \sqrt{b^2 - 4ac}}{2a}$$ Quadratic formula (see Appendix A-7)

$$x = \frac{-(5) \pm \sqrt{5^2 - 4(-1)(3)}}{2(-1)}$$

$$= \frac{-5 \pm \sqrt{37}}{-2} = -0.5414 \quad \text{or} \quad 5.5414$$

(C) Graphing in a graphing calculator:

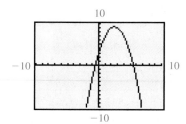

(D) Finding intercepts graphically using a graphing calculator:

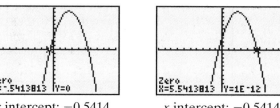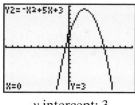

x intercept: -0.5414 x intercept: -0.5414 y intercept: 3

(E) Solving $-x^2 + 5x + 3 \geq 0$ graphically: The quadratic inequality

$$-x^2 + 5x + 3 \geq 0$$

holds for those values of x for which the graph of $f(x) = -x^2 + 5x + 3$ in the figures in parts (A) and (C) is at or above the x axis. This happens for x between the two x intercepts [found in part (B) or (D)], including the two x intercepts. Thus, the solution set for the quadratic inequality is $-0.5414 \leq x \leq 5.5414$ or $[-0.5414, 5.5414]$.

(F) Solving the equation $-x^2 + 5x + 3 = 4$ using a graphing calculator:

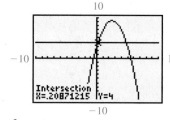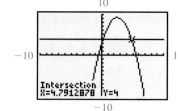

$-x^2 + 5x + 3 = 4$ at $x = 0.2087$ $-x^2 + 5x + 3 = 4$ at $x = 4.7913$ ▬

MATCHED PROBLEM 1

(A) Sketch a graph of $g(x) = 2x^2 - 5x - 5$ in a rectangular coordinate system.

(B) Find x and y intercepts algebraically to four decimal places.

(C) Graph $g(x) = 2x^2 - 5x - 5$ in a standard viewing window.

(D) Find the x and y intercepts to four decimal places using TRACE and the zero command on your graphing calculator.

(E) Solve $2x^2 - 5x - 5 \geq 0$ graphically to four decimal places using the results of parts (A) and (B) or (C) and (D).

(F) Solve the equation $2x^2 - 5x - 5 = -3$ graphically to four decimal places using an appropriate built-in routine in your graphing calculator.

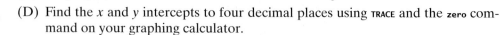

Explore & Discuss **2**   How many x intercepts can the graph of a quadratic function have? How many y intercepts? Explain your reasoning.

■ Properties of Quadratic Functions and Their Graphs

Many useful properties of the quadratic function can be uncovered by transforming

$$f(x) = ax^2 + bx + c \qquad a \neq 0$$

into the **vertex form***

$$f(x) = a(x - h)^2 + k$$

The process of *completing the square* (see Appendix A-7) is central to the transformation. We illustrate the process through a specific example and then generalize the results.

Consider the quadratic function given by

$$f(x) = -2x^2 + 16x - 24 \tag{1}$$

We use completing the square to transform this function into vertex form:

$$\begin{aligned}
f(x) &= -2x^2 + 16x - 24 & \text{Factor the coefficient of } x^2 \text{ out of} \\
&= -2(x^2 - 8x) - 24 & \text{the first two terms.} \\
&= -2(x^2 - 8x + \,?\,) - 24 \\[4pt]
&= -2(x^2 - 8x + \mathbf{16}) - 24 + \mathbf{32} & \text{Add 16 to complete the square in-} \\
& & \text{side the parentheses. Because of} \\
& & \text{the } -2 \text{ outside the parentheses, we} \\
& & \text{have actually added } -32, \text{ so we} \\
& & \text{must add 32 to the outside.} \\[4pt]
&= -2(x - 4)^2 + 8 & \text{The transformation is complete and} \\
& & \text{can be checked by multiplying out.}
\end{aligned}$$

Thus,

$$f(x) = -2(x - 4)^2 + 8 \tag{2}$$

If $x = 4$, then $-2(x - 4)^2 = 0$ and $f(4) = 8$. For any other value of x, the negative number $-2(x - 4)^2$ is added to 8, making it smaller. (Think about this.) Therefore,

$$f(4) = 8$$

is the *maximum value* of $f(x)$ for all x—a very important result! Furthermore, if we choose any two x values that are the same distance from 4, we will obtain the same function value. For example, $x = 3$ and $x = 5$ are each one unit from $x = 4$ and their function values are

$$f(3) = -2(3 - 4)^2 + 8 = 6$$
$$f(5) = -2(5 - 4)^2 + 8 = 6$$

Thus, the vertical line $x = 4$ is a line of symmetry. That is, if the graph of equation (1) is drawn on a piece of paper and the paper is folded along the line $x = 4$, then the two sides of the parabola will match exactly. All these results are illustrated by graphing equations (1) and (2) and the line $x = 4$ simultaneously in the same coordinate system (Fig. 3).

From the preceding discussion, we see that as x moves from left to right, $f(x)$ is increasing on $(-\infty, 4]$, and decreasing on $[4, \infty)$, and that $f(x)$ can assume no value greater than 8. Thus,

$$\text{Range of } f: \quad y \leq 8 \quad \text{or} \quad (-\infty, 8]$$

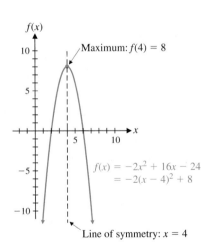

FIGURE 3 Graph of a quadratic function

Maximum: $f(4) = 8$

$f(x) = -2x^2 + 16x - 24$
$= -2(x - 4)^2 + 8$

Line of symmetry: $x = 4$

---

* This terminology is not universally agreed upon. Some call this the *standard form*.

In general, the graph of a quadratic function is a parabola with line of symmetry parallel to the vertical axis. The lowest or highest point on the parabola, whichever exists, is called the **vertex.** The maximum or minimum value of a quadratic function always occurs at the vertex of the parabola. The line of symmetry through the vertex is called the **axis** of the parabola. In the example above, $x = 4$ is the axis of the parabola and $(4, 8)$ is its vertex.

INSIGHT

Applying the graph transformation properties discussed in Section 2-2 to the transformed equation,

$$f(x) = -2x^2 + 16x - 24$$
$$= -2(x - 4)^2 + 8$$

we see that the graph of $f(x) = -2x^2 + 16x - 24$ is the graph of $g(x) = x^2$ vertically stretched by a factor of 2, reflected in the x axis, and shifted to the right 4 units and up 8 units, as shown in Figure 4.

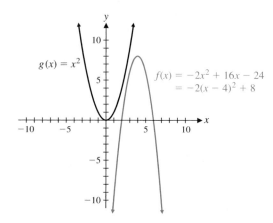

FIGURE 4 Graph of f is the graph of g transformed

Note the important results we have obtained from the vertex form of the quadratic function f:

- The vertex of the parabola
- The axis of the parabola
- The maximum value of $f(x)$
- The range of the function f
- The relationship between the graph of $g(x) = x^2$ and the graph of $f(x) = -2x^2 + 16x - 24$

Now, let us explore the effects of changing the constants a, h, and k on the graph of $f(x) = a(x - h)^2 + k$.

Explore & Discuss **3**

(A) Let $a = 1$ and $h = 5$. Graph $f(x) = a(x - h)^2 + k$ for $k = -4, 0$, and 3 simultaneously in the same coordinate system. Explain the effect of changing k on the graph of f.

(B) Let $a = 1$ and $k = 2$. Graph $f(x) = a(x - h)^2 + k$ for $h = -4, 0$, and 5 simultaneously in the same coordinate system. Explain the effect of changing h on the graph of f.

(C) Let $h = 5$ and $k = -2$. Graph $f(x) = a(x - h)^2 + k$ for $a = 0.25, 1$, and 3 simultaneously in the same coordinate system. Graph function f for

$a = 1, -1,$ and -0.25 simultaneously in the same coordinate system. Explain the effect of changing a on the graph of f.

 (D) Discuss parts (A)–(C) using a graphing calculator and a standard viewing window.

The preceding discussion is generalized for all quadratic functions in the following summary:

SUMMARY   **PROPERTIES OF A QUADRATIC FUNCTION AND ITS GRAPH**

Given a quadratic function and the vertex form obtained by completing the square

$$f(x) = ax^2 + bx + c \qquad a \neq 0 \quad \text{Standard form}$$
$$= a(x - h)^2 + k \qquad\qquad \text{Vertex form}$$

we summarize general properties as follows:

1. The graph of f is a parabola:

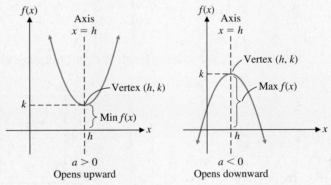

2. Vertex: (h, k) (parabola increases on one side of the vertex and decreases on the other)

3. Axis (of symmetry): $x = h$ (parallel to y axis)

4. $f(h) = k$ is the minimum if $a > 0$ and the maximum if $a < 0$

5. Domain: All real numbers
Range: $(-\infty, k]$ if $a < 0$ or $[k, \infty)$ if $a > 0$

6. The graph of f is the graph of $g(x) = ax^2$ translated horizontally h units and vertically k units.

EXAMPLE 2   **Analyzing a Quadratic Function** Given the quadratic function

$$f(x) = 0.5x^2 - 6x + 21$$

(A) Find the vertex form for f.

(B) Find the vertex and the maximum or minimum. State the range of f.

(C) Describe how the graph of function f can be obtained from the graph of $g(x) = x^2$ using transformations discussed in Section 2-2.

(D) Sketch a graph of function f in a rectangular coordinate system.

 (E) Graph function f using a suitable viewing window.

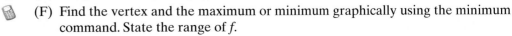 (F) Find the vertex and the maximum or minimum graphically using the minimum command. State the range of f.

SOLUTION   (A) Complete the square to find the vertex form:

$$f(x) = 0.5x^2 - 6x + 21$$
$$= 0.5(x^2 - 12x + ?) + 21$$
$$= 0.5(x^2 - 12x + 36) + 21 - 18$$
$$= 0.5(x - 6)^2 + 3$$

(B) From the vertex form, we see that $h = 6$ and $k = 3$. Thus, vertex: $(6, 3)$; minimum: $f(6) = 3$; range: $y \geq 3$ or $[3, \infty)$.

(C) The graph of $f(x) = 0.5(x - 6)^2 + 3$ is the same as the graph of $g(x) = x^2$ vertically shrunk by a factor of 0.5, and shifted to the right 6 units and up 3 units.

(D) Graph in a rectangular coordinate system:

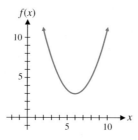

(E) Graph in a graphing calculator:

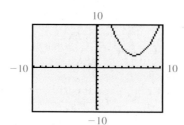

(F) Finding the vertex, minimum, and range graphically using a graphing calculator:

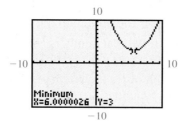

Vertex: $(6, 3)$; minimum: $f(6) = 3$; range: $y \geq 3$ or $[3, \infty)$.   ▬

MATCHED PROBLEM 2   Given the quadratic function $f(x) = -0.25x^2 - 2x + 2$

(A) Find the vertex form for f.

(B) Find the vertex and the maximum or minimum. State the range of f.

(C) Describe how the graph of function f can be obtained from the graph of $g(x) = x^2$ using transformations discussed in Section 2-2.

(D) Sketch a graph of function f in a rectangular coordinate system.

   (E) Graph function f using a suitable viewing window.

   (F) Find the vertex and the maximum or minimum graphically using the minimum command. State the range of f.

APPLICATIONS

EXAMPLE 3

Maximum Revenue This is a continuation of Example 6 in Section 2-1. Recall that the financial department in the company that produces a digital camera arrived at the following price–demand function and the corresponding revenue function:

$$p(x) = 94.8 - 5x \qquad \text{Price–demand function}$$
$$R(x) = xp(x) = x(94.8 - 5x) \qquad \text{Revenue function}$$

where $p(x)$ is the wholesale price per camera at which x million cameras can be sold and $R(x)$ is the corresponding revenue (in million dollars). Both functions have domain $1 \le x \le 15$.

(A) Find the output to the nearest thousand cameras that will produce the maximum revenue. What is the maximum revenue to the nearest thousand dollars? Solve the problem algebraically by completing the square.

(B) What is the wholesale price per camera (to the nearest dollar) that produces the maximum revenue?

 (C) Graph the revenue function using an appropriate viewing window.

 (D) Find the output to the nearest thousand cameras that will produce the maximum revenue. What is the maximum revenue to the nearest thousand dollars? Solve the problem graphically using the maximum command.

SOLUTION (A) Algebraic solution:

$$\begin{aligned}
R(x) &= x(94.8 - 5x) \\
&= -5x^2 + 94.8x \\
&= -5(x^2 - 18.96x + \;?) \\
&= -5(x^2 - 18.96x + 89.8704) + 449.352 \\
&= -5(x - 9.48)^2 + 449.352
\end{aligned}$$

The maximum revenue of 449.352 million dollars ($449,352,000) occurs when $x = 9.480$ million cameras (9,480,000 cameras).

(B) Finding the wholesale price per camera: Use the price–demand function for an output of 9.480 million cameras:

$$\begin{aligned}
p(x) &= 94.8 - 5x \\
p(9.480) &= 94.8 - 5(9.480) \\
&= \$47
\end{aligned}$$

 (C) Graph in a graphing calculator:

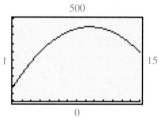

 (D) Graphical solution using a graphing calculator:

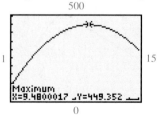

An output of 9.480 million cameras (9,480,000 cameras) will produce a maximum revenue of 449.352 million dollars ($449,352,000).

MATCHED PROBLEM 3 | The financial department in Example 3, using statistical and analytical techniques (see Matched Problem 7 in Section 2-1), arrived at the cost function

$$C(x) = 156 + 19.7x \qquad \text{Cost function}$$

where $C(x)$ is the cost (in million dollars) for manufacturing and selling x million cameras.

(A) Using the revenue function from Example 3 and the preceding cost function, write an equation for the profit function.

(B) Find the output to the nearest thousand cameras that will produce the maximum profit. What is the maximum profit to the nearest thousand dollars? Solve the problem algebraically by completing the square.

(C) What is the wholesale price per camera (to the nearest dollar) that produces the maximum profit?

 (D) Graph the profit function using an appropriate viewing window.

 (E) Find the output to the nearest thousand cameras that will produce the maximum profit. What is the maximum profit to the nearest thousand dollars? Solve the problem graphically using the maximum routine.

EXAMPLE 4 | **Break-Even Analysis** Use the revenue function from Example 3 and the cost function from Matched Problem 3:

$$R(x) = x(94.8 - 5x) \qquad \text{Revenue function}$$
$$C(x) = 156 + 19.7x \qquad \text{Cost function}$$

Both have domain $1 \le x \le 15$.

(A) Sketch the graphs of both functions in the same coordinate system.

(B) **Break-even points** are the production levels at which $R(x) = C(x)$. Find the break-even points algebraically to the nearest thousand cameras.

 (C) Plot both functions simultaneously in the same viewing window.

 (D) Use INTERSECT to find the break-even points graphically to the nearest thousand cameras.

(E) Recall that a loss occurs if $R(x) < C(x)$ and a profit occurs if $R(x) > C(x)$. For what outputs (to the nearest thousand cameras) will a loss occur? A profit?

SOLUTION (A) Sketch of functions:

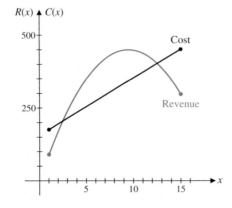

(B) Algebraic solution:
Find x such that $R(x) = C(x)$:

$$x(94.8 - 5x) = 156 + 19.7x$$
$$-5x^2 + 75.1x - 156 = 0$$
$$x = \frac{-75.1 \pm \sqrt{75.1^2 - 4(-5)(-156)}}{2(-5)}$$
$$= \frac{-75.1 \pm \sqrt{2,520.01}}{-10}$$
$$x = 2.490 \quad \text{or} \quad 12.530$$

The company breaks even at $x = 2.490$ and 12.530 million cameras.

(C) Graph in a graphing calculator:

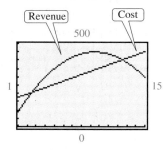

(D) Graphical solution:

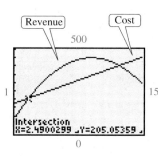 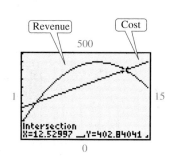

The company breaks even at $x = 2.490$ and 12.530 million cameras.

(E) Use the results from parts (A) and (B) or (C) and (D):

$$\text{Loss:} \quad 1 \le x < 2.490 \quad \text{or} \quad 12.530 < x \le 15$$
$$\text{Profit:} \quad 2.490 < x < 12.530$$

MATCHED PROBLEM 4

Use the profit equation from Matched Problem 3:

$$P(x) = R(x) - C(x)$$
$$= -5x^2 + 75.1x - 156 \quad \text{Profit function}$$
$$\text{Domain:} \quad 1 \le x \le 15$$

(A) Sketch a graph of the profit function in a rectangular coordinate system.

(B) Break-even points occur when $P(x) = 0$. Find the break-even points algebraically to the nearest thousand cameras.

 (C) Plot the profit function in an appropriate viewing window.

(D) Find the break-even points graphically to the nearest thousand cameras.

(E) A loss occurs if $P(x) < 0$, and a profit occurs if $P(x) > 0$. For what outputs (to the nearest thousand cameras) will a loss occur? A profit?

A visual inspection of the plot of a data set might indicate that a parabola would be a better model of the data than a straight line. In that case, rather than using linear regression (Section 1-3) to fit a linear model to the data, we would use **quadratic regression** on a graphing calculator to find the function of the form $y = ax^2 + bx + c$ that best fits the data.

EXAMPLE 5

Outboard Motors Table 1 gives performance data for a boat powered by an Evinrude outboard motor. Use quadratic regression to find the best model of the form $y = ax^2 + bx + c$ for fuel consumption y (in miles per gallon) as a function of speed x (in miles per hour). Estimate the fuel consumption (to one decimal place) at a speed of 12 miles per hour.

TABLE 1

rpm	mph	mpg
2,500	10.3	4.1
3,000	18.3	5.6
3,500	24.6	6.6
4,000	29.1	6.4
4,500	33.0	6.1
5,000	36.0	5.4
5,400	38.9	4.9

SOLUTION

Enter the data in a graphing calculator (Fig. 5A) and find the quadratic regression equation (Fig. 5B). The data set and the regression equation are graphed in Figure 5C. Using TRACE we see that the estimated fuel consumption at a speed of 12 mph is 4.5 mpg.

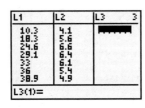

(A)　　　　　　(B)

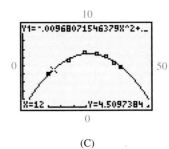

(C)

FIGURE 5

MATCHED PROBLEM 5

Refer to Table 1. Use quadratic regression to find the best model of the form $y = ax^2 + bx + c$ for boat speed y (in miles per hour) as a function of engine speed x (in revolutions per minute). Estimate the boat speed (in miles per hour, to one decimal place) at an engine speed of 3,400 rpm.

▪ More General Functions: Polynomial and Rational Functions

Linear and quadratic functions are special cases of the more general class of *polynomial functions.*

DEFINITION **Polynomial Function**

A **polynomial function** is a function that can be written in the form

$$f(x) = a_n x^n + a_{n-1} x^{n-1} + \cdots + a_1 x + a_0$$

for n a nonnegative integer, called the **degree** of the polynomial. The coefficients $a_0, a_1, \ldots, a_n$ are real numbers with $a_n \neq 0$. The **domain** of a polynomial function is the set of all real numbers.

Figure 6 shows graphs of representative polynomial functions of degrees 1 through 6. The figure suggests some general properties of graphs of polynomial functions.

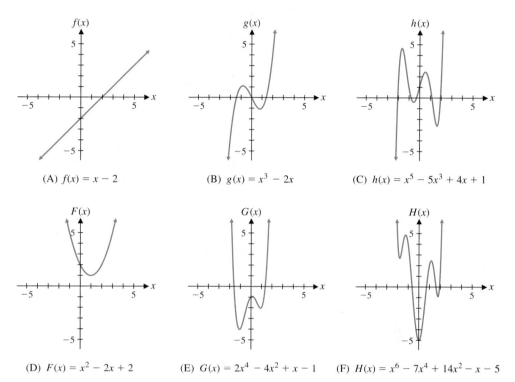

(A) $f(x) = x - 2$

(B) $g(x) = x^3 - 2x$

(C) $h(x) = x^5 - 5x^3 + 4x + 1$

(D) $F(x) = x^2 - 2x + 2$

(E) $G(x) = 2x^4 - 4x^2 + x - 1$

(F) $H(x) = x^6 - 7x^4 + 14x^2 - x - 5$

FIGURE 6 Graphs of polynomial functions

Notice that the odd-degree polynomial graphs start negative, end positive, and cross the x axis at least once. The even-degree polynomial graphs start positive, end positive, and may not cross the x axis at all. In all cases in Figure 1, the coefficient of the highest-degree term was chosen positive. If any leading coefficient had been chosen negative, then we would have a similar graph but reflected in the x axis.

A polynomial of degree n can have at most n linear factors. Therefore, the graph of a polynomial function of positive degree n can intersect the x axis at most n times. Note from Figure 6 that a polynomial of degree n may intersect the x axis fewer than n times.

The graph of a polynomial function is **continuous,** with no holes or breaks. That is, the graph can be drawn without removing a pen from the paper. Also, the graph of a polynomial has no sharp corners. Figure 7 on the next page shows the graphs of two functions, one that is not continuous, and the other that is continuous, but with a sharp corner. Neither function is a polynomial.

Just as rational numbers are defined in terms of quotients of integers, *rational functions* are defined in terms of quotients of polynomials. The following equations specify rational functions:

$$f(x) = \frac{1}{x} \quad g(x) = \frac{x - 2}{x^2 - x - 6} \quad h(x) = \frac{x^3 - 8}{x}$$
$$p(x) = 3x^2 - 5x \quad q(x) = 7 \quad r(x) = 0$$

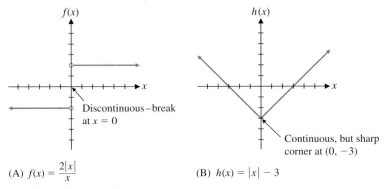

(A) $f(x) = \dfrac{2|x|}{x}$ (B) $h(x) = |x| - 3$

FIGURE 7 Discontinuous and sharp-corner functions

DEFINITION   **Rational Function**

A **rational function** is any function that can be written in the form

$$f(x) = \frac{n(x)}{d(x)} \qquad d(x) \neq 0$$

where $n(x)$ and $d(x)$ are polynomials. The **domain** is the set of all real numbers such that $d(x) \neq 0$.

Figure 8 shows the graphs of representative rational functions. Note, for example, that in Figure 8A the line $x = 2$ is a *vertical asymptote* for the function. The graph of f gets closer to this line as x gets closer to 2. The line $y = 1$ in Figure 8A is a *horizontal asymptote* for the function. The graph of f gets closer to this line as x increases or decreases without bound.

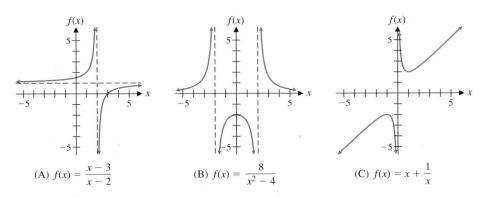

(A) $f(x) = \dfrac{x - 3}{x - 2}$ (B) $f(x) = \dfrac{8}{x^2 - 4}$ (C) $f(x) = x + \dfrac{1}{x}$

FIGURE 8 Graphs of rational functions

The number of vertical asymptotes of a rational function $f(x) = n(x)/d(x)$ is at most equal to the degree of $d(x)$. A rational function has at most one horizontal asymptote (note that the graph in Fig. 8C does not have a horizontal asymptote). Moreover, the graph of a rational function approaches the horizontal asymptote (when one exists) both as x increases and decreases without bound.

Answers to Matched Problems   **1.** (A)

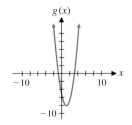

(B) x intercepts: -0.7656, 3.2656; y intercept: -5

(C)

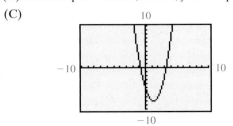

(D) x intercepts: $-0.7656, 3.2656$; y intercept: -5

(E) $x \le -0.7656$ or $x \ge 3.2656$; or $(-\infty, -0.7656]$ or $[3.2656, \infty)$

 (F) $x = -0.3508, 2.8508$

2. (A) $f(x) = -0.25(x + 4)^2 + 6$.

(B) Vertex: $(-4,6)$; maximum: $f(-4) = 6$; range: $y \le 6$ or $(-\infty, 6]$

(C) The graph of $f(x) = -0.25(x + 4)^2 + 6$ is the same as the graph of $g(x) = x^2$ vertically shrunk by a factor of 0.25, reflected in the x axis, and shifted 4 units to the left and 6 units up.

(D)

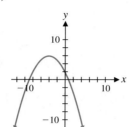

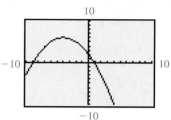 (E)

(F) Vertex: $(-4,6)$; maximum: $f(-4) = 6$; range: $y \le 6$ or $(-\infty, 6]$

3. (A) $P(x) = R(x) - C(x) = -5x^2 + 75.1x - 156$

(B) $P(x) = R(x) - C(x) = -5(x - 7.51)^2 + 126.0005$; output of 7.510 million cameras will produce a maximum profit of 126.001 million dollars.

(C) $p(7.510) = \$57$

 (D)

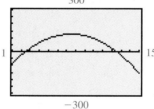

 (E)

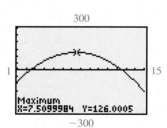

An output of 7.51 million cameras will produce a maximum profit of 126.001 million dollars. (Notice that maximum profit does not occur at the same output where maximum revenue occurs.)

4. (A)

(B) $x = 2.490$ or 12.530 million cameras

 (C)

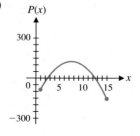

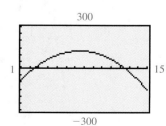

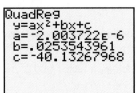

(D) $x = 2.490$ or 12.530 million cameras

(E) Loss: $1 \leq x < 2.490$ or $12.530 < x \leq 15$; profit: $2.490 < x < 12.530$

5. 22.9 mph

```
QuadReg
y=ax²+bx+c
a=-2.003722ᴇ-6
b=.0253543961
c=-40.13267968
```

Exercise 2-3

A *In Problems 1–4, complete the square and find the standard form of each quadratic function.*

1. $f(x) = x^2 - 4x + 3$ **2.** $g(x) = x^2 - 2x - 5$

3. $m(x) = -x^2 + 6x - 4$ **4.** $n(x) = -x^2 + 8x - 9$

In Problems 5–8, write a brief verbal description of the relationship between the graph of the indicated function (from Problems 1–4) and the graph of $y = x^2$.

5. $f(x) = x^2 - 4x + 3$ **6.** $g(x) = x^2 - 2x - 5$

7. $m(x) = -x^2 + 6x - 4$ **8.** $n(x) = -x^2 + 8x - 9$

9. Match each equation with a graph of one of the functions f, g, m, or n in the figure.

(A) $y = -(x + 2)^2 + 1$ (B) $y = (x - 2)^2 - 1$

(C) $y = (x + 2)^2 - 1$ (D) $y = -(x - 2)^2 + 1$

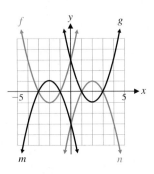

Figure for 9

10. Match each equation with a graph of one of the functions f, g, m, or n in the figure.

(A) $y = (x - 3)^2 - 4$ (B) $y = -(x + 3)^2 + 4$

(C) $y = -(x - 3)^2 + 4$ (D) $y = (x + 3)^2 - 4$

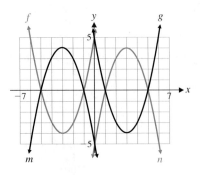

Figure for 10

For the functions indicated in Problems 11–14, find each of the following to the nearest integer by referring to the graphs for Problems 9 and 10.

(A) *Intercepts*   (B) *Vertex*

(C) *Maximum or minimum*   (D) *Range*

(E) *Increasing interval*   (F) *Decreasing interval*

11. Function n in the figure for Problem 9

12. Function m in the figure for Problem 10

13. Function f in the figure for Problem 9

14. Function g in the figure for Problem 10

In Problems 15–18, find each of the following:

(A) *Intercepts*   (B) *Vertex*

(C) *Maximum or minimum*   (D) *Range*

15. $f(x) = -(x - 3)^2 + 2$

16. $g(x) = -(x + 2)^2 + 3$

17. $m(x) = (x + 1)^2 - 2$

18. $n(x) = (x - 4)^2 - 3$

B *In Problems 19–22, write an equation for each graph in the form $y = a(x - h)^2 + k$, where a is either 1 or -1 and h and k are integers.*

19.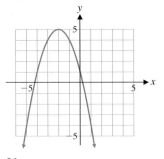

20.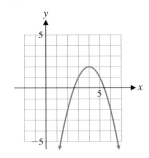

21.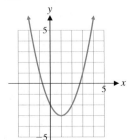

22.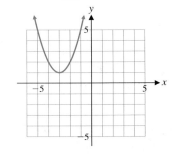

In Problems 23–28, find the vertex form for each quadratic function. Then find each of the following:

(A) *Intercepts* (B) *Vertex*

(C) *Maximum or minimum* (D) *Range*

23. $f(x) = x^2 - 8x + 12$

24. $g(x) = x^2 - 6x + 5$

25. $r(x) = -4x^2 + 16x - 15$

26. $s(x) = -4x^2 - 8x - 3$

27. $u(x) = 0.5x^2 - 2x + 5$

28. $v(x) = 0.5x^2 + 4x + 10$

29. Let $f(x) = 0.3x^2 - x - 8$. Solve each equation graphically to four decimal places.

(A) $f(x) = 4$ (B) $f(x) = -1$ (C) $f(x) = -9$

30. Let $g(x) = -0.6x^2 + 3x + 4$. Solve each equation graphically to four decimal places.

(A) $g(x) = -2$ (B) $g(x) = 5$ (C) $g(x) = 8$

31. Let $f(x) = 125x - 6x^2$. Find the maximum value of f to four decimal places graphically.

32. Let $f(x) = 100x - 7x^2 - 10$. Find the maximum value of f to four decimal places graphically.

33. Explain under what graphical conditions a quadratic function has exactly one real zero.

34. Explain under what graphical conditions a quadratic function has no real zeros.

C *In Problems 35–38, first write each function in vertex form; then find each of the following (to four decimal places):*

(A) *Intercepts*

(B) *Vertex*

(C) *Maximum or minimum*

(D) *Range*

35. $g(x) = 0.25x^2 - 1.5x - 7$

36. $m(x) = 0.20x^2 - 1.6x - 1$

37. $f(x) = -0.12x^2 + 0.96x + 1.2$

38. $n(x) = -0.15x^2 - 0.90x + 3.3$

Solve Problems 39–44 graphically to two decimal places using a graphing calculator.

39. $2 - 5x - x^2 = 0$

40. $7 + 3x - 2x^2 = 0$

41. $1.9x^2 - 1.5x - 5.6 < 0$

42. $3.4 + 2.9x - 1.1x^2 \geq 0$

43. $2.8 + 3.1x - 0.9x^2 \leq 0$

44. $1.8x^2 - 3.1x - 4.9 > 0$

45. Given that f is a quadratic function with minimum $f(x) = f(2) = 4$, find the axis, vertex, range, and x intercepts.

46. Given that f is a quadratic function with maximum $f(x) = f(-3) = -5$, find the axis, vertex, range, and x intercepts.

In Problems 47–50,

(A) *Graph f and g in the same coordinate system.*

(B) *Solve $f(x) = g(x)$ algebraically to two decimal places.*

(C) *Solve $f(x) > g(x)$ using parts (A) and (B).*

(D) *Solve $f(x) < g(x)$ using parts (A) and (B).*

47. $f(x) = -0.4x(x - 10)$
$g(x) = 0.3x + 5$
$0 \leq x \leq 10$

48. $f(x) = -0.7x(x - 7)$
$g(x) = 0.5x + 3.5$
$0 \leq x \leq 7$

49. $f(x) = -0.9x^2 + 7.2x$
$g(x) = 1.2x + 5.5$
$0 \leq x \leq 8$

50. $f(x) = -0.7x^2 + 6.3x$
$g(x) = 1.1x + 4.8$
$0 \leq x \leq 9$

51. Give a simple example of a quadratic function that has no real zeros. Explain how its graph is related to the x axis.

52. Give a simple example of a quadratic function that has exactly one real zero. Explain how its graph is related to the x axis.

Applications

53. *Tire mileage.* An automobile tire manufacturer collected the data in the table relating tire pressure x (in pounds per square inch) and mileage (in thousands of miles):

x	Mileage
28	45
30	52
32	55
34	51
36	47

A mathematical model for the data is given by

$$f(x) = -0.518x^2 + 33.3x - 481$$

(A) Complete the following table. Round values of $f(x)$ to one decimal place.

x	Mileage	$f(x)$
28	45	
30	52	
32	55	
34	51	
36	47	

(B) Sketch the graph of *f* and the mileage data in the same coordinate system.

(C) Use values of the modeling function rounded to two decimal places to estimate the mileage for a tire pressure of 31 pounds per square inch. For 35 pounds per square inch.

(D) Write a brief description of the relationship between tire pressure and mileage.

54. *Automobile production.* The table shows the retail market share of passenger cars from Ford Motor Company as a percentage of the U.S. market.

Year	Market Share
1980	17.2%
1985	18.8%
1990	20.0%
1995	20.7%
2000	20.2%
2005	17.4%

A mathematical model for this data is given by

$$f(x) = -0.0206x^2 + 0.548x + 16.9$$

where $x = 0$ corresponds to 1980.

(A) Complete the following table.

x	Market Share	f(x)
0	17.2	
5	18.8	
10	20.0	
15	20.7	
20	20.2	
25	17.4	

(B) Sketch the graph of *f* and the market share data in the same coordinate system.

(C) Use values of the modeling function *f* to estimate Ford's market share in 2010. In 2015.

(D) Write a brief verbal description of Ford's market share from 1980 to 2005.

55. *Tire mileage.* Using quadratic regression on a graphing calculator, show that the quadratic function that best fits the data on tire mileage in Problem 53 is

$$f(x) = -0.518x^2 + 33.3x - 481$$

(coefficients rounded to three significant digits).

56. *Automobile production.* Using quadratic regression on a graphing calculator, show that the quadratic function that best fits the data on market share in Problem 54 is

$$f(x) = -0.0206x^2 + 0.548x + 16.9$$

(coefficients rounded to three significant digits).

57. *Revenue.* The marketing research department for a company that manufactures and sells memory chips for microcomputers established the following price–demand and revenue functions:

$$p(x) = 75 - 3x \qquad \text{Price–demand function}$$
$$R(x) = xp(x) = x(75 - 3x) \quad \text{Revenue function}$$

where $p(x)$ is the wholesale price in dollars at which x million chips can be sold, and $R(x)$ is in millions of dollars. Both functions have domain $1 \le x \le 20$.

(A) Sketch a graph of the revenue function in a rectangular coordinate system.

(B) Find the output that will produce the maximum revenue. What is the maximum revenue?

(C) What is the wholesale price per chip (to the nearest dollar) that produces the maximum revenue?

58. *Revenue.* The marketing research department for a company that manufactures and sells "notebook" computers established the following price–demand and revenue functions:

$$p(x) = 2{,}000 - 60x \qquad \text{Price–demand function}$$
$$R(x) = xp(x) \qquad\qquad \text{Revenue function}$$
$$ = x(2{,}000 - 60x)$$

where $p(x)$ is the wholesale price in dollars at which x thousand computers can be sold, and $R(x)$ is in thousands of dollars. Both functions have domain $1 \le x \le 25$.

(A) Sketch a graph of the revenue function in a rectangular coordinate system.

(B) Find the output (to the nearest hundred computers) that will produce the maximum revenue. What is the maximum revenue to the nearest thousand dollars?

(C) What is the wholesale price per computer (to the nearest dollar) that produces the maximum revenue?

59. *Break-even analysis.* Use the revenue function from Problem 57 in this exercise and the given cost function:

$$R(x) = x(75 - 3x) \quad \text{Revenue function}$$
$$C(x) = 125 + 16x \quad \text{Cost function}$$

where x is in millions of chips, and $R(x)$ and $C(x)$ are in millions of dollars. Both functions have domain $1 \le x \le 20$.

(A) Sketch a graph of both functions in the same rectangular coordinate system.

(B) Find the break-even points to the nearest thousand chips.

(C) For what outputs will a loss occur? A profit?

60. *Break-even analysis.* Use the revenue function from Problem 58, in this exercise and the given cost function:

$$R(x) = x(2{,}000 - 60x) \quad \text{Revenue function}$$
$$C(x) = 4{,}000 + 500x \quad \text{Cost function}$$

where x is thousands of computers, and $C(x)$ and $R(x)$ are in thousands of dollars. Both functions have domain $1 \le x \le 25$.

(A) Sketch a graph of both functions in the same rectangular coordinate system.

(B) Find the break-even points.

(C) For what outputs will a loss occur? Will a profit occur?

61. *Profit–loss analysis.* Use the revenue and cost functions from Problem 59 in this exercise:

$$R(x) = x(75 - 3x) \quad \textit{Revenue function}$$

$$C(x) = 125 + 16x \quad \textit{Cost function}$$

where x is in millions of chips, and $R(x)$ and $C(x)$ are in millions of dollars. Both functions have domain $1 \le x \le 20$.

(A) Form a profit function P, and graph R, C, and P in the same rectangular coordinate system.

(B) Discuss the relationship between the intersection points of the graphs of R and C and the x intercepts of P.

(C) Find the x intercepts of P to the nearest thousand chips. Find the break-even points to the nearest thousand chips.

(D) Refer to the graph drawn in part (A). Does the maximum profit appear to occur at the same output level as the maximum revenue? Are the maximum profit and the maximum revenue equal? Explain.

(E) Verify your conclusion in part (D) by finding the output (to the nearest thousand chips) that produces the maximum profit. Find the maximum profit (to the nearest thousand dollars), and compare with Problem 57B.

62. *Profit–loss analysis.* Use the revenue and cost functions from Problem 60 in this exercise:

$$R(x) = x(2,000 - 60x) \quad \textit{Revenue function}$$

$$C(x) = 4,000 + 500x \quad \textit{Cost function}$$

where x is thousands of computers, and $R(x)$ and $C(x)$ are in thousands of dollars. Both functions have domain $1 \le x \le 25$.

(A) Form a profit function P, and graph R, C, and P in the same rectangular coordinate system.

(B) Discuss the relationship between the intersection points of the graphs of R and C and the x intercepts of P.

(C) Find the x intercepts of P to the nearest hundred computers. Find the break-even points.

(D) Refer to the graph drawn in part (A). Does the maximum profit appear to occur at the same output level as the maximum revenue? Are the maximum profit and the maximum revenue equal? Explain.

(E) Verify your conclusion in part (D) by finding the output that produces the maximum profit. Find the maximum profit and compare with Problem 58B.

63. *Medicine.* The French physician Poiseuille was the first to discover that blood flows faster near the center of an artery than near the edge. Experimental evidence has shown that the rate of flow v (in centimeters per second) at a point x centimeters from the center of an artery (see the figure) is given by

$$v = f(x) = 1,000(0.04 - x^2) \quad 0 \le x \le 0.2$$

Find the distance from the center that the rate of flow is 20 centimeters per second. Round answer to two decimal places.

Figure for 63 and 64

64. *Medicine.* Refer to Problem 63. Find the distance from the center that the rate of flow is 30 centimeters per second. Round answer to two decimal places.

65. The table gives performance data for a boat powered by an Evinrude outboard motor. Find a quadratic regression model $(y = ax^2 + bx + c)$ for boat speed y (in miles per hour) as a function of engine speed (in revolutions per minute). Estimate the boat speed at 3,100 revolutions per minute.

Table for 65 and 66

rpm	mph	mpg
1,500	4.5	8.2
2,000	5.7	6.9
2,500	7.8	4.8
3,000	9.6	4.1
3,500	13.4	3.7

66. The table gives performance data for a boat powered by an Evinrude outboard motor. Find a quadratic regression model $(y = ax^2 + bx + c)$ for fuel consumption y (in miles per gallon) as a function of engine speed (in revolutions per minute). Estimate the fuel consumption at 2,300 revolutions per minute.

Section 2-4 EXPONENTIAL FUNCTIONS

- Exponential Functions
- Base e Exponential Functions
- Growth and Decay Applications
- Compound Interest

This section introduces the important class of functions called *exponential functions*. These functions are used extensively in modeling and solving a wide variety of real-world problems, including growth of money at compound interest; growth of populations of people, animals, and bacteria; radioactive decay; and learning associated with the mastery of such devices as a new computer or an assembly process in a manufacturing plant.

■ Exponential Functions

We start by noting that

$$f(x) = 2^x \quad \text{and} \quad g(x) = x^2$$

are not the same function. Whether a variable appears as an exponent with a constant base or as a base with a constant exponent makes a big difference. The function g is a quadratic function, which we have already discussed. The function f is a new type of function called an *exponential function*. In general,

DEFINITION	Exponential Function

The equation

$$f(x) = b^x \qquad b > 0, b \neq 1$$

defines an **exponential function** for each different constant b, called the **base.** The **domain** of f is the set of all real numbers, and the **range** of f is the set of all positive real numbers.

We require the base b to be positive to avoid imaginary numbers such as $(-2)^{1/2} = \sqrt{-2} = i\sqrt{2}$. We exclude $b = 1$ as a base, since $f(x) = 1^x = 1$ is a constant function, which we have already considered.

Asked to hand-sketch graphs of equations such as $y = 2^x$ or $y = 2^{-x}$, many students would not hesitate at all. [*Note:* $2^{-x} = 1/2^x = (1/2)^x$.] They would probably make up tables by assigning integers to x, plot the resulting points, and then join these points with a smooth curve as in Figure 1. The only catch is that we have not defined 2^x for all real numbers. From Appendix A, Section A-7, we know what $2^5, 2^{-3}, 2^{2/3}, 2^{-3/5}, 2^{1.4}$, and $2^{-3.14}$ mean (that is, 2^p, where p is a rational number), but what does

$$2^{\sqrt{2}}$$

mean? The question is not easy to answer at this time. In fact, a precise definition of $2^{\sqrt{2}}$ must wait for more advanced courses, where it is shown that

$$2^x$$

names a positive real number for x any real number, and that the graph of $y = 2^x$ is indeed as indicated in Figure 1.

It is useful to compare the graphs of $y = 2^x$ and $y = 2^{-x}$ by plotting both on the same set of coordinate axes, as shown in Figure 2A. The graph of

$$f(x) = b^x \quad b > 1 \text{ (Fig. 2B)}$$

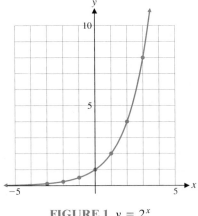

FIGURE 1 $y = 2^x$

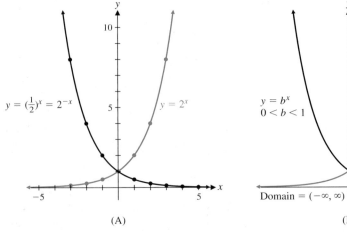

(A)   (B)

FIGURE 2 Exponential functions

looks very much like the graph of $y = 2^x$, and the graph of

$$f(x) = b^x \quad 0 < b < 1 \text{ (Fig. 2B)}$$

looks very much like the graph of $y = 2^{-x}$. Note that in both cases the x axis is a horizontal asymptote for the graphs.

The graphs in Figure 2 suggest the following important general properties of exponential functions, which we state without proof:

THEOREM 1   **BASIC PROPERTIES OF THE GRAPH OF $f(x) = b^x, b > 0, b \neq 1$**

    1. All graphs will pass through the point $(0, 1)$. $b^0 = 1$ for any
                                                                 permissible base b.

    2. All graphs are continuous curves, with no holes or jumps.

    3. The x axis is a horizontal asymptote.

    4. If $b > 1$, then b^x increases as x increases.

    5. If $0 < b < 1$, then b^x decreases as x increases.

INSIGHT

Recall that the graph of a rational function has at most one horizontal asymptote and that it approaches the horizontal asymptote (if one exists) both as $x \rightarrow \infty$ *and* as $x \rightarrow -\infty$ (see Section 2-3). The graph of an exponential function, on the other hand, approaches its horizontal asymptote as $x \rightarrow \infty$ *or* as $x \rightarrow -\infty$, but not both. In particular, there is no rational function that has the same graph as an exponential function.

The use of a calculator with the key $\boxed{y^x}$, or its equivalent, makes the graphing of exponential functions almost routine. Example 1 illustrates the process.

EXAMPLE 1   **Graphing Exponential Functions** Sketch a graph of $y = (\frac{1}{2})4^x, -2 \leq x \leq 2$.

SOLUTION   Use a calculator to create the table of values shown. Plot these points, and then join them with a smooth curve as in Figure 3.

x	y
-2	0.031
-1	0.125
0	0.50
1	2.00
2	8.00

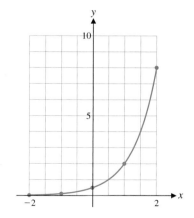

FIGURE 3 Graph of $y = (\frac{1}{2})4^x$

MATCHED PROBLEM 1   Sketch a graph of $y = (\frac{1}{2})4^{-x}, -2 \leq x \leq 2$.

Explore & Discuss **1**

Graph the functions $f(x) = 2^x$ and $g(x) = 3^x$ on the same set of coordinate axes. At which values of x do the graphs intersect? For which values of x is the graph of f above the graph of g? Below the graph of g? Are the graphs close together as x increases without bound? Are the graphs close together as x decreases without bound? Discuss.

Exponential functions, whose domains include irrational numbers, obey the familiar laws of exponents discussed in Appendix A, Section A-6 for rational exponents. We summarize these exponent laws here and add two other important and useful properties.

THEOREM 2 **PROPERTIES OF EXPONENTIAL FUNCTIONS**

For a and b positive, $a \neq 1, b \neq 1$, and x and y real,

1. Exponent laws:

$$a^x a^y = a^{x+y} \qquad \frac{a^x}{a^y} = a^{x-y} \qquad \qquad \frac{4^{2y}}{4^{5y}} = 4^{2y-5y} = 4^{-3y}$$

$$(a^x)^y = a^{xy} \qquad (ab)^x = a^x b^x \qquad \left(\frac{a}{b}\right)^x = \frac{a^x}{b^x}$$

2. $a^x = a^y$ if and only if $x = y$ If $7^{5t+1} = 7^{3t-3}$, then

$5t + 1 = 3t - 3$, and $t = -2$.

3. For $x \neq 0$,

$a^x = b^x$ if and only if $a = b$ If $a^5 = 2^5$, then $a = 2$.

■ Base *e* Exponential Functions

Of all the possible bases b we can use for the exponential function $y = b^x$, which ones are the most useful? If you look at the keys on a calculator, you will probably see $\boxed{10^x}$ and $\boxed{e^x}$. It is clear why base 10 would be important, because our number system is a base 10 system. But what is e, and why is it included as a base? It turns out that base e is used more frequently than all other bases combined. The reason for this is that certain formulas and the results of certain processes found in calculus and more advanced mathematics take on their simplest form if this base is used. This is why you will see e used extensively in expressions and formulas that model real-world phenomena. In fact, its use is so prevalent that you will often hear people refer to $y = e^x$ as *the* exponential function.

The base e is an irrational number, and like π, it cannot be represented exactly by any finite decimal fraction. However, e can be approximated as closely as we like by evaluating the expression

$$\left(1 + \frac{1}{x}\right)^x \tag{1}$$

for sufficiently large x. What happens to the value of expression (1) as x increases without bound? Think about this for a moment before proceeding. Maybe you guessed that the value approaches 1, because

$$1 + \frac{1}{x}$$

approaches 1, and 1 raised to any power is 1. Let us see if this reasoning is correct by actually calculating the value of the expression for larger and larger values of x. Table 1 summarizes the results.

TABLE 1	
x	$\left(1 + \dfrac{1}{x}\right)^x$
1	2
10	2.59374...
100	2.70481...
1,000	2.71692...
10,000	2.71814...
100,000	2.71827...
1,000,000	2.71828...

Interestingly, the value of expression (1) is never close to 1, but seems to be approaching a number close to 2.7183. In fact, as x increases without bound, the value of expression (1) approaches an irrational number that we call e. The irrational number e to 12 decimal places is

$$e = 2.718\ 281\ 828\ 459$$

Compare this value of e with the value of e^1 from a calculator. Exactly who discovered the constant e is still being debated. It is named after the great Swiss mathematician Leonhard Euler (1707–1783).

DEFINITION   **Exponential Function with Base e**

Exponential functions with base e and base $1/e$, respectively, are defined by

$$y = e^x \quad \text{and} \quad y = e^{-x}$$
$$\text{Domain:} \quad (-\infty, \infty)$$
$$\text{Range:} \quad (0, \infty)$$

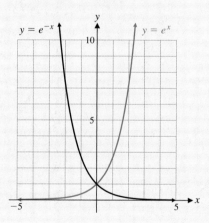

Explore & Discuss **2**   Graph the functions $f(x) = e^x, g(x) = 2^x$, and $h(x) = 3^x$ on the same set of coordinate axes. At which values of x do the graphs intersect? For positive values of x, which of the three graphs lies above the other two? Below the other two? How does your answer change for negative values of x?

■ Growth and Decay Applications

Functions of the form $y = ce^{kt}$, where c and k are constants and the independent variable t represents time, are often used to model population growth and radioactive decay. Note that if $t = 0$, then $y = c$. So the constant c represents the initial population (or initial amount). The constant k is called the **relative growth rate** and has the following interpretation: Suppose that $y = ce^{kt}$ models the population growth of a country, where y is the number of persons and t is time in years. If the relative growth rate is $k = 0.02$, then at any time t, the population is growing at a rate of $0.02y$ persons (that is, 2% of the population) per year.

We say that **population is growing continuously at relative growth rate k** to mean that the population y is given by the model $y = ce^{kt}$.

EXAMPLE 2

Exponential Growth   Cholera, an intestinal disease, is caused by a cholera bacterium that multiplies exponentially by cell division. The number of bacteria grows continuously at relative growth rate 1.386, that is,

$$N = N_0 e^{1.386t}$$

where N is the number of bacteria present after t hours and N_0 is the number of bacteria present at the start ($t = 0$). If we start with 25 bacteria, how many bacteria (to the nearest unit) will be present:

(A) In 0.6 hour?   (B) In 3.5 hours?

SOLUTION   Substituting $N_0 = 25$ into the preceding equation, we obtain

$$N = 25e^{1.386t} \qquad \text{The graph is shown in Figure 4.}$$

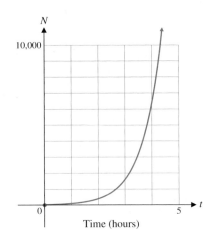

10,000

Time (hours)

FIGURE 4

(A) Solve for N when $t = 0.6$:

$$N = 25e^{1.386(0.6)} \qquad \text{Use a calculator.}$$
$$= 57 \text{ bacteria}$$

(B) Solve for N when $t = 3.5$:

$$N = 25e^{1.386(3.5)} \qquad \text{Use a calculator.}$$
$$= 3{,}197 \text{ bacteria}$$

MATCHED PROBLEM 2   Refer to the exponential growth model for cholera in Example 2. If we start with 55 bacteria, how many bacteria (to the nearest unit) will be present

(A) In 0.85 hour?   (B) In 7.25 hours?

EXAMPLE 3

Exponential Decay   Cosmic-ray bombardment of the atmosphere produces neutrons, which in turn react with nitrogen to produce radioactive carbon-14 (^{14}C). Radioactive ^{14}C enters all living tissues through carbon dioxide, which is first absorbed by plants.

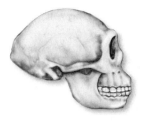

As long as a plant or animal is alive, ^{14}C is maintained in the living organism at a constant level. Once the organism dies, however, ^{14}C decays according to the equation

$$A = A_0 e^{-0.000124t}$$

where A is the amount present after t years and A_0 is the amount present at time $t = 0$. If 500 milligrams of ^{14}C is present in a sample from a skull at the time of death, how many milligrams will be present in the sample in

(A) 15,000 years?   (B) 45,000 years?

Compute answers to two decimal places.

SOLUTION   Substituting $A_0 = 500$ in the decay equation, we have

$$A = 500e^{-0.000124t} \quad \textit{See the graph in Figure 5.}$$

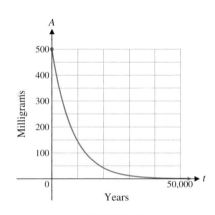

FIGURE 5

(A) Solve for A when $t = 15,000$:

$$A = 500e^{-0.000124(15,000)} \quad \textit{Use a calculator.}$$
$$= 77.84 \text{ milligrams}$$

(B) Solve for A when $t = 45,000$:

$$A = 500e^{-0.000124(45,000)} \quad \textit{Use a calculator.}$$
$$= 1.89 \text{ milligrams}$$

MATCHED PROBLEM 3   Refer to the exponential decay model in Example 3. How many milligrams of ^{14}C would have to be present at the beginning in order to have 25 milligrams present after 18,000 years? Compute the answer to the nearest milligram.

Explore & Discuss **3**

(A) On the same set of coordinate axes, graph the three decay equations $A = A_0 e^{-0.35t}, t \geq 0$, for $A_0 = 10, 20$, and 30.

(B) Identify any asymptotes for the three graphs in part (A).

(C) Discuss the long-term behavior for the equations in part (A).

If you buy a new car, it is likely to depreciate in value by several thousand dollars during the first year you own it. You would expect the value of the car to decrease in each subsequent year, but not by as much as in the previous year. If you drive the car long enough, its resale value will get close to zero. An exponential decay function will often be a good model of depreciation; a linear or quadratic function would not be suitable (why?). We can use **exponential regression** on a graphing calculator to find the function of the form $y = ab^x$ that best fits a data set.

EXAMPLE 4

Depreciation Table 2 gives the market value of a minivan (in dollars) x years after its purchase. Find an exponential regression model of the form $y = ab^x$ for this data set. Estimate the purchase price of the van. Estimate the value of the van 10 years after its purchase. Round answers to the nearest dollar.

TABLE 2

x	Value ($)
1	12,575
2	9,455
3	8,115
4	6,845
5	5,225
6	4,485

SOLUTION

Enter the data into a graphing calculator (Fig. 6A) and find the exponential regression equation (Fig. 6B). The estimated purchase price is $y_1(0) = \$14{,}910$. The data set and the regression equation are graphed in Figure 6C. Using TRACE, we see that the estimated value after 10 years is $1,959.

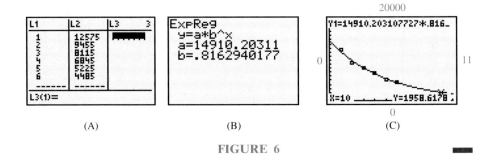

(A) (B) (C)

FIGURE 6

MATCHED PROBLEM 4

Table 3 gives the market value of a luxury sedan (in dollars) x years after its purchase. Find an exponential regression model of the form $y = ab^x$ for this data set. Estimate the purchase price of the sedan. Estimate the value of the sedan 10 years after its purchase. Round answers to the nearest dollar.

TABLE 3

x	Value ($)
1	23,125
2	19,050
3	15,625
4	11,875
5	9,450
6	7,125

■ Compound Interest

The fee paid to use another's money is called **interest.** It is usually computed as a percent (called **interest rate**) of the principal over a given period of time. If, at the end of a payment period, the interest due is reinvested at the same rate, then the interest earned as well as the principal will earn interest during the next payment period. Interest paid on interest reinvested is called **compound interest,** and may be calculated using the following compound interest formula:

If a **principal P (present value)** is invested at an annual **rate r** (expressed as a decimal) compounded m times a year, then the **amount A (future value)** in the account at the end of t years is given by

$$A = P\left(1 + \frac{r}{m}\right)^{mt} \qquad \text{Compound interest formula}$$

Note: P could be replaced by A_0, but convention dictates otherwise. For given r and m, the amount A is equal to the principal P multiplied by the exponential function b^t, where $b = (1 + r/m)^m$.

EXAMPLE 5

Compound Growth If $1,000 is invested in an account paying 10% compounded monthly, how much will be in the account at the end of 10 years? Compute the answer to the nearest cent.

SOLUTION We use the compound interest formula as follows:

$$A = P\left(1 + \frac{r}{m}\right)^{mt}$$

$$= 1,000\left(1 + \frac{0.10}{12}\right)^{(12)(10)} \qquad \text{Use a calculator.}$$

$$= \$2,707.04$$

The graph of

$$A = 1,000\left(1 + \frac{0.10}{12}\right)^{12t}$$

for $0 \le t \le 20$ is shown in Figure 7.

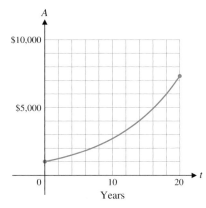

FIGURE 7

MATCHED PROBLEM 5   If you deposit $5,000 in an account paying 9% compounded daily, how much will you have in the account in 5 years? Compute the answer to the nearest cent.

Explore & Discuss **4**   Suppose that $1,000 is deposited in a savings account at an annual rate of 5%. Guess the amount in the account at the end of 1 year if interest is compounded (1) quarterly, (2) monthly, (3) daily, (4) hourly. Use the compound interest formula to compute the amounts at the end of 1 year to the nearest cent. Discuss the accuracy of your initial guesses.

The formula for compound interest is summarized on the next page for convenient reference.

SUMMARY COMPOUND INTEREST FORMULA

$$A = P\left(1 + \frac{r}{m}\right)^{mt}$$

where

A = amount (future value) at the end of t years

P = principal (present value)

r = annual rate, expressed as a decimal

m = number of compounding periods per year

t = time in years

Answers to Matched Problems

1.

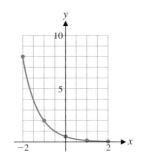

2. (A) 179 bacteria (B) 1,271,659 bacteria

3. 233 mg

4. Purchase price: $30,363; value after 10 yr: $2,864

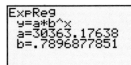

5. $7,841.13

6. (A) $9,155.23 (B) $9,156.29

Exercise 2-4

A

1. Match each equation with the graph of $f, g, h,$ or k in the figure.

(A) $y = 2^x$ (B) $y = (0.2)^x$

(C) $y = 4^x$ (D) $y = \left(\frac{1}{3}\right)^x$

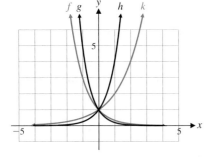

2. Match each equation with the graph of $f, g, h,$ or k in the figure.

(A) $y = \left(\frac{1}{4}\right)^x$ (B) $y = (0.5)^x$

(C) $y = 5^x$ (D) $y = 3^x$

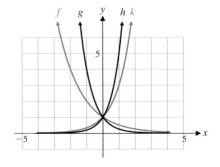

Graph each function in Problems 3–14 over the indicated interval.

3. $y = 5^x$; $[-2, 2]$

4. $y = 3^x$; $[-3, 3]$

5. $y = (\frac{1}{5})^x = 5^{-x}$; $[-2, 2]$ **6.** $y = (\frac{1}{3})^x = 3^{-x}$; $[-3, 3]$

7. $f(x) = -5^x$; $[-2, 2]$

8. $g(x) = -3^{-x}$; $[-3, 3]$

9. $y = -e^{-x}$; $[-3, 3]$

10. $y = -e^x$; $[-3, 3]$

11. $y = 100e^{0.1x}$; $[-5, 5]$

12. $y = 10e^{0.2x}$; $[-10, 10]$

13. $g(t) = 10e^{-0.2t}$; $[-5, 5]$ **14.** $f(t) = 100e^{-0.1t}$; $[-5, 5]$

Simplify each expression in Problems 15–20.

15. $(4^{3x})^{2y}$

16. $10^{3x-1}10^{4-x}$

17. $\dfrac{e^{x-3}}{e^{x-4}}$

18. $\dfrac{e^x}{e^{1-x}}$

19. $(2e^{1.2t})^3$

20. $(3e^{-1.4x})^2$

B *In Problems 21–28, describe verbally the transformations that can be used to obtain the graph of g from the graph of f (see Section 2-2).*

21. $g(x) = -2^x$; $f(x) = 2^x$

22. $g(x) = 2^{x-2}$; $f(x) = 2^x$

23. $g(x) = 3^{x+1}$; $f(x) = 3^x$

24. $g(x) = -3^x$; $f(x) = 3^x$

25. $g(x) = e^x + 1$; $f(x) = e^x$

26. $g(x) = e^x - 2$; $f(x) = e^x$

27. $g(x) = 2e^{-(x+2)}$; $f(x) = e^{-x}$

28. $g(x) = 0.5e^{-(x-1)}$; $f(x) = e^{-x}$

29. Use the graph of f shown in the figure to sketch the graph of each of the following.

(A) $y = f(x) - 1$ (B) $y = f(x + 2)$

(C) $y = 3f(x) - 2$ (D) $y = 2 - f(x - 3)$

30. Use the graph of f shown in the figure to sketch the graph of each of the following.

(A) $y = f(x) + 2$ (B) $y = f(x - 3)$

(C) $y = 2f(x) - 4$ (D) $y = 4 - f(x + 2)$

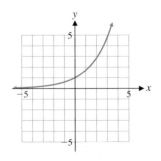

Figure for 29 and 30

In Problems 31–40, graph each function over the indicated interval.

31. $f(t) = 2^{t/10}$; $[-30, 30]$

32. $G(t) = 3^{t/100}$; $[-200, 200]$

33. $y = -3 + e^{1+x}$; $[-4, 2]$

34. $y = 2 + e^{x-2}$; $[-1, 5]$

35. $y = e^{|x|}$; $[-3, 3]$

36. $y = e^{-|x|}$; $[-3, 3]$

37. $C(x) = \dfrac{e^x + e^{-x}}{2}$; $[-5, 5]$

38. $M(x) = e^{x/2} + e^{-x/2}$; $[-5, 5]$

39. $y = e^{-x^2}$; $[-3, 3]$

40. $y = 2^{-x^2}$; $[-3, 3]$

41. Find all real numbers a such that $a^2 = a^{-2}$. Explain why this does not violate the second exponential function property in Theorem 2 on page 96.

42. Find real numbers a and b such that $a \neq b$ but $a^4 = b^4$. Explain why this does not violate the third exponential function property in Theorem 2 on page 96.

Solve each equation in Problems 43–48 for x.

43. $10^{2-3x} = 10^{5x-6}$ **44.** $5^{3x} = 5^{4x-2}$

45. $4^{5x-x^2} = 4^{-6}$ **46.** $7^{x^2} = 7^{2x+3}$

47. $5^3 = (x + 2)^3$ **48.** $(1 - x)^5 = (2x - 1)^5$

C *Solve each equation in Problems 49–52 for x. (Remember: $e^x \neq 0$ and $e^{-x} \neq 0$.)*

49. $(x - 3)e^x = 0$

50. $2xe^{-x} = 0$

51. $3xe^{-x} + x^2e^{-x} = 0$

52. $x^2e^x - 5xe^x = 0$

Graph each function in Problems 53–56 over the indicated interval.

53. $h(x) = x(2^x)$; $[-5, 0]$ **54.** $m(x) = x(3^{-x})$; $[0, 3]$

55. $N = \dfrac{100}{1 + e^{-t}}$; $[0, 5]$ **56.** $N = \dfrac{200}{1 + 3e^{-t}}$; $[0, 5]$

In Problems 57–60, approximate the real zeros of each function to two decimal places.

57. $f(x) = 4^x - 7$ **58.** $f(x) = 5 - 3^{-x}$

59. $f(x) = 2 + 3x + 10^x$ **60.** $f(x) = 7 - 2x^2 + 2^{-x}$

Applications

In all problems involving days, a 365-day year is assumed.

61. *Finance.* Suppose that $2,500 is invested at 7% compounded quarterly. How much money will be in the account in

(A) $\frac{3}{4}$ year?　　(B) 15 years?

Compute answers to the nearest cent.

62. *Finance.* Suppose that $4,000 is invested at 6% compounded weekly. How much money will be in the account in

(A) $\frac{1}{2}$ year?　　(B) 10 years?

Compute answers to the nearest cent.

63. *Finance.* A person wishes to have $15,000 cash for a new car 5 years from now. How much should be placed in an account now, if the account pays 6.75% compounded weekly? Compute the answer to the nearest dollar.

64. *Finance.* A couple just had a baby. How much should they invest now at 5.5% compounded daily in order to have $40,000 for the child's education 17 years from now? Compute the answer to the nearest dollar.

65. *Money growth.* BanxQuote operates a network of Web sites providing real-time market data from leading financial providers. The following rates for 12-month certificates of deposit were taken from the Web sites:

(A) Stonebridge Bank, 5.40% compounded monthly

(B) DeepGreen Bank, 4.95% compounded daily

(C) Provident Bank, 5.15% compounded quarterly

Compute the value of $10,000 invested in each account at the end of 1 year.

66. *Money growth.* Refer to Problem 65. The following rates for 60-month certificates of deposit were also taken from Banx-Quote Web sites:

(A) Oriental Bank & Trust, 5.50% compounded quarterly

(B) BMW Bank of North America, 5.12% compounded monthly

(C) BankFirst Corporation, 4.86% compounded daily

Compute the value of $10,000 invested in each account at the end of 5 years.

67. *Advertising.* A company is trying to introduce a new product to as many people as possible through television advertising in a large metropolitan area with 2 million possible viewers. A model for the number of people N (in millions) who are aware of the product after t days of advertising was found to be

$$N = 2(1 - e^{-0.037t})$$

Graph this function for $0 \le t \le 50$. What value does N tend to as t increases without bound?

68. *Learning curve.* People assigned to assemble circuit boards for a computer manufacturing company undergo on-the-job training. From past experience it was found that the learning curve for the average employee is given by

$$N = 40(1 - e^{-0.12t})$$

where N is the number of boards assembled per day after t days of training. Graph this function for $0 \le t \le 30$. What is the maximum number of boards an average employee can be expected to produce in 1 day?

69. *Sports salaries.* Table 4 shows the average salaries for players in Major League Baseball (MLB) and the National Basketball Association (NBA) in selected years since 1990.

(A) Let x represent the number of years since 1990 and find an exponential regression model ($y = ab^x$) for the average salary in MLB. Use the model to estimate the average salary (to the nearest thousand dollars) in 2015.

(B) The average salary in MLB in 2000 was 1.984 million. How does this compare with the value given by the model of part (A)? How would the inclusion of the year 2000 data affect the estimated average salary in 2015? Explain.

TABLE 4 Average Salary (thousand $)

Year	MLB	NBA
1990	589	750
1993	1,062	1,300
1996	1,101	2,000
1999	1,724	2,400
2002	2,300	4,500
2005	2,633	5,000

70. *Sports salaries.* Refer to Table 4.

(A) Let x represent the number of years since 1990 and find an exponential regression model ($y = ab^x$) for the average salary in the NBA. Use the model to estimate the average salary (to the nearest thousand dollars) in 2015.

(B) The average salary in the NBA in 1997 was $2.2 million. How does this compare with the value given by the model of part (A)? How would the inclusion of the year 1997 data affect the estimated average salary in 2015? Explain.

71. *Marine biology.* Marine life is dependent upon the microscopic plant life that exists in the photic zone, a zone that goes to a depth where about 1% of the surface light remains. In some waters with a great deal of sediment, the photic zone may go down only 15 to 20 feet. In some murky harbors, the intensity of light d feet below the surface is given approximately by

$$I = I_0 e^{-0.23d}$$

What percentage of the surface light will reach a depth of

(A) 10 feet?   (B) 20 feet?

72. *Marine biology.* Refer to Problem 71. Light intensity I relative to depth d (in feet) for one of the clearest bodies of water in the world, the Sargasso Sea in the West Indies, can be approximated by

$$I = I_0 e^{-0.00942d}$$

where I_0 is the intensity of light at the surface. What percentage of the surface light will reach a depth of

(A) 50 feet?   (B) 100 feet?

73. *HIV/AIDS epidemic.* The Joint United Nations Program on HIV/AIDS reported that HIV had infected 65 million people worldwide prior to 2006. Assume that number increases continuously at a relative growth rate of 2%.

(A) Write an equation that models the worldwide spread of HIV, letting 2006 be year 0.

(B) Based on the model, how many people (to the nearest million) had been infected prior to 2002? How many would be infected prior to 2014?

(C) Sketch a graph of this growth equation from 2006 to 2014.

74. *HIV/AIDS epidemic.* The Joint United Nations Program on HIV/AIDS reported that 25 million people worldwide had died of AIDS prior to 2006. Assume that number increases continuously at a relative growth rate of 1%.

(A) Write an equation that models total worldwide deaths from AIDS, letting 2006 be year 0.

(B) Based on the model, how many people (to the nearest million) had died from AIDS prior to 2002? How many would die from AIDS prior to 2014?

(C) Sketch a graph of this growth equation from 2006 to 2014.

75. *World population growth.* It took from the dawn of humanity to 1830 for the population to grow to the first billion people, just 100 more years (by 1930) for the second billion, and 3 billion more were added in only 60 more years (by 1990). In 2006, the estimated world population was 6.5 billion with a relative growth rate of 1.14%.

(A) Write an equation that models the world population growth, letting 2006 be year 0.

(B) Based on the model, what is the expected world population (to the nearest hundred million) in 2015? In 2030?

(C) Sketch a graph of the equation found in part (A). Cover the years from 2006 through 2030.

76. *Population growth in Ethiopia.* In 2006, the estimated population in Ethiopia was 75 million people with a relative growth rate of 2.3%.

(A) Write an equation that models the population growth in Ethiopia, letting 2006 be year 0.

(B) Based on the model, what is the expected population in Ethiopia (to the nearest million) in 2015? In 2030?

(C) Sketch a graph of the equation found in part (A). Cover the years from 2006 through 2030.

77. *Internet growth.* The number of Internet hosts grew very rapidly from 1994 to 2006 (Table 5).

(A) Let x represent the number of years since 1994. Find an exponential regression model ($y = ab^x$) for this data set and estimate the number of hosts in 2015 (to the nearest million).

(B) Discuss the implications of this model if the number of Internet hosts continues to grow at this rate.

TABLE 5 Internet Hosts (Millions)

Year	Hosts
1994	2.4
1997	16.1
2000	72.4
2003	171.6
2006	394.0

Source: Internet Software Consortium

78. *Life expectancy.* Table 6 shows the life expectancy (in years) at birth for residents of the United States from 1970 to 2005. Let x represent years since 1970. Find an exponential regression model for this data and use it to estimate the life expectancy for a person born in 2015.

TABLE 6

Year of Birth	Life Expectancy
1970	70.8
1975	72.6
1980	73.7
1985	74.7
1990	75.4
1995	75.9
2000	76.9
2005	77.7

Section 2-5 LOGARITHMIC FUNCTIONS

- Inverse Functions
- Logarithmic Functions
- Properties of Logarithmic Functions
- Calculator Evaluation of Logarithms
- Applications

Find the exponential function keys $\boxed{10^x}$ and $\boxed{e^x}$ on your calculator. Close to these keys you will find $\boxed{\text{LOG}}$ and $\boxed{\text{LN}}$ keys. The latter represent *logarithmic functions,* and each is closely related to the exponential function it is near. In fact, the exponential function and the corresponding logarithmic function are said to be *inverses* of each other. In this section we will develop the concept of inverse functions and use it to define a logarithmic function as the inverse of an exponential function. We will then investigate basic properties of logarithmic functions, use a calculator to evaluate them for particular values of x, and apply them to real-world problems.

Logarithmic functions are used in modeling and solving many types of problems. For example, the decibel scale is a logarithmic scale used to measure sound intensity, and the Richter scale is a logarithmic scale used to measure the strength of the force of an earthquake. An important business application has to do with finding the time it takes money to double if it is invested at a certain rate compounded a given number of times a year or compounded continuously. This requires the solution of an exponential equation, and logarithms play a central role in the process.

■ Inverse Functions

Look at the graphs of $f(x) = \dfrac{x}{2}$ and $g(x) = \dfrac{|x|}{2}$ in Figure 1:

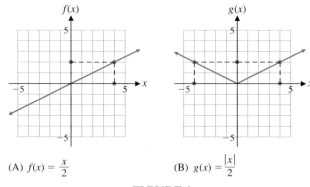

(A) $f(x) = \dfrac{x}{2}$ (B) $g(x) = \dfrac{|x|}{2}$

FIGURE 1

Because both f and g are functions, each domain value corresponds to exactly one range value. For which function does each range value correspond to exactly one domain value? This is the case only for function f. Note that for the range value 2, the corresponding domain value is 4. For function g the range value 2 corresponds to both -4 and 4. Function f is said to be *one-to-one*. In general,

---
DEFINITION   One-to-One Functions

A function f is said to be **one-to-one** if each range value corresponds to exactly one domain value.

It can be shown that any continuous function that is either increasing or decreasing for all domain values is one-to-one. If a continuous function increases for some domain values and decreases for others, it cannot be one-to-one. Figure 1 shows an example of each case.

Explore & Discuss **1**   Graph $f(x) = 2^x$ and $g(x) = x^2$. For a range value of 4, what are the corresponding domain values for each function? Which of the two functions is one-to-one? Explain why.

Starting with a one-to-one function f, we can obtain a new function called the *inverse* of f as follows:

DEFINITION **Inverse of a Function**

If f is a one-to-one function, then the **inverse** of f is the function formed by interchanging the independent and dependent variables for f. Thus, if (a, b) is a point on the graph of f, then (b, a) is a point on the graph of the inverse of f.

Note: If f is not one-to-one, then f **does not have an inverse.**

A number of important functions in any library of elementary functions are the inverses of other basic functions in the library. In this course, we are interested in the inverses of exponential functions, called *logarithmic functions.*

■ Logarithmic Functions

If we start with the exponential function f defined by

$$y = 2^x \tag{1}$$

and interchange the variables, we obtain the inverse of f:

$$x = 2^y \tag{2}$$

We call the inverse the **logarithmic function with base 2,** and write

$$y = \log_2 x \quad \text{if and only if} \quad x = 2^y$$

We can graph $y = \log_2 x$ by graphing $x = 2^y$, since they are equivalent. Any ordered pair of numbers on the graph of the exponential function will be on the graph of the logarithmic function if we interchange the order of the components. For example, $(3, 8)$ satisfies equation (1) and $(8, 3)$ satisfies equation (2). The graphs of $y = 2^x$ and $y = \log_2 x$ are shown in Figure 2. Note that if we fold the paper along the dashed line $y = x$ in Figure 2, the two graphs match exactly. The line $y = x$ is a line of symmetry for the two graphs.

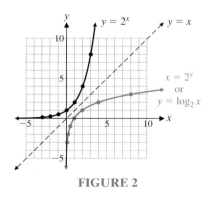

Exponential Function		Logarithmic Function	
x	$y = 2^x$	$x = 2^y$	y
-3	$\frac{1}{8}$	$\frac{1}{8}$	-3
-2	$\frac{1}{4}$	$\frac{1}{4}$	-2
-1	$\frac{1}{2}$	$\frac{1}{2}$	-1
0	1	1	0
1	2	2	1
2	4	4	2
3	8	8	3

FIGURE 2

$$\begin{bmatrix} \text{Ordered} \\ \text{pairs} \\ \text{reversed} \end{bmatrix}$$

In general, since the graphs of all exponential functions of the form $f(x) = b^x, b \neq 1, b > 0$, are either increasing or decreasing (see Section 2-2), exponential functions have inverses.

DEFINITION Logarithmic Functions

The inverse of an exponential function is called a **logarithmic function**. For $b > 0$ and $b \neq 1$,

Logarithmic form   Exponential form

$$y = \log_b x \qquad \text{is equivalent to} \qquad x = b^y$$

The **log to the base b of x** is the exponent to which b must be raised to obtain x. [*Remember:* A logarithm is an exponent.] The **domain** of the logarithmic function is the set of all positive real numbers, which is also the range of the corresponding exponential function; and the **range** of the logarithmic function is the set of all real numbers, which is also the domain of the corresponding exponential function. Typical graphs of an exponential function and its inverse, a logarithmic function, are shown in the figure in the margin.

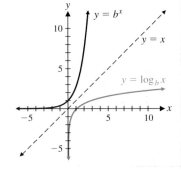

INSIGHT

Because the domain of a logarithmic function consists of the positive real numbers, the entire graph of a logarithmic function lies to the right of the y axis. In contrast, the graphs of polynomial and exponential functions intersect every vertical line, and the graphs of rational functions intersect all but a finite number of vertical lines.

The following examples involve converting logarithmic forms to equivalent exponential forms, and vice versa.

EXAMPLE 1 **Logarithmic–Exponential Conversions** Change each logarithmic form to an equivalent exponential form:

(A) $\log_5 25 = 2$ (B) $\log_9 3 = \frac{1}{2}$ (C) $\log_2(\frac{1}{4}) = -2$

SOLUTION (A) $\log_5 25 = 2$ is equivalent to $25 = 5^2$

(B) $\log_9 3 = \frac{1}{2}$ is equivalent to $3 = 9^{1/2}$

(C) $\log_2(\frac{1}{4}) = -2$ is equivalent to $\frac{1}{4} = 2^{-2}$ ▬

MATCHED PROBLEM 1 Change each logarithmic form to an equivalent exponential form:

(A) $\log_3 9 = 2$ (B) $\log_4 2 = \frac{1}{2}$ (C) $\log_3(\frac{1}{9}) = -2$

EXAMPLE 2 **Exponential–Logarithmic Conversions** Change each exponential form to an equivalent logarithmic form:

(A) $64 = 4^3$ (B) $6 = \sqrt{36}$ (C) $\frac{1}{8} = 2^{-3}$

SOLUTION (A) $64 = 4^3$ is equivalent to $\log_4 64 = 3$

(B) $6 = \sqrt{36}$ is equivalent to $\log_{36} 6 = \frac{1}{2}$

(C) $\frac{1}{8} = 2^{-3}$ is equivalent to $\log_2(\frac{1}{8}) = -3$ ▬

| MATCHED PROBLEM 2 | Change each exponential form to an equivalent logarithmic form:

(A) $49 = 7^2$ (B) $3 = \sqrt{9}$ (C) $\frac{1}{3} = 3^{-1}$

To gain a little deeper understanding of logarithmic functions and their relationship to the exponential functions, we consider a few problems where we want to find $x, b,$ or y in $y = \log_b x$, given the other two values. All values are chosen so that the problems can be solved exactly without a calculator.

| EXAMPLE 3 | **Solutions of the Equation $y = \log_b x$**   Find $y, b,$ or x, as indicated.

(A) Find y: $y = \log_4 16$ (B) Find x: $\log_2 x = -3$
(C) Find y: $y = \log_8 4$ (D) Find b: $\log_b 100 = 2$

SOLUTION | (A) $y = \log_4 16$ is equivalent to $16 = 4^y$. Thus,
$$y = 2$$
(B) $\log_2 x = -3$ is equivalent to $x = 2^{-3}$. Thus,
$$x = \frac{1}{2^3} = \frac{1}{8}$$
(C) $y = \log_8 4$ is equivalent to
$$4 = 8^y \quad \text{or} \quad 2^2 = 2^{3y}$$
Thus,
$$3y = 2$$
$$y = \tfrac{2}{3}$$
(D) $\log_b 100 = 2$ is equivalent to $100 = b^2$. Thus,
$$b = 10 \quad \text{\textit{Recall that b cannot be negative.}}$$

| MATCHED PROBLEM 3 | Find $y, b,$ or x, as indicated.

(A) Find y: $y = \log_9 27$ (B) Find x: $\log_3 x = -1$
(C) Find b: $\log_b 1,000 = 3$

■ Properties of Logarithmic Functions

Logarithmic functions have many powerful and useful properties. We list eight basic properties in Theorem 1.

THEOREM 1 | **PROPERTIES OF LOGARITHMIC FUNCTIONS**

If $b, M,$ and N are positive real numbers, $b \neq 1$, and p and x are real numbers, then

1. $\log_b 1 = 0$ 5. $\log_b MN = \log_b M + \log_b N$

2. $\log_b b = 1$ 6. $\log_b \dfrac{M}{N} = \log_b M - \log_b N$

3. $\log_b b^x = x$ 7. $\log_b M^p = p \log_b M$

4. $b^{\log_b x} = x, \quad x > 0$ 8. $\log_b M = \log_b N$ if and only if $M = N$

The first four properties in Theorem 1 follow directly from the definition of a logarithmic function. Here we will sketch a proof of property 5. The other properties are established in a similar way. Let

$$u = \log_b M \quad \text{and} \quad v = \log_b N$$

Or, in equivalent exponential form,

$$M = b^u \quad \text{and} \quad N = b^v$$

Now, see if you can provide reasons for each of the following steps:

$$\log_b MN = \log_b b^u b^v = \log_b b^{u+v} = u + v = \log_b M + \log_b N$$

EXAMPLE 4   **Using Logarithmic Properties**

(A) $\log_b \dfrac{wx}{yz}$ $\begin{aligned} &= \log_b wx - \log_b yz \\ &= \log_b w + \log_b x - (\log_b y + \log_b z) \end{aligned}$

$\qquad\qquad\qquad = \log_b w + \log_b x - \log_b y - \log_b z$

(B) $\log_b(wx)^{3/5}$ $\ = \frac{3}{5}\log_b wx\ $ $= \frac{3}{5}(\log_b w + \log_b x)$

(C) $e^{x \log_e b} = e^{\log_e b^x} = b^x$

(D) $\dfrac{\log_e x}{\log_e b} = \dfrac{\log_e(b^{\log_b x})}{\log_e b} = \dfrac{(\log_b x)(\log_e b)}{\log_e b} = \log_b x$

MATCHED PROBLEM 4   Write in simpler forms, as in Example 4.

(A) $\log_b \dfrac{R}{ST}$ (B) $\log_b\left(\dfrac{R}{S}\right)^{2/3}$ (C) $2^{u \log_2 b}$ (D) $\dfrac{\log_2 x}{\log_2 b}$

The following examples and problems, although somewhat artificial, will give you additional practice in using basic logarithmic properties.

EXAMPLE 5   **Solving Logarithmic Equations**   Find x so that

$$\tfrac{3}{2}\log_b 4 - \tfrac{2}{3}\log_b 8 + \log_b 2 = \log_b x$$

SOLUTION

$$\tfrac{3}{2}\log_b 4 - \tfrac{2}{3}\log_b 8 + \log_b 2 = \log_b x$$

$$\log_b 4^{3/2} - \log_b 8^{2/3} + \log_b 2 = \log_b x \qquad \textit{Property 7}$$

$$\log_b 8 - \log_b 4 + \log_b 2 = \log_b x$$

$$\log_b \frac{8 \cdot 2}{4} = \log_b x \qquad \textit{Properties 5 and 6}$$

$$\log_b 4 = \log_b x$$

$$x = 4 \qquad \textit{Property 8}$$

MATCHED PROBLEM 5   Find x so that $3 \log_b 2 + \tfrac{1}{2}\log_b 25 - \log_b 20 = \log_b x$.

| EXAMPLE 6 | **Solving Logarithmic Equations** Solve: $\log_{10} x + \log_{10}(x + 1) = \log_{10} 6$. |

SOLUTION

$$\log_{10} x + \log_{10}(x + 1) = \log_{10} 6$$
$$\log_{10}[x(x + 1)] = \log_{10} 6 \qquad \text{Property 5}$$
$$x(x + 1) = 6 \qquad \text{Property 8}$$
$$x^2 + x - 6 = 0 \qquad \text{Solve by factoring.}$$
$$(x + 3)(x - 2) = 0$$
$$x = -3, 2$$

We must exclude $x = -3$, since the domain of the function $\log_{10}(x + 1)$ is $x > -1$ or $(-1, \infty)$; hence, $x = 2$ is the only solution. ▬

| MATCHED PROBLEM 6 | Solve: $\log_3 x + \log_3(x - 3) = \log_3 10$. |

Explore & Discuss **2**

Discuss the relationship between each of the following pairs of expressions. If the two expressions are equivalent, explain why. If they are not, give an example.

(A) $\log_b M - \log_b N$; $\dfrac{\log_b M}{\log_b N}$

(B) $\log_b M - \log_b N$; $\log_b \dfrac{M}{N}$

(C) $\log_b M + \log_b N$; $\log_b MN$

(D) $\log_b M + \log_b N$; $\log_b(M + N)$

■ Calculator Evaluation of Logarithms

Of all possible logarithmic bases, the base e and the base 10 are used almost exclusively. Before we can use logarithms in certain practical problems, we need to be able to approximate the logarithm of any positive number either to base 10 or to base e. And conversely, if we are given the logarithm of a number to base 10 or base e, we need to be able to approximate the number. Historically, tables were used for this purpose, but now calculators make computations faster and far more accurate.

Common logarithms (also called **Briggsian logarithms**) are logarithms with base 10. **Natural logarithms** (also called **Napierian logarithms**) are logarithms with base e. Most calculators have a key labeled "log" (or "LOG") and a key labeled "ln" (or "LN"). The former represents a common (base 10) logarithm and the latter a natural (base e) logarithm. In fact, "log" and "ln" are both used extensively in mathematical literature, and whenever you see either used in this book without a base indicated, they will be interpreted as follows:

Common logarithm: $\log x$ means $\log_{10} x$
Natural logarithm: $\ln x$ means $\log_e x$

Finding the common or natural logarithm using a calculator is very easy. On some calculators, you simply enter a number from the domain of the function and press LOG or LN. On other calculators, you press either LOG or LN, enter a number from the domain, and then press ENTER. Check the user's manual for your calculator.

| EXAMPLE 7 | **Calculator Evaluation of Logarithms** Use a calculator to evaluate each to six decimal places: |

(A) $\log 3{,}184$ (B) $\ln 0.000\,349$ (C) $\log(-3.24)$

SOLUTION (A) $\log 3{,}184 = 3.502\ 973$

(B) $\ln 0.000\ 349 = -7.960\ 439$

(C) $\log(-3.24) = $ Error*   *−3.24 is not in the domain of the log function.*   ▬

MATCHED PROBLEM 7 Use a calculator to evaluate each to six decimal places:

(A) $\log 0.013\ 529$

(B) $\ln 28.693\ 28$

(C) $\ln(-0.438)$

We now turn to the second problem mentioned previously: Given the logarithm of a number, find the number. We make direct use of the logarithmic–exponential relationships, which follow from the definition of logarithmic function given at the beginning of this section.

$$\log x = y \quad \textbf{is equivalent to} \quad x = 10^y$$
$$\ln x = y \quad \textbf{is equivalent to} \quad x = e^y$$

EXAMPLE 8 **Solving $\log_b x = y$ for x** Find x to four decimal places, given the indicated logarithm:

(A) $\log x = -2.315$

(B) $\ln x = 2.386$

SOLUTION (A) $\log x = -2.315$ *Change to equivalent exponential form.*

$\quad\quad\quad x = 10^{-2.315}$ *Evaluate with a calculator.*

$\quad\quad\quad\quad = 0.0048$

(B) $\ln x = 2.386$ *Change to equivalent exponential form.*

$\quad\quad\quad x = e^{2.386}$ *Evaluate with a calculator.*

$\quad\quad\quad\quad = 10.8699$ ▬

MATCHED PROBLEM 8 Find x to four decimal places, given the indicated logarithm:

(A) $\ln x = -5.062$

(B) $\log x = 2.0821$

EXAMPLE 9 **Solving Exponential Equations** Solve for x to four decimal places:

(A) $10^x = 2$ (B) $e^x = 3$ (C) $3^x = 4$

---

* Some calculators use a more advanced definition of logarithms involving complex numbers and will display an ordered pair of real numbers as the value of $\log(-3.24)$. You should interpret such a result as an indication that the number entered is not in the domain of the logarithm function as we have defined it.

SOLUTION   (A) $10^x = 2$ *Take common logarithms of both sides.*

$\log 10^x = \log 2$ *Property 3*

$x = \log 2$ *Use a calculator.*

$= 0.3010$ *To four decimal places*

(B) $e^x = 3$ *Take natural logarithms of both sides.*

$\ln e^x = \ln 3$ *Property 3*

$x = \ln 3$ *Use a calculator.*

$= 1.0986$ *To four decimal places*

(C) $3^x = 4$ *Take either natural or common logarithms of both sides. (We choose common logarithms.)*

$\log 3^x = \log 4$ *Property 7*

$x \log 3 = \log 4$ *Solve for x.*

$x = \dfrac{\log 4}{\log 3}$ *Use a calculator.*

$= 1.2619$ *To four decimal places*   ▬

Exponential equations can also be solved graphically by graphing both sides of an equation and finding the points of intersection. Figure 3 illustrates this approach for the equations in Example 9.

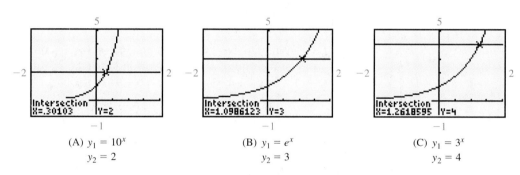

(A) $y_1 = 10^x$ (B) $y_1 = e^x$ (C) $y_1 = 3^x$
$y_2 = 2$ $y_2 = 3$ $y_2 = 4$

FIGURE 3 Graphical solution of exponential equations

MATCHED PROBLEM 9   Solve for x to four decimal places:

(A) $10^x = 7$ (B) $e^x = 6$ (C) $4^x = 5$

Explore & Discuss **3**   Discuss how you could find $y = \log_5 38.25$ using either natural or common logarithms on a calculator. [*Hint:* Start by rewriting the equation in exponential form.]

REMARK   In the usual notation for natural logarithms, the simplifications of Example 4, parts (C) and (D) on page 114, become

$$e^{x \ln b} = b^x \quad \text{and} \quad \frac{\ln x}{\ln b} = \log_b x$$

With these formulas we can change an exponential function with base b, or a logarithmic function with base b, to expressions involving exponential or logarithmic functions, respectively, to the base e. Such **change-of-base formulas** are useful in calculus.

APPLICATIONS

A convenient and easily understood way of comparing different investments is to use their **doubling times**—the length of time it takes the value of an investment to double. Logarithm properties, as you will see in Example 10, provide us with just the right tool for solving some doubling-time problems.

| EXAMPLE 10 | **Doubling Time for an Investment**   How long (to the next whole year) will it take money to double if it is invested at 20% compounded annually? |

SOLUTION   We use the compound interest formula discussed in Section 2-4:

$$A = P\left(1 + \frac{r}{m}\right)^{mt} \qquad \text{Compound interest}$$

The problem is to find t, given $r = 0.20$, $m = 1$, and $A = 2p$; that is,

$$2P = P(1 + 0.2)^t$$

$$2 = 1.2^t$$ Solve for t by taking the natural or common

$$1.2^t = 2$$ logarithm of both sides (we choose the

$$\ln 1.2^t = \ln 2$$ natural logarithm).

$$t \ln 1.2 = \ln 2$$ Property 7

$$t = \frac{\ln 2}{\ln 1.2}$$ Use a calculator.

$$\approx 3.8 \text{ years} \qquad [\text{Note: } (\ln 2)/(\ln 1.2) \neq \ln 2 - \ln 1.2]$$

$$\approx 4 \text{ years} \qquad \text{To the next whole year}$$

When interest is paid at the end of 3 years, the money will not be doubled; when paid at the end of 4 years, the money will be slightly more than doubled. ▬

 Example 10 can also be solved graphically by graphing both sides of the equation $2 = 1.2^t$, and finding the intersection point (Fig. 4).

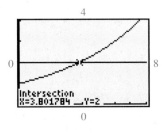

FIGURE 4 $y_1 = 1.2^x, y_2 = 2$

| MATCHED PROBLEM 10 | How long (to the next whole year) will it take money to triple if it is invested at 13% compounded annually? |

It is interesting and instructive to graph the doubling times for various rates compounded annually. We proceed as follows:

$$A = P(1 + r)^t$$

$$2P = P(1 + r)^t$$

$$2 = (1 + r)^t$$

$$(1 + r)^t = 2$$

$$\ln(1 + r)^t = \ln 2$$

$$t \ln(1 + r) = \ln 2$$

$$t = \frac{\ln 2}{\ln(1 + r)}$$

Figure 5 shows the graph of this equation (doubling time in years) for interest rates compounded annually from 1 to 70% (expressed as decimals). Note the dramatic change in doubling time as rates change from 1 to 20% (from 0.01 to 0.20).

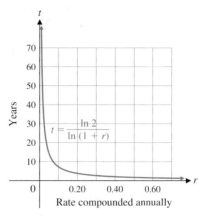

FIGURE 5

Among increasing function, the logarithmic functions (with bases $b > 1$) increase much more slowly for large values of x than either exponential or polynomial functions. When a visual inspection of the plot of a data set indicates a slowly increasing function, a logarithmic function often provides a good model. We use **logarithmic regression** on a graphing calculator to find the function of the form $y = a + b \ln x$ that best fits the data.

EXAMPLE 11

Home Ownership Rates   The U.S. Census Bureau published the data in Table 1 on home ownership rates. Let x represent time in years with $x = 0$ representing 1900. Use logarithmic regression to find the best model of the form $y = a + b \ln x$ for the home ownership rate y as a function of time x. Use the model to predict the home ownership rate in the United States in 2015 (to the nearest tenth of a percent).

TABLE 1   Home Ownership Rates

Year	Rate (%)
1950	55.0
1960	61.9
1970	62.9
1980	64.4
1990	64.2
2000	67.4

SOLUTION   Enter the data in a graphing calculator (Fig. 6A) and find the logarithmic regression equation (Fig. 6B). The data set and the regression equation are graphed in Figure 6C. Using TRACE, we predict that the home ownership rate in 2015 would be 69.4%.

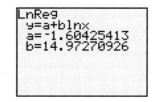

(A)

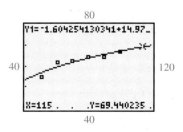

(B)   (C)

FIGURE 6

MATCHED PROBLEM 11   The home ownership rate in 2005 was 69.1%. Add this point to the data of Table 1, and, with $x = 0$ representing 1900, use logarithmic regression to find the best model of the form $y = a + b \ln x$ for the home ownership rate y as a function of time x. Use the model to predict the home ownership rate in the United States in 2015 (to the nearest tenth of a percent).

 CAUTION: Note that in Example 11 we let $x = 0$ represent 1900. If we let $x = 0$ represent 1940, for example, we would obtain a different logarithmic regression equation, but the prediction for 2015 would be the same. We would *not* let $x = 0$ represent 1950 (the first year in Table 1) or any later year, because logarithmic functions are undefined at 0.

Answers to Matched Problems

1. (A) $9 = 3^2$ (B) $2 = 4^{1/2}$ (C) $\frac{1}{9} = 3^{-2}$
2. (A) $\log_7 49 = 2$ (B) $\log_9 3 = \frac{1}{2}$ (C) $\log_3(\frac{1}{3}) = -1$
3. (A) $y = \frac{3}{2}$ (B) $x = \frac{1}{3}$ (C) $b = 10$
4. (A) $\log_b R - \log_b S - \log_b T$ (B) $\frac{2}{3}(\log_b R - \log_b S)$ (C) b^u (D) $\log_b x$
5. $x = 2$
6. $x = 5$
7. (A) $-1.868\ 734$ (B) $3.356\ 663$ (C) Not defined
8. (A) 0.0063 (B) 120.8092
9. (A) 0.8451 (B) 1.7918 (C) 1.1610
10. 9 yr
11. 69.9%

```
LnReg
y=a+blnx
a=-4.545195557
b=15.68361102
```

Exercise 2-5

A *For Problems 1–6, rewrite in equivalent exponential form.*

1. $\log_3 27 = 3$ 2. $\log_2 32 = 5$
3. $\log_{10} 1 = 0$ 4. $\log_e 1 = 0$
5. $\log_4 8 = \frac{3}{2}$ 6. $\log_9 27 = \frac{3}{2}$

For Problems 7–12, rewrite in equivalent logarithmic form.

7. $49 = 7^2$ 8. $36 = 6^2$
9. $8 = 4^{3/2}$ 10. $9 = 27^{2/3}$
11. $A = b^u$ 12. $M = b^x$

For Problems 13–24, evaluate without a calculator.

13. $\log_{10} 1$ 14. $\log_e 1$
15. $\log_e e$ 16. $\log_{10} 10$
17. $\log_{0.2} 0.2$ 18. $\log_{13} 13$
19. $\log_{10} 10^3$ 20. $\log_{10} 10^{-5}$
21. $\log_2 2^{-3}$ 22. $\log_3 3^5$
23. $\log_{10} 1,000$ 24. $\log_6 36$

For Problems 25–30, write in terms of simpler forms, as in Example 4.

25. $\log_b \dfrac{P}{Q}$ 26. $\log_b FG$
27. $\log_b L^5$ 28. $\log_b w^{15}$
29. $3^{p \log_3 q}$ 30. $\dfrac{\log_3 P}{\log_3 R}$

B *For Problems 31–42, find x, y, or b without a calculator.*

31. $\log_3 x = 2$ 32. $\log_2 x = 2$
33. $\log_7 49 = y$ 34. $\log_3 27 = y$
35. $\log_b 10^{-4} = -4$ 36. $\log_b e^{-2} = -2$
37. $\log_4 x = \frac{1}{2}$ 38. $\log_{25} x = \frac{1}{2}$
39. $\log_{1/3} 9 = y$ 40. $\log_{49}(\frac{1}{7}) = y$
41. $\log_b 1,000 = \frac{3}{2}$ 42. $\log_b 4 = \frac{2}{3}$

In Problems 43–52, discuss the validity of each statement. If the statement is always true, explain why. If not, give a counter-example.

43. Every polynomial function is one-to-one.

44. Every polynomial function of odd degree is one-to-one.

45. If g is the inverse of a function f, then g is one-to-one.

46. The graph of a one-to-one function intersects each vertical line exactly once.

47. The inverse of $f(x) = 2x$ is $g(x) = x/2$.

48. The inverse of $f(x) = x^2$ is $g(x) = \sqrt{x}$.

49. If $b > 0$ and $b \neq 1$, then the graph of $y = \log_b x$ is increasing.

50. If $b > 1$, then the graph of $y = \log_b x$ is increasing.

51. If f is one-to-one, then the domain of f is equal to the range of f.

52. If g is the inverse of a function f, then f is the inverse of g.

Find x in Problems 53–60.

53. $\log_b x = \frac{2}{3}\log_b 8 + \frac{1}{2}\log_b 9 - \log_b 6$

54. $\log_b x = \frac{2}{3}\log_b 27 + 2\log_b 2 - \log_b 3$

55. $\log_b x = \frac{3}{2}\log_b 4 - \frac{2}{3}\log_b 8 + 2\log_b 2$

56. $\log_b x = 3\log_b 2 + \frac{1}{2}\log_b 25 - \log_b 20$

57. $\log_b x + \log_b(x - 4) = \log_b 21$

58. $\log_b(x + 2) + \log_b x = \log_b 24$

59. $\log_{10}(x - 1) - \log_{10}(x + 1) = 1$

60. $\log_{10}(x + 6) - \log_{10}(x - 3) = 1$

Graph Problems 61 and 62 by converting to exponential form first.

61. $y = \log_2(x - 2)$

62. $y = \log_3(x + 2)$

63. Explain how the graph of the equation in Problem 61 can be obtained from the graph of $y = \log_2 x$ using a simple transformation (see Section 2-2).

64. Explain how the graph of the equation in Problem 62 can be obtained from the graph of $y = \log_3 x$ using a simple transformation (see Section 2-2).

65. What are the domain and range of the function defined by $y = 1 + \ln(x + 1)$?

66. What are the domain and range of the function defined by $y = \log(x - 1) - 1$?

For Problems 67 and 68, evaluate to five decimal places using a calculator.

67. (A) log 3,527.2   (B) log 0.006 913 2
    (C) ln 277.63   (D) ln 0.040 883

68. (A) log 72.604   (B) log 0.033 041
    (C) ln 40,257   (D) ln 0.005 926 3

For Problems 69 and 70, find x to four decimal places.

69. (A) $\log x = 1.1285$ (B) $\log x = -2.0497$
    (C) $\ln x = 2.7763$ (D) $\ln x = -1.8879$

70. (A) $\log x = 2.0832$ (B) $\log x = -1.1577$
    (C) $\ln x = 3.1336$ (D) $\ln x = -4.3281$

For Problems 71–78, solve each equation to four decimal places.

71. $10^x = 12$ 72. $10^x = 153$

73. $e^x = 4.304$ 74. $e^x = 0.3059$

75. $1.03^x = 2.475$ 76. $1.075^x = 1.837$

77. $1.005^{12t} = 3$ 78. $1.02^{4t} = 2$

Graph Problems 79–86 using a calculator and point-by-point plotting. Indicate increasing and decreasing intervals.

79. $y = \ln x$ 80. $y = -\ln x$

81. $y = |\ln x|$ 82. $y = \ln|x|$

83. $y = 2\ln(x + 2)$ 84. $y = 2\ln x + 2$

85. $y = 4\ln x - 3$ 86. $y = 4\ln(x - 3)$

C

87. Explain why the logarithm of 1 for any permissible base is 0.

88. Explain why 1 is not a suitable logarithmic base.

89. Write $\log_{10} y - \log_{10} c = 0.8x$ in an exponential form that is free of logarithms.

90. Write $\log_e x - \log_e 25 = 0.2t$ in an exponential form that is free of logarithms.

91. Let $p(x) = \ln x$, $q(x) = \sqrt{x}$, and $r(x) = x$. Use a graphing calculator to draw graphs of all three functions in the same viewing window for $1 \leq x \leq 16$. Discuss what it means for one function to be larger than another on an interval, and then order the three functions from largest to smallest for $1 < x \leq 16$.

92. Let $p(x) = \log x$, $q(x) = \sqrt[3]{x}$, and $r(x) = x$. Use a graphing calculator to draw graphs of all three functions in the same viewing window for $1 \leq x \leq 16$. Discuss what it means for one function to be smaller than another on an interval, and then order the three functions from smallest to largest for $1 < x \leq 16$.

Applications

93. *Doubling time.* In its first 10 years the Gabelli Growth Fund produced an average annual return of 21.36%. Assume that money invested in this fund continues to earn 21.36% compounded annually. How long will it take money invested in this fund to double?

94. *Doubling time.* In its first 10 years the Janus Flexible Income Fund produced an average annual return of 9.58%. Assume that money invested in this fund continues to earn 9.58% compounded annually. How long will it take money invested in this fund to double?

95. *Investing.* How many years (to two decimal places) will it take $1,000 to grow to $1,800 if it is invested at 6% compounded quarterly? Compounded daily?

96. *Investing.* How many years (to two decimal places) will it take $5,000 to grow to $7,500 if it is invested at 8% compounded semiannually? Compounded monthly?

97. *Supply and demand.* A cordless screwdriver is sold through a national chain of discount stores. A marketing company established price–demand and price–supply tables (Tables 1 and 2), where x is the number of screwdrivers people are willing to buy and the store is willing to sell each month at a price of p dollars per screwdriver.

(A) Find a logarithmic regression model ($y = a + b \ln x$) for the data in Table 1. Estimate the demand (to the nearest unit) at a price level of $50.

TABLE 1 Price–Demand

x	$p = D(x)$ ($)
1,000	91
2,000	73
3,000	64
4,000	56
5,000	53

(B) Find a logarithmic regression model ($y = a + b \ln x$) for the data in Table 2. Estimate the supply (to the nearest unit) at a price level of $50.

TABLE 2 Price–Supply

x	$p = S(x)$ ($)
1,000	9
2,000	26
3,000	34
4,000	38
5,000	41

(C) Does a price level of $50 represent a stable condition, or is the price likely to increase or decrease? Explain.

98. *Equilibrium point.* Use the models constructed in Problem 97 to find the equilibrium point. Write the equilibrium price to the nearest cent and the equilibrium quantity to the nearest unit.

99. *Sound intensity: decibels.* Because of the extraordinary range of sensitivity of the human ear (a range of over 1,000 million millions to 1), it is helpful to use a logarithmic scale, rather than an absolute scale, to measure sound intensity over this range. The unit of measure is called the *decibel*, after the inventor of the

telephone, Alexander Graham Bell. If we let N be the number of decibels, I the power of the sound in question (in watts per square centimeter), and I_0 the power of sound just below the threshold of hearing (approximately 10^{-16} watt per square centimeter), then

$$I = I_0 10^{N/10}$$

Show that this formula can be written in the form

$$N = 10 \log \frac{I}{I_0}$$

100. *Sound intensity: decibels.* Use the formula in Problem 99 (with $I_0 = 10^{-16}$ W/cm^2) to find the decibel ratings of the following sounds:

(A) Whisper: 10^{-13} W/cm^2

(B) Normal conversation: 3.16×10^{-10} W/cm^2

(C) Heavy traffic: 10^{-8} W/cm^2

(D) Jet plane with afterburner: 10^{-1} W/cm^2

101. *Agriculture.* Table 3 shows the yield (in bushels per acre) and the total production (in millions of bushels) for corn in the United States for selected years since 1950. Let x represent years since 1900. Find a logarithmic regression model ($y = a + b \ln x$) for the yield. Estimate (to one decimal place) the yield in 2015.

TABLE 3 United States Corn Production

Year	x	Yield (bushels per acre)	Total Production (million bushels)
1950	50	37.6	2,782
1960	60	55.6	3,479
1970	70	81.4	4,802
1980	80	97.7	6,867
1990	90	115.6	7,802
2000	100	139.6	10,192

102. *Agriculture.* Refer to Table 3. Find a logarithmic regression model ($y = a + b \ln x$) for the total production. Estimate (to the nearest million) the production in 2015.

103. *World population.* If the world population is now 6.5 billion people and if it continues to grow at an annual rate of 1.14% compounded continuously, how long (to the nearest year) will it take before there is only 1 square yard of land per person? (The Earth contains approximately 1.68×10^{14} square yards of land.)

104. *Archaeology: carbon-14 dating.* The radioactive carbon-14 (^{14}C) in an organism at the time of its death decays according to the equation

$$A = A_0 e^{-0.000124t}$$

where t is time in years and A_0 is the amount of ^{14}C present at time $t = 0$. (See Example 3 in Section 2-4.) Estimate the age of a skull uncovered in an archaeological site if 10% of the original amount of ^{14}C is still present. [*Hint:* Find t such that $A = 0.1A_0$.]

CHAPTER 2 REVIEW

Important, Terms, Symbols, and Concepts

2-1 Functions

Examples

- **Point-by-point plotting** may be used to **sketch the graph** of an equation in two variables: Plot enough points from its solution set in a rectangular cordinate system so that the total graph is apparent and then connect these points with a smooth curve.

  Ex. 1, p. 47

- A **function** is a correspondence between two sets of elements such that to each element in the first set there corresponds one and only one element in the second set. The first set is called the **domain** and the set of corresponding elements in the second set is called the **range**.

- If x is a placeholder for the elements in the domain of a function, then x is called the **independent variable** or the **input**. If y is a placeholder for the elements in the range, then y is called the **dependent variable** or the **output**.

- If in an equation in two variables we get exactly one ouput for each input, then the equation specifies a function. The graph of such a function is just the graph of the specifying equation. If we get more than one output for a given input, then the equation does not specify a function.

  Ex. 2, p. 50

- The **vertical-line test** can be used to determine whether or not an equation in two variables specifies a function (Theorem 1, p. 51).

- The functions specified by equations of the form $y = mx + b$, where $m \neq 0$, are called **linear functions**. Functions specified by equations of the form $y = b$ are called **constant functions**.

- If a function is specified by an equation and the domain is not indicated, we agree to assume that the domain is the set of all inputs that produce outputs that are real numbers.

  Ex. 3, p. 52
  Ex. 5, p. 54

- The symbol $f(x)$ represents the element in the range of f that corresponds to the element x of the domain.

  Ex. 4, p. 53
  Ex. 6, p. 55

- **Break-even** and **profit–loss** analysis use a cost function C and a revenue function R to determine when a company will have a loss ($R < C$), will break even ($R = C$), or will have a profit ($R > C$). Typical **cost**, **revenue**, **profit**, and **price–demand functions** are given on page 55.

  Ex. 7, p.56

2-2 Elementary Functions: Graphs and Transformations

- The graphs of **six basic elementary functions** (the identity function, the square and cube functions, the square root and cube root functions, and the absolute value function) are shown on page 65.

  Ex. 1, p. 64

- Performing an operation on a function produces a **transformation** of the graph of the function. The basic graph transformations, **vertical and horizontal translations** (shifts), **reflection in the x axis**, and **vertical stretches and shrinks**, are summarized on page 69.

  Ex. 2, p. 66
  Ex. 3, p. 67
  Ex. 4, p. 68
  Ex. 5, p. 69

- A **piecewise-defined function** is a function whose definition involves more than one formula.

  Ex. 6, p. 70

2-3 Quadratic Functions

- If $a, b,$ and c are real numbers with $a \neq 0$, then the function

$$f(x) = ax^2 + bx + c \qquad \textit{Standard form}$$

is a **quadratic function** in **standard form** and its graph is a **parabola**.

- The quadratic formula

  Ex. 1, p. 77

$$x = \frac{-b \pm \sqrt{b^2 - 4ac}}{2a} \qquad \textit{b}^2 - 4ac \geq 0$$

can be used to find the x intercepts of a quadratic function.

- Completing the square in the standard form of a quadratic function produces the **vertex form**

$$f(x) = a(x - h)^2 + k \qquad \textit{Vertex form}$$

- From the vertex form of a quadratic function, we can read off the vertex, axis of symmetry, maximum or minimum, and range, and can easily sketch the graph (page 81).

  Ex. 2, p. 81
  Ex. 3, p. 83

- If a revenue function $R(x)$ and a cost function $C(x)$ intersect at a point (x_0, y_0), then both this point and its x coordinate x_0 are referred to as **break-even points**.

  Ex. 4, p. 84

- **Quadratic regression** on a graphing calculator produces the function of the form $y = ax^2 + bx + c$ that best fits a data set. Ex. 5, p. 86

- A quadratic function is a special case of a **polynomial function**, that is, a function that can be written in the form

$$f(x) = a_n x^n + a_{n-1} x^{n-1} + \ldots + a_1 x + a_0$$

  for n a nonnegative integer called the **degree** of the polynomial. The coefficients $a_0, a_1, \ldots, a_n$ are real numbers with $a_n \neq 0$. The **domain** of a polynomial function is the set of all real numbers. Graphs of representative polynomial functions are shown on page 87 and inside the front cover.

- The graph of a polynomial function of degree n can intersect the x axis at most n times.

- The graph of a polynomial function has no sharp corners and is **continuous**, that is, it has no holes or breaks.

- A **rational function** is any function that can be written in the form

$$f(x) = \frac{n(x)}{d(x)} \qquad d(x) \neq 0$$

  where $n(x)$ and $d(x)$ are polynomials. The **domain** is the set of all real numbers such that $d(x) \neq 0$. Graphs of representative rational functions are shown on page 88 and inside the front cover.

- Unlike polynomial functions, a rational function can have vertical asymptotes [but not more than the degree of the denominator $d(x)$] and at most one horizontal asymptote.

2-4 Exponential Functions

- An **exponential function** is a function of the form

$$f(x) = b^x$$

  where $b \neq 1$ is a positive constant called the **base**. The **domain** of f is the set of all real numbers and the **range** is the set of positive real numbers.

- The graph of an exponential function is continuous, passes through $(0, 1)$, and has the x axis as a horizontal asymptote. If $b > 1$, then b^x increases as x increases; if $0 < b < 1$, then b^x decreases as x increases (Theorem 1, p. 95). Ex. 1, p. 95

- Exponential functions obey the familiar laws of exponents and satisfy additional properties (Theorem 2, p. 96).

- The base that is used most frequently in mathematics is the irrational number $e \approx 2.7183$. Ex. 2, p. 98

- Exponential functions can be used to model population growth and radioactive decay. Ex. 3, p. 98

- Exponential regression on a graphing calculator produces the function of the form $y = ab^x$ that best fits a data set. Ex. 4, p. 100

- Exponential functions are used in computations of **compound interest**: Ex. 5, p. 101

$$A = P\left(1 + \frac{r}{m}\right)^{mt} \qquad \text{Compound interest formula}$$

  (see summary on page 102).

2-5 Logarithmic Functions

- A function is said to be **one-to-one** if each range value corresponds to exactly one domain value.

- The **inverse** of a one-to-one function f is the function formed by interchanging the independent and dependent variables of f. That is, (a, b) is a point on the graph of f if and only if (b, a) is a point on the graph of the inverse of f. A function that is not one-to-one does not have an inverse.

- The inverse of the exponential function with base b is called the **logarithmic function with base b**, denoted $y = \log_b x$. The **domain** of $\log_b x$ is the set of all positive real numbers (which is the range of b^x), and the range of $\log_b x$ is the set of all real numbers (which is the domain of b^x).

- Because $\log_b x$ is the inverse of the function b^x, Ex. 1, p. 108
Ex. 2, p. 108
Ex. 3, p. 109

  Logarithmic form   Exponential form

$$y = \log_b x \qquad \text{is equivalent to} \qquad x = b^y$$

- Properties of logarithmic functions can be obtained from corresponding properties of expontial functions (Theorem 1, p. 109).

- Logarithms to the base 10 are called **common logarithms**, often denoted simply by log x. Logarithms to the base e are called **natural logarithms**, often denoted by ln x.

- Logarithms can be used to find an investment's **doubling time**—the length of time it takes for the value of an investment to double.

- **Logarithmic regression** on a graphing calculator produces the function of the form $y = a + b \ln x$ that best fits a data set.

Ex. 4, p.110
Ex. 5, p. 110
Ex. 6, p. 111
Ex. 7, p. 111
Ex. 8, p. 112
Ex. 9, p. 112
Ex. 10, p. 114
Ex. 11, p. 115

REVIEW EXERCISE

Work through all the problems in this chapter review and check your answers in the back of the book. Answers to all review problems are there along with section numbers in italics to indicate where each type of problem is discussed. Where weaknesses show up, review appropriate sections in the text.

A *In Problems 1–3, use point-by-point plotting to sketch the graph of each equation.*

1. $y = 5 - x^2$

2. $x^2 = y^2$

3. $y^2 = 4x^2$

4. Indicate whether each graph specifies a function:

(A) (B)

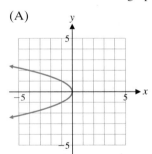

(C) (D)

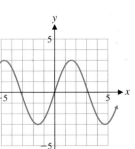

5. For $f(x) = 2x - 1$ and $g(x) = x^2 - 2x$, find:

(A) $f(-2) + g(-1)$ (B) $f(0) \cdot g(4)$

(C) $\dfrac{g(2)}{f(3)}$ (D) $\dfrac{f(3)}{g(2)}$

6. Write in logarithmic form using base e: $u = e^v$.

7. Write in logarithmic form using base 10: $x = 10^y$.

8. Write in exponential form using base e: $\ln M = N$.

9. Write in exponential form using base 10: $\log u = v$.

Simplify Problems 10 and 11.

10. $\dfrac{5^{x+4}}{5^{4-x}}$ **11.** $\left(\dfrac{e^u}{e^{-u}}\right)^u$

Solve Problems 12–14 for x exactly without using a calculator.

12. $\log_3 x = 2$ **13.** $\log_x 36 = 2$

14. $\log_2 16 = x$

Solve Problems 15–18 for x to three decimal places.

15. $10^x = 143.7$ **16.** $e^x = 503{,}000$

17. $\log x = 3.105$ **18.** $\ln x = -1.147$

19. Use the graph of function f in the figure to determine (to the nearest integer) x or y as indicated.

(A) $y = f(0)$ (B) $4 = f(x)$

(C) $y = f(3)$ (D) $3 = f(x)$

(E) $y = f(-6)$ (F) $-1 = f(x)$

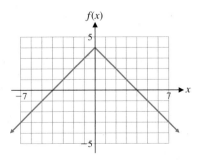

Figure for 19

20. Sketch a graph of each of the functions in parts (A)–(D) using the graph of function f in the figure below.
(A) $y = -f(x)$ (B) $y = f(x) + 4$
(C) $y = f(x - 2)$ (D) $y = -f(x + 3) - 3$

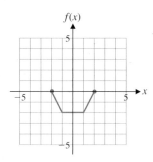

Figure for 20

21. Complete the square and find the standard form for the quadratic function
$$f(x) = -x^2 + 4x$$
Then write a brief verbal description of the relationship between the graph of f and the graph of $y = x^2$.

22. Match each equation with a graph of one of the functions $f, g, m,$ or n in the figure.

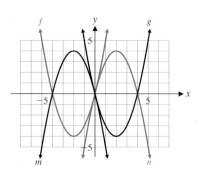

Figure for 22

(A) $y = (x - 2)^2 - 4$ (B) $y = -(x + 2)^2 + 4$
(C) $y = -(x - 2)^2 + 4$ (D) $y = (x + 2)^2 - 4$

23. Referring to the graph of function f in the figure for Problem 11 and using known properties of quadratic functions, find each of the following to the nearest integer:
(A) Intercepts   (B) Vertex
(C) Maximum or minimum   (D) Range
(E) Increasing interval   (F) Decreasing interval

24. Consider the set S of people living in the town of Newville. Which of the following correspondences specify a function? Explain.
(A) Each person in Newville (input) is paired with his or her mother (output).
(B) Each person in Newville (input) is paired with his or her children (output).

25. The three points in the table are on the graph of the indicated function f. Do these three points provide sufficient in-formation for you to sketch the graph of $y = f(x)$? Add more points to the table until you are satisfied that your sketch is a good representation of the graph of $y = f(x)$ on the interval $[-5, 5]$.

x	-2	0	2
$f(x)$	-4	0	4

$f(x) = 6x - x^3$

In Problems 26–29, each equation specifies a function. Determine whether the function is linear, quadratic, constant, or none of these.

26. $y = 4 - x + 3x^2$ **27.** $y = \dfrac{1 + 5x}{6}$

28. $y = \dfrac{7 - 4x}{2x}$ **29.** $y = 8x + 2(10 - 4x)$

Solve Problems 30–37 for x exactly without using a calculator.
30. $\log(x + 5) = \log(2x - 3)$ **31.** $2 \ln(x - 1) = \ln(x^2 - 5)$
32. $9^{x-1} = 3^{1+x}$ **33.** $e^{2x} = e^{x^2-3}$
34. $2x^2e^x = 3xe^x$ **35.** $\log_{1/3} 9 = x$
36. $\log_x 8 = -3$ **37.** $\log_9 x = \frac{3}{2}$

Solve Problems 38–47 for x to four decimal places.
38. $x = 3(e^{1.49})$ **39.** $x = 230(10^{-0.161})$
40. $\log x = -2.0144$ **41.** $\ln x = 0.3618$
42. $35 = 7(3^x)$ **43.** $0.01 = e^{-0.05x}$
44. $8,000 = 4,000(1.08^x)$ **45.** $5^{2x-3} = 7.08$
46. $x = \log_2 7$ **47.** $x = \log_{0.2} 5.321$
48. Find the domain of each function:
$$\text{(A)} \ f(x) = \frac{2x - 5}{x^2 - x - 6} \qquad \text{(B)} \ g(x) = \frac{3x}{\sqrt{5 - x}}$$

49. The function g is defined by $g(x) = 2x - 3\sqrt{x}$. Translate into a verbal definition.

50. Find the vertex form for $f(x) = 4x^2 + 4x - 3$ and then find the intercepts, the vertex, the maximum or minimum, and the range.

51. Let $f(x) = e^x - 1$ and $g(x) = \ln(x + 2)$. Find all points of intersection for the graphs of f and g. Round answers to two decimal places.

In Problems 52 and 53, use point-by-point plotting to sketch the graph of each function.

52. $f(x) = \dfrac{50}{x^2 + 1}$

53. $f(x) = \dfrac{-66}{2 + x^2}$

If $f(x) = 5x + 1$, find and simplify each of the following in Problems 54–57.
54. $f(f(0))$
55. $f(f(-1))$
56. $f(2x - 1)$
57. $f(4 - x)$
58. Let $f(x) = 3 - 2x$. Find
(A) $f(2)$ (B) $f(2 + h)$
(C) $f(2 + h) - f(2)$ (D) $\dfrac{f(2 + h) - f(2)}{h}$

59. Let $f(x) = x^2 - 3x + 1$. Find

(A) $f(a)$ (B) $f(a + h)$

(C) $f(a + h) - f(a)$ (D) $\dfrac{f(a + h) - f(a)}{h}$

60. Explain how the graph of $m(x) = -|x - 4|$ is related to the graph of $y = |x|$.

61. Explain how the graph of $g(x) = 0.3x^3 + 3$ is related to the graph of $y = x^3$.

62. The following graph is the result of applying a sequence of transformations to the graph of $y = x^2$. Describe the transformations verbally and write an equation for the graph.

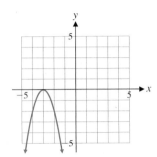

Figure for 62

63. The graph of a function f is formed by vertically stretching the graph of $y = \sqrt{x}$ by a factor of 2, and shifting it to the left 3 units and down 1 unit. Find an equation for function f and graph it for $-5 \le x \le 5$ and $-5 \le y \le 5$.

In Problems 64–71, discuss the validity of each statement. If the statement is always true, explain why. If not, give a counterexample.

64. Every polynomial function is a rational function.

65. Every rational function is a polynomial function.

66. The graph of every rational function has at least one vertical asymptote.

67. The graph of every exponential function has a horizontal asymptote.

68. The graph of every logarithmic function has a vertical asymptote.

69. There exists a logarithmic function that has both a vertical and horizontal asymptote.

70. There exists a rational function that has both a vertical and horizontal asymptote.

71. There exists an exponential function that has both a vertical and horizontal asymptote.

Graph Problems 72–74 over the indicated interval. Indicate increasing and decreasing intervals.

72. $y = 2^{x-1}$; $[-2, 4]$ **73.** $f(t) = 10e^{-0.08t}$; $t \ge 0$

74. $y = \ln(x + 1)$; $(-1, 10]$

75. Sketch the graph of f for $x \ge 0$.

$$f(x) = \begin{cases} 9 + 0.3x & \text{if } 0 \le x \le 20 \\ 5 + 0.2x & \text{if } x > 20 \end{cases}$$

76. Sketch the graph of g for $x \ge 0$.

$$g(x) = \begin{cases} 0.5x + 5 & \text{if } 0 \le x \le 10 \\ 1.2x - 2 & \text{if } 10 < x \le 30 \\ 2x - 26 & \text{if } x > 30 \end{cases}$$

77. Write an equation for the graph shown in the form $y = a(x - h)^2 + k$, where a is either -1 or $+1$ and h and k are integers.

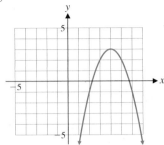

Figure for 77

78. Given $f(x) = -0.4x^2 + 3.2x + 1.2$, find the following algebraically (to three decimal places) without referring to a graph:

(A) Intercepts (B) Vertex

(C) Maximum or minimum (D) Range

79. Graph $f(x) = -0.4x^2 + 3.2x + 1.2$ in a graphing calculator and find the following (to three decimal places) using TRACE and appropriate built-in commands:

(A) Intercepts (B) Vertex

(C) Maximum or minimum (D) Range

C 80. Noting that $\pi = 3.141\,592\,654\ldots$ and $\sqrt{2} = 1.414\,213\,562\ldots$ explain why the calculator results shown here are obvious. Discuss similar connections between the natural logarithmic function and the exponential function with base e.

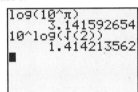

Solve Problems 81–84 exactly without using a calculator.

81. $\log x - \log 3 = \log 4 - \log(x + 4)$

82. $\ln(2x - 2) - \ln(x - 1) = \ln x$

83. $\ln(x + 3) - \ln x = 2 \ln 2$

84. $\log 3x^2 = 2 + \log 9x$

85. Write $\ln y = -5t + \ln c$ in an exponential form free of logarithms. Then solve for y in terms of the remaining variables.

86. Explain why 1 cannot be used as a logarithmic base.

87. The following graph is the result of applying a sequence of transformations to the graph of $y = \sqrt[3]{x}$. Describe the transformations verbally, and write an equation for the graph.

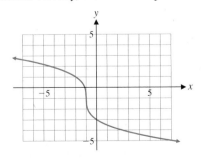

Figure for 87

88. Given $G(x) = 0.3x^2 + 1.2x - 6.9$, find the following algebraically (to three decimal places) without the use of a graph:

(A) Intercepts

(B) Vertex

(C) Maximum or minimum

(D) Range

(E) Increasing and decreasing intervals

89. Graph $G(x) = 0.3x^2 + 1.2x - 6.9$ in a standard viewing window. Then find each of the following (to three decimal places) using appropriate commands.

(A) Intercepts

(B) Vertex

(C) Maximum or minimum

(D) Range

(E) Increasing and decreasing intervals

Applications

In all problems involving days, a 365-day year is assumed.

90. *Egg consumption.* The table shows the annual per capita consumption of eggs in the United States.

Year	Number of Eggs
1980	271
1985	255
1990	233
1995	236
2000	256
2005	257

A mathematical model for this data is given by
$$f(x) = 0.181x^2 - 4.88x + 271$$
where $x = 0$ corresponds to 1980.

(A) Complete the following table.

x	Consumption	f(x)
0	271	
5	255	
10	233	
15	236	
20	256	
25	257	

(B) Graph $y = f(x)$ and the data in the same coordinate system.

(C) Use the modeling function f to estimate the per capita egg consumption in 2012.

(D) Based on the information in the table, write a brief description of egg consumption from 1980 to 2005.

91. *Egg consumption.* Using quadratic regression on a graphing calculator, show that the quadratic function that best fits the data on egg consumption in Problem 90 is
$$f(x) = 0.181x^2 - 4.88x + 271$$
(coefficients rounded to three significant digits).

92. *Electricity rates.* The table shows the electricity rates charged by Easton Utilities in the summer months.

(A) Write a piecewise definition of the monthly charge $S(x)$ (in dollars) for a customer who uses x kWh in a summer month.

(B) Graph $S(x)$.

Energy Charge (June–September)

\$3.00 for the first 20 kWh or less
5.70¢ per kWh for the next 180 kWh
3.46¢ per kWh for the next 800 kWh
2.17¢ per kWh for all over 1,000 kWh

93. *Money growth.* Provident Bank of Cincinnati, Ohio recently offered a certificate of deposit that paid 5.35% compounded quarterly. If a \$5,000 CD earns this rate for 5 years, how much will it be worth?

94. *Money growth.* Capital One Bank of Glen Allen, Virginia recently offered a certificate of deposit that paid 4.82% compounded daily. If a \$5,000 CD earns this rate for 5 years, how much will it be worth?

95. *Money growth.* How long will it take for money invested at 6.59% compounded monthly to triple?

96. *Money growth.* How long will it take for money invested at 7.39% compounded daily to double?

97. *Break–even analysis.* The research department in a company that manufactures AM/FM clock radios established the following price–demand, cost, and revenue functions:

$$p(x) = 50 - 1.25x \quad \text{Price–demand function}$$
$$C(x) = 160 + 10x \quad \text{Cost function}$$
$$R(x) = xp(x)$$
$$\quad\quad = x(50 - 1.25x) \quad \text{Revenue function}$$

where x is in thousands of units, and $C(x)$ and $R(x)$ are in thousands of dollars. All three functions have domain $1 \le x \le 40$.

(A) Graph the cost function and the revenue function simultaneously in the same coordinate system.

(B) Determine algebraically when $R = C$. Then, with the aid of part (A), determine when $R < C$ and $R > C$ to the nearest unit.

(C) Determine algebraically the maximum revenue (to the nearest thousand dollars) and the output (to the nearest unit) that produces the maximum revenue. What is the wholesale price of the radio (to the nearest dollar) at this output?

98. *Profit–loss analysis.* Use the cost and revenue functions from Problem 97.

(A) Write a profit function and graph it in a graphing calculator.

(B) Determine graphically when $P = 0, P < 0,$ and $P > 0$ to the nearest unit.

(C) Determine graphically the maximum profit (to the nearest thousand dollars) and the output (to the nearest unit) that produces the maximum profit. What is the wholesale price of the radio (to the nearest dollar) at this output? [Compare with Problem 97C.]

99. *Construction.* A construction company has 840 feet of chain-link fence that is used to enclose storage areas for equipment and materials at construction sties. The supervisor wants to set up two identical rectangular storage areas sharing a common fence (see the figure).

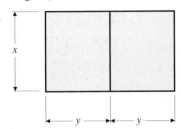

Assuming that all fencing is used,
(A) Express the total area $A(x)$ enclosed by both pens as a function of x.
(B) From physical considerations, what is the domain of the function A?
(C) Graph function A in a rectangular coordinate system.
(D) Use the graph to discuss the number and approximate locations of values of x that would produce storage areas with a combined area of 25,000 square feet.
(E) Approximate graphically (to the nearest foot) the values of x that would produce storage areas with a combined area of 25,000 square feet.
(F) Determine algebraically the dimensions of the storage areas that have the maximum total combined area. What is the maximum area?

100. *Equilibrium point.* A company is planning to introduce a 10-piece set of nonstick cookware. A marketing company established price–demand and price–supply tables for selected prices (Tables 1 and 2), where x is the number of cookware sets people are willing to buy and the company is willing to sell each month at a price of p dollars per set.

TABLE 1 Price–Demand

x	$p = D(x)$ ($)
985	330
2,145	225
2,950	170
4,225	105
5,100	50

TABLE 2 Price–Supply

x	$p = S(x)$ ($)
985	30
2,145	75
2,950	110
4,225	155
5,100	190

(A) Find a quadratic regression model for the data in Table 1. Estimate the demand at a price level of $180.
(B) Find a linear regression model for the data in Table 2. Estimate the supply at a price level of $180.
(C) Does a price level of $180 represent a stable condition, or is the price likely to increase or decrease? Explain.
(D) Use the models in parts (A) and (B) to find the equilibrium point. Write the equilibrium price to the nearest cent and the equilibrium quantity to the nearest unit.

101. *Telecommunications.* According to the Telecommunications Industry Association, wireless telephone subscriptions grew from about 4 million in 1990 to over 180 million in 2005 (Table 3). Let x represent years since 1990.

TABLE 3 Wireless Telephone Subscribers

Year	Million Subscribers
1990	4
1993	13
1996	38
1999	76
2002	143
2005	180

(A) Find an exponential regression model ($y = ab^x$) for this data. Estimate (to the nearest million) the number of subscribers in 2015.
(B) Some analysts estimate the number of wireless subscribers in 2018 to be 300 million. How does this compare with the prediction of the model of part (A)? Explain why the model will not give reliable predictions far into the future.

102. *Medicine.* One leukemic cell injected into a healthy mouse will divide into 2 cells in about $\frac{1}{2}$ day. At the end of the day these 2 cells will divide into 4. This doubling continues until 1 billion cells are formed; then the animal dies with leukemic cells in every part of the body.
(A) Write an equation that will give the number N of leukemic cells at the end of t days.
(B) When, to the nearest day, will the mouse die?

103. *Marine biology.* The intensity of light entering water is reduced according to the exponential equation

$$I = I_0 e^{-kd}$$

where I is the intensity d feet below the surface, I_0 is the intensity at the surface, and k is the coefficient of extinction. Measurements in the Sargasso Sea in the West Indies have indicated that half of the surface light reaches a depth of

73.6 feet. Find k (to five decimal places), and find the depth (to the nearest foot) at which 1% of the surface light remains.

104. *Agriculture.* The number of dairy cows on farms in the United States is shown in Table 4 for selected years since 1950. Let 1940 be year 0.

TABLE 4 Dairy Cows on Farms in the United States	
Year	**Dairy Cows (thousands)**
1950	23,853
1960	19,527
1970	12,091
1980	10,758
1990	10,015
2000	9,190

(A) Find a logarithmic regression model ($y = a + b \ln x$) for the data. Estimate (to the nearest thousand) the number of dairy cows in 2015.

(B) Explain why it is not a good idea to let 1950 be year 0.

105. *Population growth.* Some countries have a relative growth rate of 3% (or more) per year. At this rate, how many years (to the nearest tenth of a year) will it take a population to double?

106. *Medicare.* The annual expenditures for Medicare (in billions of dollars) by the U.S. government for selected years since 1980 are shown in Table 5. Let x represent years since 1980.

TABLE 5 Medicare Expenditures	
Year	**Billion $**
1980	37
1985	72
1990	111
1995	181
2000	197
2005	330

(A) Find an exponential regression model ($y = ab^x$) for the data. Estimate (to the nearest billion) the annual expenditures in 2015.

(B) When will the annual expenditures reach one trillion dollars?

PART TWO

Calculus

Limits and the Derivative

CHAPTER 3

3-1 Introduction to Limits

3-2 Continuity

3-3 Infinite Limits and Limits at Infinity

3-4 The Derivative

3-5 Basic Differentiation Properties

3-6 Differentials

3-7 Marginal Analysis in Business and Economics

Chapter 3 Review

Review Exercise

INTRODUCTION

How do algebra and calculus differ? The two words *static* and *dynamic* probably come as close as any to expressing the difference between the two disciplines. In algebra, we solve equations for a particular value of a variable—a static notion. In calculus, we are interested in how a change in one variable affects another variable—a dynamic notion.

Parts (A)–(C) of the figure at the top of the next page illustrate three basic problems in calculus. It may surprise you to learn that all three problems—as different as they appear—are related mathematically. The solutions to these problems and the discovery of their relationship required the creation of a new kind of mathematics. Isaac Newton (1642–1727) of England and Gottfried Wilhelm von Leibniz (1646–1716) of Germany developed this new mathematics, called **the calculus,** simultaneously and independently. It was an idea whose time had come.

In addition to solving the problems described in the figure, calculus will enable us to solve many other important problems. Until fairly recently, calculus was used primarily in the physical sciences, but now people in many other disciplines are finding it a useful tool as well.

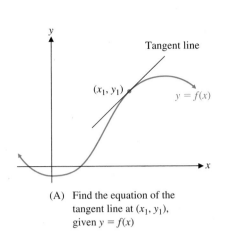

(A) Find the equation of the tangent line at (x_1, y_1), given $y = f(x)$

(B) Find the instantaneous velocity of a falling object

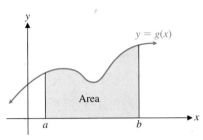

(C) Find the indicated area bounded by $y = g(x)$, $x = a$, $x = b$, and the x axis

Section 3-1 INTRODUCTION TO LIMITS

- Functions and Graphs: Brief Review
- Limits: A Graphical Approach
- Limits: An Algebraic Approach
- Limits of Difference Quotients

Basic to the study of calculus is the concept of a *limit*. This concept helps us to describe, in a precise way, the behavior of $f(x)$ when x is close, but not equal, to a particular value c. In this section, we develop an intuitive and informal approach to evaluating limits. Our discussion concentrates on developing and understanding concepts rather than on presenting formal mathematical details.

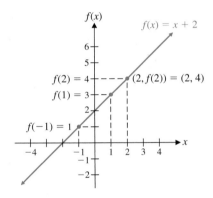

FIGURE 1

■ Functions and Graphs: Brief Review

The graph of the function $y = f(x) = x + 2$ is the graph of the set of all ordered pairs $(x, f(x))$. For example, if $x = 2$, then $f(2) = 4$ and $(2, f(2)) = (2, 4)$ is a point on the graph of f. Figure 1 shows $(-1, f(-1))$, $(1, f(1))$, and $(2, f(2))$ plotted on the graph of f. Notice that the domain values -1, 1, and 2 are associated with the x axis and the range values $f(-1) = 1$, $f(1) = 3$, and $f(2) = 4$ are associated with the y axis.

Given x, it is sometimes useful to be able to read $f(x)$ directly from the graph of f. Example 1 reviews this process.

EXAMPLE 1   **Finding Values of a Function from Its Graph** Complete the following table, using the given graph of the function g:

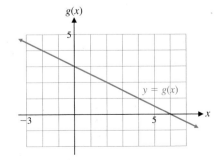

x	$g(x)$
-2	
1	
3	
4	

SOLUTION To determine $g(x)$, proceed vertically from the x value on the x axis to the graph of g and then horizontally to the corresponding y value $g(x)$ on the y axis (as indicated by the dashed lines):

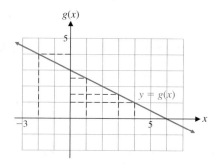

x	$g(x)$
-2	4.0
1	2.5
3	1.5
4	1.0

MATCHED PROBLEM 1   Complete the following table, using the given graph of the function h:

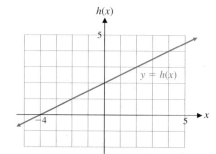

x	$h(x)$
-2	
-1	
0	
1	
2	
3	
4	

■ Limits: A Graphical Approach

We introduce the important concept of a *limit* through an example, which will lead to an intuitive definition of the concept.

EXAMPLE 2   **Analyzing a Limit** Let $f(x) = x + 2$. Discuss the behavior of the values of $f(x)$ when x is close to 2.

SOLUTION   We begin by drawing a graph of f that includes the domain value $x = 2$ (Fig. 2).

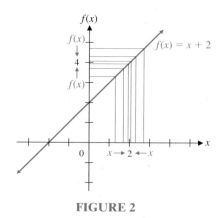

FIGURE 2

In Figure 2, we are using a static drawing to describe a dynamic process. This requires careful interpretation. The thin vertical lines in Figure 2 represent values of x that are close to 2. The corresponding horizontal lines identify the value of $f(x)$ associated with each value of x. [Example 1 dealt with the relationship between x and $f(x)$ on a graph.] The graph in Figure 2 indicates that as the values of x get closer and closer to 2 on either side of 2, the corresponding values of $f(x)$ get closer and closer to 4. Symbolically, we write

$$\lim_{x \to 2} f(x) = 4$$

This equation is read as "The limit of $f(x)$ as x approaches 2 is 4." Note that $f(2) = 4$. That is, the value of the function at 2 and the limit of the function as x approaches 2 are the same. This relationship can be expressed as

$$\lim_{x \to 2} f(x) = f(2) = 4$$

Graphically, this means that there is no hole or break in the graph of f at $x = 2$. ▬

MATCHED PROBLEM 2   Let $f(x) = x + 1$. Discuss the behavior of the values of $f(x)$ when x is close to 1.

We now present an informal definition of the important concept of a limit. A precise definition is not needed for our discussion, but one is given in a footnote.*

DEFINITION   **Limit**

We write

$$\lim_{x \to c} f(x) = L \qquad \text{or} \qquad f(x) \to L \quad \text{as} \quad x \to c$$

if the functional value $f(x)$ is close to the single real number L whenever x is close, but not equal, to c (on either side of c).

Note: The existence of a limit at c has nothing to do with the value of the function at c. In fact, c may not even be in the domain of f. However, the function must be defined on both sides of c.

The next example involves the **absolute value function:**

$$f(x) = |x| = \begin{cases} -x & \text{if } x < 0 \\ x & \text{if } x \geq 0 \end{cases} \quad \begin{array}{l} f(-2) = |-2| = -(-2) = 2 \\ f(3) = |3| = 3 \end{array}$$

The graph of f is shown in Figure 3.

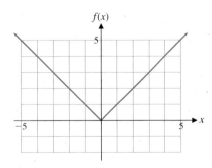

FIGURE 3 $f(x) = |x|$

---

* To make the informal definition of *limit* precise, the use of the word *close* must be made more precise. This is done as follows: We write $\lim_{x \to c} f(x) = L$ if, for each $e > 0$, there exists a $d > 0$ such that $|f(x) - L| < e$ whenever $0 < |x - c| < d$. This definition is used to establish particular limits and to prove many useful properties of limits that will be helpful to us in finding particular limits. [Even though intuitive notions of a limit existed for a long time, it was not until the nineteenth century that a precise definition was given by the German mathematician Karl Weierstrass (1815–1897).]

EXAMPLE 3

Analyzing a Limit Let $h(x) = |x|/x$. Explore the behavior of $h(x)$ for x near, but not equal, to 0. Find $\lim_{x\to 0} h(x)$ if it exists.

SOLUTION

The function h is defined for all real numbers except 0. For example,

$$h(-2) = \frac{|-2|}{-2} = \frac{2}{-2} = -1$$

$$h(0) = \frac{|0|}{0} = \frac{0}{0} \qquad \text{Not defined}$$

$$h(2) = \frac{|2|}{2} = \frac{2}{2} = 1$$

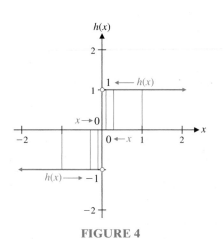

In general, $h(x)$ is -1 for all negative x and 1 for all positive x. Figure 4 illustrates the behavior of $h(x)$ for x near 0. Note that the absence of a solid dot on the vertical axis indicates that h is not defined when $x = 0$.

When x is near 0 (on either side of 0), is $h(x)$ near one specific number? The answer is "No," because $h(x)$ is -1 for $x < 0$ and 1 for $x > 0$. Consequently, we say that

$$\lim_{x\to 0}\frac{|x|}{x} \text{ does not exist}$$

Thus, neither $h(x)$ nor the limit of $h(x)$ exists at $x = 0$. However, the limit from the left and the limit from the right both exist at 0, but they are not equal. ▬

FIGURE 4

MATCHED PROBLEM 3

Graph

$$h(x) = \frac{x - 2}{|x - 2|}$$

and find $\lim_{x\to 2} h(x)$ if it exists.

In Example 3, we saw that the values of the function $h(x)$ approached two different numbers, depending on the direction of approach, and it was natural to refer to these values as "the limit from the left" and "the limit from the right." These experiences suggest that the notion of **one-sided limits** will be very useful in discussing basic limit concepts.

DEFINITION **One-Sided Limits**

We write

$$\lim_{x\to c^-} f(x) = K \quad \text{$x \to c^-$ is read "x approaches c from the left" and means $x\to c$ and $x < c$.}$$

and call K the **limit from the left** or the **left-hand limit** if $f(x)$ is close to K whenever x is close to, but to the left of, c on the real number line. We write

$$\lim_{x\to c^+} f(x) = L \quad \text{$x \to c^+$ is read "x approaches c from the right" and means $x\to c$ and $x > c$.}$$

and call L the **limit from the right** or the **right-hand limit** if $f(x)$ is close to L whenever x is close to, but to the right of, c on the real number line.

If no direction is specified in a limit statement, we will always assume that the limit is **two sided** or **unrestricted.** Theorem 1 states an important relationship between one-sided limits and unrestricted limits.

THEOREM 1 ON THE EXISTENCE OF A LIMIT

For a (two-sided) limit to exist, the limit from the left and the limit from the right must exist and be equal. That is,

$$\lim_{x \to c} f(x) = L \quad \text{if and only if} \quad \lim_{x \to c^-} f(x) = \lim_{x \to c^+} f(x) = L$$

In Example 3,

$$\lim_{x \to 0^-} \frac{|x|}{x} = -1 \quad \text{and} \quad \lim_{x \to 0^+} \frac{|x|}{x} = 1$$

Since the left- and right-hand limits are *not* the same,

$$\lim_{x \to 0} \frac{|x|}{x} \text{ does not exist}$$

EXAMPLE 4 **Analyzing Limits Graphically** Given the graph of the function f shown in Figure 5, discuss the behavior of $f(x)$ for x near (A) -1, (B) 1, and (C) 2.

SOLUTION (A) Since we have only a graph to work with, we use vertical and horizontal lines to relate the values of x and the corresponding values of $f(x)$. For any x near -1 on either side of -1, we see that the corresponding value of $f(x)$, determined by a horizontal line, is close to 1.

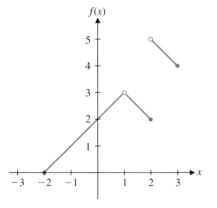

FIGURE 5

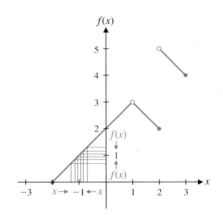

$$\lim_{x \to -1^-} f(x) = 1$$
$$\lim_{x \to -1^+} f(x) = 1$$
$$\lim_{x \to -1} f(x) = 1$$
$$f(-1) = 1$$

(B) Again, for any x near, but not equal to, 1, the vertical and horizontal lines indicate that the corresponding value of $f(x)$ is close to 3. The open dot at $(1, 3)$, together with the absence of a solid dot anywhere on the vertical line through $x = 1$, indicates that $f(1)$ is not defined.

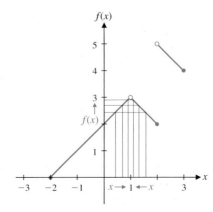

$$\lim_{x \to 1^-} f(x) = 3$$
$$\lim_{x \to 1^+} f(x) = 3$$
$$\lim_{x \to 1} f(x) = 3$$
$$f(1) \text{ not defined}$$

(C) The abrupt break in the graph at $x = 2$ indicates that the behavior of the graph near $x = 2$ is more complicated than in the two preceding cases. If x is close to 2 on the left side of 2, the corresponding horizontal line intersects the y axis at a point close to 2. If x is close to 2 on the right side of 2, the corresponding horizontal line intersects the y axis at a point close to 5. Thus, this is a case where the one-sided limits are different.

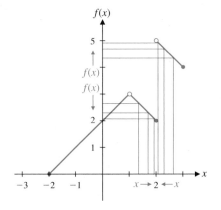

$$\lim_{x \to 2^-} f(x) = 2$$

$$\lim_{x \to 2^+} f(x) = 5$$

$$\lim_{x \to 2} f(x) \text{ does not exist}$$

$$f(2) = 2$$

MATCHED PROBLEM 4 Given the graph of the function f shown in Figure 6, discuss the following, as we did in Example 4:

(A) Behavior of $f(x)$ for x near 0

(B) Behavior of $f(x)$ for x near 1

(C) Behavior of $f(x)$ for x near 3

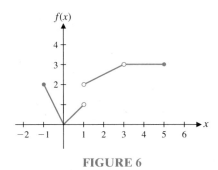

FIGURE 6

INSIGHT

In Example 4B, note that $\lim_{x \to 1} f(x)$ exists even though f is not defined at $x = 1$ and the graph has a hole at $x = 1$. In general, the value of a function at $x = c$ has no effect on the limit of the function as x approaches c.

■ Limits: An Algebraic Approach

Graphs are very useful tools for investigating limits, especially if something unusual happens at the point in question. However, many of the limits encountered in calculus are routine and can be evaluated quickly with a little algebraic simplification, some intuition, and basic properties of limits. The following list of properties of limits forms the basis for this approach:

THEOREM 2 PROPERTIES OF LIMITS

Let f and g be two functions, and assume that

$$\lim_{x \to c} f(x) = L \qquad \lim_{x \to c} g(x) = M$$

where L and M are real numbers (both limits exist). Then

1. $\lim_{x \to c} k = k$ for any constant k

2. $\lim_{x \to c} x = c$

3. $\lim_{x \to c}[f(x) + g(x)] = \lim_{x \to c} f(x) + \lim_{x \to c} g(x) = L + M$

4. $\lim_{x \to c}[f(x) - g(x)] = \lim_{x \to c} f(x) - \lim_{x \to c} g(x) = L - M$

5. $\lim_{x \to c} kf(x) = k \lim_{x \to c} f(x) = kL$ for any constant k

6. $\lim_{x \to c}[f(x) \cdot g(x)] = [\lim_{x \to c} f(x)][\lim_{x \to c} g(x)] = LM$

7. $\lim_{x \to c} \dfrac{f(x)}{g(x)} = \dfrac{\lim_{x \to c} f(x)}{\lim_{x \to c} g(x)} = \dfrac{L}{M}$ if $M \neq 0$

8. $\lim_{x \to c} \sqrt[n]{f(x)} = \sqrt[n]{\lim_{x \to c} f(x)} = \sqrt[n]{L}$ $L > 0$ for n even

Each property in Theorem 2 is also valid if $x \to c$ is replaced everywhere by $x \to c^-$ or replaced everywhere by $x \to c^+$.

Explore & Discuss **1**   The properties listed in Theorem 2 can be paraphrased in brief verbal statements. For example, property 3 simply states that *the limit of a sum is equal to the sum of the limits*. Write brief verbal statements for the remaining properties in Theorem 2.

| EXAMPLE 5 | **Using Limit Properties** Find $\lim_{x \to 3}(x^2 - 4x)$. |

SOLUTION

$$\lim_{x \to 3}(x^2 - 4x) = \lim_{x \to 3} x^2 - \lim_{x \to 3} 4x \qquad \textit{Property 4}$$

$$= \left(\lim_{x \to 3} x\right) \cdot \left(\lim_{x \to 3} x\right) - 4\lim_{x \to 3} x \qquad \textit{Properties 5 and 6}$$

$$= 3 \cdot 3 - 4 \cdot 3 = -3 \qquad \textit{Property 2}$$

With a little practice, you will soon be able to omit the steps in the dashed boxes and simply write

$$\lim_{x \to 3}(x^2 - 4x) = 3 \cdot 3 - 4 \cdot 3 = -3$$

MATCHED PROBLEM 5   Find $\lim_{x \to -2}(x^2 + 5x)$.

What happens if we try to evaluate a limit like the one in Example 5, but with x approaching an unspecified number, such as c? Proceeding as we did in Example 5, we have

$$\lim_{x \to c}(x^2 - 4x) = c \cdot c - 4 \cdot c = c^2 - 4c$$

If we let $f(x) = x^2 - 4x$, we have

$$\lim_{x \to c} f(x) = \lim_{x \to c}(x^2 - 4x) = c^2 - 4c = f(c)$$

That is, this limit can be evaluated simply by evaluating the function f at c. It would certainly simplify the process of evaluating limits if we could identify the functions for which

$$\lim_{x \to c} f(x) = f(c) \qquad (1)$$

since we could use this fact to evaluate the limit. It turns out that there are many functions that satisfy equation (1). We postpone a detailed discussion of these functions until the next section. For now, we note that if

$$f(x) = a_n x^n + a_{n-1}x^{n-1} + \cdots + a_0$$

is a polynomial function, then, by the properties in Theorem 1,

$$\lim_{x \to c} f(x) = \lim_{x \to c}(a_n x^n + a_{n-1}x^{n-1} + \cdots + a_0)$$
$$= a_n c^n + a_{n-1}c^{n-1} + \cdots + a_0 = f(c)$$

and if

$$r(x) = \frac{n(x)}{d(x)}$$

is a rational function, where $n(x)$ and $d(x)$ are polynomials with $d(c) \neq 0$, then, by property 7 and the fact that polynomials $n(x)$ and $d(x)$ satisfy equation (1),

$$\lim_{x \to c} r(x) = \lim_{x \to c}\frac{n(x)}{d(x)} = \frac{\lim_{x \to c} n(x)}{\lim_{x \to c} d(x)} = \frac{n(c)}{d(c)} = r(c)$$

These results are summarized in Theorem 3.

THEOREM 3 LIMITS OF POLYNOMIAL AND RATIONAL FUNCTIONS

1. $\lim_{x \to c} f(x) = f(c)$ f any polynomial function

2. $\lim_{x \to c} r(x) = r(c)$ r any rational function with a nonzero denominator at $x = c$

EXAMPLE 6   **Evaluating Limits** Find each limit:

(A) $\lim_{x \to 2}(x^3 - 5x - 1)$ (B) $\lim_{x \to -1}\sqrt{2x^2 + 3}$ (C) $\lim_{x \to 4}\frac{2x}{3x + 1}$

SOLUTION   (A) $\lim_{x \to 2}(x^3 - 5x - 1) = 2^3 - 5 \cdot 2 - 1 = -3$ *Theorem 3*

(B) $\lim_{x \to -1}\sqrt{2x^2 + 3} = \sqrt{\lim_{x \to -1}(2x^2 + 3)}$ *Property 8*

$\qquad\qquad\qquad = \sqrt{2(-1)^2 + 3}$ *Theorem 3*

$\qquad\qquad\qquad = \sqrt{5}$

(C) $\lim_{x \to 4}\frac{2x}{3x + 1} = \frac{2 \cdot 4}{3 \cdot 4 + 1}$ *Theorem 3*

$\qquad\qquad\quad = \frac{8}{13}$

MATCHED PROBLEM 6   Find each limit:

(A) $\lim_{x \to -1}(x^4 - 2x + 3)$ (B) $\lim_{x \to 2}\sqrt{3x^2 - 6}$ (C) $\lim_{x \to -2}\frac{x^2}{x^2 + 1}$

EXAMPLE 7   **Evaluating Limits**   Let

$$f(x) = \begin{cases} x^2 + 1 & \text{if } x < 2 \\ x - 1 & \text{if } x > 2 \end{cases}$$

Find

(A) $\lim\limits_{x \to 2^-} f(x)$ (B) $\lim\limits_{x \to 2^+} f(x)$ (C) $\lim\limits_{x \to 2} f(x)$ (D) $f(2)$

SOLUTION   (A) $\lim\limits_{x \to 2^-} f(x) = \lim\limits_{x \to 2^-} (x^2 + 1)$ If $x < 2$, $f(x) = x^2 + 1$.

$$= 2^2 + 1 = 5$$

(B) $\lim\limits_{x \to 2^+} f(x) = \lim\limits_{x \to 2^+} (x - 1)$ If $x > 2$, $f(x) = x - 1$.

$$= 2 - 1 = 1$$

(C) Since the one-sided limits are not equal, $\lim\limits_{x \to 2} f(x)$ does not exist.

(D) Because the definition of f does not assign a value to f for $x = 2$, only for $x < 2$ and $x > 2$, $f(2)$ does not exist.   ■

MATCHED PROBLEM 7   Let

$$f(x) = \begin{cases} 2x + 3 & \text{if } x < 5 \\ -x + 12 & \text{if } x > 5 \end{cases}$$

Find each limit:

(A) $\lim\limits_{x \to 5^-} f(x)$ (B) $\lim\limits_{x \to 5^+} f(x)$ (C) $\lim\limits_{x \to 5} f(x)$ (D) $f(5)$

It is important to note that there are restrictions on some of the limit properties. In particular, if

$$\lim\limits_{x \to c} f(x) = 0 \quad \text{and} \quad \lim\limits_{x \to c} g(x) = 0, \quad \text{then finding } \lim\limits_{x \to c} \frac{f(x)}{g(x)}$$

may present some difficulties, since limit property 7 (the limit of a quotient) does not apply when $\lim\limits_{x \to c} g(x) = 0$. The next example illustrates some techniques that can be useful in this situation.

EXAMPLE 8   **Evaluating Limits**   Find each limit:

(A) $\lim\limits_{x \to 2} \dfrac{x^2 - 4}{x - 2}$

(B) $\lim\limits_{x \to -1} \dfrac{x|x + 1|}{x + 1}$

SOLUTION   (A) Algebraic simplification is often useful when the numerator and denominator are both approaching 0.

$$\lim\limits_{x \to 2} \frac{x^2 - 4}{x - 2} = \lim\limits_{x \to 2} \frac{(x - 2)(x + 2)}{x - 2} = \lim\limits_{x \to 2} (x + 2) = 4$$

(B) One-sided limits are helpful for limits involving the absolute value function.

$$\lim\limits_{x \to -1^+} \frac{x|x + 1|}{x + 1} = \lim\limits_{x \to -1^+} (x) = -1 \quad \text{If } x > -1, \text{ then } \frac{|x + 1|}{x + 1} = 1.$$

$$\lim\limits_{x \to -1^-} \frac{x|x + 1|}{x + 1} = \lim\limits_{x \to -1^-} (-x) = 1 \quad \text{If } x < -1, \text{ then } \frac{|x + 1|}{x + 1} = -1.$$

Since the limit from the left and the limit from the right are not the same, we conclude that

$$\lim_{x \to -1} \frac{x|x + 1|}{x + 1} \quad \text{does not exist}$$

MATCHED PROBLEM 8   Find each limit:

(A) $\lim_{x \to -3} \dfrac{x^2 + 4x + 3}{x + 3}$ (B) $\lim_{x \to 4} \dfrac{x^2 - 16}{|x - 4|}$

INSIGHT

In the solution to part A of Example 8, we used the following algebraic identity:

$$\frac{x^2 - 4}{x - 2} = \frac{(x - 2)(x + 2)}{x - 2} = x + 2 \qquad x \neq 2$$

The restriction $x \neq 2$ is necessary here because the first two expressions are not defined at $x = 2$. Why didn't we include this restriction in the solution in part A? When x approaches 2 in a limit problem, it is assumed that x is close, but not equal, to 2. It is important that you understand that both of the following statements are valid:

$$\lim_{x \to 2} \frac{x^2 - 4}{x - 2} = \lim_{x \to 2}(x + 2) \qquad \frac{x^2 - 4}{x - 2} = x + 2, \quad x \neq 2$$

Limits like those in Example 8 occur so frequently in calculus that they are given a special name.

DEFINITION   **Indeterminate Form**

If $\lim\limits_{x \to c} f(x) = 0$ and $\lim\limits_{x \to c} g(x) = 0$, then $\lim\limits_{x \to c} \dfrac{f(x)}{g(x)}$ is said to be **indeterminate,** or, more specifically, a **0/0 indeterminate form.**

The term *indeterminate* is used because the limit of an indeterminate form may or may not exist (see parts A and B of Example 8).

 CAUTION The expression 0/0 does not represent a real number and should never be used as the value of a limit. If a limit is a 0/0 indeterminate form, further investigation is always required to determine whether the limit exists and to find its value if it does exist.

If the denominator of a quotient approaches 0 and the numerator approaches a nonzero number, the limit of the quotient is not an indeterminate form. In fact, a limit of this form never exists, as Theorem 4 states.

THEOREM 4   **LIMIT OF A QUOTIENT**

If $\lim\limits_{x \to c} f(x) = L, L \neq 0$, and $\lim\limits_{x \to c} g(x) = 0$,

then

$$\lim_{x \to c} \frac{f(x)}{g(x)} \quad \text{does not exist.}$$

Explore & Discuss **2**   Use algebraic and/or graphical techniques to analyze each of the following indeterminate forms:

(A) $\lim\limits_{x \to 1} \dfrac{x - 1}{x^2 - 1}$ *DNE*   (B) $\lim\limits_{x \to 2} \dfrac{(x - 1)(x - 1)}{(x^2 - 1)(x + 1)}$ (C) $\lim\limits_{x \to 1} \dfrac{x^2(x - 1)(x + 1)}{(x - 1)(x - 1)} = \dfrac{2}{0} \cdot DNE$

$(x + 1)(x - 1)$ $\dfrac{x^2 - 2x - 2}{x^2 - 1}$

■ Limits of Difference Quotients

Let the function f be defined in an open interval containing the number a. One of the most important limits in calculus is the limit of the **difference quotient:**

$$\lim_{h \to 0} \frac{f(a + h) - f(a)}{h} \qquad (3)$$

If

$$\lim_{h \to 0} [f(a + h) - f(a)] = 0$$

as it often does, then limit (3) is an indeterminate form. The examples that follow illustrate some useful techniques for evaluating limits of difference quotients.

EXAMPLE 9   **Limit of a Difference Quotient** Find the following limit for $f(x) = 4x - 5$:

$$\lim_{h \to 0} \frac{f(3 + h) - f(3)}{h}$$

SOLUTION

$$\begin{aligned}
\lim_{h \to 0} \frac{f(3 + h) - f(3)}{h} &= \lim_{h \to 0} \frac{[4(\mathbf{3 + h}) - 5] - [4(\mathbf{3}) - 5]}{h} \\
&= \lim_{h \to 0} \frac{12 + 4h - 5 - 12 + 5}{h} \\
&= \lim_{h \to 0} \frac{4h}{h} = \lim_{h \to 0} 4 = 4
\end{aligned}$$

Since this is a 0/0 indeterminate form and property 7 in Theorem 2 does not apply, we proceed with algebraic simplification. ■

MATCHED PROBLEM 9   Find the following limit for $f(x) = 7 - 2x$: $\lim\limits_{h \to 0} \dfrac{f(4 + h) - f(4)}{h}$.

Explore & Discuss **3**   The following is an incorrect solution to Example 9, with the invalid statements indicated by $\neq$:

$$\begin{aligned}
\lim_{h \to 0} \frac{f(3 + h) - f(3)}{h} &\neq \lim_{h \to 0} \frac{4(3 + h) - 5 - 4(3) - 5}{h} \\
&= \lim_{h \to 0} \frac{-10 + 4h}{h} \\
&\neq \lim_{h \to 0} \frac{-10 + 4}{1} = -6
\end{aligned}$$

Explain why each $\neq$ is used.

EXAMPLE 10   **Limit of a Difference Quotient** Find the following limit for $f(x) = |x + 5|$:

$$\lim_{h \to 0} \frac{f(-5 + h) - f(-5)}{h}$$

SOLUTION

$$\lim_{h \to 0} \frac{f(-5 + h) - f(-5)}{h} = \lim_{h \to 0} \frac{|(-5 + h) + 5| - |-5 + 5|}{h}$$

Since this is a 0/0 indeterminate form and property 7 in Theorem 2 does not apply, we proceed with algebraic simplification.

$$= \lim_{h \to 0} \frac{|h|}{h} \quad \text{does not exist}$$

MATCHED PROBLEM 10   Find the following limit for $f(x) = |x - 1|$: $\displaystyle\lim_{h \to 0} \frac{f(1 + h) - f(1)}{h}$.

EXAMPLE 11   **Limit of a Difference Quotient**   Find the following limit for $f(x) = \sqrt{x}$:

$$\lim_{h \to 0} \frac{f(2 + h) - f(2)}{h}$$

SOLUTION

$$\lim_{h \to 0} \frac{f(2 + h) - f(2)}{h} = \lim_{h \to 0} \frac{\sqrt{2 + h} - \sqrt{2}}{h}$$

This is a 0/0 indeterminate form, so property 7 in Theorem 2 does not apply. Rationalizing the numerator will be of help.

$$= \lim_{h \to 0} \frac{\sqrt{2 + h} - \sqrt{2}}{h} \cdot \frac{\sqrt{2 + h} + \sqrt{2}}{\sqrt{2 + h} + \sqrt{2}}$$

$(A - B)(A + B) = A^2 - B^2$

$$= \lim_{h \to 0} \frac{2 + h - 2}{h(\sqrt{2 + h} + \sqrt{2})}$$

$$= \lim_{h \to 0} \frac{1}{\sqrt{2 + h} + \sqrt{2}}$$

$$= \frac{1}{\sqrt{2} + \sqrt{2}} = \frac{1}{2\sqrt{2}}$$

MATCHED PROBLEM 11   Find the following limit for $f(x) = \sqrt{x}$: $\displaystyle\lim_{h \to 0} \frac{f(3 + h) - f(3)}{h}$.

Answers to Matched Problem

1.

x	-2	-1	0	1	2	3	4
$h(x)$	1.0	1.5	2.0	2.5	3.0	3.5	4.0

2. $\displaystyle\lim_{x \to 1} f(x) = 2$

3.

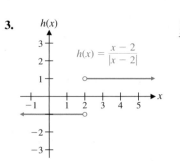

$\displaystyle\lim_{x \to 2} \frac{x - 2}{|x - 2|}$ does not exist

4. (A) $\displaystyle\lim_{x \to 0^-} f(x) = 0$ (B) $\displaystyle\lim_{x \to 1^-} f(x) = 1$

$\displaystyle\lim_{x \to 0^+} f(x) = 0$ $\displaystyle\lim_{x \to 1^+} f(x) = 2$

$\displaystyle\lim_{x \to 0} f(x) = 0$ $\displaystyle\lim_{x \to 1} f(x)$ does not exist

$f(0) = 0$ $f(1)$ not defined

(C) $\lim_{x \to 3^-} f(x) = 3$

$\lim_{x \to 3^+} f(x) = 3$

$\lim_{x \to 3} f(x) = 3$

$f(3)$ not defined

5. -6

6. (A) 6

(B) $\sqrt{6}$

(C) $\frac{4}{5}$

7. (A) 13

(B) 7

(C) Does not exist

(D) Not defined

8. (A) -2

(B) Does not exist

9. -2 **10.** Does not exist   **11.** $1/(2\sqrt{3})$

Exercise 3-1

A   In Problems 1–4, use the graph of the function f shown to estimate the indicated limits and function values.

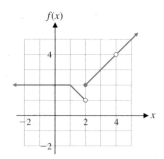

f(x)

1. (A) $\lim_{x \to 0^-} f(x)$ (B) $\lim_{x \to 0^+} f(x)$

(C) $\lim_{x \to 0} f(x)$ (D) $f(0)$

2. (A) $\lim_{x \to 1^-} f(x)$ (B) $\lim_{x \to 1^+} f(x)$

(C) $\lim_{x \to 1} f(x)$ (D) $f(1)$

3. (A) $\lim_{x \to 2^-} f(x)$ (B) $\lim_{x \to 2^+} f(x)$

(C) $\lim_{x \to 2} f(x)$ (D) $f(2)$

(E) Is it possible to redefine $f(2)$ so that $\lim_{x \to 2} f(x) = f(2)$? Explain.

4. (A) $\lim_{x \to 4^-} f(x)$ (B) $\lim_{x \to 4^+} f(x)$

(C) $\lim_{x \to 4} f(x)$ (D) $f(4)$

(E) Is it possible to define $f(4)$ so that $\lim_{x \to 4} f(x) = f(4)$? Explain.

In Problems 5–8, use the graph of the function g shown to estimate the indicated limits and function values.

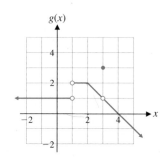

g(x)

5. (A) $\lim_{x \to 1^-} g(x)$ (B) $\lim_{x \to 1^+} g(x)$

(C) $\lim_{x \to 1} g(x)$ (D) $g(1)$

(E) Is it possible to define $g(1)$ so that $\lim_{x \to 1} g(x) = g(1)$? Explain.

6. (A) $\lim_{x \to 2^-} g(x)$ (B) $\lim_{x \to 2^+} g(x)$

(C) $\lim_{x \to 2} g(x)$ (D) $g(2)$

7. (A) $\lim_{x \to 3^-} g(x)$ (B) $\lim_{x \to 3^+} g(x)$

(C) $\lim_{x \to 3} g(x)$ (D) $g(3)$

(E) Is it possible to redefine $g(3)$ so that $\lim_{x \to 3} g(x) = g(3)$? Explain.

8. (A) $\lim_{x \to 4^-} g(x)$ (B) $\lim_{x \to 4^+} g(x)$

(C) $\lim_{x \to 4} g(x)$ (D) $g(4)$

In Problems 9–12, use the graph of the function f shown to estimate the indicated limits and function values.

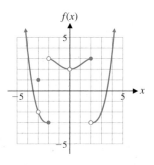

f(x)

9. (A) $\lim_{x \to -3^+} f(x)$ (B) $\lim_{x \to -3^-} f(x)$

(C) $\lim_{x \to -3} f(x)$ (D) $f(-3)$

(E) Is it possible to redefine $f(-3)$ so that $\lim_{x \to -3} f(x) = f(-3)$? Explain.

10. (A) $\lim_{x \to -2^+} f(x)$ (B) $\lim_{x \to -2^-} f(x)$

(C) $\lim_{x \to -2} f(x)$ (D) $f(-2)$

(E) Is it possible to define $f(-2)$ so that $\lim_{x \to -2} f(x) = f(-2)$? Explain.

11. (A) $\lim_{x \to 0^+} f(x)$ (B) $\lim_{x \to 0^-} f(x)$

    (C) $\lim_{x \to 0} f(x)$ (D) $f(0)$

    (E) Is it possible to redefine $f(0)$ so that $\lim_{x \to 0} f(x) = f(0)$? Explain.

12. (A) $\lim_{x \to 2^+} f(x)$ (B) $\lim_{x \to 2^-} f(x)$

    (C) $\lim_{x \to 2} f(x)$ (D) $f(2)$

    (E) Is it possible to redefine $f(2)$ so that $\lim_{x \to 2} f(x) = f(2)$? Explain.

In Problems 13–22, find each limit if it exists.

13. $\lim_{x \to 3} 4x$ 14. $\lim_{x \to -2} 3x$

15. $\lim_{x \to -4} (x + 5)$ 16. $\lim_{x \to 5}(x - 3)$

17. $\lim_{x \to 2} x(x - 4)$ 18. $\lim_{x \to -1} x(x + 3)$

19. $\lim_{x \to -3} \dfrac{x}{x + 5}$ 20. $\lim_{x \to 4} \dfrac{x - 2}{x}$

21. $\lim_{x \to 1} \sqrt{5x + 4}$ 22. $\lim_{x \to 0} \sqrt{16 - 7x}$

Given that $\lim_{x \to 1} f(x) = -5$ and $\lim_{x \to 1} g(x) = 4$, find the indicated limits in Problems 23–34.

23. $\lim_{x \to 1}(-3)f(x)$ 24. $\lim_{x \to 1} 2g(x)$

25. $\lim_{x \to 1}[2f(x) + g(x)]$ 26. $\lim_{x \to 1}[g(x) - 3f(x)]$

27. $\lim_{x \to 1} \dfrac{2 - f(x)}{x + g(x)}$ 28. $\lim_{x \to 1} \dfrac{3 - f(x)}{1 - 4g(x)}$

29. $\lim_{x \to 1} f(x)[2 - g(x)]$ 30. $\lim_{x \to 1}[f(x) - 7x]g(x)$

31. $\lim_{x \to 1} \sqrt{g(x) - f(x)}$ 32. $\lim_{x \to 1} \sqrt[3]{2x + 2f(x)}$

33. $\lim_{x \to 1}[f(x) + 1]^2$ 34. $\lim_{x \to 1}[2 - g(x)]^3$

In Problems 35–38, sketch a possible graph of a function that satisfies the given conditions.

35. $f(0) = 1$; $\lim_{x \to 0^-} f(x) = 3$; $\lim_{x \to 0^+} f(x) = 1$

36. $f(1) = -2$; $\lim_{x \to 1^-} f(x) = 2$; $\lim_{x \to 1^+} f(x) = -2$

37. $f(-2) = 2$; $\lim_{x \to -2^-} f(x) = 1$; $\lim_{x \to -2^+} f(x) = 1$

38. $f(0) = -1$; $\lim_{x \to 0^-} f(x) = 2$; $\lim_{x \to 0^+} f(x) = 2$

B *In Problems 39–54, find each indicated quantity if it exists.*

39. Let $f(x) = \begin{cases} 1 - x^2 & \text{if } x \le 0 \\ 1 + x^2 & \text{if } x > 0 \end{cases}$. Find

    (A) $\lim_{x \to 0^+} f(x)$ (B) $\lim_{x \to 0^-} f(x)$

    (C) $\lim_{x \to 0} f(x)$ (D) $f(0)$

40. Let $f(x) = \begin{cases} 2 + x & \text{if } x \le 0 \\ 2 - x & \text{if } x > 0 \end{cases}$. Find

    (A) $\lim_{x \to 0^+} f(x)$ (B) $\lim_{x \to 0^-} f(x)$

    (C) $\lim_{x \to 0} f(x)$ (D) $f(0)$

41. Let $f(x) = \begin{cases} x^2 & \text{if } x < 1 \\ 2x & \text{if } x > 1 \end{cases}$. Find

    (A) $\lim_{x \to 1^+} f(x)$ (B) $\lim_{x \to 1^-} f(x)$

    (C) $\lim_{x \to 1} f(x)$ (D) $f(1)$

42. Let $f(x) = \begin{cases} x + 3 & \text{if } x < -2 \\ \sqrt{x + 2} & \text{if } x > -2 \end{cases}$. Find

    (A) $\lim_{x \to -2^+} f(x)$ (B) $\lim_{x \to -2^-} f(x)$

    (C) $\lim_{x \to -2} f(x)$ (D) $f(-2)$

43. Let $f(x) = \begin{cases} \dfrac{x^2 - 9}{x + 3} & \text{if } x < 0 \\ \dfrac{x^2 - 9}{x - 3} & \text{if } x > 0 \end{cases}$. Find

    (A) $\lim_{x \to -3} f(x)$ (B) $\lim_{x \to 0} f(x)$

    (C) $\lim_{x \to 3} f(x)$

44. Let $f(x) = \begin{cases} \dfrac{x}{x + 3} & \text{if } x < 0 \\ \dfrac{x}{x - 3} & \text{if } x > 0 \end{cases}$. Find

    (A) $\lim_{x \to -3} f(x)$ (B) $\lim_{x \to 0} f(x)$

    (C) $\lim_{x \to 3} f(x)$

45. Let $f(x) = \dfrac{|x - 1|}{x - 1}$. Find

    (A) $\lim_{x \to 1^+} f(x)$ (B) $\lim_{x \to 1^-} f(x)$

    (C) $\lim_{x \to 1} f(x)$ (D) $f(1)$

46. Let $f(x) = \dfrac{x - 3}{|x - 3|}$. Find

    (A) $\lim_{x \to 3^+} f(x)$ (B) $\lim_{x \to 3^-} f(x)$

    (C) $\lim_{x \to 3} f(x)$ (D) $f(3)$

47. Let $f(x) = \dfrac{x - 2}{x^2 - 2x}$. Find

    (A) $\lim_{x \to 0} f(x)$ (B) $\lim_{x \to 2} f(x)$

    (C) $\lim_{x \to 4} f(x)$

48. Let $f(x) = \dfrac{x + 3}{x^2 + 3x}$. Find

    (A) $\lim_{x \to -3} f(x)$ (B) $\lim_{x \to 0} f(x)$

    (C) $\lim_{x \to 3} f(x)$

49. Let $f(x) = \dfrac{x^2 - x - 6}{x + 2}$. Find

    (A) $\lim_{x \to -2} f(x)$ (B) $\lim_{x \to 0} f(x)$

    (C) $\lim_{x \to 3} f(x)$

50. Let $f(x) = \dfrac{x^2 + x - 6}{x + 3}$. Find

    (A) $\lim_{x \to -3} f(x)$ (B) $\lim_{x \to 0} f(x)$

    (C) $\lim_{x \to 2} f(x)$

51. Let $f(x) = \dfrac{(x+2)^2}{x^2-4}$. Find

  (A) $\lim\limits_{x \to -2} f(x)$ (B) $\lim\limits_{x \to 0} f(x)$

  (C) $\lim\limits_{x \to 2} f(x)$

52. Let $f(x) = \dfrac{x^2-1}{(x+1)^2}$. Find

  (A) $\lim\limits_{x \to -1} f(x)$ (B) $\lim\limits_{x \to 0} f(x)$

  (C) $\lim\limits_{x \to 1} f(x)$

53. Let $f(x) = \dfrac{2x^2-3x-2}{x^2+x-6}$. Find

  (A) $\lim\limits_{x \to 2} f(x)$ (B) $\lim\limits_{x \to 0} f(x)$

  (C) $\lim\limits_{x \to 1} f(x)$

54. Let $f(x) = \dfrac{3x^2+2x-1}{x^2+3x+2}$. Find

  (A) $\lim\limits_{x \to -3} f(x)$ (B) $\lim\limits_{x \to -1} f(x)$

  (C) $\lim\limits_{x \to 2} f(x)$

Compute the following limit for each function in Problems 55–64:

$$\lim_{h \to 0} \frac{f(2+h) - f(2)}{h}$$

55. $f(x) = 3x + 1$ **56.** $f(x) = 5x - 1$

57. $f(x) = x^2 + 1$ **58.** $f(x) = x^2 - 2$

59. $f(x) = \sqrt{x} - 2$ **60.** $f(x) = 1 + \sqrt{x}$

61. $f(x) = |x - 2| - 3$ **62.** $f(x) = 2 + |x - 2|$

63. $f(x) = \dfrac{2}{x-1}$ **64.** $f(x) = \dfrac{1}{x+2}$

65. Let f be defined by

$$f(x) = \begin{cases} 1 + mx & \text{if } x \le 1 \\ 4 - mx & \text{if } x > 1 \end{cases}$$

where m is a constant.

  (A) Graph f for $m = 1$, and find

$$\lim_{x \to 1^-} f(x) \quad \text{and} \quad \lim_{x \to 1^+} f(x)$$

  (B) Graph f for $m = 2$, and find

$$\lim_{x \to 1^-} f(x) \quad \text{and} \quad \lim_{x \to 1^+} f(x)$$

  (C) Find m so that

$$\lim_{x \to 1^-} f(x) = \lim_{x \to 1^+} f(x)$$

and graph f for this value of m.

  (D) Write a brief verbal description of each graph. How does the graph in part (C) differ from the graphs in parts (A) and (B)?

66. Let f be defined by

$$f(x) = \begin{cases} -3m + 0.5x & \text{if } x \le 2 \\ 3m - x & \text{if } x > 2 \end{cases}$$

where m is a constant.

  (A) Graph f for $m = 0$, and find

$$\lim_{x \to 2^-} f(x) \quad \text{and} \quad \lim_{x \to 2^+} f(x)$$

  (B) Graph f for $m = 1$, and find

$$\lim_{x \to 2^-} f(x) \quad \text{and} \quad \lim_{x \to 2^+} f(x)$$

  (C) Find m so that

$$\lim_{x \to 2^-} f(x) = \lim_{x \to 2^+} f(x)$$

and graph f for this value of m.

  (D) Write a brief verbal description of each graph. How does the graph in part (C) differ from the graphs in parts (A) and (B)?

Find each limit in Problems 67–70, where a is a real constant.

67. $\lim\limits_{h \to 0} \dfrac{(a+h)^2 - a^2}{h}$

68. $\lim\limits_{h \to 0} \dfrac{[3(a+h) - 2] - (3a - 2)}{h}$

69. $\lim\limits_{h \to 0} \dfrac{\sqrt{a+h} - \sqrt{a}}{h}, \quad a > 0$

70. $\lim\limits_{h \to 0} \dfrac{\dfrac{1}{a+h} - \dfrac{1}{a}}{h}, \quad a \ne 0$

Applications

71. *Telephone rates.* A long-distance telephone service charges $0.99 for the first 20 minutes or less of a call and $0.07 per minute for each additional minute or fraction thereof.

  (A) Write a piecewise definition of the charge $F(x)$ for a long-distance call lasting x minutes.

  (B) Graph $F(x)$ for $0 < x \le 40$.

  (C) Find $\lim\limits_{x \to 20^-} F(x)$, $\lim\limits_{x \to 20^+} F(x)$, and $\lim\limits_{x \to 20} F(x)$, whichever exist.

72. *Telephone rates.* A second long-distance telephone service charges $0.09 per minute or fraction thereof for calls lasting 10 minutes or more and $0.18 per minute or fraction thereof for calls lasting less than 10 minutes.

  (A) Write a piecewise definition of the charge $G(x)$ for a long-distance call lasting x minutes.

  (B) Graph $G(x)$ for $0 < x \le 40$.

  (C) Find $\lim\limits_{x \to 10^-} G(x)$, $\lim\limits_{x \to 10^+} G(x)$, and $\lim\limits_{x \to 10} G(x)$, whichever exist.

73. *Telephone rates.* Refer to Problems 71 and 72. Write a brief verbal comparison of the two services described for calls lasting 20 minutes or less.

74. *Telephone rates.* Refer to Problems 71 and 72. Write a brief verbal comparison of the two services described for calls lasting more than 20 minutes.

A company sells custom embroidered apparel and promotional products. The table that follows shows the volume discounts offered by the company, where x is the volume of a purchase in dollars. Problems 75 and 76 deal with two different interpretations of this discount method.

Volume Discount (Excluding Tax)	
Volume (x)	**Discount Amount**
$300 \leq x < \$1,000$	3%
$\$1,000 \leq x < \$3,000$	5%
$\$3,000 \leq x < \$5,000$	7%
$\$5,000 \leq x$	10%

75. *Volume discount.* Assume that the volume discounts in the table apply to the entire purchase. That is, if the volume x satisfies $\$300 \leq x < \$1,000$, the entire purchase is discounted 3%. If the volume x satisfies $\$1,000 \leq x < \$3,000$, the entire purchase is discounted 5%, and so on.

(A) If x is the volume of a purchase before the discount is applied, write a piecewise definition for the discounted price $D(x)$ of this purchase.

(B) Use one-sided limits to investigate the limit of $D(x)$ as x approaches $\$1,000$; as x approaches $\$3,000$.

76. *Volume discount.* Assume that the volume discounts in the table apply only to that portion of the volume in each interval in the table. That is, the discounted price for a $4,000 purchase would be computed as follows:

$$300 + 0.97(700) + 0.95(2,000) + 0.93(1,000) = 3,809$$

(A) If x is the volume of a purchase before the discount is applied, write a piecewise definition for the discounted price $P(x)$ of this purchase.

(B) Use one-sided limits to investigate the limit of $P(x)$ as x approaches $\$1,000$; as x approaches $\$3,000$.

(C) Compare this discount method with the one in Problem 75. Does one always produce a lower price than the other? Discuss.

77. *Pollution.* A state charges polluters an annual fee of $20 per ton for each ton of pollutant emitted into the atmosphere, up to a maximum of 4,000 tons. No fees are charged for emissions beyond the 4,000-ton limit. Write a piecewise definition of the fees $F(x)$ charged for the emission of x tons of pollutant in a year. What is the limit of $F(x)$ as x approaches 4,000 tons? As x approaches 8,000 tons?

78. *Pollution.* Refer to Problem 77. The fee per ton of pollution is given by $A(x) = F(x)/x$. Write a piecewise definition of $A(x)$. What is the limit of $A(x)$ as x approaches 4,000 tons? As x approaches 8,000 tons?

79. *Voter turnout.* Statisticians often use piecewise-defined functions to predict outcomes of elections. For the following functions f and g, find the limit of each function as x approaches 5 and as x approaches 10:

$$f(x) = \begin{cases} 0 & \text{if } x \leq 5 \\ 0.8 - 0.08x & \text{if } 5 < x < 10 \\ 0 & \text{if } 10 \leq x \end{cases}$$

$$g(x) = \begin{cases} 0 & \text{if } x \leq 5 \\ 0.8x - 0.04x^2 - 3 & \text{if } 5 < x < 10 \\ 1 & \text{if } 10 \leq x \end{cases}$$

Section 3-2 CONTINUITY

- Continuity
- Continuity Properties
- Solving Inequalities by Using Continuity Properties

Theorem 3 in Section 3-1 states that if f is a polynomial function or a rational function with a nonzero denominator at $x = c$, then

$$\lim_{x \to c} f(x) = f(c) \tag{1}$$

Functions that satisfy equation (1) are said to be *continuous* at $x = c$. A firm understanding of continuous functions is essential for sketching and analyzing graphs. We will also see that continuity properties provide a simple and efficient method for solving inequalities—a tool that we will use extensively in later sections.

■ Continuity

Compare the graphs shown in Figure 1. Notice that two of the graphs are broken; that is, they cannot be drawn without lifting a pen off the paper. Informally, a function is *continuous over an interval* if its graph over the interval can be drawn without removing a pen from the paper. A function whose graph is broken (disconnected) at

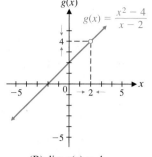

(A) $\lim_{x \to 2} f(x) = 4$
$f(2) = 4$

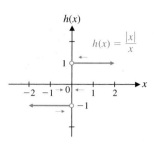

(B) $\lim_{x \to 2} g(x) = 4$
$g(2)$ is not defined

(C) $\lim_{x \to 0} h(x)$ does not exist
$h(0)$ is not defined

FIGURE 1

$x = c$ is said to be *discontinuous* at $x = c$. Function f (Fig. 1A) is continuous for all x. Function g (Fig. 1B) is discontinuous at $x = 2$, but is continuous over any interval that does not include 2. Function h (Fig. 1C) is discontinuous at $x = 0$, but is continuous over any interval that does not include 0.

Most graphs of natural phenomena are continuous, whereas many graphs in business and economics applications have discontinuities. Figure 2A illustrates temperature variation over a 24-hour period—a continuous phenomenon. Figure 2B illustrates warehouse inventory over a 1-week period—a discontinuous phenomenon.

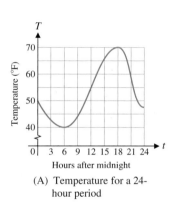

(A) Temperature for a 24-hour period

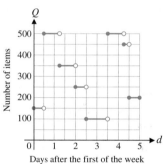

(B) Inventory in a warehouse during 1 week

FIGURE 2

Explore & Discuss **1**
(A) Write a brief verbal description of the temperature variation illustrated in Figure 2A, including estimates of the high and low temperatures during the period shown and the times at which they occurred.
(B) Write a brief verbal description of the changes in inventory illustrated in Figure 2B, including estimates of the changes in inventory and the times at which those changes occurred.

The preceding discussion leads to the following formal definition of continuity:

DEFINITION Continuity

A function f is **continuous at the point $x = c$ if**

1. $\lim_{x \to c} f(x)$ exists   **2.** $f(c)$ exists   **3.** $\lim_{x \to c} f(x) = f(c)$

A function is **continuous on the open interval* (a, b)** if it is continuous at each point on the interval.

---

* See Section 1-1 for a review of interval notation.

If one or more of the three conditions in the definition fails, the function is **discontinuous** at $x = c$.

Explore & Discuss **2**   Sketch a graph of a function that is discontinuous at a point because it fails to satisfy condition 1 in the definition of continuity. Repeat for conditions 2 and 3.

EXAMPLE 1   **Continuity of a Function Defined by a Graph**   Use the definition of continuity to discuss the continuity of the function whose graph is shown in Figure 3.

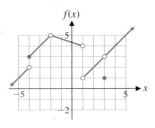

FIGURE 3

SOLUTION   We begin by identifying the points of discontinuity. Examining the graph, we see breaks or holes at $x = -4, -2, 1$, and 3. Now we must determine which conditions in the definition of continuity are not satisfied at each of these points. In each case, we find the value of the function and the limit of the function at the point in question.

Discontinuity at $x = -4$:

$$\lim_{x \to -4^-} f(x) = 2$$
$$\lim_{x \to -4^+} f(x) = 3$$
$$\lim_{x \to -4} f(x) \text{ does not exist}$$
$$f(-4) = 3$$

Since the one-sided limits are different, the limit does not exist (Section 3-1).

Thus, f is not continuous at $x = -4$ because condition 1 is not satisfied.

Discontinuity at $x = -2$:

$$\lim_{x \to -2^-} f(x) = 5$$
$$\lim_{x \to -2^+} f(x) = 5$$
$$\lim_{x \to -2} f(x) = 5$$
$$f(-2) \text{ does not exist}$$

The hole at $(-2, 5)$ indicates that 5 is not the value of f at -2. Since there is no solid dot elsewhere on the vertical line $x = -2$, $f(-2)$ is not defined.

Thus, f is not continuous at $x = -2$ because condition 2 is not satisfied.

Discontinuity at $x = 1$:

$$\lim_{x \to 1^-} f(x) = 4$$
$$\lim_{x \to 1^+} f(x) = 1$$
$$\lim_{x \to 1} f(x) \text{ does not exist}$$
$$f(1) \text{ does not exist}$$

This time, f is not continuous at $x = 1$ because neither of conditions 1 and 2 is satisfied.

Discontinuity at $x = 3$:

$$\lim_{x \to 3^-} f(x) = 3$$
$$\lim_{x \to 3^+} f(x) = 3$$
$$\lim_{x \to 3} f(x) = 3$$
$$f(3) = 1$$

The solid dot at $(3, 1)$ indicates that $f(3) = 1$.

Conditions 1 and 2 are satisfied, but f is not continuous at $x = 3$ because condition 3 is not satisfied.

Having identified and discussed all points of discontinuity, we can now conclude that f is continuous except at $x = -4, -2, 1$, and 3.

Rather than list the points where a function is discontinuous, sometimes it is useful to state the intervals on which the function is continuous. Using the set operation **union,** denoted by $\cup$, we can express the set of points where the function in Example 1 is continuous as follows:

$$(-\infty, -4) \cup (-4, -2) \cup (-2, 1) \cup (1, 3) \cup (3, \infty)$$

MATCHED PROBLEM 1 Use the definition of continuity to discuss the continuity of the function whose graph is shown in Figure 4.

$X = -3, -1, 2, 4$

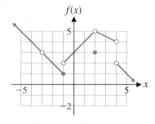

FIGURE 4

For functions defined by equations, it is important to be able to locate points of discontinuity by examining the equation.

EXAMPLE 2 **Continuity of Functions Defined by Equations** Using the definition of continuity, discuss the continuity of each function at the indicated point(s).

(A) $f(x) = x + 2$ at $x = 2$ 　　(B) $g(x) = \dfrac{x^2 - 4}{x - 2}$ at $x = 2$

(C) $h(x) = \dfrac{|x|}{x}$ at $x = 0$ and at $x = 1$

SOLUTION (A) f is continuous at $x = 2$, since

$$\lim_{x \to 2} f(x) = 4 = f(2) \quad \text{See Figure 1A.}$$

(B) g is not continuous at $x = 2$, since $g(2) = 0/0$ is not defined (see Fig. 1B).

(C) h is not continuous at $x = 0$, since $h(0) = |0|/0$ is not defined; also, $\lim_{x \to 0} h(x)$ does not exist.

h is continuous at $x = 1$, since

$$\lim_{x \to 1} \frac{|x|}{x} = 1 = h(1) \quad \text{See Figure 1C.}$$

MATCHED PROBLEM 2 Using the definition of continuity, discuss the continuity of each function at the indicated point(s).

(A) $f(x) = x + 1$ at $x = 1$ 　　(B) $g(x) = \dfrac{x^2 - 1}{x - 1}$ at $x = 1$

(C) $h(x) = \dfrac{x - 2}{|x - 2|}$ at $x = 2$ and at $x = 0$

We can also talk about one-sided continuity, just as we talked about one-sided limits. For example, a function is said to be **continuous on the right** at $x = c$ if $\lim_{x \to c^+} f(x) = f(c)$ and **continuous on the left** at $x = c$ if $\lim_{x \to c^-} f(x) = f(c)$. A function is **continuous on the closed interval [a, b]** if it is continuous on the open interval (a, b) and is continuous both on the right at a and on the left at b.

Figure 5A illustrates a function that is continuous on the closed interval $[-1, 1]$. Figure 5B illustrates a function that is continuous on the half-closed interval $[0, \infty)$.

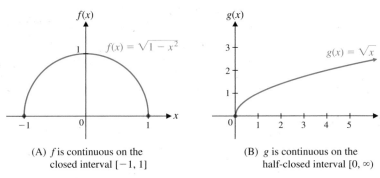

FIGURE 5 Continuity on closed and half-closed intervals

■ Continuity Properties

Functions have some useful **general continuity properties:**

> If two functions are continuous on the same interval, then their sum, difference, product, and quotient are continuous on the same interval, except for values of x that make a denominator 0.

These properties, along with Theorem 1 below, enable us to determine intervals of continuity for some important classes of functions without having to look at their graphs or use the three conditions in the definition.

THEOREM 1   **CONTINUITY PROPERTIES OF SOME SPECIFIC FUNCTIONS**

(A) A constant function $f(x) = k$, where k is a constant, is continuous for all x.

   $f(x) = 7$ is continuous for all x.

(B) For n a positive integer, $f(x) = x^n$ is continuous for all x.

   $f(x) = x^5$ is continuous for all x.

(C) A polynomial function is continuous for all x.

   $2x^3 - 3x^2 + x - 5$ is continuous for all x.

(D) A rational function is continuous for all x except those values that make a denominator 0.

   $\dfrac{x^2 + 1}{x - 1}$ is continuous for all x except x = 1, a value that makes the denominator 0.

(E) For n an odd positive integer greater than 1, $\sqrt[n]{f(x)}$ is continuous wherever $f(x)$ is continuous.

   $\sqrt[3]{x^2}$ is continuous for all x.

(F) For n an even positive integer, $\sqrt[n]{f(x)}$ is continuous wherever $f(x)$ is continuous and nonnegative.

   $\sqrt[4]{x}$ is continuous on the interval $[0, \infty)$.

Parts C and D of Theorem 1 are the same as Theorem 3 in Section 3-1. They are repeated here to emphasize their importance.

EXAMPLE 3   **Using Continuity Properties**   Using Theorem 1 and the general properties of continuity, determine where each function is continuous.

(A) $f(x) = x^2 - 2x + 1$

(B) $f(x) = \dfrac{x}{(x + 2)(x - 3)}$

(C) $f(x) = \sqrt[3]{x^2 - 4}$

(D) $f(x) = \sqrt{x - 2}$

SOLUTION

(A) Since f is a polynomial function, f is continuous for all x.

(B) Since f is a rational function, f is continuous for all x except -2 and 3 (values that make the denominator 0).

(C) The polynomial function $x^2 - 4$ is continuous for all x. Since $n = 3$ is odd, f is continuous for all x.

(D) The polynomial function $x - 2$ is continuous for all x and nonnegative for $x \geq 2$. Since $n = 2$ is even, f is continuous for $x \geq 2$, or on the interval $[2, \infty)$. ■

MATCHED PROBLEM 3

Using Theorem 1 and the general properties of continuity, determine where each function is continuous.

(A) $f(x) = x^4 + 2x^2 + 1$ (B) $f(x) = \dfrac{x^2}{(x + 1)(x - 4)}$

(C) $f(x) = \sqrt{x - 4}$ (D) $f(x) = \sqrt[3]{x^3 + 1}$

■ Solving Inequalities by Using Continuity Properties

One of the basic tools for analyzing graphs in calculus is a special line graph called a *sign chart*. We will make extensive use of this type of chart in later sections. In the discussion that follows, we use continuity properties to develop a simple and efficient procedure for constructing sign charts.

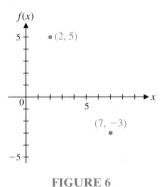

FIGURE 6

Suppose that a function f is continuous over the interval $(1, 8)$ and $f(x) \neq 0$ for any x in $(1, 8)$. Suppose also that $f(2) = 5$, a positive number. Is it possible for $f(x)$ to be negative for any x in the interval $(1, 8)$? The answer is "no." If $f(7)$ were -3, for example, as shown in Figure 6, how would it be possible to join the points $(2, 5)$ and $(7, -3)$ with the graph of a continuous function without crossing the x axis between 1 and 8 at least once? [Crossing the x axis would violate our assumption that $f(x) \neq 0$ for any x in $(1, 8)$.] Thus, we conclude that $f(x)$ must be positive for all x in $(1, 8)$. If $f(2)$ were negative, then, using the same type of reasoning, $f(x)$ would have to be negative over the entire interval $(1, 8)$.

In general, **if f is continuous and $f(x) \neq 0$ on the interval (a, b), then $f(x)$ cannot change sign on (a, b).** This is the essence of Theorem 2.

THEOREM 2   SIGN PROPERTIES ON AN INTERVAL (*a*, *b*)

If f is continuous on (a, b) and $f(x) \neq 0$ for all x in (a, b), then either $f(x) > 0$ for all x in (a, b) or $f(x) < 0$ for all x in (a, b).

Theorem 2 provides the basis for an effective method of solving many types of inequalities. Example 4 illustrates the process.

EXAMPLE 4   **Solving an Inequality**   Solve $\dfrac{x + 1}{x - 2} > 0$.

SOLUTION   We start by using the left side of the inequality to form the function f:

$$f(x) = \frac{x + 1}{x - 2}$$

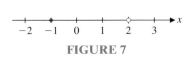

FIGURE 7

The rational function f is discontinuous at $x = 2$, and $f(x) = 0$ for $x = -1$ (a fraction is 0 when the numerator is 0 and the denominator is not 0). We plot $x = 2$ and $x = -1$, which we call *partition numbers,* on a real-number line (Fig. 7). (Note that the dot at 2 is open, because the function is not defined at $x = 2$.) The partition numbers 2 and -1 determine three open intervals: $(-\infty, -1)$, $(-1, 2)$, and $(2, \infty)$. The function f is continuous and nonzero on each of these intervals. From Theorem 2, we know that $f(x)$ does not change sign on any of these intervals. Thus, we can find

Test Numbers	
x	**f(x)**
−2	$\frac{1}{4}$ (+)
0	$-\frac{1}{2}$ (−)
3	4 (+)

the sign of $f(x)$ on each of the intervals by selecting a **test number** in each interval and evaluating $f(x)$ at that number. Since any number in each subinterval will do, we choose test numbers that are easy to evaluate: −2, 0, and 3. The table in the margin shows the results.

The sign of $f(x)$ at each test number is the same as the sign of $f(x)$ over the interval containing that test number. Using this information, we construct a **sign chart** for $f(x)$:

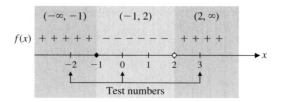

From the sign chart, we can easily write the solution of the given nonlinear inequality:

$$f(x) > 0 \quad \text{for} \quad \begin{array}{ll} x < -1 \quad \text{or} \quad x > 2 & \text{Inequality notation} \\ (-\infty, -1) \cup (2, \infty) & \text{Interval notation} \end{array}$$

Most of the inequalities we encounter will involve strict inequalities ($>$ or $<$). If it is necessary to solve inequalities of the form $\geq$ or $\leq$, we simply include the endpoint x of any interval if f is defined at x and $f(x)$ satisfies the given inequality. For example, from the sign chart in Example 4, the solution of the inequality

$$\frac{x + 1}{x - 2} \geq 0 \quad \text{is} \quad \begin{array}{ll} x \leq -1 \quad \text{or} \quad x > 2 & \text{Inequality notation} \\ (-\infty, -1] \cup (2, \infty) & \text{Interval notation} \end{array}$$

In general, given a function f, a **partition number** is a value of x such that f is discontinuous at x or $f(x) = 0$. **Partition numbers determine open intervals in which $f(x)$ does not change sign.** By using a test number from each interval, we can construct a sign chart for $f(x)$ on the real-number line. It is then an easy matter to determine where $f(x) < 0$ or $f(x) > 0$; that is, it is easy to solve the inequality $f(x) < 0$ or $f(x) > 0$.

We summarize the procedure for constructing sign charts in the following box:

PROCEDURE

Constructing Sign Charts

Given a function f,

Step 1. Find all partition numbers. That is,

(A) Find all numbers such that f is discontinuous. (Rational functions are discontinuous for values of x that make a denominator 0.)

(B) Find all numbers such that $f(x) = 0$. (For a rational function, this occurs where the numerator is 0 and the denominator is not 0.)

Step 2. Plot the numbers found in step 1 on a real-number line, dividing the number line into intervals.

Step 3. Select a test number in each open interval determined in step 2, and evaluate $f(x)$ at each test number to determine whether $f(x)$ is positive (+) or negative (−) in each interval.

Step 4. Construct a sign chart, using the real-number line in step 2. This will show the sign of $f(x)$ on each open interval.

Note: From the sign chart, it is easy to find the solution of the inequality $f(x) < 0$ or $f(x) > 0$.

MATCHED PROBLEM 4   Solve $\dfrac{x^2 - 1}{x - 3} < 0$.

Answers to Matched Problems

1. f is not continuous at $x = -3, -1, 2,$ and 4.

$x = -3$: $\lim\limits_{x \to -3} f(x) = 3$, but $f(-3)$ does not exist

$x = -1$: $f(-1) = 1$, but $\lim\limits_{x \to -1} f(x)$ does not exist

$x = 2$: $\lim\limits_{x \to 2} f(x) = 5$, but $f(2) = 3$

$x = 4$: $\lim\limits_{x \to 4} f(x)$ does not exist, and $f(4)$ does not exist

2. (A) f is continuous at $x = 1$, since $\lim\limits_{x \to 1} f(x) = 2 = f(1)$.

(B) g is not continuous at $x = 1$, since $g(1)$ is not defined.

(C) h is not continuous at $x = 2$ for two reasons: $h(2)$ does not exist and $\lim\limits_{x \to 2} h(x)$ does not exist.

h is continuous at $x = 0$, since $\lim\limits_{x \to 0} h(x) = -1 = h(0)$.

3. (A) Since f is a polynomial function, f is continuous for all x.

(B) Since f is a rational function, f is continuous for all x except -1 and 4 (values that make the denominator 0).

(C) The polynomial function $x - 4$ is continuous for all x and nonnegative for $x \geq 4$. Since $n = 2$ is even, f is continuous for $x \geq 4$, or on the interval $[4, \infty)$.

(D) The polynomial function $x^3 + 1$ is continuous for all x. Since $n = 3$ is odd, f is continuous for all x.

4. $-\infty < x < -1$ or $1 < x < 3$; $(-\infty, -1) \cup (1, 3)$

Exercise 3-2

A *In Problems 1–6, sketch a possible graph of a function that satisfies the given conditions at $x = 1$, and discuss the continuity of f at $x = 1$.*

1. $f(1) = 2$ and $\lim\limits_{x \to 1} f(x) = 2$

2. $f(1) = -2$ and $\lim\limits_{x \to 1} f(x) = 2$

3. $f(1) = 2$ and $\lim\limits_{x \to 1} f(x) = -2$

4. $f(1) = -2$ and $\lim\limits_{x \to 1} f(x) = -2$

5. $f(1) = -2$, $\lim\limits_{x \to 1^-} f(x) = 2$, and $\lim\limits_{x \to 1^+} f(x) = -2$

6. $f(1) = 2$, $\lim\limits_{x \to 1^-} f(x) = 2$, and $\lim\limits_{x \to 1^+} f(x) = -2$

Problems 7–10 refer to the function f shown in the figure. Use the graph to estimate the indicated quantities to the nearest integer.

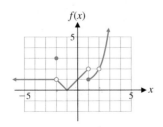

7. (A) $\lim\limits_{x \to 1^-} f(x)$ (B) $\lim\limits_{x \to 1^+} f(x)$

    (C) $\lim\limits_{x \to 1} f(x)$ (D) $f(1)$

    (E) Is f continuous at $x = 1$? Explain.

8. (A) $\lim\limits_{x \to 2^-} f(x)$ (B) $\lim\limits_{x \to 2^+} f(x)$

    (C) $\lim\limits_{x \to 2} f(x)$ (D) $f(2)$

    (E) Is f continuous at $x = 2$? Explain.

9. (A) $\lim\limits_{x \to -2^-} f(x)$ (B) $\lim\limits_{x \to -2^+} f(x)$

    (C) $\lim\limits_{x \to -2} f(x)$ (D) $f(-2)$

    (E) Is f continuous at $x = -2$? Explain.

10. (A) $\lim\limits_{x \to -1^-} f(x)$ (B) $\lim\limits_{x \to -1^+} f(x)$

    (C) $\lim\limits_{x \to -1} f(x)$ (D) $f(-1)$

    (E) Is f continuous at $x = -1$? Explain.

Problems 11–14 refer to the function g shown in the figure. Use the graph to estimate the indicated quantities to the nearest integer.

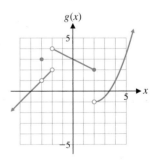

11. (A) $\lim\limits_{x \to -3^-} g(x)$ (B) $\lim\limits_{x \to -3^+} g(x)$

    (C) $\lim\limits_{x \to -3} g(x)$ (D) $g(-3)$

    (E) Is g continuous at $x = -3$? Explain.

12. (A) $\lim\limits_{x \to -2^-} g(x)$ (B) $\lim\limits_{x \to -2^+} g(x)$

    (C) $\lim\limits_{x \to -2} g(x)$ (D) $g(-2)$

    (E) Is g continuous at $x = -2$? Explain.

13. (A) $\lim\limits_{x \to 2^-} g(x)$ (B) $\lim\limits_{x \to 2^+} g(x)$

    (C) $\lim\limits_{x \to 2} g(x)$ (D) $g(2)$

    (E) Is g continuous at $x = 2$? Explain.

14. (A) $\lim\limits_{x \to 4^-} g(x)$ (B) $\lim\limits_{x \to 4^+} g(x)$

    (C) $\lim\limits_{x \to 4} g(x)$ (D) $g(4)$

    (E) Is g continuous at $x = 4$? Explain.

Use Theorem 1 to determine where each function in Problems 15–24 is continuous.

15. $f(x) = 3x - 4$ 16. $h(x) = 4 - 2x$

17. $g(x) = \dfrac{3x}{x + 2}$ 18. $k(x) = \dfrac{2x}{x - 4}$

19. $m(x) = \dfrac{x + 1}{(x - 1)(x + 4)}$

20. $n(x) = \dfrac{x - 2}{(x - 3)(x + 1)}$

21. $F(x) = \dfrac{2x}{x^2 + 9}$ 22. $G(x) = \dfrac{1 - x^2}{x^2 + 1}$

23. $M(x) = \dfrac{x - 1}{4x^2 - 9}$ 24. $N(x) = \dfrac{x^2 + 4}{4 - 25x^2}$

B

25. Given the function

$$f(x) = \begin{cases} 2 & \text{if } x \text{ is an integer} \\ 1 & \text{if } x \text{ is not an integer} \end{cases}$$

    (A) Graph f.

    (B) $\lim\limits_{x \to 2} f(x) = ?$

    (C) $f(2) = ?$

    (D) Is f continuous at $x = 2$?

    (E) Where is f discontinuous?

26. Given the function

$$g(x) = \begin{cases} -1 & \text{if } x \text{ is an even integer} \\ 1 & \text{if } x \text{ is not an even integer} \end{cases}$$

    (A) Graph g.

    (B) $\lim\limits_{x \to 1} g(x) = ?$

    (C) $g(1) = ?$

    (D) Is g continuous at $x = 1$?

    (E) Where is g discontinuous?

In Problems 27–34, use a sign chart to solve each inequality. Express answers in inequality and interval notation.

27. $x^2 - x - 12 < 0$ 28. $x^2 - 2x - 8 < 0$

29. $x^2 + 21 > 10x$ 30. $x^2 + 7x > -10$

31. $x^3 < 4x$ 32. $x^4 - 9x^2 > 0$

33. $\dfrac{x^2 + 5x}{x - 3} > 0$ 34. $\dfrac{x - 4}{x^2 + 2x} < 0$

35. Use the graph of f to determine where

    (A) $f(x) > 0$ (B) $f(x) < 0$

Express answers in interval notation.

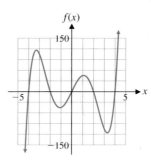

36. Use the graph of g to determine where

    (A) $g(x) > 0$ (B) $g(x) < 0$

Express answers in interval notation.

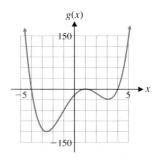

 In Problems 37–40, use a graphing calculator to approximate the partition numbers of each function $f(x)$ to four decimal places. Then solve the following inequalities:

    (A) $f(x) > 0$ (B) $f(x) < 0$

Express answers in interval notation.

37. $f(x) = x^4 - 6x^2 + 3x + 5$

38. $f(x) = x^4 - 4x^2 - 2x + 2$

39. $f(x) = \dfrac{3 + 6x - x^3}{x^2 - 1}$

40. $f(x) = \dfrac{x^3 - 5x + 1}{x^2 - 1}$

Use Theorem 1 to determine where each function in Problems 41–48 is continuous. Express the answer in interval notation.

41. $\sqrt{x - 6}$ 42. $\sqrt{7 - x}$

43. $\sqrt[3]{5 - x}$ 44. $\sqrt[3]{x - 8}$

45. $\sqrt{x^2 - 9}$ 46. $\sqrt{4 - x^2}$

47. $\sqrt{x^2 + 1}$ 48. $\sqrt[3]{x^2 + 2}$

In Problems 49–54, graph f, locate all points of discontinuity, and discuss the behavior of f at these points.

49. $f(x) = \begin{cases} 1 + x & \text{if } x < 1 \\ 5 - x & \text{if } x \ge 1 \end{cases}$

50. $f(x) = \begin{cases} x^2 & \text{if } x \le 1 \\ 2x & \text{if } x > 1 \end{cases}$

51. $f(x) = \begin{cases} 1 + x & \text{if } x \le 2 \\ 5 - x & \text{if } x > 2 \end{cases}$

52. $f(x) = \begin{cases} x^2 & \text{if } x \le 2 \\ 2x & \text{if } x > 2 \end{cases}$

53. $f(x) = \begin{cases} -x & \text{if } x < 0 \\ 1 & \text{if } x = 0 \\ x & \text{if } x > 0 \end{cases}$

54. $f(x) = \begin{cases} 1 & \text{if } x < 0 \\ 0 & \text{if } x = 0 \\ 1 + x & \text{if } x > 0 \end{cases}$

 In Problems 55–58, use a graphing calculator to locate all points of discontinuity of f, and discuss the behavior of f at these points. [Hint: Select Xmin and Xmax so that the suspected point of discontinuity is the midpoint of the graphing interval (Xmin, Xmax).]

55. $f(x) = x + \dfrac{|2x - 4|}{x - 2}$ **56.** $f(x) = x + \dfrac{|3x + 9|}{x + 3}$

57. $f(x) = \dfrac{x^2 - 1}{|x| - 1}$ **58.** $f(x) = \dfrac{x^3 - 8}{|x| - 2}$

C

59. Use the graph of the function g to answer the following questions:

(A) Is g continuous on the open interval $(-1, 2)$?

(B) Is g continuous from the right at $x = -1$? That is, does $\lim_{x \to -1^+} g(x) = g(-1)$?

(C) Is g continuous from the left at $x = 2$? That is, does $\lim_{x \to 2^-} g(x) = g(2)$?

(D) Is g continuous on the closed interval $[-1, 2]$?

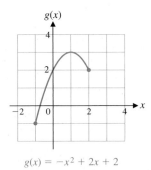

$g(x) = -x^2 + 2x + 2$

Figure for 59

60. Use the graph of the function f to answer the following questions:

(A) Is f continuous on the open interval $(0, 3)$?

(B) Is f continuous from the right at $x = 0$? That is, does $\lim_{x \to 0^+} f(x) = f(0)$?

(C) Is f continuous from the left at $x = 3$? That is, does $\lim_{x \to 3^-} f(x) = f(3)$?

(D) Is f continuous on the closed interval $[0, 3]$?

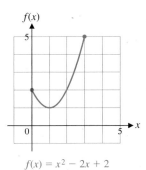

$f(x) = x^2 - 2x + 2$

Figure for 60

*Problems 61 and 62 refer to the **greatest integer function,** which is denoted by $[\![x]\!]$ and is defined as*

$$[\![x]\!] = \text{greatest integer} \le x$$

For example,

$$[\![-3.6]\!] = \text{greatest integer} \le -3.6 = -4$$
$$[\![2]\!] = \text{greatest integer} \le 2 = 2$$
$$[\![2.5]\!] = \text{greatest integer} \le 2.5 = 2$$

The graph of $f(x) = [\![x]\!]$ is shown. There, we can see that

$$\begin{aligned} [\![x]\!] &= -2 & for & \quad -2 \le x < -1 \\ [\![x]\!] &= -1 & for & \quad -1 \le x < 0 \\ [\![x]\!] &= 0 & for & \quad 0 \le x < 1 \\ [\![x]\!] &= 1 & for & \quad 1 \le x < 2 \\ [\![x]\!] &= 2 & for & \quad 2 \le x < 3 \end{aligned}$$

and so on.

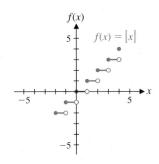

Figure for 61 and 62

61. (A) Is f continuous from the right at $x = 0$?

(B) Is f continuous from the left at $x = 0$?

(C) Is f continuous on the open interval $(0, 1)$?

(D) Is f continuous on the closed interval $[0, 1]$?

(E) Is f continuous on the half-closed interval $[0, 1)$?

62. (A) Is f continuous from the right at $x = 2$?

(B) Is f continuous from the left at $x = 2$?

(C) Is f continuous on the open interval $(1, 2)$?

(D) Is f continuous on the closed interval $[1, 2]$?

(E) Is f continuous on the half-closed interval $[1, 2)$?

In Problems 63–66, sketch a possible graph of a function f that is continuous for all real numbers and satisfies the given conditions. Find the x intercepts of f.

63. $f(x) < 0$ on $(-\infty, -5)$ and $(2, \infty)$; $f(x) > 0$ on $(-5, 2)$

64. $f(x) > 0$ on $(-\infty, -4)$ and $(3, \infty)$; $f(x) < 0$ on $(-4, 3)$

65. $f(x) < 0$ on $(-\infty, -6)$ and $(-1, 4)$; $f(x) > 0$ on $(-6, -1)$ and $(4, \infty)$

66. $f(x) > 0$ on $(-\infty, -3)$ and $(2, 7)$; $f(x) < 0$ on $(-3, 2)$ and $(7, \infty)$

67. The function $f(x) = 2/(1 - x)$ satisfies $f(0) = 2$ and $f(2) = -2$. Is f equal to 0 anywhere on the interval $(-1, 3)$? Does this contradict Theorem 2? Explain.

68. The function $f(x) = 6/(x - 4)$ satisfies $f(2) = -3$ and $f(7) = 2$. Is f equal to 0 anywhere on the interval $(0, 9)$? Does this contradict Theorem 2? Explain.

69. The function f is continuous and never 0 on the interval $(0, 4)$ and is continuous and never 0 on the interval $(4, 8)$. Also, $f(2) = 3$ and $f(6) = -3$. Discuss the validity of the following statement, and illustrate your conclusions with graphs: Either $f(4) = 0$ or f is discontinuous at $x = 4$.

70. The function f is continuous and never 0 on the interval $(-3, 1)$ and is continuous and never 0 on the interval $(1, 4)$. Also, $f(-2) = -3$ and $f(3) = 4$. Discuss the validity of the following statement and illustrate your conclusions with graphs: Either $f(1) = 0$ or f is discontinuous at $x = 1$.

Applications

71. *Postal rates.* First-class postage in 2006 was \$0.39 for the first ounce (or any fraction thereof) and \$0.24 for each additional ounce (or fraction thereof).

(A) Write a piecewise definition of the first-class postage $P(x)$ for a letter weighing x ounces.

(B) Graph $P(x)$ for $0 < x \le 5$.

(C) Is $P(x)$ continuous at $x = 4.5$? At $x = 4$? Explain.

72. *Telephone rates.* A long-distance telephone service charges \$0.07 for the first minute (or any fraction thereof) and \$0.05 for each additional minute (or fraction thereof).

(A) Write a piecewise definition of the charge $R(x)$ for a long-distance call lasting x minutes.

(B) Graph $R(x)$ for $0 < x \le 6$.

(C) Is $R(x)$ continuous at $x = 3.5$? At $x = 3$? Explain.

73. *Postal rates.* Discuss the differences between the function $Q(x) = 0.39 + 0.24[\![x]\!]$ and the function $P(x)$ defined in Problem 71.

74. *Telephone rates.* Discuss the differences between the function $S(x) = 0.07 + 0.05[\![x]\!]$ and the function $R(x)$ defined in Problem 72.

75. *Natural-gas rates.* Table 1 shows the rates for natural gas charged by the Middle Tennessee Natural Gas Utility District during the summer months. The customer charge is a fixed monthly charge, independent of the amount of gas used during the month.

(A) Write a piecewise definition of the monthly charge $S(x)$ for a customer who uses x therms* in a summer month.

(B) Graph $S(x)$.

(C) Is $S(x)$ continuous at $x = 50$? Explain.

TABLE 1 Summer (May–September)	
Base charge	\$5.00
First 50 therms	0.63 per therm
Over 50 therms	0.45 per therm

76. *Natural-gas rates.* Table 2 shows the rates for natural gas charged by the Middle Tennessee Natural Gas Utility District during the winter months. The customer charge is a fixed monthly charge, independent of the amount of gas used during the month.

(A) Write a piecewise definition of the monthly charge $S(x)$ for a customer who uses x therms in a winter month.

(B) Graph $S(x)$.

(C) Is $S(x)$ continuous at $x = 5$? At $x = 50$? Explain.

TABLE 2 Winter (October– April)	
Base charge	\$5.00
First 5 therms	0.69 per therm
Next 45 therms	0.65 per therm
Over 50 therms	0.63 per therm

77. *Income.* A personal-computer salesperson receives a base salary of \$1,000 per month and a commission of 5% of all sales over \$10,000 during the month. If the monthly sales are \$20,000 or more, the salesperson is given an additional \$500 bonus. Let $E(s)$ represent the person's earnings during the month as a function of the monthly sales s.

(A) Graph $E(s)$ for $0 \le s \le 30,000$.

(B) Find $\lim_{s \to 10,000} E(s)$ and $E(10,000)$.

(C) Find $\lim_{s \to 20,000} E(s)$ and $E(20,000)$.

(D) Is E continuous at $s = 10,000$? At $s = 20,000$?

78. *Equipment rental.* An office equipment rental and leasing company rents copiers for \$10 per day (and any

---

*A British thermal unit (Btu) is the amount of heat required to raise the temperature of 1 pound of water 1 degree Fahrenheit, and a therm is 100,000 Btu.

fraction thereof) or for $50 per 7-day week. Let $C(x)$ be the cost of renting a copier for x days.

(A) Graph $C(x)$ for $0 \le x \le 10$.

(B) Find $\lim_{x \to 4.5} C(x)$ and $C(4.5)$.

(C) Find $\lim_{x \to 8} C(x)$ and $C(8)$.

(D) Is C continuous at $x = 4.5$? At $x = 8$?

79. *Animal supply.* A medical laboratory raises its own rabbits. The number of rabbits $N(t)$ available at any time t depends on the number of births and deaths. When a birth or death occurs, the function N generally has a discontinuity, as shown in the figure.

80. *Learning.* The graph shown might represent the history of a particular person learning the material on limits and continuity in this book. At time t_2, the student's mind goes blank during a quiz. At time t_4, the instructor explains a concept particularly well, and suddenly a big jump in understanding takes place.

(A) Where is the function p discontinuous?

(B) $\lim_{t \to t_1} p(t) = ?$; $p(t_1) = ?$

(C) $\lim_{t \to t_2} p(t) = ?$; $p(t_2) = ?$

(D) $\lim_{t \to t_4} p(t) = ?$; $p(t_4) = ?$

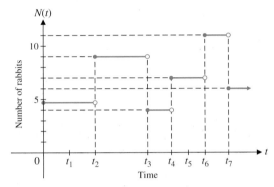

Figure for 79

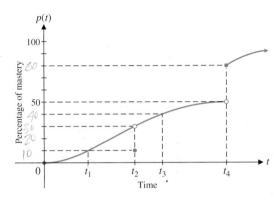

Figure for 80

(A) Where is the function N discontinuous?

(B) $\lim_{t \to t_5} N(t) = ?$; $N(t_5) = ?$

(C) $\lim_{t \to t_3} N(t) = ?$; $N(t_3) = ?$

Section 3-3   INFINITE LIMITS AND LIMITS AT INFINITY

- Infinite Limits
- Locating Vertical Asymptotes
- Limits at Infinity
- Finding Horizontal Asymptotes

In this section, we consider two new types of limits: infinite limits and limits at infinity. Infinite limits and vertical asymptotes are used to analyze the behavior of a function that is unbounded near $x = a$. Limits at infinity and horizontal asymptotes are used to describe the behavior of a function as x assumes arbitrarily large positive values or arbitrarily large negative values. Although we will include graphs to illustrate basic concepts, we postpone a discussion of graphing techniques until Chapter 5.

■ Infinite Limits

A function is discontinuous at any point a where $\lim_{x \to a} f(x)$ does not exist. For example, if the one-sided limits are different at $x = a$, then the limit does not exist and the function is discontinuous at $x = a$. Another situation in which a limit may fail to exist involves functions whose values become very large as x approaches a. The special symbol ∞ is often used to describe this type of behavior. To illustrate, consider the function $f(x) = 1/(x - 1)$, which is discontinuous only at $x = 1$. As x approaches 1 from the right, the values of $f(x)$ are positive and become larger and larger; that is, $f(x)$ increases without bound (Table 1). We express this behavior symbolically as

$$f(x) = \frac{1}{x - 1} \to \infty \quad \text{as} \quad x \to 1^+ \qquad (1)$$

Since ∞ is a not a real number, *the limit in (1) does not exist.* We are using the symbol ∞ to describe the manner in which the limit fails to exist, and we call this situation an **infinite limit.** If, now, x approaches 1 from the left, the values of $f(x)$ are negative and become larger and larger in absolute value; that is, $f(x)$ decreases through negative values without bound (see Table 2). We express this behavior symbolically as

$$f(x) = \frac{1}{x - 1} \to -\infty \quad \text{as} \quad x \to 1^- \qquad (2)$$

From (1) and (2), we have

$$\lim_{x \to 1^+} \frac{1}{x - 1} = \infty \quad \text{and} \quad \lim_{x \to 1^-} \frac{1}{x - 1} = -\infty$$

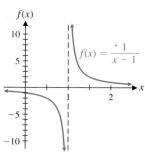

FIGURE 1

TABLE 1	
x	$f(x) = \dfrac{1}{x - 1}$
1.1	10
1.01	100
1.001	1,000
1.0001	10,000
1.00001	100,000
1.000001	1,000,000

TABLE 2	
x	$f(x) = \dfrac{1}{x - 1}$
0.9	−10
0.99	−100
0.999	−1,000
0.9999	−10,000
0.99999	−100,000
0.999999	−1,000,000

The one-sided limits in (1) and (2) also illustrate the behavior of the graph near $x \to 1$ (Fig. 1). Does the unrestricted limit of $f(x)$ as $x = 1$ exist? No, because neither of the one-sided limits exist. Also, there is no reasonable way to use the symbol ∞ to describe the behavior of $f(x)$ as $x \to 1$ on both sides of 1. Thus, we say simply that

$$\lim_{x \to 1} \frac{1}{x - 1} \text{ does not exist}$$

Explore & Discuss 1

Let $g(x) = \dfrac{1}{(x - 1)^2}$

Construct tables for $g(x)$ as $x \to 1^+$ and as $x \to 1^-$. Use these tables and infinite limits to discuss the behavior of $g(x)$ near $x = 1$.

We used the dashed vertical line $x = 1$ in Figure 1 to illustrate the infinite limits as x approaches 1 from the right and from the left. We call this line a *vertical asymptote*.

DEFINITION **Infinite Limits and Vertical Asymptotes**

The vertical line $x = a$ is a **vertical asymptote** for the graph of $y = f(x)$ if

$$f(x) \to \infty \quad \text{or} \quad f(x) \to -\infty \quad \text{as} \quad x \to a^+ \quad \text{or} \quad x \to a^-$$

[that is, if $f(x)$ either increases or decreases without bound as x approaches a from the right or from the left].

■ Locating Vertical Asymptotes

How do we locate vertical asymptotes? If a function f is continuous at $x = a$, then

$$\lim_{x \to a} f(x) = \lim_{x \to a^+} f(x) = \lim_{x \to a^-} f(x) = f(a) \tag{3}$$

Since all of the limits in (3) exist and are finite, f cannot have a vertical asymptote at $x = a$. In order for f to have a vertical asymptote at $x = a$, at least one of the limits in (3) must be an infinite limit and f must be discontinuous at $x = a$. We know that polynomial functions are continuous for all real numbers, so *a polynomial has no vertical asymptotes*. Since a rational function is discontinuous at the zeros of its denominator, *a vertical asymptote of a rational function can occur only at a zero of its denominator*. Theorem 1 provides a simple procedure for locating the vertical asymptotes of a rational function.

THEOREM 1 **LOCATING VERTICAL ASYMPTOTES OF RATIONAL FUNCTIONS**

If $f(x) = n(x)/d(x)$ is a rational function, $d(c) = 0$ and $n(c) \neq 0$, then the line $x = c$ is a vertical asymptote of the graph of f.

If $f(x) = n(x)/d(x)$ and both $n(c) = 0$ and $d(c) = 0$, then the limit of $f(x)$ as x approaches c involves an indeterminate form and Theorem 1 does not apply:

$$\lim_{x \to c} f(x) = \lim_{x \to c} \frac{n(x)}{d(x)} \quad \frac{0}{0} \text{ indeterminate form}$$

Algebraic simplification is often useful in this situation.

EXAMPLE 1 **Locating Vertical Asymptotes** Let $f(x) = \dfrac{x^2 + x - 2}{x^2 - 1}$

Describe the behavior of f at each point of discontinuity. Use ∞ and $-\infty$ when appropriate. Identify all vertical asymptotes.

SOLUTION Let $n(x) = x^2 + x - 2$ and $d(x) = x^2 - 1$. Factoring the denominator, we see that

$$d(x) = x^2 - 1 = (x - 1)(x + 1)$$

has two zeros: $x = -1$ and $x = 1$. These are the points of discontinuity of f.

First we consider $x = -1$. Since $d(-1) = 0$ and $n(-1) = -2 \neq 0$, Theorem 1 tells us that the line $x = -1$ is a vertical asymptote. So at least one of the one-sided limits at $x = -1$ must be either ∞ or $-\infty$. Examining tables of values of f for x near -1 or a graph on a graphing calculator will show which is the case. From Tables 3 and 4, we see that

$$\lim_{x \to -1^-} \frac{x^2 + x - 2}{x^2 - 1} = -\infty \quad \text{and} \quad \lim_{x \to -1^+} \frac{x^2 + x - 2}{x^2 - 1} = \infty$$

TABLE 3	
x	$f(x) = \dfrac{x^2 + x - 2}{x^2 - 1}$
-1.1	-9
-1.01	-99
-1.001	-999
-1.0001	$-9{,}999$
-1.00001	$-99{,}999$

TABLE 4	
x	$f(x) = \dfrac{x^2 + x - 2}{x^2 - 1}$
-0.9	11
-0.99	101
-0.999	$1{,}001$
-0.9999	$10{,}001$
-0.99999	$100{,}001$

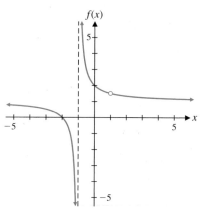

FIGURE 2 $f(x) = \dfrac{x^2 + x - 2}{x^2 - 1}$

Now we consider the other zero of $d(x)$, $x = 1$. This time $n(1) = 0$ and Theorem 1 does not apply. We use algebraic simplification to investigate the behavior of the function at $x = 1$:

$$\lim_{x \to 1} f(x) = \lim_{x \to 1} \frac{x^2 + x - 2}{x^2 - 1} \quad \frac{0}{0} \text{ indeterminate form}$$

$$= \lim_{x \to 1} \frac{(x - 1)(x + 2)}{(x - 1)(x + 1)}$$

$$= \lim_{x \to 1} \frac{x + 2}{x + 1} \quad \text{Reduced to lowest terms (see Appendix A, Section A-4)}$$

$$= \frac{3}{2}$$

Since the limit exists as x approaches 1, f does not have a vertical asymptote at $x = 1$. The graph of f (Fig. 2) shows the behavior at the vertical asymptote $x = -1$ and also at $x = 1$.

MATCHED PROBLEM 1   Let $f(x) = \dfrac{x - 3}{x^2 - 4x + 3}$

Describe the behavior of f at each discontinuity. Use ∞ and $-\infty$ when appropriate. Identify all vertical asymptotes.

EXAMPLE 2   **Locating Vertical Asymptotes**   Let $f(x) = \dfrac{x^2 + 20}{5(x - 2)^2}$

Describe the behavior of f at each discontinuity. Use ∞ and $-\infty$ when appropriate. Identify all vertical asymptotes.

SOLUTION   Let $n(x) = x^2 + 20$ and $d(x) = 5(x - 2)^2$. Since $d = 0$ only at $x = 2$, f is discontinuous only at 2. Since $n(2) = 24 \neq 0$, f has a vertical asymptote at $x = 2$ (Theorem 1). Tables 5 and 6 show that $f(x) \to \infty$ as $x \to 2$ from either side, and we have

$$\lim_{x \to 2^+} \frac{x^2 + 20}{5(x - 2)^2} = \infty \quad \text{and} \quad \lim_{x \to 2^-} \frac{x^2 + 20}{5(x - 2)^2} = \infty$$

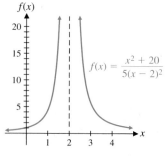

FIGURE 3

TABLE 5	
x	$f(x) = \dfrac{x^2 + 20}{5(x - 2)^2}$
2.1	488.2
2.01	48,080.02
2.001	4,800,800.2

TABLE 6	
x	$f(x) = \dfrac{x^2 + 20}{5(x - 2)^2}$
1.9	472.2
1.99	47,920.02
1.999	4,799,200.2

The denominator d has no other zeros, so f does not have any other vertical asymptotes. The graph of f (Fig. 3) shows the behavior at the vertical asymptote $x = 2$. With the left- and right-hand limits both infinite, we have

$$\lim_{x \to 2} \frac{x^2 + 20}{5(x - 2)^2} = \infty$$

MATCHED PROBLEM 2 Let $f(x) = \dfrac{x - 1}{(x + 3)^2}$

Describe the behavior of f at each discontinuity. Use ∞ and $-\infty$ when appropriate. Identify all vertical asymptotes.

INSIGHT

When is it correct to say that a limit does not exist, and when is it correct to use $\pm\infty$? It depends on the situation. Table 7 lists the infinite limits we discussed in Examples 1 and 2.

TABLE 7

Right-hand limit	Left-hand limit	Unrestricted limit
$\displaystyle\lim_{x\to-1^+}\frac{x^2 + x - 2}{x^2 - 1} = \infty$	$\displaystyle\lim_{x\to-1^-}\frac{x^2 + x - 2}{x^2 - 1} = -\infty$	$\displaystyle\lim_{x\to-1}\frac{x^2 + x - 2}{x^2 - 1}$ does not exist
$\displaystyle\lim_{x\to-2^+}\frac{x^2 + 20}{5(x - 2)^2} = \infty$	$\displaystyle\lim_{x\to2^-}\frac{x^2 + 20}{5(x - 2)^2} = \infty$	$\displaystyle\lim_{x\to2}\frac{x^2 + 20}{5(x - 2)^2} = \infty$

The instructions in Examples 1 and 2 said that we should use infinite limits to describe the behavior at vertical asymptotes. If we had been asked to *evaluate* the limits, with no mention of ∞ or asymptotes, then the correct answer would be that **all of these limits do not exist**. Remember, ∞ is a symbol used to describe the behavior of limits that do not exist.

■ Limits at Infinity

The symbol ∞ can also be used to indicate that an independent variable is increasing or decreasing without bound. We will write $x \to \infty$ to indicate that x is increasing without bound through positive values and $x \to -\infty$ to indicate that x is decreasing without bound through negative values. We begin by considering power functions of the form x^p and $1/x^p$ where p is a positive real number.

Explore & Discuss **2**

(A) Complete the following table:

x	100	1,000	10,000	100,000	1,000,000
x^2					
$1/x^2$					

(B) Describe verbally the behavior of x^2 as x increases without bound. Then use limit notation to describe this behavior.

(C) Repeat part B for $1/x^2$.

If p is a positive real number, then x^p increases as x increases and it can be shown that there is no upper bound on the values of x^p. We indicate this behavior by writing

$$x^p \to \infty \quad \text{as} \quad x \to \infty \quad \text{or} \quad \lim_{x\to\infty} x^p = \infty$$

Since the reciprocals of very large numbers are very small numbers, it follows that $1/x^p$ approaches 0 as x increases without bound. We indicate this behavior by writing

$$\frac{1}{x^p} \to 0 \quad \text{as} \quad x \to \infty \quad \text{or} \quad \lim_{x\to\infty}\frac{1}{x^p} = 0$$

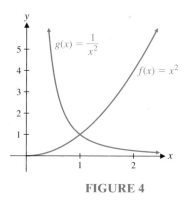

FIGURE 4

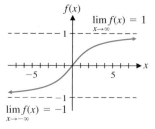

FIGURE 5

Figure 4 illustrates the preceding behavior for $f(x) = x^2$ and $g(x) = 1/x^2$, and we say that

$$\lim_{x \to \infty} f(x) = \infty \quad \text{and} \quad \lim_{x \to \infty} g(x) = 0$$

Limits of power forms as x decreases without bound behave in a similar manner, with two important differences. First, if x is negative, then x^p is not defined for all values of p. For example, $x^{1/2} = \sqrt{x}$ is not defined for negative values of x. Second, if x^p is defined, then it may approach ∞ or $-\infty$, depending on the value of p. For example,

$$\lim_{x \to \infty} x^2 = \infty \quad \text{but} \quad \lim_{x \to -\infty} x^3 = -\infty$$

For the function g in Figure 4, the line $y = 0$ (the x axis) is called a *horizontal asymptote*. In general, a line $y = b$ is a **horizontal asymptote** of the graph of $y = f(x)$ if $f(x)$ approaches b as either x increases without bound or x decreases without bound. Symbolically, $y = b$ is a horizontal asymptote if either

$$\lim_{x \to -\infty} f(x) = b \quad \text{or} \quad \lim_{x \to \infty} f(x) = b$$

In the first case, the graph of f will be close to the horizontal line $y = b$ for large (in absolute value) negative x. In the second case, the graph will be close to the horizontal line $y = b$ for large positive x. Figure 5 shows the graph of a function with two horizontal asymptotes: $y = 1$ and $y = -1$.

Theorem 2 summarizes the various possibilities for limits of power functions as x increases or decreases without bound.

THEOREM 2   LIMITS OF POWER FUNCTIONS AT INFINITY

If p is a positive real number and k is any real constant, then

1. $\displaystyle\lim_{x \to -\infty} \frac{k}{x^p} = 0$ **2.** $\displaystyle\lim_{x \to \infty} \frac{k}{x^p} = 0$

3. $\displaystyle\lim_{x \to -\infty} kx^p = \pm\infty$ **4.** $\displaystyle\lim_{x \to \infty} kx^p = \pm\infty$

provided that x^p is a real number for negative values of x. The limits in 3 and 4 will be either $-\infty$ or ∞, depending on k and p.

How can we use Theorem 2 to evaluate limits at infinity? It turns out that the limit properties listed in Theorem 2, Section 3-1, are also valid if we replace the statement $x \to c$ with $x \to \infty$ or $x \to -\infty$.

EXAMPLE 3   **Limit of a Polynomial Function at Infinity**   Let $p(x) = 2x^3 - x^2 - 7x + 3$. Find the limit of $p(x)$ as x approaches ∞ and as x approaches $-\infty$.

SOLUTION   Since limits of power functions of the form $1/x^p$ approach 0 as x approaches ∞ or $-\infty$, it is convenient to work with these reciprocal forms whenever possible. If we factor out the term involving the highest power of x, we can write $p(x)$ as

$$p(x) = 2x^3\left(1 - \frac{1}{2x} - \frac{7}{2x^2} + \frac{3}{2x^3}\right)$$

Using Theorem 2 in Section 3-3 and Theorem 2 in Section 3-1, we can write

$$\lim_{x \to \infty}\left(1 - \frac{1}{2x} - \frac{7}{2x^2} + \frac{3}{2x^3}\right) = 1 - 0 - 0 + 0 = 1$$

For large values of x,

$$\left(1 - \frac{1}{2x} - \frac{7}{2x^2} + \frac{3}{2x^3}\right) \approx 1$$

and

$$p(x) = 2x^3\left(1 - \frac{1}{2x} - \frac{7}{2x^2} + \frac{3}{2x^3}\right) \approx 2x^3$$

Since $2x^3 \to \infty$ as $x \to \infty$, it follows that

$$\lim_{x \to \infty} p(x) = \lim_{x \to \infty} 2x^3 = \infty$$

Similarly, $2x^3 \to -\infty$ as $x \to -\infty$ implies

$$\lim_{x \to -\infty} p(x) = \lim_{x \to -\infty} 2x^3 = -\infty$$

So the behavior of $p(x)$ for large values is the same as the behavior of the highest-degree term, $2x^3$.   ■

MATCHED PROBLEM 3   Let $p(x) = -4x^4 + 2x^3 + 3x$. Find the limit of $p(x)$ as x approaches ∞ and as x approaches $-\infty$.

The term with highest degree in a polynomial is called the **leading term.** In the solution in Example 3, the limits at infinity of $p(x) = 2x^3 - x^2 - 7x + 3$ were the same as the limits of the leading term $2x^3$. Theorem 3 states that this is true for any polynomial of degree greater than or equal to 1.

THEOREM 3   **LIMITS OF POLYNOMIAL FUNCTIONS AT INFINITY**

If

$$p(x) = a_n x^n + a_{n-1} x^{n-1} + \cdots + a_1 x + a_0, a_n \neq 0, n \geq 1$$

then

$$\lim_{x \to \infty} p(x) = \lim_{x \to \infty} a_n x^n = \pm\infty$$

and

$$\lim_{x \to -\infty} p(x) = \lim_{x \to -\infty} a_n x^n = \pm\infty$$

Each limit will be either $-\infty$ or ∞, depending on a_n and n.

A polynomial of degree 0 is a constant function $p(x) = a_0$, and its limit as x approaches ∞ or $-\infty$ is the number a_0. For any polynomial of degree 1 or greater, Theorem 3 states that the limit as x approaches ∞ or $-\infty$ cannot be equal to a number. This means that **polynomials of degree 1 or greater never have horizontal asymptotes.**

A description of the behavior of the limits of any function f at infinity is called the **end behavior** of f. $\lim_{x \to \infty} f(x)$ determines the right end behavior and $\lim_{x \to -\infty} f(x)$ determines the left end behavior. According to Theorem 3, the right and left end behavior of a nonconstant polynomial $p(x)$ is always an infinite limit.

EXAMPLE   4   **End Behavior of a Polynomial** Describe the end behavior of each polynomial:

(A) $p(x) = 3x^3 - 500x^2$ (B) $p(x) = 3x^3 - 500x^4$

SOLUTION (A) According to Theorem 3, as x increases without bound to the right, the right end behavior is

$$\lim_{x \to \infty} (3x^3 - 500x^2) = \lim_{x \to \infty} 3x^3 = \infty$$

and as x increases without bound to the left, the end behavior is

$$\lim_{x \to -\infty} (3x^3 - 500x^2) = \lim_{x \to -\infty} 3x^3 = -\infty$$

(B) As x increases without bound to the right, the end behavior is

$$\lim_{x \to \infty} (3x^3 - 500x^4) = \lim_{x \to \infty} (-500x^4) = -\infty$$

and as x increases without bound to the left, the end behavior is

$$\lim_{x \to -\infty} (3x^3 - 500x^4) = \lim_{x \to -\infty} (-500x^4) = -\infty$$

MATCHED PROBLEM 4 Describe the end behavior of each polynomial:

(A) $p(x) = 300x^2 - 4x^5$ (B) $p(x) = 300x^6 - 4x^5$

Explore & Discuss **3** Let $p(x)$ be a polynomial of degree $n > 0$ with $a_n \neq 0$. Complete the following table:

n	a_n	Left End Behavior	Right End Behavior
Even	Positive	$p(x) \to \infty$	$p(x) \to \infty$
Even	Negative		
Odd	Positive		
Odd	Negative		

▪ Finding Horizontal Asymptotes

Since a rational function is the ratio of two polynomials, it is not surprising that reciprocals of powers of x can be used to analyze limits of rational functions at infinity. For example, consider the rational function

$$f(x) = \frac{3x^2 - 5x + 9}{2x^2 + 7}$$

Factoring the highest-degree term out of the numerator and the denominator, we can write

$$f(x) = \frac{3x^2}{2x^2} \cdot \frac{1 - \dfrac{5}{3x} + \dfrac{3}{x^2}}{1 + \dfrac{7}{2x^2}}$$

$$\lim_{x \to \infty} f(x) = \lim_{x \to \infty} \frac{3x^2}{2x^2} \cdot \lim_{x \to \infty} \frac{1 - \dfrac{5}{3x} + \dfrac{3}{x^2}}{1 + \dfrac{7}{2x^2}} = \frac{3}{2} \cdot \frac{1 - 0 + 0}{1 + 0} = \frac{3}{2}$$

Thus, the behavior of this rational function as x approaches infinity is determined by the ratio of the highest-degree term in the numerator $(3x^2)$ to the highest-degree term in the denominator $(2x^2)$. Theorem 4 generalizes this result to any rational function and lists the three possible outcomes.

THEOREM 4   **LIMITS OF RATIONAL FUNCTIONS AT INFINITY AND HORIZONTAL ASYMPTOTES OF RATIONAL FUNCTIONS**

(A) If $f(x) = \dfrac{a_m x^m + a_{m-1} x^{m-1} + \cdots + a_1 x + a_0}{b_n x^n + b_{n-1} x^{n-1} + \cdots + b_1 x + b_0}$, $a_m \neq 0, b_n = 0$

then $\displaystyle\lim_{x \to \infty} f(x) = \lim_{x \to \infty} \dfrac{a_m x^m}{b_n x^n}$ and $\displaystyle\lim_{x \to -\infty} f(x) = \lim_{x \to -\infty} \dfrac{a_m x^m}{b_n x^n}$

(B) There are three possible cases for these limits:

1. If $m < n$, then $\displaystyle\lim_{x \to \infty} f(x) = \lim_{x \to -\infty} f(x) = 0$, and the line $y = 0$ (the x axis) is a horizontal asymptote of $f(x)$.

2. If $m = n$, then $\displaystyle\lim_{x \to \infty} f(x) = \lim_{x \to -\infty} f(x) = \dfrac{a_m}{b_n}$, and the line $y = \dfrac{a_m}{b_n}$ is a horizontal asymptote of $f(x)$.

3. If $m > n$, then each limit will be ∞ or $-\infty$, depending on m, n, a_m, and b_n, and $f(x)$ does not have a horizontal asymptote.

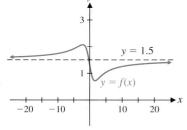

FIGURE 6 $f(x) = \dfrac{3x^2 - 5x + 9}{2x^2 + 7}$

Notice in cases 1 and 2 of Theorem 4 that the limit is the same if x approaches ∞ or $-\infty$. Thus, **a rational function can have at most one horizontal asymptote** (see Fig. 6).

INSIGHT

The graph of f in Figure 6 dispels the misconception that the graph of a function cannot cross a horizontal asymptote. Horizontal asymptotes give us information about the graph of a function only as $x \to \infty$ and $x \to -\infty$, not at any specific value of x.

EXAMPLE 5   **Finding Horizontal Asymptotes** Find all horizontal asymptotes, if any, of each function.

(A) $f(x) = \dfrac{5x^3 - 2x^2 + 1}{4x^3 + 2x - 7}$ (B) $f(x) = \dfrac{3x^4 - x^2 + 1}{8x^6 - 10}$

(C) $f(x) = \dfrac{2x^5 - x^3 - 1}{6x^3 + 2x^2 - 7}$

SOLUTION   We will make use of part A of Theorem 3:

(A) $\displaystyle\lim_{x \to \infty} f(x) = \lim_{x \to \infty} \dfrac{5x^3 - 2x^2 + 1}{4x^3 + 2x - 7} = \lim_{x \to \infty} \dfrac{5x^3}{4x^3} = \dfrac{5}{3}$

The line $y = 5/3$ is a horizontal asymptote of $f(x)$.

(B) $\displaystyle\lim_{x \to \infty} f(x) = \lim_{x \to \infty} \dfrac{3x^4 - x^2 + 1}{8x^6 - 10} = \lim_{x \to \infty} \dfrac{3x^4}{8x^6} = \lim_{x \to \infty} \dfrac{3}{8x^2} = 0$

The line $y = 0$ (the x axis) is a horizontal asymptote of $f(x)$.

(C) $\displaystyle\lim_{x \to \infty} f(x) = \lim_{x \to \infty} \dfrac{2x^5 - x^3 - 1}{6x^3 + 2x^2 - 7} = \lim_{x \to \infty} \dfrac{2x^5}{6x^3} = \lim_{x \to \infty} \dfrac{x^2}{3} = \infty$

The function $f(x)$ has no horizontal asymptotes.

MATCHED PROBLEM 5   Find the horizontal asymptotes of each function.

$$\text{(A) } f(x) = \frac{4x^3 - 5x + 8}{2x^4 - 7} \qquad \text{(B) } f(x) = \frac{5x^6 + 3x}{2x^5 - x - 5}$$

$$\text{(C) } f(x) = \frac{2x^3 - x + 7}{4x^3 + 3x^2 - 100}$$

An accurate sketch of the graph of a rational function requires knowledge of both vertical and horizontal asymptotes. As we mentioned earlier, we are postponing a detailed discussion of graphing techniques until Section 5-4.

EXAMPLE 6   **Finding the Asymptotes of a Rational Function**   Find all the asymptotes of the function

$$f(x) = \frac{2x^2 - 5}{x^2 + 4x + 4}$$

SOLUTION   Let $n(x) = 2x^2 - 5$ and $d(x) = x^2 + 4x + 4 = (x + 2)^2$. Since $d(x) = 0$ only at $x = 2$ and $n(2) = 3$, the only vertical asymptote of f is the line $x = 2$ (Theorem 1). Since

$$\lim_{x \to \infty} f(x) = \lim_{x \to \infty} \frac{2x^2 - 5}{x^2 + 4x + 4} = \lim_{x \to \infty} \frac{2x^2}{x^2} = 2$$

the horizontal asymptote is the line $y = 2$ (Theorem 3).

MATCHED PROBLEM 6   Find all the asymptotes of the function $f(x) = \dfrac{x^2 - 9}{x^2 - 4}$

Answers to Matched Problems   **1.** Vertical asymptote: $x = 1$; $\lim_{x \to 1^+} f(x) = \infty$, $\lim_{x \to 1^-} f(x) = -\infty$

2. Vertical asymptote: $x = -3$; $\lim_{x \to -3^+} f(x) = \lim_{x \to -3^-} f(x) = -\infty$

3. $\lim_{x \to \infty} p(x) = \lim_{x \to -\infty} p(x) = -\infty$

4. (A) $\lim_{x \to \infty} p(x) = -\infty$, $\lim_{x \to -\infty} p(x) = \infty$

(B) $\lim_{x \to \infty} p(x) = \infty$, $\lim_{x \to -\infty} p(x) = \infty$

5. (A) $y = 0$ (B) No horizontal asymptotes   (C) $y = 1/2$

6. Vertical asymptotes: $x = -2$, $x = 4$; horizontal asymptote: $y = 1$

Exercise 3-3

A *Problems 1–8 refer to the following graph of* $y = f(x)$:

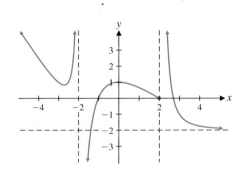

1. $\lim\limits_{x \to \infty} f(x) = ?$

2. $\lim\limits_{x \to -\infty} f(x) = ?$

3. $\lim\limits_{x \to -2^+} f(x) = ?$

4. $\lim\limits_{x \to -2^-} f(x) = ?$

5. $\lim\limits_{x \to -2} f(x) = ?$

6. $\lim\limits_{x \to 2^+} f(x) = ?$

7. $\lim\limits_{x \to 2^-} f(x) = ?$

8. $\lim\limits_{x \to 2} f(x) = ?$

In Problems 9–16, find each limit. Use $-\infty$ *and* ∞ *when appropriate.*

9. $f(x) = \dfrac{x}{x - 5}$

   (A) $\lim\limits_{x \to 5^-} f(x)$ (B) $\lim\limits_{x \to 5^+} f(x)$ (C) $\lim\limits_{x \to 5} f(x)$

10. $f(x) = \dfrac{x^2}{x + 3}$

    (A) $\lim\limits_{x \to -3^-} f(x)$ (B) $\lim\limits_{x \to -3^+} f(x)$ (C) $\lim\limits_{x \to -3} f(x)$

11. $f(x) = \dfrac{2x - 4}{(x - 4)^2}$

    (A) $\lim\limits_{x \to 4^-} f(x)$ (B) $\lim\limits_{x \to 4^+} f(x)$ (C) $\lim\limits_{x \to 4} f(x)$

12. $f(x) = \dfrac{2x + 2}{(x + 2)^2}$

    (A) $\lim\limits_{x \to -2^-} f(x)$ (B) $\lim\limits_{x \to -2^+} f(x)$ (C) $\lim\limits_{x \to -2} f(x)$

13. $f(x) = \dfrac{x^2 + x - 2}{(x - 1)}$

    (A) $\lim\limits_{x \to 1^-} f(x)$ (B) $\lim\limits_{x \to 1^+} f(x)$ (C) $\lim\limits_{x \to 1} f(x)$

14. $f(x) = \dfrac{x^2 + x + 2}{(x - 1)}$

    (A) $\lim\limits_{x \to 1^-} f(x)$ (B) $\lim\limits_{x \to 1^+} f(x)$ (C) $\lim\limits_{x \to 1} f(x)$

15. $f(x) = \dfrac{x^2 - 3x + 2}{(x + 2)}$

    (A) $\lim\limits_{x \to -2^-} f(x)$ (B) $\lim\limits_{x \to -2^+} f(x)$ (C) $\lim\limits_{x \to -2} f(x)$

16. $f(x) = \dfrac{x^2 + x - 2}{(x + 2)}$

    (A) $\lim\limits_{x \to -2^-} f(x)$ (B) $\lim\limits_{x \to -2^+} f(x)$ (C) $\lim\limits_{x \to -2} f(x)$

In Problems 17–20, find the limit of each polynomial $p(x)$
(A) as x approaches ∞ *(B) as x approaches* $-\infty$

17. $p(x) = 4x^5 - 3x^4 + 1$

18. $p(x) = 4x^3 - 3x^4 + x^2$

19. $p(x) = 2x^5 - 2x^6 - 11$

20. $p(x) = 2x^4 - 2x^3 + 9x$

B *In Problems 21–30, use* $-\infty$ *or* ∞ *where appropriate to describe the behavior at each discontinuity and identify all vertical asymptotes.*

21. $f(x) = \dfrac{1}{x + 3}$

22. $g(x) = \dfrac{x}{4 - x}$

23. $h(x) = \dfrac{x^2 + 4}{x^2 - 4}$

24. $k(x) = \dfrac{x^2 - 9}{x^2 + 9}$

25. $F(x) = \dfrac{x^2 - 4}{x^2 + 4}$

26. $G(x) = \dfrac{x^2 + 9}{9 - x^2}$

27. $H(x) = \dfrac{x^2 - 2x - 3}{x^2 - 4x + 3}$

28. $K(x) = \dfrac{x^2 + 2x - 3}{x^2 - 4x + 3}$

29. $T(x) = \dfrac{8x - 16}{x^4 - 8x^3 + 16x^2}$

30. $S(x) = \dfrac{6x + 9}{x^4 + 6x^3 + 9x^2}$

In Problems 31–36, evaluate the indicated limit. Use $-\infty$ *or* ∞ *where appropriate.*

31. $\lim\limits_{x \to \infty} \dfrac{4x + 7}{5x - 9}$

32. $\lim\limits_{x \to \infty} \dfrac{2 - 3x^3}{7 + 4x^3}$

33. $\lim_{x \to \infty} \dfrac{5x^2 + 11}{7x - 2}$

34. $\lim_{x \to \infty} \dfrac{5x + 11}{7x^3 - 2}$

35. $\lim_{x \to \infty} \dfrac{7x^4 - 14x^2}{6x^5 + 3}$

36. $\lim_{x \to \infty} \dfrac{4x^7 - 8x}{6x^4 + 9x^2}$

In Problems 37–50, find all horizontal and vertical asymptotes.

37. $f(x) = \dfrac{2x}{x + 2}$

38. $f(x) = \dfrac{3x + 2}{x - 4}$

39. $f(x) = \dfrac{x^2 + 1}{x^2 - 1}$

40. $f(x) = \dfrac{x^2 - 1}{x^2 + 2}$

41. $f(x) = \dfrac{x^3}{x^2 + 6}$

42. $f(x) = \dfrac{x}{x^2 - 4}$

43. $f(x) = \dfrac{x}{x^2 + 4}$

44. $f(x) = \dfrac{x^2 + 9}{x}$

45. $f(x) = \dfrac{x^2}{x - 3}$

46. $f(x) = \dfrac{x + 5}{x^2}$

47. $f(x) = \dfrac{2x^2 + 3x - 2}{x^2 - x - 2}$

48. $f(x) = \dfrac{2x^2 + 7x + 12}{2x^2 + 5x - 12}$

49. $f(x) = \dfrac{2x^2 - 5x + 2}{x^2 - x - 2}$

50. $f(x) = \dfrac{x^2 - x - 12}{2x^2 + 5x - 12}$

C *In Problems 51–54, determine a so that f is continuous on* $(-\infty, \infty)$.

51. $f(x) = \begin{cases} \dfrac{x^2 + 2x - 3}{x - 1} & \text{if } x \neq 1 \\ a & \text{if } x = 1 \end{cases}$

52. $f(x) = \begin{cases} \dfrac{x^2 + 3x + 2}{x + 2} & \text{if } x \neq -2 \\ a & \text{if } x = -2 \end{cases}$

53. $f(x) = \begin{cases} \dfrac{x^2 + 4x + 3}{x + 3} & \text{if } x \neq -3 \\ a & \text{if } x = -3 \end{cases}$

54. $f(x) = \begin{cases} \dfrac{x^2 - 5x + 6}{x - 2} & \text{if } x \neq 2 \\ a & \text{if } x = 2 \end{cases}$

55. Theorem 3 states that
$$\lim_{x \to \infty} (a_n x^n + a_{n-1} x^{n-1} + \cdots + a^0) = \pm\infty.$$ What conditions must n and a_n satisfy for the limit to be ∞? For the limit to be $-\infty$?

56. Theorem 3 also states that
$$\lim_{x \to -\infty} (a_n x^n + a_{n-1} x^{n-1} + \cdots + a_0) = \pm\infty.$$ What conditions must n and a_n satisfy for the limit to be ∞? For the limit to be $-\infty$?

Describe the end behavior of each function in Problems 57–64.

57. $f(x) = 2x^4 - 5x + 11$

58. $f(x) = 3x - 7x^5 - 2$

59. $f(x) = 7x^2 + 9x^3 + 5x$

60. $f(x) = 4x^3 - 5x^2 - 6x^6$

61. $f(x) = \dfrac{x^2 - 5x - 7}{x + 11}$

62. $f(x) = \dfrac{5x^4 + 7x^7 - 10}{x^2 + 6x^4 + 3}$

63. $f(x) = \dfrac{5x^5 + 7x^4 - 10}{-x^3 + 6x^2 + 3}$

64. $f(x) = \dfrac{-3x^6 + 4x^4 + 2}{2x^2 - 2x - 3}$

Applications

65. *Average cost.* A company manufacturing snowboards has fixed costs of $200 per day and total costs of $3,800 per day at a daily output of 20 boards.

(A) Assuming that the total cost per day $C(x)$ is linearly related to the total output per day x, write an equation for the cost function.

(B) The average cost per board for an output of x boards is given by $\overline{C}(x) = C(x)/x$. Find the average cost function.

(C) Sketch a graph of the average cost function, including any asymptotes, for $1 \leq x \leq 30$.

(D) What does the average cost per board tend to as production increases?

66. *Average cost.* A company manufacturing surfboards has fixed costs of $300 per day and total costs of $5,100 per day at a daily output of 20 boards.

(A) Assuming that the total cost per day $C(x)$ is linearly related to the total output per day x, write an equation for the cost function.

(B) The average cost per board for an output of x boards is given by $\overline{C}(x) = C(x)/x$. Find the average cost function.

(C) Sketch a graph of the average cost function, including any asymptotes, for $1 \leq x \leq 30$.

(D) What does the average cost per board tend to as production increases?

67. *Energy costs.* Most appliance manufacturers produce both conventional models and energy-efficient models. The energy-efficient models are more expensive to make, but cheaper to operate. The costs of purchasing and operating a 23-cubic-foot refrigerator of each type are given in Table 8. These costs do not include maintenance charges or changes in the price of electricity.

TABLE 8 Refrigerators

	Energy-Efficient Model	Conventional Model
Initial cost	$950	$900
Total volume	23 ft^3	23 ft^3
Annual cost of electricity	$56	$66

(A) Express the total cost $C_e(x)$ and the average cost $\overline{C}_e(x) = C_e(x)/x$ of purchasing and operating an energy-efficient refrigerator for x years.

(B) Express the total cost $C_c(x)$ and the average cost $\overline{C}_c(x) = C_c(x)/x$ of purchasing and operating a conventional refrigerator for x years.

(C) Are the total cost for an energy-efficient model and the total cost for a conventional model ever the same? If so, when?

(D) Are the average cost for an energy-efficient model and the average cost for a conventional model ever the same? If so, when?

(E) Find the limit of each average cost function as $x \to \infty$, and discuss the implications of the results.

68. *Energy costs.* Most appliance manufacturers produce both conventional models and energy-efficient models. The energy-efficient models are more expensive to make, but cheaper to operate. The costs of purchasing and operating a 36,000-Btu central air-conditioner of each type are given in Table 9. These costs do not include maintenance charges or changes in the price of electricity.

TABLE 9 Central Air-Conditioner

	Energy Efficient Model	Conventional Model
Initial cost	$4,000	$2,700
Total capacity	36,000 Btu	36,000 Btu
Annual cost of electricity	$932	$1,332

(A) Express the total cost $C_e(x)$ and the average cost $\overline{C}_e(x) = C_e(x)/x$ of purchasing and operating an energy-efficient central air-conditioner for x years.

(B) Express the total cost $C_c(x)$ and the average cost $\overline{C}_c(x) = C_c(x)/x$ of purchasing and operating a conventional central air-conditioner for x years.

(C) Are the total cost for an energy-efficient model and the total cost for a conventional model ever the same? If so, when?

(D) Are the average cost for an energy-efficient model and the average cost for a conventional model ever the same? If so, when?

(E) Find the limit of each average cost function as $x \to \infty$, and discuss the implications of the results.

69. *Drug concentration.* A drug is administered to a patient through an injection. The concentration of the drug (in milligrams/milliliter) in the bloodstream t hours after the injection is given by $C(t) = \dfrac{5t^2(t + 50)}{t^3 + 100}$. Find and interpret $\lim\limits_{t \to \infty} C(t)$.

70. *Drug concentration.* A drug is administered to a patient through an IV drip. The concentration of the drug (in milligrams/milliliter) in the bloodstream t hours after the drip was started is given by $C(t) = \dfrac{5t(t + 50)}{t^3 + 100}$. Find and interpret $\lim\limits_{t \to \infty} C(t)$.

71. *Pollution.* In Silicon Valley (California), a number of computer-related manufacturing firms were found to be contaminating underground water supplies with toxic chemicals stored in leaking underground containers. A water quality control agency ordered the companies to take immediate corrective action and to contribute to a monetary pool for testing and cleanup of the underground contamination. Suppose that the required monetary pool (in millions of dollars) for the testing and cleanup is estimated to be given by

$$P(x) = \frac{2x}{1 - x} \quad 0 \le x < 1$$

where x is the percentage (expressed as a decimal fraction) of the total contaminant removed.

(A) How much must be in the pool to remove 90% of the contaminant?

(B) How much must be in the pool to remove 95% of the contaminant?

(C) Find $\lim\limits_{x \to 1^-} P(x)$ and discuss the implications of this limit.

72. *Employee training.* A company producing computer components has established that, on the average, a new employee can assemble $N(t)$ components per day after t days of on-the-job training, as given by

$$N(t) = \frac{100t}{t + 9} \quad t \ge 0$$

(A) How many components per day can a new employee assemble after 6 days of on-the-job training?

(B) How many days of on-the-job training will a new employee need to reach the level of 70 components per day?

(C) Find $\lim\limits_{t \to \infty} N(t)$ and discuss the implications of this limit.

73. *Biochemistry.* In 1913, biochemists Leonor Michaelis (1875–1949) and Maude Menten (1879–1960) proposed the rational function model (see figure)

$$v(s) = \frac{V_{\max}\, s}{K_M + s}$$

for the velocity of the enzymatic reaction v, where s is the substrate concentration. The constants $V_{\max}$ and K_M are determined from experimental data.

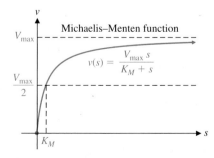

Figure for 73

(A) Show that $\lim\limits_{s\to\infty} v(s) = V_{\max}$.

(B) Show that $v(K_M) = \dfrac{V_{\max}}{2}$.

(C) Table 10* lists data for an enzyme treated with the substrate saccharose.

TABLE 10

s	v
5.2	0.866
10.4	1.466
20.8	2.114
41.6	2.666
83.3	3.236
167	3.636
333	3.636

Plot the preceding points on graph paper and estimate $V_{\max}$ to the nearest integer. To estimate K_M, add the horizontal line $v = \dfrac{V_{\max}}{2}$ to your graph, connect successive points on the graph with straight-line segments, and estimate the value of s (to the nearest multiple of 10) that satisfies $v(s) = \dfrac{V_{\max}}{2}$.

(D) Use the constants $V_{\max}$ and K_M from part C to form a Michaelis–Menten function for the data in Table 10.

(E) Use the function from part D to estimate the velocity of the enzyme reaction when the saccharose is 15 and to estimate the saccharose when the velocity is 3.

74. *Biochemistry.* Table 11[†] lists data for the enzyme invertase treated with the substrate sucrose. We want to model these data with a Michaelis–Menten function.

TABLE 11

s	v
2.92	18.2
5.84	26.5
8.76	31.1
11.7	33
14.6	34.9
17.5	37.2
23.4	37.1

(A) Plot the points in Table 11 on graph paper and estimate $V_{\max}$ to the nearest integer. To estimate K_M, add the horizontal line $v = \dfrac{V_{\max}}{2}$ to your graph,

connect successive points on the graph with straight-line segments, and estimate the value of s (to the nearest integer) that satisfies $v(s) = \dfrac{V_{\max}}{2}$.

(B) Use the constants $V_{\max}$ and K_M from part A to form a Michaelis–Menten function for the data in Table 11.

(C) Use the function from part B to estimate the velocity of the enzyme reaction when the sucrose is 9 and to estimate the sucrose when the velocity is 32.

75. *Physics: Thermal expansion.* The coefficient of thermal expansion (CTE) is a measure of the expansion of an object subjected to extreme temperatures. We want to use a Michaelis–Menten function of the form

$$C(T) = \frac{C_{\max}\, T}{M + T} \qquad \text{(Problem 73)}$$

where $C = $ CTE, T is temperature in K (degrees Kelvin), and $C_{\max}$ and M are constants, to model this coefficient. Table 12* lists the coefficients of thermal expansion for nickel and for copper at various temperatures.

TABLE 12 Coefficients of thermal expansion

T (K)	Nickel	Copper
100	6.6	10.3
200	11.3	15.2
293	13.4	16.5
500	15.3	18.3
800	16.8	20.3
1,100	17.8	23.7

(A) Plot the points in columns 1 and 2 of Table 12 on graph paper and estimate $C_{\max}$ to the nearest integer. To estimate M, add the horizontal line CTE $= \dfrac{C_{\max}}{2}$ to your graph, connect successive points on the graph with straight-line segments, and estimate the value of T (to the nearest multiple of fifty) that satisfies $C(T) = \dfrac{C_{\max}}{2}$.

(B) Use the constants $\dfrac{C_{\max}}{2}$ and M from part A to form a Michaelis–Menten function for the CTE of nickel.

(C) Use the function from part B to estimate the CTE of nickel at 600 K and to estimate the temperature when the CTE of nickel is 12.

76. *Physics: Thermal expansion.* Repeat Problem 75 for the CTE of copper (column 3 of Table 12).

---

*Michaelis and Menten (1913) *Biochem. Z.* 49, 333–369.

[†]Institute of Chemistry, Macedonia.

*National Physical Laboratory

Section 3-4   THE DERIVATIVE

- Rate of Change
- Slope of the Tangent Line
- The Derivative
- Nonexistence of the Derivative

We will now make use of the limit concepts developed in Sections 3-1, 3-2, and 3-3 to solve the two important problems illustrated in Figure 1. The solution of each of these apparently unrelated problems involves a common concept called the *derivative*.

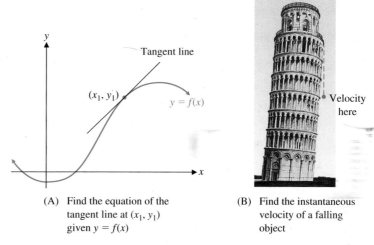

(A) Find the equation of the tangent line at (x_1, y_1) given $y = f(x)$

(B) Find the instantaneous velocity of a falling object

FIGURE 1 Two basic problems of calculus

■ Rate of Change

Let us start by considering a simple example.

EXAMPLE 1   **Revenue Analysis** The revenue (in dollars) from the sale of x plastic planter boxes is given by

$$R(x) = 20x - 0.02x^2 \qquad 0 \le x \le 1,000$$

which is graphed in Figure 2.

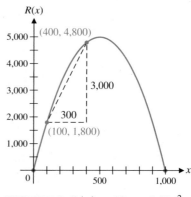

FIGURE 2 $R(x) = 20x - 0.02x^2$

(A) What is the change in revenue if production is changed from 100 planters to 400 planters?

(B) What is the average change in revenue for this change in production?

SOLUTION (A) The change in revenue is given by

$$R(400) - R(100) = 20(400) - 0.02(400)^2 - [20(100) - 0.02(100)^2]$$
$$= 4,800 - 1,800 = \$3,000$$

Thus, increasing production from 100 planters to 400 planters will increase revenue by $3,000.

(B) To find the average change in revenue, we divide the change in revenue by the change in production:

$$\frac{R(400) - R(100)}{400 - 100} = \frac{3,000}{300} = \$10$$

Thus, the average change in revenue is $10 per planter when production is increased from 100 to 400 planters. ▬

MATCHED PROBLEM 1   Refer to the revenue function in Example 1.

(A) What is the change in revenue if production is changed from 600 planters to 800 planters?

(B) What is the average change in revenue for this change in production?

In general, if we are given a function $y = f(x)$ and if x is changed from a to $a + h$, then y will change from $f(a)$ to $f(a + h)$. The *average rate of change is the ratio of the change in y to the change in x.*

DEFINITION   **Average Rate of Change**

For $y = f(x)$, the **average rate of change from $x = a$ to $x = a + h$** is

$$\frac{f(a + h) - f(a)}{(a + h) - a} = \frac{f(a + h) - f(a)}{h} \qquad h \neq 0 \qquad (1)$$

As we noted in Section 3-1, the mathematical expression (1) is called the **difference quotient.** The preceding discussion shows that the difference quotient can be interpreted as an average rate of change. The next example illustrates another interpretation of this quotient: the velocity of a moving object.

EXAMPLE 2   **Velocity**   A small steel ball dropped from a tower will fall a distance of y feet in x seconds, as given approximately by the physics formula

$$y = f(x) = 16x^2$$

Figure 3 shows the position of the ball on a coordinate line (positive direction down) at the end of 0, 1, 2, and 3 seconds.

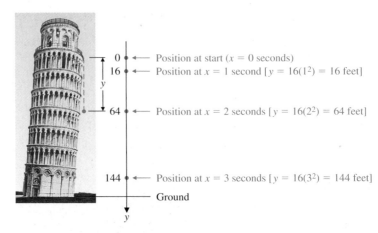

0 ◄—— Position at start ($x = 0$ seconds)
16 ◄—— Position at $x = 1$ second [$y = 16(1^2) = 16$ feet]
64 ◄—— Position at $x = 2$ seconds [$y = 16(2^2) = 64$ feet]
144 ◄—— Position at $x = 3$ seconds [$y = 16(3^2) = 144$ feet]
Ground

FIGURE 3 *Note:* Positive y direction is down.

(A) Find the average velocity from $x = 2$ seconds to $x = 3$ seconds.

(B) Find and simplify the average velocity from $x = 2$ seconds to $x = 2 + h$ seconds, $h \neq 0$.

(C) Find the limit of the expression from part B as $h \to 0$ if that limit exists.

(D) Discuss possible interpretations of the limit from part C.

SOLUTION   (A) Recall the formula $d = rt$, which can be written in the form

$$r = \frac{d}{t} = \frac{\text{Distance covered}}{\text{Elapsed time}} = \text{Average velocity}$$

For example, if a person drives from San Francisco to Los Angeles (a distance of about 420 miles) in 7 hours, then the average velocity is

$$r = \frac{d}{t} = \frac{420}{7} = 60 \text{ miles per hour}$$

Sometimes the person will be traveling faster and sometimes slower, but the average velocity is 60 miles per hour. In our present problem, the average velocity of the steel ball from $x = 2$ seconds to $x = 3$ seconds is

$$\begin{aligned}
\text{Average velocity} &= \frac{\text{Distance covered}}{\text{Elapsed time}} \\
&= \frac{f(3) - f(2)}{3 - 2} \\
&= \frac{16(3)^2 - 16(2)^2}{1} = 80 \text{ feet per second}
\end{aligned}$$

Thus, we see that if $y = f(x)$ is the position of the falling ball, then the average velocity is simply the average rate of change of $f(x)$ with respect to time x, and we have another interpretation of the difference quotient (1).

(B) Proceeding as in part A, we have

$$\begin{aligned}
\text{Average velocity} &= \frac{\text{Distance covered}}{\text{Elapsed time}} \\
&= \frac{f(2 + h) - f(2)}{h} &&\textit{Difference quotient} \\
&= \frac{16(2 + h)^2 - 16(2)^2}{h} &&\textit{Simplify this 0/0} \\
& &&\textit{indeterminate form.} \\
&= \frac{64 + 64h + 16h^2 - 64}{h} \\
&= \frac{h(64 + 16h)}{h} = 64 + 16h \qquad h \neq 0
\end{aligned}$$

Notice that if $h = 1$, the average velocity is 80 feet per second, our result in part A.

(C) The limit of the average velocity expression from part B as $h \to 0$ is

$$\lim_{h \to 0} \frac{f(2 + h) - f(2)}{h} = \lim_{h \to 0} (64 + 16h)$$
$$= 64 \text{ feet per second}$$

(D) The average velocity over smaller and smaller time intervals approaches 64 feet per second. This limit can be interpreted as the velocity of the ball at the *instant* the ball has been falling for exactly 2 seconds. Thus, 64 feet per second is referred to as the **instantaneous velocity** at $x = 2$ seconds, and we have solved the one of the basic problems of calculus (see Fig. 1B).

MATCHED PROBLEM 2   For the falling steel ball in Example 2, find

(A) The average velocity from $x = 1$ second to $x = 2$ seconds

(B) The average velocity (in simplified form) from $x = 1$ second to $x = 1 + h$ seconds, $h \neq 0$

(C) The instantaneous velocity at $x = 1$ second

Explore & Discuss **1**   Recall the revenue function in Example 1: $R(x) = 20x - 0.02x^2$. Find

$$\lim_{h \to 0} \frac{R(100 + h) - R(100)}{h}$$

Discuss possible interpretations of this limit.

The ideas introduced in Example 2 are not confined to average velocity, but can be applied to the average rate of change of any function.

DEFINITION   **Instantaneous Rate of Change**

For $y = f(x)$, the **instantaneous rate of change at $x = a$** is

$$\lim_{h \to 0} \frac{f(a + h) - f(a)}{h} \qquad (2)$$

if the limit exists.

The adjective *instantaneous* is often omitted with the understanding that the phrase **rate of change** always refers to the instantaneous rate of change and not the average rate of change. Similarly, **velocity** always refers to the instantaneous rate of change of distance with respect to time.

■ Slope of the Tangent Line

So far, our interpretations of the difference quotient have been numerical in nature. Now we want to consider a geometric interpretation. A line through two points on the graph of a function is called a **secant line**. If $(a, f(a))$ and $(a + h, f(a + h))$ are two points on the graph of $y = f(x)$, we can use the slope formula from Section 1-2 to find the slope of the secant line through these points (see Fig. 4).

$$\textbf{Slope of secant line} = \frac{f(a + h) - f(a)}{(a + h) - a}$$

$$= \frac{f(a + h) - f(a)}{h} \qquad \textit{Difference quotient}$$

Thus, the difference quotient can be interpreted as both the average rate of change and the slope of the secant line.

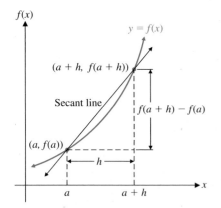

FIGURE 4 Secant line

INSIGHT

If (x_1, y_1) and (x_2, y_2) are two points in the plane with $x_1 \neq x_2$ and L is the line passing through these two points (see Section 1-2), then

$$\text{Slope of } L \qquad\qquad \text{Point–slope form for } L$$

$$m = \frac{y_2 - y_1}{x_2 - x_1} \qquad\qquad y - y_1 = m(x - x_1)$$

These formulas will be used extensively in the remainder of this chapter.

EXAMPLE 3   **Slope of a Secant Line**   Given $f(x) = x^2$,

(A) Find the slope of the secant line for $a = 1$ and $h = 2$ and 1, respectively. Graph $y = f(x)$ and the two secant lines.

(B) Find and simplify the slope of the secant line for $a = 1$ and h any nonzero number.

(C) Find the limit of the expression in part B.

(D) Discuss possible interpretations of the limit in part C.

SOLUTION   (A) For $a = 1$ and $h = 2$, the secant line goes through $(1, f(1)) = (1, 1)$ and $(3, f(3)) = (3, 9)$, and its slope is

$$\frac{f(1 + 2) - f(1)}{2} = \frac{3^2 - 1^2}{2} = 4$$

For $a = 1$ and $h = 1$, the secant line goes through $(1, f(1)) = (1, 1)$ and $(2, f(2)) = (2, 4)$, and its slope is

$$\frac{f(1 + 1) - f(1)}{1} = \frac{2^2 - 1^2}{1} = 3$$

The graphs of $y = f(x)$ and the two secant lines are shown in Figure 5.

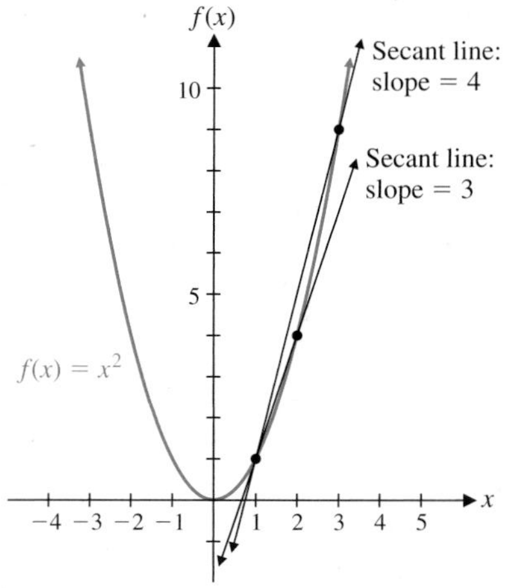

(B) For $a = 1$ and h any nonzero number, the secant line goes through $(1, f(1)) = (1, 1)$ and $(1 + h, f(1 + h)) = (1 + h, (1 + h)^2)$, and its slope is

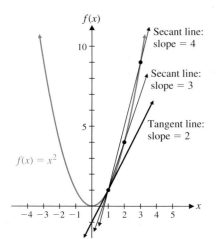

FIGURE 5 Secant lines

$$\frac{f(1 + h) - f(1)}{h} = \frac{(1 + h)^2 - 1^2}{h} \qquad \text{Square the binomial.}$$

$$= \frac{1 + 2h + h^2 - 1}{h} \qquad \text{Combine like terms and factor the numerator.}$$

$$= \frac{h(2 + h)}{h} \qquad \text{Cancel.}$$

$$= 2 + h \qquad h \neq 0$$

(C) The limit of the secant line slope from part B is

$$\lim_{h \to 0} \frac{f(1 + h) - f(1)}{h} = \lim_{h \to 0} (2 + h)$$
$$= 2$$

(D) In part C, we saw that the limit of the slopes of the secant lines through the point $(1, f(1))$ is 2. If we graph the line through $(1, f(1))$ with slope 2 (Fig. 6), this line appears to be the limit of the secant lines. The slope obtained from the limit of slopes of secant lines is called the *slope of the graph* at $x = 1$, the line through the point $(1, f(1))$ with this slope is called the *tangent line,* and we have solved another basic problem of calculus (see Fig. 1A).

FIGURE 6 Tangent line

MATCHED PROBLEM 3   Given $f(x) = x^2$,

(A) Find the slope of the secant line for $a = 2$ and $h = 2$ and 1, respectively.

(B) Find and simplify the slope of the secant line for $a = 2$ and h any nonzero number.

(C) Find the limit of the expression in part B.

(D) Find the slope of the graph and the slope of the tangent line at $a = 2$.

The ideas introduced in the preceding example are summarized next:

DEFINITION **Slope of a Graph**

Given $y = f(x)$, the **slope of the graph** at the point $(a, f(a))$ is given by

$$\lim_{h \to 0} \frac{f(a + h) - f(a)}{h} \tag{3}$$

provided that the limit exists. The slope of the graph is also the **slope of the tangent line** at the point $(a, f(a))$.

> ### INSIGHT
>
> If the function f is continuous at a, then
>
> $$\lim_{h \to 0} f(a + h) = f(a)$$
>
> and limit (3) will be a 0/0 indeterminate form. As we saw in Examples 2 and 3, evaluation of this type of limit typically involves algebraic simplification.

From plane geometry, we know that a line tangent to a circle is a line that passes through one and only one point of the circle (see Fig. 7A). Although this definition cannot be extended to graphs of functions in general, the visual relationship between graphs of functions and their tangent lines (see Fig. 7B) is similar to the circle case. Limit (3) provides both a mathematically sound definition of a tangent line and a method for approximating the slope of the tangent line.

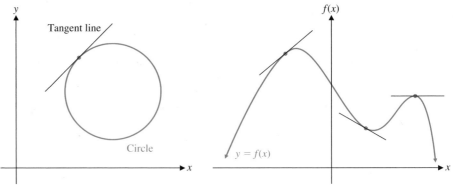

(A) Tangent line for a circle   (B) Tangent lines for the graph of a function

FIGURE 7

■ The Derivative

We have seen that the limit of a difference quotient can be interpreted as a rate of change, as a velocity, or as the slope of a tangent line. In addition, this limit provides solutions to two of the three basic problems stated at the beginning of the chapter. We are now ready to introduce some terms that are used to refer to that limit. To follow customary practice, we use x in place of a and think of the difference quotient

$$\frac{f(x + h) - f(x)}{h}$$

as a function of h, with x held fixed as h tends to 0.

DEFINITION **The Derivative**

For $y = f(x)$, we define the **derivative of f at x,** denoted by $f'(x)$, to be

$$f'(x) = \lim_{h \to 0} \frac{f(x + h) - f(x)}{h} \text{ if the limit exists}$$

If $f'(x)$ exists for each x in the open interval (a, b), then f is said to be **differentiable** over (a, b).

(Differentiability from the left or from the right is defined by using $h \rightarrow 0^-$ or $h \rightarrow 0^+$, respectively, in place of $h \rightarrow 0$ in the preceding definition.)

The process of finding the derivative of a function is called **differentiation.** That is, the derivative of a function is obtained by **differentiating** the function.

SUMMARY | **Interpretations of the Derivative**

The derivative of a function f is a new function f'. The domain of f' is a subset of the domain of f. The derivative has various applications and interpretations, including the following:

1. *Slope of the tangent line.* For each x in the domain of f', $f'(x)$ is the slope of the line tangent to the graph of f at the point $(x, f(x))$.
2. *Instantaneous rate of change.* For each x in the domain of f', $f'(x)$ is the instantaneous rate of change of $y = f(x)$ with respect to x.
3. *Velocity.* If $f(x)$ is the position of a moving object at time x, then $v = f'(x)$ is the velocity of the object at that time.

Example 4 illustrates the **four-step process** that we use to find derivatives in this section. In subsequent sections, we develop rules for finding derivatives that do not involve limits. However, it is important that you master the limit process in order to fully comprehend and appreciate the various applications we will consider.

EXAMPLE 4 | **Finding a Derivative** Find $f'(x)$, the derivative of f at x, for $f(x) = 4x - x^2$.

SOLUTION | To find $f'(x)$, we use a four-step process.

Step 1. Find $f(x + h)$.

$$f(x + h) = 4(x + h) - (x + h)^2$$
$$= 4x + 4h - x^2 - 2xh - h^2$$

Step 2. Find $f(x + h) - f(x)$.

$$f(x + h) - f(x) = 4x + 4h - x^2 - 2xh - h^2 - (4x - x^2)$$
$$= 4h - 2xh - h^2$$

Step 3. Find $\dfrac{f(x + h) - f(x)}{h}$.

$$\frac{f(x + h) - f(x)}{h} = \frac{4h - 2xh - h^2}{h} = \frac{h(4 - 2x - h)}{h}$$
$$= 4 - 2x - h, \quad h \neq 0$$

Step 4. Find $f'(x) = \lim\limits_{h \to 0} \dfrac{f(x + h) - f(x)}{h}$.

$$f'(x) = \lim_{h \to 0} \frac{f(x + h) - f(x)}{h} = \lim_{h \to 0}(4 - 2x - h) = 4 - 2x$$

Thus, if $f(x) = 4x - x^2$, then $f'(x) = 4 - 2x$. The function f' is a new function derived from the function f. ▬

MATCHED PROBLEM 4 | Find $f'(x)$, the derivative of f at x, for $f(x) = 8x - 2x^2$.

The four-step process used in Example 4 is summarized as follows for easy reference:

PROCEDURE | The four-step process for finding the derivative of a function f:

Step 1. Find $f(x + h)$.

Step 2. Find $f(x + h) - f(x)$.

Step 3. Find $\dfrac{f(x + h) - f(x)}{h}$.

Step 4. Find $\lim\limits_{h \to 0} \dfrac{f(x + h) - f(x)}{h}$.

EXAMPLE 5 **Finding Tangent Line Slopes** In Example 4, we started with the function specified by $f(x) = 4x - x^2$ and found the derivative of f at x to be $f'(x) = 4 - 2x$. Thus, the slope of a line tangent to the graph of f at any point $(x, f(x))$ on the graph is

$$m = f'(x) = 4 - 2x$$

(A) Find the slope of the graph of f at $x = 0$, $x = 2$, and $x = 3$.

(B) Graph $y = f(x) = 4x - x^2$, and use the slopes found in part (A) to make a rough sketch of the lines tangent to the graph at $x = 0$, $x = 2$, and $x = 3$.

SOLUTION (A) Using $f'(x) = 4 - 2x$, we have

$$f'(0) = 4 - 2(0) = 4 \qquad \text{Slope at } x = 0$$
$$f'(2) = 4 - 2(2) = 0 \qquad \text{Slope at } x = 2$$
$$f'(3) = 4 - 2(3) = -2 \qquad \text{Slope at } x = 3$$

(B)

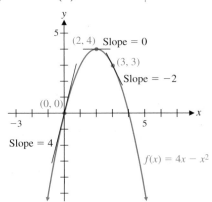

MATCHED PROBLEM 5 In Matched Problem 4, we started with the function specified by $f(x) = 8x - 2x^2$. Using the derivative found there,

(A) Find the slope of the graph of f at $x = 1$, $x = 2$, and $x = 4$.

(B) Graph $y = f(x) = 8x - 2x^2$, and use the slopes from part (A) to make a rough sketch of the lines tangent to the graph at $x = 1$, $x = 2$, and $x = 4$.

Explore & Discuss **2** In Example 4 we found that the derivative of $f(x) = 4x - x^2$ is $f'(x) = 4 - 2x$, and in Example 5 we graphed $f(x)$ and several tangent lines.

(A) Graph f and f' on the same set of axes.

(B) The graph of f' is a straight line. Is it a tangent line for the graph of f? Explain.

(C) Find the x intercept for the graph of f'. What is the slope of the line tangent to the graph of f for this value of x? Write a verbal description of the relationship between the slopes of the tangent lines of a function and the x intercepts of the derivative of the function.

EXAMPLE 6   **Finding a Derivative** Find $f'(x)$, the derivative of f at x, for $f(x) = \sqrt{x} + 2$.

SOLUTION   We use the four-step process to find $f'(x)$.

Step 1. Find $f(x + h)$.

$$f(x + h) = \sqrt{x + h} + 2$$

Step 2. Find $f(x + h) - f(x)$.

$$f(x + h) - f(x) = \sqrt{x + h} + 2 - (\sqrt{x} + 2) \quad \text{Combine like terms.}$$

$$= \sqrt{x + h} - \sqrt{x}$$

Step 3. Find $\dfrac{f(x + h) - f(x)}{h}$.

$$\frac{f(x + h) - f(x)}{h} = \frac{\sqrt{x + h} - \sqrt{x}}{h}$$

We rationalize the numerator (Appendix A, Section A-6) to change the form of this fraction. Combine like terms.

$$= \frac{\sqrt{x + h} - \sqrt{x}}{h} \cdot \frac{\sqrt{x + h} + \sqrt{x}}{\sqrt{x + h} + \sqrt{x}}$$

$$= \frac{x + h - x}{h(\sqrt{x + h} + \sqrt{x})}$$

Cancel.

$$= \frac{h}{h(\sqrt{x + h} + \sqrt{x})}$$

$$= \frac{1}{\sqrt{x + h} + \sqrt{x}} \quad h \neq 0$$

Step 4. Find $f'(x) = \lim\limits_{h \to 0} \dfrac{f(x + h) - f(x)}{h}$.

$$\lim_{h \to 0} \frac{f(x + h) - f(x)}{h} = \lim_{h \to 0} \frac{1}{\sqrt{x + h} + \sqrt{x}}$$

$$= \frac{1}{\sqrt{x} + \sqrt{x}} = \frac{1}{2\sqrt{x}} \quad x > 0$$

Thus, the derivative of $f(x) = \sqrt{x} + 2$ is $f'(x) = 1/(2\sqrt{x})$, a new function. The domain of f is $[0, \infty)$. Since $f'(0)$ is not defined, the domain of f' is $(0, \infty)$, a subset of the domain of f. ▬

MATCHED PROBLEM 6   Find $f'(x)$ for $f(x) = \sqrt{x} + 4$.

EXAMPLE 7   **Sales Analysis** The total sales of a company (in millions of dollars) t months from now are given by $S(t) = \sqrt{t} + 2$. Find and interpret $S(25)$ and $S'(25)$. Use these results to estimate the total sales after 26 months and after 27 months.

SOLUTION The total sales function S has the same form as the function f in Example 6; only the letters used to represent the function and the independent variable have been changed. It follows that S' and f' also have the same form:

$$S(t) = \sqrt{t} + 2 \qquad f(x) = \sqrt{x} + 2$$

$$S'(t) = \frac{1}{2\sqrt{t}} \qquad f'(x) = \frac{1}{2\sqrt{x}}$$

Evaluating S and S' at $t = 25$, we have

$$S(25) = \sqrt{25} + 2 = 7 \qquad S'(25) = \frac{1}{2\sqrt{25}} = 0.1$$

Thus, 25 months from now the total sales will be $7 million and will be increasing at the rate of $0.1 million ($100,000) per month. If this instantaneous rate of change of sales remained constant, the sales would grow to $7.1 million after 26 months, $7.2 million after 27 months, and so on. Even though $S'(t)$ is not a constant function in this case, these values provide useful estimates of the total sales. ■

MATCHED PROBLEM 7 The total sales of a company (in millions of dollars) t months from now are given by $S(t) = \sqrt{t} + 4$. Find and interpret $S(12)$ and $S'(12)$. Use these results to estimate the total sales after 13 months and after 14 months. (Use the derivative found in Matched Problem 6.)

In Example 7, it is instructive to compare the estimates of total sales obtained by using the derivative with the corresponding exact values of $S(t)$:

	Exact values	Estimated values
$S(26) =$	$\sqrt{26} + 2 = 7.099\ldots$	≈ 7.1
$S(27) =$	$\sqrt{27} + 2 = 7.196\ldots$	≈ 7.2

For this function, the estimated values provide very good approximations to the exact values of $S(t)$. For other functions, the approximation might not be as accurate.

Using the instantaneous rate of change of a function at a point to estimate values of the function at nearby points is a simple, but important, application of the derivative.

■ Nonexistence of the Derivative

The existence of a derivative at $x = a$ depends on the existence of a limit at $x = a$, that is, on the existence of

$$f'(a) = \lim_{h \to 0} \frac{f(a + h) - f(a)}{h} \tag{4}$$

If the limit does not exist at $x = a$, we say that the function f is **nondifferentiable at** $x = a$, or $f'(a)$ **does not exist.**

Explore & Discuss 3 Let $f(x) = |x - 1|$.

(A) Graph f.

(B) Complete the following table:

h	-0.1	-0.01	$-0.001 \to 0 \leftarrow$	0.001	0.01	0.1
$\dfrac{f(1 + h) - f(1)}{h}$	?	?	? $\;\to$? $\leftarrow$	?	?	?

(C) Find the following limit if it exists:

$$\lim_{h \to 0} \frac{f(1 + h) - f(1)}{h}$$

(D) Use the results of parts (A)–(C) to discuss the existence of $f'(1)$.

Repeat parts (A)–(D) for $g(x) = \sqrt[3]{x - 1}$.

How can we recognize the points on the graph of f where $f'(a)$ does not exist? It is impossible to describe all the ways that the limit of a difference quotient can fail to exist. However, we can illustrate some common situations where $f'(a)$ does fail to exist (see Fig. 8):

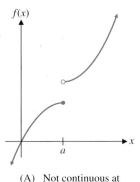

(A) Not continuous at $x = a$

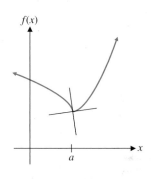

(B) Graph has sharp corner at $x = a$

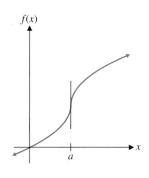

(C) Vertical tangent at $x = a$

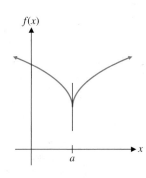

(D) Vertical tangent at $x = a$

FIGURE 8 The function f is nondifferentiable at $x = a$.

1. If the graph of f has a hole or a break at $x = a$, then $f'(a)$ does not exist (Fig. 8A).

2. If the graph of f has a sharp corner at $x = a$, then $f'(a)$ does not exist and the graph has no tangent line at $x = a$ (Fig. 8B). (In Fig. 8B, the left- and right-hand derivatives exist, but are not equal.)

3. If the graph of f has a vertical tangent line at $x = a$, then $f'(a)$ does not exist (Fig. 8C and D).

Answers to Matched Problems

1. (A) −$1,600   (B) −$8 per planter
2. (A) 48 ft/s   (B) $32 + 16h$ (C) 32 ft/s
3. (A) 6, 5   (B) $4 + h$
   (C) 4   (D) Both are 4
4. $f'(x) = 8 - 4x$
5. (A) $f'(1) = 4, f'(2) = 0, f'(4) = -8$
   (B)

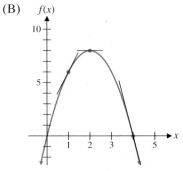

6. $f'(x) = 1/(2\sqrt{x} + 4)$
7. $S(12) = 4, S'(12) = 0.125$; 12 months from now, the total sales will be $4 million and will be increasing at the rate of $0.125 million ($125,000) per month. The estimated total sales are $4.125 million after 13 months and $4.25 million after 14 months.

Exercise 3-4

A *In Problems 1 and 2, find the indicated quantity for*
$y = f(x) = 5 - x^2$, *and interpret that quantity in terms of the*
following graph.

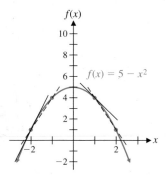

1. (A) $\dfrac{f(2) - f(1)}{2 - 1}$ (B) $\dfrac{f(1 + h) - f(1)}{h}$

  (C) $\lim\limits_{h \to 0} \dfrac{f(1 + h) - f(1)}{h}$

2. (A) $\dfrac{f(-1) - f(-2)}{-1 - (-2)}$ (B) $\dfrac{f(-2 + h) - f(-2)}{h}$

  (C) $\lim\limits_{h \to 0} \dfrac{f(-2 + h) - f(-2)}{h}$

3. Find the indicated quantities for $f(x) = 3x^2$.

  (A) The average rate of change of $f(x)$ if x changes from 1 to 4.

  (B) The slope of the secant line through the points $(1, f(1))$ and $(4, f(4))$ on the graph of $y = f(x)$.

  (C) The slope of the secant line through the points $(1, f(1))$ and $(1 + h, f(1 + h))$, $h \neq 0$. Simplify your answer.

  (D) The slope of the graph at $(1, f(1))$.

  (E) The instantaneous rate of change of $y = f(x)$ with respect to x at $x = 1$.

  (F) The slope of the tangent line at $(1, f(1))$.

  (G) The equation of the tangent line at $(1, f(1))$.

4. Find the indicated quantities for $f(x) = 3x^2$.

  (A) The average rate of change of $f(x)$ if x changes from 2 to 5.

  (B) The slope of the secant line through the points $(2, f(2))$ and $(5, f(5))$ on the graph of $y = f(x)$.

  (C) The slope of the secant line through the points $(2, f(2))$ and $(2 + h, f(2 + h))$, $h \neq 0$. Simplify your answer.

  (D) The slope of the graph at $(2, f(2))$.

  (E) The instantaneous rate of change of $y = f(x)$ with respect to x at $x = 2$.

  (F) The slope of the tangent line at $(2, f(2))$.

  (G) The equation of the tangent line at $(2, f(2))$.

In Problems 5–26, use the four-step process to find $f'(x)$ and
then find $f'(1), f'(2)$, and $f'(3)$.

5. $f(x) = -5$ **6.** $f(x) = 9$

7. $f(x) = 3x - 7$ **8.** $f(x) = 4 - 6x$

9. $f(x) = 2 - 3x^2$ **10.** $f(x) = 2x^2 + 8$

11. $f(x) = x^2 + 6x - 10$ **12.** $f(x) = x^2 + 4x + 7$

13. $f(x) = 2x^2 - 7x + 3$ **14.** $f(x) = 2x^2 + 5x + 1$

15. $f(x) = -x^2 + 4x - 9$ **16.** $f(x) = -x^2 + 9x - 2$

17. $f(x) = 2x^3 + 1$ **18.** $f(x) = -2x^3 + 5$

19. $f(x) = 4 + \dfrac{4}{x}$ **20.** $f(x) = \dfrac{6}{x} - 2$

21. $f(x) = 5 + 3\sqrt{x}$ **22.** $f(x) = 3 - 7\sqrt{x}$

23. $f(x) = 10\sqrt{x + 5}$ **24.** $f(x) = 16\sqrt{x + 9}$

25. $f(x) = \dfrac{3x}{x + 2}$ **26.** $f(x) = \dfrac{5x}{3 + x}$

B *Problems 27 and 28 refer to the graph of $y = f(x) = x^2 + x$*
shown.

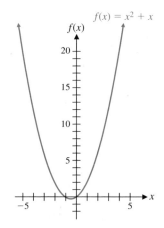

27. (A) Find the slope of the secant line joining $(1, f(1))$ and $(3, f(3))$.

  (B) Find the slope of the secant line joining $(1, f(1))$ and $(1 + h, f(1 + h))$.

  (C) Find the slope of the tangent line at $(1, f(1))$.

  (D) Find the equation of the tangent line at $(1, f(1))$.

28. (A) Find the slope of the secant line joining $(2, f(2))$ and $(4, f(4))$.

  (B) Find the slope of the secant line joining $(2, f(2))$ and $(2 + h, f(2 + h))$.

  (C) Find the slope of the tangent line at $(2, f(2))$.

  (D) Find the equation of the tangent line at $(2, f(2))$.

In Problems 29 and 30, suppose an object moves along the y
axis so that its location is $y = f(x) = x^2 + x$ at time x (y is in
meters and x is in seconds). Find

29. (A) The average velocity (the average rate of change of y with respect to x) for x changing from 1 to 3 seconds

  (B) The average velocity for x changing from 1 to $1 + h$ seconds

  (C) The instantaneous velocity at $x = 1$ second

30. (A) The average velocity (the average rate of change of y with respect to x) for x changing from 2 to 4 seconds

  (B) The average velocity for x changing from 2 to $2 + h$ seconds

  (C) The instantaneous velocity at $x = 2$ seconds

Problems 31–38 refer to the function F in the graph shown. Use the graph to determine whether F′(x) exists at each indicated value of x.

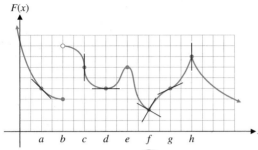

31. $x = a$

32. $x = b$

33. $x = c$

34. $x = d$

35. $x = e$

36. $x = f$

37. $x = g$

38. $x = h$

39. Given $f(x) = x^2 - 4x$,

(A) Find $f'(x)$.

(B) Find the slopes of the lines tangent to the graph of f at $x = 0, 2,$ and 4.

(C) Graph f, and sketch in the tangent lines at $x = 0, 2,$ and 4.

40. Given $f(x) = x^2 + 2x$,

(A) Find $f'(x)$.

(B) Find the slopes of the lines tangent to the graph of f at $x = -2, -1,$ and 1.

(C) Graph f, and sketch in the tangent lines at $x = -2, -1,$ and 1.

41. If an object moves along a line so that it is at $y = f(x) = 4x^2 - 2x$ at time x (in seconds), find the instantaneous velocity function $v = f'(x)$, and find the velocity at times $x = 1, 3,$ and 5 seconds (y is measured in feet).

42. Repeat Problem 41 with $f(x) = 8x^2 - 4x$.

43. Let $f(x) = x^2$, $g(x) = x^2 - 1$, and $h(x) = x^2 + 2$.

(A) How are the graphs of these functions related? How would you expect the derivatives of these functions to be related?

(B) Use the four-step process to find the derivative of $m(x) = x^2 + C$, where C is any real constant.

44. Let $f(x) = -x^2$, $g(x) = -x^2 - 1$, and $h(x) = -x^2 + 2$.

(A) How are the graphs of these functions related? How would you expect the derivatives of these functions to be related?

(B) Use the four-step process to find the derivative of $m(x) = -x^2 + C$, where C is any real constant.

45. (A) Give a geometric explanation of the following statement: If $f(x) = C$ is a constant function, then $f'(x) = 0$.

(B) Use the four-step process to verify the statement in part (A).

46. (A) Give a geometric explanation of the following statement: If $f(x) = mx + b$ is a linear function, then $f'(x) = m$.

(B) Use the four-step process to verify the statement in part (A).

C *In Problems 47–50, sketch the graph of f and determine where f is nondifferentiable.*

47. $f(x) = \begin{cases} 2x & \text{if } x < 1 \\ 2 & \text{if } x \geq 1 \end{cases}$

48. $f(x) = \begin{cases} 2x & \text{if } x < 2 \\ 6 - x & \text{if } x \geq 2 \end{cases}$

49. $f(x) = \begin{cases} x^2 + 1 & \text{if } x < 0 \\ 1 & \text{if } x \geq 0 \end{cases}$

50. $f(x) = \begin{cases} 2 - x^2 & \text{if } x \leq 0 \\ 2 & \text{if } x > 0 \end{cases}$

In Problems 51–56, determine whether f is differentiable at $x = 0$ by considering

$$\lim_{h \to 0} \frac{f(0 + h) - f(0)}{h}$$

51. $f(x) = |x|$

52. $f(x) = 1 - |x|$

53. $f(x) = x^{1/3}$

54. $f(x) = x^{2/3}$

55. $f(x) = \sqrt{1 - x^2}$

56. $f(x) = \sqrt{1 + x^2}$

57. A ball dropped from a balloon falls $y = 16x^2$ feet in x seconds. If the balloon is 576 feet above the ground when the ball is dropped, when does the ball hit the ground? What is the velocity of the ball at the instant it hits the ground?

58. Repeat Problem 57 if the balloon is 1,024 feet above the ground when the ball is dropped.

Applications

59. *Revenue.* The revenue (in dollars) from the sale of x car seats for infants is given by

$$R(x) = 60x - 0.025x^2 \qquad 0 \leq x \leq 2,400$$

(A) Find the average change in revenue if production is changed from 1,000 car seats to 1,050 car seats.

(B) Use the four-step process to find $R'(x)$.

(C) Find the revenue and the instantaneous rate of change of revenue at a production level of 1,000 car seats, and write a brief verbal interpretation of these results.

60. *Profit.* The profit (in dollars) from the sale of x car seats for infants is given by

$$P(x) = 45x - 0.025x^2 - 5,000 \qquad 0 \leq x \leq 2,400$$

(A) Find the average change in profit if production is changed from 800 car seats to 850 car seats.

(B) Use the four-step process to find $P'(x)$.

(C) Find the profit and the instantaneous rate of change of profit at a production level of 800 car seats, and write a brief verbal interpretation of these results.

61. *Sales analysis.* The total sales of a company (in millions of dollars) t months from now are given by

$$S(t) = 2\sqrt{t} + 10$$

(A) Use the four-step process to find $S'(t)$.

(B) Find $S(15)$ and $S'(15)$. Write a brief verbal interpretation of these results.

(C) Use the results in part (B) to estimate the total sales after 16 months and after 17 months.

62. *Sales analysis.* The total sales of a company (in millions of dollars) t months from now are given by

$$S(t) = 2\sqrt{t} + 6$$

(A) Use the four-step process to find $S'(t)$.

(B) Find $S(10)$ and $S'(10)$. Write a brief verbal interpretation of these results.

(C) Use the results in part (B) to estimate the total sales after 11 months and after 12 months.

63. *Mineral production.* The U.S. production of zinc (in thousands of metric tons) is given approximately by

$$p(t) = 14t^2 - 6.6t + 602.4$$

where t is time in years and $t = 0$ corresponds to 1995.

(A) Use the four-step process to find $p'(t)$.

(B) Find the annual production in 2010 and the instantaneous rate of change of production in 2010, and write a brief verbal interpretation of these results.

64. *Mineral consumption.* The U.S. consumption of copper (in thousands of metric tons) is given approximately by

$$p(t) = 27t^2 - 75t + 6{,}015$$

where t is time in years and $t = 0$ corresponds to 1990.

(A) Use the four-step process to find $p'(t)$.

(B) Find the annual production in 2010 and the instantaneous rate of change of production in 2010, and write a brief verbal interpretation of these results.

65. *Electricity consumption.* Table 1 gives the retail sales of electricity (in billions of kilowatt-hours) for the residential and commercial sectors in the United States during the 1990s. (*Source:* Energy Information Administration)

TABLE 1 Electricity Sales

Year	Residential	Commercial
1996	1,083	887
1997	1,076	926
1998	1,130	979
1999	1,145	1,002
2000	1,192	1,055
2001	1,202	1,083
2002	1,265	1,104
2003	1,276	1,199
2004	1,292	1,230
2005	1,359	1,275

(A) Let x represent time (in years) with $x = 0$ corresponding to 1996, and let y represent the corresponding residential sales. Enter the appropriate data set in a graphing calculator and find a quadratic regression equation for the data.

(B) If $y = R(x)$ denotes the regression equation found in part (A), find $R(20)$ and $R'(20)$, and write a brief verbal interpretation of these results. Round answers to the nearest tenth of a billion.

 66. *Electricity consumption.* Refer to the data in Table 1.

(A) Let x represent time (in years) with $x = 0$ corresponding to 1996, and let y represent the corresponding commercial sales. Enter the appropriate data set in a graphing calculator and find a quadratic regression equation for the data.

(B) If $y = C(x)$ denotes the regression equation found in part (A), find $C(20)$ and $C'(20)$, and write a brief verbal interpretation of these results. Round answers to the nearest tenth of a billion.

67. *Air pollution.* The ozone level (in parts per billion) on a summer day in a metropolitan area is given by

$$P(t) = 80 + 12t - t^2$$

where t is time in hours and $t = 0$ corresponds to 9 A.M.

(A) Use the four-step process to find $P'(t)$.

(B) Find $P(3)$ and $P'(3)$. Write a brief verbal interpretation of these results.

68. *Medicine.* The body temperature (in degrees Fahrenheit) of a patient t hours after being given a fever-reducing drug is given by

$$F(t) = 98 + \frac{4}{t + 1} \, .$$

(A) Use the four-step process to find $F'(t)$.

(B) Find $F(3)$ and $F'(3)$. Write a brief verbal interpretation of these results.

69. *Infant mortality.* The number of infant deaths per 100,000 births for males in the United States is given approximately by

$$f(t) = 0.008t^2 - 0.5t + 14.4$$

where t is time in years and $t = 0$ corresponds to 1980.

(A) Use the four-step process to find $f'(t)$.

(B) Find the number of male infant deaths in 2010 and the instantaneous rate of change of the number of male infant deaths in 2010, and write a brief verbal interpretation of these results.

70. *Infant mortality.* The number of infant deaths per 100,000 births for females in the United States is given approximately by

$$f(t) = 0.007t^2 - 0.4t + 11.6$$

where t is time in years and $t = 0$ corresponds to 1980.

(A) Use the four-step process to find $f'(t)$.

(B) Find the number of female infant deaths in 2010 and the instantaneous rate of change of the number of female infant deaths in 2010, and write a brief verbal interpretation of these results.

Section 3-5   BASIC DIFFERENTIATION PROPERTIES

- Constant Function Rule
- Power Rule
- Constant Multiple Property
- Sum and Difference Properties
- Applications

In the preceding section, we defined the derivative of f at x as

$$f'(x) = \lim_{h \to 0} \frac{f(x + h) - f(x)}{h}$$

if the limit exists, and we used this definition and a four-step process to find the derivatives of several functions. Now we want to develop some rules of differentiation. These rules will enable us to find the derivative of a great many functions without using the four-step process.

Before starting on these rules, we list some symbols that are widely used to represent derivatives.

NOTATION   **The Derivative**

If $y = f(x)$, then

$$f'(x) \qquad y' \qquad \frac{dy}{dx}$$

all represent the derivative of f at x.

Each of these symbols for the derivative has its particular advantage in certain situations. All of them will become familiar to you after a little experience.

■ Constant Function Rule

If $f(x) = C$ is a constant function, then the four-step process can be used to show that $f'(x) = 0$ (see Problem 45 in Exercise 3-4). Thus,

The derivative of any constant function is 0.

THEOREM 1   **CONSTANT FUNCTION RULE**

If $y = f(x) = C$, then

$$f'(x) = 0$$

Also, $y' = 0$ and $dy/dx = 0$.

Note: When we write $C' = 0$ or $\dfrac{d}{dx}C = 0$, we mean that $y' = \dfrac{dy}{dx} = 0$ when $y = C$.

INSIGHT

The graph of $f(x) = C$ is a horizontal line with slope 0 (see Fig. 1), so we would expect that $f'(x) = 0$.

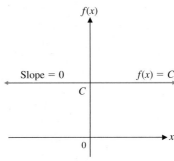

FIGURE 1

| EXAMPLE 1 | **Differentiating Constant Functions** |

(A) If $f(x) = 3$, then $f'(x) = 0$.   (B) If $y = -1.4$, then $y' = 0$.

(C) If $y = \pi$, then $\dfrac{dy}{dx} = 0$.   (D) $\dfrac{d}{dx}23 = 0$

| MATCHED PROBLEM 1 | Find |

(A) $f'(x)$ for $f(x) = -24$ (B) y' for $y = 12$

(C) $\dfrac{dy}{dx}$ for $y = -\sqrt{7}$ (D) $\dfrac{d}{dx}(-\pi)$

■ Power Rule

A function of the form $f(x) = x^k$, where k is a real number, is called a **power function.** The following elementary functions (see the inside front cover) are examples of power functions:

$$f(x) = x \qquad h(x) = x^2 \qquad m(x) = x^3$$
$$n(x) = \sqrt{x} \qquad p(x) = \sqrt[3]{x} \tag{1}$$

Explore & Discuss 1

(A) It is clear that the functions f, h, and m in (1) are power functions. Explain why the functions n and p are also power functions.

(B) The domain of a power function depends on the power. Discuss the domain of each of the following power functions:

$$r(x) = x^4 \qquad s(x) = x^{-4} \qquad t(x) = x^{1/4}$$
$$u(x) = x^{-1/4} \qquad v(x) = x^{1/5} \qquad w(x) = x^{-1/5}$$

The definition of the derivative and the four-step process introduced in the preceding section can be used to find the derivatives of many power functions. For example, it can be shown that

If	$f(x) = x^2$,	then	$f'(x) = 2x$.
If	$f(x) = x^3$,	then	$f'(x) = 3x^2$.
If	$f(x) = x^4$,	then	$f'(x) = 4x^3$.
If	$f(x) = x^5$,	then	$f'(x) = 5x^4$.

Notice the pattern in these derivatives. In each case, the power in f becomes the coefficient in f' and the power in f' is 1 less than the power in f. In general, for any positive integer n,

$$\text{If} \quad f(x) = x^n, \quad \text{then} \quad f'(x) = nx^{n-1}. \tag{2}$$

In fact, more advanced techniques can be used to show that (2) holds for *any* real number n. We will assume this general result for the remainder of the book.

| THEOREM 2 | **POWER RULE** |

If $y = f(x) = x^n$, where n is a real number, then

$$f'(x) = nx^{n-1}$$

Also, $y' = nx^{n-1}$ and $dy/dx = nx^{n-1}$.

Explore & Discuss 2

(A) Write a verbal description of the power rule.

(B) If $f(x) = x$, what is $f'(x)$? Discuss how this derivative can be obtained from the power rule.

EXAMPLE 2

Differentiating Power Functions

(A) If $f(x) = x^5$, then $f'(x) = 5x^{5-1} = 5x^4$.

(B) If $y = x^{25}$, then $y' = 25x^{25-1} = 25x^{24}$.

(C) If $y = t^{-3}$, then $\dfrac{dy}{dt} = -3t^{-3-1} = -3t^{-4} = -\dfrac{3}{t^4}$.

(D) $\dfrac{d}{dx} x^{5/3} = \dfrac{5}{3} x^{(5/3)-1} = \dfrac{5}{3} x^{2/3}$.

MATCHED PROBLEM 2

Find

(A) $f'(x)$ for $f(x) = x^6$ (B) y' for $y = x^{30}$

(C) $\dfrac{dy}{dt}$ for $y = t^{-2}$ (D) $\dfrac{d}{dx} x^{3/2}$

In some cases, properties of exponents must be used to rewrite an expression before the power rule is applied.

EXAMPLE 3

Differentiating Power Functions

(A) If $f(x) = 1/x^4$, we can write $f(x) = x^{-4}$ and

$$f'(x) = -4x^{-4-1} = -4x^{-5}, \quad \text{or} \quad \dfrac{-4}{x^5}$$

(B) If $y = \sqrt{u}$, we can write $y = u^{1/2}$ and

$$y' = \dfrac{1}{2} u^{(1/2)-1} = \dfrac{1}{2} u^{-1/2}, \quad \text{or} \quad \dfrac{1}{2\sqrt{u}}$$

(C) $\dfrac{d}{dx} \dfrac{1}{\sqrt[3]{x}} = \dfrac{d}{dx} x^{-1/3} = -\dfrac{1}{3} x^{(-1/3)-1} = -\dfrac{1}{3} x^{-4/3}, \quad \text{or} \quad \dfrac{-1}{3\sqrt[3]{x^4}}$

MATCHED PROBLEM 3

Find

(A) $f'(x)$ for $f(x) = \dfrac{1}{x}$ (B) y' for $y = \sqrt[3]{u^2}$ (C) $\dfrac{d}{dx} \dfrac{1}{\sqrt{x}}$

■ Constant Multiple Property

Let $f(x) = ku(x)$, where k is a constant and u is differentiable at x. Then, using the four-step process, we have the following:

Step 1. $f(x + h) = ku(x + h)$

Step 2. $f(x + h) - f(x) = ku(x + h) - ku(x) = k[u(x + h) - u(x)]$

Step 3. $\dfrac{f(x + h) - f(x)}{h} = \dfrac{k[u(x + h) - u(x)]}{h} = k\left[\dfrac{u(x + h) - u(x)}{h}\right]$

Step 4. $f'(x) = \displaystyle\lim_{h \to 0} \dfrac{f(x + h) - f(x)}{h}$

$\qquad\qquad = \displaystyle\lim_{h \to 0} k\left[\dfrac{u(x + h) - u(x)}{h}\right]$ $\displaystyle\lim_{x \to c} kg(x) = k \lim_{x \to c} g(x)$

$\qquad\qquad = k \displaystyle\lim_{h \to 0}\left[\dfrac{u(x + h) - u(x)}{h}\right]$ Definition of $u'(x)$

$\qquad\qquad = ku'(x)$

Thus,

> The derivative of a constant times a differentiable function is the constant times the derivative of the function.

THEOREM 3 CONSTANT MULTIPLE PROPERTY

If $y = f(x) = ku(x)$, then

$$f'(x) = ku'(x)$$

Also,

$$y' = ku' \qquad \frac{dy}{dx} = k\frac{du}{dx}$$

EXAMPLE 4 **Differentiating a Constant Times a Function**

(A) If $f(x) = 3x^2$, then $f'(x) = 3 \cdot 2x^{2-1} = 6x$.

(B) If $y = \dfrac{t^3}{6} = \dfrac{1}{6}t^3$, then $\dfrac{dy}{dt} = \dfrac{1}{6} \cdot 3t^{3-1} = \dfrac{1}{2}t^2$.

(C) If $y = \dfrac{1}{2x^4} = \dfrac{1}{2}x^{-4}$, then $y' = \dfrac{1}{2}(-4x^{-4-1}) = -2x^{-5}$, or $\dfrac{-2}{x^5}$.

(D) $\dfrac{d}{dx}\dfrac{0.4}{\sqrt{x^3}} = \dfrac{d}{dx}\dfrac{0.4}{x^{3/2}} = \dfrac{d}{dx}0.4x^{-3/2} = 0.4\left[-\dfrac{3}{2}x^{(-3/2)-1}\right]$

$$= -0.6x^{-5/2}, \quad \text{or} \quad -\dfrac{0.6}{\sqrt{x^5}}$$

MATCHED PROBLEM 4 Find

(A) $f'(x)$ for $f(x) = 4x^5$ (B) $\dfrac{dy}{dt}$ for $y = \dfrac{t^4}{12}$

(C) y' for $y = \dfrac{1}{3x^3}$ (D) $\dfrac{d}{dx}\dfrac{0.9}{\sqrt[3]{x}}$

■ Sum and Difference Properties

Let $f(x) = u(x) + v(x)$, where $u'(x)$ and $v'(x)$ exist. Then, using the four-step process, we have the following:

Step 1. $f(x + h) = u(x + h) + v(x + h)$

Step 2. $f(x + h) - f(x) = u(x + h) + v(x + h) - [u(x) + v(x)]$

$$= u(x + h) - u(x) + v(x + h) - v(x)$$

Step 3. $\dfrac{f(x + h) - f(x)}{h} = \dfrac{u(x + h) - u(x) + v(x + h) - v(x)}{h}$

$$= \dfrac{u(x + h) - u(x)}{h} + \dfrac{v(x + h) - v(x)}{h}$$

Step 4. $f'(x) = \lim\limits_{h \to 0} \dfrac{f(x + h) - f(x)}{h}$

$$= \lim_{h \to 0}\left[\dfrac{u(x + h) - u(x)}{h} + \dfrac{v(x + h) - v(x)}{h}\right]$$

$$\lim_{x \to c}[g(x) + h(x)] = \lim_{x \to c} g(x) + \lim_{x \to c} h(x)$$

$$= \lim_{h \to 0} \dfrac{u(x + h) - u(x)}{h} + \lim_{h \to 0} \dfrac{v(x + h) - v(x)}{h}$$

$$= u'(x) + v'(x)$$

Thus,

> The derivative of the sum of two differentiable functions is the sum of the derivatives of the functions.

Similarly, we can show that

> The derivative of the difference of two differentiable functions is the difference of the derivatives of the functions.

Together, we then have the **sum and difference property** for differentiation:

THEOREM 4   SUM AND DIFFERENCE PROPERTY

If $y = f(x) = u(x) \pm v(x)$, then

$$f'(x) = u'(x) \pm v'(x)$$

Also,

$$y' = u' \pm v' \qquad \frac{dy}{dx} = \frac{du}{dx} \pm \frac{dv}{dx}$$

Note: This rule generalizes to the sum and difference of any given number of functions.

With the preceding rule and the other rules stated previously, we will be able to compute the derivatives of all polynomials and a variety of other functions.

EXAMPLE 5   **Differentiating Sums and Differences**

(A) If $f(x) = 3x^2 + 2x$, then

$$f'(x) = (3x^2)' + (2x)' = 3(2x) + 2(1) = 6x + 2$$

(B) If $y = 4 + 2x^3 - 3x^{-1}$, then

$$y' = (4)' + (2x^3)' - (3x^{-1})' = 0 + 2(3x^2) - 3(-1)x^{-2} = 6x^2 + 3x^{-2}$$

(C) If $y = \sqrt[3]{w} - 3w$, then

$$\frac{dy}{dw} = \frac{d}{dw}w^{1/3} - \frac{d}{dw}3w = \frac{1}{3}w^{-2/3} - 3 = \frac{1}{3w^{2/3}} - 3$$

(D) $\dfrac{d}{dx}\left(\dfrac{5}{3x^2} - \dfrac{2}{x^4} + \dfrac{x^3}{9} \right) = \dfrac{d}{dx}\dfrac{5}{3}x^{-2} - \dfrac{d}{dx}2x^{-4} + \dfrac{d}{dx}\dfrac{1}{9}x^3$

$$= \frac{5}{3}(-2)x^{-3} - 2(-4)x^{-5} + \frac{1}{9}\cdot 3x^2$$

$$= -\frac{10}{3x^3} + \frac{8}{x^5} + \frac{1}{3}x^2$$

MATCHED PROBLEM 5   Find

(A) $f'(x)$ for $f(x) = 3x^4 - 2x^3 + x^2 - 5x + 7$

(B) y' for $y = 3 - 7x^{-2}$

(C) $\dfrac{dy}{dv}$ for $y = 5v^3 - \sqrt[4]{v}$

(D) $\dfrac{d}{dx}\left(-\dfrac{3}{4x} + \dfrac{4}{x^3} - \dfrac{x^4}{8} \right)$

APPLICATIONS

| EXAMPLE 6 | **Instantaneous Velocity** An object moves along the y axis (marked in feet) so that its position at time x (in seconds) is |

$$f(x) = x^3 - 6x^2 + 9x$$

(A) Find the instantaneous velocity function v.

(B) Find the velocity at $x = 2$ and $x = 5$ seconds.

(C) Find the time(s) when the velocity is 0.

| SOLUTION | (A) $v = f'(x) = (x^3)' - (6x^2)' + (9x)' = 3x^2 - 12x + 9$ |

(B) $f'(2) = 3(2)^2 - 12(2) + 9 = -3$ feet per second

$f'(5) = 3(5)^2 - 12(5) + 9 = 24$ feet per second

(C) $v = f'(x) = 3x^2 - 12x + 9 = 0$ *Factor 3 out of each term.*

$3(x^2 - 4x + 3) = 0$ *Factor the quadratic term.*

$3(x - 1)(x - 3) = 0$ *Use the zero properly.*

$x = 1, 3$

Thus, $v = 0$ at $x = 1$ and $x = 3$ seconds.

| MATCHED PROBLEM 6 | Repeat Example 6 for $f(x) = x^3 - 15x^2 + 72x$. |

| EXAMPLE 7 | **Tangents** Let $f(x) = x^4 - 6x^2 + 10$. |

(A) Find $f'(x)$.

(B) Find the equation of the tangent line at $x = 1$.

(C) Find the values of x where the tangent line is horizontal.

| SOLUTION | (A) $f'(x) = (x^4)' - (6x^2)' + (10)'$ |

$= 4x^3 - 12x$

(B) $y - y_1 = m(x - x_1)$ $y_1 = f(x_1) = f(1) = (1)^4 - 6(1)^2 + 10 = 5$

$y - 5 = -8(x - 1)$ $m = f'(x_1) = f'(1) = 4(1)^3 - 12(1) = -8$

$y = -8x + 13$ *Tangent line at* $x = 1$

(C) Since a horizontal line has 0 slope, we must solve $f'(x) = 0$ for x:

$f'(x) = 4x^3 - 12x = 0$ *Factor 4x out of each term.*

$4x(x^2 - 3) = 0$ *Factor the difference of two squares.*

$4x(x + \sqrt{3})(x - \sqrt{3}) = 0$ *Use the zero properly.*

$x = 0, -\sqrt{3}, \sqrt{3}$

| MATCHED PROBLEM 7 | Repeat Example 7 for $f(x) = x^4 - 8x^3 + 7$. |

Answers to Matched Problems

1. All are 0.

2. (A) $6x^5$ (B) $30x^{29}$ (C) $-2t^{-3} = -2/t^3$ (D) $\frac{3}{2}x^{1/2}$

3. (A) $-x^{-2}$, or $-1/x^2$ (B) $\frac{2}{3}u^{-1/3}$, or $2/(3\sqrt[3]{u})$ (C) $-\frac{1}{2}x^{-3/2}$, or $-1/(2\sqrt{x^3})$

4. (A) $20x^4$ (B) $t^3/3$ (C) $-x^{-4}$, or $-1/x^4$ (D) $-0.3x^{-4/3}$, or $-0.3/\sqrt[3]{x^4}$

5. (A) $12x^3 - 6x^2 + 2x - 5$ (B) $14x^{-3}$, or $14/x^3$

     (C) $15v^2 - \frac{1}{4}v^{-3/4}$, or $15v^2 - 1/(4v^{3/4})$ (D) $3/(4x^2) - (12/x^4) - (x^3/2)$

6. (A) $v = 3x^2 - 30x + 72$

(B) $f'(2) = 24$ ft/s; $f'(5) = -3$ ft/s

(C) $x = 4$ and $x = 6$ seconds

7. (A) $f'(x) = 4x^3 - 24x^2$

(B) $y = -20x + 20$

(C) $x = 0$ and $x = 6$

Exercise 3-5

A *Find the indicated derivatives in Problems 1–18.*

1. $f'(x)$ for $f(x) = 7$

2. $\dfrac{d}{dx}3$

3. $\dfrac{dy}{dx}$ for $y = x^9$

4. y' for $y = x^6$

5. $\dfrac{d}{dx}x^3$

6. $g'(x)$ for $g(x) = x^5$

7. y' for $y = x^{-4}$

8. $\dfrac{dy}{dx}$ for $y = x^{-8}$

9. $g'(x)$ for $g(x) = x^{8/3}$

10. $f'(x)$ for $f(x) = x^{9/2}$

11. $\dfrac{dy}{dx}$ for $y = \dfrac{1}{x^{10}}$

12. y' for $y = \dfrac{1}{x^{12}}$

13. $f'(x)$ for $f(x) = 5x^2$

14. $\dfrac{d}{dx}(-2x^3)$

15. y' for $y = 0.4x^7$

16. $f'(x)$ for $f(x) = 0.8x^4$

17. $\dfrac{d}{dx}\left(\dfrac{x^3}{18}\right)$

18. $\dfrac{dy}{dx}$ for $y = \dfrac{x^5}{25}$

Problems 19–24 refer to functions f and g that satisfy $f'(2) = 3$ and $g'(2) = -1$. In each problem, find $h'(2)$ for the indicated function h.

19. $h(x) = 4f(x)$

20. $h(x) = 5g(x)$

21. $h(x) = f(x) + g(x)$

22. $h(x) = g(x) - f(x)$

23. $h(x) = 2f(x) - 3g(x) + 7$

24. $h(x) = -4f(x) + 5g(x) - 9$

B *Find the indicated derivatives in Problems 25–48.*

25. $\dfrac{d}{dx}(2x - 5)$

26. $\dfrac{d}{dx}(-4x + 9)$

27. $f'(t)$ if $f(t) = 2t^2 - 3t + 1$

28. $\dfrac{dy}{dt}$ if $y = 2 + 5t - 8t^3$

29. y' for $y = 5x^{-2} + 9x^{-1}$

30. $g'(x)$ if $g(x) = 5x^{-7} - 2x^{-4}$

31. $\dfrac{d}{du}(5u^{0.3} - 4u^{2.2})$

32. $\dfrac{d}{du}(2u^{4.5} - 3.1u + 13.2)$

33. $h'(t)$ if $h(t) = 2.1 + 0.5t - 1.1t^3$

34. $F'(t)$ if $F(t) = 0.2t^3 - 3.1t + 13.2$

35. y' if $y = \dfrac{2}{5x^4}$

36. w' if $w = \dfrac{7}{5u^2}$

37. $\dfrac{d}{dx}\left(\dfrac{3x^2}{2} - \dfrac{7}{5x^2}\right)$

38. $\dfrac{d}{dx}\left(\dfrac{5x^3}{4} - \dfrac{2}{5x^3}\right)$

39. $G'(w)$ if $G(w) = \dfrac{5}{9w^4} + 5\sqrt[3]{w}$

40. $H'(w)$ if $H(w) = \dfrac{5}{w^6} - 2\sqrt{w}$

41. $\dfrac{d}{du}(3u^{2/3} - 5u^{1/3})$

42. $\dfrac{d}{du}(8u^{3/4} + 4u^{-1/4})$

43. $h'(t)$ if $h(t) = \dfrac{3}{t^{3/5}} - \dfrac{6}{t^{1/2}}$

44. $F'(t)$ if $F(t) = \dfrac{5}{t^{1/5}} - \dfrac{8}{t^{3/2}}$

45. y' if $y = \dfrac{1}{\sqrt[3]{x}}$

46. w' if $w = \dfrac{10}{\sqrt[5]{u}}$

47. $\dfrac{d}{dx}\left(\dfrac{1.2}{\sqrt{x}} - 3.2x^{-2} + x\right)$

48. $\dfrac{d}{dx}\left(2.8x^{-3} - \dfrac{0.6}{\sqrt[3]{x^2}} + 7\right)$

For Problems 49–52, find

(A) $f'(x)$

(B) *The slope of the graph of f at $x = 2$ and $x = 4$*

(C) *The equations of the tangent lines at $x = 2$ and $x = 4$*

(D) *The value(s) of x where the tangent line is horizontal*

49. $f(x) = 6x - x^2$

50. $f(x) = 2x^2 + 8x$

51. $f(x) = 3x^4 - 6x^2 - 7$

52. $f(x) = x^4 - 32x^2 + 10$

If an object moves along the y axis (marked in feet) so that its position at time x (in seconds) is given by the indicated functions in Problems 53–56, find

(A) *The instantaneous velocity function* $v = f'(x)$

(B) *The velocity when* $x = 0$ *and* $x = 3$ *seconds*

(C) *The time(s) when* $v = 0$

53. $f(x) = 176x - 16x^2$

54. $f(x) = 80x - 10x^2$

55. $f(x) = x^3 - 9x^2 + 15x$

56. $f(x) = x^3 - 9x^2 + 24x$

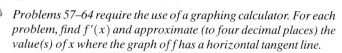

Problems 57–64 require the use of a graphing calculator. For each problem, find $f'(x)$ and approximate (to four decimal places) the value(s) of x where the graph of f has a horizontal tangent line.

57. $f(x) = x^2 - 3x - 4\sqrt{x}$

58. $f(x) = x^2 + x - 10\sqrt{x}$

59. $f(x) = 3\sqrt[3]{x^4} - 1.5x^2 - 3x$

60. $f(x) = 3\sqrt[3]{x^4} - 2x^2 + 4x$

61. $f(x) = 0.05x^4 + 0.1x^3 - 1.5x^2 - 1.6x + 3$

62. $f(x) = 0.02x^4 - 0.06x^3 - 0.78x^2 + 0.94x + 2.2$

63. $f(x) = 0.2x^4 - 3.12x^3 + 16.25x^2 - 28.25x + 7.5$

64. $f(x) = 0.25x^4 - 2.6x^3 + 8.1x^2 - 10x + 9$

65. Let $f(x) = ax^2 + bx + c, a \neq 0$. Recall that the graph of $y = f(x)$ is a parabola. Use the derivative $f'(x)$ to derive a formula for the x coordinate of the vertex of this parabola.

66. Now that you know how to find derivatives, explain why it is no longer necessary for you to memorize the formula for the x coordinate of the vertex of a parabola.

67. Give an example of a cubic polynomial function that has

(A) No horizontal tangents

(B) One horizontal tangent

(C) Two horizontal tangents

68. Can a cubic polynomial function have more than two horizontal tangents? Explain.

C *Find the indicated derivatives in Problems 69–76.*

69. $f'(x)$ if $f(x) = (2x - 1)^2$

70. y' if $y = (2x - 5)^2$

71. $\dfrac{d}{dx} \dfrac{10x + 20}{x}$

72. $\dfrac{dy}{dx}$ if $y = \dfrac{x^2 + 25}{x^2}$

73. $\dfrac{dy}{dx}$ if $y = \dfrac{3x - 4}{12x^2}$

74. $\dfrac{d}{dx} \dfrac{5x - 3}{15x^6}$

75. y' if $y = \dfrac{x^4 - 3x^3 + 5}{x^2}$

76. $f'(x)$ if $f(x) = \dfrac{2x^5 - 4x^3 + 2x}{x^3}$

In Problems 77 and 78, use the definition of the derivative and the four-step process to verify each statement.

77. $\dfrac{d}{dx} x^3 = 3x^2$

78. $\dfrac{d}{dx} x^4 = 4x^3$

79. The domain of the power function $f(x) = x^{1/3}$ is the set of all real numbers. Find the domain of the derivative $f'(x)$. Discuss the nature of the graph of $y = f(x)$ for any x values excluded from the domain of $f'(x)$.

80. The domain of the power function $f(x) = x^{2/3}$ is the set of all real numbers. Find the domain of the derivative $f'(x)$. Discuss the nature of the graph of $y = f(x)$ for any x values excluded from the domain of $f'(x)$.

Applications

81. *Sales analysis.* The total sales of a company (in millions of dollars) t months from now are given by

$$S(t) = 0.03t^3 + 0.5t^2 + 2t + 3$$

(A) Find $S'(t)$.

(B) Find $S(5)$ and $S'(5)$ (to two decimal places). Write a brief verbal interpretation of these results.

(C) Find $S(10)$ and $S'(10)$ (to two decimal places). Write a brief verbal interpretation of these results.

82. *Sales analysis.* The total sales of a company (in millions of dollars) t months from now are given by

$$S(t) = 0.015t^4 + 0.4t^3 + 3.4t^2 + 10t - 3$$

(A) Find $S'(t)$.

(B) Find $S(4)$ and $S'(4)$ (to two decimal places). Write a brief verbal interpretation of these results.

(C) Find $S(8)$ and $S'(8)$ (to two decimal places). Write a brief verbal interpretation of these results.

83. *Advertising.* From past records, it is estimated that a marine manufacturer will sell $N(x)$ power boats after spending $\$x$ thousand on advertising, as given by

$$N(x) = 1,000 - \dfrac{3,780}{x} \qquad 5 \leq x \leq 30$$

(see the figure).

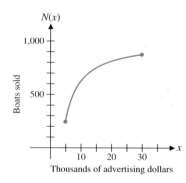

Figure for 83

(A) Find $N'(x)$.

(B) Find $N'(10)$ and $N'(20)$. Write a brief verbal interpretation of these results.

84. *Price–demand equation.* Suppose that, in a given gourmet food store, people are willing to buy x pounds of chocolate candy per day at $\$p$ per quarter pound, as given by the price–demand equation

$$x = 10 + \frac{180}{p} \qquad 2 \le p \le 10$$

This function is graphed in the accompanying figure. Find the demand and the instantaneous rate of change of demand with respect to price when the price is $5. Write a brief verbal interpretation of these results.

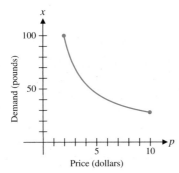

Figure for 84

 85. *Motor vehicle production.* Annual limousine production in the United States for selected years is given in Table 1.

TABLE 1

Year	Limousine Production
1980	2,500
1985	6,500
1990	4,400
1995	3,400
2000	5,700

(A) Let x represent time (in years) since 1980, and let y represent the corresponding U.S. production of limousines. Enter the data in a graphing calculator and find a cubic regression equation for the data.

(B) If $y = L(x)$ denotes the regression equation found in part (A), find $L(12)$ and $L'(12)$ (to the nearest hundred), and write a brief verbal interpretation of these results.

(C) Repeat part (B) for $L(18)$ and $L'(18)$.

86. *Motor vehicle operation.* The total number of registered limousine operators in the United States for selected years is given in Table 2.

TABLE 2

Year	Limousine Operators
1985	4,000
1990	7,000
1995	8,900
2000	11,000

(A) Let x represent time (in years) since 1980, and let y represent the corresponding number of registered limousine operators in the United States. Enter the data in a graphing calculator and find a quadratic regression equation for the data.

(B) If $y = L(x)$ denotes the regression equation found in part (A), find $L(18)$ and $L'(18)$ (to the nearest hundred), and write a brief verbal interpretation of these results.

87. *Medicine.* A person x inches tall has a pulse rate of y beats per minute, as given approximately by

$$y = 590x^{-1/2} \qquad 30 \le x \le 75$$

What is the instantaneous rate of change of pulse rate at the

(A) 36-inch level?

(B) 64-inch level?

88. *Ecology.* A coal-burning electrical generating plant emits sulfur dioxide into the surrounding air. The concentration $C(x)$, in parts per million, is given approximately by

$$C(x) = \frac{0.1}{x^2}$$

where x is the distance from the plant in miles. Find the instantaneous rate of change of concentration at

(A) $x = 1$ mile

(B) $x = 2$ miles

89. *Learning.* Suppose that a person learns y items in x hours, as given by

$$y = 50\sqrt{x} \qquad 0 \le x \le 9$$

(see the figure). Find the rate of learning at the end of

(A) 1 hour   (B) 9 hours

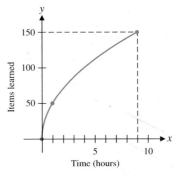

Figure for 89

90. *Learning.* If a person learns y items in x hours, as given by

$$y = 21\sqrt[3]{x^2} \qquad 0 \le x \le 8$$

find the rate of learning at the end of

(A) 1 hour   (B) 8 hours

Section 3-6 DIFFERENTIALS

- Increments
- Differentials
- Approximations Using Differentials

In this section, we introduce increments and differentials. Increments are useful in their own right, and they also provide an alternative notation for defining the derivative. Differentials are often easier to compute than increments and can be used to approximate increments.

■ Increments

In Section 3-4, we defined the derivative of f at x as the limit of the difference quotient

$$f'(x) = \lim_{h \to 0} \frac{f(x + h) - f(x)}{h}$$

We considered various interpretations of this limit, including slope, velocity, and instantaneous rate of change. Increment notation enables us to interpret the numerator and the denominator of the difference quotient separately.

Given $y = f(x) = x^3$, if x changes from 2 to 2.1, then y will change from $y = f(2) = 2^3 = 8$ to $y = f(2.1) = 2.1^3 = 9.261$. The change in x is called the *increment in x* and is denoted by Δx (read "delta x"), since Δ is the Greek letter delta. Similarly, the change in y is called the *increment in y* and is denoted by Δy. In terms of the given example, we write

$$\Delta x = 2.1 - 2 = 0.1 \qquad \textit{Change in x}$$
$$\Delta y = f(2.1) - f(2) \qquad \textit{f(x) = x}^3$$
$$= 2.1^3 - 2^3 \qquad \textit{Use a calculator.}$$
$$= 9.261 - 8$$
$$= 1.261 \qquad \textit{Corresponding change in y}$$

INSIGHT

The symbol Δx does not represent the product of Δ and x, but is the symbol for a single quantity: the *change in x*. Likewise, the symbol Δy represents a single quantity: the *change in y*.

The definition of these quantities is given next.

DEFINITION **Increments**

For $y = f(x)$, $\Delta x = x_2 - x_1$, so $x_2 = x_1 + \Delta x$, and

$$\Delta y = y_2 - y_1$$
$$= f(x_2) - f(x_1)$$
$$= f(x_1 + \Delta x) - f(x_1)$$

Δy represents the change in y corresponding to a change Δx in x.

Δx can be either positive or negative.

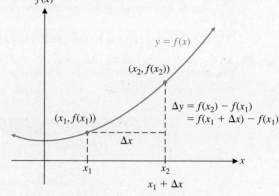

[*Note*: Δy depends on the function f, the input x_1, and the increment Δx.]

EXAMPLE 1 **Increments** Given the function $y = f(x) = \dfrac{x^2}{2}$,

(A) Find Δx, Δy, and $\Delta y / \Delta x$ for $x_1 = 1$ and $x_2 = 2$.

(B) Find $\dfrac{f(x_1 + \Delta x) - f(x_1)}{\Delta x}$ for $x_1 = 1$ and $\Delta x = 2$.

SOLUTION (A) $\Delta x = x_2 - x_1 = 2 - 1 = 1$

$$\Delta y = f(x_2) - f(x_1)$$

$$= f(2) - f(1) = \frac{4}{2} - \frac{1}{2} = \frac{3}{2}$$

$$\frac{\Delta y}{\Delta x} = \frac{f(x_2) - f(x_1)}{x_2 - x_1} = \frac{\frac{3}{2}}{1} = \frac{3}{2}$$

(B) $\dfrac{f(x_1 + \Delta x) - f(x_1)}{\Delta x} = \dfrac{f(1 + 2) - f(1)}{2}$

$$= \frac{f(3) - f(1)}{2} = \frac{\frac{9}{2} - \frac{1}{2}}{2} = \frac{4}{2} = 2$$

MATCHED PROBLEM 1 Given the function $y = f(x) = x^2 + 1$,

(A) Find Δx, Δy, and $\Delta y / \Delta x$ for $x_1 = 2$ and $x_2 = 3$.

(B) Find $\dfrac{f(x_1 + \Delta x) - f(x_1)}{\Delta x}$ for $x_1 = 1$ and $\Delta x = 2$.

In Example 1, we observe another notation for the familiar difference quotient

$$\frac{f(x + h) - f(x)}{h} \tag{1}$$

It is common to refer to h, the change in x, as Δx. Then the difference quotient (1) takes on the form

$$\frac{f(x + \Delta x) - f(x)}{\Delta x} \quad \text{or} \quad \frac{\Delta y}{\Delta x} \qquad \Delta y = f(x + \Delta x) - f(x)$$

and the derivative is defined by

$$f'(x) = \lim_{\Delta x \to 0} \frac{f(x + \Delta x) - f(x)}{\Delta x}$$

or

$$f'(x) = \lim_{\Delta x \to 0} \frac{\Delta y}{\Delta x} \tag{2}$$

if the limit exists.

Explore & Discuss 1 Suppose that $y = f(x)$ defines a function whose domain is the set of all real numbers. If every increment Δy is equal to 0, what is the range of f?

■ Differentials

Assume that the limit in equation (2) exists. Then, for small Δx, the difference quotient $\Delta y / \Delta x$ provides a good approximation for $f'(x)$. Also, $f'(x)$ provides a good approximation for $\Delta y / \Delta x$. We indicate the latter by writing

$$\frac{\Delta y}{\Delta x} \approx f'(x) \qquad \Delta x \text{ is small, but} \neq 0 \tag{3}$$

Multiplying both sides of (3) by Δx gives us

$$\Delta y \approx f'(x)\,\Delta x \qquad \text{Δx is small, but $\neq 0$} \qquad (4)$$

From equation (4), we see that $f'(x)\Delta x$ provides a good approximation for Δy when Δx is small.

Because of the practical and theoretical importance of $f'(x)\,\Delta x$, we give it the special name **differential** and represent it by the special symbol dy or df:

$$dy = f'(x)\Delta x \qquad \text{or} \qquad df = f'(x)\Delta x$$

For example,

$$d(2x^3) = (2x^3)'\,\Delta x = 6x^2\,\Delta x$$
$$d(x) = (x)'\,\Delta x = 1\,\Delta x = \Delta x$$

In the second example, we usually drop the parentheses in $d(x)$ and simply write

$$dx = \Delta x$$

In summary, we have the following:

DEFINITION **Differentials**

If $y = f(x)$ defines a differential function, then the **differential dy, or df,** is defined as the product of $f'(x)$ and dx, where $dx = \Delta x$. Symbolically,

$$dy = f'(x)\,dx, \qquad \text{or} \qquad df = f'(x)\,dx$$

where

$$dx = \Delta x$$

[*Note*: The differential dy (or df) is actually a function involving two independent variables: x and dx; a change in either one or both will affect dy (or df).]

EXAMPLE 2 **Differentials** Find dy for $f(x) = x^2 + 3x$. Evaluate dy for

(A) $x = 2$ and $dx = 0.1$ (B) $x = 3$ and $dx = 0.1$

(C) $x = 1$ and $dx = 0.02$

SOLUTION $dy = f'(x)\,dx$
$$= (2x + 3)\,dx$$

(A) When $x = 2$ and $dx = 0.1$,   (B) When $x = 3$ and $dx = 0.1$,
$$dy = [2(2) + 3]0.1 = 0.7 \qquad\qquad dy = [2(3) + 3]0.1 = 0.9$$

(C) When $x = 1$ and $dx = 0.02$,
$$dy = [2(1) + 3]0.02 = 0.1$$

MATCHED PROBLEM 2 Find dy for $f(x) = \sqrt{x} + 3$. Evaluate dy for

(A) $x = 4$ and $dx = 0.1$

(B) $x = 9$ and $dx = 0.12$

(C) $x = 1$ and $dx = 0.01$

We now have two interpretations of the symbol dy/dx. Referring to the function $y = f(x) = x^2 + 3x$ in Example 2 with $x = 2$ and $dx = 0.1$, we have

$$\frac{dy}{dx} = f'(2) = 7 \quad \textit{Derivative}$$

and

$$\frac{dy}{dx} = \frac{0.7}{0.1} = 7 \quad \textit{Ratio of differentials}$$

INSIGHT

Finding a differential is not a new operation; it is new notation and a new interpretation of the familiar operation of differentiation.

■ Approximations Using Differentials

Earlier, we noted that for small Δx,

$$\frac{\Delta y}{\Delta x} \approx f'(x) \quad \text{and} \quad \Delta y \approx f'(x)\Delta x$$

Also, since

$$dy = f'(x)\, dx$$

it follows that

$$\Delta y \approx dy$$

and dy can be used to approximate Δy.

To interpret this result geometrically, we need to recall a basic property of the slope. The vertical change in a line is equal to the product of the slope and the horizontal change, as shown in Figure 1.

Now consider the line tangent to the graph of $y = f(x)$, as shown in Figure 2. Since $f'(x)$ is the slope of the tangent line and dx is the horizontal change in the tangent line, it follows that the vertical change in the tangent line is given by $dy = f'(x)\, dx$, as indicated in Figure 2.

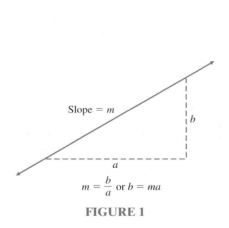

$$m = \frac{b}{a} \text{ or } b = ma$$

FIGURE 1

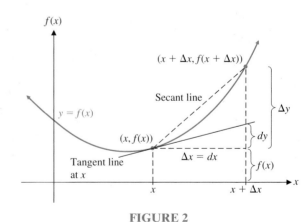

FIGURE 2

Explore & Discuss **2** Suppose that $y = f(x)$ defines a differentiable function. Explain why the graph of the function dy (of the independent variable $\Delta x = dx$) is a straight line through the origin for a fixed value of x.

EXAMPLE 3 **Comparing Increments and Differentials** Let $y = f(x) = 6x - x^2$.

(A) Find Δy and dy when $x = 2$.

(B) Graph Δy and dy from part A for $-1 \le \Delta x \le 1$.

(C) Compare Δy and dy from part A for $\Delta x = 0.1, 0.2$, and 0.3.

SOLUTION

(A) $\Delta y = f(2 + \Delta x) - f(2)$
$= 6(2 + \Delta x) - (2 + \Delta x)^2 - (6 \cdot 2 - 2^2)$ Remove parentheses.
$= 12 + 6\Delta x - 4 - 4\Delta x - \Delta x^2 - 12 + 4$ Collect like terms.
$= 2\Delta x - \Delta x^2$

Since $f'(x) = 6 - 2x$, $f'(2) = 2$, and $dx = \Delta x$, $dy = f'(2)\,dx = 2\Delta x$

(B) The graphs of $\Delta y = 2\Delta x - \Delta x^2$ and $dy = 2\Delta x$ are shown in Figure 3. Examining the graphs, we conclude that the differential provides a linear approximation of the increments and that the approximation is better for values of $\Delta x = dx$ close to 0.

(C) The table on the left compares the values of Δx and dy for the indicated values of Δx.

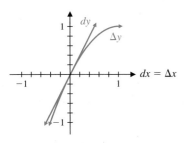

FIGURE 3
$\Delta y = 2\Delta x - \Delta x^2$, $dy = 2\Delta x$

MATCHED PROBLEM 3 Repeat Example 3 for $x = 4$ and $\Delta x = dx = -0.1, -0.2$, and -0.3.

Δx	Δy	dy
0.1	0.19	0.2
0.2	0.36	0.4
0.3	0.51	0.6

INSIGHT

The error in the approximation $\Delta y \approx dy$ is usually small when $\Delta x = dx$ is small (see Example 3C), but can be quite substantial in certain cases.

EXAMPLE 4

Cost–Revenue A company manufactures and sells x transistor radios per week. If the weekly cost and revenue equations are

$$C(x) = 5,000 + 2x \qquad R(x) = 10x - \frac{x^2}{1,000} \qquad 0 \le x \le 8,000$$

find the approximate changes in revenue and profit if production is increased from 2,000 to 2,010 units per week.

SOLUTION

We will approximate ΔR and ΔP with dR and dP, respectively, using $x = 2,000$ and $dx = 2,010 - 2,000 = 10$.

$$R(x) = 10x - \frac{x^2}{1,000} \qquad\qquad P(x) = R(x) - C(x) = 10x - \frac{x^2}{1,000} - 5,000 - 2x$$

$$dR = R'(x)\,dx \qquad\qquad\qquad\qquad = 8x - \frac{x^2}{1,000} - 5,000$$

$$= \left(10 - \frac{x}{500}\right)dx \qquad dP = P'(x)\,dx$$

$$= \left(10 - \frac{2,000}{500}\right)10 \qquad = \left(8 - \frac{x}{500}\right)dx$$

$$= \$60 \text{ per week} \qquad\qquad = \left(8 - \frac{2,000}{500}\right)10$$

$$\qquad\qquad\qquad\qquad\qquad = \$40 \text{ per week}$$

MATCHED PROBLEM 4 Repeat Example 4 with production increasing from 6,000 to 6,010.

Comparing the results in Example 4 and Matched Problem 4, we see that an increase in production results in a revenue and profit increase at the 2,000 production level, but a revenue and profit loss at the 6,000 production level.

Answers to Matched Problems

1. (A) $\Delta x = 1$, $\Delta y = 5$, $\Delta y/\Delta x = 5$ (B) 4

2. $dy = \dfrac{1}{2\sqrt{x}}\,dx$ (A) 0.025 (B) 0.02 (C) 0.005

3. (A) $\Delta y = -2\Delta x - \Delta x^2$; $dy = -2\Delta x$

(B)

(C)

Δx	Δy	dy
-0.1	0.19	0.2
-0.2	0.36	0.4
-0.3	0.51	0.6

4. $dR = -\$20/\text{wk}; dP = -\$40/\text{wk}$

Exercise 3-6

A *In Problems 1–6, find the indicated quantities for*
$y = f(x) = 3x^2$.

1. $\Delta x, \Delta y,$ and $\Delta y/\Delta x$; given $x_1 = 1$ and $x_2 = 4$

2. $\Delta x, \Delta y,$ and $\Delta y/\Delta x$; given $x_1 = 2$ and $x_2 = 5$

3. $\dfrac{f(x_1 + \Delta x) - f(x_1)}{\Delta x}$; given $x_1 = 1$ and $\Delta x = 2$

4. $\dfrac{f(x_1 + \Delta x) - f(x_1)}{\Delta x}$; given $x_1 = 2$ and $\Delta x = 1$

5. $\Delta y/\Delta x$; given $x_1 = 1$ and $x_2 = 3$

6. $\Delta y/\Delta x$; given $x_1 = 2$ and $x_2 = 3$

In Problems 7–12, find dy for each function.

7. $y = 30 + 12x^2 - x^3$

8. $y = 200x - \dfrac{x^2}{30}$

9. $y = x^2\left(1 - \dfrac{x}{9}\right)$

10. $y = x^3(60 - x)$

11. $y = \dfrac{590}{\sqrt{x}}$

12. $y = 52\sqrt{x}$

B *In Problems 13 and 14, find the indicated quantities for*
$y = f(x) = 3x^2$.

13. (A) $\dfrac{f(2 + \Delta x) - f(2)}{\Delta x}$ (simplify)

(B) What does the quantity in part (A) approach as Δx
approaches 0?

14. (A) $\dfrac{f(3 + \Delta x) - f(3)}{\Delta x}$ (simplify)

(B) What does the quantity in part (A) approach as Δx
approaches 0?

In Problems 15–18, find dy for each function.

15. $y = (2x + 1)^3$

16. $y = (3x + 5)^5$

17. $y = \dfrac{x}{x^2 + 9}$

18. $y = \dfrac{x^2}{(x + 1)^2}$

*In Problems 19–22, evaluate dy and Δy for each function for
the indicated values.*

19. $y = f(x) = x^2 - 3x + 2; x = 5, dx = \Delta x = 0.2$

20. $y = f(x) = 30 + 12x^2 - x^3; x = 2, dx = \Delta x = 0.1$

21. $y = f(x) = 75\left(1 - \dfrac{2}{x}\right); x = 5, dx = \Delta x = -0.5$

22. $y = f(x) = 100\left(x - \dfrac{4}{x^2}\right); x = 2, dx = \Delta x = -0.1$

23. A cube with sides 10 inches long is covered with a coat of
fiberglass 0.2 inch thick. Use differentials to estimate the
volume of the fiberglass shell.

24. A sphere with a radius of 5 centimeters is coated with ice
0.1 centimeter thick. Use differentials to estimate the volume of the ice. [Recall that $V = \frac{4}{3}\pi r^3$.]

C *In Problems 25–28,*

(A) Find Δy and dy for the function f at the indicated value
of x.

(B) Graph Δy and dy from part A.

(C) Compare the values of Δy and dy from part A at the indicated values of Δx.

25. $f(x) = x^2 + 2x + 3; x = -0.5, \Delta x = dx = 0.1, 0.2, 0.3$

26. $f(x) = x^2 + 2x + 3;$
$x = -2, \Delta x = dx = -0.1, -0.2, -0.3$

27. $f(x) = x^3 - 2x^2; x = 1, \Delta x = dx = 0.05, 0.10, 0.15$

28. $f(x) = x^3 - 2x^2;$
$x = 2, \Delta x = dx = -0.05, -0.10, -0.15$

*In Problems 29–32, discuss the validity of each statement.
If the statement is always true, explain why. If not, give a
counterexample.*

29. If the graph of the function $y = f(x)$ is a line, then the
functions Δy and dy (of the independent variable
$\Delta x = dx$) for $f(x)$ at $x = 3$ are identical.

30. If the graph of the function $y = f(x)$ is a parabola, then
the functions Δy and dy (of the independent variable
$\Delta x = dx$) for $f(x)$ at $x = 0$ are identical.

31. Suppose that $y = f(x)$ defines a differentiable function
whose domain is the set of all real numbers. If every
differential dy at $x = 2$ is equal to 0, then $f(x)$ is a
constant function.

32. Suppose that $y = f(x)$ defines a function whose
domain is the set of all real numbers. If every

increment at $x = 2$ is equal to 0, then $f(x)$ is a constant function.

33. Find dy if $y = \sqrt[3]{3x^2 - 2x + 1}$.

34. Find dy if $y = (2x^2 - 4)\sqrt{x + 2}$.

35. Find dy and Δy for $y = 52\sqrt{x}$, $x = 4$, and $\Delta x = dx = 0.3$.

36. Find dy and Δy for $y = 590/\sqrt{x}$, $x = 64$, and $\Delta x = dx = 1$.

Applications

Use differential approximations in the following problems.

37. *Advertising.* From past records, it is estimated that a company will sell N units of a product after spending $\$x$ thousand in advertising, as given by

$$N = 60x - x^2 \qquad 5 \le x \le 30$$

Approximately what increase in sales will result by increasing the advertising budget from \$10,000 to \$11,000? From \$20,000 to \$21,000?

38. *Price–demand.* Suppose, in a grocery chain, the daily demand (in pounds) for chocolate candy at $\$x$ per pound is given by

$$D = 1,000 - 40x^2 \qquad 1 \le x \le 5$$

If the price is increased from \$3.00 per pound to \$3.20 per pound, what is the approximate change in demand?

39. *Average cost.* For a company that manufactures tennis rackets, the average cost per racket $\overline{C}$ is found to be

$$\overline{C} = \frac{400}{x} + 5 + \frac{1}{2}x \qquad x \ge 1$$

where x is the number of rackets produced per hour. What will the approximate change in average cost per racket be if production is increased from 20 per hour to 25 per hour? From 40 per hour to 45 per hour?

40. *Revenue and profit.* A company manufactures and sells x televisions per month. If the cost and revenue equations are

$$C(x) = 72,000 + 60x$$

$$R(x) = 200x - \frac{x^2}{30} \qquad 0 \le x \le 6,000$$

what will the approximate changes in revenue and profit be if production is increased from 1,500 to 1,510? From 4,500 to 4,510?

41. *Pulse rate.* The average pulse rate y (in beats per minute) of a healthy person x inches tall is given approximately by

$$y = \frac{590}{\sqrt{x}} \qquad 30 \le x \le 75$$

Approximately how will the pulse rate change for a change in height from 36 to 37 inches? From 64 to 65 inches?

42. *Measurement.* An egg of a particular bird is very nearly spherical. If the radius to the inside of the shell is 5 millimeters and the radius to the outside of the shell is 5.3 millimeters, approximately what is the volume of the shell? [Remember that $V = \frac{4}{3}\pi r^3$.]

43. *Medicine.* A drug is given to a patient to dilate her arteries. If the radius of an artery is increased from 2 to 2.1 millimeters, approximately how much is the cross-sectional area increased? [Assume that the cross section of the artery is circular; that is, $A = \pi r^2$.]

44. *Drug sensitivity.* One hour after x milligrams of a particular drug are given to a person, the change in body temperature T (in degrees Fahrenheit) is given by

$$T = x^2\left(1 - \frac{x}{9}\right) \qquad 0 \le x \le 6$$

Approximate the changes in body temperature produced by the following changes in drug dosages:

(A) From 2 to 2.1 milligrams

(B) From 3 to 3.1 milligrams

(C) From 4 to 4.1 milligrams

45. *Learning.* A particular person learning to type has an achievement record given approximately by

$$N = 75\left(1 - \frac{2}{t}\right) \qquad 3 \le t \le 20$$

where N is the number of words per minute typed after t weeks of practice. What is the approximate improvement from 5 to 5.5 weeks of practice?

46. *Learning.* If a person learns y items in x hours, as given approximately by

$$y = 52\sqrt{x} \qquad 0 \le x \le 9$$

what is the approximate increase in the number of items learned when x changes from 1 to 1.1 hours? From 4 to 4.1 hours?

47. *Politics.* In a newly incorporated city, it is estimated that the voting population (in thousands) will increase according to

$$N(t) = 30 + 12t^2 - t^3 \qquad 0 \le t \le 8$$

where t is time in years. Find the approximate change in votes for the following changes in time:

(A) From 1 to 1.1 years

(B) From 4 to 4.1 years

(C) From 7 to 7.1 years

Section 3-7   MARGINAL ANALYSIS IN BUSINESS AND ECONOMICS

- Marginal Cost, Revenue, and Profit
- Application
- Marginal Average Cost, Revenue, and Profit

■ Marginal Cost, Revenue, and Profit

One important use of calculus in business and economics is in *marginal analysis*. In economics, the word *marginal* refers to a rate of change—that is, to a derivative. Thus, if $C(x)$ is the total cost of producing x items, then $C'(x)$ is called the *marginal cost* and represents the instantaneous rate of change of total cost with respect to the number of items produced. Similarly, the *marginal revenue* is the derivative of the total revenue function and the *marginal profit* is the derivative of the total profit function.

DEFINITION   **Marginal Cost, Revenue, and Profit**

If x is the number of units of a product produced in some time interval, then

$$\text{total cost} = C(x)$$
$$\textbf{marginal cost} = C'(x)$$
$$\text{total revenue} = R(x)$$
$$\textbf{marginal revenue} = R'(x)$$
$$\text{total profit} = P(x) = R(x) - C(x)$$
$$\textbf{marginal profit} = P'(x) = R'(x) - C'(x)$$
$$= (\text{marginal revenue}) - (\text{marginal cost})$$

Marginal cost (or revenue or profit) is the instantaneous rate of change of cost (or revenue or profit) relative to production at a given production level.

To begin our discussion, we consider a cost function $C(x)$. It is important to remember that $C(x)$ represents the *total* cost of producing x items, not the cost of producing a *single* item. To find the cost of producing a single item, we use the difference of two successive values of $C(x)$:

$$\text{Total cost of producing } x + 1 \text{ items} = C(x + 1)$$
$$\text{Total cost of producing } x \text{ items} = C(x)$$
$$\text{Exact cost of producing the } (x + 1)\text{st item} = C(x + 1) - C(x)$$

EXAMPLE 1   **Cost Analysis**   A company manufactures fuel tanks for automobiles. The total weekly cost (in dollars) of producing x tanks is given by

$$C(x) = 10,000 + 90x - 0.05x^2$$

(A) Find the marginal cost function.

(B) Find the marginal cost at a production level of 500 tanks per week, and interpret the results.

(C) Find the exact cost of producing the 501st item.

SOLUTION   (A) $C'(x) = 90 - 0.1x$

(B) $C'(500) = 90 - 0.1(500) = \$40$ *Marginal cost*

At a production level of 500 tanks per week, the total production costs are increasing at the rate of $40 tank.

(C) $C(501) = 10{,}000 + 90(501) - 0.05(501)^2$

$\qquad = \$42{,}539.95$ *Total cost of producing 501 tanks per week*

$\qquad C(500) = 10{,}000 + 90(500) - 0.05(500)^2$

$\qquad = \$42{,}500.00$ *Total cost of producing 500 tanks per week*

$\qquad C(501) - C(500) = 42{,}539.95 - 42{,}500.00$

$\qquad = \$39.95$ *Exact cost of producing the 501st tank*

MATCHED PROBLEM 1 A company manufactures automatic transmissions for automobiles. The total weekly cost (in dollars) of producing x transmissions is given by

$$C(x) = 50{,}000 + 600x - 0.75x^2$$

(A) Find the marginal cost function.

(B) Find the marginal cost at a production level of 200 transmissions per week, and interpret the results.

(C) Find the exact cost of producing the 201st transmission.

In Example 1, we found that the cost of the 501st tank and the marginal cost at a production level of 500 tanks differ by only a nickel. Increments and differentials will help us understand the relationship between marginal cost and the cost of a single item. If $C(x)$ is any total cost function, then

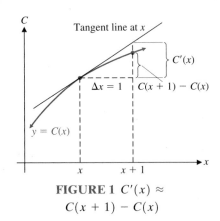

FIGURE 1 $C'(x) \approx$ $C(x + 1) - C(x)$

$$C'(x) \approx \frac{C(x + \Delta x) - C(x)}{\Delta x} \qquad \textit{See Section 3-6}$$

$$C'(x) \approx C(x + 1) - C(x) \qquad \Delta x = 1$$

Thus, we see that the marginal cost $C'(x)$ approximates $C(x + 1) - C(x)$, the exact cost of producing the $(x + 1)$st item. These observations are summarized next and are illustrated in Figure 1.

THEOREM 1 MARGINAL COST AND EXACT COST

If $C(x)$ is the total cost of producing x items, the marginal cost function approximates the exact cost of producing the $(x + 1)$st item:

$$\underset{\text{Marginal cost}}{} \quad \underset{\text{Exact cost}}{}$$
$$C'(x) \approx C(x + 1) - C(x)$$

Similar statements can be made for total revenue functions and total profit functions.

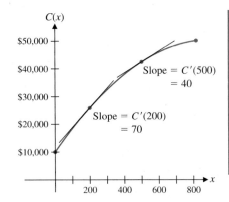

FIGURE 2 $C(x) = 10{,}000 + 90x - 0.05x^2$

INSIGHT

Theorem 1 states that the marginal cost at a given production level x approximates the cost of producing the $(x + 1)$st, or *next*, item. In practice, the marginal cost is used more frequently than the exact cost. One reason for this is that the marginal cost is easily visualized when one is examining the graph of the total cost function. Figure 2 shows the graph of the cost function discussed in Example 1, with tangent lines added at $x = 200$ and $x = 500$. The graph clearly shows that as production increases, the slope of the tangent line decreases. Thus, the cost of producing the next tank also decreases, a desirable characteristic of a total cost function. We will have much more to say about graphical analysis in Chapter 5.

APPLICATION

We next discuss how price, demand, revenue, cost, and profit are tied together in typical applications. Although either price or demand can be used as the independent variable in a price–demand equation, it is common practice to use demand as the independent variable when marginal revenue, cost, and profit are also involved.

Explore & Discuss **1**

The market research department of a company used test marketing to determine the demand for a new headphone set (Table 1).

(A) Assuming that the relationship between price p and demand x is linear, find the price–demand equation and write the result in the form $x = f(p)$. Graph the equation and find the domain of f. Discuss the effect of price increases and decreases on demand.

(B) Solve the equation found in part (A) for p, obtaining an equation of the form $p = g(x)$. Graph this equation and find the domain of g. Discuss the effect of price increases and decreases on demand.

TABLE 1

Demand x	Price p
3,000	$7
6,000	$4

EXAMPLE 2

Production Strategy The market research department of a company recommends that the company manufacture and market a new headphone set for MP3 players. After suitable test marketing, the research department presents the following **price–demand equation:**

$$x = 10{,}000 - 1{,}000p \quad \text{\small x is demand at price p.} \tag{1}$$

Solving (1) for p gives

$$p = 10 - 0.001x \tag{2}$$

where x is the number of headphones retailers are likely to buy at $\$p$ per set.

The financial department provides the **cost function**

$$C(x) = 7{,}000 + 2x \tag{3}$$

where $\$7{,}000$ is the estimate of fixed costs (tooling and overhead) and $\$2$ is the estimate of variable costs per headphone set (materials, labor, marketing, transportation, storage, etc.).

(A) Find the domain of the function defined by the price–demand equation (2).

(B) Find and interpret the marginal cost function $C'(x)$.

(C) Find the revenue function as a function of x, and find its domain.

(D) Find the marginal revenue at $x = 2{,}000$, $5{,}000$, and $7{,}000$. Interpret these results.

(E) Graph the cost function and the revenue function in the same coordinate system, find the intersection points of these two graphs, and interpret the results.

(F) Find the profit function and its domain, and sketch the graph of the function.

(G) Find the marginal profit at $x = 1{,}000$, $4{,}000$, and $6{,}000$. Interpret these results.

SOLUTION

(A) Since price p and demand x must be nonnegative, we have $x \geq 0$ and

$$p = 10 - 0.001x \geq 0$$
$$10 \geq 0.001x$$
$$10{,}000 \geq x$$

Thus, the permissible values of x are $0 \leq x \leq 10{,}000$.

(B) The marginal cost is $C'(x) = 2$. Since this is a constant, it costs an additional $\$2$ to produce one more headphone set at any production level.

(C) The **revenue** is the amount of money R received by the company for manufacturing and selling x headphone sets at $\$p$ per set and is given by

$$R = (\text{number of headphone sets sold})(\text{price per headphone set}) = xp$$

In general, the revenue R can be expressed as a function of p by using equation (1) or as a function of x by using equation (2). As we mentioned earlier, when using marginal functions, we will always use the number of items x as the independent variable. Thus, the **revenue function** is

$$R(x) = xp = x(10 - 0.001x) \quad \text{Using equation (2)} \qquad (4)$$
$$= 10x - 0.001x^2$$

Since equation (2) is defined only for $0 \le x \le 10{,}000$, it follows that the domain of the revenue function is $0 \le x \le 10{,}000$.

(D) The **marginal revenue** is

$$R'(x) = 10 - 0.002x$$

For production levels of $x = 2{,}000, 5{,}000,$ and $7{,}000$, we have

$$R'(2{,}000) = 6 \qquad R'(5{,}000) = 0 \qquad R'(7{,}000) = -4$$

This means that at production levels of 2,000, 5,000, and 7,000, the respective approximate changes in revenue per unit change in production are $6, $0, and −$4. That is, at the 2,000 output level, revenue increases as production increases; at the 5,000 output level, revenue does not change with a "small" change in production; and at the 7,000 output level, revenue decreases with an increase in production.

(E) When we graph $R(x)$ and $C(x)$ in the same coordinate system, we obtain Figure 3. The intersection points are called the **break-even points,** because revenue equals cost at these production levels: The company neither makes nor loses money, but just breaks even. The break-even points are obtained as follows:

$$C(x) = R(x)$$
$$7{,}000 + 2x = 10x - 0.001x^2$$
$$0.001x^2 - 8x + 7{,}000 = 0 \quad \text{Solve by the quadratic formula}$$
$$x^2 - 8{,}000x + 7{,}000{,}000 = 0 \quad \text{(see Appendix A-7)}.$$

$$x = \frac{8{,}000 \pm \sqrt{8{,}000^2 - 4(7{,}000{,}000)}}{2}$$
$$= \frac{8{,}000 \pm \sqrt{36{,}000{,}000}}{2}$$
$$= \frac{8{,}000 \pm 6{,}000}{2}$$
$$= 1{,}000, \quad 7{,}000$$

$$R(1{,}000) = 10(1{,}000) - 0.001(1{,}000)^2 = 9{,}000$$
$$C(1{,}000) = 7{,}000 + 2(1{,}000) = 9{,}000$$
$$R(7{,}000) = 10(7{,}000) - 0.001(7{,}000)^2 = 21{,}000$$
$$C(7{,}000) = 7{,}000 + 2(7{,}000) = 21{,}000$$

Thus, the break-even points are (1,000, 9,000) and (7,000, 21,000), as shown in Figure 3. Further examination of the figure shows that cost is greater than revenue for production levels between 0 and 1,000 and also between 7,000 and 10,000. Consequently, the company incurs a loss at these levels. By contrast, for production levels between 1,000 and 7,000, revenue is greater than cost and the company makes a profit.

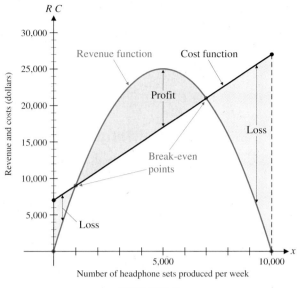

FIGURE 3

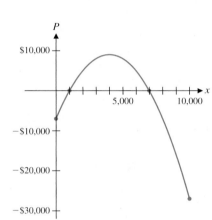

FIGURE 4

(F) The **profit function** is

$$P(x) = R(x) - C(x)$$
$$= (10x - 0.001x^2) - (7,000 + 2x)$$
$$= -0.001x^2 + 8x - 7,000$$

The domain of the cost function is $x \geq 0$ and the domain of the revenue function is $0 \leq x \leq 10,000$. Thus, the domain of the profit function is the set of x values for which both functions are defined—that is, $0 \leq x \leq 10,000$. The graph of the profit function is shown in Figure 4. Notice that the x coordinates of the break-even points in Figure 3 are the x intercepts of the profit function. Furthermore, the intervals on which cost is greater than revenue and on which revenue is greater than cost correspond, respectively, to the intervals on which profit is negative and the intervals on which profit is positive.

(G) The **marginal profit** is

$$P'(x) = -0.002x + 8$$

For production levels of 1,000, 4,000, and 6,000, we have

$$P'(1,000) = 6 \qquad P'(4,000) = 0 \qquad P'(6,000) = -4$$

This means that at production levels of 1,000, 4,000, and 6,000, the respective approximate changes in profit per unit change in production are $6, $0, and $-$4. That is, at the 1,000 output level, profit will be increased if production is increased; at the 4,000 output level, profit does not change for "small" changes in production; and at the 6,000 output level, profits will decrease if production is increased. It seems that the best production level to produce a maximum profit is 4,000.

Example 2 warrants careful study, since a number of important ideas in economics and calculus are involved. In the next chapter, we will develop a systematic procedure for finding the production level (and, using the demand equation, the selling price) that will maximize profit.

MATCHED PROBLEM 2   Refer to the revenue and profit functions in Example 2.

(A) Find $R'(3,000)$ and $R'(6,000)$, and interpret the results.

(B) Find $P'(2,000)$ and $P'(7,000)$, and interpret the results.

Explore & Discuss **2**   Let

$$C(x) = 12,000 + 5x \qquad \text{and} \qquad R(x) = 9x - 0.002x^2$$

Then

$$P(x) = R(x) - C(x) \neq 9x - 0.002x^2 - 12,000 + 5x$$

Explain why $\neq$ is used in the preceding equation. Then find the correct expression for the profit function.

■ Marginal Average Cost, Revenue, and Profit

Sometimes it is desirable to carry out marginal analysis relative to **average cost (cost per unit)**, **average revenue (revenue per unit)**, and **average profit (profit per unit)**. The relevant definitions are summarized as follows:

DEFINITION   **Marginal Average Cost, Revenue, and Profit**

If x is the number of units of a product produced in some time interval, then

Cost per unit:

$$\text{average cost} = \overline{C}(x) = \frac{C(x)}{x}$$

$$\text{marginal average cost} = \overline{C}'(x) = \frac{d}{dx}\overline{C}(x)$$

Revenue per unit:

$$\text{average revenue} = \overline{R}(x) = \frac{R(x)}{x}$$

$$\text{marginal average revenue} = \overline{R}'(x) = \frac{d}{dx}\overline{R}(x)$$

Profit per unit:

$$\text{average profit} = \overline{P}(x) = \frac{P(x)}{x}$$

$$\text{marginal average profit} = \overline{P}'(x) = \frac{d}{dx}\overline{P}(x)$$

EXAMPLE 3   **Cost Analysis** A small machine shop manufactures drill bits used in the petroleum industry. The shop manager estimates that the total daily cost (in dollars) of producing x bits is

$$C(x) = 1,000 + 25x - 0.1x^2$$

(A) Find $\overline{C}(x)$ and $\overline{C}'(x)$.

(B) Find $\overline{C}(10)$ and $\overline{C}'(10)$, and interpret these quantities.

(C) Use the results in part (B) to estimate the average cost per bit at a production level of 11 bits per day.

SOLUTION   (A) $\overline{C}(x) = \dfrac{C(x)}{x} = \dfrac{1,000 + 25x - 0.1x^2}{x}$

$$= \frac{1,000}{x} + 25 - 0.1x \qquad \textit{Average cost function}$$

$$\overline{C}'(x) = \frac{d}{dx}\overline{C}(x) = -\frac{1,000}{x^2} - 0.1 \qquad \textit{Marginal average cost function}$$

(B) $\overline{C}(10) = \dfrac{1,000}{10} + 25 - 0.1(10) = \$124$

$$\overline{C}'(10) = -\frac{1,000}{10^2} - 0.1 = -\$10.10$$

At a production level of 10 bits per day, the average cost of producing a bit is $124, and this cost is decreasing at the rate of $10.10 per bit.

(C) If production is increased by 1 bit, then the average cost per bit will decrease by approximately $10.10. Thus, the average cost per bit at a production level of 11 bits per day is approximately $124 − $10.10 = $113.90. ▬

MATCHED PROBLEM 3   Consider the cost function for the production of headphone sets from Example 2:

$$C(x) = 7{,}000 + 2x$$

(A) Find $\overline{C}(x)$ and $\overline{C}'(x)$.

(B) Find $\overline{C}(100)$ and $\overline{C}'(100)$, and interpret these quantities.

(C) Use the results in part (B) to estimate the average cost per headphone set at a production level of 101 headphone sets. ▬▬

Explore & Discuss **3**   A student produced the following solution to Matched Problem 3:

$$C(x) = 7{,}000 + 2x \quad \textit{Cost}$$
$$C'(x) = 2 \qquad\qquad \textit{Marginal cost}$$
$$\frac{C'(x)}{x} = \frac{2}{x} \qquad \textit{"Average" of the marginal cost}$$

Explain why the last function is not the same as the marginal average cost function.

 CAUTION

1. The marginal average cost function must be computed by first finding the average cost function and then finding its derivative. As Explore & Discuss 3 illustrates, reversing the order of these two steps produces a different function that does not have any useful economic interpretations.

2. Recall that the marginal cost function has two interpretations: the usual interpretation of any derivative as an instantaneous rate of change and the special interpretation as an approximation to the exact cost of the $(x + 1)$st item. This special interpretation does not apply to the marginal average cost function. Referring to Example 3, we would be incorrect to interpret $\overline{C}'(10) = -\$10.10$ to mean that the average cost of the next bit is approximately −$10.10. In fact, the phrase "average cost of the next bit" does not even make sense. Averaging is a concept applied to a collection of items, not to a single item.

These remarks also apply to revenue and profit functions.

Answers to Matched Problems   **1.** (A) $C'(x) = 600 − 1.5x$

(B) $C'(200) = 300$. At a production level of 200 transmissions, total costs are increasing at the rate of $300 per transmission.

(C) $C(201) − C(200) = \$299.25$

2. (A) $R'(3{,}000) = 4$. At a production level of 3,000, a unit increase in production will increase revenue by approximately $4.
    $R'(6{,}000) = −2$. At a production level of 6,000, a unit increase in production will decrease revenue by approximately $2.

(B) $P'(2{,}000) = 4$. At a production level of 2,000, a unit increase in production will increase profit by approximately $4.
    $P'(7{,}000) = −6$. At a production level of 7,000, a unit increase in production will decrease profit by approximately $6.

3. (A) $\overline{C}(x) = \dfrac{7{,}000}{x} + 2; \overline{C}'(x) = -\dfrac{7{,}000}{x^2}$

(B) $\overline{C}(100) = \$72; \overline{C}'(100) = −\$0.70$. At a production level of 100 headphone sets, the average cost per headphone set is $72, and this average cost is decreasing at a rate of $0.70 per headphone set.

(C) Approx. $71.30.

Exercise 3-7

Applications

1. *Cost analysis.* The total cost (in dollars) of producing x food processors is

$$C(x) = 2,000 + 50x - 0.5x^2$$

(A) Find the exact cost of producing the 21st food processor.

(B) Use the marginal cost to approximate the cost of producing the 21st food processor.

2. *Cost analysis.* The total cost (in dollars) of producing x electric guitars is

$$C(x) = 1,000 + 100x - 0.25x^2$$

(A) Find the exact cost of producing the 51st guitar.

(B) Use the marginal cost to approximate the cost of producing the 51st guitar.

3. *Cost analysis.* The total cost (in dollars) of manufacturing x auto body frames is

$$C(x) = 60,000 + 300x$$

(A) Find the average cost per unit if 500 frames are produced.

(B) Find the marginal average cost at a production level of 500 units, and interpret the results.

(C) Use the results from parts (A) and (B) to estimate the average cost per frame if 501 frames are produced.

4. *Cost analysis.* The total cost (in dollars) of printing x dictionaries is

$$C(x) = 20,000 + 10x$$

(A) Find the average cost per unit if 1,000 dictionaries are produced.

(B) Find the marginal average cost at a production level of 1,000 units, and interpret the results.

(C) Use the results from parts (A) and (B) to estimate the average cost per dictionary if 1,001 dictionaries are produced.

5. *Profit analysis.* The total profit (in dollars) from the sale of x skateboards is

$$P(x) = 30x - 0.3x^2 - 250 \qquad 0 \le x \le 100$$

(A) Find the exact profit from the sale of the 26th skateboard.

(B) Use the marginal profit to approximate the profit from the sale of the 26th skateboard.

6. *Profit analysis.* The total profit (in dollars) from the sale of x portable stereos is

$$P(x) = 22x - 0.2x^2 - 400 \qquad 0 \le x \le 100$$

(A) Find the exact profit from the sale of the 41st stereo.

(B) Use the marginal profit to approximate the profit from the sale of the 41st stereo.

7. *Profit analysis.* The total profit (in dollars) from the sale of x videocassettes is

$$P(x) = 5x - 0.005x^2 - 450 \qquad 0 \le x \le 1,000$$

Evaluate the marginal profit at the given values of x, and interpret the results.

(A) $x = 450$ \qquad (B) $x = 750$

8. *Profit analysis.* The total profit (in dollars) from the sale of x cameras is

$$P(x) = 12x - 0.02x^2 - 1,000 \qquad 0 \le x \le 600$$

Evaluate the marginal profit at the given values of x, and interpret the results.

(A) $x = 200$ \qquad (B) $x = 350$

9. *Profit analysis.* The total profit (in dollars) from the sale of x lawn mowers is

$$P(x) = 30x - 0.03x^2 - 750 \qquad 0 \le x \le 1,000$$

(A) Find the average profit per mower if 50 mowers are produced.

(B) Find the marginal average profit at a production level of 50 mowers, and interpret the results.

(C) Use the results from parts (A) and (B) to estimate the average profit per mower if 51 mowers are produced.

10. *Profit analysis.* The total profit (in dollars) from the sale of x charcoal grills is

$$P(x) = 20x - 0.02x^2 - 320 \qquad 0 \le x \le 1,000$$

(A) Find the average profit per grill if 40 grills are produced.

(B) Find the marginal average profit at a production level of 40 grills, and interpret the results.

(C) Use the results from parts (A) and (B) to estimate the average profit per grill if 41 grills are produced.

11. *Revenue analysis.* The price p (in dollars) and the demand x for a particular clock radio are related by the equation

$$x = 4,000 - 40p$$

(A) Express the price p in terms of the demand x, and find the domain of this function.

(B) Find the revenue $R(x)$ from the sale of x clock radios. What is the domain of R?

(C) Find the marginal revenue at a production level of 1,600 clock radios, and interpret the results.

(D) Find the marginal revenue at a production level of 2,500 clock radios, and interpret the results.

12. *Revenue analysis.* The price p (in dollars) and the demand x for a particular steam iron are related by the equation

$$x = 1{,}000 - 20p$$

(A) Express the price p in terms of the demand x, and find the domain of this function.

(B) Find the revenue $R(x)$ from the sale of x clock radios. What is the domain of R?

(C) Find the marginal revenue at a production level of 400 steam irons, and interpret the results.

(D) Find the marginal revenue at a production level of 650 steam irons, and interpret the results.

13. *Revenue, cost and profit.* The price–demand equation and the cost function for the production of table saws are given, respectively, by

$$x = 6{,}000 - 30p \quad \text{and} \quad C(x) = 72{,}000 + 60x$$

where x is the number of saws that can be sold at a price of \$$p$ per saw and $C(x)$ is the total cost (in dollars) of producing x saws.

(A) Express the price p as a function of the demand x, and find the domain of this function.

(B) Find the marginal cost.

(C) Find the revenue function and state its domain.

(D) Find the marginal revenue.

(E) Find $R'(1{,}500)$ and $R'(4{,}500)$, and interpret these quantities.

(F) Graph the cost function and the revenue function on the same coordinate system for $0 \le x \le 6{,}000$. Find the break-even points, and indicate regions of loss and profit.

(G) Find the profit function in terms of x.

(H) Find the marginal profit.

(I) Find $P'(1{,}500)$ and $P'(3{,}000)$, and interpret these quantities.

14. *Revenue, cost, and profit.* The price–demand equation and the cost function for the production of television sets are given, respectively, by

$$x = 9{,}000 - 30p \quad \text{and} \quad C(x) = 150{,}000 + 30x$$

where x is the number of sets that can be sold at a price of \$$p$ per set and $C(x)$ is the total cost (in dollars) of producing x sets.

(A) Express the price p as a function of the demand x, and find the domain of this function.

(B) Find the marginal cost.

(C) Find the revenue function and state its domain.

(D) Find the marginal revenue.

(E) Find $R'(3{,}000)$ and $R'(6{,}000)$, and interpret these quantities.

(F) Graph the cost function and the revenue function on the same coordinate system for $0 \le x \le 9{,}000$. Find the break-even points, and indicate regions of loss and profit.

(G) Find the profit function in terms of x.

(H) Find the marginal profit.

(I) Find $P'(1{,}500)$ and $P'(4{,}500)$, and interpret these quantities.

15. *Revenue, cost, and profit.* A company is planning to manufacture and market a new two-slice electric toaster. After conducting extensive market surveys, the research department provides the following estimates: a weekly demand of 200 toasters at a price of \$16 per toaster and a weekly demand of 300 toasters at a price of \$14 per toaster. The financial department estimates that weekly fixed costs will be \$1,400 and variable costs (cost per unit) will be \$4.

(A) Assume that the relationship between the price p and the demand x is linear. Use the research department's estimates to express p as a function of x, and find the domain of this function.

(B) Find the revenue function in terms of x and state its domain.

(C) Assume that the cost function is linear. Use the financial department's estimates to express the cost function in terms of x.

(D) Graph the cost function and the revenue function on the same coordinate system for $0 \le x \le 1{,}000$. Find the break-even points, and indicate regions of loss and profit.

(E) Find the profit function in terms of x.

(F) Evaluate the marginal profit at $x = 250$ and $x = 475$, and interpret the results.

16. *Revenue, cost, and profit.* The company in Problem 15 is also planning to manufacture and market a four-slice toaster. For this toaster, the research department's estimates are a weekly demand of 300 toasters at a price of \$25 per toaster and a weekly demand of 400 toasters at a price of \$20. The financial department's estimates are fixed weekly costs of \$5,000 and variable costs of \$5 per toaster.

(A) Assume that the relationship between the price p and the demand x is linear. Use the research department's estimates to express p as a function of x, and find the domain of this function.

(B) Find the revenue function in terms of x and state its domain.

(C) Assume that the cost function is linear. Use the financial department's estimates to express the cost function in terms of x.

(D) Graph the cost function and the revenue function on the same coordinate system for $0 \leq x \leq 800$. Find the break-even points, and indicate regions of loss and profit.

(E) Find the profit function in terms of x.

(F) Evaluate the marginal profit at $x = 325$ and $x = 425$, and interpret the results.

17. *Revenue, cost, and profit.* The total cost and the total revenue (in dollars) for the production and sale of x ski jackets are given, respectively, by

$$C(x) = 24x + 21{,}900 \quad \text{and} \quad R(x) = 200x - 0.2x^2$$

$$0 \leq x \leq 1{,}000$$

(A) Find the value of x where the graph of $R(x)$ has a horizontal tangent line.

(B) Find the profit function $P(x)$.

(C) Find the value of x where the graph of $P(x)$ has a horizontal tangent line.

(D) Graph $C(x)$, $R(x)$, and $P(x)$ on the same coordinate system for $0 \leq x \leq 1{,}000$. Find the break-even points. Find the x intercepts of the graph of $P(x)$.

18. *Revenue, cost, and profit.* The total cost and the total revenue (in dollars) for the production and sale of x hair dryers are given, respectively, by

$$C(x) = 5x + 2{,}340 \quad \text{and} \quad R(x) = 40x - 0.1x^2$$

$$0 \leq x \leq 400$$

(A) Find the value of x where the graph of $R(x)$ has a horizontal tangent line.

(B) Find the profit function $P(x)$.

(C) Find the value of x where the graph of $P(x)$ has a horizontal tangent line.

(D) Graph $C(x)$, $R(x)$, and $P(x)$ on the same coordinate system for $0 \leq x \leq 400$. Find the break-even points. Find the x intercepts of the graph of $P(x)$.

19. *Break-even analysis.* The price–demand equation and the cost function for the production of garden hoses are given, respectively, by

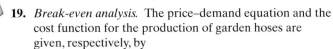

$$p = 20 - \sqrt{x} \quad \text{and} \quad C(x) = 500 + 2x$$

where x is the number of garden hoses that can be sold at a price of $\$p$ per unit and $C(x)$ is the total cost (in dollars) of producing x garden hoses.

(A) Express the revenue function in terms of x.

(B) Graph the cost function and the revenue function in the same viewing window for $0 \leq x \leq 400$. Use approximation techniques to find the break-even points correct to the nearest unit.

20. *Break-even analysis.* The price–demand equation and the cost function for the production of handwoven silk scarves are given, respectively, by

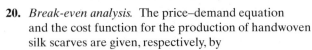

$$p = 60 - 2\sqrt{x} \quad \text{and} \quad C(x) = 3{,}000 + 5x$$

where x is the number of scarves that can be sold at a price of $\$p$ per unit and $C(x)$ is the total cost (in dollars) of producing x scarves.

(A) Express the revenue function in terms of x.

(B) Graph the cost function and the revenue function in the same viewing window for $0 \leq x \leq 900$. Use approximation techniques to find the break-even points correct to the nearest unit.

21. *Break-even analysis.* Table 2 contains price–demand and total cost data for the production of overhead projectors, where p is the wholesale price (in dollars) of a projector for an annual demand of x projectors and C is the total cost (in dollars) of producing x projectors.

TABLE 2		
x	$p(\$)$	$C(\$)$
3,190	581	1,130,000
4,570	405	1,241,000
5,740	181	1,410,000
7,330	85	1,620,000

(A) Find a quadratic regression equation for the price–demand data, using x as the independent variable.

(B) Find a linear regression equation for the cost data, using x as the independent variable. Use this equation to estimate the fixed costs and the variable costs per projector. Round answers to the nearest dollar.

(C) Find the break-even points. Round answers to the nearest integer.

(D) Find the price range for which the company will make a profit. Round answers to the nearest dollar.

22. *Break-even analysis.* Table 3 contains price–demand and total cost data for the production of treadmills, where p is the wholesale price (in dollars) of a treadmill for an annual demand of x treadmills and C is the total cost (in dollars) of producing x treadmills.

TABLE 3		
x	$p(\$)$	$C(\$)$
2,910	1,435	3,650,000
3,415	1,280	3,870,000
4,645	1,125	4,190,000
5,330	910	4,380,000

(A) Find a linear regression equation for the price–demand data, using x as the independent variable.

(B) Find a linear regression equation for the cost data, using x as the independent variable. Use this equation to estimate the fixed costs and the variable costs per treadmill. Round answers to the nearest dollar.

(C) Find the break-even points. Round answers to the nearest integer.

(D) Find the price range for which the company will make a profit. Round answers to the nearest dollar.

CHAPTER 3 REVIEW

Important Terms, Symbols, and Concepts

3-1 Introduction to Limits

Examples

- The graph of the function $y = f(x)$ is the graph of the set of all ordered pairs $(x, f(x))$.
- The limit of the function $y = f(x)$ as x approaches c is L, written $\lim_{x \to c} f(x) = L$, if the functional value $f(x)$ is close to the single real number L whenever x is close, but not equal, to c (on either side of c).
- The limit of the function $y = f(x)$ as x approaches c from the left is K, written $\lim_{x \to c^-} f(x) = K$, if $f(x)$ is close to K whenever x is close to, but to the left of, c on the real-number line.
- The limit of the function $y = f(x)$ as x approaches c from the right is L, written $\lim_{x \to c^+} f(x) = L$, if $f(x)$ is close to L whenever x is close to, but to the right of, c on the real-number line.
- Limit properties are useful for evaluating limits.
- The limit of the difference quotient $[f(a + h) - f(a)]/h$ always results in a 0/0 indeterminate form. Algebraic simplification is often required to evaluate this type of limit.

Ex. 1, p. 129
Ex. 2, p. 130
Ex. 3, p. 132
Ex. 4, p. 133
Ex. 5, p. 135
Ex. 6, p. 136
Ex. 7, p. 137
Ex. 8, p. 137
Ex. 9, p. 139
Ex. 10, p. 139
Ex. 11, p. 140

3-2 Continuity

- Intuitively, the graph of a continuous function can be drawn without lifting a pen off the paper. Algebraically, a function f is **continuous at c** if

  1. $\lim_{x \to c} f(x)$ exists   2. $f(c)$ exists   3. $\lim_{x \to c} f(x) = f(c)$

- Continuity properties are useful for determining where a function is continuous and where it is not.
- Continuity properties are also useful for solving inequalities.

Ex. 1, p. 146
Ex. 2, p. 147
Ex. 3, p. 148
Ex. 4, p. 149

3-3 Infinite Limits and Limits at Infinity

- If $f(x)$ increases or decreases without bound as x approaches a from either side of a, then the line $x = a$ is a **vertical asymptote** of the graph of $y = f(x)$.
- If $f(x)$ gets close to L as x increases without bound or decreases without bound, then L is called the limit of f at ∞ or $-\infty$.
- The end behavior of a polynomial is described in terms of limits at infinity.
- If $f(x)$ approaches L as $x \to \infty$ or as $x \to -\infty$, then the line $y = L$ is a **horizontal asymptote** of the graph of $y = f(x)$. Polynomial functions never have horizontal asymptotes. A rational function can have at most one.

Ex. 1, p. 157
Ex. 2, p. 158
Ex. 3, p. 160
Ex. 4, p. 161
Ex. 5, p. 163
Ex. 6, p. 164

3-4 The Derivative

- Given a function $y = f(x)$, the **average rate of change** is the ratio of the change in y to the change in x.
- The **instantaneous rate of change** is the limit of the average rate of change as the change in x approaches 0.
- The slope of the secant line through two points on the graph of a function $y = f(x)$ is the ratio of the change in y to the change in x. The slope of the tangent line at the point $(a, f(a))$ is the limit of the slope of the secant line through the points $(a, f(a))$ and $(a + h, f(a + h))$ as h approaches 0.
- The **derivative of $y = f(x)$ at x,** denoted $f'(x)$, is the limit of the difference quotient $[f(x + h) - f(x)]/h$ as $h \to 0$ (if the limit exists).
- The four-step method is used to find derivatives.
- If the limit of the difference quotient does not exist at $x = a$, then f is nondifferentiable at a and $f'(a)$ does not exist.

Ex. 1, p. 169
Ex. 2, p. 170
Ex. 3, p. 173
Ex. 4, p. 175
Ex. 5, p. 176
Ex. 6, p. 177

3-5 Basic Differentiation Properties

- The derivative of a constant function is 0.
- For any real number n, the derivative of $f(x) = x^n$ is nx^{n-1}.
- If f is a differential function, then the derivative of $kf(x)$ is $kf'(x)$.
- The derivative of the sum or difference of two differential functions is the sum or difference of the derivatives of the functions.

Ex. 1, p. 184
Ex. 2, p. 185
Ex. 3, p. 185
Ex. 4, p. 186
Ex. 5, p. 187

3-6 Differentials

- Given the function $y = f(x)$, the change in x is also called the **increment of x** and is denoted by Δx. The corresponding change in y is called the **increment of y** and is given by $\Delta y = f(x + \Delta x) - f(x)$.
- If $y = f(x)$ is differentiable at x, then the **differential of x** is $dx = \Delta x$ and the **differential of $y = f(x)$** is $dy = f'(x)dx$, or $df = f'(x)dx$. In this context, x and dx are both independent variables.

Ex. 1, p. 193
Ex. 2, p. 194
Ex. 3, p. 195

3-7 Marginal Analysis in Business and Economics

- If $y = C(x)$ is the total cost of producing x items, then $y = C'(x)$ is the **marginal cost** and $C(x + 1) - C(x)$ is the exact cost of producing item $x + 1$. Furthermore, $C'(x) \approx C(x + 1) - C(x)$. Similar statements can be made regarding total revenue and total profit functions.

  Ex. 1, p. 199
  Ex. 2, p. 201

- If $y = C(x)$ is the total cost of producing x items, then the **average cost,** or cost per unit, is $\overline{C}(x) = \dfrac{C(x)}{x}$

  Ex. 3, p. 204

  and the **marginal average cost** is $\overline{C}'(x) = \dfrac{d}{dx}\overline{C}(x)$. Similar statements can be made regarding total revenue and total profit functions.

REVIEW EXERCISE

Work through all the problems in this chapter review, and check your answers in the back of the book. Answers to all review problems are there, along with section numbers in italics to indicate where each type of problem is discussed. Where weaknesses show up, review appropriate sections of the text.

Many of the problems in this exercise set ask you to find a derivative. Most of the answers to these problems contain both an unsimplified form and a simplified form of the derivative. When checking your work, first check that you applied the rules correctly, and then check that you performed the algebraic simplification correctly.

A **1.** Find the indicated quantities for $y = f(x) = 2x^2 + 5$:

(A) The change in y if x changes from 1 to 3

(B) The average rate of change of y with respect to x if x changes from 1 to 3

(C) The slope of the secant line through the points $(1, f(1))$ and $(3, f(3))$ on the graph of $y = f(x)$

(D) The instantaneous rate of change of y with respect to x at $x = 1$

(E) The slope of the line tangent to the graph of $y = f(x)$ at $x = 1$

(F) $f'(1)$

2. Use the two-step limiting process to find $f'(x)$ for $f(x) = -3x + 2$.

3. If $\lim\limits_{x \to 1} f(x) = 2$ and $\lim\limits_{x \to 1} g(x) = 4$, find

(A) $\lim\limits_{x \to 1}(5f(x) + 3g(x))$ (B) $\lim\limits_{x \to 1}[f(x)g(x)]$

(C) $\lim\limits_{x \to 1}\dfrac{g(x)}{f(x)}$ (D) $\lim\limits_{x \to 1}[5 + 2x - 3g(x)]$

In Problems 4–6, use the graph of f to estimate the indicated limits and function values.

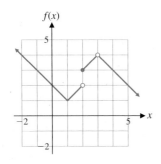

4. (A) $\lim\limits_{x \to 1^-} f(x)$

(B) $\lim\limits_{x \to 1^+} f(x)$

(C) $\lim\limits_{x \to 1} f(x)$

(D) $f(1)$

5. (A) $\lim\limits_{x \to 2^-} f(x)$

(B) $\lim\limits_{x \to 2^+} f(x)$

(C) $\lim\limits_{x \to 2} f(x)$

(D) $f(2)$

6. (A) $\lim\limits_{x \to 3^-} f(x)$

(B) $\lim\limits_{x \to 3^+} f(x)$

(C) $\lim\limits_{x \to 3} f(x)$

(D) $f(3)$

In Problems 7–9, use the graph of the function f shown in the figure to answer each question.

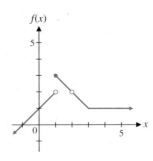

7. (A) $\lim\limits_{x \to 1} f(x) = ?$

(B) $f(1) = ?$

(C) Is f continuous at $x = 1$?

8. (A) $\lim\limits_{x \to 2} f(x) = ?$

(B) $f(2) = ?$

(C) Is f continuous at $x = 2$?

9. (A) $\lim\limits_{x \to 3} f(x) = ?$

(B) $f(3) = ?$

(C) Is f continuous at $x = 3$?

In Problems 10–19, refer to the following graph of $y = f(x)$:

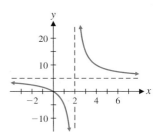

10. $\lim\limits_{x \to \infty} f(x) = ?$

11. $\lim\limits_{x \to -\infty} f(x) = ?$

12. $\lim\limits_{x \to 2^+} f(x) = ?$

13. $\lim\limits_{x \to 2^-} f(x) = ?$

14. $\lim\limits_{x \to 0^-} f(x) = ?$

15. $\lim\limits_{x \to 0^+} f(x) = ?$

16. $\lim\limits_{x \to 0} f(x) = ?$

17. Identify any vertical asymptotes.

18. Identify any horizontal asymptotes.

19. Where is $y = f(x)$ discontinuous?

20. Use the four-step limiting process to find $f'(x)$ for $f(x) = 5x^2$.

21. If $f(5) = 4$, $f'(5) = -1$, $g(5) = 2$, and $g'(5) = -3$, find $h'(5)$ for each of the following functions:

 (A) $h(x) = 3f(x)$

 (B) $h(x) = -2g(x)$

 (C) $h(x) = 2f(x) + 5$

 (D) $h(x) = -g(x) - 1$

 (E) $h(x) = 2f(x) + 3g(x)$

In Problems 22–27, find $f'(x)$ *and simplify.*

22. $f(x) = \dfrac{1}{3}x^3 - 5x^2 + 1$

23. $f(x) = 2x^{1/2} - 3x$

24. $f(x) = 5$

25. $f(x) = \dfrac{3}{2x} + \dfrac{5x^3}{4}$

26. $f(x) = \dfrac{0.5}{x^4} + 0.25x^4$

27. $f(x) = (3x^3 - 2)(x + 1)$ (*Hint:* Multiply and then differentiate.)

In Problems 28–31, find the indicated quantities for $y = f(x) = x^2 + x$.

28. Δx, Δy, and $\Delta y / \Delta x$ for $x_1 = 1$ and $x_2 = 3$.

29. $[f(x_1 + \Delta x) - f(x_1)]/\Delta x$ for $x_1 = 1$ and $\Delta x = 2$.

30. dy for $x_1 = 1$ and $x_2 = 3$.

31. Δy and dy for $x = 1$, $\Delta x = dx = 0.2$.

B *Problems 32–34 refer to the function f in the figure.*

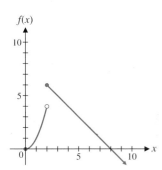

Figure for 32–34:

$$f(x) = \begin{cases} x^2 & \text{if } 0 \le x < 2 \\ 8 - x & \text{if } x \ge 2 \end{cases}$$

32. (A) $\lim\limits_{x \to 2^-} f(x) = ?$

 (B) $\lim\limits_{x \to 2^+} f(x) = ?$

 (C) $\lim\limits_{x \to 2} f(x) = ?$

 (D) $f(2) = ?$

 (E) Is f continuous at $x = 2$?

33. (A) $\lim\limits_{x \to 5^-} f(x) = ?$

 (B) $\lim\limits_{x \to 5^+} f(x) = ?$

 (C) $\lim\limits_{x \to 5} f(x) = ?$

 (D) $f(5) = ?$

 (E) Is f continuous at $x = 5$?

34. Solve each inequality. Express answers in interval notation.

 (A) $f(x) < 0$

 (B) $f(x) \ge 0$

In Problems 35–37, solve each inequality. Express the answer in interval notation. Use a graphing calculator in Problem 37 to approximate partition numbers to four decimal places.

35. $x^2 - x < 12$

36. $\dfrac{x - 5}{x^2 + 3x} > 0$

37. $x^3 + x^2 - 4x - 2 > 0$

38. Let $f(x) = 0.5x^2 - 5$.

  (A) Find the slope of the secant line through $(2, f(2))$ and $(4, f(4))$.

  (B) Find the slope of the secant line through $(2, f(2))$ and $(2 + h, f(2 + h))$, $h \neq 0$.

  (C) Find the slope of the tangent line at $x = 2$.

In Problems 39–42, find the indicated derivative and simplify.

39. $\dfrac{dy}{dx}$ for $y = \dfrac{1}{3}x^{-3} - 5x^{-2} + 1$

40. y' for $y = \dfrac{3\sqrt{x}}{2} + \dfrac{5}{3\sqrt{x}}$

41. $g'(x)$ for $g(x) = 1.8\sqrt[3]{x} + \dfrac{0.9}{\sqrt[3]{x}}$

42. $\dfrac{dy}{dx}$ for $y = \dfrac{2x^3 - 3}{5x^3}$

43. For $y = f(x) = x^2 + 4$, find

  (A) The slope of the graph at $x = 1$

  (B) The equation of the tangent line at $x = 1$ in the form $y = mx + b$

In Problems 44 and 45, find the value(s) of x where the tangent line is horizontal.

44. $f(x) = 10x - x^2$

45. $f(x) = x^3 + 3x^2 - 45x - 135$

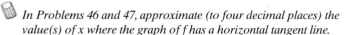 *In Problems 46 and 47, approximate (to four decimal places) the value(s) of x where the graph of f has a horizontal tangent line.*

46. $f(x) = x^4 - 2x^3 - 5x^2 + 7x$

47. $f(x) = x^3 - 10x^3 - 5x + 10$

48. If an object moves along the y axis (scale in feet) so that it is at $y = f(x) = 8x^2 - 4x + 1$ at time x (in seconds), find

  (A) The instantaneous velocity function

  (B) The velocity at time $x = 3$ seconds

49. An object moves along the y axis (scale in feet) so that at time x (in seconds) it is at $y = f(x) = -5x^2 + 16x + 3$. Find

  (A) The instantaneous velocity function

  (B) The time(s) when the velocity is 0

50. Let $f(x) = x^3$, $g(x) = (x - 4)^3$, and $h(x) = (x + 3)^3$.

  (A) How are the graphs of f, g, and h related? Illustrate your conclusion by graphing f, g, and h on the same coordinate axes.

  (B) How would you expect the graphs of the derivatives of these functions to be related? Illustrate your conclusion by graphing f', g', and h' on the same coordinate axes.

51. Let $f(x)$ be a differentiable function and let k be a nonzero constant. For each function g, write a brief verbal description of the relationship between the graphs of f and g. Do the same for the graphs of f' and g'.

  (A) $g(x) = kf(x)$, $k > 1$

  (B) $g(x) = f(x) + k$

52. Find dy and Δy for $y = 12\sqrt{x}$, $x = 3$, and $\Delta x = dx = 0.2$, (correct to four decimal places).

53. If $f(x) = x^{1/3}$, $x = 8$, and $\Delta x = dx = 0.1$, use differentials to approximate $\sqrt[3]{8.1}$, (correct to four decimal places). Use a calculator to find $\sqrt[3]{8.1}$, (correct to four decimal places) and compare your approximation with the calculator value.

In Problems 54–58, determine where f is continuous. Express the answer in interval notation.

54. $f(x) = x^2 - 4$

55. $f(x) = \dfrac{x + 1}{x - 2}$

56. $f(x) = \dfrac{x + 4}{x^2 + 3x - 4}$

57. $f(x) = \sqrt[3]{4 - x^2}$

58. $f(x) = \sqrt{4 - x^2}$

In Problems 59–68, evaluate the indicated limits if they exist.

59. Let $f(x) = \dfrac{2x}{x^2 - 3x}$. Find

  (A) $\lim\limits_{x \to 1} f(x)$ (B) $\lim\limits_{x \to 3} f(x)$ (C) $\lim\limits_{x \to 0} f(x)$

60. Let $f(x) = \dfrac{x + 1}{(3 - x)^2}$. Find

  (A) $\lim\limits_{x \to 1} f(x)$ (B) $\lim\limits_{x \to -1} f(x)$ (C) $\lim\limits_{x \to 3} f(x)$

61. Let $f(x) = \dfrac{|x - 4|}{x - 4}$. Find

  (A) $\lim\limits_{x \to 4^-} f(x)$ (B) $\lim\limits_{x \to 4^+} f(x)$ (C) $\lim\limits_{x \to 4} f(x)$

62. Let $f(x) = \dfrac{x - 3}{9 - x^2}$. Find

  (A) $\lim\limits_{x \to 3} f(x)$ (B) $\lim\limits_{x \to -3} f(x)$ (C) $\lim\limits_{x \to 0} f(x)$

63. Let $f(x) = \dfrac{x^2 - x - 2}{x^2 - 7x + 10}$. Find

  (A) $\lim\limits_{x \to -1} f(x)$ (B) $\lim\limits_{x \to 2} f(x)$ (C) $\lim\limits_{x \to 5} f(x)$

64. Let $f(x) = \dfrac{2x}{3x - 6}$. Find

  (A) $\lim\limits_{x \to \infty} f(x)$

  (B) $\lim\limits_{x \to -\infty} f(x)$

  (C) $\lim\limits_{x \to 2} f(x)$

65. Let $f(x) = \dfrac{2x^3}{3(x - 2)^2}$. Find

  (A) $\lim\limits_{x \to \infty} f(x)$

  (B) $\lim\limits_{x \to -\infty} f(x)$

  (C) $\lim\limits_{x \to 2} f(x)$

66. Let $f(x) = \dfrac{2x}{3(x-2)^3}$. Find

   (A) $\lim\limits_{x \to \infty} f(x)$

   (B) $\lim\limits_{x \to -\infty} f(x)$

   (C) $\lim\limits_{x \to 2} f(x)$

67. $\lim\limits_{h \to 0} \dfrac{f(2+h) - f(2)}{h}$ for $f(x) = x^2 + 4$

68. $\lim\limits_{h \to 0} \dfrac{f(x+h) - f(x)}{h}$ for $f(x) = \dfrac{1}{x+2}$

69. Let $f(x) = \dfrac{x^3 - 4x^2 - 4x + 16}{|x^2 - 4|}$

Graph f and use zoom and trace to investigate the left- and right-hand limits at the indicated values of c.

   (A) $c = -2$ (B) $c = 0$ (C) $c = 2$

In Problems 70 and 71, use the definition of the derivative and the four-step process to find $f'(x)$.

70. $f(x) = x^2 - x$ **71.** $f(x) = \sqrt{x} - 3$

C   *Problems 72–75 refer to the function f in the figure. Determine whether f is differentiable at the indicated value of x.*

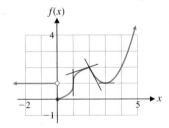

72. $x = 0$ **73.** $x = 1$ **74.** $x = 2$ **75.** $x = 3$

In Problems 76–80, find all horizontal and vertical asymptotes.

76. $f(x) = \dfrac{5x}{x-7}$

77. $f(x) = \dfrac{-2x+5}{(x-4)^2}$

78. $f(x) = \dfrac{x^2+9}{x-3}$

79. $f(x) = \dfrac{x^2 - 9}{x^2 + x - 2}$

80. $f(x) = \dfrac{x^3 - 1}{x^3 - x^2 - x + 1}$

81. The domain of the power function $f(x) = x^{1/5}$ is the set of all real numbers. Find the domain of the derivative $f'(x)$. Discuss the nature of the graph of $y = f(x)$ for any x values excluded from the domain of $f'(x)$.

82. Let f be defined by

$$f(x) = \begin{cases} x^2 - m & \text{if } x \le 1 \\ -x^2 + m & \text{if } x > 1 \end{cases}$$

where m is a constant.

   (A) Graph f for $m = 0$, and find

$$\lim\limits_{x \to 1^-} f(x) \quad \text{and} \quad \lim\limits_{x \to 1^+} f(x)$$

   (B) Graph f for $m = 2$, and find

$$\lim\limits_{x \to 1^-} f(x) \quad \text{and} \quad \lim\limits_{x \to 1^+} f(x)$$

   (C) Find m so that

$$\lim\limits_{x \to 1^-} f(x) = \lim\limits_{x \to 1^+} f(x)$$

   and graph f for this value of m.

   (D) Write a brief verbal description of each graph. How does the graph in part (C) differ from the graphs in parts (A) and (B)?

83. Let $f(x) = 1 - |x - 1|, 0 \le x \le 2$ (see the figure).

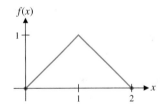

Figure for 83

   (A) $\lim\limits_{h \to 0^-} \dfrac{f(1+h) - f(1)}{h} = ?$

   (B) $\lim\limits_{h \to 0^+} \dfrac{f(1+h) - f(1)}{h} = ?$

   (C) $\lim\limits_{h \to 0} \dfrac{f(1+h) - f(1)}{h} = ?$

   (D) Does $f'(1)$ exist?

APPLICATIONS

84. *Natural-gas rates.* The accompanying table shows the winter rates for natural gas charged by the Bay State Gas Company. The customer charge is a fixed monthly charge, independent of the amount of gas used during the month.

   (A) Write a piecewise definition of the monthly charge $S(x)$ for a customer who uses x therms in a winter month.

   (B) Graph $S(x)$.

   (C) Is $S(x)$ continuous at $x = 90$? Explain.

Natural Gas Rates	
Monthly customer charge	$7.47
First 90 therms	$0.4000 per therm
All usage over 90 therms	$0.2076 per therm

85. *Cost analysis.* The total cost (in dollars) of producing x television sets is

$$C(x) = 10,000 + 200x - 0.1x^2$$

(A) Find the exact cost of producing the 101st television set.

(B) Use the marginal cost to approximate the cost of producing the 101st television set.

86. *Cost analysis.* The total cost (in dollars) of producing x bicycles is

$$C(x) = 5,000 + 40x + 0.05x^2$$

(A) Find the total cost and the marginal cost at a production level of 100 bicycles, and interpret the results.

(B) Find the average cost and the marginal average cost at a production level of 100 bicycles, and interpret the results.

87. *Cost analysis.* The total cost (in dollars) of producing x laser printers per week is shown in the accompanying figure. Which is greater, the approximate cost of producing the 201st printer or the approximate cost of producing the 601st printer? Does this graph represent a manufacturing process that is becoming more efficient or less efficient as production levels increase? Explain.

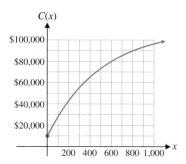

Figure for 87

88. *Cost analysis.* Let

$$p = 25 - 0.01x \quad \text{and} \quad C(x) = 2x + 9,000$$
$$0 \le x \le 2,500$$

be the price–demand equation and the cost function, respectively, for the manufacture of umbrellas.

(A) Find the marginal cost, average cost, and marginal average cost functions.

(B) Express the revenue in terms of x, and find the marginal revenue, average revenue, and marginal average revenue functions.

(C) Find the profit, marginal profit, average profit, and marginal average profit functions.

(D) Find the break-even point(s).

(E) Evaluate the marginal profit at $x = 1,000$, $1,150$, and $1,400$, and interpret the results.

(F) Graph $R = R(x)$ and $C = C(x)$ on the same coordinate system, and locate regions of profit and loss.

89. *Employee training.* A company producing computer components has established that, on the average, a new employee can assemble $N(t)$ components per day after t days of on-the-job training, as given by

$$N(t) = \frac{40t - 80}{t}, t \ge 2$$

(A) Find the average rate of change of $N(t)$ from 2 days to 5 days.

(B) Find the instantaneous rate of change of $N(t)$ at 2 days.

90. *Sales analysis.* Past sales records for a swimming-pool manufacturer indicate that the total number of swimming pools, N (in thousands), sold during a year is given by

$$N(t) = 2t + \frac{1}{3}t^{3/2}$$

where t is the number of months since the beginning of the year. Find $N(9)$ and $N'(9)$, and interpret these quantities.

91. *Natural-gas consumption.* The data in Table 1 give the U.S. consumption of natural gas in trillions of cubic feet.

TABLE 1	
Year	Natural-Gas Consumption
1960	12.0
1970	21.1
1980	19.9
1990	18.7
2000	21.9

(A) Let x represent time (in years), with $x = 0$ corresponding to 1960, and let y represent the corresponding U.S. consumption of natural gas. Enter the data set in a graphing calculator and find a cubic regression equation for the data.

(B) If $y = N(x)$ denotes the regression equation found in part (A), find $N(50)$ and $N'(50)$, and write a brief verbal interpretation of these results.

92. *Break-even analysis.* Table 2 contains price–demand and total cost data from a bakery for the production of kringles (a Danish pastry), where p is the price (in dollars) of a kringle for a daily demand of x kringles and C is the total cost (in dollars) of producing x kringles.

TABLE 2		
x	$p(\$)$	$C(\$)$
125	9	740
140	8	785
170	7	850
200	6	900

(A) Find a linear regression equation for the price–demand data, using x as the independent variable.

(B) Find a linear regression equation for the cost data, using x as the independent variable. Use this equation to estimate the fixed costs and the variable costs per kringle.

(C) Find the break-even points.

(D) Find the price range for which the bakery will make a profit.

In all answers, round dollar amounts to the nearest cent.

93. *Pollution.* A sewage treatment plant disposes of its effluent through a pipeline that extends 1 mile toward the center of a large lake. The concentration of effluent $C(x)$ in parts per million, x meters from the end of the pipe is given approximately by

$$C(x) = \frac{500}{x^2}, x \geq 1$$

What is the instantaneous rate of change of concentration at 10 meters? At 100 meters?

94. *Medicine.* The body temperature (in degrees Fahrenheit) of a patient t hours after being given a fever-reducing drug is given by

$$F(t) = 0.16t^2 - 1.6t + 102$$

Find $F(4)$ and $F'(4)$. Write a brief verbal interpretation of these quantities.

95. *Learning.* If a person learns N items in t hours, as given by

$$N(t) = 20\sqrt{t}$$

find the rate of learning after

(A) 1 hour   (B) 4 hours

96. *Physics: Thermal expansion.* The coefficient of thermal expansion (CTE) is a measure of the expansion of an object subjected to extreme temperatures. We want to use a Michaelis–Menten function of the form

$$C(T) = \frac{C_{max}T}{M + T}$$

where C = CTE, T is temperature in K (degrees Kelvin), and C_{max} and M are constants. Table 3 lists the coefficients of thermal expansion for titanium at various temperatures.

TABLE 3 Coefficients of Thermal Expansion	
T (K)	Titanium
100	4.5
200	7.4
293	8.6
500	9.9
800	11.1
1100	11.7

(A) Plot the points in columns 1 and 2 of Table 3 on graph paper and estimate C_{max} to the nearest integer. To estimate M, add the horizontal line CTE $= \dfrac{C_{max}}{2}$ to your graph, connect successive points on the graph with straight-line segments, and estimate the value of T (to the nearest multiple of fifty) that satisfies

$$C(T) = \frac{C_{max}}{2}.$$

(B) Use the constants $\dfrac{C_{max}}{2}$ and M from part A to form a Michaelis–Menten function for the CTE of germanium.

(C) Use the function from part B to estimate the CTE of germanium at 600 K and to estimate the temperature when the CTE of germanium is 10.

Additional Derivative Topics

CHAPTER 4

4-1   The Constant e and Continuous Compound Interest

4-2   Derivatives of Exponential and Logarithmic Functions

4-3   Derivatives of Products and Quotients

4-4   The Chain Rule

4-5   Implicit Differentiation

4-6   Related Rates

4-7   Elasticity of Demand

Chapter 4 Review

Review Exercise

INTRODUCTION

In this chapter, we complete our discussion of derivatives by first looking at the differentiation of exponential forms. Next, we will consider the differentiation of logarithmic forms. Finally, we will examine some additional topics and applications involving all the different types of functions we have encountered thus far, including the general chain rule. You will probably find it helpful to review some of the important properties of the exponential and logarithmic functions given in Chapter 2 before proceeding further.

Section 4-1 THE CONSTANT *e* AND CONTINUOUS COMPOUND INTEREST

- The Constant *e*
- Continuous Compound Interest

In Chapter 2, both the exponential function with base *e* and continuous compound interest were introduced informally. Now, with limit concepts at our disposal, we can give precise definitions of *e* and continuous compound interest.

■ The Constant *e*

The special irrational number *e* is a particularly suitable base for both exponential and logarithmic functions. The reasons for choosing this number as a base will become clear as we develop differentiation formulas for the exponential function e^x and the natural logarithmic function ln *x*.

In precalculus treatments (Chapter 2), the number *e* is informally defined as the irrational number that can be approximated by the expression $[1 + (1/n)]^n$ by taking *n* sufficiently large. Now we will use the limit concept to formally define *e* as either of the following two limits:

DEFINITION **The Number *e***

$$e = \lim_{n \to \infty} \left(1 + \frac{1}{n}\right)^n \qquad \text{or, alternatively,} \qquad e = \lim_{s \to 0}(1 + s)^{1/s}$$

$$e = 2.718\ 281\ 828\ 459\ldots$$

We will use both of these limit forms. [*Note:* If $s = 1/n$, then, as $n \to \infty$, $s \to 0$.]

The proof that the indicated limits exist and represent an irrational number between 2 and 3 is not easy and is omitted here.

INSIGHT

The two limits used to define *e* are unlike any we have encountered thus far. Some people reason (incorrectly) that both limits are 1, since $1 + s \to 1$ as $s \to 0$ and 1 to any power is 1. A calculator experiment on an ordinary scientific calculator with a y^x key can convince you otherwise. Consider the following table of values for *s* and $f(s) = (1 + s)^{1/s}$ and the graph shown in Figure 1 for *s* close to 0.

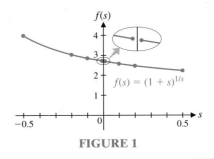

FIGURE 1

---

s approaches 0 from the left → 0 ← *s* approaches 0 from the right

s	−0.5	−0.2	−0.1	−0.01 →0← 0.01	0.1	0.2	0.5
$(1 + s)^{1/s}$	4.0000	3.0518	2.8680	2.7320 → *e* ← 2.7048	2.5937	2.4883	2.2500

Compute some of the table values with a calculator yourself, and also try several values of s even closer to 0. Note that the function is discontinuous at $s = 0$.

Exactly who discovered e is still debated. It is named after the great mathematician Leonhard Euler (1707–1783), who computed e to 23 decimal places, using $[1 + (1/n)]^n$.

■ Continuous Compound Interest

Now we will see how e appears quite naturally in the important application of compound interest. Let us start with simple interest, move on to compound interest, and then proceed on to continuous compound interest.

On the one hand, if a principal P is borrowed at an annual rate r,* then after t years at simple interest, the borrower will owe the lender an amount A given by

$$A = P + Prt = P(1 + rt) \quad \text{Simple interest} \qquad (1)$$

On the other hand, if interest is compounded n times a year, the borrower will owe the lender an amount A given by

$$A = P\left(1 + \frac{r}{n}\right)^{nt} \quad \text{Compound interest} \qquad (2)$$

where r/n is the interest rate per compounding period and nt is the number of compounding periods. Suppose that P, r, and t in equation (2) are held fixed and n is increased. Will the amount A increase without bound, or will it tend to approach some limiting value?

Let us perform a calculator experiment before we attack the general limit problem. If $P = \$100$, $r = 0.06$, and $t = 2$ years, then

$$A = 100\left(1 + \frac{0.06}{n}\right)^{2n}$$

We compute A for several values of n in Table 1. The biggest gain appears in the first step; then the gains slow down as n increases. In fact, it appears that A tends to approach $\$112.75$ as n gets larger and larger.

TABLE 1

Compounding Frequency	n	$A = 100\left(1 + \dfrac{0.06}{n}\right)^{2n}$
Annually	1	$112.3600
Semiannually	2	112.5509
Quarterly	4	112.6493
Monthly	12	112.7160
Weekly	52	112.7419
Daily	365	112.7486
Hourly	8,760	112.7496

Explore & Discuss 1

(A) Suppose that $\$1,000$ is deposited in a savings account that earns 6% simple interest. How much will be in the account after 2 years?

(B) Suppose that $\$1,000$ is deposited in a savings account that earns compound interest at a rate of 6% per year. How much will be in the account after 2 years if interest is compounded annually? Semiannually? Quarterly? Weekly?

(C) How frequently must interest be compounded at the 6% rate in order to have $\$1,150$ in the account after 2 years?

---

* If r is the interest rate written as a decimal, then $100r\%$ is the rate in percent. For example, if $r = 0.12$, we have $100r\% = 100(0.12)\% = 12\%$. The expressions 0.12 and 12% are therefore equivalent. Unless stated otherwise, all formulas in this book use r in decimal form.

Now we turn back to the general problem for a moment. Keeping P, r, and t fixed in equation (2), we compute the following limit and observe an interesting and useful result:

$$\lim_{n \to \infty} P\left(1 + \frac{r}{n}\right)^{nt} = P \lim_{n \to \infty} \left(1 + \frac{r}{n}\right)^{(n/r)rt} \quad \text{Insert } r/r \text{ in the exponent and}$$
$$\text{let } s = r/n. \text{ Note that}$$
$$n \to \infty \text{ implies } s \to 0.$$
$$= P \lim_{s \to 0}[(1 + s)^{1/s}]^{rt} \quad \text{Use a limit property.*}$$
$$= P[\lim_{s \to 0}(1 + s)^{1/s}]^{rt} \quad \lim_{s \to 0}(1 + s)^{1/s} = e$$
$$= Pe^{rt}$$

The resulting formula is called the **continuous compound interest formula,** a very important and widely used formula in business and economics.

Continuous Compound Interest

$$A = Pe^{rt}$$

where

$$P = \text{principal}$$
$$r = \text{annual nominal interest rate compounded continuously}$$
$$t = \text{time in years}$$
$$A = \text{amount at time } t$$

EXAMPLE 1   **Computing Continuously Compounded Interest** If \$100 is invested at 6% compounded continuously,[†] what amount will be in the account after 2 years? How much interest will be earned?

SOLUTION
$$A = Pe^{rt}$$
$$= 100e^{(0.06)(2)} \quad \text{6\% is equivalent to } r = 0.06.$$
$$\approx \$112.7497$$

(Compare this result with the values calculated in Table 1.) The interest earned is \$112.7497 − \$100 = \$12.7497.   ■

MATCHED PROBLEM 1   What amount (to the nearest cent) will an account have after 5 years if \$100 is invested at an annual nominal rate of 8% compounded annually? Semiannually? Continuously?

EXAMPLE 2   **Graphing the Growth of an Investment** Recently, Union Savings Bank offered a 5-year certificate of deposit (CD) that earned 5.75% compounded continuously. If \$1,000 is invested in one of these CDs, graph the amount in the account as a function of time for a period of 5 years.

SOLUTION   We want to graph

$$A = 1{,}000e^{0.0575t} \quad 0 \le t \le 5$$

Using a calculator, we construct a table of values (Table 2). Then we graph the points from the table and join the points with a smooth curve (Fig. 2).

---

* The following new limit property is used: If $\lim_{x \to c} f(x)$ exists, then $\lim_{x \to c}[f(x)]^p = [\lim_{x \to c} f(x)]^p$, provided that the last expression names a real number.

† Following common usage, we will often write "at 6% compounded continuously," understanding that this means "at an annual nominal rate of 6% compounded continuously."

TABLE 2	
t	*A*($)
0	1,000
1	1,059
2	1,122
3	1,188
4	1,259
5	1,333

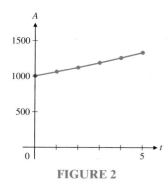

FIGURE 2

INSIGHT

Depending on the domain, the graph of an exponential function can appear to be linear. Table 3 shows the slopes of the line segments connecting the dots in Figure 2. Since these slopes are not identical, this graph is not the graph of a straight line.

TABLE 3		
t	*A*	*m*
0	1,000	
		} 59
1	1,059	
		} 63
2	1,122	
		} 66
3	1,188	
		} 71
4	1,259	
		} 74
5	1,333	

MATCHED PROBLEM 2 If $5,000 is invested in a Union Savings Bank 4-year CD that earns 5.61% compounded continuously, graph the amount in the account as a function of time for a period of 4 years.

EXAMPLE 3 **Computing Growth Time** How long will it take an investment of $5,000 to grow to $8,000 if it is invested at 5% compounded continuously?

SOLUTION Starting with the continous compound interest formula $A = Pe^{rt}$, we must solve for t:

$$A = Pe^{rt}$$
$$8{,}000 = 5{,}000e^{0.05t} \quad \text{Divide both sides by 5,000 and reverse the equation.}$$
$$e^{0.05t} = 1.6 \quad \text{Take the natural logarithm of both sides—recall that}$$
$$\ln e^{0.05t} = \ln 1.6 \quad \log_b b^x = x.$$
$$0.05t = \ln 1.6$$
$$t = \frac{\ln 1.6}{0.05}$$
$$t \approx 9.4 \text{ years}$$

Intersection
X=9.4000726 Y=8000

FIGURE 3
$y_1 = 5{,}000e^{0.05x}$
$y_2 = 8{,}000$

  Figure 3 shows an alternative method for solving Example 3 on a graphing calculator.

MATCHED PROBLEM 3   How long will it take an investment of $10,000 to grow to $15,000 if it is invested at 9% compounded continuously?

EXAMPLE 4   **Computing Doubling Time**   How long will it take money to double if it is invested at 6.5% compounded continuously?

SOLUTION   Starting with the continuous compound interest formula $A = Pe^{rt}$, we must solve for t, given $A = 2P$ and $r = 0.065$:

$$2P = Pe^{0.065t} \qquad \text{Divide both sides by } P \text{ and reverse the equation.}$$
$$e^{0.065t} = 2 \qquad \text{Take the natural logarithm of both sides.}$$
$$\ln e^{0.065t} = \ln 2$$
$$0.065t = \ln 2$$
$$t = \frac{\ln 2}{0.065}$$
$$t \approx 10.66 \text{ years}$$

MATCHED PROBLEM 4   How long will it take money to triple if it is invested at 5.5% compounded continuously?

Explore & Discuss **2**   You are considering three options for investing $10,000: at 7% compounded annually, at 6% compounded monthly, and at 5% compounded continuously.

(A) Which option would be the best for investing $10,000 for 8 years?

(B) How long would you need to invest your money for the third option to be the best?

Answers to Matched Problems   **1.** $146.93; $148.02; $149.18

2. $A = 5,000e^{0.0561t}$ **3.** 4.51 yr   **4.** 19.97 yr

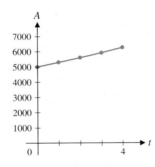

t	$A(\$)$
0	5,000
1	5,289
2	5,594
3	5,916
4	6,258

Exercise 4-1

A *Use a calculator to evaluate A to the nearest cent in Problems 1 and 2.*

1. $A = \$1,000e^{0.1t}$ for $t = 2, 5,$ and 8

2. $A = \$5,000e^{0.08t}$ for $t = 1, 4,$ and 10

3. If $6,000 is invested at 10% compounded continuously, graph the amount in the account as a function of time for a period of 8 years.

4. If $4,000 is invested at 8% compounded continuously, graph the amount in the account as a function of time for a period of 6 years.

B *In Problems 5–10, solve for t or r to two decimal places.*

5. $2 = e^{0.06t}$ **6.** $2 = e^{0.03t}$ **7.** $3 = e^{0.1r}$

8. $3 = e^{0.25t}$ **9.** $2 = e^{5r}$ **10.** $3 = e^{10r}$

C *In Problems 11 and 12, use a calculator to complete each table to five decimal places.*

11.

n	$[1 + (1/n)]^n$
10	2.593 74
100	
1,000	
10,000	
100,000	
1,000,000	
10,000,000	
↓	↓
∞	$e = 2.718\,281\,828\,459\ldots$

12.

s	$(1 + s)^{1/s}$
0.01	2.704 81
−0.01	
0.001	
−0.001	
0.000 1	
−0.000 1	
0.000 01	
−0.000 01	
↓	↓
0	$e = 2.718\,281\,828\,459\ldots$

13. Use a calculator and a table of values to investigate

$$\lim_{n\to\infty} (1 + n)^{1/n}$$

Do you think this limit exists? If so, what do you think it is?

14. Use a calculator and a table of values to investigate

$$\lim_{s\to 0^+} \left(1 + \frac{1}{s}\right)^s$$

Do you think this limit exists? If so, what do you think it is?

 15. It can be shown that the number e satisfies the inequality

$$\left(1 + \frac{1}{n}\right)^n < e < \left(1 + \frac{1}{n}\right)^{n+1} \qquad n \geq 1$$

Illustrate this condition by graphing

$$y_1 = (1 + 1/n)^n$$
$$y_2 = 2.718\,281\,828 \approx e$$
$$y_3 = (1 + 1/n)^{n+1}$$

in the same viewing window, for $1 \leq n \leq 20$.

16. It can be shown that

$$e^s = \lim_{n\to\infty} \left(1 + \frac{s}{n}\right)^n$$

for any real number s. Illustrate this equation graphically for $s = 2$ by graphing

$$y_1 = (1 + 2/n)^n$$
$$y_2 = 7.389\,056\,099 \approx e^2$$

in the same viewing window, for $1 \leq n \leq 50$.

Applications

17. *Continuous compound interest.* Recently, Provident Bank offered a 10-year CD that earns 5.51% compounded continuously.

(A) If $10,000 is invested in this CD, how much will it be worth in 10 years?

(B) How long will it take for the account to be worth $15,000?

18. *Continuous compound interest.* Provident Bank also offers a 3-year CD that earns 5.28% compounded continuously.

(A) If $10,000 is invested in this CD, how much will it be worth in 3 years?

(B) How long will it take for the account to be worth $11,000?

19. *Present value.* A note will pay $20,000 at maturity 10 years from now. How much should you be willing to pay for the note now if money is worth 5.2% compounded continuously?

20. *Present value.* A note will pay $50,000 at maturity 5 years from now. How much should you be willing to pay for the note now if money is worth 6.4% compounded continuously?

21. *Continuous compound interest.* An investor bought stock for $20,000. Five years later, the stock was sold for $30,000. If interest is compounded continuously, what annual nominal rate of interest did the original $20,000 investment earn?

22. *Continuous compound interest.* A family paid $40,000 cash for a house. Fifteen years later, the house was sold for $100,000. If interest is compounded continuously, what annual nominal rate of interest did the original $40,000 investment earn?

23. *Present value.* Solving $A = Pe^{rt}$ for P, we obtain

$$P = Ae^{-rt}$$

which is the present value of the amount A due in t years if money earns interest at an annual nominal rate r compounded continuously.

(A) Graph $P = 10,000e^{-0.08t}$, $0 \leq t \leq 50$.

(B) $\lim_{t\to\infty} 10,000e^{-0.08t} = $? [Guess, using part (A).]

[*Conclusion:* The longer the time until the amount A is due, the smaller is its present value, as we would expect.]

24. *Present value.* Referring to Problem 23, in how many years will the $10,000 have to be due in order for its present value to be $5,000?

25. *Doubling time.* How long will it take money to double if it is invested at 7% compounded continuously?

26. *Doubling time.* How long will it take money to double if it is invested at 5% compounded continuously?

27. *Doubling rate.* At what nominal rate compounded continuously must money be invested to double in 8 years?

28. *Doubling rate.* At what nominal rate compounded continuously must money be invested to double in 10 years?

29. *Growth time.* A man with $20,000 to invest decides to diversify his investments by placing $10,000 in an account that earns 7.2% compounded continuously and $10,000 in an account that earns 8.4% compounded annually. Use graphical approximation methods to determine how long it will take for his total investment in the two accounts to grow to $35,000.

30. *Growth time.* A woman invests $5,000 in an account that earns 8.8% compounded continuously and $7,000 in an account that earns 9.6% compounded annually. Use graphical approximation methods to determine how long it will take for her total investment in the two accounts to grow to $20,000.

31. *Doubling times*

(A) Show that the doubling time t (in years) at an annual rate r compounded continuously is given by

$$t = \frac{\ln 2}{r}$$

(B) Graph the doubling-time equation from part (A) for $0.02 \le r \le 0.30$. Is this restriction on r reasonable? Explain.

(C) Determine the doubling times (in years, to two decimal places) for $r = 5\%, 10\%, 15\%, 20\%, 25\%$, and 30%.

32. *Doubling rates*

(A) Show that the rate r that doubles an investment at continuously compounded interest in t years is given by

$$r = \frac{\ln 2}{t}$$

(B) Graph the doubling-rate equation from part (A) for $1 \le t \le 20$. Is this restriction on t reasonable? Explain.

(C) Determine the doubling rates for $t = 2, 4, 6, 8, 10$, and 12 years.

33. *Radioactive decay.* A mathematical model for the decay of radioactive substances is given by

$$Q = Q_0 e^{rt}$$

where
$Q_0 =$ amount of the substance at time $t = 0$
$r =$ continuous compound rate of decay
$t =$ time in years
$Q =$ amount of the substance at time t

If the continuous compound rate of decay of radium per year is $r = -0.000\ 433\ 2$, how long will it take a certain amount of radium to decay to half the original amount? (This period is the *half-life* of the substance.)

34. *Radioactive decay.* The continuous compound rate of decay of carbon-14 per year is $r = -0.000\ 123\ 8$. How long will it take a certain amount of carbon-14 to decay to half the original amount? (Use the radioactive decay model in Problem 33.)

35. *Radioactive decay.* A cesium isotope has a half-life of 30 years. What is the continuous compound rate of decay? (Use the radioactive decay model in Problem 33.)

36. *Radioactive decay.* A strontium isotope has a half-life of 90 years. What is the continuous compound rate of decay? (Use the radioactive decay model in Problem 33.)

37. *World population.* A mathematical model for world population growth over short intervals is given by

$$P = P_0 e^{rt}$$

where
$P_0 =$ population at time $t = 0$
$r =$ continuous compound rate of growth
$t =$ time in years
$P =$ population at time t

How long will it take world population to double if it continues to grow at its current continuous compound rate of 1.3% per year?

38. *U.S. population.* How long will it take for the U.S. population to double if it continues to grow at a rate of 0.85% per year?

39. *Population growth.* Some underdeveloped nations have population doubling times of 50 years. At what continuous compound rate is the population growing? (Use the population growth model in Problem 37.)

40. *Population growth.* Some developed nations have population doubling times of 200 years. At what continuous compound rate is the population growing? (Use the population growth model in Problem 37.)

Section 4-2 DERIVATIVES OF EXPONENTIAL AND LOGARITHMIC FUNCTIONS

- The Derivative of e^x
- The Derivative of $\ln x$
- Other Logarithmic and Exponential Functions
- Exponential and Logarithmic Models

In this section, we obtain formulas for the derivatives of logarithmic and exponential functions. A review of Sections 2–4 and 2–5 may prove helpful. In particular, recall that $f(x) = e^x$ is the exponential function with base $e \approx 2.718$ and the inverse of the function e^x is the natural logarithm function $\ln x$. More generally, if b is a positive real number, $b \neq 1$, then the exponential function b^x with base b, and the logarithmic function $\log_b x$ with base b, are inverses of each other.

■ The Derivative of e^x

In the process of finding the derivative of e^x, we will use (without proof) the fact that

$$\lim_{h \to 0} \frac{e^h - 1}{h} = 1 \tag{1}$$

Explore & Discuss **1**   Complete Table 1.

TABLE 1

h	−0.1	−0.01	−0.001 → 0 ← 0.001	0.01	0.1
$\dfrac{e^h - 1}{h}$					

Do your calculations make it reasonable to conclude that

$$\lim_{h \to 0} \frac{e^h - 1}{h} = 1?$$

Discuss.

We now apply the four-step process (Section 3-4) to the exponential function $f(x) = e^x$.

Step 1. Find $f(x + h)$.

$$f(x + h) = e^{x+h} = e^x e^h \quad \text{See Section 2-4.}$$

Step 2. Find $f(x + h) - f(x)$.

$$f(x + h) - f(x) = e^x e^h - e^x \quad \text{Factor out } e^x.$$
$$= e^x(e^h - 1)$$

Step 3. Find $\dfrac{f(x + h) - f(x)}{h}$.

$$\frac{f(x + h) - f(x)}{h} = \frac{e^x(e^h - 1)}{h} = e^x\left(\frac{e^h - 1}{h}\right)$$

Step 4. Find $f'(x) = \lim_{h \to 0} \dfrac{f(x + h) - f(x)}{h}$.

$$f'(x) = \lim_{h \to 0} \frac{f(x + h) - f(x)}{h}$$

$$= \lim_{h \to 0} e^x \left(\frac{e^h - 1}{h} \right)$$

$$= e^x \lim_{h \to 0} \left(\frac{e^h - 1}{h} \right) \qquad \text{Use the limit in (1).}$$

$$= e^x \cdot 1 = e^x$$

Thus,

$$\frac{d}{dx} e^x = e^x \qquad \text{The derivative of the exponential function is the exponential function.}$$

EXAMPLE 1   **Finding Derivatives** Find $f'(x)$ for

(A) $f(x) = 5e^x - 3x^4 + 9x + 16$

(B) $f(x) = -7x^e + 2e^x + e^2$

SOLUTIONS   (A) $f'(x) = 5e^x - 12x^3 + 9$

(B) $f'(x) = -7ex^{e-1} + 2e^x$

Remember that e is a real number, so the power rule (Section 3–5) is used to find the derivative of x^e. The derivative of the exponential function e^x, however, is e^x. Note that $e^2 \approx 7.389$ is a constant, so its derivative is 0.

MATCHED PROBLEM 1   Find $f'(x)$ for

(A) $f(x) = 4e^x + 8x^2 + 7x - 14$

(B) $f(x) = x^7 - x^5 + e^3 - x + e^x$

 CAUTION:

$$\frac{d}{dx} e^x \neq xe^{x-1} \qquad \frac{d}{dx} e^x = e^x$$

The power rule cannot be used to differentiate the exponential function. The power rule applies to exponential forms x^n where the exponent is a constant and the base is a variable. In the exponential form e^x, the base is a constant and the exponent is a variable.

■ The Derivative of ln x

We summarize some important facts about logarithmic functions from Section 2–5:

SUMMARY   Recall that the inverse of an exponential function is called a **logarithmic function.** For $b > 0$ and $b \neq 1$,

Logarithmic form		Exponential form
$y = \log_b x$	is equivalent to	$x = b^y$
Domain: $(0, \infty)$		Domain: $(-\infty, \infty)$
Range: $(-\infty, \infty)$		Range: $(0, \infty)$

The graphs of $y = \log_b x$ and $y = b^x$ are symmetric with respect to the line $y = x$. (See Figure 1.)

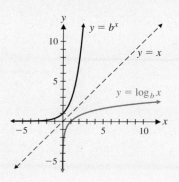

FIGURE 1

Of all the possible bases for logarithmic functions, the two most widely used are

$$\log x = \log_{10} x \quad \textit{Common logarithm (base 10)}$$
$$\ln x = \log_e x \quad \textit{Natural logarithm (base e)}$$

We are now ready to use the definition of the derivative and the four-step process discussed in Section 3-4 to find a formula for the derivative of $\ln x$. Later we will extend this formula to include $\log_b x$ for any base b.

Let $f(x) = \ln x$, $x > 0$.

Step 1. Find $f(x + h)$.

$$f(x + h) = \ln(x + h) \quad \textit{ln(x + h) cannot be simplified.}$$

Step 2. Find $f(x + h) - f(x)$.

$$f(x + h) - f(x) = \ln(x + h) - \ln x \quad \textit{Use ln A} - \textit{ln B} = \textit{ln} \dfrac{A}{B}.$$
$$= \ln \frac{x + h}{x}$$

Step 3. Find $\dfrac{f(x + h) - f(x)}{h}$.

$$\frac{f(x + h) - f(x)}{h} = \frac{\ln(x + h) - \ln x}{h}$$
$$= \frac{1}{h} \ln \frac{x + h}{x} \qquad \textit{Multiply by 1} = \textit{x/x to change form.}$$
$$= \frac{x}{x} \cdot \frac{1}{h} \ln \frac{x + h}{x}$$
$$= \frac{1}{x} \left[\frac{x}{h} \ln \left(1 + \frac{h}{x} \right) \right] \qquad \textit{Use p ln A} = \textit{ln A}^p.$$
$$= \frac{1}{x} \ln \left(1 + \frac{h}{x} \right)^{x/h}$$

Step 4. Find $f'(x) = \lim\limits_{h \to 0} \dfrac{f(x + h) - f(x)}{h}$.

$$f'(x) = \lim_{h \to 0} \frac{f(x + h) - f(x)}{h}$$
$$= \lim_{h \to 0} \left[\frac{1}{x} \ln \left(1 + \frac{h}{x} \right)^{x/h} \right] \qquad \textit{Let s} = \textit{h/x. Note that h} \to 0$$
$$\textit{implies s} \to 0.$$

$$= \frac{1}{x} \lim_{s \to 0} [\ln(1 + s)^{1/s}] \qquad \text{Use a new limit property.*}$$

$$= \frac{1}{x} \ln \left[\lim_{s \to 0} (1 + s)^{1/s} \right] \qquad \text{Use the definition of } e.$$

$$= \frac{1}{x} \ln e \qquad\qquad \ln e = \log_e e = 1$$

$$= \frac{1}{x}$$

Thus,

$$\frac{d}{dx} \ln x = \frac{1}{x}$$

INSIGHT

In the derivation of the derivative of $\ln x$, we used the following properties of logarithms:

$$\ln \frac{A}{B} = \ln A - \ln B \qquad \ln A^p = p \ln A$$

We also noted that there is no property that simplifies $\ln(A + B)$. (See Theorem 1 in Section 2-5 for a list of properties of logarithms.)

EXAMPLE 2

Finding Derivatives Find y' for

(A) $y = 3e^x + 5 \ln x$

(B) $y = x^4 - \ln x^4$

SOLUTIONS (A) $y' = 3e^x + \dfrac{5}{x}$

(B) Before taking the derivative, we use a property of logarithms (see Theorem 1, Section 2-5) to rewrite y.

$$y = x^4 - \ln x^4 \qquad \text{Use ln } M^p = p \text{ ln } M.$$
$$y = x^4 - 4 \ln x \qquad \text{Now take the derivative of both sides.}$$
$$y' = 4x^3 - \frac{4}{x}$$

MATCHED PROBLEM 2

Find y' for

(A) $y = 10x^3 - 100 \ln x$

(B) $y = \ln x^5 + e^x - \ln e^2$

■ Other Logarithmic and Exponential Functions

In most applications involving logarithmic or exponential functions, the number e is the preferred base. However, in some situations it is convenient to use a base other than e. Derivatives of $y = \log_b x$ and $y = b^x$ can be obtained by expressing these functions in terms of the natural logarithmic and exponential functions.

---

* The following new limit property is used: If $\lim_{x \to c} f(x)$ exists and is positive, then $\lim_{x \to c} [\ln f(x)] = \ln[\lim_{x \to c} f(x)]$.

We begin by finding a relationship between $\log_b x$ and $\ln x$ for any base $b, b > 0$ and $b \neq 1$. Some of you may prefer to remember the process, others the formula.

$$y = \log_b x \qquad \textit{Change to exponential form.}$$
$$b^y = x \qquad \textit{Take the natural logarithm of both sides.}$$
$$\ln b^y = \ln x \qquad \textit{Recall that } \ln b^y = y \ln b.$$
$$y \ln b = \ln x \qquad \textit{Solve for y.}$$
$$y = \frac{1}{\ln b} \ln x$$

Thus,

$$\log_b x = \frac{1}{\ln b} \ln x \qquad \textit{Change-of-base formula for logarithms*} \qquad (2)$$

Similarly, we can find a relationship between b^x and e^x for any base $b, b > 0, b \neq 1$.

$$y = b^x \qquad \textit{Take the natural logarithm of both sides.}$$
$$\ln y = \ln b^x \qquad \textit{Recall that } \ln b^x = x \ln b.$$
$$\ln y = x \ln b \qquad \textit{Take the exponential function of both sides.}$$
$$y = e^{x \ln b}$$

Thus,

$$b^x = e^{x \ln b} \qquad \textit{Change-of-base formula for exponential functions} \qquad (3)$$

Differentiating both sides of equation (2) gives

$$\frac{d}{dx} \log_b x = \frac{1}{\ln b} \frac{d}{dx} \ln x = \frac{1}{\ln b}\left(\frac{1}{x}\right)$$

It can be shown that the derivative of the function e^{cx}, where c is a constant, is the function ce^{cx} (see Problems 49–50 in Exercise 4-2 or the more general results of Section 4-4). Therefore, differentiating both sides of equation (3), we have

$$\frac{d}{dx} b^x = e^{x \ln b} \ln b = b^x \ln b$$

For convenient reference, we list the derivative formulas that we have obtained for exponential and logarithmic functions:

DERIVATIVES OF EXPONENTIAL AND LOGARITHMIC FUNCTIONS

For $b > 0, b \neq 1$,

$$\frac{d}{dx} e^x = e^x \qquad\qquad \frac{d}{dx} b^x = b^x \ln b$$

$$\frac{d}{dx} \ln x = \frac{1}{x} \qquad\qquad \frac{d}{dx} \log_b x = \frac{1}{\ln b}\left(\frac{1}{x}\right)$$

EXAMPLE 3 **Finding Derivatives** Find $g'(x)$ for

(A) $g(x) = 2^x - 3^x$

(B) $g(x) = \log_4 x^5$

---

*Equation (2) is a special case of the **general change-of-base formula** for logarithms (which can be derived in the same way): $\log_b x = (\log_a x)/(\log_a b)$.

SOLUTIONS   (A) $g'(x) = 2^x \ln 2 - 3^x \ln 3$

(B) First use a property of logarithms to rewrite $g(x)$.

$$g(x) = \log_4 x^5 \qquad \text{Use } \log_b M^p = p \log_b M.$$
$$g(x) = 5 \log_4 x \qquad \text{Take the derivative of both sides.}$$
$$g'(x) = \frac{5}{\ln 4}\left(\frac{1}{x}\right)$$

MATCHED PROBLEM 3   Find $g'(x)$ for

(A) $g(x) = x^{10} + 10^x$

(B) $g(x) = \log_2 x - 6 \log_5 x$

Explore & Discuss 2   (A) The graphs of $f(x) = \log_2 x$ and $g(x) = \log_4 x$ are shown in Figure 2. Which graph belongs to which function?

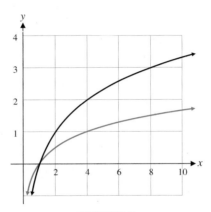

FIGURE 2

(B) Sketch graphs of $f'(x)$ and $g'(x)$.

(C) The function $f(x)$ is related to $g(x)$ in the same way that $f'(x)$ is related to $g'(x)$. What is that relationship?

■ Exponential and Logarithmic Models

EXAMPLE 4   **Price–Demand Model** An Internet store sells blankets made of the finest Australian wool. If the store sells x blankets at a price of $\$p$ per blanket, then the price–demand equation is $p = 350(0.999)^x$. Find the rate of change of price with respect to demand when the demand is 800 blankets, and interpret the result.

SOLUTION $\dfrac{dp}{dx} = 350(0.999)^x \ln 0.999$

If $x = 800$, then

$$\frac{dp}{dx} = 350(0.999)^{800} \ln 0.999 \approx -0.157, \text{ or } -\$0.16$$

When the demand is 800 blankets, the price is decreasing about $0.16 per blanket.

MATCHED PROBLEM 4   The store in Example 4 also sells a reversible fleece blanket. If the price–demand equation for reversible fleece blankets is $p = 200(0.998)^x$, find the rate of change of price with respect to demand when the demand is 400 blankets and interpret the result.

EXAMPLE 5   **Cable TV Subscribers** A statistician used data from the U.S. census to construct the model

$$S(t) = 21 \ln t + 2$$

where $S(t)$ is the number of cable TV subscribers (in millions) in year t ($t = 0$ corresponds to 1980). Use this model to estimate the number of cable TV subscribers in 2010 and the rate of change of the number of subscribers in 2010 (round both to the nearest tenth of a million). Interpret these results.

SOLUTION   Since 2010 corresponds to $t = 30$, we must find $S(30)$ and $S'(30)$.

$$S(30) = 21 \ln 30 + 2 = 73.4 \text{ million}$$

$$S'(t) = 21\frac{1}{t} = \frac{21}{t}$$

$$S'(30) = \frac{21}{30} = 0.7 \text{ million}$$

In 2010 there will be approximately 73.4 million subscribers, and this number is growing at the rate of 0.7 million per year.   ■

MATCHED PROBLEM 5   A model for newspaper circulation is

$$C(t) = 83 - 9 \ln t$$

where $C(t)$ is newspaper circulation (in millions) in year t ($t = 0$ corresponds to 1980). Use this model to estimate the circulation and the rate of change of circulation in 2010 (round both to the nearest tenth of a million). Interpret these results.

INSIGHT

On most graphing calculators, exponential regression produces a function of the form $y = a \cdot b^x$. Formula (3) enables you to change the base b (chosen by the graphing calculator) to the more familiar base e:

$$y = a \cdot b^x = a \cdot e^{x \ln b}$$

On most graphing calculators, logarithmic regression produces a function of the form $y = a + b \ln x$. Formula (2) enables you to write the function in terms of logarithms to any base d that you may prefer:

$$y = a + b \ln x = a + b(\ln d) \log_d x$$

Answers to Matched Problems   **1.** (A) $4e^x + 16x + 7$

(B) $7x^6 - 5x^4 - 1 + e^x$

2. (A) $30x^2 - \dfrac{100}{x}$

(B) $\dfrac{5}{x} + e^x$

3. (A) $10x^9 + 10^x \ln 10$

(B) $\left(\dfrac{1}{\ln 2} - \dfrac{6}{\ln 5} \right) \dfrac{1}{x}$

4. The price is decreasing at the rate of $0.18 per blanket.

5. The circulation in 2010 is approximately 52.4 million and is decreasing at the rate of 0.3 million per year.

Exercise 4-2

A *In Problems 1–14, find $f'(x)$.*

1. $f(x) = 5e^x + 3x + 1$

2. $f(x) = -7e^x - 2x + 5$

3. $f(x) = -2 \ln x + x^2 - 4$

4. $f(x) = 6 \ln x - x^3 + 2$

5. $f(x) = x^3 - 6e^x$

6. $f(x) = 9e^x + 2x^2$

7. $f(x) = e^x + x - \ln x$

8. $f(x) = \ln x + 2e^x - 3x^2$

9. $f(x) = \ln x^3$

10. $f(x) = \ln x^8$

11. $f(x) = 5x - \ln x^5$

12. $f(x) = 4 + \ln x^9$

13. $f(x) = \ln x^2 + 4e^x$

14. $f(x) = \ln x^{10} + 2 \ln x$

B *In Problems 15–22, find the equation of the line tangent to the graph of f at the indicated value of x.*

15. $f(x) = 3 + \ln x; \; x = 1$

16. $f(x) = 2 \ln x; \; x = 1$

17. $f(x) = 3e^x; \; x = 0$

18. $f(x) = e^x + 1; \; x = 0$

19. $f(x) = \ln x^3; \; x = e$

20. $f(x) = 1 + \ln x^4; \; x = e$

21. $f(x) = 2 + e^x; \; x = 1$

22. $f(x) = 5e^x; \; x = 1$

23. A student claims that the line tangent to the graph of $f(x) = e^x$ at $x = 3$ passes through the point $(2, 0)$ (see the figure). Is she correct? Will the line tangent at $x = 4$ pass through $(3, 0)$? Explain.

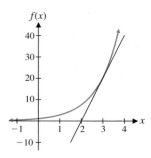

Figure for 23

24. Refer to Problem 23. Does the line tangent to the graph of $f(x) = e^x$ at $x = 1$ pass through the origin? Are there any other lines tangent to the graph of f that pass through the origin? Explain.

25. A student claims that the line tangent to the graph of $g(x) = \ln x$ at $x = 3$ passes through the origin (see the figure). Is he correct? Will the line tangent at $x = 4$ pass through the origin? Explain.

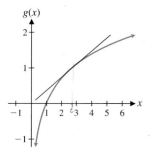

Figure for 25

26. Refer to Problem 25. Does the line tangent to the graph of $f(x) = \ln x$ at $x = e$ pass through the origin? Are there any other lines tangent to the graph of f that pass through the origin? Explain.

In Problems 27–30, first use appropriate properties of logarithms to rewrite f(x), and then find $f'(x)$.

27. $f(x) = 10x + \ln 10x$

28. $f(x) = 2 + 3 \ln \dfrac{1}{x}$

29. $f(x) = \ln \dfrac{4}{x^3}$

30. $f(x) = x + 5 \ln 6x$

C *In Problems 31–42, find $\dfrac{dy}{dx}$ for the indicated function y.*

31. $y = \log_2 x$

32. $y = 3 \log_5 x$

33. $y = 3^x$

34. $y = 4^x$

35. $y = 2x - \log x$

36. $y = \log x + 4x^2 + 1$

37. $y = 10 + x + 10^x$

38. $y = x^5 - 5^x$

39. $y = 3 \ln x + 2 \log_3 x$

40. $y = -\log_2 x + 10 \ln x$

41. $y = 2^x + e^2$

42. $y = e^3 - 3^x$

In Problems 43–48, use graphical approximation methods to find the points of intersection of f(x) and g(x) (to two decimal places).

43. $f(x) = e^x; g(x) = x^4$

   [Note that there are three points of intersection and that e^x is greater than x^4 for large values of x.]

44. $f(x) = e^x$; $g(x) = x^5$

  [Note that there are two points of intersection and that e^x is greater than x^5 for large values of x.]

45. $f(x) = (\ln x)^2$; $g(x) = x$

46. $f(x) = (\ln x)^3$; $g(x) = \sqrt{x}$

47. $f(x) = \ln x$; $g(x) = x^{1/5}$

48. $f(x) = \ln x$; $g(x) = x^{1/4}$

49. Explain why $\lim\limits_{h \to 0} \dfrac{e^{ch} - 1}{h} = c$.

50. Use the result of Problem 49 and the four-step process to show that if $f(x) = e^{cx}$, then $f'(x) = ce^{cx}$.

Applications

51. *Salvage value.* The salvage value S (in dollars) of a company airplane after t years is estimated to be given by

$$S(t) = 300,000(0.9)^t$$

What is the rate of depreciation (in dollars per year) after 1 year? 5 years? 10 years?

52. *Resale value.* The resale value R (in dollars) of a company car after t years is estimated to be given by

$$R(t) = 20,000(0.86)^t$$

What is the rate of depreciation (in dollars per year) after 1 year? 2 years? 3 years?

53. *Bacterial growth.* A single cholera bacterium divides every 0.5 hour to produce two complete cholera bacteria. If we start with a colony of 5,000 bacteria, after t hours there will be

$$A(t) = 5,000 \cdot 2^{2t} = 5,000 \cdot 4^t$$

bacteria. Find $A'(t)$, $A'(1)$, and $A'(5)$, and interpret the results.

54. *Bacterial growth.* Repeat Problem 53 for a starting colony of 1,000 bacteria such that a single bacterium divides every 0.25 hour.

55. *Blood pressure.* An experiment was set up to find a relationship between weight and systolic blood pressure in normal children. Using hospital records for 5,000 normal children, the experimenters found that the systolic blood pressure was given approximately by

$$P(x) = 17.5(1 + \ln x) \qquad 10 \le x \le 100$$

where $P(x)$ is measured in millimeters of mercury and x is measured in pounds. What is the rate of change of blood pressure with respect to weight at the 40-pound weight level? At the 90-pound weight level?

56. Refer to Problem 55. Find the weight (to the nearest pound) at which the rate of change of blood pressure with respect to weight is 0.3 millimeter of mercury per pound.

57. *Psychology: stimulus/response.* In psychology, the Weber–Fechner law for the response to a stimulus is

$$R = k \ln \frac{S}{S_0}$$

where R is the response, S is the stimulus, and S_0 is the lowest level of stimulus that can be detected. Find dR/dS.

58. *Psychology: learning.* A mathematical model for the average of a group of people learning to type is given by

$$N(t) = 10 + 6 \ln t \qquad t \ge 1$$

where $N(t)$ is the number of words per minute typed after t hours of instruction and practice (2 hours per day, 5 days per week). What is the rate of learning after 10 hours of instruction and practice? After 100 hours?

Section 4-3   DERIVATIVES OF PRODUCTS AND QUOTIENTS

  ■ Derivatives of Products
  ■ Derivatives of Quotients

The derivative properties discussed in Section 3-5 added substantially to our ability to compute and apply derivatives to many practical problems. In this and the next two sections, we add a few more properties that will increase this ability even further.

■ Derivatives of Products

In Section 3-5, we found that the derivative of a sum is the sum of the derivatives. Is the derivative of a product the product of the derivatives?

Explore & Discuss 1   Let $F(x) = x^2$, $S(x) = x^3$, and $f(x) = F(x)S(x) = x^5$. Which of the following is $f'(x)$?

(A) $F'(x)S'(x)$

(B) $F(x)S'(x)$

(C) $F'(x)S(x)$

(D) $F(x)S'(x) + F'(x)S(x)$

Comparing the various expressions computed in Explore & Discuss 1, we see that the derivative of a product is not the product of the derivatives, but appears to involve a slightly more complicated form.

Using the definition of the derivative and the four-step process, we can show that

The derivative of the product of two functions is the first function times the derivative of the second function, plus the second function times the derivative of the first function.

That is,

THEOREM 1   PRODUCT RULE

If

$$y = f(x) = F(x)S(x)$$

and if $F'(x)$ and $S'(x)$ exist, then

$$f'(x) = F(x)S'(x) + S(x)F'(x)$$

Also,

$$y' = FS' + SF' \qquad \frac{dy}{dx} = F\frac{dS}{dx} + S\frac{dF}{dx}$$

EXAMPLE 1   **Differentiating a Product** Use two different methods to find $f'(x)$ for

$$f(x) = 2x^2(3x^4 - 2).$$

SOLUTION   **Method 1.** Use the product rule:

$$\begin{aligned}
f'(x) &= 2x^2(3x^4 - 2)' + (3x^4 - 2)(2x^2)' \\
&= 2x^2(12x^3) + (3x^4 - 2)(4x) \\
&= 24x^5 + 12x^5 - 8x \\
&= 36x^5 - 8x
\end{aligned}$$

First times derivative of second, plus second times derivative of first

Method 2. Multiply first; then take derivatives:

$$f(x) = 2x^2(3x^4 - 2) = 6x^6 - 4x^2$$
$$f'(x) = 36x^5 - 8x$$

MATCHED PROBLEM 1   Use two different methods to find $f'(x)$ for $f(x) = 3x^3(2x^2 - 3x + 1)$.

Some of the products we encounter can be differentiated by either of the methods illustrated in Example 1. In other situations, the product rule *must* be used. Unless instructed otherwise, you should use the product rule to differentiate all products in this section in order to gain experience with the use of this important differentiation rule.

EXAMPLE 2 **Tangent Lines** Let $f(x) = (2x - 9)(x^2 + 6)$.

(A) Find the equation of the line tangent to the graph of $f(x)$ at $x = 3$.

(B) Find the value(s) of x where the tangent line is horizontal.

SOLUTION (A) First, find $f'(x)$:

$$f'(x) = (2x - 9)(x^2 + 6)' + (x^2 + 6)(2x - 9)'$$
$$= (2x - 9)(2x) + (x^2 + 6)(2)$$

Then, find $f(3)$ and $f'(3)$:

$$f(3) = [2(3) - 9](3^2 + 6) = (-3)(15) = -45$$
$$f'(3) = [2(3) - 9]2(3) + (3^2 + 6)(2) = -18 + 30 = 12$$

Now, find the equation of the tangent line at $x = 3$:

$$y - y_1 = m(x - x_1) \quad \text{\small $y_1 = f(x_1) = f(3) = -45$}$$
$$y - (-45) = 12(x - 3) \quad \text{\small $m = f'(x_1) = f'(3) = 12$}$$
$$y = 12x - 81 \quad \text{\small Tangent line at } x = 3$$

(B) The tangent line is horizontal at any value of x such that $f'(x) = 0$, so

$$f'(x) = (2x - 9)2x + (x^2 + 6)2 = 0$$
$$6x^2 - 18x + 12 = 0$$
$$x^2 - 3x + 2 = 0$$
$$(x - 1)(x - 2) = 0$$
$$x = 1, 2$$

The tangent line is horizontal at $x = 1$ and at $x = 2$. ▬

MATCHED PROBLEM 2 Repeat Example 2 for $f(x) = (2x + 9)(x^2 - 12)$.

INSIGHT

As Example 2 illustrates, the way we write $f'(x)$ depends on what we want to do with it. If we are interested only in evaluating $f'(x)$ at specified values of x, the form in part (A) is sufficient. However, if we want to solve $f'(x) = 0$, we must multiply and collect like terms, as we did in part (B).

EXAMPLE 3 **Finding Derivatives** Find $f'(x)$ for

(A) $f(x) = 2x^3 e^x$

(B) $f(x) = 6x^4 \ln x$

SOLUTIONS

(A) $f'(x) = 2x^3(e^x)' + e^x(2x^3)'$
$$= 2x^3 e^x + e^x(6x^2)$$
$$= 2x^2 e^x(x + 3)$$

(B) $f'(x) = 6x^4 (\ln x)' + (\ln x)(6x^4)'$
$$= 6x^4 \frac{1}{x} + (\ln x)(24x^3)$$
$$= 6x^3 + 24x^3 \ln x$$
$$= 6x^3 (1 + 4 \ln x)$$

MATCHED PROBLEM 3   Find $f'(x)$ for

(A) $f(x) = 5x^8 e^x$

(B) $f(x) = x^7 \ln x$

■ Derivatives of Quotients

As is the case with a product, the derivative of a quotient of two functions is not the quotient of the derivatives of the two functions.

Explore & Discuss **2**   Let $T(x) = x^5$, $B(x) = x^2$, and

$$f(x) = \frac{T(x)}{B(x)} = \frac{x^5}{x^2} = x^3$$

Which of the following is $f'(x)$?

(A) $\dfrac{T'(x)}{B'(x)}$ (B) $\dfrac{T'(x)B(x)}{[B(x)]^2}$ (C) $\dfrac{T(x)B'(x)}{[B(x)]^2}$

(D) $\dfrac{T'(x)B(x)}{[B(x)]^2} - \dfrac{T(x)B'(x)}{[B(x)]^2} = \dfrac{T'(x)B(x) - T(x)B'(x)}{[B(x)]^2}$

The expressions in Explore & Discuss 2 suggest that the derivative of a quotient leads to a more complicated quotient than you might expect.

In general, if $T(x)$ and $B(x)$ are any two differentiable functions and

$$f(x) = \frac{T(x)}{B(x)}$$

then

$$f'(x) = \frac{B(x)T'(x) - T(x)B'(x)}{[B(x)]^2}$$

Thus,

> The derivative of the quotient of two functions is the denominator function times the derivative of the numerator function, minus the numerator function times the derivative of the denominator function, divided by the denominator function squared.

THEOREM 2   QUOTIENT RULE

If

$$y = f(x) = \frac{T(x)}{B(x)}$$

and if $T'(x)$ and $B'(x)$ exist, then

$$f'(x) = \frac{B(x)T'(x) - T(x)B'(x)}{[B(x)]^2}$$

Also,

$$y' = \frac{BT' - TB'}{B^2} \qquad \frac{dy}{dx} = \frac{B\dfrac{dT}{dx} - T\dfrac{dB}{dx}}{B^2}$$

| EXAMPLE 4 | **Differentiating Quotients** |

(A) If $f(x) = \dfrac{x^2}{2x - 1}$, find $f'(x)$. (B) If $y = \dfrac{t^2 - t}{t^3 + 1}$, find y'.

(C) Find $\dfrac{d}{dx} \dfrac{x^2 - 3}{x^2}$ by using the quotient rule and also by splitting the fraction into two fractions.

SOLUTION (A) $f'(x) = \dfrac{(2x - 1)(x^2)' - x^2(2x - 1)'}{(2x - 1)^2}$

The denominator times the derivative of the numerator, minus the numerator times the derivative of the denominator, divided by the square of the denominator

$= \dfrac{(2x - 1)(2x) - x^2(2)}{(2x - 1)^2}$

$= \dfrac{4x^2 - 2x - 2x^2}{(2x - 1)^2}$

$= \dfrac{2x^2 - 2x}{(2x - 1)^2}$

(B) $y' = \dfrac{(t^3 + 1)(t^2 - t)' - (t^2 - t)(t^3 + 1)'}{(t^3 + 1)^2}$

$= \dfrac{(t^3 + 1)(2t - 1) - (t^2 - t)(3t^2)}{(t^3 + 1)^2}$

$= \dfrac{2t^4 - t^3 + 2t - 1 - 3t^4 + 3t^3}{(t^3 + 1)^2}$

$= \dfrac{-t^4 + 2t^3 + 2t - 1}{(t^3 + 1)^2}$

(C) **Method 1.** Use the quotient rule:

$\dfrac{d}{dx} \dfrac{x^2 - 3}{x^2} = \dfrac{x^2 \dfrac{d}{dx}(x^2 - 3) - (x^2 - 3)\dfrac{d}{dx}x^2}{(x^2)^2}$

$= \dfrac{x^2(2x) - (x^2 - 3)2x}{x^4}$

$= \dfrac{2x^3 - 2x^3 + 6x}{x^4} = \dfrac{6x}{x^4} = \dfrac{6}{x^3}$

Method 2. Split into two fractions:

$\dfrac{x^2 - 3}{x^2} = \dfrac{x^2}{x^2} - \dfrac{3}{x^2} = 1 - 3x^{-2}$

$\dfrac{d}{dx}(1 - 3x^{-2}) = 0 - 3(-2)x^{-3} = \dfrac{6}{x^3}$

Comparing methods 1 and 2, we see that it often pays to change an expression algebraically before blindly using a differentiation formula. ▰

| MATCHED PROBLEM 4 | Find |

(A) $f'(x)$ for $f(x) = \dfrac{2x}{x^2 + 3}$

(B) y' for $y = \dfrac{t^3 - 3t}{t^2 - 4}$

(C) $\dfrac{d}{dx} \dfrac{2 + x^3}{x^3}$ in two ways

Explore & Discuss **3**   Explain why ≠ is used in the following expression, and then find the correct derivative:

$$\frac{d}{dx}\frac{x^3}{x^2 + 3x + 4} \neq \frac{3x^2}{2x + 3}$$

EXAMPLE 5   **Finding Derivatives** Find $f'(x)$ for

(A) $f(x) = \dfrac{3e^x}{1 + e^x}$

(B) $f(x) = \dfrac{\ln x}{2x + 5}$

SOLUTIONS   (A) $f'(x) = \dfrac{(1 + e^x)(3e^x)' - 3e^x(1 + e^x)'}{(1 + e^x)^2}$

$$= \frac{(1 + e^x)3e^x - 3e^x e^x}{(1 + e^x)^2}$$

$$= \frac{3e^x}{(1 + e^x)^2}$$

(B) $f'(x) = \dfrac{(2x + 5)(\ln x)' - (\ln x)(2x + 5)'}{(2x + 5)^2}$

$$= \frac{(2x + 5) \cdot \dfrac{1}{x} - (\ln x)2}{(2x + 5)^2} \qquad \text{Multiply by } \frac{x}{x}$$

$$= \frac{2x + 5 - 2x \ln x}{x(2x + 5)^2}$$

MATCHED PROBLEM 5   Find $f'(x)$ for

(A) $f'(x) = \dfrac{x^3}{e^x + 2}$

(B) $f(x) = \dfrac{4x}{1 + \ln x}$

EXAMPLE 6   **Sales Analysis** The total sales S (in thousands of games) of a home video game t months after the game is introduced are given by

$$S(t) = \frac{125t^2}{t^2 + 100}$$

(A) Find $S'(t)$.

(B) Find $S(10)$ and $S'(10)$. Write a brief verbal interpretation of these results.

(C) Use the results from part (B) to estimate the total sales after 11 months.

SOLUTION (A) $S'(t) = \dfrac{(t^2 + 100)(125t^2)' - 125t^2(t^2 + 100)'}{(t^2 + 100)^2}$

$= \dfrac{(t^2 + 100)(250t) - 125t^2(2t)}{(t^2 + 100)^2}$

$= \dfrac{250t^3 + 25{,}000t - 250t^3}{(t^2 + 100)^2}$

$= \dfrac{25{,}000t}{(t^2 + 100)^2}$

(B) $S(10) = \dfrac{125(10)^2}{10^2 + 100} = 62.5$ and $S'(10) = \dfrac{25{,}000(10)}{(10^2 + 100)^2} = 6.25$

The total sales after 10 months are 62,500 games, and sales are increasing at the rate of 6,250 games per month.

(C) The total sales will increase by approximately 6,250 games during the next month. Thus, the estimated total sales after 11 months are 62,500 + 6,250 = 68,750 games. ∎

MATCHED PROBLEM 6 Refer to Example 6. Suppose that the total sales S (in thousands of games) t months after the game is introduced are given by

$$S(t) = \frac{150t}{t + 3}$$

(A) Find $S'(t)$.

(B) Find $S(12)$ and $S'(12)$. Write a brief verbal interpretation of these results.

(C) Use the results from part (B) to estimate the total sales after 13 months.

Answers to Matched Problems

1. $30x^4 - 36x^3 + 9x^2$

2. (A) $y = 84x - 297$

   (B) $x = -4, x = 1$

3. (A) $5x^8 e^x + e^x(40x^7) = 5x^7(x + 8)e^x$

   (B) $x^7 \cdot \dfrac{1}{x} + \ln x\,(7x^6) = x^6\,(1 + 7\ln x)$

4. (A) $\dfrac{(x^2 + 3)2 - (2x)(2x)}{(x^2 + 3)^2} = \dfrac{6 - 2x^2}{(x^2 + 3)^2}$

   (B) $\dfrac{(t^2 - 4)(3t^2 - 3) - (t^3 - 3t)(2t)}{(t^2 - 4)^2} = \dfrac{t^4 - 9t^2 + 12}{(t^2 - 4)^2}$

   (C) $-\dfrac{6}{x^4}$

5. (A) $\dfrac{(e^x + 2)\,3x^2 - x^3 e^x}{(e^x + 2)^2}$

   (B) $\dfrac{(1 + \ln x)\,4 - 4x\,\dfrac{1}{x}}{(1 + \ln x)^2} = \dfrac{4\ln x}{(1 + \ln x)^2}$

6. (A) $S'(t) = \dfrac{450}{(t + 3)^2}$

   (B) $S(12) = 120;\ S'(12) = 2$. After 12 months, the total sales are 120,000 games, and sales are increasing at the rate of 2,000 games per month.

   (C) 122,000 games

Exercise 4-3

The answers to most of the problems in this exercise set contain both an unsimplified form and a simplified form of the derivative. When checking your work, first check that you applied the rules correctly, and then check that you performed the algebraic simplification correctly. Unless instructed otherwise, when differentiating a product, use the product rule rather than performing the multiplication first.

A *In Problems 1–26, find $f'(x)$ and simplify.*

1. $f(x) = 2x^3(x^2 - 2)$ **2.** $f(x) = 5x^2(x^3 + 2)$

3. $f(x) = (x - 3)(2x - 1)$

4. $f(x) = (3x + 2)(4x - 5)$ **5.** $f(x) = \dfrac{x}{x - 3}$

6. $f(x) = \dfrac{3x}{2x + 1}$ **7.** $f(x) = \dfrac{2x + 3}{x - 2}$

8. $f(x) = \dfrac{3x - 4}{2x + 3}$ **9.** $f(x) = 3xe^x$

10. $f(x) = x^2 e^x$ **11.** $f(x) = x^3 \ln x$

12. $f(x) = 5x \ln x$

13. $f(x) = (x^2 + 1)(2x - 3)$

14. $f(x) = (3x + 5)(x^2 - 3)$

15. $f(x) = (0.4x + 2)(0.5x - 5)$

16. $f(x) = (0.5x - 4)(0.2x + 1)$

17. $f(x) = \dfrac{x^2 + 1}{2x - 3}$ **18.** $f(x) = \dfrac{3x + 5}{x^2 - 3}$

19. $f(x) = (x^2 + 2)(x^2 - 3)$

20. $f(x) = (x^2 - 4)(x^2 + 5)$

21. $f(x) = \dfrac{x^2 + 2}{x^2 - 3}$ **22.** $f(x) = \dfrac{x^2 - 4}{x^2 + 5}$

23. $f(x) = \dfrac{e^x}{x^2 + 1}$ **24.** $f(x) = \dfrac{1 - e^x}{1 + e^x}$

25. $f(x) = \dfrac{\ln x}{1 + x}$ **26.** $f(x) = \dfrac{2x}{1 + \ln x}$

In Problems 27–38, find $h'(x)$, where $f(x)$ is an unspecified differentiable function.

27. $h(x) = xf(x)$ **28.** $h(x) = x^2 f(x)$

29. $h(x) = x^3 f(x)$ **30.** $h(x) = \dfrac{f(x)}{x}$

31. $h(x) = \dfrac{f(x)}{x^2}$ **32.** $h(x) = \dfrac{f(x)}{x^3}$

33. $h(x) = \dfrac{x}{f(x)}$ **34.** $h(x) = \dfrac{x^2}{f(x)}$

35. $h(x) = e^x f(x)$ **36.** $h(x) = \dfrac{e^x}{f(x)}$

37. $h(x) = \dfrac{\ln x}{f(x)}$ **38.** $h(x) = \dfrac{f(x)}{\ln x}$

B *In Problems 39–48, find the indicated derivatives and simplify.*

39. $f'(x)$ for $f(x) = (2x + 1)(x^2 - 3x)$

40. y' for $y = (x^3 + 2x^2)(3x - 1)$

41. $\dfrac{dy}{dt}$ for $y = (2.5t - t^2)(4t + 1.4)$

42. $\dfrac{d}{dt}[(3 - 0.4t^3)(0.5t^2 - 2t)]$

43. y' for $y = \dfrac{5x - 3}{x^2 + 2x}$

44. $f'(x)$ for $f(x) = \dfrac{3x^2}{2x - 1}$

45. $\dfrac{d}{dw} \dfrac{w^2 - 3w + 1}{w^2 - 1}$

46. $\dfrac{dy}{dw}$ for $y = \dfrac{w^4 - w^3}{3w - 1}$

47. y' for $y = (1 + x - x^2)e^x$

48. $\dfrac{dy}{dt}$ for $y = (1 + e^t)\ln t$

In Problems 49–54, find $f'(x)$ and find the equation of the line tangent to the graph of f at $x = 2$.

49. $f(x) = (1 + 3x)(5 - 2x)$

50. $f(x) = (7 - 3x)(1 + 2x)$

51. $f(x) = \dfrac{x - 8}{3x - 4}$ **52.** $f(x) = \dfrac{2x - 5}{2x - 3}$

53. $f(x) = \dfrac{x}{2^x}$ **54.** $f(x) = (x - 2)\ln x$

In Problems 55–58, find $f'(x)$ and find the value(s) of x where $f'(x) = 0$.

55. $f(x) = (2x - 15)(x^2 + 18)$

56. $f(x) = (2x - 3)(x^2 - 6)$

57. $f(x) = \dfrac{x}{x^2 + 1}$ **58.** $f(x) = \dfrac{x}{x^2 + 9}$

In Problems 59–62, find $f'(x)$ in two ways: by using the product or quotient rule and by simplifying first.

59. $f(x) = x^3(x^4 - 1)$ **60.** $f(x) = x^4(x^3 - 1)$

61. $f(x) = \dfrac{x^3 + 9}{x^3}$ **62.** $f(x) = \dfrac{x^4 + 4}{x^4}$

C *In Problems 63–82, find each indicated derivative and simplify.*

63. $f(w) = (w + 1)2^w$

64. $g(w) = (w - 5)\log_3 w$ **65.** $\dfrac{d}{dx} \dfrac{3x^2 - 2x + 3}{4x^2 + 5x - 1}$

66. y' for $y = \dfrac{x^3 - 3x + 4}{2x^2 + 3x - 2}$

67. $\dfrac{dy}{dx}$ for $y = 9x^{1/3}(x^3 + 5)$

68. $\dfrac{d}{dx}[(4x^{1/2} - 1)(3x^{1/3} + 2)]$

69. y' for $y = \dfrac{\log_2 x}{1 + x^2}$ **70.** $\dfrac{dy}{dx}$ for $y = \dfrac{10^x}{1 + x^4}$

71. $f'(x)$ for $f(x) = \dfrac{6\sqrt[3]{x}}{x^2 - 3}$

72. y' for $y = \dfrac{2\sqrt{x}}{x^2 - 3x + 1}$

73. $g'(t)$ if $g(t) = \dfrac{0.2t}{3t^2 - 1}$

74. $h'(t)$ if $h(t) = \dfrac{-0.05t^2}{2t + 1}$

75. $\dfrac{d}{dx}\,[4x \log x^5]$

76. $\dfrac{d}{dt}\,[10^t \log t]$

77. $\dfrac{d}{dx}\,\dfrac{x^3 - 2x^2}{\sqrt[3]{x^2}}$

78. $\dfrac{dy}{dx}$ for $y = \dfrac{x^2 - 3x + 1}{\sqrt[4]{x}}$

79. $f'(x)$ for $f(x) = \dfrac{(2x^2 - 1)(x^2 + 3)}{x^2 + 1}$

80. y' for $y = \dfrac{2x - 1}{(x^3 + 2)(x^2 - 3)}$

81. $\dfrac{dy}{dt}$ for $y = \dfrac{t \ln t}{e^t}$

82. $\dfrac{dy}{du}$ for $y = \dfrac{u^2 e^u}{1 + \ln u}$

Applications

83. *Sales analysis.* The total sales S (in thousands of CDs) of a compact disk are given by

$$S(t) = \frac{90t^2}{t^2 + 50}$$

where t is the number of months since the release of the CD.

(A) Find $S'(t)$.

(B) Find $S(10)$ and $S'(10)$. Write a brief verbal interpretation of these results.

(C) Use the results from part (B) to estimate the total sales after 11 months.

84. *Sales analysis.* A communications company has installed a cable television system in a city. The total number N (in thousands) of subscribers t months after the installation of the system is given by

$$N(t) = \frac{180t}{t + 4}$$

(A) Find $N'(t)$.

(B) Find $N(16)$ and $N'(16)$. Write a brief verbal interpretation of these results.

(C) Use the results from part (B) to estimate the total number of subscribers after 17 months.

85. *Price–demand equation.* According to classical economic theory, the demand x for a quantity in a free market decreases as the price p increases (see the figure). Suppose that the number x of CD players people are willing to buy per week from a retail chain at a price of \$$p$ is given by

$$x = \frac{4{,}000}{0.1p + 1} \qquad 10 \le p \le 70$$

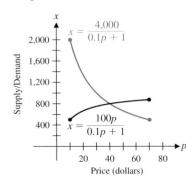

Figure for 85 and 86

(A) Find dx/dp.

(B) Find the demand and the instantaneous rate of change of demand with respect to price when the

price is \$40. Write a brief verbal interpretation of these results.

(C) Use the results from part (B) to estimate the demand if the price is increased to \$41.

86. *Price–supply equation.* Also according to classical economic theory, the supply x of a quantity in a free market increases as the price p increases (see the figure). Suppose that the number x of CD players a retail chain is willing to sell per week at a price of \$$p$ is given by

$$x = \frac{100p}{0.1p + 1} \qquad 10 \le p \le 70$$

(A) Find dx/dp.

(B) Find the supply and the instantaneous rate of change of supply with respect to price when the price is \$40. Write a brief verbal interpretation of these results.

(C) Use the results from part (B) to estimate the supply if the price is increased to \$41.

87. *Medicine.* A drug is injected into the bloodstream of a patient through her right arm. The concentration of the drug (in milligrams per cubic centimeter) in the bloodstream of the left arm t hours after the injection is given by

$$C(t) = \frac{0.14t}{t^2 + 1}$$

(A) Find $C'(t)$.

(B) Find $C'(0.5)$ and $C'(3)$, and interpret the results.

88. *Drug sensitivity.* One hour after a dose of x milligrams of a particular drug is administered to a person, the change in body temperature $T(x)$, in degrees Fahrenheit, is given approximately by

$$T(x) = x^2\left(1 - \frac{x}{9}\right) \qquad 0 \le x \le 7$$

The rate $T'(x)$ at which T changes with respect to the size of the dosage, x, is called the *sensitivity* of the body to the dosage.

(A) Use the product rule to find $T'(x)$.

(B) Find $T'(1)$, $T'(3)$, and $T'(6)$.

89. *Learning.* In the early days of quantitative learning theory (around 1917), L. L. Thurstone found that a given person successfully accomplished $N(x)$ acts after x practice acts, as given by

$$N(x) = \frac{100x + 200}{x + 32}$$

(A) Find the instantaneous rate of change of learning, $N'(x)$, with respect to the number of practice acts, x.

(B) Find $N'(4)$ and $N'(68)$.

Section 4-4   THE CHAIN RULE

- Composite Functions
- General Power Rule
- The Chain Rule

The word *chain* in the name "chain rule" comes from the fact that a function formed by composition involves a chain of functions—that is, a function of a function. The *chain rule* enables us to compute the derivative of a composite function in terms of the derivatives of the functions making up the composition. In this section, we review composite functions, introduce the chain rule by means of the special case known as the *general power rule,* and then discuss the chain rule itself.

■ Composite Functions

The function $m(x) = (x^2 + 4)^3$ is a combination of a quadratic function and a cubic function. To see this more clearly, let

$$y = f(u) = u^3 \qquad \text{and} \qquad u = g(x) = x^2 + 4$$

Then we can express y as a function of x as follows:

$$y = f(u) = f[g(x)] = [x^2 + 4]^3 = m(x)$$

The function m is said to be the *composite* of the two functions f and g. In general, we have the following:

DEFINITION   **Composite Functions**

A function m is a **composite** of functions f and g if

$$m(x) = f[g(x)]$$

The domain of m is the set of all numbers x such that x is in the domain of g and $g(x)$ is in the domain of f.

EXAMPLE 1   **Composite Functions**   Let $f(u) = e^u$ and $g(x) = -3x$. Find $f[g(x)]$ and $g[f(u)]$.

SOLUTION

$$f[g(x)] = f(-3x) = e^{-3x}$$
$$g[f(u)] = g(e^u) = -3e^u$$

MATCHED PROBLEM 1   Let $f(u) = 2u$ and $g(x) = e^x$. Find $f[g(x)]$ and $g[f(u)]$.

EXAMPLE 2   **Composite Functions**   Write each function as a composition of two simpler functions.

(A) $y = 100e^{0.04x}$

(B) $y = \sqrt{4 - x^2}$

SOLUTION   (A) Let

$$y = f(u) = 100e^u$$
$$u = g(x) = 0.04x$$

Check: $y = f[g(x)] = 100e^{g(x)} = 100e^{0.04x}$

(B) Let

$$y = f(u) = \sqrt{u}$$

$$u = g(x) = 4 - x^2$$

Check: $y = f[g(x)] = \sqrt{g(x)} = \sqrt{4 - x^2}$ ▬

| MATCHED PROBLEM 2 | Write each function as a composition of two simpler functions.

(A) $y = 50e^{-2x}$ (B) $y = \sqrt[3]{1 + x^3}$

INSIGHT

There can be more than one way to express a function as a composition of simpler functions. Choosing $y = f(u) = 100u$ and $u = g(x) = e^{0.04x}$ in Example 2A produces the same result:

$$y = f[g(x)] = 100g(x) = 100e^{0.04x}$$

Since we will be using composition as a means to an end (finding a derivative), usually it will not matter what choices you make for the functions in the composition.

■ General Power Rule

We have already made extensive use of the power rule,

$$\frac{d}{dx}x^n = nx^{n-1} \tag{1}$$

Now we want to generalize this rule so that we can differentiate composite functions of the form $[u(x)]^n$, where $u(x)$ is a differentiable function. Is rule (1) still valid if we replace x with a function $u(x)$?

Explore & Discuss **1**   Let $u(x) = 2x^2$ and $f(x) = [u(x)]^3 = 8x^6$. Which of the following is $f'(x)$?

(A) $3[u(x)]^2$

(B) $3[u'(x)]^2$

(C) $3[u(x)]^2u'(x)$

The calculations in Explore & Discuss 1 show that we cannot generalize the power rule simply by replacing x with $u(x)$ in equation (1).

How can we find a formula for the derivative of $[u(x)]^n$, where $u(x)$ is an arbitrary differentiable function? Let's begin by considering the derivatives of $[u(x)]^2$ and $[u(x)]^3$ to see if a general pattern emerges. Since $[u(x)]^2 = u(x)u(x)$, we use the product rule to write

$$\frac{d}{dx}[u(x)]^2 = \frac{d}{dx}[u(x)u(x)]$$

$$= u(x)u'(x) + u(x)u'(x)$$

$$= 2u(x)u'(x) \tag{2}$$

Because $[u(x)]^3 = [u(x)]^2u(x)$, we now use the product rule and the result in equation (2) to write

$$\frac{d}{dx}[u(x)]^3 = \frac{d}{dx}\{[u(x)]^2u(x)\}$$
Use equation (2) to substitute for $\frac{d}{dx}[u(x)]^2$.

$$= [u(x)]^2\frac{d}{dx}u(x) + u(x)\frac{d}{dx}[u(x)]^2$$

$$= [u(x)]^2u'(x) + u(x)[2u(x)u'(x)]$$

$$= 3[u(x)]^2u'(x)$$

Continuing in this fashion, we can show that

$$\frac{d}{dx}[u(x)]^n = n[u(x)]^{n-1}u'(x) \qquad n \text{ a positive integer} \qquad (3)$$

Using more advanced techniques, we can establish formula (3) for all real numbers n. Thus, we have the **general power rule:**

THEOREM 1   GENERAL POWER RULE

If $u(x)$ is a differentiable function, n is any real number, and

$$y = f(x) = [u(x)]^n$$

then

$$f'(x) = n[u(x)]^{n-1}u'(x)$$

This rule is often written more compactly as

$$y' = nu^{n-1}u' \qquad \text{or} \qquad \frac{d}{dx}u^n = nu^{n-1}\frac{du}{dx} \qquad \text{where } u = u(x)$$

EXAMPLE 3   **Using the General Power Rule**   Find the indicated derivatives:

(A) $f'(x)$ if $f(x) = (3x + 1)^4$

(B) y' if $y = (x^3 + 4)^7$

(C) $\dfrac{d}{dt}\dfrac{1}{(t^2 + t + 4)^3}$

(D) $\dfrac{dh}{dw}$ if $h(w) = \sqrt{3 - w}$

SOLUTION   (A) $f(x) = (3x + 1)^4$ Let $u = 3x + 1, n = 4$.

$$f'(x) = 4(3x + 1)^3(3x + 1)' \qquad nu^{n-1}\frac{du}{dx}$$

$$= 4(3x + 1)^3\, 3 \qquad \frac{du}{dx} = 3$$

$$= 12(3x + 1)^3$$

(B) $y = (x^3 + 4)^7$ Let $u = (x^3 + 4), n = 7$.

$$y' = 7(x^3 + 4)^6(x^3 + 4)' \qquad nu^{n-1}\frac{du}{dx}$$

$$= 7(x^3 + 4)^6\, 3x^2 \qquad \frac{du}{dx} = 3x^2$$

$$= 21x^2(x^3 + 4)^6$$

(C) $\dfrac{d}{dt}\dfrac{1}{(t^2 + t + 4)^3}$

$$= \frac{d}{dt}(t^2 + t + 4)^{-3} \qquad \text{Let } u = t^2 + t + 4, n = -3.$$

$$= -3(t^2 + t + 4)^{-4}(t^2 + t + 4)' \qquad nu^{n-1}\frac{du}{dt}$$

$$= -3(t^2 + t + 4)^{-4}(2t + 1) \qquad \frac{du}{dt} = 2t + 1$$

$$= \frac{-3(2t + 1)}{(t^2 + t + 4)^4}$$

(D) $h(w) = \sqrt{3 - w} = (3 - w)^{1/2}$ Let $u = 3 - w$, $n = \dfrac{1}{2}$

$$\dfrac{dh}{dw} = \boxed{\dfrac{1}{2}(3 - w)^{-1/2}(3 - w)'}\quad nu^{n-1}\dfrac{du}{dw}$$

$$= \dfrac{1}{2}(3 - w)^{-1/2}(-1)\qquad \dfrac{du}{dw} = -1$$

$$= -\dfrac{1}{2(3 - w)^{1/2}}\quad\text{or}\quad -\dfrac{1}{2\sqrt{3 - w}}$$

MATCHED PROBLEM 3

Find the indicated derivatives:

(A) $h'(x)$ if $h(x) = (5x + 2)^3$

(B) y' if $y = (x^4 - 5)^5$

(C) $\dfrac{d}{dt}\dfrac{1}{(t^2 + 4)^2}$

(D) $\dfrac{dg}{dw}$ if $g(w) = \sqrt{4 - w}$

Notice that we used two steps to differentiate each function in Example 3. First, we applied the general power rule; then we found du/dx. As you gain experience with the general power rule, you may want to combine these two steps. If you do this, be certain to multiply by du/dx. For example,

$$\dfrac{d}{dx}(x^5 + 1)^4 = 4(x^5 + 1)^3 5x^4 \quad \text{Correct}$$

$$\dfrac{d}{dx}(x^5 + 1)^4 \neq 4(x^5 + 1)^3 \qquad du/dx = 5x^4 \text{ is missing}$$

INSIGHT

If we let $u(x) = x$, then $du/dx = 1$, and the general power rule reduces to the (ordinary) power rule discussed in Section 3-5. Compare the following:

$$\dfrac{d}{dx}x^n = nx^{n-1} \qquad \text{Yes—power rule}$$

$$\dfrac{d}{dx}u^n = nu^{n-1}\dfrac{du}{dx} \qquad \text{Yes—general power rule}$$

$$\dfrac{d}{dx}u^n \neq nu^{n-1} \qquad \text{Unless } u(x) = x + k, \text{ so that } du/dx = 1$$

■ The Chain Rule

We have used the general power rule to find derivatives of composite functions of the form $f(g(x))$, where $f(u) = u^n$ is a power function. But what if f is not a power function? Then a more general rule, the *chain rule,* enables us to compute the derivatives of many composite functions of the form $f(g(x))$.

Suppose that

$$y = m(x) = f[g(x)]$$

is a composite of f and g, where

$$y = f(u) \qquad \text{and} \qquad u = g(x)$$

We would like to express the derivative dy/dx in terms of the derivatives of f and g. From the definition of a derivative (see Section 3-4), we have

$$\frac{dy}{dx} = \lim_{h \to 0} \frac{m(x + h) - m(x)}{h} \qquad \text{Substitute } m(x + h) = f[g(x + h)] \text{ and } m(x) = f[g(x)].$$

$$= \lim_{h \to 0} \frac{f[g(x + h)] - f[g(x)]}{h} \qquad \text{Multiply by } 1 = \frac{g(x + h) - g(x)}{g(x + h) - g(x)}.$$

$$= \lim_{h \to 0} \left[\frac{f[g(x + h)] - f[g(x)]}{h} \cdot \frac{g(x + h) - g(x)}{g(x + h) - g(x)} \right]$$

$$= \lim_{h \to 0} \left[\frac{f[g(x + h)] - f[g(x)]}{g(x + h) - g(x)} \cdot \frac{g(x + h) - g(x)}{h} \right] \tag{4}$$

We recognize the second factor in equation (4) as the difference quotient for $g(x)$. To interpret the first factor as the difference quotient for $f(u)$, we let $k = g(x + h) - g(x)$. Since $u = g(x)$, we can write

$$u + k \; \fbox{$= g(x) + g(x + h) - g(x)$} = g(x + h)$$

Substituting in equation (1), we now have

$$\frac{dy}{dx} = \lim_{h \to 0} \left[\frac{f(u + k) - f(u)}{k} \cdot \frac{g(x + h) - g(x)}{h} \right] \tag{5}$$

If we assume that $k = [g(x + h) - g(x)] \to 0$ as $h \to 0$, we can find the limit of each difference quotient in equation (5):

$$\frac{dy}{dx} = \left[\lim_{k \to 0} \frac{f(u + k) - f(u)}{k} \right] \left[\lim_{h \to 0} \frac{g(x + h) - g(x)}{h} \right]$$

$$= f'(u)g'(x)$$

$$= \frac{dy}{du} \frac{du}{dx}$$

This result is correct under rather general conditions and is called the *chain rule,* but our "derivation" is superficial, because it ignores a number of hidden problems. Since a formal proof of the chain rule is beyond the scope of this book, we simply state it as follows:

THEOREM 2   CHAIN RULE

If $y = f(u)$ and $u = g(x)$ define the composite function

$$y = m(x) = f[g(x)]$$

then

$$\frac{dy}{dx} = \frac{dy}{du} \frac{du}{dx} \qquad \text{provided that } \frac{dy}{du} \text{ and } \frac{du}{dx} \text{ exist}$$

or, equivalently,

$$m'(x) = f'[g(x)]g'(x) \qquad \text{provided that } f'[g(x)] \text{ and } g'(x) \text{ exist}$$

EXAMPLE 4

Using the Chain Rule   Find dy/du, du/dx, and dy/dx (express dy/dx as a function of x) for

(A) $y = u^{3/2}$ and $u = 3x^2 + 1$

(B) $y = e^u$ and $u = 2x^3 + 5$

(C) $y = \ln u$ and $u = x^2 - 4x + 2$

SOLUTION (A) $\dfrac{dy}{du} = \dfrac{3}{2}u^{1/2}$ and $\dfrac{du}{dx} = 6x$ *Basic derivative rules*

$$\frac{dy}{dx} = \frac{dy}{du}\frac{du}{dx} \qquad \text{Chain rule}$$

$$= \frac{3}{2}u^{1/2}(6x) = 9x(3x^2 + 1)^{1/2} \quad \text{Since } u = 3x^2 + 1$$

(B) $\dfrac{dy}{du} = e^u$ and $\dfrac{du}{dx} = 6x^2$ *Basic derivative rules*

$$\frac{dy}{dx} = \frac{dy}{du}\frac{du}{dx} \qquad \text{Chain rule}$$

$$= e^u(6x^2) = 6x^2 e^{2x^3+5} \qquad \text{Since } u = 2x^3 + 5$$

(C) $\dfrac{dy}{du} = \dfrac{1}{u}$ and $\dfrac{du}{dx} = 2x - 4$ *Basic derivative rules*

$$\frac{dy}{dx} = \frac{dy}{du}\frac{du}{dx} \qquad \text{Chain rule}$$

$$= \frac{1}{u}(2x - 4) = \frac{2x - 4}{x^2 - 4x + 2} \qquad \text{Since } u = x^2 - 4x + 2$$

MATCHED PROBLEM 4   Find dy/du, du/dx, and dy/dx (express dy/dx as a function of x) for

(A) $y = u^{-5}$ and $u = 2x^3 + 4$

(B) $y = e^u$ and $u = 3x^4 + 6$

(C) $y = \ln u$ and $u = x^2 + 9x + 4$

Explore & Discuss **2**   Let $m(x) = f[g(x)]$. Use the chain rule and the graphs in Figures 1 and 2 to find

   (A) $f(4)$ (B) $g(6)$ (C) $m(6)$

   (D) $f'(4)$ (E) $g'(6)$ (F) $m'(6)$

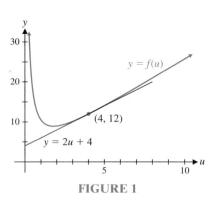

FIGURE 1

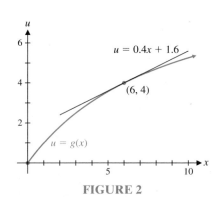

FIGURE 2

The chain rule can be extended to compositions of three or more functions. For example, if $y = f(w)$, $w = g(u)$, and $u = h(x)$, then

$$\frac{dy}{dx} = \frac{dy}{dw}\frac{dw}{du}\frac{du}{dx}$$

EXAMPLE 5   **Using the Chain Rule** For $y = h(x) = e^{1+(\ln x)^2}$, find dy/dx.

SOLUTION   Note that h is of the form $y = e^w$, where $w = 1 + u^2$ and $u = \ln x$. Thus,

$$\frac{dy}{dx} = \frac{dy}{dw}\frac{dw}{du}\frac{du}{dx}$$

$$= e^w(2u)\left(\frac{1}{x}\right)$$

$$= e^{1+u^2}(2u)\left(\frac{1}{x}\right) \qquad \text{Since } w = 1 + u^2$$

$$= e^{1+(\ln x)^2}(2\ln x)\left(\frac{1}{x}\right) \qquad \text{Since } u = \ln x$$

$$= \frac{2}{x}(\ln x)e^{1+(\ln x)^2}$$

MATCHED PROBLEM 5   For $y = h(x) = [\ln(1 + e^x)]^3$, find dy/dx.

The chain rule generalizes basic derivative rules. We list three general derivative rules here for convenient reference [the first, equation (6), is the general power rule of Theorem 1].

General Derivative Rules

$$\frac{d}{dx}[f(x)]^n = n[f(x)]^{n-1}f'(x) \tag{6}$$

$$\frac{d}{dx}\ln[f(x)] = \frac{1}{f(x)}f'(x) \tag{7}$$

$$\frac{d}{dx}e^{f(x)} = e^{f(x)}f'(x) \tag{8}$$

Unless directed otherwise, you now have a choice between the chain rule and the general derivative rules. However, practicing with the chain rule will help prepare you for concepts that appear later in the text. Examples 4 and 5 illustrate the chain rule method, and the next example illustrates the general derivative rules method.

EXAMPLE 6   **Using General Derivative Rules**

(A) $\dfrac{d}{dx}e^{2x} = e^{2x}\dfrac{d}{dx}2x$ Using equation (5)

$\qquad = e^{2x}(2) = 2e^{2x}$

(B) $\dfrac{d}{dx}\ln(x^2 + 9) = \dfrac{1}{x^2 + 9}\dfrac{d}{dx}(x^2 + 9)$ Using equation (4)

$\qquad = \dfrac{1}{x^2 + 9}2x = \dfrac{2x}{x^2 + 9}$

(C) $\dfrac{d}{dx}(1 + e^{x^2})^3 = 3(1 + e^{x^2})^2\dfrac{d}{dx}(1 + e^{x^2})$ Using equation (3)

$\qquad = 3(1 + e^{x^2})^2 e^{x^2}\dfrac{d}{dx}x^2$ Using equation (5)

$\qquad = 3(1 + e^{x^2})^2 e^{x^2}(2x)$

$\qquad = 6xe^{x^2}(1 + e^{x^2})^2$

MATCHED PROBLEM 6   Find

$$\text{(A) } \frac{d}{dx}\ln(x^3 + 2x) \qquad \text{(B) } \frac{d}{dx}e^{3x^2+2} \qquad \text{(C) } \frac{d}{dx}(2 + e^{-x^2})^4$$

Answers to Matched Problems

1. $f[g(x)] = 2e^x$, $g[f(u)] = e^{2u}$

2. (A) $f(u) = 50e^u$, $u = -2x$ (B) $f(u) = \sqrt[3]{u}$, $u = 1 + x^3$
[*Note:* There are other correct answers.]

3. (A) $15(5x + 2)^2$

(B) $20x^3(x^4 - 5)^4$

(C) $-4t/(t^2 + 4)^3$

(D) $-1/(2\sqrt{4 - w})$

4. (A) $\dfrac{dy}{dx} = -5u^{-4}, \dfrac{du}{dx} = 6x^2, \dfrac{dy}{dx} = -30x^2(2x^3 + 4)^{-6}$

(B) $\dfrac{dy}{du} = e^u, \dfrac{du}{dx} = 12x^3, \dfrac{dy}{dx} = 12x^3e^{3x^4+6}$

(C) $\dfrac{dy}{du} = \dfrac{1}{u}, \dfrac{du}{dx} = 2x + 9, \dfrac{dy}{dx} = \dfrac{2x + 9}{x^2 + 9x + 4}$

5. $\dfrac{3e^x[\ln(1 + e^x)]^2}{1 + e^x}$

6. (A) $\dfrac{3x^2 + 2}{x^3 + 2x}$ (B) $6xe^{3x^2+2}$ (C) $-8xe^{-x^2}(2 + e^{-x^2})^3$

Exercise 4-4

For many of the problems in this exercise set, the answers in the back of the book include both an unsimplified form and a simplified form. When checking your work, first check that you applied the rules correctly, and then check that you performed the algebraic simplification correctly.

A *In Problems 1–4, find f[g(x)].*

1. $f(u) = u^3$; $g(x) = 3x^2 + 2$

2. $f(u) = u^4$; $g(x) = 1 - 4x^3$

3. $f(u) = e^u$; $g(x) = -x^2$

4. $g(u) = e^u$; $g(x) = 3x^3$

In Problems 5–8, write each composite function in the form y = f(u) and u = g(x).

5. $y = (3x^2 - x + 5)^4$

6. $y = (2x^3 + x + 3)^5$

7. $y = e^{1+x+x^2}$

8. $y = e^{x^4+2x^2+5}$

In Problems 9–16, replace the ? with an expression that will make the indicated equation valid.

9. $\dfrac{d}{dx}(3x + 4)^4 = 4(3x + 4)^3$?

10. $\dfrac{d}{dx}(5 - 2x)^6 = 6(5 - 2x)^5$?

11. $\dfrac{d}{dx}(4 - 2x^2)^3 = 3(4 - 2x^2)^2$?

12. $\dfrac{d}{dx}(3x^2 + 7)^5 = 5(3x^2 + 7)^4$?

13. $\dfrac{d}{dx}e^{x^2+1} = e^{x^2+1}$?

14. $\dfrac{d}{dx}e^{4x-2} = e^{4x-2}$?

15. $\dfrac{d}{dx}\ln(x^4 + 1) = \dfrac{1}{x^4 + 1}$?

16. $\dfrac{d}{dx}\ln(x - x^3) = \dfrac{1}{x - x^3}$?

In Problems 17–36, find f'(x) and simplify.

17. $f(x) = (2x + 5)^3$

18. $f(x) = (3x - 7)^5$

19. $f(x) = (5 - 2x)^4$

20. $f(x) = (9 - 5x)^2$

21. $f(x) = (4 + 0.2x)^5$

22. $f(x) = (6 - 0.5x)^4$

23. $f(x) = (3x^2 + 5)^5$

24. $f(x) = (5x^2 - 3)^6$

25. $f(x) = e^{5x}$

26. $f(x) = 6e^{-2x}$

27. $f(x) = 3e^{-6x}$

28. $f(x) = e^{x^2 + 3x + 1}$

29. $f(x) = (2x - 5)^{1/2}$

30. $f(x) = (4x + 3)^{1/2}$

31. $f(x) = (x^4 + 1)^{-2}$

32. $f(x) = (x^5 + 2)^{-3}$

33. $f(x) = 3\ln(1 + x^2)$

34. $f(x) = 2\ln(x^2 - 3x + 4)$

35. $f(x) = (1 + \ln x)^3$

36. $f(x) = (x - 2\ln x)^4$

In Problems 37–42, find f'(x) and the equation of the line tangent to the graph of f at the indicated value of x. Find the value(s) of x where the tangent line is horizontal.

37. $f(x) = (2x - 1)^3$; $x = 1$

38. $f(x) = (3x - 1)^4$; $x = 1$

39. $f(x) = (4x - 3)^{1/2};\quad x = 3$

40. $f(x) = (2x + 8)^{1/2};\quad x = 4$

41. $f(x) = 5e^{x^2 - 4x + 1};\quad x = 0$

42. $f(x) = \ln(1 - x^2 + 2x^4);\quad x = 1$

B *In Problems 43–62, find the indicated derivative and simplify.*

43. y' if $y = 3(x^2 - 2)^4$

44. y' if $y = 2(x^3 + 6)^5$

45. $\dfrac{d}{dt} 2(t^2 + 3t)^{-3}$

46. $\dfrac{d}{dt} 3(t^3 + t^2)^{-2}$

47. $\dfrac{dh}{dw}$ if $h(w) = \sqrt{w^2 + 8}$

48. $\dfrac{dg}{dw}$ if $g(w) = \sqrt[3]{3w - 7}$

49. $g'(x)$ if $g(x) = 4xe^{3x}$

50. $h'(x)$ if $h(x) = \dfrac{e^{2x}}{x^2 + 9}$

51. $\dfrac{d}{dx} \dfrac{\ln(1 + x)}{x^3}$

52. $\dfrac{d}{dx} [x^4 \ln(1 + x^4)]$

53. $F'(t)$ if $F(t) = (e^{t^2 + 1})^3$

54. $G'(t)$ if $G(t) = (1 - e^{2t})^2$

55. y' if $y = \ln(x^2 + 3)^{3/2}$

56. y' if $y = [\ln(x^2 + 3)]^{3/2}$

57. $\dfrac{d}{dw} \dfrac{1}{(w^3 + 4)^5}$

58. $\dfrac{d}{dw} \dfrac{1}{(w^2 - 2)^6}$

59. $\dfrac{dy}{dx}$ if $y = (3\sqrt{x} - 1)^5$

60. $\dfrac{dy}{dx}$ if $y = \left(\dfrac{1}{x^2} - 5\right)^{-2}$

61. $f'(t)$ if $f(t) = \dfrac{4}{\sqrt{t^2 - 3t}}$

62. $g'(t)$ if $g(t) = \dfrac{3}{\sqrt[3]{t - t^2}}$

In Problems 63–68, find $f'(x)$ and find the equation of the line tangent to the graph of f at the indicated value of x.

63. $f(x) = x(4 - x)^3;\quad x = 2$

64. $f(x) = x^2(1 - x)^4;\quad x = 2$

65. $f(x) = \dfrac{x}{(2x - 5)^3};\quad x = 3$

66. $f(x) = \dfrac{x^4}{(3x - 8)^2};\quad x = 4$

67. $f(x) = \sqrt{\ln x};\quad x = e$

68. $f(x) = e^{\sqrt{x}};\quad x = 1$

In Problems 69–74, find $f'(x)$ and find the value(s) of x where the tangent line is horizontal.

69. $f(x) = x^2(x - 5)^3$

70. $f(x) = x^3(x - 7)^4$

71. $f(x) = \dfrac{x}{(2x + 5)^2}$

72. $f(x) = \dfrac{x - 1}{(x - 3)^3}$

73. $f(x) = \sqrt{x^2 - 8x + 20}$

74. $f(x) = \sqrt{x^2 + 4x + 5}$

75. Suppose a student reasons that the functions $f(x) = \ln[5(x^2 + 3)^4]$ and $g(x) = 4\ln(x^2 + 3)$ must have the same derivative, since he has entered $f(x)$, $g(x)$, $f'(x)$, and $g'(x)$ into a graphing calculator, but only three graphs appear (see the figure). Is his reasoning correct? Are $f'(x)$ and $g'(x)$ the same function? Explain.

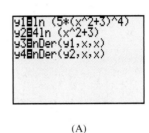
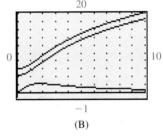

(A)   (B)

Figure for 75

76. Suppose a student reasons that the functions $f(x) = (x + 1)\ln(x + 1) - x$ and $g(x) = (x + 1)^{1/3}$ must have the same derivative, since she has entered $f(x)$, $g(x)$, $f'(x)$, and $g'(x)$ into a graphing calculator, but only three graphs appear (see the figure). Is her reasoning correct? Are $f'(x)$ and $g'(x)$ the same function? Explain.

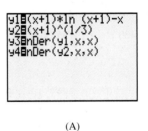
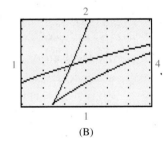

(A)   (B)

Figure for 76

C *In Problems 77–92, find each derivative and simplify.*

77. $\dfrac{d}{dx}[3x(x^2 + 1)^3]$

78. $\dfrac{d}{dx}[2x^2(x^3 - 3)^4]$

79. $\dfrac{d}{dx} \dfrac{(x^3 - 7)^4}{2x^3}$

80. $\dfrac{d}{dx} \dfrac{3x^2}{(x^2 + 5)^3}$

81. $\dfrac{d}{dx} \log_2(3x^2 - 1)$

82. $\dfrac{d}{dx} \log(x^3 - 1)$

83. $\dfrac{d}{dx} 10^{x^2 + x}$

84. $\dfrac{d}{dx} 8^{1 - 2x^2}$

85. $\dfrac{d}{dx} \log_3(4x^3 + 5x + 7)$

86. $\dfrac{d}{dx} \log_5(5^{x^2 - 1})$

87. $\dfrac{d}{dx} 2^{x^3 - x^2 + 4x + 1}$

88. $\dfrac{d}{dx} 10^{\ln x}$

89. $\dfrac{d}{dx} \dfrac{2x}{\sqrt{x - 3}}$

90. $\dfrac{d}{dx} \dfrac{x^2}{\sqrt{x^2 + 1}}$

91. $\dfrac{d}{dx} \sqrt{(2x - 1)^3(x^2 + 3)^4}$

92. $\dfrac{d}{dx} \sqrt{\dfrac{4x + 1}{2x^2 + 1}}$

Applications

93. *Cost function.* The total cost (in hundreds of dollars) of producing x calculators per day is

$$C(x) = 10 + \sqrt{2x + 16} \qquad 0 \le x \le 50$$

(see the figure).

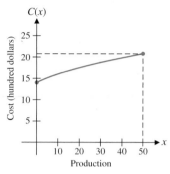

Figure for 93

(A) Find $C'(x)$.

(B) Find $C'(24)$ and $C'(42)$, and interpret the results.

94. *Cost function.* The total cost (in hundreds of dollars) of producing x cameras per week is

$$C(x) = 6 + \sqrt{4x + 4} \qquad 0 \le x \le 30$$

(A) Find $C'(x)$.

(B) Find $C'(15)$ and $C'(24)$, and interpret the results.

95. *Price–supply equation.* The number x of stereo speakers a retail chain is willing to sell per week at a price of \$$p$ is given by

$$x = 80\sqrt{p + 25} - 400 \qquad 20 \le p \le 100$$

(see the figure).

(A) Find dx/dp.

(B) Find the supply and the instantaneous rate of change of supply with respect to price when the price is \$75. Write a brief verbal interpretation of these results.

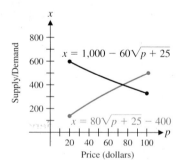

Figure for 95 and 96

96. *Price–demand equation.* The number x of stereo speakers people are willing to buy per week from a retail chain at a price of \$$p$ is given by

$$x = 1{,}000 - 60\sqrt{p + 25} \qquad 20 \le p \le 100$$

(see the previous figure).

(A) Find dx/dp.

(B) Find the demand and the instantaneous rate of change of demand with respect to price when the price is \$75. Write a brief verbal interpretation of these results.

97. *Drug concentration.* The concentration of a drug in the bloodstream t hours after injection is given approximately by

$$C(t) = 4.35e^{-t} \qquad 0 \le t \le 5$$

where $C(t)$ is concentration in milligrams per milliliter.

(A) What is the rate of change of concentration after 1 hour? After 4 hours?

(B) Graph C.

98. *Water pollution.* The use of iodine crystals is a popular way of making small quantities of nonpotable water safe to drink. Crystals placed in a 1-ounce bottle of water will dissolve until the solution is saturated. After saturation, half of the solution is poured into a quart container of nonpotable water, and after about an hour, the water is usually safe to drink. The half-empty 1-ounce bottle is then refilled, to be used again in the same way. Suppose that the concentration of iodine in the 1-ounce bottle t minutes after the crystals are introduced can be approximated by

$$C(t) = 250(1 - e^{-t}) \qquad t \ge 0$$

where $C(t)$ is the concentration of iodine in micrograms per milliliter.

(A) What is the rate of change of the concentration after 1 minute? After 4 minutes?

(B) Graph C for $0 \le t \le 5$.

99. *Blood pressure and age.* A research group using hospital records developed the following approximate mathematical model relating systolic blood pressure and age:

$$P(x) = 40 + 25 \ln(x + 1) \qquad 0 \le x \le 65$$

Here, $P(x)$ is pressure, measured in millimeters of mercury, and x is age in years. What is the rate of change of pressure at the end of 10 years? At the end of 30 years? At the end of 60 years?

100. *Biology.* A yeast culture at room temperature (68°F) is placed in a refrigerator maintaining a constant temperature of 38°F. After t hours, the temperature T of the culture is given approximately by

$$T = 30e^{-0.58t} + 38 \qquad t \ge 0$$

What is the rate of change of temperature of the culture at the end of 1 hour? At the end of 4 hours?

101. *Learning.* In 1930, L. L. Thurstone developed the following formula to indicate how learning time T depends on the length of a list n:

$$T = f(n) = \frac{c}{k} n \sqrt{n - a}$$

Here, a, c, and k are empirical constants. Suppose that, for a particular person, the time T (in minutes) required to learn a list of length n is

$$T = f(n) = 2n\sqrt{n - 2}$$

(A) Find dT/dn.

(B) Find $f'(11)$ and $f'(27)$, and interpret the results.

Section 4-5 IMPLICIT DIFFERENTIATION

- Special Function Notation
- Implicit Differentiation

Special Function Notation

The equation

$$y = 2 - 3x^2 \tag{1}$$

defines a function f with y as a dependent variable and x as an independent variable. Using function notation, we would write

$$y = f(x) \quad \text{or} \quad f(x) = 2 - 3x^2$$

In order to reduce to a minimum the number of symbols involved in a discussion, we will often write equation (1) in the form

$$y = 2 - 3x^2 = y(x)$$

where y is *both* a dependent variable and a function symbol. This is a convenient notation, and no harm is done as long as one is aware of the double role of y. Other examples are

$$x = 2t^2 - 3t + 1 = x(t)$$

$$z = \sqrt{u^2 - 3u} = z(u)$$

$$r = \frac{1}{(s^2 - 3s)^{2/3}} = r(s)$$

This type of notation will simplify much of the discussion and work that follows.

Until now, we have considered functions involving only one independent variable. There is no reason to stop there: The concept can be generalized to functions involving two or more independent variables, and this will be done in detail in Chapter 8. For now, we will "borrow" the notation for a function involving two independent variables. For example,

$$F(x, y) = x^2 - 2xy + 3y^2 - 5$$

specifies a function F involving two independent variables.

Implicit Differentiation

Consider the equation

$$3x^2 + y - 2 = 0 \tag{2}$$

and the equation obtained by solving equation (2) for y in terms of x,

$$y = 2 - 3x^2 \tag{3}$$

Both equations define the same function with x as the independent variable and y as the dependent variable. On the one hand, we can write, for equation (3),

$$y = f(x)$$

where

$$f(x) = 2 - 3x^2 \tag{4}$$

and we have an **explicit** (directly stated) rule that enables us to determine y for each value of x. On the other hand, the y in equation (2) is the same y as in equation (3), and equation (2) **implicitly** gives (implies, though does not directly express) y as a function of x. Thus, we say that equations (3) and (4) define the function f explicitly and equation (2) defines f implicitly.

The direct use of an equation that defines a function implicitly to find the derivative of the dependent variable with respect to the independent variable is called

implicit differentiation. Let us differentiate equation (2) implicitly and equation (3) directly, and compare results.

Starting with

$$3x^2 + y - 2 = 0$$

we think of y as a function of x—that is, $y = y(x)$—and write

$$3x^2 + y(x) - 2 = 0$$

Then we differentiate both sides with respect to x:

$$\frac{d}{dx}[(3x^2 + y(x) - 2)] = \frac{d}{dx}0$$ *Since y is a function of x, but is not*
explicitly given, we simply write

$$\frac{d}{dx}3x^2 + \frac{d}{dx}y(x) - \frac{d}{dx}2 = 0$$

$$6x + y' - 0 = 0$$ $$\frac{d}{dx}y(x) = y' \text{ to indicate its derivative.}$$

Now we solve for y':

$$y' = -6x$$

Note that we get the same result if we start with equation (3) and differentiate directly:

$$y = 2 - 3x^2$$
$$y' = -6x$$

Why are we interested in implicit differentiation? In general, why do we not solve for y in terms of x and differentiate directly? The answer is that there are many equations of the form

$$F(x, y) = 0 \tag{5}$$

that are either difficult or impossible to solve for y explicitly in terms of x (try it for $x^2 y^5 - 3xy + 5 = 0$ or for $e^y - y = 3x$, for example). But it can be shown that, under fairly general conditions on F, equation (5) will define one or more functions in which y is a dependent variable and x is an independent variable. To find y' under these conditions, we differentiate equation (5) implicitly.

Explore & Discuss 1

(A) How many tangent lines are there to the graph in Figure 1 when $x = 0$? When $x = 1$? When $x = 2$? When $x = 4$? When $x = 6$?

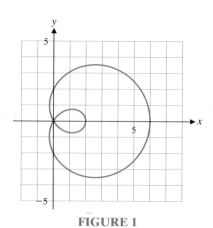

FIGURE 1

(B) Sketch the tangent lines referred to in part (A), and estimate each of their slopes.

(C) Explain why the graph in Figure 1 is not the graph of a function.

EXAMPLE 1   **Differentiating Implicitly** Given

$$F(x, y) = x^2 + y^2 - 25 = 0 \qquad (6)$$

find y' and the slope of the graph at $x = 3$.

SOLUTION   We start with the graph of $x^2 + y^2 - 25 = 0$ (a circle, as shown in Fig. 2) so that we can interpret our results geometrically. From the graph, it is clear that equation (6) does not define a function. But with a suitable restriction on the variables, equation (6) can define two or more functions. For example, the upper half and the lower half of the circle each define a function. On each half-circle, a point that corresponds to $x = 3$ is found by substituting $x = 3$ into equation (6) and solving for y:

$$x^2 + y^2 - 25 = 0$$
$$(3)^2 + y^2 = 25$$
$$y^2 = 16$$
$$y = \pm 4$$

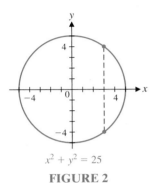

$x^2 + y^2 = 25$

FIGURE 2

Thus, the point $(3, 4)$ is on the upper half-circle, and the point $(3, -4)$ is on the lower half-circle. We will use these results in a moment. We now differentiate equation (6) implicitly, treating y as a function of x [i.e., $y = y(x)$]:

$$x^2 + y^2 - 25 = 0$$

$$x^2 + [y(x)]^2 - 25 = 0$$

$$\frac{d}{dx}\{x^2 + [y(x)]^2 - 25\} = \frac{d}{dx}0$$

$$\frac{d}{dx}x^2 + \frac{d}{dx}[y(x)]^2 - \frac{d}{dx}25 = 0 \qquad \text{Use the chain rule.}$$

$$2x + 2[y(x)]^{2-1}y'(x) - 0 = 0$$

$$2x + 2yy' = 0 \qquad \text{Solve for } y' \text{ in terms of } x \text{ and } y.$$

$$y' = -\frac{2x}{2y}$$

$$y' = -\frac{x}{y} \qquad \text{Leave the answer in terms of } x \text{ and } y.$$

We have found y' without first solving $x^2 + y^2 - 25 = 0$ for y in terms of x. And by leaving y' in terms of x and y, we can use $y' = -x/y$ to find y' for *any* point on the graph of $x^2 + y^2 - 25 = 0$ (except where $y = 0$). In particular, for $x = 3$, we found that $(3, 4)$ and $(3, -4)$ are on the graph; thus, the slope of the graph at $(3, 4)$ is

$$y'\big|_{(3, 4)} = -\frac{3}{4} \qquad \text{The slope of the graph at } (3, 4)$$

and the slope at $(3, -4)$ is

$$y'\big|_{(3, -4)} = -\frac{3}{-4} = \frac{3}{4} \qquad \text{The slope of the graph at } (3, -4)$$

The symbol

$$y'\big|_{(a, b)}$$

is used to indicate that we are evaluating y' at $x = a$ and $y = b$.
   The results are interpreted geometrically in Figure 3 on the original graph. ▬

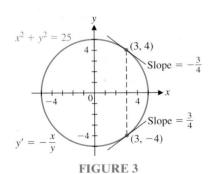

$x^2 + y^2 = 25$

Slope $= -\frac{3}{4}$

Slope $= \frac{3}{4}$

$y' = -\dfrac{x}{y}$

FIGURE 3

In Example 1, the fact that y' is given in terms of both x and y is not a great disadvantage. We have only to make certain that when we want **to evaluate y' for a particular value of x and y, say, (x_0, y_0), the ordered pair must satisfy the original equation.**

INSIGHT

In the solution for Example 1, notice that the derivative of y^2 with respect to x is $2yy'$, not just $2y$. When first learning about implicit differentiation, it is a good idea to replace y with $y(x)$ in the original equation to emphasize that y is a function of x.

MATCHED PROBLEM 1   Graph $x^2 + y^2 - 169 = 0$, find y' by implicit differentiation, and find the slope of the graph when $x = 5$.

EXAMPLE 2   **Differentiating Implicitly** Find the equation(s) of the tangent line(s) to the graph of

$$y - xy^2 + x^2 + 1 = 0 \tag{7}$$

at the point(s) where $x = 1$.

SOLUTION   We first find y when $x = 1$:

$$y - xy^2 + x^2 + 1 = 0$$
$$y - (1)y^2 + (1)^2 + 1 = 0$$
$$y - y^2 + 2 = 0$$
$$y^2 - y - 2 = 0$$
$$(y - 2)(y + 1) = 0$$
$$y = -1, 2$$

Thus, there are two points on the graph of (7) where $x = 1$, namely, $(1, -1)$ and $(1, 2)$. We next find the slope of the graph at these two points by differentiating equation (7) implicitly:

$$y - xy^2 + x^2 + 1 = 0 \qquad \text{Use the product rule and the}$$
$$\frac{d}{dx}y - \frac{d}{dx}xy^2 + \frac{d}{dx}x^2 + \frac{d}{dx}1 = \frac{d}{dx}0 \qquad \text{chain rule for } \frac{d}{dx}xy^2.$$
$$y' - (x \cdot 2yy' + y^2) + 2x = 0$$
$$y' - 2xyy' - y^2 + 2x = 0 \qquad \text{Solve for } y' \text{ by getting all}$$
$$y' - 2xyy' = y^2 - 2x \qquad \text{terms involving } y' \text{ on one side.}$$
$$(1 - 2xy)y' = y^2 - 2x$$
$$y' = \frac{y^2 - 2x}{1 - 2xy}$$

Now find the slope at each point:

$$y'|_{(1,-1)} = \frac{(-1)^2 - 2(1)}{1 - 2(1)(-1)} = \frac{1 - 2}{1 + 2} = \frac{-1}{3} = -\frac{1}{3}$$

$$y'|_{(1,2)} = \frac{(2)^2 - 2(1)}{1 - 2(1)(2)} = \frac{4 - 2}{1 - 4} = \frac{2}{-3} = -\frac{2}{3}$$

Equation of tangent line at $(1, -1)$:   Equation of tangent line at $(1, 2)$:

$$y - y_1 = m(x - x_1) \qquad\qquad y - y_1 = m(x - x_1)$$
$$y + 1 = -\tfrac{1}{3}(x - 1) \qquad\qquad y - 2 = -\tfrac{2}{3}(x - 1)$$
$$y + 1 = -\tfrac{1}{3}x + \tfrac{1}{3} \qquad\qquad y - 2 = -\tfrac{2}{3}x + \tfrac{2}{3}$$
$$y = -\tfrac{1}{3}x - \tfrac{2}{3} \qquad\qquad y = -\tfrac{2}{3}x + \tfrac{8}{3}$$

MATCHED PROBLEM 2   Repeat Example 2 for $x^2 + y^2 - xy - 7 = 0$ at $x = 1$.

Explore & Discuss **2**   The slopes of the tangent lines to $y^2 + 3xy + 4x = 9$ when $x = 0$ can be found either (1) by differentiating the equation implicitly or (2) by solving for y explicitly in terms of x (using the quadratic formula) and then computing the derivative. Which of the two methods is more efficient? Explain.

EXAMPLE 3   **Differentiating Implicitly** Find x' for $x = x(t)$ defined implicitly by

$$t \ln x = xe^t - 1$$

and evaluate x' at $(t, x) = (0, 1)$.

SOLUTION   It is important to remember that x is the dependent variable and t is the independent variable. Therefore, we differentiate both sides of the equation with respect to t (using product and chain rules where appropriate) and then solve for x':

$$t \ln x = xe^t - 1 \qquad \text{Differentiate implicitly with respect to } t.$$

$$\frac{d}{dt}(t \ln x) = \frac{d}{dt}(xe^t) - \frac{d}{dt} 1$$

$$t\frac{x'}{x} + \ln x = xe^t + x'e^t \qquad \text{Clear fractions.}$$

$$\boxed{x \cdot t\frac{x'}{x} + x \cdot \ln x = x \cdot xe^t + x \cdot e^t x'} \quad x \neq 0$$

$$tx' + x \ln x = x^2 e^t + xe^t x' \qquad \text{Solve for } x'.$$

$$tx' - xe^t x' = x^2 e^t - x \ln x \qquad \text{Factor out } x'.$$

$$(t - xe^t)x' = x^2 e^t - x \ln x$$

$$x' = \frac{x^2 e^t - x \ln x}{t - xe^t}$$

Now we evaluate x' at $(t, x) = (0, 1)$, as requested:

$$x'|_{(0,1)} = \frac{(1)^2 e^0 - 1 \ln 1}{0 - 1e^0}$$

$$= \frac{1}{-1} = -1$$

MATCHED PROBLEM 3   Find x' for $x = x(t)$ defined implicitly by

$$1 + x \ln t = te^x$$

and evaluate x' at $(t, x) = (1, 0)$.

Answers to Matched Problems   **1.** $y' = -x/y.$ When $x = 5$, $y = \pm 12$; thus, $y'|_{(5,12)} = -\frac{5}{12}$ and $y'|_{(5,-12)} = \frac{5}{12}$

2. $y' = \dfrac{y - 2x}{2y - x};$ $y = \frac{4}{5}x - \frac{14}{5}, y = \frac{1}{5}x + \frac{14}{5}$ **3.** $x' = \dfrac{te^x - x}{t \ln t - t^2 e^x}; x'|_{(1,0)} = -1$

Exercise 4-5

A *In Problems 1–4, find y′ in two ways:*

(A) *Differentiate the given equation implicitly and then solve for y′.*

(B) *Solve the given equation for y and then differentiate directly.*

1. $3x + 5y + 9 = 0$

2. $-2x + 6y - 4 = 0$

3. $3x^2 - 4y - 18 = 0$

4. $2x^3 + 5y - 2 = 0$

In Problems 5–22, use implicit differentiation to find y′ and evaluate y′ at the indicated point.

5. $y - 5x^2 + 3 = 0; (1, 2)$

6. $5x^3 - y - 1 = 0; (1, 4)$

7. $x^2 - y^3 - 3 = 0; (2, 1)$

8. $y^2 + x^3 + 4 = 0; (-2, 2)$

9. $y^2 + 2y + 3x = 0; (-1, 1)$

10. $y^2 - y - 4x = 0; (0, 1)$

B

11. $xy - 6 = 0; (2, 3)$

12. $3xy - 2x - 2 = 0; (2, 1)$

13. $2xy + y + 2 = 0; (-1, 2)$

14. $2y + xy - 1 = 0; (-1, 1)$

15. $x^2y - 3x^2 - 4 = 0; (2, 4)$

16. $2x^3y - x^3 + 5 = 0; (-1, 3)$

17. $e^y = x^2 + y^2; (1, 0)$

18. $x^2 - y = 4e^y; (2, 0)$

19. $x^3 - y = \ln y; (1, 1)$

20. $\ln y = 2y^2 - x; (2, 1)$

21. $x \ln y + 2y = 2x^3; (1, 1)$

22. $xe^y - y = x^2 - 2; (2, 0)$

In Problems 23 and 24, find x′ for x = x(t) defined implicitly by the given equation. Evaluate x′ at the indicated point.

23. $x^2 - t^2x + t^3 + 11 = 0; (-2, 1)$

24. $x^3 - tx^2 - 4 = 0; (-3, -2)$

Problems 25 and 26 refer to the equation and graph shown in the figure.

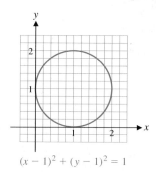

$(x - 1)^2 + (y - 1)^2 = 1$

25. Use implicit differentiation to find the slopes of the tangent lines at the points on the graph where $x = 1.6$. Check your answers by visually estimating the slopes on the graph in the figure.

26. Find the slopes of the tangent lines at the points on the graph where $x = 0.2$. Check your answers by visually estimating the slopes on the graph in the figure.

In Problems 27–30, find the equation(s) of the tangent line(s) to the graphs of the indicated equations at the point(s) with the given value of x.

27. $xy - x - 4 = 0; x = 2$

28. $3x + xy + 1 = 0; x = -1$

29. $y^2 - xy - 6 = 0; x = 1$

30. $xy^2 - y - 2 = 0; x = 1$

31. If $xe^y = 1$, find $y′$ in two ways, first by differentiating implicitly and then by solving for y explicitly in terms of x. Which method do you prefer? Explain.

32. Explain the difficulty that arises in solving $x^3 + y + xe^y = 1$ for y as an explicit function of x. Find the slope of the tangent line to the graph of the equation at the point $(0, 1)$.

C *In Problems 33–40, find y′ and the slope of the tangent line to the graph of each equation at the indicated point.*

33. $(1 + y)^3 + y = x + 7; (2, 1)$

34. $(y - 3)^4 - x = y; (-3, 4)$

35. $(x - 2y)^3 = 2y^2 - 3; (1, 1)$

36. $(2x - y)^4 - y^3 = 8; (-1, -2)$

37. $\sqrt{7 + y^2} - x^3 + 4 = 0; (2, 3)$

38. $6\sqrt{y^3 + 1} - 2x^{3/2} - 2 = 0; (4, 2)$

39. $\ln(xy) = y^2 - 1; (1, 1)$

40. $e^{xy} - 2x = y + 1; (0, 0)$

41. Find the equation(s) of the tangent line(s) at the point(s) on the graph of the equation

$$y^3 - xy - x^3 = 2$$

where $x = 1$. Round all approximate values to two decimal places.

42. Refer to the equation in Problem 41. Find the equation(s) of the tangent line(s) at the point(s) on the graph where $y = -1$. Round all approximate values to two decimal places.

Applications

For the demand equations in Problems 43–46, find the rate of change of p with respect to x by differentiating implicitly (x is the number of items that can be sold at a price of $p).

43. $x = p^2 - 2p + 1,000$

44. $x = p^3 - 3p^2 + 200$

45. $x = \sqrt{10,000 - p^2}$

46. $x = \sqrt[3]{1,500 - p^3}$

47. *Biophysics.* In biophysics, the equation
$$(L + m)(V + n) = k$$
is called the *fundamental equation of muscle contraction,* where $m, n,$ and k are constants and V is the velocity of the shortening of muscle fibers for a muscle subjected to a load L. Find dL/dV by implicit differentiation.

48. *Biophysics.* In Problem 47, find dV/dL by implicit differentiation.

Section 4-6   RELATED RATES

The workers in a union are concerned that the rate at which wages are increasing is lagging behind the rate of increase in the company's profits. An automobile dealer wants to predict how badly an anticipated increase in interest rates will decrease his rate of sales. An investor is studying the connection between the rate of increase in the Dow Jones average and the rate of increase in the gross domestic product over the past 50 years.

In each of these situations, there are two quantities—wages and profits in the first instance, for example—that are changing with respect to time. We would like to discover the precise relationship between the rates of increase (or decrease) of the two quantities. We will begin our discussion of such *related rates* by considering some familiar situations in which the two quantities are distances and the two rates are velocities.

EXAMPLE 1

Related Rates and Motion A 26-foot ladder is placed against a wall (Fig. 1). If the top of the ladder is sliding down the wall at 2 feet per second, at what rate is the bottom of the ladder moving away from the wall when the bottom of the ladder is 10 feet away from the wall?

SOLUTION Many people reason that since the ladder is of constant length, the bottom of the ladder will move away from the wall at the rate that the top of the ladder is moving down the wall. This is not the case, as we will see.

At any moment in time, let x be the distance of the bottom of the ladder from the wall and let y be the distance of the top of the ladder from the ground (see Fig. 1). Both x and y are changing with respect to time and can be thought of as functions of time; that is, $x = x(t)$ and $y = y(t)$. Furthermore, x and y are related by the Pythagorean relationship:

$$x^2 + y^2 = 26^2 \tag{1}$$

Differentiating equation (1) implicitly with respect to time t and using the chain rule where appropriate, we obtain

$$2x\frac{dx}{dt} + 2y\frac{dy}{dt} = 0 \tag{2}$$

The rates dx/dt and dy/dt are related by equation (2); hence, this type of problem is referred to as a **related-rates problem.**

Now, our problem is to find dx/dt when $x = 10$ feet, given that $dy/dt = -2$ (y is decreasing at a constant rate of 2 feet per second). We have all the quantities we need in equation (2) to solve for dx/dt, except y. When $x = 10$, y can be found from equation (1):

$$10^2 + y^2 = 26^2$$

$$y = \sqrt{26^2 - 10^2} = 24 \text{ feet}$$

26 ft

FIGURE 1

Substitute $dy/dt = -2$, $x = 10$, and $y = 24$ into (2); then solve for dx/dt:

$$2(10)\frac{dx}{dt} + 2(24)(-2) = 0$$

$$\frac{dx}{dt} = \frac{-2(24)(-2)}{2(10)} = 4.8 \text{ feet per second}$$

Thus, the bottom of the ladder is moving away from the wall at a rate of 4.8 feet per second.

INSIGHT

In the solution to Example 1, we used equation (1) in two ways: first, to find an equation relating dy/dt and dx/dt, and second, to find the value of y when $x = 10$. These steps must be done in this order; substituting $x = 10$ and then differentiating does not produce any useful results:

$$x^2 + y^2 = 26^2 \qquad \text{Substituting 10 for x has the}$$
$$100 + y^2 = 26^2 \qquad \text{effect of stopping the ladder.}$$
$$0 + 2yy' = 0 \qquad \text{The rate of change of a stationary}$$
$$y' = 0 \qquad \text{object is always 0, but that is not the}$$
$$\text{rate of change of the moving ladder.}$$

MATCHED PROBLEM 1 Again, a 26-foot ladder is placed against a wall (Fig. 1). If the bottom of the ladder is moving away from the wall at 3 feet per second, at what rate is the top moving down when the top of the ladder is 24 feet up the wall?

Explore & Discuss **1** (A) For which values of x and y in Example 1 is dx/dt equal to 2 (i.e., the same rate at which the ladder is sliding down the wall)?

(B) When is dx/dt greater than 2? Less than 2?

DEFINITION **Suggestions for Solving Related-Rates Problems**

Step 1. Sketch a figure if helpful.

Step 2. Identify all relevant variables, including those whose rates are given and those whose rates are to be found.

Step 3. Express all given rates and rates to be found as derivatives.

Step 4. Find an equation connecting the variables identified in step 2.

Step 5. Implicitly differentiate the equation found in step 4, using the chain rule where appropriate, and substitute in all given values.

Step 6. Solve for the derivative that will give the unknown rate.

EXAMPLE 2 **Related Rates and Motion** Suppose that two motorboats leave from the same point at the same time. If one travels north at 15 miles per hour and the other travels east at 20 miles per hour, how fast will the distance between them be changing after 2 hours?

SOLUTION First, draw a picture, as shown in Figure 2.

All variables, x, y, and z, are changing with time. Hence, they can be thought of as functions of time: $x = x(t)$, $y = y(t)$, and $z = z(t)$, given implicitly. It now makes sense to take derivatives of each variable with respect to time. From the Pythagorean theorem,

$$z^2 = x^2 + y^2 \tag{3}$$

We also know that

$$\frac{dx}{dt} = 20 \text{ miles per hour} \qquad \text{and} \qquad \frac{dy}{dt} = 15 \text{ miles per hour}$$

FIGURE 2

We would like to find dz/dt at the end of 2 hours—that is, when $x = 40$ miles and $y = 30$ miles. To do this, we differentiate both sides of equation (3) with respect to t and solve for dz/dt:

$$2z\frac{dz}{dt} = 2x\frac{dx}{dt} + 2y\frac{dy}{dt} \tag{4}$$

We have everything we need except z. From equation (3), when $x = 40$ and $y = 30$, we find z to be 50. Substituting the known quantities into equation (4), we obtain

$$2(50)\frac{dz}{dt} = 2(40)(20) + 2(30)(15)$$

$$\frac{dz}{dt} = 25 \text{ miles per hour}$$

Thus, the boats will be separating at a rate of 25 miles per hour. ▬

MATCHED PROBLEM 2 Repeat Example 2 for the situation at the end of 3 hours.

EXAMPLE 3 **Related Rates and Motion** Suppose a point is moving along the graph of $x^2 + y^2 = 25$ (Fig. 3). When the point is at $(-3, 4)$, its x coordinate is increasing at the rate of 0.4 unit per second. How fast is the y coordinate changing at that moment?

SOLUTION Since both x and y are changing with respect to time, we can think of each as a function of time, namely,

$$x = x(t) \qquad \text{and} \qquad y = y(t)$$

but restricted so that

$$x^2 + y^2 = 25 \tag{5}$$

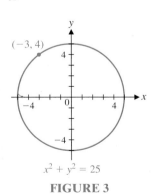

$(-3, 4)$

$x^2 + y^2 = 25$

FIGURE 3

Our problem now is to find dy/dt, given $x = -3$, $y = 4$, and $dx/dt = 0.4$. Implicitly differentiating both sides of equation (5) with respect to t, we have

$$x^2 + y^2 = 25$$

$$2x\frac{dx}{dt} + 2y\frac{dy}{dt} = 0 \qquad \text{Divide both sides by 2.}$$

$$x\frac{dx}{dt} + y\frac{dy}{dt} = 0 \qquad \text{Substitute } x = -3, y = 4, \text{ and } dx/dt = 0.4, \text{ and solve for } dy/dt.$$

$$(-3)(0.4) + 4\frac{dy}{dt} = 0$$

$$\frac{dy}{dt} = 0.3 \text{ unit per second}$$

| MATCHED PROBLEM 3 | A point is moving on the graph of $y^3 = x^2$. When the point is at $(-8, 4)$, its y coordinate is decreasing by 2 units per second. How fast is the x coordinate changing at that moment? |

| EXAMPLE 4 | **Related Rates and Business** Suppose that for a company manufacturing transistor radios, the cost, revenue, and profit equations are given by |

$$C = 5,000 + 2x \qquad \text{Cost equation}$$
$$R = 10x - 0.001x^2 \quad \text{Revenue equation}$$
$$P = R - C \qquad\qquad \text{Profit equation}$$

where the production output in 1 week is x radios. If production is increasing at the rate of 500 radios per week when production is 2,000 radios, find the rate of increase in

(A) Cost   (B) Revenue   (C) Profit

SOLUTION   If production x is a function of time (it must be, since it is changing with respect to time), then C, R, and P must also be functions of time. These functions are implicitly (rather than explicitly) given. Letting t represent time in weeks, we differentiate both sides of each of the preceding three equations with respect to t and then substitute $x = 2,000$ and $dx/dt = 500$ to find the desired rates.

(A) $C = 5,000 + 2x$ Think: $C = C(t)$ and $x = x(t)$.

$$\frac{dC}{dt} = \frac{d}{dt}(5,000) + \frac{d}{dt}(2x) \quad \text{Differentiate both sides with respect to } t.$$

$$\frac{dC}{dt} = 0 + 2\frac{dx}{dt} = 2\frac{dx}{dt}$$

Since $dx/dt = 500$ when $x = 2,000$,

$$\frac{dC}{dt} = 2(500) = \$1,000 \text{ per week}$$

Cost is increasing at a rate of $1,000 per week.

(B) $R = 10x - 0.001x^2$

$$\frac{dR}{dt} = \frac{d}{dt}(10x) - \frac{d}{dt}0.001x^2$$

$$\frac{dR}{dt} = 10\frac{dx}{dt} - 0.002x\frac{dx}{dt}$$

$$\frac{dR}{dt} = (10 - 0.002x)\frac{dx}{dt}$$

Since $dx/dt = 500$ when $x = 2,000$,

$$\frac{dR}{dt} = [10 - 0.002(2,000)](500) = \$3,000 \text{ per week}$$

Revenue is increasing at a rate of $3,000 per week.

(C) $P = R - C$

$$\frac{dP}{dt} = \frac{dR}{dt} - \frac{dC}{dt}$$

$$= \$3,000 - \$1,000 \quad \text{Results from parts (A) and (B)}$$

$$= \$2,000 \text{ per week}$$

Profit is increasing at a rate of $2,000 per week.

MATCHED PROBLEM 4   Repeat Example 4 for a production level of 6,000 radios per week.

Explore & Discuss **2**   (A) In Example 4, suppose that $x(t) = 500t + 500$. Find the time and production level at which the profit is maximized.

(B) Suppose that $x(t) = t^2 + 492t + 16$. Find the time and production level at which the profit is maximized.

(C) Explain why it is unnecessary to know a formula for $x(t)$ in order to determine the production level at which the profit is maximized.

Answers to Matched Problems   1. $dy/dt = -1.25$ ft/sec
2. $dz/dt = 25$ mi/hr
3. $dx/dt = 6$ units/sec
4. (A) $dC/dt = \$1,000$/wk
(B) $dR/dt = -\$1,000$/wk
(C) $dP/dt = -\$2,000$/wk

Exercise 4-6

A *In Problems 1–6, assume that $x = x(t)$ and $y = y(t)$. Find the indicated rate, given the other information.*

1. $y = x^2 + 2$; $dx/dt = 3$ when $x = 5$; find dy/dt

2. $y = x^3 - 3$; $dx/dt = -2$ when $x = 2$; find dy/dt

3. $x^2 + y^2 = 1$; $dy/dt = -4$ when $x = -0.6$ and $y = 0.8$; find dx/dt

4. $x^2 + y^2 = 4$; $dy/dt = 5$ when $x = 1.2$ and $y = -1.6$; find dx/dt

5. $x^2 + 3xy + y^2 = 11$; $dx/dt = 2$ when $x = 1$ and $y = 2$; find dy/dt

6. $x^2 - 2xy - y^2 = 7$; $dy/dt = -1$ when $x = 2$ and $y = -1$; find dx/dt

B

7. A point is moving on the graph of $xy = 36$. When the point is at $(4, 9)$, its x coordinate is increasing by 4 units per second. How fast is the y coordinate changing at that moment?

8. A point is moving on the graph of $4x^2 + 9y^2 = 36$. When the point is at $(3, 0)$, its y coordinate is decreasing by 2 units per second. How fast is its x coordinate changing at that moment?

9. A boat is being pulled toward a dock as indicated in the figure. If the rope is being pulled in at 3 feet per second, how fast is the distance between the dock and the boat decreasing when it is 30 feet from the dock?

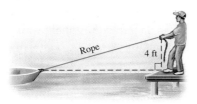

Figure for 9 and 10

10. Refer to Problem 9. Suppose that the distance between the boat and the dock is decreasing by 3.05 feet per second. How fast is the rope being pulled in when the boat is 10 feet from the dock?

11. A rock thrown into a still pond causes a circular ripple. If the radius of the ripple is increasing by 2 feet per second, how fast is the area changing when the radius is 10 feet? [Use $A = \pi R^2$, $\pi \approx 3.14$.]

12. Refer to Problem 11. How fast is the circumference of a circular ripple changing when the radius is 10 feet? [Use $C = 2\pi R$, $\pi \approx 3.14$.]

13. The radius of a spherical balloon is increasing at the rate of 3 centimeters per minute. How fast is the volume changing when the radius is 10 centimeters? [Use $V = \frac{4}{3}\pi R^3$, $\pi \approx 3.14$.]

14. Refer to Problem 13. How fast is the surface area of the sphere increasing when the radius is 10 centimeters? [Use $S = 4\pi R^2$, $\pi \approx 3.14$.]

15. Boyle's law for enclosed gases states that if the volume is kept constant, the pressure P and temperature T are related by the equation

$$\frac{P}{T} = k$$

where k is a constant. If the temperature is increasing at 3 kelvins per hour, what is the rate of change of pressure when the temperature is 250 kelvins and the pressure is 500 pounds per square inch?

16. Boyle's law for enclosed gases states that if the temperature is kept constant, the pressure P and volume V of a gas are related by the equation

$$VP = k$$

where k is a constant. If the volume is decreasing by 5 cubic inches per second, what is the rate of change of pressure when the volume is 1,000 cubic inches and the pressure is 40 pounds per square inch?

17. A 10-foot ladder is placed against a vertical wall. Suppose the bottom of the ladder slides away from the wall at a constant rate of 3 feet per second. How fast is the top of the ladder sliding down the wall (negative rate) when the bottom is 6 feet from the wall? [*Hint:* Use the Pythagorean theorem, $a^2 + b^2 = c^2$, where c is the length of the hypotenuse of a right triangle and a and b are the lengths of the two shorter sides.]

18. A weather balloon is rising vertically at the rate of 5 meters per second. An observer is standing on the ground 300 meters from the point where the balloon was released. At what rate is the distance between the observer and the balloon changing when the balloon is 400 meters high?

19. A streetlight is on top of a 20-foot pole. A person who is 5 feet tall walks away from the pole at the rate of 5 feet per second. At what rate is the tip of the person's shadow moving away from the pole when he is 20 feet from the pole?

20. Refer to Problem 19. At what rate is the person's shadow growing when he is 20 feet from the pole?

21. Helium is pumped into a spherical balloon at a constant rate of 4 cubic feet per second. How fast is the radius increasing after 1 minute? After 2 minutes? Is there any time at which the radius is increasing at a rate of 100 feet per second? Explain.

22. A point is moving along the x axis at a constant rate of 5 units per second. At which point is its distance from $(0, 1)$ increasing at a rate of 2 units per second? At 4 units per second? At 5 units per second? At 10 units per second? Explain.

23. A point is moving on the graph of $y = e^x + x + 1$ in such a way that its x coordinate is always increasing at a rate of 3 units per second. How fast is the y coordinate changing when the point crosses the x axis?

24. A point is moving on the graph of $x^3 + y^2 = 1$ in such a way that its y coordinate is always increasing at a rate of 2 units per second. At which point(s) is the x coordinate increasing at a rate of 1 unit per second?

Applications

25. *Cost, revenue, and profit rates.* Suppose that for a company manufacturing calculators, the cost, revenue, and profit equations are given by

$$C = 90,000 + 30x \qquad R = 300x - \frac{x^2}{30}$$

$$P = R - C$$

where the production output in 1 week is x calculators. If production is increasing at a rate of 500 calculators per week when production output is 6,000 calculators, find the rate of increase (decrease) in

(A) Cost   (B) Revenue   (C) Profit

26. *Cost, revenue, and profit rates.* Repeat Problem 25 for

$$C = 72,000 + 60x \qquad R = 200x - \frac{x^2}{30}$$

$$P = R - C$$

where production is increasing at a rate of 500 calculators per week at a production level of 1,500 calculators.

27. *Advertising.* A retail store estimates that weekly sales s and weekly advertising costs x (both in dollars) are related by

$$s = 60,000 - 40,000e^{-0.0005x}$$

The current weekly advertising costs are $2,000, and these costs are increasing at the rate of $300 per week. Find the current rate of change of sales.

28. *Advertising.* Repeat Problem 27 for

$$s = 50,000 - 20,000e^{-0.0004x}$$

29. *Price–demand.* The price p (in dollars) and demand x for a product are related by

$$2x^2 + 5xp + 50p^2 = 80,000$$

(A) If the price is increasing at a rate of $2 per month when the price is $30, find the rate of change of the demand.

(B) If the demand is decreasing at a rate of 6 units per month when the demand is 150 units, find the rate of change of the price.

30. *Price–demand.* Repeat Problem 29 for

$$x^2 + 2xp + 25p^2 = 74,500$$

31. *Pollution.* An oil tanker aground on a reef is leaking oil that forms a circular oil slick about 0.1 foot thick (see the figure). To estimate the rate dV/dt (in cubic feet per minute) at which the oil is leaking from the tanker, it was found that the radius of the slick was increasing at 0.32 foot per minute ($dR/dt = 0.32$) when the radius R was 500 feet. Find dV/dt, using $\pi \approx 3.14$.

$$A = \pi R^2$$
$$V = 0.1 A$$

Figure for 31

32. *Learning.* A person who is new on an assembly line performs an operation in T minutes after x performances of the operation, as given by

$$T = 6\left(1 + \frac{1}{\sqrt{x}}\right)$$

If $dx/dt = 6$ operations per hour, where t is time in hours, find dT/dt after 36 performances of the operation.

Section 4-7   ELASTICITY OF DEMAND

- Relative Rate of Change
- Elasticity of Demand

When will a price increase lead to an increase in revenue? To answer this question and, more generally, to study relationships among price, demand, and revenue, economists use the notion of *elasticity of demand.* In this section, we define the concepts of *relative rate of change* and *percentage rate of change,* and we give an introduction to elasticity of demand.

Relative Rate of Change

Explore & Discuss **1**

A broker is trying to sell you two stocks: Biotech and Comstat. The broker estimates that Biotech's earnings will increase $2 per year over the next several years, while Comstat's earnings will increase only $1 per year. Is this sufficient information for you to choose between the two stocks? What other information might you request from the broker to help you decide?

Interpreting rates of change is a fundamental application of calculus. In Explore & Discuss 1, Biotech's earnings are increasing at twice the rate of Comstat's, but that does not automatically make Biotech the better buy. The obvious information that is missing is the cost of each stock. If Biotech costs $100 a share and Comstat costs $25 share, then which stock is the better buy? To answer this question, we introduce two new concepts: *relative rate of change* and *percentage rate of change.*

DEFINITION   **Relative and Percentage Rates of Change**

The **relative rate of change** of a function $f(x)$ is $\dfrac{f'(x)}{f(x)}$.

The **percentage rate of change** is $100 \times \dfrac{f'(x)}{f(x)}$.

Because

$$\frac{d}{dx} \ln f(x) = \frac{f'(x)}{f(x)}$$

the relative rate of change of $f(x)$ is the derivative of the logarithm of $f(x)$. This is also referred to as the **logarithmic derivative** of $f(x)$. Returning to Explore & Discuss 1, we can now write

	Relative rate of change		Percentage rate of change
Biotech	$\dfrac{2}{100} = 0.02$	or	2%
Comstat	$\dfrac{1}{25} = 0.04$	or	4%

EXAMPLE 1   **Relative Rate of Change**   Table 1 at the top of the next page lists the real GDP (gross domestic product expressed in billions of 1996 dollars) and population of the United States from 1995 to 2002. A model for the GDP is

$$f(t) = 300t + 6{,}000$$

where t is years since 1990. Find and graph the percentage rate of change of $f(t)$ for $5 \le t \le 12$.

TABLE 1		
Year	Real GDP (billions of 1996 dollars)	Population (in millions)
1995	$7,540	262.765
1996	$7,810	265.19
1997	$8,150	267.744
1998	$8,500	270.299
1999	$8,850	272.82
2000	$9,190	275.306
2001	$9,210	277.803
2002	$9,440	280.306

SOLUTION If $p(t)$ is the percentage rate of change of $f(t)$, then

$$p(t) = 100 \times \frac{d}{dx}\ln(300t + 6,000)$$

$$= \frac{30,000}{300t + 6,000}$$

$$= \frac{100}{t + 20}$$

The graph of $p(t)$ is shown in Figure 1 (graphing details omitted). Notice that $p(t)$ is decreasing, even though the GDP is increasing.

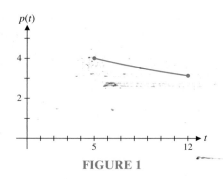

FIGURE 1

MATCHED PROBLEM 1 A model for the population data in Table 1 is

$$f(t) = 2.5t + 250$$

where t is years since 1990. Find and graph $p(t)$, the percentage rate of change of $f(t)$ for $5 \leq t \leq 12$.

■ Elasticity of Demand

Logarithmic derivatives and relative rates of change are used by economists to study the relationship among price changes, demand, and revenue. Suppose the price $\$p$ and the demand x for a certain product are related by the price–demand equation

$$x + 500p = 10,000 \tag{1}$$

In problems involving revenue, cost, and profit, it is customary to use the demand equation to express price as a function of demand. Since we are now interested in the effects that changes in price have on demand, it will be more convenient to express demand as a function of price. Solving (1) for x, we have

$$x = 10,000 - 500p$$

$$= 500(20 - p) \qquad \text{Demand as a function of price}$$

or

$$x = f(p) = 500(20 - p) \qquad 0 \le p \le 20 \tag{2}$$

Since x and p both represent nonnegative quantities, we must restrict p so that $0 \le p \le 20$. For most products, demand is assumed to be a decreasing function of price. That is, price increases result in lower demand, and price decreases result in higher demand (see Figure 2).

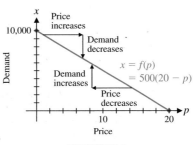

FIGURE 2

Economists use the *elasticity of demand* to study the relationship between changes in price and changes in demand. The **elasticity of demand** is the negative of the ratio of the relative rate of change of demand to the relative rate of change of price. If price and demand are related by a price–demand equation of the form $x = f(p)$, then the elasticity of demand can be expressed as

$$-\frac{\text{relative rate of change of demand}}{\text{relative rate of change of price}} = -\frac{\dfrac{d}{dp} \ln f(p)}{\dfrac{d}{dp} \ln p}$$

$$= -\frac{\dfrac{f'(p)}{f(p)}}{\dfrac{1}{p}}$$

$$= -\frac{pf'(p)}{f(p)} \tag{3}$$

INSIGHT

Since p and $f(p)$ are nonnegative and $f'(p)$ is negative (remember, demand is usually a decreasing function of price), (3) is always nonnegative. This is the reason that elasticity of demand was defined as the negative of a ratio.

Summarizing the preceding discussion, we have Theorem 1.

THEOREM 1 **Elasticity of Demand**

If price and demand are related by $x = f(p)$, then the elasticity of demand is given by

$$E(p) = -\frac{pf'(p)}{f(p)}$$

The next example illustrates interpretations of the elasticity of demand.

EXAMPLE 2

Elasticity of Demand Find $E(p)$ for the price–demand equation

$$x = f(p) = 500(20 - p)$$

Find and interpret each of the following:

(A) $E(4)$ (B) $E(16)$ (C) $E(10)$

SOLUTION

$$E(p) = -\frac{pf'(p)}{f(p)} = -\frac{p(-500)}{500(20 - p)} = \frac{p}{20 - p}$$

In order to interpret values of $E(p)$, we must recall the definition of elasticity:

$$E(p) = -\frac{\text{relative rate of change of demand}}{\text{relative rate of change of price}}$$

or

$$-\left(\begin{array}{c}\text{relative rate of}\\\text{change of demand}\end{array}\right) \approx E(p)\left(\begin{array}{c}\text{relative rate of}\\\text{change of price}\end{array}\right)$$

(A) $E(4) = \frac{4}{16} = 0.25 < 1$. If the \$4 price changes by 10%, then the demand will change by approximately $0.25(10\%) = 2.5\%$.

(B) $E(16) = \frac{16}{4} = 4 > 1$. If the \$16 price changes by 10%, then the demand will change by approximately $4(10\%) = 40\%$.

(C) $E(10) = \frac{10}{10} = 1$. If the \$10 price changes by 10%, then the demand will also change by approximately 10%. ▬

INSIGHT

Do not be concerned with the omission of any negative signs in these interpretations. We already know that if price increases, then demand decreases, and vice versa (Fig. 2). $E(p)$ is a measure of *how much* the demand changes for a given change in price.

MATCHED PROBLEM 2

Find $E(p)$ for the price–demand equation

$$x = f(p) = 1,000(40 - p)$$

Find and interpret each of the following:

(A) $E(8)$ (B) $E(30)$ (C) $E(20)$

The three cases illustrated in the solution to Example 2 are referred to as **inelastic demand, elastic demand,** and **unit elasticity,** as indicated in Table 2.

TABLE 2		
$E(p)$	Demand	Interpretation
$0 < E(p) < 1$	Inelastic	Demand is not sensitive to changes in price. A change in price produces a smaller change in demand.
$E(p) > 1$	Elastic	Demand is sensitive to changes in price. A change in price produces a larger change in demand.
$E(p) = 1$	Unit	A change in price produces the same change in demand.

Now we want to see how revenue and elasticity are related. We have used the following model for revenue many times before:

$$\text{Revenue} = (\text{demand}) \times (\text{price})$$

Since we are looking for a connection between $E(p)$ and revenue, we will use a price–demand equation that has been written in the form $x = f(p)$, where x is demand and p is price.

$$R(p) = xp = pf(p) \qquad \textit{Revenue as a function of price}$$
$$R'(p) = pf'(p) + f(p)$$
$$= f(p)\left[\frac{pf'(p)}{f(p)} + 1\right] \qquad E(p) = -\frac{pf'(p)}{f(p)}$$
$$= f(p)[1 - E(p)] \tag{4}$$

Since $x = f(p) > 0$, it follows from (4) that $R'(p)$ and $[1 - E(p)]$ always have the same sign (see Table 3).

TABLE 3 Revenue and Elasticity of Demand

All Are True or All Are False	All Are True or All Are False
$R'(p) > 0$	$R'(p) < 0$
$E(p) < 1$	$E(p) > 1$
Demand is inelastic	Demand is elastic

These facts are interpreted in the following summary and in Figure 3:

Revenue and Elasticity of Demand

Demand is inelastic:

A price increase will increase revenue.
A price decrease will decrease revenue.

Demand is elastic:

A price increase will decrease revenue.
A price decrease will increase revenue.

INSIGHT

We know that a price increase will decrease demand at all price levels (Fig. 2). As Figure 3 illustrates, the effects of a price increase on revenue depends on the price level. If demand is elastic, a price increase decreases revenue. If demand is inelastic, a price increase increases revenue.

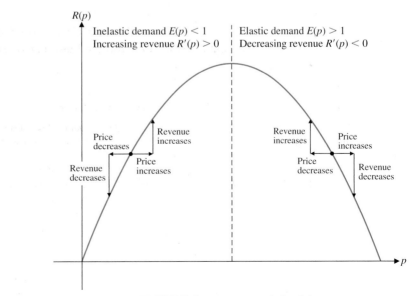

FIGURE 3 Revenue and elasticity

EXAMPLE 3

Elasticity and Revenue A manufacturer of sunglasses currently sells one type for $4 a pair. The price p and the demand x for these glasses are related by

$$x = f(p) = 7,000 - 500p$$

If the current price is increased, will revenue increase or decrease?

SOLUTION

$$E(p) = -\frac{pf'(p)}{f(p)}$$

$$= -\frac{p(-500)}{7,000 - 500p}$$

$$= \frac{p}{14 - p}$$

$$E(4) = \frac{4}{10} = 0.4$$

At the $4 price level, demand is inelastic and a price increase will increase revenue. ▬▬▬

MATCHED PROBLEM 3   Repeat Example 3 if the current price for sunglasses is $10 a pair.

Answers to Matched Problems   **1.** $p(t) = \dfrac{100}{t + 100}$

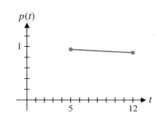

2. $E(p) = \dfrac{p}{20 - p}$

(A) $E(8) = 0.25$; demand is inelastic.

(B) $E(30) = 3$; demand is elastic.

(C) $E(20) = 1$; demand has unit elasticity.

3. $E(10) = 2.5$; demand is elastic. Increasing price will decrease revenue.

Exercise 4-7

A *In Problems 1–12, find the relative rate of change of f(x).*

1. $f(x) = 25$ **2.** $f(x) = 150$

3. $f(x) = 30x$ **4.** $f(x) = 4x$

5. $f(x) = 10x + 500$ **6.** $f(x) = 5x + 200$

7. $f(x) = 100x - 0.5x^2$ **8.** $f(x) = 50x - 0.01x^2$

9. $f(x) = 4 + 2e^{-2x}$ **10.** $f(x) = 5 - 3e^{-x}$

11. $f(x) = 25x + 3x \ln x$ **12.** $f(x) = 15x + 2x \ln x$

B *In Problems 13–16, use the price–demand equation to determine whether demand is elastic, is inelastic, or has unit elasticity at the indicated values of p.*

13. $x = f(p) = 12,000 - 10p^2$

   (A) $p = 10$ (B) $p = 20$ (C) $p = 30$

14. $x = f(p) = 1,875 - p^2$

   (A) $p = 15$ (B) $p = 25$ (C) $p = 40$

15. $x = f(p) = 950 - 2p - 0.1p^2$

    (A) $p = 30$ (B) $p = 50$ (C) $p = 70$

16. $x = f(p) = 875 - p - 0.05p^2$

    (A) $p = 50$ (B) $p = 70$ (C) $p = 100$

17. Given the price–demand equation

$$p + 0.005x = 30$$

    (A) Express the demand x as a function of the price p.

    (B) Find the elasticity of demand, $E(p)$.

    (C) What is the elasticity of demand when $p = \$10$? If this price is increased by 10%, what is the approximate change in demand?

    (D) What is the elasticity of demand when $p = \$25$? If this price is increased by 10%, what is the approximate change in demand?

    (E) What is the elasticity of demand when $p = \$15$? If this price is increased by 10%, what is the approximate change in demand?

18. Given the price–demand equation

$$p + 0.01x = 50$$

    (A) Express the demand x as a function of the price p.

    (B) Find the elasticity of demand, $E(p)$.

    (C) What is the elasticity of demand when $p = \$10$? If this price is decreased by 5%, what is the approximate change in demand?

    (D) What is the elasticity of demand when $p = \$45$? If this price is decreased by 5%, what is the approximate change in demand?

    (E) What is the elasticity of demand when $p = \$25$? If this price is decreased by 5%, what is the approximate change in demand?

19. Given the price–demand equation

$$0.02x + p = 60$$

    (A) Express the demand x as a function of the price p.

    (B) Express the revenue R as a function of the price p.

    (C) Find the elasticity of demand, $E(p)$.

    (D) For which values of p is demand elastic? Inelastic?

    (E) For which values of p is revenue increasing? Decreasing?

    (F) If $p = \$10$ and the price is decreased, will revenue increase or decrease?

    (G) If $p = \$40$ and the price is decreased, will revenue increase or decrease?

20. Repeat Problem 19 for the price–demand equation

$$0.025x + p = 50$$

In Problems 21–26, use the price–demand equation to find the values of p for which demand is elastic and the values for which demand is inelastic.

21. $x = f(p) = 10(p - 30)^2$

22. $x = f(p) = 5(p - 60)^2$

23. $x = f(p) = \sqrt{144 - 2p}$

24. $x = f(p) = \sqrt{324 - 2p}$

25. $x = f(p) = \sqrt{2{,}500 - 2p^2}$

26. $x = f(p) = \sqrt{3{,}600 - 2p^2}$

In Problems 27–32, use the demand equation to find the revenue function. Sketch the graph of the revenue function, and indicate the regions of inelastic and elastic demand on the graph.

27. $x = f(p) = 20(10 - p)$

28. $x = f(p) = 10(16 - p)$

29. $x = f(p) = 40(p - 15)^2$

30. $x = f(p) = 10(p - 9)^2$

31. $x = f(p) = 30 - 10\sqrt{p}$

32. $x = f(p) = 30 - 5\sqrt{p}$

C *If a price–demand equation is solved for p, then price is expressed as $p = g(x)$ and x becomes the independent variable. In this case, it can be shown that the elasticity of demand is given by*

$$E(x) = -\frac{g(x)}{xg'(x)}$$

In Problems 33–36, use the price–demand equation to find $E(x)$ at the indicated value of x.

33. $p = g(x) = 50 - 0.1x$, $x = 200$

34. $p = g(x) = 30 - 0.05x$, $x = 400$

35. $p = g(x) = 50 - 2\sqrt{x}$, $x = 400$

36. $p = g(x) = 20 - \sqrt{x}$, $x = 100$

37. Find $E(p)$ for $x = f(p) = Ap^{-k}$, where A and k are positive constants.

38. Find $E(p)$ for $x = f(p) = Ae^{-kp}$, where A and k are positive constants.

Applications

39. *Rate of change of cost.* A fast-food restaurant can produce a hamburger for $1.25. If the restaurant's daily sales are increasing at the rate of 20 hamburgers per day, how fast is its daily cost for hamburgers increasing?

40. *Rate of change of cost.* The fast-food restaurant in Problem 39 can produce an order of fries for $0.40. If the restaurant's daily sales are increasing at the

rate of 15 per day, how fast is its daily cost for fries increasing?

41. *Revenue and elasticity.* The price–demand equation for hamburgers at a fast-food restaurant is

$$x + 400p = 2{,}000$$

Currently, the price of a hamburger is $2.00. If the price is increased by 10%, will revenue increase or decrease?

42. *Revenue and elasticity.* Refer to Problem 41. If the current price of a hamburger is $3.00, will a 10% price increase cause revenue to increase or decrease?

43. *Revenue and elasticity.* The price–demand equation for an order of fries at a fast-food restaurant is

$$x + 1,000p = 1,000$$

Currently, the price of an order of fries is $0.30. If the price is decreased by 10%, will revenue increase or decrease?

44. *Revenue and elasticity.* Refer to Problem 43. If the current price of an order of fries is $0.60, will a 10% price decrease cause revenue to increase or decrease?

45. *Maximum revenue.* Refer to Problem 41. What price will maximize the revenue from selling hamburgers?

46. *Maximum revenue.* Refer to Problem 43. What price will maximize the revenue from selling fries?

47. *Population growth.* A model for Canada's population growth (Table 4) is

$$f(t) = 0.34t + 14.6$$

where t is years since 1950. Find and graph the percentage rate of change of $f(t)$ for $0 \le t \le 50$.

TABLE 4 Population Growth

Year	Canada (millions)	Mexico (millions)
1950	14	28
1960	18	39
1970	22	53
1980	25	68
1990	28	85
2000	31	100

48. *Population growth.* A model for Mexico's population growth (Table 4) is

$$f(t) = 1.47t + 25.5$$

where t is years since 1950. Find and graph the percentage rate of change of $f(t)$ for $0 \le t \le 50$.

49. *Crime.* A model for the number of robberies in the United States (Table 5) is

$$r(t) = 11.3 - 3.6 \ln t$$

where t is years since 1990. Find the relative rate of change for robberies in 2002.

TABLE 5 Number of Victimizations per 1,000 Population Age 12 and Over

	Robbery	Aggravated Assault
1995	5.4	9.5
1996	5.2	8.8
1997	4.3	8.6
1998	4.0	7.5
1999	3.6	6.7
2000	3.2	5.7
2001	2.8	5.3
2002	2.2	4.3

50. *Crime.* A model for the number of assaults in the United States (Table 5) is

$$a(t) = 19.6 - 6.0 \ln t$$

where t is years since 1990. Find the relative rate of change for assaults in 2002.

CHAPTER 4 REVIEW

Important Terms, Symbols, and Concepts

4-1 The Constant e and Continuous Compound Interest   Examples
- The number e is defined as

$$\lim_{x \to \infty} \left(1 + \frac{1}{n}\right)^n = \lim_{x \to 0}(1 + s)^{1/s} = 2.718\ 281\ 828\ 459 \ldots$$

If the number of compounding periods in one year is increased without limit, we obtain the **compound interest formula**

$$A = Pe^{rt}$$

where

P = principal
r = annual nominal interest rate compounded continuously
t = time in years
A = amount at time t

Ex. 1, p. 219
Ex. 2, p. 219
Ex. 3, p. 220
Ex. 4, p. 221

4-2 Derivatives of Exponential and Logarithmic Functions
- For $b > 0, b \neq 1$,

Ex. 1, p. 225
Ex. 2, p. 227
Ex. 3, p. 228
Ex. 4, p. 229
Ex. 5, p. 230

$$\frac{d}{dx} e^x = e^x \qquad \frac{d}{dx} b^x = b^x \ln b$$

$$\frac{d}{dx} \ln x = \frac{1}{x} \qquad \frac{d}{dx} \log_b x = \frac{1}{\ln b}\frac{1}{x}$$

- The **change-of-base formulas** allow conversion from base e to any base $b, b > 0, b \neq 1$:

$$b^x = e^{x \ln b} \qquad \log_b x = \frac{\ln x}{\ln b}$$

4-3 Derivatives of Products and Quotients
- Product rule. If $y = f(x) = F(x)\, S(x)$, then $f'(x) = F(x)S'(x) + S(x)F'(x)$, provided that both $F'(x)$ and $S'(x)$ exist.

Ex. 1, p. 233
Ex. 2, p. 234
Ex. 3, p. 234
Ex. 4, p. 236
Ex. 5, p. 237
Ex. 6, p. 237

- Quotient rule. If $y = f(x) = \dfrac{T(x)}{B(x)}$, then $f'(x) = \dfrac{B(x)\,T'(x) - T(x)\,B'(x)}{\left[B(x)\right]^2}$ provided that both $T'(x)$ and $B'(x)$ exist.

4-4 The Chain Rule
- A function m is a **composite** of functions f and g if $m(x) = f[g(x)]$.
- The **chain rule** gives a formula for the derivative of the composite function $m(x) = f[g(x)]$:

Ex. 1, p. 241
Ex. 2, p. 241
Ex. 4, p. 245
Ex. 5, p. 247
Ex. 3, p. 243

$$m'(x) = f'[g(x)]g'(x)$$

- A special case of the chain rule is the **general power rule:**

$$\frac{d}{dx}[f(x)]^n = n[f(x)]^{n-1}f'(x)$$

- Other special cases of the chain rule are the following **general derivative rules:**

Ex. 6, p. 247

$$\frac{d}{dx}\ln[f(x)] = \frac{1}{f(x)}f'(x)$$

$$\frac{d}{dx}e^{f(x)} = e^{f(x)}f'(x)$$

4-5 Implicit Differentiation
- If $y = y(x)$ is a function defined by the equation $F(x, y) = 0$, we use **implicit differentiation** to find an equation in $x, y,$ and y'.

Ex. 1, p. 253
Ex. 2, p. 254
Ex. 3, p. 255

4-6 Related Rates
- If x and y represent quantities that are changing with respect to time and are related by the equation $F(x, y) = 0$, then implicit differentiation produces an equation that relates $x, y, dy/dt,$ and dx/dt. Problems of this type are called **related-rates problems.**
- Suggestions for solving related-rates problems are given on page 258.

Ex. 1, p. 257
Ex. 2, p. 258
Ex. 3, p. 259

4-7 Elasticity of Demand
- The **relative rate of change,** or the **logarithmic derivative,** of a function $f(x)$ is $f'(x)/f(x)$, and the **percentage rate of change** is $100 \times [f'(x)/f(x)]$.
- If price and demand are related by $x = f(p)$, then the **elasticity of demand** is given by

Ex. 1, p. 263

Ex. 2, p. 265

$$E(p) = -\frac{pf'(p)}{f(p)} = -\frac{\text{relative rate of change of demand}}{\text{relative rate of change of price}}$$

- **Demand is inelastic** if $0 < E(p) < 1$. (Demand is not sensitive to changes in price; a change in price produces a smaller change in demand.) **Demand is elastic** if $E(p) > 1$. (Demand is sensitive to changes in price; a change in price produces a larger change in demand.) **Demand has unit elasticity** if $E(p) = 1$. (A change in price produces the same change in demand.)

Ex. 3, p. 268

- If $R(p) = pf(p)$ is the revenue function, then $R'(p)$ and $[1 - E(p)]$ always have the same sign.

REVIEW EXERCISE

Work through all the problems in this chapter review, and check your answers in the back of the book. Answers to all review problems are there, along with section numbers in italics to indicate where each type of problem is discussed. Where weaknesses show up, review appropriate sections of the text.

A

1. Use a calculator to evaluate $A = 2,000e^{0.09t}$ to the nearest cent for $t = 5, 10$, and 20.

Find the indicated derivatives in Problems 2–4.

2. $\dfrac{d}{dx}(2 \ln x + 3e^x)$ 3. $\dfrac{d}{dx}e^{2x-3}$

4. y' for $y = \ln(2x + 7)$

5. Let $y = \ln u$ and $u = 3 + e^x$.

   (A) Express y in terms of x.

   (B) Use the chain rule to find dy/dx, and then express dy/dx in terms of x.

6. Find y' for $y = y(x)$ defined implicity by the equation $2y^2 - 3x^3 - 5 = 0$, and evaluate at $(x, y) = (1, 2)$.

7. For $y = 3x^2 - 5$, where $x = x(t)$ and $y = y(t)$, find dy/dt if $dx/dt = 3$ when $x = 12$.

8. Given the demand equation $25p + x = 1,000$,

   (A) Express the demand x as a function of the price p.

   (B) Find the elasticity of demand, $E(p)$.

   (C) Find $E(15)$ and interpret.

   (D) Express the revenue function as a function of price p.

   (E) If $p = \$25$, what is the effect of a price cut on revenue?

B

9. Find the slope of the line tangent to $y = 100e^{-0.1x}$ when $x = 0$.

10. Use a calculator and a table of values to investigate

$$\lim_{n \to \infty} \left(1 + \frac{2}{n}\right)^n$$

   Do you think the limit exists? If so, what do you think it is?

Find the indicated derivatives in Problems 11–16.

11. $\dfrac{d}{dz}[(\ln z)^7 + \ln z^7]$ 12. $\dfrac{d}{dx}(x^6 \ln x)$

13. $\dfrac{d}{dx}\dfrac{e^x}{x^6}$ 14. y' for $y = \ln(2x^3 - 3x)$

15. $f'(x)$ for $f(x) = e^{x^3 - x^2}$

16. dy/dx for $y = e^{-2x} \ln 5x$

17. Find the equation of the line tangent to the graph of $y = f(x) = 1 + e^{-x}$ at $x = 0$. At $x = -1$.

18. Find y' for $y = y(x)$ defined implicitly by the equation $x^2 - 3xy + 4y^2 = 23$, and find the slope of the graph at $(-1, 2)$.

19. Find x' for $x = x(t)$ defined implicitly by $x^3 - 2t^2x + 8 = 0$, and evaluate at $(t, x) = (-2, 2)$.

20. Find y' for $y = y(x)$ defined implicitly by $x - y^2 = e^y$, and evaluate at $(1, 0)$.

21. Find y' for $y = y(x)$ defined implicitly by $\ln y = x^2 - y^2$, and evaluate at $(1, 1)$.

22. A point is moving on the graph of $y^2 - 4x^2 = 12$ so that its x coordinate is decreasing by 2 units per second when $(x, y) = (1, 4)$. Find the rate of change of the y coordinate.

23. A 17-foot ladder is placed against a wall. If the foot of the ladder is pushed toward the wall at 0.5 foot per second, how fast is the top of the ladder rising when the foot is 8 feet from the wall?

24. Water from a water heater is leaking onto a floor. A circular pool is created whose area is increasing at the rate of 24 square inches per minute. How fast is the radius R of the pool increasing when the radius is 12 inches? $[A = \pi R^2]$

C

25. Find the values of p for which demand is elastic and the values for which demand is inelastic if the price–demand equation is

$$x = f(p) = 20(p - 15)^2 \qquad 0 \le p \le 15$$

26. Graph the revenue function and indicate the regions of inelastic and elastic demand of the graph if the price–demand equation is

$$x = f(p) = 5(20 - p) \qquad 0 \le p \le 20$$

27. Let $y = w^3$, $w = \ln u$, and $u = 4 - e^x$.

   (A) Express y in terms of x.

   (B) Use the chain rule to find dy/dx, and then express dy/dx in terms of x.

Find the indicated derivatives in Problems 28–30.

28. y' for $y = 5^{x^2 - 1}$

29. $\dfrac{d}{dx}\log_5(x^2 - x)$

30. $\dfrac{d}{dx}\sqrt{\ln(x^2 + x)}$

31. Find y' for $y = y(x)$ defined implicitly by the equation $e^{xy} = x^2 + y + 1$, and evaluate at $(0, 0)$.

32. A rock thrown into a still pond causes a circular ripple. Suppose the radius is increasing at a constant rate of 3 feet per second. Show that the area does not increase at a constant rate. When is the rate of increase of the area the smallest? The largest? Explain.

33. A point moves along the graph of $y = x^3$ in such a way that its y coordinate is increasing at a constant rate of 5 units per second. Does the x coordinate ever increase at a faster rate than the y coordinate? Explain.

APPLICATIONS

34. *Doubling time.* How long will it take money to double if it is invested at 5% interest compounded

(A) Annually?   (B) Continuously?

35. *Continuous compound interest.* If $100 is invested at 10% interest compounded continuously, the amount (in dollars) at the end of t years is given by

$$A = 100e^{0.1t}$$

Find $A'(t)$, $A'(1)$, and $A'(10)$.

36. *Marginal analysis.* The price–demand equation for 14-cubic-foot refrigerators at an appliance store is

$$p(x) = 1,000e^{-0.02x}$$

where x is the monthly demand and p is the price in dollars. Find the marginal revenue equation.

37. *Demand equation.* Given the demand equation

$$x = \sqrt{5,000 - 2p^3}$$

find the rate of change of p with respect to x by implicit differentiation (x is the number of items that can be sold at a price of $\$p$ per item).

38. *Rate of change of revenue.* A company is manufacturing a new video game and can sell all that it manufactures. The revenue (in dollars) is given by

$$R = 36x - \frac{x^2}{20}$$

where the production output in 1 day is x games. If production is increasing at 10 games per day when production is 250 games per day, find the rate of increase in revenue.

39. *Revenue and elasticity.* The price–demand equation for home-delivered 12-inch pizzas is

$$p = 16.8 - 0.002x$$

where x is the number of pizzas delivered weekly. The current price of one pizza is $8. In order to generate additional revenue from the sale of 12-inch pizzas, would you recommend a price increase or a price decrease?

40. *Average income.* A model for the average income per household before taxes are paid is

$$f(t) = 1,700t + 20,500$$

where t is years since 1980. Find the relative rate of change of household income in 2010.

41. *Drug concentration.* The concentration of a drug in the bloodstream t hours after injection is given approximately by

$$C(t) = 5e^{-0.3t}$$

where $C(t)$ is concentration in milligrams per milliliter. What is the rate of change of concentration after 1 hour? After 5 hours?

42. *Wound healing.* A circular wound on an arm is healing at the rate of 45 square millimeters per day (the area of the wound is decreasing at this rate). How fast is the radius R of the wound decreasing when $R = 15$ millimeters? $[A = \pi R^2]$

43. *Psychology: learning.* In a computer assembly plant, a new employee, on the average, is able to assemble

$$N(t) = 10(1 - e^{-0.4t})$$

units after t days of on-the-job training.

(A) What is the rate of learning after 1 day? After 5 days?

(B) Find the number of days (to the nearest day) after which the rate of learning is less than 0.25 unit per day.

44. *Learning.* A new worker on the production line performs an operation in T minutes after x performances of the operation, as given by

$$T = 2\left(1 + \frac{1}{x^{3/2}}\right)$$

If, after performing the operation 9 times, the rate of improvement is $dx/dt = 3$ operations per hour, find the rate of improvement in time dT/dt in performing each operation.

Graphing and Optimization

CHAPTER 5

5-1 First Derivative and Graphs

5-2 Second Derivative and Graphs

5-3 L' Hôpital's Rule

5-4 Curve-Sketching Techniques

5-5 Absolute Maxima and Minima

5-6 Optimization

Chapter 5 Review

Review Exercise

INTRODUCTION

Since the derivative is associated with the slope of the graph of a function at a point, we might expect that it is also associated with other properties of a graph. As we will see in this and the next section, the derivative can tell us a great deal about the shape of the graph of a function. In addition, our investigation will lead to methods for finding absolute maximum and minimum values for functions that do not require graphing. Manufacturing companies can use these methods to find production levels that will minimize cost or maximize profit, pharmacologists can use them to find levels of drug dosages that will produce maximum sensitivity, and so on.

Section 5-1   FIRST DERIVATIVE AND GRAPHS

- Increasing and Decreasing Functions
- Local Extrema
- First-Derivative Test
- Applications to Economics

■ Increasing and Decreasing Functions

Sign charts (Section 3-2) will be used throughout this chapter. You will find it helpful to review the terminology and techniques for constructing sign charts now.

Explore & Discuss **1**

Figure 1 shows the graph of $y = f(x)$ and a sign chart for $f'(x)$, where

$$f(x) = x^3 - 3x$$

and

$$f'(x) = 3x^2 - 3 = 3(x + 1)(x - 1)$$

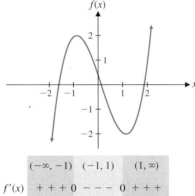

FIGURE 1

Discuss the relationship between the graph of f and the sign of $f'(x)$ over each interval on which $f'(x)$ has a constant sign. Also, describe the behavior of the graph of f at each partition number for f'.

As they are scanned from left to right, graphs of functions generally have rising and falling sections. If you scan the graph of $f(x) = x^3 - 3x$ in Figure 1 from left to right, you will see that

- On the interval $(-\infty, -1)$, the graph of f is rising, $f(x)$ is increasing,* and the slope of the graph is positive [$f'(x) > 0$].
- On the interval $(-1, 1)$, the graph of f is falling, $f(x)$ is decreasing, and the slope of the graph is negative [$f'(x) < 0$].
- On the interval $(1, \infty)$, the graph of f is rising, $f(x)$ is increasing, and the slope of the graph is positive [$f'(x) > 0$].
- At $x = -1$ and $x = 1$, the slope of the graph is 0 [$f'(x) = 0$].

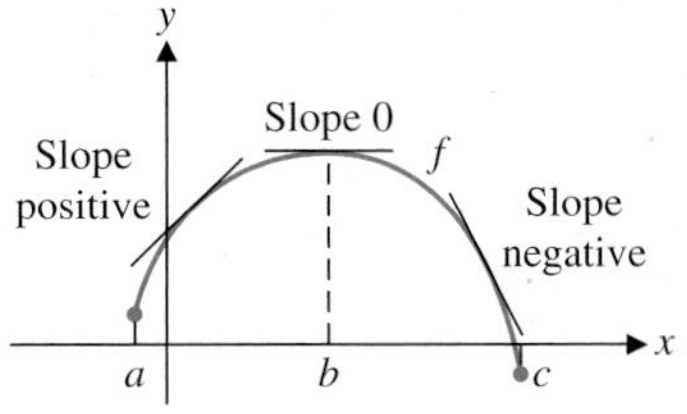

In general, if $f'(x) > 0$ (is positive) on the interval (a, b) (Fig. 2), then $f(x)$ increases (↗) and the graph of f rises as we move from left to right over the interval; if $f'(x) < 0$ (is negative) on an interval (a, b), then $f(x)$ decreases (↘) and the graph of f falls as we move from left to right over the interval. We summarize these important results in Theorem 1.

y

Slope positive

Slope 0

f

Slope negative

a   *b*   *c*   x

FIGURE 2

---

* Formally, we say that the function f is **increasing** on an interval (a, b) if $f(x_2) > f(x_1)$ whenever $a < x_1 < x_2 < b$, and f is **decreasing** on (a, b) if $f(x_2) < f(x_1)$ whenever $a < x_1 < x_2 < b$.

THEOREM 1   INCREASING AND DECREASING FUNCTIONS

For the interval (a, b),

$f'(x)$	$f(x)$	Graph of f	Examples
+	Increases ↗	Rises ↗	
−	Decreases ↘	Falls ↘	

Explore & Discuss **2**   The graphs of $f(x) = x^2$ and $g(x) = |x|$ are shown in Figure 3. Both functions change from decreasing to increasing at $x = 0$. Discuss the relationship between the graph of each function at $x = 0$ and the derivative of the function at $x = 0$.

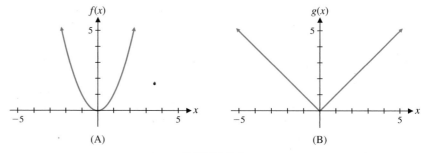

(A)   (B)

FIGURE 3

EXAMPLE 1   **Finding Intervals on Which a Function Is Increasing or Decreasing** Given the function $f(x) = 8x - x^2$,

(A) Which values of x correspond to horizontal tangent lines?

(B) For which values of x is $f(x)$ increasing? Decreasing?

(C) Sketch a graph of f. Add any horizontal tangent lines.

SOLUTION   (A) $f'(x) = 8 - 2x = 0$

$$x = 4$$

Thus, a horizontal tangent line exists at $x = 4$ only.

(B) We will construct a sign chart for $f'(x)$ to determine which values of x make $f'(x) > 0$ and which values make $f'(x) < 0$. Recall from Section 3-2 that the partition numbers for a function are the points where the function is 0 or discontinuous. Thus, when constructing a sign chart for $f'(x)$, we must locate all points where $f'(x) = 0$ or $f'(x)$ is discontinuous. From part (A), we know that $f'(x) = 8 - 2x = 0$ at $x = 4$. Since $f'(x) = 8 - 2x$ is a polynomial, it is continuous for all x. Thus, 4 is the only partition number. We construct a sign chart for the intervals $(-\infty, 4)$ and $(4, \infty)$, using test numbers 3 and 5:

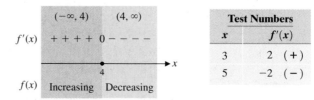

Hence, $f(x)$ is increasing on $(-\infty, 4)$ and decreasing on $(4, \infty)$.

(C)

x	f(x)
0	0
2	12
4	16
6	12
8	0

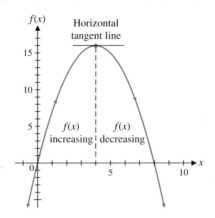

MATCHED PROBLEM 1   Repeat Example 1 for $f(x) = x^2 - 6x + 10$.

As Example 1 illustrates, the construction of a sign chart will play an important role in using the derivative to analyze and sketch the graph of a function f. The partition numbers for f' are central to the construction of these sign charts and also to the analysis of the graph of $y = f(x)$. We already know that if $f'(c) = 0$, then the graph of $y = f(x)$ will have a horizontal tangent line at $x = c$. But the partition numbers for f' also include the numbers c such that $f'(c)$ does not exist.* There are two possibilities at this type of number: $f(c)$ does not exist; or $f(c)$ exists, but the slope of the tangent line at $x = c$ is undefined.

DEFINITION   Critical Values

The values of x in the domain of f where $f'(x) = 0$ or where $f'(x)$ does not exist are called the **critical values** of f.

> ### INSIGHT
>
> The critical values of f are always in the domain of f and are also partition numbers for f', but f' may have partition numbers that are not critical values.
>
> If f is a polynomial, then both the partition numbers for f' and the critical values of f are the solutions of $f'(x) = 0$.

We will illustrate the process for locating critical values with examples.

EXAMPLE 2   **Partition Numbers and Critical Values**   Find the critical values of f, the intervals on which f is increasing, and those on which f is decreasing, for $f(x) = 1 + x^3$.

SOLUTION   Begin by finding the partition number for $f'(x)$:

$$f'(x) = 3x^2 = 0, \quad \text{only at } x = 0$$

The partition number 0 is in the domain of f, so 0 is the only critical value of f.

---

* We are assuming that $f'(c)$ does not exist at any point of discontinuity of f'. There do exist functions f such that f' is discontinuous at $x = c$, yet $f'(c)$ exists. However, we do not consider such functions in this book.

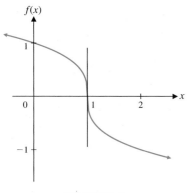

FIGURE 4

The sign chart for $f'(x) = 3x^2$ (partition number is 0) is

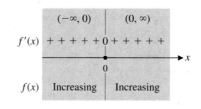

	(−∞, 0)	(0, ∞)
$f'(x)$	+ + + + + 0 + + + + +	
$f(x)$	Increasing	Increasing

Test Numbers	
x	$f'(x)$
−1	3 (+)
1	3 (+)

The sign chart indicates that $f(x)$ is increasing on $(-\infty, 0)$ and $(0, \infty)$. Since f is continuous at $x = 0$, it follows that $f(x)$ is increasing for all x. The graph of f is shown in Figure 4.

MATCHED PROBLEM 2 Find the critical values of f, the intervals on which f is increasing, and those on which f is decreasing, for $f(x) = 1 - x^3$.

EXAMPLE 3 **Partition Numbers and Critical Values** Find the critical values of f, the intervals on which f is increasing, and those on which f is decreasing, for $f(x) = (1 - x)^{1/3}$.

SOLUTION
$$f'(x) = -\frac{1}{3}(1 - x)^{-2/3} = \frac{-1}{3(1 - x)^{2/3}}$$

To find partition numbers for f', we note that f' is continuous for all x, except for values of x for which the denominator is 0; that is, $f'(1)$ does not exist and f' is discontinuous at $x = 1$. Since the numerator is the constant -1, $f'(x) \neq 0$ for any value of x. Thus, $x = 1$ is the only partition number for f'. Since 1 is in the domain of f, $x = 1$ is also the only critical value of f. When constructing the sign chart for f' we use the abbreviation ND to note the fact that $f'(x)$ is *not defined* at $x = 1$.

The sign chart for $f'(x) = -1/[3(1 - x)^{2/3}]$ (partition number is 1) is

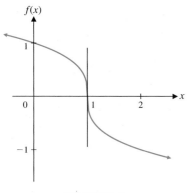

FIGURE 5

	(−∞, 1)	(1, ∞)
$f'(x)$	− − − − ND − − − −	
$f(x)$	Decreasing	Decreasing

Test Numbers	
x	$f'(x)$
0	$-\frac{1}{3}$ (−)
2	$-\frac{1}{3}$ (−)

The sign chart indicates that f is decreasing on $(-\infty, 1)$ and $(1, \infty)$. Since f is continuous at $x = 1$, it follows that $f(x)$ is decreasing for all x. Thus, **a continuous function can be decreasing (or increasing) on an interval containing values of x where $f'(x)$ does not exist.** The graph of f is shown in Figure 5. Notice that the undefined derivative at $x = 1$ results in a vertical tangent line at $x = 1$. In general, **a vertical tangent will occur at $x = c$ if f is continuous at $x = c$ and if $|f'(x)|$ becomes larger and larger as x approaches c.**

MATCHED PROBLEM 3 Find the critical values of f, the intervals on which f is increasing, and those on which f is decreasing, for $f(x) = (1 + x)^{1/3}$.

EXAMPLE 4 **Partition Numbers and Critical Values** Find the critical values of f, the intervals on which f is increasing, and those on which f is decreasing, for $f(x) = \frac{1}{x - 2}$.

SOLUTION

$$f(x) = \frac{1}{x - 2} = (x - 2)^{-1}$$

$$f'(x) = -(x - 2)^{-2} = \frac{-1}{(x - 2)^2}$$

To find the partition numbers for f', note that $f'(x) \neq 0$ for any x and f' is not defined at $x = 2$. Thus, $x = 2$ is the only partition number for f'. However, $x = 2$ is *not* in the domain of f. Consequently, $x = 2$ is not a critical value of f. This function has no critical values.

The sign chart for $f'(x) = -1/(x - 2)^2$ (partition number is 2) is

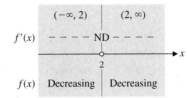

	Test Numbers	
	x	$f'(x)$
	1	-1 $(-)$
	3	-1 $(-)$

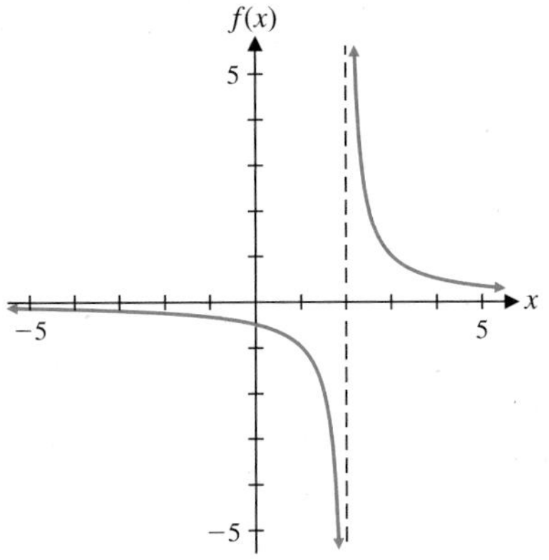

Thus, f is decreasing on $(-\infty, 2)$ and $(2, \infty)$. See the graph of f in Figure 6.   ▬

FIGURE 6

MATCHED PROBLEM 4   Find the critical values for f, the intervals on which f is increasing, and those on which f is decreasing, for $f(x) = \dfrac{1}{x}$.

EXAMPLE 5   **Partition Numbers and Critical Values**   Find the critical values of f, the intervals on which f is increasing, and those on which f is decreasing, for $f(x) = 8 \ln x - x^2$.

SOLUTION   The natural logarithm function $\ln x$ is defined on $(0, \infty)$, or $x > 0$, so $f(x)$ is defined only for $x > 0$. We have

$$f(x) = 8 \ln x - x^2, x > 0$$

$$f'(x) = \frac{8}{x} - 2x \qquad \textit{Find a common denominator.}$$

$$= \frac{8}{x} - \frac{2x^2}{x} \qquad \textit{Subtract numerators.}$$

$$= \frac{8 - 2x^2}{x} \qquad \textit{Factor numerator.}$$

$$= \frac{2(2 - x)(2 + x)}{x}, \quad x > 0$$

Note that $f'(x) = 0$ at -2 and at 2 and $f'(x)$ is discontinuous at 0. These are the partition numbers for $f'(x)$. Since the domain of f is $(0, \infty)$, 0 and -2 are not critical values. The remaining partition number, 2, is the only critical value for $f(x)$.

The sign chart for $f'(x) = \dfrac{2(2 - x)(2 + x)}{x}$, $x > 0$ (partition number is 2), is

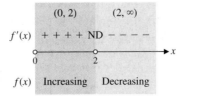

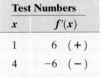

	Test Numbers	
	x	$f'(x)$
	1	6 $(+)$
	4	-6 $(-)$

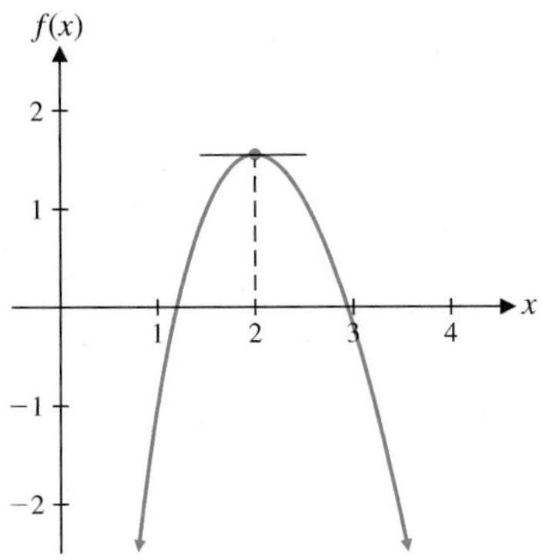

Thus, f is increasing on $(0, 2)$ and decreasing on $(2, \infty)$. See the graph of f in Figure 7.   ▬

FIGURE 7

MATCHED PROBLEM 5   Find the critical values of f, the intervals on which f is increasing, and those on which f is decreasing, for $f(x) = 5 \ln x - x$.

Explore & Discuss **3**   A student examined the sign chart in Example 4 and concluded that $f(x) = 1/(x - 2)$ is decreasing for all x except $x = 2$. However, $f(1) = -1 < f(3) = 1$, which seems to indicate that f is increasing. Discuss the difference between the correct answer in Example 4 and the student's answer. Explain why the student's description of where f is decreasing is unacceptable.

INSIGHT

Example 4 illustrates two important ideas:

1. Do not assume that all partition numbers for the derivative f' are critical values of the function f. To be a critical value, a partition number must also be in the domain of f.
2. The values for which a function is increasing or decreasing must always be expressed in terms of open intervals that are subsets of the domain of the function.

■ Local Extrema

When the graph of a continuous function changes from rising to falling, a high point, or *local maximum,* occurs; when the graph changes from falling to rising, a low point, or *local minimum,* occurs. In Figure 8, high points occur at c_3 and c_6, and low points occur at c_2 and c_4. In general, we call $f(c)$ a **local maximum** if there exists an interval (m, n) containing c such that

$$f(x) \leq f(c) \qquad \text{for all } x \text{ in } (m, n)$$

Note that this inequality need hold only for values of x near c—hence the use of the term *local.*

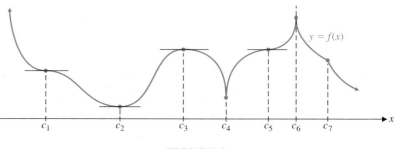

FIGURE 8

The quantity $f(c)$ is called a **local minimum** if there exists an interval (m, n) containing c such that

$$f(x) \geq f(c) \qquad \text{for all } x \text{ in } (m, n)$$

The quantity $f(c)$ is called a **local extremum** if it is either a local maximum or a local minimum. A point on a graph where a local extremum occurs is also called a **turning point.** Thus, in Figure 8 we see that local maxima occur at c_3 and c_6, local minima occur at c_2 and c_4, and all four values produce local extrema. Also, the local maximum $f(c_3)$ is not the highest point on the graph in Figure 8. Later in this chapter, we consider the problem of finding the highest and lowest points on a graph, or absolute extrema. For now, we are concerned only with locating *local* extrema.

EXAMPLE 6   **Analyzing a graph**   Use the graph of f in Figure 9 to find the intervals on which f is increasing, those on which f is decreasing, any local maxima, and any local minima.

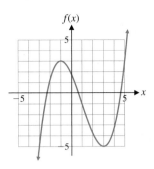

FIGURE 9

SOLUTION   The function f is increasing (the graph is rising) on $(-\infty, -1)$ and on $(3, \infty)$ and is decreasing (the graph is falling) on $(-1, 3)$. Because the graph changes from rising to falling at $x = -1$, $f(-1) = 3$ is a local maximum. Because the graph changes from falling to rising at $x = 3$, $f(3) = -5$ is a local minimum. ▬

MATCHED PROBLEM 6   Use the graph of g in Figure 10 to find the intervals on which g is increasing, those on which g is decreasing, any local maxima, and any local minima.

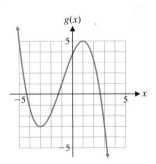

FIGURE 10

How can we locate local maxima and minima if we are given the equation of a function and not its graph? The key is to examine the critical values of the function. The local extrema of the function f in Figure 8 occur either at points where the derivative is 0 (c_2 and c_3) or at points where the derivative does not exist (c_4 and c_6). In other words, local extrema occur only at critical values of f. Theorem 2 shows that this is true in general.

THEOREM 2   **EXISTENCE OF LOCAL EXTREMA**

If f is continuous on the interval (a, b), c is a number in (a, b), and $f(c)$ is a local extremum, then either $f'(c) = 0$ or $f'(c)$ does not exist (is not defined).

Theorem 2 states that a local extremum can occur only at a critical value, but it does not imply that every critical value produces a local extremum. In Figure 8, c_1 and c_5 are critical values (the slope is 0), but the function does not have a local maximum or local minimum at either of these values.

Our strategy for finding local extrema is now clear: We find all critical values of f and test each one to see if it produces a local maximum, a local minimum, or neither.

■ **First-Derivative Test**

If $f'(x)$ exists on both sides of a critical value c, the sign of $f'(x)$ can be used to determine whether the point $(c, f(c))$ is a local maximum, a local minimum, or neither. The various possibilities are summarized in the following box and are illustrated in Figure 11:

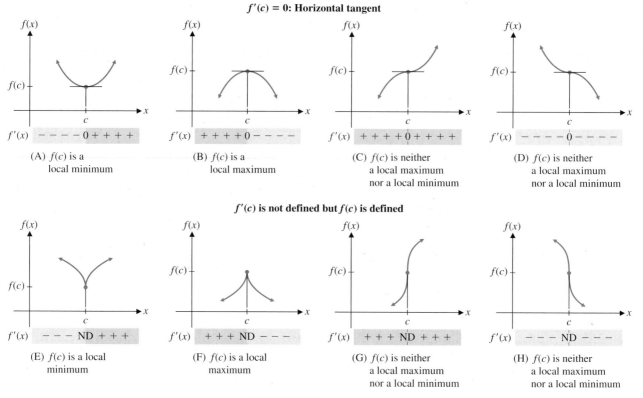

FIGURE 11 Local extrema

PROCEDURE | **First-Derivative Test for Local Extrema**

Let c be a critical value of f [$f(c)$ is defined and either $f'(c) = 0$ or $f'(c)$ is not defined]. Construct a sign chart for $f'(x)$ close to and on either side of c.

Sign Chart	$f(c)$
$f'(x)$ $---$ $+++$ m c n $f(x)$ Decreasing Increasing	$f(c)$ is local minimum. If $f'(x)$ changes from negative to positive at c, then $f(c)$ is a local minimum.
$f'(x)$ $+++$ $---$ m c n $f(x)$ Increasing Decreasing	$f(c)$ is local maximum. If $f'(x)$ changes from positive to negative at c, then $f(c)$ is a local maximum.
$f'(x)$ $---$ $\mid$ $---$ m c n $f(x)$ Decreasing Decreasing	$f(c)$ is not a local extremum. If $f'(x)$ does not change sign at c, then $f(c)$ is neither a local maximum nor a local minimum.
$f'(x)$ $+++$ $\mid$ $+++$ m c n $f(x)$ Increasing Increasing	$f(c)$ is not a local extremum. If $f'(x)$ does not change sign at c, then $f(c)$ is neither a local maximum nor a local minimum.

EXAMPLE 7   **Locating Local Extrema** Given $f(x) = x^3 - 6x^2 + 9x + 1$,

(A) Find the critical values of f.

(B) Find the local maxima and minima.

(C) Sketch the graph of f.

SOLUTION   (A) Find all numbers x in the domain of f where $f'(x) = 0$ or $f'(x)$ does not exist.

$$f'(x) = 3x^2 - 12x + 9 = 0$$
$$3(x^2 - 4x + 3) = 0$$
$$3(x - 1)(x - 3) = 0$$
$$x = 1 \quad \text{or} \quad x = 3$$

$f'(x)$ exists for all x; the critical values are $x = 1$ and $x = 3$.

(B) The easiest way to apply the first-derivative test for local maxima and minima is to construct a sign chart for $f'(x)$ for all x. Partition numbers for $f'(x)$ are $x = 1$ and $x = 3$ (which also happen to be critical values of f).

Sign chart for $f'(x) = 3(x - 1)(x - 3)$:

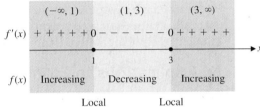

Test Numbers	
x	$f'(x)$
0	9 (+)
2	−3 (−)
4	9 (+)

The sign chart indicates that f increases on $(-\infty, 1)$, has a local maximum at $x = 1$, decreases on $(1, 3)$, has a local minimum at $x = 3$, and increases on $(3, \infty)$. These facts are summarized in the following table:

x	$f'(x)$	$f(x)$	Graph of f
$(-\infty, 1)$	+	Increasing	Rising
$x = 1$	0	Local maximum	Horizontal tangent
$(1, 3)$	−	Decreasing	Falling
$x = 3$	0	Local minimum	Horizontal tangent
$(3, \infty)$	+	Increasing	Rising

(C) We sketch a graph of f, using the information from part (B) and point-by-point plotting.

x	$f(x)$
0	1
1	5
2	3
3	1
4	5

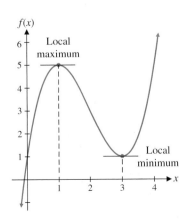

MATCHED PROBLEM 7 | Given $f(x) = x^3 - 9x^2 + 24x - 10$,

(A) Find the critical values of f.

(B) Find the local maxima and minima.

(C) Sketch a graph of f.

How can you tell if you have found all the local extrema of a function? In general, this can be a difficult question to answer. However, in the case of a polynomial function, there is an easily determined upper limit on the number of local extrema. Since the local extrema are the x intercepts of the derivative, this limit is a consequence of the number of x intercepts of a polynomial. The relevant information is summarized in the following theorem, which is stated without proof:

THEOREM 3   INTERCEPTS AND LOCAL EXTREMA OF POLYNOMIAL FUNCTIONS

If $f(x) = a_n x^n + a_{n-1} x^{n-1} + \cdots + a_1 x + a_0, a_n \neq 0$, is an nth-degree polynomial, then f has at most n x intercepts and at most $n - 1$ local extrema.

Theorem 3 does not guarantee that every nth-degree polynomial has exactly $n - 1$ local extrema; it says only that there can never be more than $n - 1$ local extrema. For example, the third-degree polynomial in Example 7 has two local extrema, while the third-degree polynomial in Example 2 does not have any.

■ Applications to Economics

In addition to providing information for hand-sketching graphs, the derivative is an important tool for analyzing graphs and discussing the interplay between a function and its rate of change. The next two examples illustrate this process in the context of some applications to economics.

EXAMPLE 8   **Agricultural Exports and Imports** Over the past several decades, the United States has exported more agricultural products than it has imported, maintaining a positive balance of trade in this area. However, the trade balance fluctuated considerably during that period. The graph in Figure 12 approximates the rate of change of the balance of trade over a 15-year period, where $B(t)$ is the balance of trade (in billions of dollars) and t is time (in years).

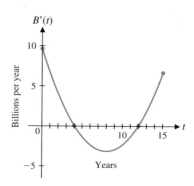

FIGURE 12 Rate of change of the balance of trade

(A) Write a brief verbal description of the graph of $y = B(t)$, including a discussion of any local extrema.

(B) Sketch a possible graph of $y = B(t)$.

SOLUTION (A) The graph of the derivative $y = B'(t)$ contains the same essential information as a sign chart. That is, we see that $B'(t)$ is positive on $(0, 4)$, 0 at $t = 4$, negative on $(4, 12)$, 0 at $t = 12$, and positive on $(12, 15)$. Hence, the trade balance increases for the first 4 years to a local maximum, decreases for the next 8 years to a local minimum, and then increases for the final 3 years.

(B) Without additional information concerning the actual values of $y = B(t)$, we cannot produce an accurate graph. However, we can sketch a possible graph that illustrates the important features, as shown in Figure 13. The absence of a scale on the vertical axis is a consequence of the lack of information about the values of $B(t)$.

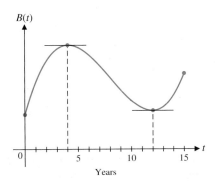

FIGURE 13 Balance of trade

MATCHED PROBLEM 8   The graph in Figure 14 approximates the rate of change of the U.S. share of the total world production of motor vehicles over a 20-year period, where $S(t)$ is the U.S. share (as a percentage) and t is time (in years).

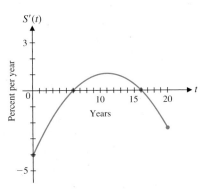

FIGURE 14

(A) Write a brief verbal description of the graph of $y = S(t)$, including a discussion of any local extrema.

(B) Sketch a possible graph of $y = S(t)$.

EXAMPLE 9   **Revenue Analysis** The graph of the total revenue $R(x)$ (in dollars) from the sale of x bookcases is shown in Figure 15.

(A) Write a brief verbal description of the graph of the marginal revenue function $y = R'(x)$, including a discussion of any x intercepts.

(B) Sketch a possible graph of $y = R'(x)$.

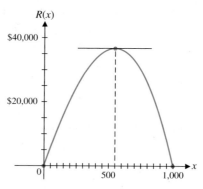

FIGURE 15 Revenue

SOLUTION   (A) The graph of $y = R(x)$ indicates that $R(x)$ increases on $(0, 550)$, has a local maximum at $x = 550$, and decreases on $(550, 1,000)$. Consequently, the marginal revenue function $R'(x)$ must be positive on $(0, 550)$, 0 at $x = 550$, and negative on $(550, 1,000)$.

(B) A possible graph of $y = R'(x)$ illustrating the information summarized in part (A) is shown in Figure 16.

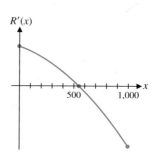

FIGURE 16 Marginal revenue

MATCHED PROBLEM 9   The graph of the total revenue $R(x)$ (in dollars) from the sale of x desks is shown in Figure 17.

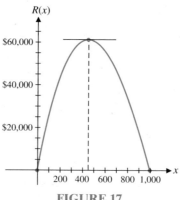

FIGURE 17

(A) Write a brief verbal description of the graph of the marginal revenue function $y = R'(x)$, including a discussion of any x intercepts.

(B) Sketch a possible graph of $y = R'(x)$.

Comparing Examples 8 and 9, we see that we were able to obtain more information about the function from the graph of its derivative (Example 8) than we were when the process was reversed (Example 9). In the next section, we introduce some ideas that will enable us to extract additional information about the derivative from the graph of the function.

Answers to Matched Problems

1. (A) Horizontal tangent line at $x = 3$.
   (B) Decreasing on $(-\infty, 3)$;
       increasing on $(3, \infty)$

(C)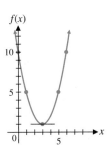

$f(x)$

2. Partition number: $x = 0$; critical value: $x = 0$; decreasing for all x
3. Partition number: $x = -1$; critical value: $x = -1$; increasing for all x
4. Partition number: $x = 0$; no critical values; decreasing on $(-\infty, 0)$ and $(0, \infty)$
5. Partition number: $x = 5$; critical value: $x = 5$; increasing on $(0, 5)$; decreasing on $(5, \infty)$
6. Increasing on $(-3, 1)$; decreasing on $(-\infty, -3)$ and $(1, \infty)$; local maximum at $x = 1$; local minimum at $x = -3$
7. (A) Critical values: $x = 2, x = 4$
   (B) Local maximum at $x = 2$;
       local minimum at $x = 4$

(C) $f(x)$

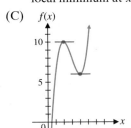

8. (A) The U.S. share of the world market decreases for 6 years to a local minimum, increases for the next 10 years to a local maximum, and then decreases for the final 4 years.

   (B) $S(t)$

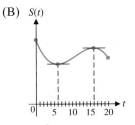

9. (A) The marginal revenue is positive on $(0, 450)$, 0 at $x = 450$, and negative on $(450, 1{,}000)$.

   (B) $R'(x)$

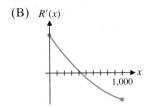

Exercise 5-1

A *Problems 1–8 refer to the following graph of $y = f(x)$:*

$f(x)$

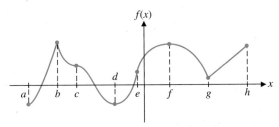

1. Identify the intervals on which $f(x)$ is increasing.
2. Identify the intervals on which $f(x)$ is decreasing.

3. Identify the intervals on which $f'(x) < 0$.
4. Identify the intervals on which $f'(x) > 0$.
5. Identify the x coordinates of the points where $f'(x) = 0$.
6. Identify the x coordinates of the points where $f'(x)$ does not exist.
7. Identify the x coordinates of the points where $f(x)$ has a local maximum.
8. Identify the x coordinates of the points where $f(x)$ has a local minimum.

In Problems 9 and 10, $f(x)$ is continuous on $(-\infty, \infty)$ and has critical values at $x = a, b, c,$ and d. Use the sign chart for $f'(x)$ to determine whether f has a local maximum, a local minimum, or neither at each critical value.

9.

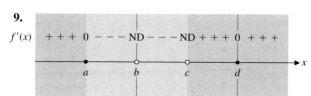

$f'(x)$ $+ + + 0 - - -$ ND $- - -$ ND $+ + + 0 + + +$

10.

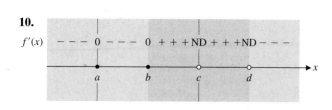

$f'(x)$ $- - - 0 - - - 0 + + +$ ND $+ + +$ ND $- - -$

In Problems 11–18, match the graph of f with one of the sign charts a–h in the figure.

11.

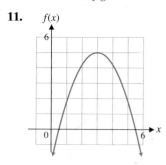

12.

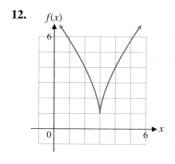

13.

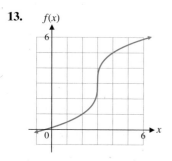

14.

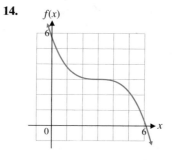

15.

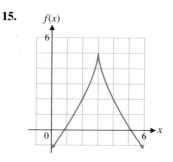

16.

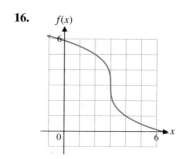

17.

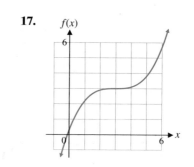

18.

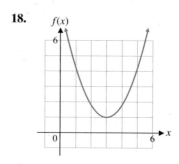

(a)
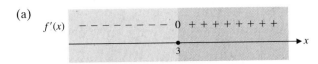

(b)
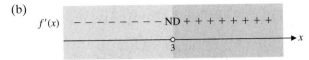

(c)
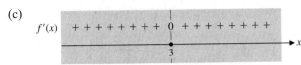

(d)
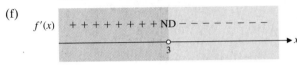

(e)
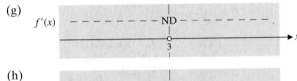

(f)

$f'(x)$ + + + + + + + + ND − − − − − − − −
→ x
3

(g)

$f'(x)$ − − − − − − − − ND − − − − − − − −
→ x
3

(h)

$f'(x)$ − − − − − − − − 0 − − − − − − − −
→ x
3

Figure for 11–18

35. $f(x) = (x^2 - 3x - 4)^{4/3}$

36. $f(x) = x^{4/3} - 7x^{1/3}$

 *In Problems 37–46, use a graphing calculator to approximate the critical values of $f(x)$ to two decimal places. Find the intervals on which $f(x)$ is increasing, the intervals on which $f(x)$ is decreasing, and the local extrema.*

37. $f(x) = x^4 + x^2 + x$

38. $f(x) = x^4 + x^2 - 9x$

39. $f(x) = x^4 - 4x^3 + 9x$

40. $f(x) = x^4 + 5x^3 - 15x$

41. $f(x) = x \ln x - (x - 2)^3$

42. $f(x) = 3x - x^{1/3} - x^{4/3}$

43. $f(x) = e^x - 2x^2$

44. $f(x) = e^{-x} - 3x^2$

45. $f(x) = x^{1/3} + x^{4/3} - 2x$

46. $f(x) = \dfrac{\ln x}{x} - 5x + x^2$

In Problems 47–54, find the intervals on which $f(x)$ is increasing and the intervals on which $f(x)$ is decreasing. Then sketch the graph. Add horizontal tangent lines.

47. $f(x) = 4 + 8x - x^2$

48. $f(x) = 2x^2 - 8x + 9$

49. $f(x) = x^3 - 3x + 1$

50. $f(x) = x^3 - 12x + 2$

51. $f(x) = 10 - 12x + 6x^2 - x^3$

52. $f(x) = x^3 + 3x^2 + 3x$

53. $f(x) = x^4 - 18x^2$

54. $f(x) = -x^4 + 50x^2$

B *In Problems 19–36, find the intervals on which $f(x)$ is increasing, the intervals on which $f(x)$ is decreasing, and the local extrema.*

19. $f(x) = 2x^2 - 4x$

20. $f(x) = -3x^2 - 12x$

21. $f(x) = -2x^2 - 16x - 25$

22. $f(x) = -3x^2 + 12x - 5$

23. $f(x) = x^3 + 4x - 5$

24. $f(x) = -x^3 - 4x + 8$

25. $f(x) = 2x^3 - 3x^2 - 36x$

26. $f(x) = -2x^3 + 3x^2 + 120x$

27. $f(x) = 3x^4 - 4x^3 + 5$

28. $f(x) = x^4 + 2x^3 + 5$

29. $f(x) = (x - 1)e^{-x}$

30. $f(x) = x \ln x - x$

31. $f(x) = 4x^{1/3} - x^{2/3}$

32. $f(x) = (x^2 - 9)^{2/3}$

33. $f(x) = 2x - x \ln x$

34. $f(x) = (x + 2)e^x$

In Problems 55–62, $f(x)$ is continuous on $(-\infty, \infty)$. Use the given information to sketch the graph of f.

55.

$f'(x)$ + + + 0 + + + 0 − − −
→ x
−1   1

x	−2	−1	0	1	2
$f(x)$	−1	1	2	3	1

56.

$f'(x)$ + + + 0 − − − 0 − − −
→ x
−1   1

x	−2	−1	0	1	2
$f(x)$	1	3	2	1	−1

57.

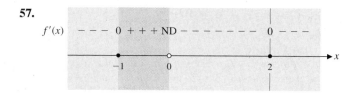

x	-2	-1	0	2	4
f(x)	2	1	2	1	0

58.

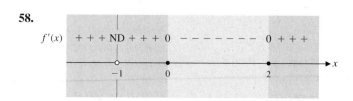

x	-2	-1	0	2	3
f(x)	-3	0	2	-1	0

59. $f(-2) = 4, f(0) = 0, f(2) = -4$;
$f'(-2) = 0, f'(0) = 0, f'(2) = 0$;
$f'(x) > 0$ on $(-\infty, -2)$ and $(2, \infty)$;
$f'(x) < 0$ on $(-2, 0)$ and $(0, 2)$

60. $f(-2) = -1, f(0) = 0, f(2) = 1$;
$f'(-2) = 0, f'(2) = 0$;
$f'(x) > 0$ on $(-\infty, -2), (-2, 2)$, and $(2, \infty)$

61. $f(-1) = 2, f(0) = 0, f(1) = -2$;
$f'(-1) = 0, f'(1) = 0, f'(0)$ is not defined;
$f'(x) > 0$ on $(-\infty, -1)$ and $(1, \infty)$;
$f'(x) < 0$ on $(-1, 0)$ and $(0, 1)$

62. $f(-1) = 2, f(0) = 0, f(1) = 2$;
$f'(-1) = 0, f'(1) = 0, f'(0)$ is not defined;
$f'(x) > 0$ on $(-\infty, -1)$ and $(0, 1)$;
$f'(x) < 0$ on $(-1, 0)$ and $(1, \infty)$

Problems 63–68 involve functions f_1–f_6 and their derivatives, g_1–g_6. Use the graphs shown in figures (A) and (B) to match each function f_i with its derivative g_j.

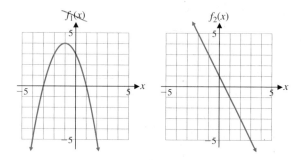

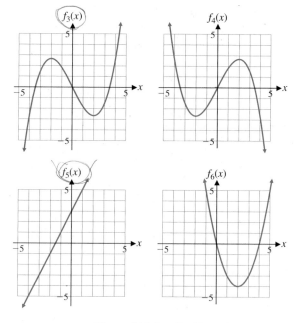

Figure (A) for 63–68

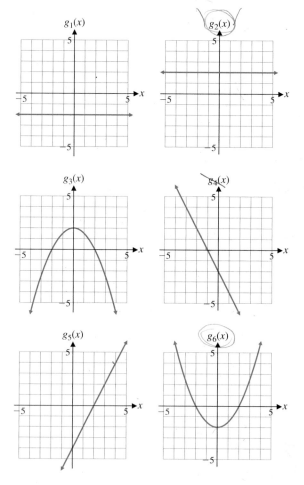

Figure (B) for 63–68

63. f_1

64. f_2

65. f_3

66. f_4

67. f_5

68. f_6

In Problems 69–74, use the given graph of $y = f'(x)$ to find the intervals on which f is increasing, the intervals on which f is decreasing, and the local extrema. Sketch a possible graph of $y = f(x)$.

69.
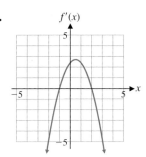

70.
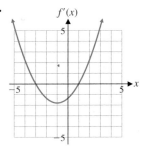

71.
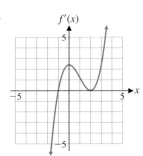

72.
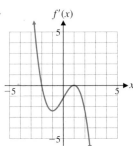

73.
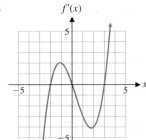

74.
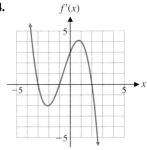

In Problems 75–78, use the given graph of $y = f(x)$ to find the intervals on which $f'(x) > 0$, the intervals on which $f'(x) < 0$, and the values of x for which $f'(x) = 0$. Sketch a possible graph of $y = f'(x)$.

75.
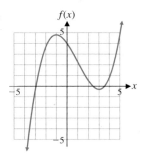

76.
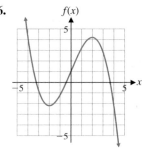

77.
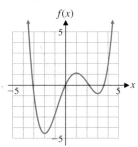

78.
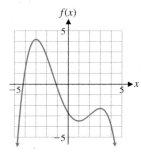

C *In Problems 79–90, find the critical values, the intervals on which $f(x)$ is increasing, the intervals on which $f(x)$ is decreasing, and the local extrema. Do not graph.*

79. $f(x) = x + \dfrac{4}{x}$

80. $f(x) = \dfrac{9}{x} + x$

81. $f(x) = 1 + \dfrac{1}{x} + \dfrac{1}{x^2}$

82. $f(x) = 3 - \dfrac{4}{x} - \dfrac{2}{x^2}$

83. $f(x) = \dfrac{x^2}{x - 2}$

84. $f(x) = \dfrac{x^2}{x + 1}$

85. $f(x) = x^4(x - 6)^2$

86. $f(x) = x^3(x - 5)^2$

87. $f(x) = 3(x - 2)^{2/3} + 4$

88. $f(x) = 6(4 - x)^{2/3} + 4$

89. $f(x) = \dfrac{2x^2}{x^2 + 1}$

90. $f(x) = \dfrac{-3x}{x^2 + 4}$

91. Let $f(x) = x^3 + kx$, where k is a constant. Discuss the number of local extrema and the shape of the graph of f if

　(A) $k > 0$　　　(B) $k < 0$　　　(C) $k = 0$

92. Let $f(x) = x^4 + kx^2$, where k is a constant. Discuss the number of local extrema and the shape of the graph of f if

　(A) $k > 0$　　　(B) $k < 0$　　　(C) $k = 0$

Applications

93. *Profit analysis.* The graph of the total profit $P(x)$ (in dollars) from the sale of x cordless electric screwdrivers is shown in the figure.

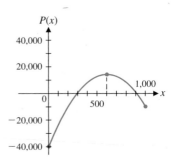

Figure for 93

(A) Write a brief verbal description of the graph of the marginal profit function $y = P'(x)$, including a discussion of any x intercepts.

(B) Sketch a possible graph of $y = P'(x)$.

94. *Revenue analysis.* The graph of the total revenue $R(x)$ (in dollars) from the sale of x cordless electric screwdrivers is shown in the figure.

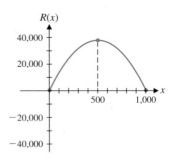

Figure for 94

(A) Write a brief verbal description of the graph of the marginal revenue function $y = R'(x)$, including a discussion of any x intercepts.

(B) Sketch a possible graph of $y = R'(x)$.

95. *Price analysis.* The graph in the figure approximates the rate of change of the price of bacon over a 70-month period, where $B(t)$ is the price of a pound of sliced bacon (in dollars) and t is time (in months).

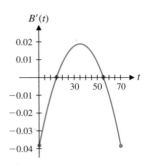

Figure for 95

(A) Write a brief verbal description of the graph of $y = B(t)$, including a discussion of any local extrema.

(B) Sketch a possible graph of $y = B(t)$.

96. *Price analysis.* The graph in the figure approximates the rate of change of the price of eggs over a 70-month period, where $E(t)$ is the price of a dozen eggs (in dollars) and t is time (in months).

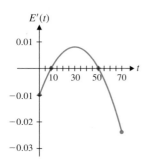

Figure for 96

(A) Write a brief verbal description of the graph of $y = E(t)$, including a discussion of any local extrema.

(B) Sketch a possible graph of $y = E(t)$.

97. *Average cost.* A manufacturer incurs the following costs in producing x toasters in one day, for $0 < x < 150$: fixed costs, \$320; unit production cost, \$20 per toaster; equipment maintenance and repairs, $0.05x^2$ dollars. Thus, the cost of manufacturing x toasters in one day is given by

$$C(x) = 0.05x^2 + 20x + 320 \qquad 0 < x < 150$$

(A) What is the average cost $\overline{C}(x)$ per toaster if x toasters are produced in one day?

(B) Find the critical values of $\overline{C}(x)$, the intervals on which the average cost per toaster is decreasing, the intervals on which the average cost per toaster is increasing, and the local extrema. Do not graph.

98. *Average cost.* A manufacturer incurs the following costs in producing x blenders in one day for $0 < x < 200$: fixed costs, \$450; unit production cost, \$30 per blender; equipment maintenance and repairs, $0.08x^2$ dollars.

(A) What is the average cost $\overline{C}(x)$ per blender if x blenders are produced in one day?

(B) Find the critical values of $\overline{C}(x)$, the intervals on which the average cost per blender is decreasing, the intervals on which the average cost per blender is increasing, and the local extrema. Do not graph.

99. *Marginal analysis.* Show that profit will be increasing over production intervals (a, b) for which marginal revenue is greater than marginal cost. [*Hint:* $P(x) = R(x) - C(x)$]

100. *Marginal analysis.* Show that profit will be decreasing over production intervals (a, b) for which marginal revenue is less than marginal cost.

101. *Medicine.* A drug is injected into the bloodstream of a patient through the right arm. The concentration of the drug in the bloodstream of the left arm t hours after the injection is approximated by

$$C(t) = \frac{0.28t}{t^2 + 4} \qquad 0 < t < 24$$

Find the critical values of $C(t)$, the intervals on which the concentration of the drug is increasing, the intervals on which the concentration of the drug is decreasing, and the local extrema. Do not graph.

102. *Medicine.* The concentration $C(t)$, in milligrams per cubic centimeter, of a particular drug in a patient's bloodstream is given by

$$C(t) = \frac{0.3t}{t^2 + 6t + 9} \qquad 0 < t < 12$$

where t is the number of hours after the drug is taken orally. Find the critical values of $C(t)$, the intervals on which the concentration of the drug is increasing, the intervals on which the concentration of the drug is decreasing, and the local extrema. Do not graph.

103. *Politics.* Public awareness of a congressional candidate before and after a successful campaign was approximated by

$$P(t) = \frac{8.4t}{t^2 + 49} + 0.1 \qquad 0 < t < 24$$

where t is time (in months) after the campaign started and $P(t)$ is the fraction of people in the congressional district who could recall the candidate's (and later, congressman's) name. Find the critical values of $P(t)$, the time intervals on which the fraction is increasing, the time intervals on which the fraction is decreasing, and the local extrema. Do not graph.

Section 5-2   SECOND DERIVATIVE AND GRAPHS

■ Using Concavity as a Graphing Tool
■ Finding Inflection Points
■ Analyzing Graphs
■ Curve Sketching
■ Point of Diminishing Returns

In Section 5-1, we saw that the derivative can be used when a graph is rising and falling. Now we want to see what the *second derivative* (the derivative of the derivative) can tell us about the shape of a graph.

■ Using Concavity as a Graphing Tool

Consider the functions

$$f(x) = x^2 \qquad \text{and} \qquad g(x) = \sqrt{x}$$

for x in the interval $(0, \infty)$. Since

$$f'(x) = 2x > 0 \qquad \text{for } 0 < x < \infty$$

and

$$g'(x) = \frac{1}{2\sqrt{x}} > 0 \qquad \text{for } 0 < x < \infty$$

both functions are increasing on $(0, \infty)$.

Explore & Discuss **1**

(A) Discuss the difference in the shapes of the graphs of *f* and *g* shown in Figure 1.

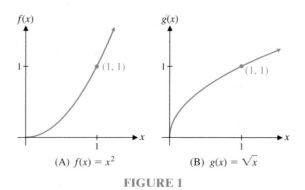

FIGURE 1

(B) Complete the following table, and discuss the relationship between the values of the derivatives of *f* and *g* and the shapes of their graphs:

x	0.25	0.5	0.75	1
$f'(x)$				
$g'(x)$				

We use the term *concave upward* to describe a graph that opens upward and *concave downward* to describe a graph that opens downward. Thus, the graph of *f* in Figure 1A is concave upward, and the graph of *g* in Figure 1B is concave downward. Finding a mathematical formulation of concavity will help us sketch and analyze graphs.

It will be instructive to examine the slopes of *f* and *g* at various points on their graphs (see Fig. 2). We can make two observations about each graph:

1. Looking at the graph of *f* in Figure 2A, we see that $f'(x)$ (the slope of the tangent line) is *increasing* and that the graph lies *above* each tangent line;

2. Looking at Figure 2B, we see that $g'(x)$ is *decreasing* and that the graph lies *below* each tangent line.

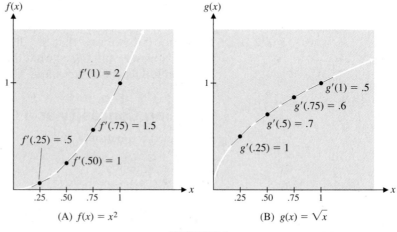

FIGURE 2

With these ideas in mind, we state the general definition of concavity.

DEFINITION Concavity

The graph of a function *f* is **concave upward** on the interval (a, b) if $f'(x)$ is *increasing* on (a, b) and is **concave downward** on the interval (a, b) if $f'(x)$ is *decreasing* on (a, b).

Geometrically, the graph is concave upward on (a, b) if it lies above its tangent lines in (a, b) and is concave downward on (a, b) if it lies below its tangent lines in (a, b).

How can we determine when $f'(x)$ is increasing or decreasing? In Section 5-1, we used the derivative of a function to determine when that function is increasing or decreasing. Thus, to determine when the function $f'(x)$ is increasing or decreasing, we use the derivative of $f'(x)$. The derivative of the derivative of a function is called the *second derivative* of the function. Various notations for the second derivative are given in the following box:

NOTATION

Second Derivative

For $y = f(x)$, the **second derivative** of f, provided that it exists, is

$$f''(x) = \frac{d}{dx} f'(x)$$

Other notations for $f''(x)$ are

$$\frac{d^2 y}{dx^2} \qquad y''$$

Returning to the functions f and g discussed at the beginning of this section, we have

$$f(x) = x^2 \qquad\qquad g(x) = \sqrt{x} = x^{1/2}$$

$$f'(x) = 2x \qquad\qquad g'(x) = \frac{1}{2} x^{-1/2} = \frac{1}{2\sqrt{x}}$$

$$f''(x) = \frac{d}{dx} 2x = 2 \qquad g''(x) = \frac{d}{dx} \frac{1}{2} x^{-1/2} = -\frac{1}{4} x^{-3/2} = -\frac{1}{4\sqrt{x^3}}$$

For $x > 0$, we see that $f''(x) > 0$; thus, $f'(x)$ is increasing and the graph of f is concave upward (see Fig. 2A). For $x > 0$, we also see that $g''(x) < 0$; thus, $g'(x)$ is decreasing and the graph of g is concave downward (see Fig. 2B). These ideas are summarized in the following box:

SUMMARY   **CONCAVITY**

For the interval (a, b),

$f''(x)$	$f'(x)$	Graph of $y = f(x)$	Examples
+	Increasing	Concave upward	$\smile \quad \smile \quad \diagdown \quad \diagup$
−	Decreasing	Concave downward	$\frown \quad \frown \quad \diagup$

INSIGHT

Be careful not to confuse concavity with falling and rising. A graph that is concave upward on an interval may be falling, rising, or both falling and rising on that interval. A similar statement holds for a graph that is concave downward. See Figure 3 on the next page.

$f''(x) > 0$ on (a, b)
Concave upward

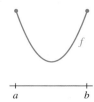

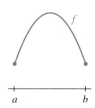

(A) $f'(x)$ is negative
and increasing.
Graph of f is falling.

(B) $f'(x)$ increases from
negative to positive.
Graph of f falls, then rises.

(C) $f'(x)$ is positive
and increasing.
Graph of f is rising.

$f''(x) < 0$ on (a, b)
Concave downward

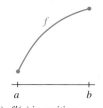

(D) $f'(x)$ is positive
and decreasing.
Graph of f is rising.

(E) $f'(x)$ decreases from
positive to negative.
Graph of f rises, then falls.

(F) $f'(x)$ is negative
and decreasing.
Graph of f is falling.

FIGURE 3 Concavity

EXAMPLE 1 **Concavity of Graphs** Determine the intervals on which the graph of each function is concave upward and the intervals on which it is concave downward. Sketch a graph of each function.

(A) $f(x) = e^x$ (B) $g(x) = \ln x$ (C) $h(x) = x^3$

SOLUTION

(A) $f(x) = e^x$

$f'(x) = e^x$

$f''(x) = e^x$

Since $f''(x) > 0$ on $(-\infty, \infty)$, the graph of $f(x) = e^x$ [Fig. 4(A)] is concave upward on $(-\infty, \infty)$.

(B) $g(x) = \ln x$

$g'(x) = \dfrac{1}{x}$

$g''(x) = -\dfrac{1}{x^2}$

The domain of $g(x) = \ln x$ is $(0, \infty)$ and $g''(x) < 0$ on this interval, so the graph of $g(x) = \ln x$ [Fig. 4(B)] is concave downward on $(0, \infty)$.

(C) $h(x) = x^3$

$h'(x) = 3x^2$

$h''(x) = 6x$

Since $h''(x) < 0$ when $x < 0$ and $h''(x) > 0$ when $x > 0$, the graph of $h(x) = x^3$ [Fig. 4(C)] is concave downward on $(-\infty, 0)$ and concave upward on $(0, \infty)$.

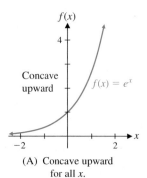

(A) Concave upward
for all x.

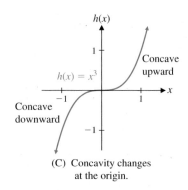

(B) Concave downward
for $x > 0$.

(C) Concavity changes
at the origin.

FIGURE 4

| MATCHED PROBLEM 1 | Determine the intervals on which the graph of each function is concave upward and the intervals on which it is cancave downward. Sketch a graph of each function. |

$$(A)\ f(x) = -e^{-x} \qquad (B)\ g(x) = \ln\frac{1}{x} \qquad (C)\ h(x) = x^{1/3}$$

Refer to Example 1. The graphs of $f(x) = e^x$ and $g(x) = \ln x$ never change concavity. But the graph of $h(x) = x^3$ changes concavity at $(0, 0)$. This point is called an *inflection point*.

■ Finding Inflection Points

Explore & Discuss **2**　Discuss the relationship between the change in concavity of each of the following functions at $x = 0$ and the second derivative at and near 0:

$$(A)\ f(x) = x^3 \qquad (B)\ g(x) = x^{4/3} \qquad (C)\ h(x) = x^4$$

In general, an **inflection point** is a point on the graph of the function where the concavity changes (from upward to downward or from downward to upward). For the concavity to change at a point, $f''(x)$ must change sign at that point. But in Section 3-2, we saw that the partition numbers* identify the points where a function can change sign. Thus, we have the following theorem:

THEOREM 1　**INFLECTION POINTS**

If $y = f(x)$ is continuous on (a, b) and has an inflection point at $x = c$, then either $f''(c) = 0$ or $f''(c)$ does not exist.

Note that inflection points can occur only at partition numbers of f'', but not every partition number of f'' produces an inflection point. Two additional requirements must be satisfied for an inflection point to occur:

A partition number c for f'' produces an inflection point for the graph of f only if

1. $f''(x)$ changes sign at c and
2. c is in the domain of f.

Figure 5 illustrates several typical cases.

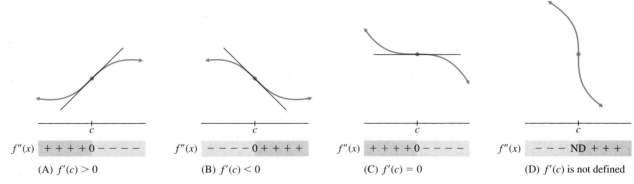

FIGURE 5　Inflection points

If $f'(c)$ exists and $f''(x)$ changes sign at $x = c$, the tangent line at an inflection point $(c, f(c))$ will always lie below the graph on the side that is concave upward and above the graph on the side that is concave downward (see Fig. 5A, B, and C).

---

*As we did with the first derivative, we assume that if f'' is discontinuous at c, then $f''(c)$ does not exist.

EXAMPLE 2 **Locating Inflection Points** Find the inflection point(s) of

$$f(x) = x^3 - 6x^2 + 9x + 1$$

SOLUTION Since inflection point(s) occur at values of x where $f''(x)$ changes sign, we construct a sign chart for $f''(x)$. We have

$$f(x) = x^3 - 6x^2 + 9x + 1$$
$$f'(x) = 3x^2 - 12x + 9$$
$$f''(x) = 6x - 12 = 6(x - 2)$$

The sign chart for $f''(x) = 6(x - 2)$ (partition number is 2) is

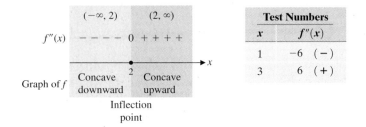

From the sign chart, we see that the graph of f has an inflection point at $x = 2$. That is, the point

$$(2, f(2)) = (2, 3) \qquad f(2) = 2^3 - 6 \cdot 2^2 + 9 \cdot 2 + 1 = 3$$

is an inflection point on the graph of f.

MATCHED PROBLEM 2 Find the inflection point(s) of

$$f(x) = x^3 - 9x^2 + 24x - 10$$

EXAMPLE 3 **Locating Inflection Points** Find the inflection point(s) of

$$f(x) = \ln(x^2 - 4x + 5)$$

SOLUTION First we find the domain of f (a good first step for most calculus problems involving properties of the graph of a function). Since $\ln x$ is defined only for $x > 0$, f is defined only for

$$x^2 - 4x + 5 > 0 \quad \text{Use completing the square (Section 2-3).}$$
$$(x - 2)^2 + 1 > 0 \quad \text{True for all x.}$$

So the domain of f is $(-\infty, \infty)$. Now we find $f''(x)$ and construct a sign chart for it. We have

$$f(x) = \ln(x^2 - 4x + 5)$$

$$f'(x) = \frac{2x - 4}{x^2 - 4x + 5}$$

$$f''(x) = \frac{(x^2 - 4x + 5)(2x - 4)' - (2x - 4)(x^2 - 4x + 5)'}{(x^2 - 4x + 5)^2}$$

$$= \frac{(x^2 - 4x + 5)2 - (2x - 4)(2x - 4)}{(x^2 - 4x + 5)^2}$$

$$= \frac{2x^2 - 8x + 10 - 4x^2 + 16x - 16}{(x^2 - 4x + 5)^2}$$

$$= \frac{-2x^2 + 8x - 6}{(x^2 - 4x + 5)^2}$$

$$= \frac{-2(x - 1)(x - 3)}{(x^2 - 4x + 5)^2}$$

The partition numbers for $f''(x)$ are $x = 1$ and $x = 3$.

Sign chart for $f''(x)$

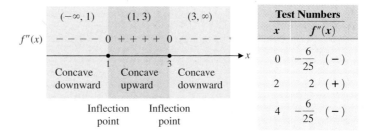

	Test Numbers	
	x	$f''(x)$
	0	$-\frac{6}{25}$ $(-)$
	2	2 $(+)$
	4	$-\frac{6}{25}$ $(-)$

The sign chart shows that the graph of f has inflection points at $x = 1$ and $x = 3$.

MATCHED PROBLEM 3   Find the inflection point(s) of

$$f(x) = \ln(x^2 - 2x + 5)$$

INSIGHT

It is important to remember that the partition numbers for f'' are only *candidates* for inflection points. The function f must be defined at $x = c$, and the second derivative must change sign at $x = c$ in order for the graph to have an inflection point at $x = c$. For example, consider

$$f(x) = x^4 \qquad g(x) = \frac{1}{x}$$

$$f'(x) = 4x^3 \qquad g'(x) = -\frac{1}{x^2}$$

$$f''(x) = 12x^2 \qquad g''(x) = \frac{2}{x^3}$$

In each case, $x = 0$ is a partition number for the second derivative, but neither the graph of $f(x)$ nor the graph of $g(x)$ has an inflection point at $x = 0$. Function f does not have an inflection point at $x = 0$ because $f''(x)$ does not change sign at $x = 0$ (see Fig. 6A). Function g does not have an inflection point at $x = 0$ because $g(0)$ is not defined (see Fig. 6B).

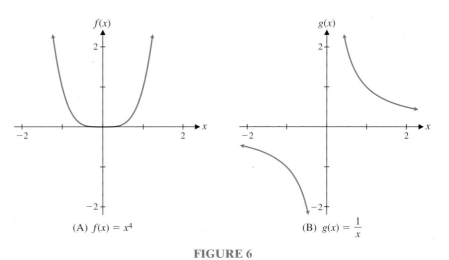

(A) $f(x) = x^4$

(B) $g(x) = \frac{1}{x}$

FIGURE 6

■ Analyzing Graphs

In the next example, we combine increasing/decreasing properties with concavity properties to analyze the graph of a function.

EXAMPLE 4

Analyzing a Graph Figure 7 shows the graph of the derivative of a function f. Use this graph to discuss the graph of f. Include a sketch of a possible graph of f.

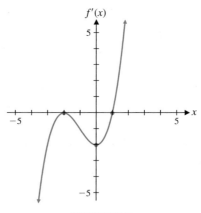

FIGURE 7

SOLUTION The sign of the derivative determines where the original function is increasing and decreasing, and the increasing/decreasing properties of the derivative determine the concavity of the original function. The relevant information obtained from the graph of f' is summarized in Table 1, and a possible graph of f is shown in Figure 8.

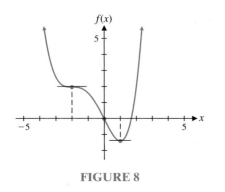

FIGURE 8

TABLE 1		
x	$f'(x)$ (Fig. 7)	$f(x)$ (Fig. 8)
$-\infty < x < -2$	Negative and increasing	Decreasing and concave upward
$x = -2$	Local maximum	Inflection point
$-2 < x < 0$	Negative and decreasing	Decreasing and concave downward
$x = 0$	Local minimum	Inflection point
$0 < x < 1$	Negative and increasing	Decreasing and concave upward
$x = 1$	x intercept	Local minimum
$1 < x < \infty$	Positive and increasing	Increasing and concave upward

MATCHED PROBLEM 4 Figure 9 shows the graph of the derivative of a function f. Use this graph to discuss the graph of f. Include a sketch of a possible graph of f.

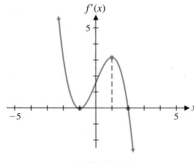

FIGURE 9

■ Curve Sketching

Graphing calculators and computers produce the graph of a function by plotting many points. Although the technology is quite accurate, important points on a plot many be difficult to identify. Using information gained from the function $f(x)$ and its derivatives, and plotting the important points—intercepts, local extrema, and inflection points—we can sketch by hand a very good representation of the graph of $f(x)$. This graphing process is called **curve sketching** and is summarized next.

PROCEDURE

Graphing Strategy (First Version)*

Step 1. *Analyze f(x).* Find the domain and the intercepts. The x intercepts are the solutions of $f(x) = 0$ and the y intercept is $f(0)$.

Step 2. *Analyze f'(x).* Find the partition numbers for, and critical values of, $f'(x)$. Construct a sign chart for $f'(x)$, determine the intervals on which f is increasing and decreasing, and find local maxima and minima.

Step 3. *Analyze f''(x).* Find the partition numbers for $f''(x)$. Construct a sign chart for $f''(x)$, determine the intervals on which the graph of f is concave upward and concave downward, and find inflection points.

Step 4. *Sketch the graph of f.* Locate intercepts, local maxima and minima, and inflection points. Sketch in what you know from steps 1–3. Plot additional points as needed and complete the sketch.

Example will illustrate the use of this strategy.

EXAMPLE 5

Using the Graphing Strategy Follow the graphing strategy and analyze the function

$$f(x) = x^4 - 2x^3$$

State all the pertinent information and sketch the graph of f.

SOLUTION

Step 1. *Analyze f(x).* Since f is a polynomial, its domain is $(-\infty, \infty)$.

x intercept: $f(x) = 0$
$$x^4 - 2x^3 = 0$$
$$x^3(x - 2) = 0$$
$$x = 0, 2$$

y intercept: $f(0) = 0$

Step 2. *Analyze f'(x).* $f'(x) = 4x^3 - 6x^2 = 4x^2\left(x - \frac{3}{2}\right)$

Critical values of $f(x)$: 0 and $\frac{3}{2}$

Partition numbers for $f'(x)$: 0 and $\frac{3}{2}$

Sign chart for $f'(x)$:

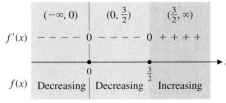

	Test Numbers	
x	$f'(x)$	
-1	-10	$(-)$
1	-2	$(-)$
2	8	$(+)$

---

*We will modify this summary in Section 5-4 to include some additional information about the graph of f.

Thus, $f(x)$ is decreasing on $(-\infty, \frac{3}{2})$, is increasing on $(\frac{3}{2}, \infty)$, and has a local minimum at $x = \frac{3}{2}$.

Step 3. *Analyze* $f''(x)$. $f''(x) = 12x^2 - 12x = 12x(x - 1)$

Partition numbers for $f''(x)$: 0 and 1

Sign chart for $f''(x)$:

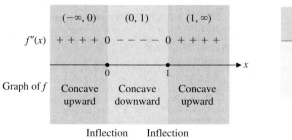

	Test Numbers	
x	**$f''(x)$**	
-1	24	(+)
$\frac{1}{2}$	-3	(−)
2	24	(+)

Thus, the graph of f is concave upward on $(-\infty, 0)$ and $(1, \infty)$, is concave downward on $(0, 1)$, and has inflection points at $x = 0$ and $x = 1$.

Step 4. *Sketch the graph of* f.

x	f(x)
0	0
1	-1
$\frac{3}{2}$	$-\frac{27}{16}$
2	0

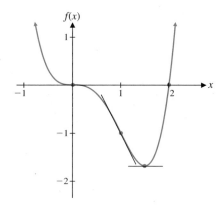

MATCHED PROBLEM 5   Follow the graphing strategy and analyze the function $f(x) = x^4 + 4x^3$. State all the pertinent information and sketch the graph of f.

INSIGHT

Refer to the solution to Example 5. Combining the sign charts for $f'(x)$ and $f''(x)$ (Fig. 10) partitions the real-number line into intervals on which neither $f'(x)$ nor $f''(x)$ changes sign. On each of these intervals, the graph of $f(x)$ must have one of four basic shapes (see also Fig. 3, parts A, C, D, and F). This reduces sketching the graph of a function to plotting the points identified in the graphing strategy and connecting them with one of the basic shapes.

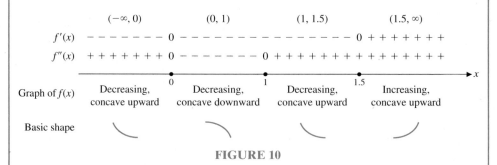

FIGURE 10

| EXAMPLE 6 | **Using the Graphing Strategy** Follow the graphing strategy and analyze the function |

$$f(x) = 3x^{5/3} - 20x$$

State all the pertinent information and sketch the graph of f. Round any decimal values to two decimal places.

SOLUTION **Step 1.** *Analyze* $f(x)$. $f(x) = 3x^{5/3} - 20x$

Since x^p is defined for any x and any positive p, the domain of f is $(-\infty, \infty)$.

$$x \text{ intercepts: Solve } f(x) = 0$$

$$3x^{5/3} - 20x = 0$$

$$3x\left(x^{2/3} - \frac{20}{3}\right) = 0 \qquad (a^2 - b^2) = (a - b)(a + b)$$

$$3x\left(x^{1/3} - \sqrt{\frac{20}{3}}\right)\left(x^{1/3} + \sqrt{\frac{20}{3}}\right) = 0$$

The x intercepts of f are

$$x = 0, \quad x = \left(\sqrt{\frac{20}{3}}\right)^3 \approx 17.21, \quad x = \left(-\sqrt{\frac{20}{3}}\right)^3 \approx -17.21$$

y intercept: $f(0) = 0$.

Step 2. *Analyze* $f'(x)$.

$$f'(x) = 5x^{2/3} - 20$$

$$= 5(x^{2/3} - 4) \qquad \text{Again, } a^2 - b^2 = (a - b)(a + b)$$

$$= 5(x^{1/3} - 2)(x^{1/3} + 2)$$

Critical values of f: $x = 2^3 = 8$ and $x = (-2)^3 = -8$.

Partition numbers for f: $-8, 8$

Sign chart for $f'(x)$:

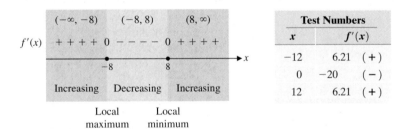

| (−∞, −8) | (−8, 8) | (8, ∞) |

$f'(x)$ + + + + 0 − − − − 0 + + + +

Increasing Decreasing Increasing

Local maximum Local minimum

Test Numbers	
x	$f'(x)$
−12	6.21 (+)
0	−20 (−)
12	6.21 (+)

So f is increasing on $(-\infty, -8)$ and $(8, \infty)$ and decreasing on $(-8, 8)$, $f(-8)$ is a local maximum, and $f(8)$ is a local minimum.

Step 3. *Analyze* $f''(x)$.

$$f'(x) = 5x^{2/3} - 20$$

$$f''(x) = \frac{10}{3}x^{-1/3} = \frac{10}{3x^{1/3}}$$

Partition number for f'': 0

Sign chart for $f''(x)$:

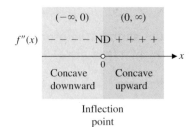

	Test Numbers	
	x	**f''(x)**
	-8	-1.67 $(-)$
	8	1.67 $(+)$

So f is concave downward on $(-\infty, 0)$, is concave upward on $(0, \infty)$, and has an inflection point at $x = 0$.

Step 4. *Sketch the graph of f.*

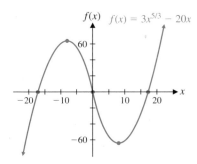

x	**f(x)**
-17.21	0
-8	64
0	0
8	-64
17.21	0

MATCHED PROBLEM 6 Follow the graphing strategy and analyze the function $f(x) = 3x^{2/3} - x$. State all the pertinent information and sketch the graph of f. Round any decimal values to two decimal places.

■ Point of Diminishing Returns

If a company decides to increase spending on advertising, it would expect sales to increase. At first, sales will increase at an increasing rate and then increase at a decreasing rate. The value of x where the rate of change of sales changes from increasing to decreasing is called the **point of diminishing returns.** This is also the point where the rate of change has a maximum value. Money spent after this point may increase sales, but at a lower rate. The next example illustrates this concept.

EXAMPLE 7 **Maximum Rate of Change** Currently, a discount appliance store is selling 200 large-screen television sets monthly. If the store invests x thousand in an advertising campaign, the ad company estimates that sales will increase to

$$N(x) = 3x^3 - 0.25x^4 + 200 \qquad 0 \le x \le 9$$

When is rate of change of sales increasing and when is it decreasing? What is the point of diminishing returns and the maximum rate of change of sales? Graph N and N' on the same coordinate system.

SOLUTION The rate of change of sales with respect to advertising expenditures is

$$N'(x) = 9x^2 - x^3 = x^2(9 - x)$$

To determine when $N'(x)$ is increasing and decreasing, we find $N''(x)$, the derivative of $N'(x)$:

$$N''(x) = 18x - 3x^2 = 3x(6 - x)$$

The information obtained by analyzing the signs of $N'(x)$ and $N''(x)$ is summarized in Table 2 (sign charts are omitted).

TABLE 2				
x	$N''(x)$	$N'(x)$	$N'(x)$	$N(x)$
$0 < x < 6$	+	+	Increasing	Increasing, concave upward
$x = 6$	0	+	Local maximum	Inflection point
$6 < x < 9$	−	+	Decreasing	Increasing, concave downward

Examining Table 2, we see that $N'(x)$ is increasing on $(0, 6)$ and decreasing on $(6, 9)$. The point of diminishing returns is $x = 6$ and the maximum rate of change is $N'(6) = 108$. Note that $N'(x)$ has a local maximum and $N(x)$ has an inflection point at $x = 6$.

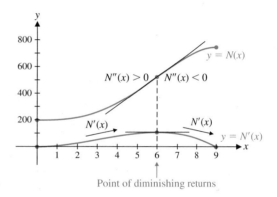

Point of diminishing returns

MATCHED PROBLEM 7   Repeat Example 7 for

$$N(x) = 4x^3 - 0.25x^4 + 500 \qquad 0 \le x \le 12$$

Answers to Matched Problems   **1. (A)** Concave downward on $(-\infty, \infty)$

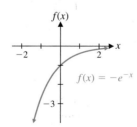

(B) Concave upward on $(0, \infty)$

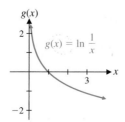

(C) Concave upward on $(-\infty, 0)$ and concave downward on $(0, \infty)$

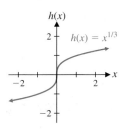

2. The only inflection point is $(3, f(3)) = (3, 8)$.

3. The inflection points are $(-1, f(-1)) = (-1, \ln 8)$ and $(3, f(3)) = (3, \ln 8)$.

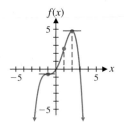

4.

x	$f'(x)$	$f(x)$
$-\infty < x < -1$	Positive and decreasing	Increasing and concave downward
$x = -1$	Local minimum	Inflection point
$-1 < x < 1$	Positive and increasing	Increasing and concave upward
$x = 1$	Local maximum	Inflection point
$1 < x < 2$	Positive and decreasing	Increasing and concave downward
$x = 2$	x intercept	Local maximum
$2 < x < \infty$	Negative and decreasing	Decreasing and concave downward

5. x intercepts: $-4, 0$; y intercept: $f(0) = 0$

Decreasing on $(-\infty, -3)$; increasing on $(-3, \infty)$; local minimum at $x = -3$

Concave upward on $(-\infty, -2)$ and $(0, \infty)$; concave downward on $(-2, 0)$

Inflection points at $x = -2$ and $x = 0$

x	$f(x)$
-4	0
-3	-27
-2	-16
0	0

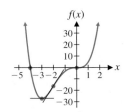

6. x intercepts: $0, 27$; y intercept: $f(0) = 0$

Decreasing on $(-\infty, 0)$ and $(8, \infty)$; increasing on $(0, 8)$; local minimum: $f(0) = 0$; local maximum: $f(8) = 4$

Concave downward on $(-\infty, 0)$ and $(0, \infty)$; no inflection points

x	$f(x)$
0	0
8	4
27	0

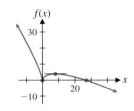

7. $N'(x)$ is increasing on $(0, 8)$ and decreasing on $(8, 12)$. The point of diminishing returns is $x = 8$ and the maximum rate of change is $N'(8) = 256$.

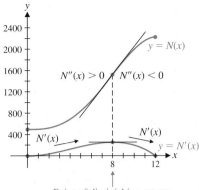

Point of diminishing returns

Exercise 5-2

A

1. Use the graph of $y = f(x)$ to identify

(A) Intervals on which the graph of f is concave upward
(B) Intervals on which the graph of f is concave downward
(C) Intervals on which $f''(x) < 0$
(D) Intervals on which $f''(x) > 0$
(E) Intervals on which $f'(x)$ is increasing
(F) Intervals on which $f'(x)$ is decreasing
(G) The x coordinates of inflection points
(H) The x coordinates of local extrema for $f'(x)$

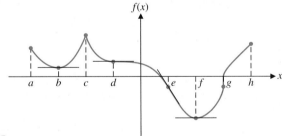

2. Use the graph of $y = g(x)$ to identify

(A) Intervals on which the graph of g is concave upward
(B) Intervals on which the graph of g is concave downward
(C) Intervals on which $g''(x) < 0$
(D) Intervals on which $g''(x) > 0$
(E) Intervals on which $g'(x)$ is increasing
(F) Intervals on which $g'(x)$ is decreasing
(G) The x coordinates of inflection points
(H) The x coordinates of local extrema for $g'(x)$

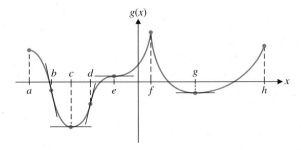

In Problems 3–6, match the indicated conditions with one of the graphs (A)–(D) shown in the figure.

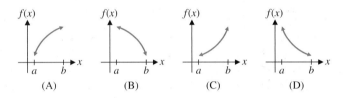

3. $f'(x) > 0$ and $f''(x) > 0$ on (a, b)
4. $f'(x) > 0$ and $f''(x) < 0$ on (a, b)
5. $f'(x) < 0$ and $f''(x) > 0$ on (a, b)
6. $f'(x) < 0$ and $f''(x) < 0$ on (a, b)

In Problems 7–18, find the indicated derivative for each function.

7. $f''(x)$ for $f(x) = 2x^3 - 4x^2 + 5x - 6$
8. $g''(x)$ for $g(x) = -x^3 + 2x^2 - 3x + 9$
9. $h''(x)$ for $h(x) = 2x^{-1} - 3x^{-2}$
10. $k''(x)$ for $k(x) = -6x^{-2} + 12x^{-3}$
11. d^2y/dx^2 for $y = x^2 - 18x^{1/2}$
12. d^2y/dx^2 for $y = x^3 - 24x^{1/3}$
13. y'' for $y = (x^2 + 9)^4$ $4(x^2+9)^{\xi 3}$
14. y'' for $y = (x^2 - 16)^5$ $20(x^2+9)^{\xi 2}$
15. $f''(x)$ for $f(x) = e^{-x^2}$
16. $f''(x)$ for $f(x) = xe^{-x}$
17. y'' for $y = \dfrac{\ln x}{x^2}$
18. y'' for $y = x^2 \ln x$

In Problems 19–28, find the intervals on which the graph of f is concave upward, the intervals on which the graph of f is concave downward, and the inflection points.

19. $f(x) = x^4 + 6x^2$
20. $f(x) = x^4 + 6x$

21. $f(x) = x^3 - 4x^2 + 5x - 2$

22. $f(x) = -x^3 - 5x^2 + 4x - 3$

23. $f(x) = -x^4 + 12x^3 - 12x + 24$

24. $f(x) = x^4 - 2x^3 - 36x + 12$

25. $f(x) = \ln(x^2 - 2x + 10)$

26. $f(x) = \ln(x^2 + 6x + 13)$

27. $f(x) = 8e^x - e^{2x}$

28. $f(x) = e^{3x} - 9e^x$

In Problems 29–36, f(x) is continuous on $(-\infty, \infty)$. Use the given information to sketch the graph of f.

29.

x	-4	-2	-1	0	2	4
f(x)	0	3	1.5	0	-1	-3

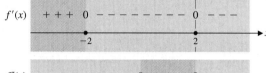

30.

x	-4	-2	-1	0	2	4
f(x)	0	-2	-1	0	1	3

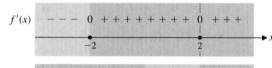

31.

x	-3	0	1	2	4	5
f(x)	-4	0	2	1	-1	0

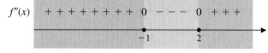

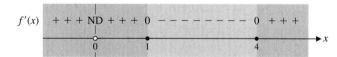

32.

x	-4	-2	0	2	4	6
f(x)	0	3	0	-2	0	3

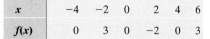

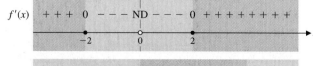

33. $f(0) = 2, f(1) = 0, f(2) = -2;$
$f'(0) = 0, f'(2) = 0;$
$f'(x) > 0$ on $(-\infty, 0)$ and $(2, \infty);$
$f'(x) < 0$ on $(0, 2);$
$f''(1) = 0;$
$f''(x) > 0$ on $(1, \infty);$
$f''(x) < 0$ on $(-\infty, 1)$

34. $f(-2) = -2, f(0) = 1, f(2) = 4;$
$f'(-2) = 0, f'(2) = 0;$
$f'(x) > 0$ on $(-2, 2);$
$f'(x) < 0$ on $(-\infty, -2)$ and $(2, \infty);$
$f''(0) = 0;$
$f''(x) > 0$ on $(-\infty, 0);$
$f''(x) < 0$ on $(0, \infty)$

35. $f(-1) = 0, f(0) = -2, f(1) = 0;$
$f'(0) = 0, f'(-1)$ and $f'(1)$ are not defined;
$f'(x) > 0$ on $(0, 1)$ and $(1, \infty);$
$f'(x) < 0$ on $(-\infty, -1)$ and $(-1, 0);$
$f''(-1)$ and $f''(1)$ are not defined;
$f''(x) > 0$ on $(-1, 1);$
$f''(x) < 0$ on $(-\infty, -1)$ and $(1, \infty)$

36. $f(0) = -2, f(1) = 0, f(2) = 4;$
$f'(0) = 0, f'(2) = 0, f'(1)$ is not defined;
$f'(x) > 0$ on $(0, 1)$ and $(1, 2);$
$f'(x) < 0$ on $(-\infty, 0)$ and $(2, \infty);$
$f''(1)$ is not defined;
$f''(x) > 0$ on $(-\infty, 1);$
$f''(x) < 0$ on $(1, \infty)$

B *In Problems 37–58, summarize the pertinent information obtained by applying the graphing strategy and sketch the graph of $y = f(x)$.*

37. $f(x) = (x - 2)(x^2 - 4x - 8)$

38. $f(x) = (x - 3)(x^2 - 6x - 3)$

39. $f(x) = (x + 1)(x^2 - x + 2)$

40. $f(x) = (1 - x)(x^2 + x + 4)$

41. $f(x) = -0.25x^4 + x^3$

42. $f(x) = 0.25x^4 - 2x^3$

43. $f(x) = 16x(x - 1)^3$

44. $f(x) = -4x(x + 2)^3$

45. $f(x) = (x^2 + 3)(9 - x^2)$

46. $f(x) = (x^2 + 3)(x^2 - 1)$

47. $f(x) = (x^2 - 4)^2$

48. $f(x) = (x^2 - 1)(x^2 - 5)$

49. $f(x) = 2x^6 - 3x^5$

50. $f(x) = 3x^5 - 5x^4$

51. $f(x) = 1 - e^{-x}$

52. $f(x) = 2 - 3e^{-2x}$

53. $f(x) = e^{0.5x} + 4e^{-0.5x}$

54. $f(x) = 2e^{0.5x} + e^{-0.5x}$

55. $f(x) = -4 + 2 \ln x$

56. $f(x) = 5 - 3 \ln x$

57. $f(x) = \ln(x + 4) - 2$

58. $f(x) = 1 - \ln(x - 3)$

In Problems 59–62, use the graph of $y = f'(x)$ to discuss the graph of $y = f(x)$. Organize your conclusions in a table (see Example 4), and sketch a possible graph of $y = f(x)$.

59.

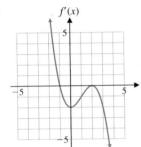

60.

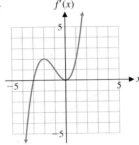

61.

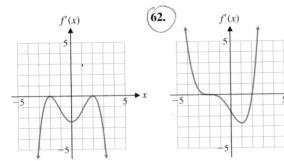

62.

In Problems 63–70, apply steps 1–3 of the graphing strategy to $f(x)$. Use a graphing calculator to approximate (to two decimal places) x intercepts, critical values, and x coordinates of inflection points. Summarize all the pertinent information.

63. $f(x) = x^4 - 5x^3 + 3x^2 + 8x - 5$

64. $f(x) = x^4 + 2x^3 - 5x^2 - 4x + 4$

65. $f(x) = x^4 - 21x^3 + 100x^2 + 20x + 100$

66. $f(x) = x^4 - 12x^3 + 28x^2 + 76x - 50$

67. $f(x) = -x^4 - x^3 + 2x^2 - 2x + 3$

68. $f(x) = -x^4 + x^3 + x^2 + 6$

69. $f(x) = 0.1x^5 + 0.3x^4 - 4x^3 - 5x^2 + 40x + 30$

70. $f(x) = x^5 + 4x^4 - 7x^3 - 20x^2 + 20x - 20$

C *In Problems 71–74, assume that f is a polynomial.*

71. Explain how you can locate inflection points for the graph of $y = f(x)$ by examining the graph of $y = f'(x)$.

72. Explain how you can determine where $f'(x)$ is increasing or decreasing by examining the graph of $y = f(x)$.

73. Explain how you can locate local maxima and minima for the graph of $y = f'(x)$ by examining the graph of $y = f(x)$.

74. Explain how you can locate local maxima and minima for the graph of $y = f(x)$ by examining the graph of $y = f'(x)$.

Applications

75. *Inflation.* One commonly used measure of inflation is the annual rate of change of the Consumer Price Index (CPI). A newspaper headline proclaims that the rate of change of inflation for consumer prices is increasing. What does this say about the shape of the graph of the CPI?

76. *Inflation.* Another commonly used measure of inflation is the annual rate of change of the Producer Price Index (PPI). A government report states that the rate of change of inflation for producer prices is decreasing. What does this say about the shape of the graph of the PPI?

77. *Cost analysis.* A company manufactures a variety of lighting fixtures at different locations. The total cost $C(x)$ (in dollars) of producing x desk lamps per week at plant A is shown in the figure. Discuss the graph of the marginal cost function $C'(x)$ and interpret the graph of $C'(x)$ in terms of the efficiency of the production process at this plant.

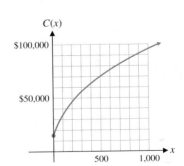

Figure for 77   Production costs at plant A

78. *Cost analysis.* The company in Problem 77 produces the same lamp at another plant. The total cost $C(x)$ (in dollars) of producing x desk lamps per week at plant B is shown in the figure. Discuss the graph of the marginal cost function $C'(x)$ and interpret the graph of $C'(x)$ in terms of the efficiency of the production process at plant B. Compare the production processes at the two plants.

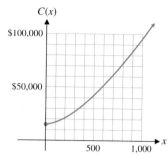

Figure for 78   Production costs at plant B

79. *Revenue.* The marketing research department of a computer company used a large city to test market the firm's new product. The department found that the relationship between price p (dollars per unit) and the demand x (units per week) was given approximately by

$$p = 1,296 - 0.12x^2 \qquad 0 < x < 80$$

Thus, weekly revenue can be approximated by

$$R(x) = xp = 1,296x - 0.12x^3 \qquad 0 < x < 80$$

(A) Find the local extrema for the revenue function.

(B) On which intervals is the graph of the revenue function concave upward? Concave downward?

80. *Profit.* Suppose that the cost equation for the company in Problem 79 is

$$C(x) = 830 + 396x$$

(A) Find the local extrema for the profit function.

(B) On which intervals is the graph of the profit function concave upward? Concave downward?

81. *Revenue.* A cosmetics company is planning to introduce and promote a new lipstick line. After test marketing the new line in a carefully selected large city, the marketing research department found that the demand in that city is given approximately by

$$p = 10e^{-x} \qquad 0 \le x \le 5$$

where x thousand lipsticks were sold per week at a price of $\$p$ each.

(A) Find the local extrema for the revenue function.

(B) On which intervals is the graph of the revenue function concave upward? Concave downward?

82. *Revenue.* A national food service runs food concessions for sporting events throughout the country. The company's marketing research department chose a particular football stadium to test market a new jumbo hot dog.

It was found that the demand for the new hot dog is given approximately by

$$P = 8 - 2 \ln x \qquad 5 \le x \le 50$$

where x is the number of hot dogs (in thousands) that can be sold during one game at a price of $\$p$.

(A) Find the local extrema for the revenue function.

(B) On which intervals is the graph of the revenue function concave upward? Concave downward?

83. *Production: point of diminishing returns.* A T-shirt manufacturer is planning to expand its workforce. It estimates that the number of T-shirts produced by hiring x new workers is given by

$$T(x) = -0.25x^4 + 5x^3 \qquad 0 \le x \le 15$$

When is the rate of change of T-shirt production increasing and when is it decreasing? What is the point of diminishing returns and the maximum rate of change of T-shirt production? Graph T and T' on the same coordinate system.

84. *Production: point of diminishing returns.* A baseball cap manufacturer is planning to expand its workforce. It estimates that the number of baseball caps produced by hiring x new workers is given by

$$T(x) = -0.25x^4 + 6x^3 \qquad 0 \le x \le 18$$

When is the rate of change of baseball cap production increasing and when is it decreasing? What is the point of diminishing returns and the maximum rate of change of baseball cap production? Graph T and T' on the same coordinate system.

85. *Advertising: point of diminishing returns.* A company estimates that it will sell $N(x)$ units of a product after spending $\$x$ thousand on advertising, as given by

$$N(x) = -0.25x^4 + 23x^3 - 540x^2 + 80,000 \qquad 24 \le x \le 45$$

When is the rate of change of sales increasing and when is it decreasing? What is the point of diminishing returns and the maximum rate of change of sales? Graph N and N' on the same coordinate system.

86. *Advertising: point of diminishing returns.* A company estimates that it will sell $N(x)$ units of a product after spending $\$x$ thousand on advertising, as given by

$$N(x) = -0.25x^4 + 13x^3 - 180x^2 + 10,000 \qquad 15 \le x \le 24$$

When is the rate of change of sales increasing and when is it decreasing? What is the point of diminishing returns and the maximum rate of change of sales? Graph N and N' on the same coordinate system.

87. *Advertising.* An automobile dealer uses television advertising to promote car sales. On the basis of past records, the dealer arrived at the following data, where x is the number of ads placed monthly and y is the number of cars sold that month:

Number of Ads x	Number of Cars y
10	325
12	339
20	417
30	546
35	615
40	682
50	795

(A) Enter the data in a graphing calculator and find a cubic regression equation for the number of cars sold monthly as a function of the number of ads.

(B) How many ads should the dealer place each month to maximize the rate of change of sales with respect to the number of ads, and how many cars can the dealer expect to sell with this number of ads? Round answers to the nearest integer.

88. *Advertising.* A music store advertises on the radio to promote sales of CDs. The store manager used past records to determine the following data, where x is the number of ads placed monthly and y is the number of CDs sold that month.

Number of Ads x	Number of CDs y
10	345
14	488
20	746
30	1,228
40	1,671
50	1,955

(A) Enter the data in a graphing calculator and find a cubic regression equation for the number of CDs sold monthly as a function of the number of ads.

(B) How many ads should the store manager place each month to maximize the rate of change of sales with respect to the number of ads, and how many CDs can the manager expect to sell with this number of ads? Round answers to the nearest integer.

89. *Population growth: bacteria.* A drug that stimulates reproduction is introduced into a colony of bacteria. After t minutes, the number of bacteria is given approximately by

$$N(t) = 1,000 + 30t^2 - t^3 \qquad 0 \le t \le 20$$

(A) When is the rate of growth, $N'(t)$, increasing? Decreasing?

(B) Find the inflection points for the graph of N.

(C) Sketch the graphs of N and N' on the same coordinate system.

(D) What is the maximum rate of growth?

90. *Drug sensitivity.* One hour after x milligrams of a particular drug are given to a person, the change in body temperature $T(x)$, in degrees Fahrenheit, is given by

$$T(x) = x^2\left(1 - \frac{x}{9}\right) \qquad 0 \le x \le 6$$

The rate $T'(x)$ at which $T(x)$ changes with respect to the size of the dosage x is called the *sensitivity* of the body to the dosage.

(A) When is $T'(x)$ increasing? Decreasing?

(B) Where does the graph of T have inflection points?

(C) Sketch the graphs of T and T' on the same coordinate system.

(D) What is the maximum value of $T'(x)$?

91. *Learning.* The time T (in minutes) it takes a person to learn a list of length n is

$$T(n) = 0.08n^3 - 1.2n^2 + 6n \qquad n \ge 0$$

(A) When is the rate of change of T with respect to the length of the list increasing? Decreasing?

(B) Where does the graph of T have inflection points? Graph T and T' on the same coordinate system.

(C) What is the minimum value of $T'(n)$?

Section 5-3 L'HÔPITAL'S RULE

- Introduction
- L'Hôpital's Rule and the Indeterminate Form 0/0
- One-Sided Limits and Limits at ∞
- L'Hôpital's Rule and the Indeterminate Form ∞/∞

■ Introduction

The ability to evaluate a wide variety of different types of limits is one of the skills that are necessary to apply the techniques of calculus successfully. Limits play a fundamental role in the development of the derivative and are an important graphing tool. In order to deal effectively with graphs, we need to develop some additional methods for evaluating limits.

In this section, we discuss a powerful technique for evaluating limits of quotients called *L'Hôpital's rule*. The rule is named after the French mathematician Marquis de L'Hôpital (1661–1704), who is generally credited with writing the first calculus textbook. To use L'Hôpital's rule, it is necessary to be familiar with the limit properties of some basic functions. Figure 1 reviews some limits involving powers of x that we have discussed earlier.

The limits in Figure 1 are easily extended to functions of the form $f(x) = (x - c)^n$ and $g(x) = 1/(x - c)^n$. In general, if n is an odd integer, limits involving $(x - c)^n$ or $1/(x - c)^n$ as x approaches c (or $\pm\infty$) behave, respectively, like the limits of x and $1/x$ as x approaches 0 (or $\pm\infty$). If n is an even integer, limits involving these expressions behave, respectively, like the limits of x^2 and $1/x^2$ as x approaches 0 (or $\pm\infty$).

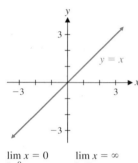

$$\lim_{x \to 0} x = 0 \qquad \lim_{x \to \infty} x = \infty$$
$$\lim_{x \to -\infty} x = -\infty$$

(A) $y = x$

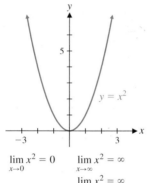

$$\lim_{x \to 0} x^2 = 0 \qquad \lim_{x \to \infty} x^2 = \infty$$
$$\lim_{x \to -\infty} x^2 = \infty$$

(B) $y = x^2$

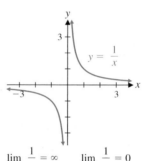

$$\lim_{x \to 0^+} \frac{1}{x} = \infty \qquad \lim_{x \to \infty} \frac{1}{x} = 0$$
$$\lim_{x \to 0^-} \frac{1}{x} = -\infty \qquad \lim_{x \to -\infty} \frac{1}{x} = 0$$
$$\lim_{x \to 0} \frac{1}{x} \quad \text{Does not exist}$$

(C) $y = \dfrac{1}{x}$

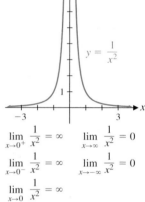

$$\lim_{x \to 0^+} \frac{1}{x^2} = \infty \qquad \lim_{x \to \infty} \frac{1}{x^2} = 0$$
$$\lim_{x \to 0^-} \frac{1}{x^2} = \infty \qquad \lim_{x \to -\infty} \frac{1}{x^2} = 0$$
$$\lim_{x \to 0} \frac{1}{x^2} = \infty$$

(D) $y = \dfrac{1}{x^2}$

FIGURE 1 Limits involving powers of x

EXAMPLE 1 **Limits Involving Powers of $x - c$**

(A) $\displaystyle\lim_{x \to 2} \frac{5}{(x - 2)^4} = \infty$ *Compare with* $\displaystyle\lim_{x \to 0} \frac{1}{x^2}$ *in Figure 1.*

(B) $\displaystyle\lim_{x \to -1^-} \frac{4}{(x + 1)^3} = -\infty$ *Compare with* $\displaystyle\lim_{x \to 0^-} \frac{1}{x}$ *in Figure 1.*

(C) $\displaystyle\lim_{x \to \infty} \frac{4}{(x - 9)^6} = 0$ *Compare with* $\displaystyle\lim_{x \to \infty} \frac{1}{x^2}$ *in Figure 1.*

(D) $\displaystyle\lim_{x \to -\infty} 3x^3 = -\infty$ *Compare with* $\displaystyle\lim_{x \to -\infty} x$ *in Figure 1.*

MATCHED PROBLEM 1 Evaluate each limit.

(A) $\displaystyle\lim_{x \to 3^+} \frac{7}{(x - 3)^5}$ (B) $\displaystyle\lim_{x \to -4} \frac{6}{(x + 4)^6}$

(C) $\displaystyle\lim_{x \to -\infty} \frac{3}{(x + 2)^3}$ (D) $\displaystyle\lim_{x \to \infty} 5x^4$

Figure 2 reviews some limits of exponential and logarithmic functions.

The limits in Figure 2 also generalize to other simple exponential and logarithmic forms.

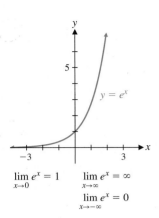

$$\lim_{x \to 0} e^x = 1 \qquad \lim_{x \to \infty} e^x = \infty$$
$$\lim_{x \to -\infty} e^x = 0$$

(A) $y = e^x$

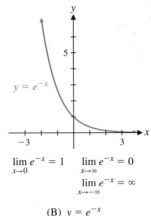

$$\lim_{x \to 0} e^{-x} = 1 \qquad \lim_{x \to \infty} e^{-x} = 0$$
$$\lim_{x \to -\infty} e^{-x} = \infty$$

(B) $y = e^{-x}$

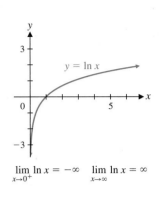

$$\lim_{x \to 0^+} \ln x = -\infty \qquad \lim_{x \to \infty} \ln x = \infty$$

(C) $y = \ln x$

FIGURE 2 Limits involving exponential and logarithmic functions

EXAMPLE 2   **Limits Involving Exponential and Logarithmic Forms**

(A) $\displaystyle\lim_{x \to \infty} 2e^{3x} = \infty$ *Compare with* $\displaystyle\lim_{x \to \infty} e^x$ *in Figure 2.*

(B) $\displaystyle\lim_{x \to \infty} 4e^{-5x} = 0$ *Compare with* $\displaystyle\lim_{x \to \infty} e^{-x}$ *in Figure 2.*

(C) $\displaystyle\lim_{x \to \infty} \ln(x + 4) = \infty$ *Compare with* $\displaystyle\lim_{x \to \infty} \ln x$ *in Figure 2.*

(D) $\displaystyle\lim_{x \to 2^+} \ln(x - 2) = -\infty$ *Compare with* $\displaystyle\lim_{x \to 0^+} \ln x$ *in Figure 2.*

MATCHED PROBLEM 2   Evaluate each limit.

(A) $\displaystyle\lim_{x \to -\infty} 2e^{-6x}$ (B) $\displaystyle\lim_{x \to -\infty} 3e^{2x}$

(C) $\displaystyle\lim_{x \to -4^+} \ln(x + 4)$ (D) $\displaystyle\lim_{x \to \infty} \ln(x - 10)$

Now that we have reviewed the limit properties of some basic functions, we are ready to consider the main topic of this section: L'Hôpital's rule—a powerful tool for evaluating certain types of limits.

Explore & Discuss **1**   It follows from Figure 2 that

$$\lim_{x \to \infty} e^{-x} = 0 \qquad \text{and} \qquad \lim_{x \to \infty} \ln(1 + 3e^{-x}) = 0$$

Does

$$\lim_{x \to \infty} \frac{\ln(1 + 3e^{-x})}{e^{-x}}$$

exist? If so, what is its value? Use graphical methods to support your answers.

■ L'Hôpital's Rule and the Indeterminate Form 0/0

Recall that the limit

$$\lim_{x \to c} \frac{f(x)}{g(x)}$$

is a 0/0 indeterminate form if

$$\lim_{x \to c} f(x) = 0 \qquad \text{and} \qquad \lim_{x \to c} g(x) = 0$$

The quotient property for limits in Section 3-1 does not apply, since $\lim\limits_{x \to c} g(x) = 0$.

If we are dealing with a 0/0 indeterminate form, the limit may or may not exist, and we cannot tell which is true without further investigation.

Each of the following is a 0/0 indeterminate form:

$$\lim_{x \to 2} \frac{x^2 - 4}{x - 2} \quad \text{and} \quad \lim_{x \to 1} \frac{e^x - e}{x - 1}$$

The first limit can be evaluated by performing some algebraic simplifications, such as

$$\lim_{x \to 2} \frac{x^2 - 4}{x - 2} = \lim_{x \to 2} \frac{(x - 2)(x + 2)}{x - 2} = \lim_{x \to 2}(x + 2) = 4$$

The second cannot. Instead, we turn to the powerful **L'Hôpital's rule,** which we now state without proof. This rule can be used whenever a limit is a 0/0 indeterminate form.

THEOREM 1   **L'HÔPITAL'S RULE FOR 0/0 INDETERMINATE FORMS: VERSION 1**

For c a real number,
if $\lim\limits_{x \to c} f(x) = 0$ and $\lim\limits_{x \to c} g(x) = 0$, then

$$\lim_{x \to c} \frac{f(x)}{g(x)} = \lim_{x \to c} \frac{f'(x)}{g'(x)}$$

provided that the second limit exists or is $+\infty$ or $-\infty$.

The use of L'Hôpital's rule is best illustrated through examples.

EXAMPLE 3   **L'Hôpital's Rule** Evaluate $\lim\limits_{x \to 1} \dfrac{e^x - e}{x - 1}$.

SOLUTION   **Step 1.** *Check to see if L'Hôpital's rule applies:*

$$\lim_{x \to 1}(e^x - e) = e^1 - e = 0 \quad \text{and} \quad \lim_{x \to 1}(x - 1) = 1 - 1 = 0$$

L'Hôpital's rule does apply.

Step 2. *Apply L'Hôpital's rule:*

0/0 form

$$\lim_{x \to 1} \frac{e^x - e}{x - 1} = \lim_{x \to 1} \frac{\dfrac{d}{dx}(e^x - e)}{\dfrac{d}{dx}(x - 1)} \qquad \text{Apply L'Hôpital's rule.}$$

$$= \lim_{x \to 1} \frac{e^x}{1} \qquad e^x \text{ is continuous at } x = 1.$$

$$= \frac{e^1}{1} = e$$

MATCHED PROBLEM 3   Evaluate $\lim\limits_{x \to 4} \dfrac{e^x - e^4}{x - 4}$.

In L'Hôpital's rule, the symbol $f'(x)/g'(x)$ represents the derivative of $f(x)$ divided by the derivative of $g(x)$, not the derivative of the quotient $f(x)/g(x)$.

When applying L'Hôpital's rule to a 0/0 indeterminate form, be certain that you differentiate the numerator and denominator separately.

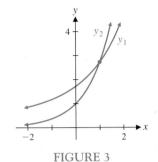

FIGURE 3

The functions

$$y_1 = \frac{e^x - e}{x - 1} \quad \text{and} \quad y_2 = \frac{e^x}{1}$$

of Example 3 are different functions (see Fig. 3), but both functions have the same limit e as x approaches 1. Although y_1 is undefined at $x = 1$, the graph of y_1 provides a check of the answer to Example 3.

EXAMPLE 4   **L'Hôpital's Rule** Evaluate $\displaystyle\lim_{x \to 0} \frac{\ln(1 + x^2)}{x^4}$.

SOLUTION   **Step 1.**   *Check to see if L'Hôpital's rule applies:*

$$\lim_{x \to 0} \ln(1 + x^2) = \ln 1 = 0 \quad \text{and} \quad \lim_{x \to 0} x^4 = 0$$

L'Hôpital's rule does apply.

Step 2.   *Apply L'Hôpital's rule:*

$$\lim_{x \to 0} \overset{0/0 \text{ form}}{\frac{\ln(1 + x^2)}{x^4}} = \lim_{x \to 0} \frac{\dfrac{d}{dx} \ln(1 + x^2)}{\dfrac{d}{dx} x^4} \qquad \text{Apply L'Hôpital's rule.}$$

$$\lim_{x \to 0} \frac{\ln(1 + x^2)}{x^4} = \lim_{x \to 0} \frac{\dfrac{2x}{1 + x^2}}{4x^3} \qquad \text{Multiply numerator and denominator by } 1/4x^3.$$

$$= \lim_{x \to 0} \frac{\dfrac{2x}{1 + x^2} \cdot \dfrac{1}{4x^3}}{\dfrac{4x^3}{1} \cdot \dfrac{1}{4x^3}} \qquad \text{Simplify.}$$

$$= \lim_{x \to 0} \frac{1}{2x^2(1 + x^2)} \qquad \begin{array}{l}\text{Apply Theorem 1 in Section 3-3 and} \\ \text{compare with Fig. 1(D).}\end{array}$$

$$= \infty$$

MATCHED PROBLEM 4   Evaluate $\displaystyle\lim_{x \to 1} \frac{\ln x}{(x - 1)^3}$.

EXAMPLE 5   **L'Hôpital's Rule May Not Be Applicable** Evaluate $\displaystyle\lim_{x \to 1} \frac{\ln x}{x}$.

SOLUTION   **Step 1.**   *Check to see if L'Hôpital's rule applies:*

$$\lim_{x \to 1} \ln x = \ln 1 = 0, \quad \text{but} \quad \lim_{x \to 1} x = 1 \neq 0$$

L'Hôpital's rule does not apply.

Step 2.   *Evaluate by another method.* The quotient property for limits from Section 3-1 does apply, and we have

$$\lim_{x \to 1} \frac{\ln x}{x} = \frac{\lim\limits_{x \to 1} \ln x}{\lim\limits_{x \to 1} x} = \frac{\ln 1}{1} = \frac{0}{1} = 0$$

Note that applying L'Hôpital's rule would give us an incorrect result:

$$\lim_{x \to 1} \frac{\ln x}{x} \neq \lim_{x \to 1} \frac{\dfrac{d}{dx} \ln x}{\dfrac{d}{dx} x} = \lim_{x \to 1} \frac{1/x}{1} = 1$$

MATCHED PROBLEM 5 Evaluate $\displaystyle\lim_{x \to 0} \frac{x}{e^x}$.

INSIGHT

As Example 5 illustrates, all limits involving quotients are not 0/0 indeterminate forms. You must always check to see if L'Hôpital's rule applies before you use it.

EXAMPLE 6 **Repeated Application of L'Hôpital's Rule** Evaluate

$$\lim_{x \to 0} \frac{x^2}{e^x - 1 - x}$$

SOLUTION **Step 1.** *Check to see if L'Hôpital's rule applies:*

$$\lim_{x \to 0} x^2 = 0 \qquad \text{and} \qquad \lim_{x \to 0}(e^x - 1 - x) = 0$$

L'Hôpital's rule does apply.

Step 2. *Apply L'Hôpital's rule:*

$$\underset{0/0 \text{ form}}{\lim_{x \to 0} \frac{x^2}{e^x - 1 - x}} = \lim_{x \to 0} \frac{\dfrac{d}{dx} x^2}{\dfrac{d}{dx}(e^x - 1 - x)} = \lim_{x \to 0} \frac{2x}{e^x - 1}$$

Since $\lim_{x \to 0} 2x = 0$ and $\lim_{x \to 0}(e^x - 1) = 0$, the new limit obtained is also a 0/0 indeterminate form, and L'Hôpital's rule can be applied again.

Step 3. *Apply L'Hôpital's rule again:*

$$\underset{0/0 \text{ form}}{\lim_{x \to 0} \frac{2x}{e^x - 1}} = \lim_{x \to 0} \frac{\dfrac{d}{dx} 2x}{\dfrac{d}{dx}(e^x - 1)} = \lim_{x \to 0} \frac{2}{e^x} = \frac{2}{e^0} = 2$$

Thus,

$$\lim_{x \to 0} \frac{x^2}{e^x - 1 - x} = \lim_{x \to 0} \frac{2x}{e^x - 1} = \lim_{x \to 0} \frac{2}{e^x} = 2$$

MATCHED PROBLEM 6 Evaluate: $\displaystyle\lim_{x \to 0} \frac{e^{2x} - 1 - 2x}{x^2}$

■ One-Sided Limits and Limits at ∞

In addition to examining the limit as x approaches c, we have discussed one-sided limits and limits at ∞ in Chapter 3. L'Hôpital's rule is valid in these cases also.

THEOREM 2   L'HÔPITAL'S RULE FOR 0/0 INDETERMINATE FORMS: VERSION 2 (FOR ONE-SIDED LIMITS AND LIMITS AT INFINITY)

The first version of L'Hôpital's rule (Theorem 1) remains valid if the symbol $x \to c$ is replaced everywhere it occurs with one of the following symbols:

$$x \to c^+ \qquad x \to c^- \qquad x \to \infty \qquad x \to -\infty$$

For example, if $\lim_{x \to \infty} f(x) = 0$ and $\lim_{x \to \infty} g(x) = 0$, then

$$\lim_{x \to \infty} \frac{f(x)}{g(x)} = \lim_{x \to \infty} \frac{f'(x)}{g'(x)}$$

provided that the second limit exists or is $+\infty$ or $-\infty$. Similar rules can be written for $x \to c^+$, $x \to c^-$, and $x \to -\infty$.

EXAMPLE 7   **L'Hôpital's Rule for One-Sided Limits**   Evaluate $\lim_{x \to 1^+} \dfrac{\ln x}{(x-1)^2}$.

SOLUTION   **Step 1.**   *Check to see if L'Hôpital's rule applies:*

$$\lim_{x \to 1^+} \ln x = 0 \qquad \text{and} \qquad \lim_{x \to 1^+} (x-1)^2 = 0$$

L'Hôpital's rule does apply.

Step 2.   *Apply L'Hôpital's rule:*

$$\underset{\text{0/0 form}}{\lim_{x \to 1^+} \frac{\ln x}{(x-1)^2}} = \lim_{x \to 1^+} \frac{\dfrac{d}{dx}(\ln x)}{\dfrac{d}{dx}(x-1)^2} \qquad \text{Apply L'Hôpital's rule.}$$

$$= \lim_{x \to 1^+} \frac{1/x}{2(x-1)} \qquad \text{Simplify.}$$

$$= \lim_{x \to 1^+} \frac{1}{2x(x-1)}$$

$$= \infty$$

The limit as $x \to 1^+$ is ∞ because $1/2x(x-1)$ has a vertical asymptote at $x = 1$ (Theorem 1, Section 3-3) and $x(x-1) > 0$ for $x > 1$.   ∎

MATCHED PROBLEM 7   Evaluate $\lim_{x \to 1^-} \dfrac{\ln x}{(x-1)^2}$.

EXAMPLE 8   **L'Hôpital's Rule for Limits at Infinity**   Evaluate $\lim_{x \to \infty} \dfrac{\ln(1 + e^{-x})}{e^{-x}}$.

SOLUTION   **Step 1.**   *Check to see if L'Hôpital's rule applies:*

$$\lim_{x \to \infty} \ln(1 + e^{-x}) = \ln(1 + 0) = \ln 1 = 0 \qquad \text{and} \qquad \lim_{x \to \infty} e^{-x} = 0$$

L'Hôpital's rule does apply.

Step 2. *Apply L'Hôpital's rule:*

$$\lim_{x \to \infty} \frac{\ln(1 + e^{-x})}{e^{-x}} = \lim_{x \to \infty} \frac{\dfrac{d}{dx}[\ln(1 + e^{-x})]}{\dfrac{d}{dx}e^{-x}}$$

0/0 form (above left)

Apply L'Hôpital's rule.

$$= \lim_{x \to \infty} \frac{-e^{-x}/(1 + e^{-x})}{-e^{-x}}$$

Multiply numerator and denominator by $-e^x$.

$$\lim_{x \to \infty} \frac{\ln(1 + e^{-x})}{e^{-x}} = \lim_{x \to \infty} \frac{1}{1 + e^{-x}}$$

$\lim_{x \to \infty} e^{-x} = 0$

$$= \frac{1}{1 + 0} = 1$$

MATCHED PROBLEM 8   Evaluate $\displaystyle \lim_{x \to -\infty} \frac{\ln(1 + 2e^x)}{e^x}$.

■ L'Hôpital's Rule and the Indeterminate Form ∞/∞

In Section 3-3, we discussed techniques for evaluating limits of rational functions such as

$$\lim_{x \to \infty} \frac{2x^2}{x^3 + 3} \qquad \lim_{x \to \infty} \frac{4x^3}{2x^2 + 5} \qquad \lim_{x \to \infty} \frac{3x^3}{5x^3 + 6} \qquad (1)$$

Each of these limits is an ∞/∞ *indeterminate form*. In general, if $\lim_{x \to c} f(x) = \pm\infty$ and $\lim_{x \to c} g(x) = \pm\infty$, then

$$\lim_{x \to c} \frac{f(x)}{g(x)}$$

is called an **∞/∞ indeterminate form**. Furthermore, $x \to c$ can be replaced in all three limits above with $x \to c^+$, $x \to c^-$, $x \to \infty$, or $x \to -\infty$. It can be shown that L'Hôpital's rule also applies to these ∞/∞ indeterminate forms.

THEOREM 3   L'HÔPITAL'S RULE FOR THE INDETERMINATE FORM ∞/∞: VERSION 3

Versions 1 and 2 of L'Hôpital's rule for the indeterminate form 0/0 are also valid if the limit of f and the limit of g are both infinite; that is, both $+\infty$ and $-\infty$ are permissible for either limit.

For example, if $\lim_{x \to c^+} f(x) = \infty$ and $\lim_{x \to c^+} g(x) = -\infty$, then L'Hôpital's rule can be applied to $\lim_{x \to c^+} [f(x)/g(x)]$.

Explore & Discuss **2**

Evaluate each of the limits in (1) in two ways:

1. Use Theorem 4 in Section 3-3.

2. Use L'Hôpital's rule.

Given a choice, which method would you choose? Why?

EXAMPLE 9   L'Hôpital's Rule for the Indeterminate Form ∞/∞

Evaluate $\displaystyle \lim_{x \to \infty} \frac{\ln x}{x^2}$.

SOLUTION   **Step 1.** *Check to see if L'Hôpital's rule applies:*

$$\lim_{x \to \infty} \ln x = \infty \qquad \text{and} \qquad \lim_{x \to \infty} x^2 = \infty$$

L'Hôpital's rule does apply.

Step 2.   *Apply L'Hôpital's rule:*

∞/∞ form

$$\lim_{x \to \infty} \frac{\ln x}{x^2} = \lim_{x \to \infty} \frac{\dfrac{d}{dx}(\ln x)}{\dfrac{d}{dx}x^2}$$ Apply L'Hôpital's rule.

$$= \lim_{x \to \infty} \frac{1/x}{2x}$$ Simplify.

$$\lim_{x \to \infty} \frac{\ln x}{x^2} = \lim_{x \to \infty} \frac{1}{2x^2}$$ See Figure 1(D).

$$= 0$$

MATCHED PROBLEM 9   Evaluate $\displaystyle\lim_{x \to \infty} \frac{\ln x}{x}$.

EXAMPLE 10   L'Hôpital's Rule for the Indeterminate Form ∞/∞

Evaluate $\displaystyle\lim_{x \to \infty} \frac{e^x}{x^2}$.

SOLUTION   Step 1.   *Check to see if L'Hôpital's rule applies:*

$$\lim_{x \to \infty} e^x = \infty \qquad \text{and} \qquad \lim_{x \to \infty} x^2 = \infty$$

L'Hôpital's rule does apply.

Step 2.   *Apply L'Hôpital's rule:*

∞/∞ form

$$\lim_{x \to \infty} \frac{e^x}{x^2} = \lim_{x \to \infty} \frac{\dfrac{d}{dx}e^x}{\dfrac{d}{dx}x^2} = \lim_{x \to \infty} \frac{e^x}{2x}$$

Since $\lim_{x \to \infty} e^x = \infty$ and $\lim_{x \to \infty} 2x = \infty$, this limit is an ∞/∞ indeterminate form and L'Hôpital's rule can be applied again.

Step 3.   *Apply L'Hôpital's rule again:*

∞/∞ form

$$\lim_{x \to \infty} \frac{e^x}{2x} = \lim_{x \to \infty} \frac{\dfrac{d}{dx}e^x}{\dfrac{d}{dx}2x} = \lim_{x \to \infty} \frac{e^x}{2} = \infty$$

Thus,

$$\lim_{x \to \infty} \frac{e^x}{x^2} = \lim_{x \to \infty} \frac{e^x}{2x} = \lim_{x \to \infty} \frac{e^x}{2} = \infty$$

MATCHED PROBLEM 10   Evaluate $\displaystyle\lim_{x \to \infty} \frac{e^{2x}}{x^2}$.

Explore & Discuss **3**   Let *n* be a positive integer. Explain how L'Hôpital's rule can be used to show that

$$\lim_{x \to \infty} \frac{e^x}{x^n} = \infty$$

Does this imply that e^x increases more rapidly than any power of x?

Answers to Matched Problems
  1. (A) ∞ (B) ∞ (C) 0 (D) ∞ **2.** (A) ∞ (B) 0 (C) $-\infty$ (D) ∞
  3. e^4 **4.** ∞ **5.** 0 **6.** 2 **7.** $-\infty$ **8.** 2 **9.** 0 **10.** ∞

EXERCISE 5-3

A Use L'Hôpital's rule to find each limit in Problems 1–14.

1. $\lim\limits_{x \to 2} \dfrac{x^4 - 16}{x^3 - 8}$

2. $\lim\limits_{x \to 1} \dfrac{x^6 - 1}{x^5 - 1}$

3. $\lim\limits_{x \to 2} \dfrac{x^2 + x - 6}{x^2 + 6x - 16}$

4. $\lim\limits_{x \to 4} \dfrac{x^2 - 8x + 16}{x^2 - 5x + 4}$

5. $\lim\limits_{x \to 0} \dfrac{e^x - 1}{x}$

6. $\lim\limits_{x \to 1} \dfrac{\ln x}{x - 1}$

7. $\lim\limits_{x \to 0} \dfrac{\ln(1 + 4x)}{x}$

8. $\lim\limits_{x \to 0} \dfrac{e^{2x} - 1}{x}$

9. $\lim\limits_{x \to \infty} \dfrac{2x^2 + 7}{5x^3 + 9}$

10. $\lim\limits_{x \to \infty} \dfrac{3x^4 + 6}{2x^2 + 5}$

11. $\lim\limits_{x \to \infty} \dfrac{e^{3x}}{x}$

12. $\lim\limits_{x \to \infty} \dfrac{x}{e^{4x}}$

13. $\lim\limits_{x \to \infty} \dfrac{x^2}{\ln x}$

14. $\lim\limits_{x \to \infty} \dfrac{\ln x}{x^4}$

In Problems 15–18, explain why L'Hôpital's rule does not apply. If the limit exists, find it by other means.

15. $\lim\limits_{x \to 1} \dfrac{x^2 + 5x + 4}{x^3 + 1}$

16. $\lim\limits_{x \to \infty} \dfrac{e^{-x}}{\ln x}$

17. $\lim\limits_{x \to 2} \dfrac{x + 2}{(x - 2)^4}$

18. $\lim\limits_{x \to -3} \dfrac{x^2}{(x + 3)^5}$

B Find each limit in Problems 19–42. Note that L'Hôpital's rule does not apply to every problem, and some problems will require more than one application of L'Hôpital's rule.

19. $\lim\limits_{x \to 0} \dfrac{e^{4x} - 1 - 4x}{x^2}$

20. $\lim\limits_{x \to 0} \dfrac{3x + 1 - e^{3x}}{x^2}$

21. $\lim\limits_{x \to 2} \dfrac{\ln(x - 1)}{x - 1}$

22. $\lim\limits_{x \to -1} \dfrac{\ln(x + 2)}{x + 2}$

23. $\lim\limits_{x \to 0^+} \dfrac{\ln(1 + x^2)}{x^3}$

24. $\lim\limits_{x \to 0^-} \dfrac{\ln(1 + 2x)}{x^2}$

25. $\lim\limits_{x \to 0^+} \dfrac{\ln(1 + \sqrt{x})}{x}$

26. $\lim\limits_{x \to 0^+} \dfrac{\ln(1 + x)}{\sqrt{x}}$

27. $\lim\limits_{x \to -2} \dfrac{x^2 + 2x + 1}{x^2 + x + 1}$

28. $\lim\limits_{x \to 1} \dfrac{2x^3 - 3x^2 + 1}{x^3 - 3x + 2}$

29. $\lim\limits_{x \to -1} \dfrac{x^3 + x^2 - x - 1}{x^3 + 4x^2 + 5x + 2}$

30. $\lim\limits_{x \to 3} \dfrac{x^3 + 3x^2 - x - 3}{x^2 + 6x + 9}$

31. $\lim\limits_{x \to 2^-} \dfrac{x^3 - 12x + 16}{x^3 - 6x^2 + 12x - 8}$

32. $\lim\limits_{x \to 1^+} \dfrac{x^3 - x^2 - x + 1}{x^3 - 3x^2 + 3x - 1}$

33. $\lim\limits_{x \to \infty} \dfrac{3x^2 + 5x}{4x^3 + 7}$

34. $\lim\limits_{x \to \infty} \dfrac{4x^2 + 9x}{5x^2 + 8}$

35. $\lim\limits_{x \to \infty} \dfrac{x^2}{e^{2x}}$

36. $\lim\limits_{x \to \infty} \dfrac{e^{3x}}{x^3}$

37. $\lim\limits_{x \to \infty} \dfrac{1 + e^{-x}}{1 + x^2}$

38. $\lim\limits_{x \to -\infty} \dfrac{1 + e^{-x}}{1 + x^2}$

39. $\lim\limits_{x \to \infty} \dfrac{e^{-x}}{\ln(1 + 4e^{-x})}$

40. $\lim\limits_{x \to \infty} \dfrac{\ln(1 + 2e^{-x})}{\ln(1 + e^{-x})}$

41. $\lim\limits_{x \to 0} \dfrac{e^x - e^{-x} - 2x}{x^3}$

42. $\lim\limits_{x \to 0} \dfrac{e^{2x} - 1 - 2x - 2x^2}{x^3}$

C **43.** Find $\lim\limits_{x \to 0^+} (x \ln x)$.
    [*Hint:* Write $x \ln x = (\ln x)/x^{-1}$.]

44. Find $\lim\limits_{x \to 0^+} (\sqrt{x} \ln x)$.
    [*Hint:* Write $\sqrt{x} \ln x = (\ln x)/x^{-1/2}$.]

In Problems 45–48, n is a positive integer. Find each limit.

45. $\lim\limits_{x \to \infty} \dfrac{\ln x}{x^n}$

46. $\lim\limits_{x \to \infty} \dfrac{x^n}{\ln x}$

47. $\lim\limits_{x \to \infty} \dfrac{e^x}{x^n}$

48. $\lim\limits_{x \to \infty} \dfrac{x^n}{e^x}$

In Problems 49–52, show that the repeated application of L'Hôpital's rule does not lead to a solution. Then use algebraic manipulation to evaluate each limit. [Hint: For $x > 0$ and $n > 0$, $\sqrt[n]{x^n} = x$.]

49. $\lim\limits_{x \to \infty} \dfrac{\sqrt{1 + x^2}}{x}$

50. $\lim\limits_{x \to -\infty} \dfrac{x}{\sqrt{4 + x^2}}$

51. $\lim\limits_{x \to -\infty} \dfrac{\sqrt[3]{x^3 + 1}}{x}$

52. $\lim\limits_{x \to \infty} \dfrac{x^2}{\sqrt[3]{(x^3 + 1)^2}}$

Section 5-4   CURVE-SKETCHING TECHNIQUES

- Modifying the Graphing Strategy
- Using the Graphing Strategy
- Modeling Average Cost

When we summarized the graphing strategy in Section 5-2, we omitted one important topic: asymptotes. Polynomial functions do not have any asymptotes. Asymptotes of rational functions were discussed in Section 3-3, but what about all the other functions, such as logarithmic and exponential functions? Since investigating asymptotes always involves limits, we can now use L'Hôpital's rule (Section 5-3) as a tool for finding asymptotes of many different types of functions.

■ Modifying the Graphing Strategy

The first version of the graphing strategy in Section 5-2 made no mention of asymptotes. Including information about asymptotes produces the following (and final) version of the graphing strategy:

PROCEDURE

Graphing Strategy (Final Version)

Step 1. *Analyze $f(x)$.*

   (A) Find the domain of f.

   (B) Find the intercepts.

   (C) Find asymptotes.

Step 2. *Analyze $f'(x)$.* Find the partition numbers for, and critical values of, $f'(x)$. Construct a sign chart for $f'(x)$, determine the intervals on which f is increasing and decreasing, and find local maxima and minima.

Step 3. *Analyze $f''(x)$.* Find the partition numbers of $f''(x)$. Construct a sign chart for $f''(x)$, determine the intervals on which the graph of f is concave upward and concave downward, and find inflection points.

Step 4. *Sketch the graph of f.* Draw asymptotes and locate intercepts, local maxima and minima, and inflection points. Sketch in what you know from steps 1–3. Plot additional points as needed and complete the sketch.

■ Using the Graphing Strategy

INSIGHT

From now on, you should always use the final version of the graphing strategy. If a function does not have any asymptotes, simply state this fact.

We will illustrate the graphing strategy with several examples:

EXAMPLE 1   **Using the Graphing Strategy**   Use the graphing strategy to analyze the function $f(x) = (x - 1)/(x - 2)$. State all the pertinent information and sketch the graph of f.

SOLUTION   **Step 1.** *Analyze $f(x)$.* $f(x) = \dfrac{x - 1}{x - 2}$

   (A) Domain: All real x, except $x = 2$

   (B) y intercept: $f(0) = \dfrac{0 - 1}{0 - 2} = \dfrac{1}{2}$

   x intercepts: Since a fraction is 0 when its numerator is 0 and its denominator is not 0, the x intercept is $x = 1$.

(C) Horizontal asymptote: $\dfrac{a_m x^m}{b_n x^n} = \dfrac{x}{x} = 1$

Thus, the line $y = 1$ is a horizontal asymptote.

Vertical asymptote: The denominator is 0 for $x = 2$, and the numerator is not 0 for this value. Therefore, the line $x = 2$ is a vertical asymptote.

Step 2. *Analyze $f'(x)$.* $f'(x) = \dfrac{(x-2)(1) - (x-1)(1)}{(x-2)^2} = \dfrac{-1}{(x-2)^2}$

Critical values of $f(x)$:   None

Partition number for $f'(x)$: $x = 2$

Sign chart for $f'(x)$:

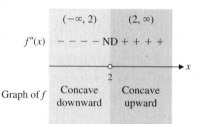

	Test Numbers	
x	**f'(x)**	
1	−1	(−)
3	−1	(−)

Thus, $f(x)$ is decreasing on $(-\infty, 2)$ and $(2, \infty)$. There are no local extrema.

Step 3. *Analyze $f''(x)$.* $f''(x) = \dfrac{2}{(x-2)^3}$

Partition number for $f''(x)$: $x = 2$

Sign chart for $f''(x)$:

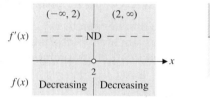

	Test Numbers	
x	**f''(x)**	
1	−2	(−)
3	2	(+)

Thus, the graph of f is concave downward on $(-\infty, 2)$ and concave upward on $(2, \infty)$. Since $f(2)$ is not defined, there is no inflection point at $x = 2$, even though $f''(x)$ changes sign at $x = 2$.

Step 4. *Sketch the graph of f.*   Insert intercepts and asymptotes, and plot a few additional points (for functions with asymptotes, plotting additional points is often helpful). Then sketch the graph:

x	f(x)
−2	$\frac{3}{4}$
0	$\frac{1}{2}$
1	0
$\frac{3}{2}$	−1
$\frac{5}{2}$	3
3	2
4	$\frac{3}{2}$

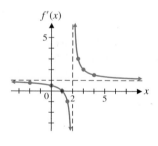

MATCHED PROBLEM 1   Follow the graphing strategy and analyze the function $f(x) = 2x/(1-x)$. State all the pertinent information and sketch the graph of f.

EXAMPLE 2 **Using the Graphing Strategy** Use the graphing strategy to analyze the function

$$g(x) = \frac{2x - 1}{x^2}$$

State all pertinent information and sketch the graph of g.

SOLUTION **Step 1.** *Analyze $g(x)$.*

(A) Domain: All real x, except $x = 0$

(B) x intercept: $x = \frac{1}{2} = 0.5$

y intercept: Since 0 is not in the domain of g, there is no y intercept.

(C) Horizontal asymptote: $y = 0$ (the x axis)

Vertical asymptote: The denominator of $g(x)$ is 0 at $x = 0$ and the numerator is not. So the line $x = 0$ (the y axis) is a vertical asymptote.

Step 2. *Analyze $g'(x)$.*

$$g(x) = \frac{2x - 1}{x^2} = \frac{2}{x} - \frac{1}{x^2} = 2x^{-1} - x^{-2}$$

$$g'(x) = -2x^{-2} + 2x^{-3} = -\frac{2}{x^2} + \frac{2}{x^3} = \frac{-2x + 2}{x^3}$$

$$= \frac{2(1 - x)}{x^3}$$

Critical values of $g(x)$: $x = 1$

Partition numbers for $g'(x)$: $x = 0, x = 1$

Sign chart for $g'(x)$:

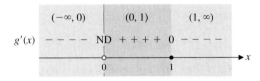

Function $f(x)$ is decreasing on $(-\infty, 0)$ and $(1, \infty)$, is increasing on $(0, 1)$, and has a local maximum at $x = 1$.

Step 3. *Analyze $g''(x)$.*

$$g'(x) = -2x^{-2} + 2x^{-3}$$

$$g''(x) = 4x^{-3} - 6x^{-4} = \frac{4}{x^3} - \frac{6}{x^4} = \frac{4x - 6}{x^4} = \frac{2(2x - 3)}{x^4}$$

Partition numbers for $g''(x)$: $x = 0, x = \frac{3}{2} = 1.5$

Sign chart for $g''(x)$:

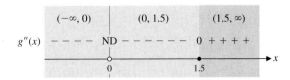

Function $g(x)$ is concave downward on $(-\infty, 0)$ and $(0, 1.5)$, is concave upward on $(1.5, \infty)$, and has an inflection point at $x = 1.5$.

Step 4. *Sketch the graph of g.* Plot key points, note that the coordinate axes are asymptotes, and sketch the graph.

x	g(x)
−10	−0.21
−1	−3
0.5	0
1	1
1.5	0.89
10	0.19

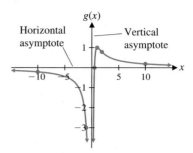

MATCHED PROBLEM 2 | Use the graphing strategy to analyze the function

$$h(x) = \frac{4x + 3}{x^2}$$

State all pertinent information and sketch the graph of h.

EXAMPLE 3 | **Graphing Strategy** Follow the steps of the graphing strategy and analyze the function $f(x) = xe^x$. State all the pertinent information and sketch the graph of f.

SOLUTION | Step 1. Analyze f(x): $f(x) = xe^x$.

(A) Domain: All real numbers

(B) y intercept: $f(0) = 0$

x intercept: $xe^x = 0$ for $x = 0$ only, since $e^x > 0$ for all x.

(C) Vertical asymptotes: None

Horizontal asymptotes: We use tables to determine the nature of the graph of f as $x \to \infty$ and $x \to -\infty$:

x	1	5	10	$\to \infty$
f(x)	2.72	742.07	220,264.66	$\to \infty$

x	−1	−5	−10	$\to -\infty$
f(x)	−0.37	−0.03	−0.000 45	$\to 0$

Step 2. *Analyze f'(x):*

$$f'(x) = x\frac{d}{dx}e^x + e^x\frac{d}{dx}x$$

$$= xe^x + e^x = e^x(x + 1)$$

Critical value of f(x):   −1
Partition number for f'(x):   −1
Sign chart for f'(x):

	$(-\infty, -1)$	$(-1, \infty)$
f'(x)	− − − − 0	+ + + +
		−1
f(x)	Decreasing	Increasing

Test Numbers	
x	f'(x)
−2	$-e^{-2}$ (−)
0	1 (+)

Thus, f(x) decreases on $(-\infty, -1)$, has a local minimum at $x = -1$, and increases on $(-1, \infty)$. (Since $e^x > 0$ for all x, we do not have to evaluate e^{-2} to conclude that $-e^{-2} < 0$ when using the test number −2.)

Step 3. *Analyze* $f''(x)$:

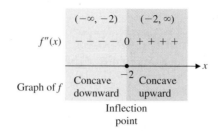

$$f''(x) = e^x \frac{d}{dx}(x+1) + (x+1)\frac{d}{dx}e^x$$

$$= e^x + (x+1)e^x = e^x(x+2)$$

Sign chart for $f''(x)$ (partition number is -2):

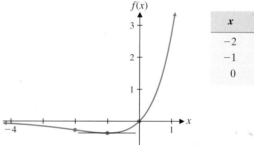

	$(-\infty, -2)$	$(-2, \infty)$
$f''(x)$	$----$ 0	$++++$

Graph of f — Concave downward -2 Concave upward

Inflection point

Test Numbers

x	$f''(x)$
-3	$-e^{-3}$ $(-)$
-1	e^{-1} $(+)$

Thus, the graph of f is concave downward on $(-\infty, -2)$, has an inflection point at $x = -2$, and is concave upward on $(-2, \infty)$.

Step 4. *Sketch the graph of f, using the information from steps 1 to 3:*

x	$f(x)$
-2	-0.27
-1	-0.37
0	0

MATCHED PROBLEM 3   Analyze the function $f(x) = xe^{-0.5x}$. State all the pertinent information and sketch the graph of f.

INSIGHT

Remember, if p is any real number, then $e^p > 0$. The sign of the exponent p does not matter. This is a useful fact to remember when you work with sign charts involving exponential forms.

Explore & Discuss **1**   Refer to the discussion of vertical asymptotes in the solution of Example 3. We used tables of values to estimate limits at infinity and determine horizontal asymptotes. In some cases, the functions involved in these limits can be written in a form that allows us to use L'Hôpital's rule.

$$\lim_{x \to -\infty} f(x) = \lim_{x \to -\infty} xe^x \qquad -\infty \cdot 0 \text{ form}$$
Rewrite as a fraction.

$$= \lim_{x \to -\infty} \frac{x}{e^{-x}} \qquad -\infty/\infty \text{ form}$$
Apply L'Hôpital's rule.

$$= \lim_{x \to -\infty} \frac{1}{-e^{-x}} \qquad \text{Simplify.}$$

$$= \lim_{x \to -\infty} (-e^x) \qquad \text{Property of } e^x$$

$$= 0$$

Use algebraic manipulation and L'Hôpital's rule to verify the value of each of the following limits:

(A) $\lim\limits_{x \to \infty} xe^{-0.5x} = 0$

(B) $\lim\limits_{x \to 0^+} x^2(\ln x - 0.5) = 0$

(C) $\lim\limits_{x \to 0^+} x \ln x = 0$

EXAMPLE 4

Graphing Strategy Let $f(x) = x^2 \ln x - 0.5x^2$. Follow the steps in the graphing strategy and analyze this function. State all the pertinent information and sketch the graph of f.

SOLUTION **Step 1.** *Analyze* $f(x)$: $f(x) = x^2 \ln x - 0.5x^2 = x^2(\ln x - 0.5)$.

(A) Domain: $(0, \infty)$

(B) y intercept: None [$f(0)$ is not defined.]

x intercept: Solve $x^2(\ln x - 0.5) = 0$

$$\ln x - 0.5 = 0 \qquad \text{or} \qquad x^2 = 0 \qquad \text{\small\textit{Discard, since 0 is not in the domain of f.}}$$

$$\ln x = 0.5 \quad \text{\small \textit{ln x = a if and only if x = } } e^a.$$

$$x = e^{0.5} \quad \text{\small \textit{x intercept}}$$

(C) Asymptotes: None. The following tables suggest the nature of the graph as $x \to 0^+$ and as $x \to \infty$:

x	0.1	0.01	0.001	$\to 0^+$
$f(x)$	−0.0280	−0.00051	−0.00007	$\to 0$

See Explore & Discuss 1(B).

x	10	100	1,000	$\to \infty$
$f(x)$	180	41,000	6,400,000	$\to \infty$

Step 2. *Analyze* $f'(x)$:

$$f'(x) = x^2 \frac{d}{dx} \ln x + (\ln x)\frac{d}{dx} x^2 - 0.5\frac{d}{dx} x^2$$

$$= x^2 \frac{1}{x} + (\ln x)\, 2x - 0.5(2x)$$

$$= x + 2x \ln x - x$$

$$= 2x \ln x$$

Critical value of $f(x)$: 1
Partition number for $f'(x)$: 1
Sign chart for $f'(x)$:

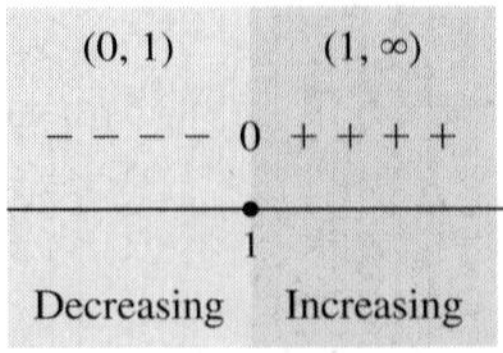

	Test Numbers	
x	$f'(x)$	
0.5	−0.2983	$(-)$
2	0.7726	$(+)$

The function $f(x)$ decreases on $(0, 1)$, has a local minimum at $x = 1$, and increases on $(1, \infty)$.

Step 3. *Analyze $f''(x)$:*

$$f''(x) \boxed{= 2x \frac{d}{dx}(\ln x) + (\ln x)\frac{d}{dx}(2x)}$$

$$= 2x\frac{1}{x} + (\ln x)\,2$$

$$= 2 + 2\ln x = 0$$

$$2\ln x = -2$$

$$\ln x = -1$$

$$x = e^{-1} \approx 0.3679$$

Sign chart for $f'(x)$ (partition number is e^{-1}):

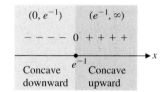

	Test Numbers	
x	$f'(x)$	
0.2	−1.2189	(−)
1	2	(+)

The graph of $f(x)$ is concave downward on $(0, e^{-1})$, has an inflection point at $x = e^{-1}$, and is concave upward on (e^{-1}, ∞).

Step 4. *Sketch the graph of f, using the information from steps 1 to 3:*

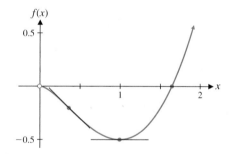

x	$f(x)$
e^{-1}	$-1.5e^{-2}$
1	−0.5
$e^{0.5}$	0

MATCHED PROBLEM 4   Analyze the function $f(x) = x \ln x$. State all pertinent information and sketch the graph of f.

■ Modeling Average Cost

EXAMPLE 5   **Average Cost**   Given the cost function $C(x) = 5{,}000 + 0.5x^2$, where x is the number of items produced, use the graphing strategy to analyze the graph of the average cost function. State all the pertinent information and sketch the graph of the average cost function. Find the marginal cost function and graph it on the same set of coordinate axes.

SOLUTION   The average cost function is

$$\overline{C}(x) = \frac{5{,}000 + 0.5x^2}{x} = \frac{5{,}000}{x} + 0.5x$$

Step 1. *Analyze $\overline{C}(x)$.*

(A) Domain: Since negative values of x do not make sense and $\overline{C}(0)$ is not defined, the domain is the set of positive real numbers.

(B) Intercepts: None

(C) Horizontal asymptote: $\dfrac{a_m x^m}{b_n x^n} = \dfrac{0.5x^2}{x} = 0.5x$

Thus, there is no horizontal asymptote.

Vertical asymptote: The line $x = 0$ is a vertical asymptote, since the denominator is 0 and the numerator is not 0 for $x = 0$.

Oblique asymptotes: If a graph approaches a line that is neither horizontal nor vertical as x approaches ∞ or $-\infty$, that line is called an **oblique asymptote**. If x is a large positive number, then $5{,}000/x$ is very small and

$$\overline{C}(x) = \frac{5{,}000}{x} + 0.5x \approx 0.5x$$

That is,

$$\lim_{x \to \infty} [\overline{C}(x) - 0.5x] = \lim_{x \to \infty} \frac{5{,}000}{x} = 0$$

This implies that the graph of $y = \overline{C}(x)$ approaches the line $y = 0.5x$ as x approaches ∞. That line is an oblique asymptote for the graph of $y = \overline{C}(x)$. [More generally, whenever $f(x) = n(x)/d(x)$ is a rational function for which the degree of $n(x)$ is 1 more than the degree of $d(x)$, we can use polynomial long division to write $f(x) = mx + b + r(x)/d(x)$, where the degree of $r(x)$ is less than the degree of $d(x)$. The line $y = mx + b$ is then an oblique asymptote for the graph of $y = f(x)$.]

Step 2. *Analyze* $\overline{C}'(x)$.

$$\begin{aligned}\overline{C}'(x) &= -\frac{5{,}000}{x^2} + 0.5 \\[1mm] &= \frac{0.5x^2 - 5{,}000}{x^2} \\[1mm] &= \frac{0.5(x - 100)(x + 100)}{x^2}\end{aligned}$$

Critical value for $\overline{C}(x)$: 100

Partition numbers for $\overline{C}'(x)$: 0 and 100

Sign chart for $\overline{C}'(x)$:

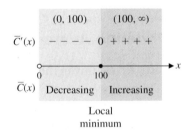

Test Numbers	
x	$\overline{C}'(x)$
50	-1.5 $(-)$
125	0.18 $(+)$

Thus, $\overline{C}(x)$ is decreasing on $(0, 100)$, is increasing on $(100, \infty)$, and has a local minimum at $x = 100$.

Step 3. *Analyze* $\overline{C}''(x)$: $\overline{C}''(x) = \dfrac{10{,}000}{x^3}$.

$\overline{C}''(x)$ is positive for all positive x; therefore, the graph of $y = \overline{C}(x)$ is concave upward on $(0, \infty)$.

Step 4. *Sketch the graph of* $\overline{C}$. The graph of $\overline{C}$ is shown in Figure 1.

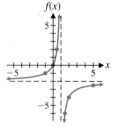

FIGURE 1

The marginal cost function is $C'(x) = x$. The graph of this linear function is also shown in Figure 1.

The graph in Figure 1 illustrates an important principle in economics:

The minimum average cost occurs when the average cost is equal to the marginal cost.

MATCHED PROBLEM 5   Given the cost function $C(x) = 1,600 + 0.25x^2$, where x is the number of items produced,

(A) Use the graphing strategy to analyze the graph of the average cost function. State all the pertinent information and sketch the graph of the average cost function. Find the marginal cost function and graph it on the same set of coordinate axes. Include any oblique asymptotes.

(B) Find the minimum average cost.

Answers to Matched Problems   **1.** Domain: All real x, except $x = 1$
y intercept: $f(0) = 0$; x intercept: 0
Horizontal asymptote: $y = -2$;   vertical asymptote: $x = 1$
Increasing on $(-\infty, 1)$ and $(1, \infty)$
Concave upward on $(-\infty, 1)$; concave downward on $(1, \infty)$

x	$f(x)$
-1	-1
0	0
$\frac{1}{2}$	2
$\frac{3}{2}$	-6
2	-4
5	$-\frac{5}{2}$

2. Domain: All real x, except $x = 0$
x intercept: $= -\dfrac{3}{4} = -0.75$

$h(0)$ is not defined
Vertical asymptote: $x = 0$ (the y axis)
Horizontal asymptote: $y = 0$ (the x axis)
Increasing on $(-1.5, 0)$
Decreasing on $(-\infty, -1.5)$ and $(0, \infty)$
Local minimum at $x = 1.5$
Concave upward on $(-2.25, 0)$ and $(0, \infty)$

Concave downward on $(-\infty, -2.25)$
Inflection point at $x = -2.25$

x	$h(x)$
-10	-0.37
-2.25	-1.19
-1.5	-1.33
-0.75	0
2	2.75
10	0.43

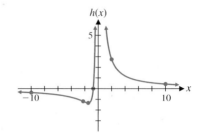

3. Domain: $(-\infty, \infty)$
y intercept: $f(0) = 0$
x intercept: $x = 0$
Horizontal asymptote: $y = 0$ (the x axis)
Increasing on $(-\infty, 2)$
Decreasing on $(2, \infty)$
Local maximum at $x = 2$
Concave downward on $(-\infty, 4)$
Concave upward on $(4, \infty)$
Inflection point at $x = 4$

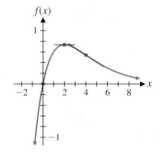

4. Domain: $(0, \infty)$
y intercept: None [$f(0)$ is not defined]
x intercept: $x = 1$
Increasing on (e^{-1}, ∞)
Decreasing on $(0, e^{-1})$
Local minimum at $x = e^{-1} \approx 0.368$
Concave upward on $(0, \infty)$

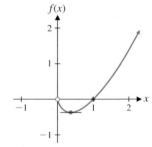

x	5	10	100	$\to \infty$
$f(x)$	8.05	23.03	460.52	$\to \infty$

x	0.1	0.01	0.001	$0.000\ 1$	$\to 0$
$f(x)$	-0.23	-0.046	$-0.006\ 9$	$-0.000\ 92$	$\to 0$

5. (A) Domain: $(0, \infty)$
Intercepts: None
Vertical asymptote: $x = 0$; oblique asymptote: $y = 0.25x$
Decreasing on $(0, 80)$; increasing on $(80, \infty)$; local minimum at $x = 80$
Concave upward on $(0, \infty)$

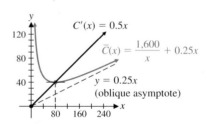

(B) Minimum average cost is 40 at $x = 80$.

Exercise 5-4

A

1. Use the graph of *f* to identify

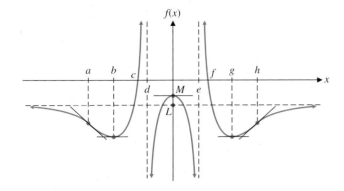

(A) the intervals on which $f'(x) < 0$
(B) the intervals on which $f'(x) > 0$
(C) the intervals on which $f(x)$ is increasing
(D) the intervals on which $f(x)$ is decreasing
(E) the *x* coordinate(s) of the point(s) where $f(x)$ has a local maximum
(F) the *x* coordinate(s) of the point(s) where $f(x)$ has a local minimum
(G) the intervals on which $f''(x) < 0$
(H) the intervals on which $f''(x) > 0$
(I) the intervals on which the graph of *f* is concave upward
(J) the intervals on which the graph of *f* is concave downward
(K) the *x* coordinate(s) of the inflection point(s)
(L) the horizontal asymptote(s)
(M) the vertical asymptote(s)

2. Repeat Problem 1 for the following graph of *f*:

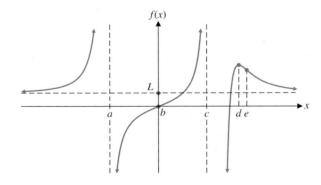

In Problems 3–10, use the given information to sketch the graph of f. Assume that f is continuous on its domain and that all intercepts are included in the table of values.

3. Domain: All real *x*; $\lim\limits_{x \to \pm\infty} f(x) = 2$

x	−4	−2	0	2	4
f(x)	0	−2	0	−2	0

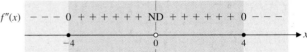

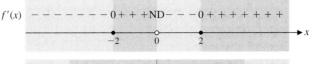

4. Domain: All real *x*; $\lim\limits_{x \to -\infty} f(x) = -3$; $\lim\limits_{x \to \infty} f(x) = 3$

x	−2	−1	0	1	2
f(x)	0	2	0	−2	0

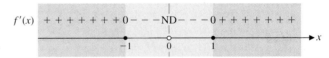

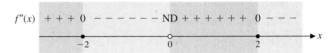

5. Domain: All real *x*, except $x = -2$; $\lim\limits_{x \to -2^-} f(x) = \infty$; $\lim\limits_{x \to -2^+} f(x) = -\infty$; $\lim\limits_{x \to \infty} f(x) = 1$

x	−4	0	4	6
f(x)	0	0	3	2

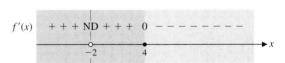

6. Domain: All real *x*, except $x = 1$; $\lim\limits_{x \to 1^-} f(x) = \infty$; $\lim\limits_{x \to 1^+} f(x) = \infty$; $\lim\limits_{x \to \infty} f(x) = -2$

x	−4	−2	0	2
f(x)	0	−2	0	0

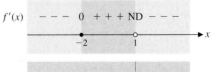

7. Domain: All real x, except $x = -1$;
$f(-3) = 2, f(-2) = 3, f(0) = -1, f(1) = 0$;
$f'(x) > 0$ on $(-\infty, -1)$ and $(-1, \infty)$;
$f''(x) > 0$ on $(-\infty, -1)$; $f''(x) < 0$ on $(-1, \infty)$;
vertical asymptote: $x = -1$;
horizontal asymptote: $y = 1$

8. Domain: All real x, except $x = 1$;
$f(0) = -2, f(2) = 0$;
$f'(x) < 0$ on $(-\infty, 1)$ and $(1, \infty)$;
$f''(x) < 0$ on $(-\infty, 1)$;
$f''(x) > 0$ on $(1, \infty)$;
vertical asymptote: $x = 1$;
horizontal asymptote: $y = -1$

9. Domain: All real x, except $x = -2$ and $x = 2$;
$f(-3) = -1, f(0) = 0, f(3) = 1$;
$f'(x) < 0$ on $(-\infty, -2)$ and $(2, \infty)$;
$f'(x) > 0$ on $(-2, 2)$;
$f''(x) < 0$ on $(-\infty, -2)$ and $(-2, 0)$;
$f''(x) > 0$ on $(0, 2)$ and $(2, \infty)$;
vertical asymptotes: $x = -2$ and $x = 2$;
horizontal asymptote: $y = 0$

10. Domain: All real x, except $x = -1$ and $x = 1$;
$f(-2) = 1, f(0) = 0, f(2) = 1$;
$f'(x) > 0$ on $(-\infty, -1)$ and $(0, 1)$;
$f'(x) < 0$ on $(-1, 0)$ and $(1, \infty)$;
$f''(x) > 0$ on $(-\infty, -1), (-1, 1)$, and $(1, \infty)$;
vertical asymptotes: $x = -1$ and $x = 1$;
horizontal asymptote: $y = 0$

B *In Problems 11–56, summarize the pertinent information obtained by applying the graphing strategy and sketch the graph of $y = f(x)$.*

11. $f(x) = \dfrac{x + 3}{x - 3}$

12. $f(x) = \dfrac{2x - 4}{x + 2}$

13. $f(x) = \dfrac{x}{x - 2}$

14. $f(x) = \dfrac{2 + x}{3 - x}$

15. $f(x) = 5 + 5e^{-0.1x}$

16. $f(x) = 3 + 7e^{-0.2x}$

17. $f(x) = 5xe^{-0.2x}$

18. $f(x) = 10xe^{-0.1x}$

19. $f(x) = \ln(1 - x)$

20. $f(x) = \ln(2x + 4)$

21. $f(x) = x - \ln x$

22. $f(x) = \ln(x^2 + 4)$

23. $f(x) = \dfrac{x}{x^2 - 4}$

24. $f(x) = \dfrac{1}{x^2 - 4}$

25. $f(x) = \dfrac{1}{1 + x^2}$

26. $f(x) = \dfrac{x^2}{1 + x^2}$

27. $f(x) = \dfrac{2x}{1 - x^2}$

28. $f(x) = \dfrac{2x}{x^2 - 9}$

29. $f(x) = \dfrac{-5x}{(x - 1)^2}$

30. $f(x) = \dfrac{x}{(x - 2)^2}$

31. $f(x) = \dfrac{x^2 + 2}{x}$

32. $f(x) = \dfrac{3 - 2x}{x^2}$

33. $f(x) = \dfrac{x^2 + x - 2}{x^2}$

34. $f(x) = \dfrac{x^2 - 5x - 6}{x^2}$

35. $f(x) = \dfrac{x^2}{x - 1}$

36. $f(x) = \dfrac{x^2}{2 + x}$

37. $f(x) = \dfrac{3x^2 + 2}{x^2 - 9}$

38. $f(x) = \dfrac{2x^2 + 5}{4 - x^2}$

39. $f(x) = \dfrac{x^3}{x - 2}$

40. $f(x) = \dfrac{x^3}{4 - x}$

41. $f(x) = (3 - x)e^x$

42. $f(x) = (x - 2)e^x$

43. $f(x) = e^{-0.5x^2}$

44. $f(x) = e^{-2x^2}$

45. $f(x) = x^2 \ln x$

46. $f(x) = \dfrac{\ln x}{x}$

47. $f(x) = (\ln x)^2$

48. $f(x) = \dfrac{x}{\ln x}$

49. $f(x) = \dfrac{1}{x^2 + 2x - 8}$

50. $f(x) = \dfrac{1}{3 - 2x - x^2}$

51. $f(x) = \dfrac{x}{x^2 - 4}$

52. $f(x) = \dfrac{x}{x^2 - 36}$

53. $f(x) = \dfrac{x^3}{(x - 4)^2}$

54. $f(x) = \dfrac{x^3}{(x + 2)^2}$

55. $f(x) = \dfrac{x^3}{3 - x^2}$

56. $f(x) = \dfrac{x^3}{x^2 - 12}$

C *In Problems 57–64, show that the line $y = x$ is an oblique asymptote for the graph of $y = f(x)$, summarize all pertinent information obtained by applying the graphing strategy, and sketch the graph of $y = f(x)$.*

57. $f(x) = x + \dfrac{4}{x}$

58. $f(x) = x - \dfrac{9}{x}$

59. $f(x) = x - \dfrac{4}{x^2}$

60. $f(x) = x + \dfrac{32}{x^2}$

61. $f(x) = x - \dfrac{9}{x^3}$

62. $f(x) = x + \dfrac{27}{x^3}$

63. $f(x) = x + \dfrac{1}{x} + \dfrac{4}{x^3}$

64. $f(x) = x - \dfrac{16}{x^3}$

In Problems 65–72, summarize all pertinent information obtained by applying the graphing strategy, and sketch the graph of $y = f(x)$. [Note: These rational functions are not reduced to lowest terms.]

65. $f(x) = \dfrac{x^2 + x - 6}{x^2 - 6x + 8}$

66. $f(x) = \dfrac{x^2 + x - 6}{x^2 - x - 12}$

67. $f(x) = \dfrac{2x^2 + x - 15}{x^2 - 9}$

68. $f(x) = \dfrac{2x^2 + 11x + 14}{x^2 - 4}$

69. $f(x) = \dfrac{x^3 - 5x^2 + 6x}{x^2 - x - 2}$

70. $f(x) = \dfrac{x^3 - 5x^2 - 6x}{x^2 + 3x + 2}$

71. $f(x) = \dfrac{x^2 + x - 2}{x^2 - 2x + 1}$

72. $f(x) = \dfrac{x^2 + x - 2}{x^2 + 4x + 4}$

Applications

73. *Revenue.* The marketing research department for a computer company used a large city to test market the firm's new product. The department found that the relationship between price p (dollars per unit) and demand x (units sold per week) was given approximately by

$$p = 1{,}296 - 0.12x^2 \qquad 0 \le x \le 80$$

Thus, weekly revenue can be approximated by

$$R(x) = xp = 1{,}296x - 0.12x^3 \qquad 0 \le x \le 80$$

Graph the revenue function R.

74. *Profit.* Suppose that the cost function $C(x)$ (in dollars) for the company in Problem 73 is

$$C(x) = 830 + 396x$$

(A) Write an equation for the profit $P(x)$.

(B) Graph the profit function P.

75. *Pollution.* In Silicon Valley (California), a number of computer-related manufacturing firms were found to be contaminating underground water supplies with toxic chemicals stored in leaking underground containers. A water quality control agency ordered the companies to take immediate corrective action and to contribute to a monetary pool for testing and cleanup of the underground contamination. Suppose that the required monetary pool (in millions of dollars) for the testing and cleanup is estimated to be given by

$$P(x) = \frac{2x}{1 - x} \qquad 0 \le x < 1$$

where x is the percentage (expressed as a decimal fraction) of the total contaminant removed.

(A) Where is $P(x)$ increasing? Decreasing?

(B) Where is the graph of P concave upward? Downward?

(C) Find any horizontal and vertical asymptotes.

(D) Find the x and y intercepts.

(E) Sketch a graph of P.

76. *Employee training.* A company producing computer components has established that, on the average, a new employee can assemble $N(t)$ components per day after t days of on-the-job training, as given by

$$N(t) = \frac{100t}{t + 9} \qquad t \ge 0$$

(A) Where is $N(t)$ increasing? Decreasing?

(B) Where is the graph of N concave upward? Downward?

(C) Find any horizontal and vertical asymptotes.

(D) Find the intercepts.

(E) Sketch a graph of N.

77. *Replacement time.* An office copier has an initial price of $3,200. A maintenance/service contract costs $300 for the first year and increases $100 per year thereafter. It can be shown that the total cost of the copier (in dollars) after n years is given by

$$C(n) = 3{,}200 + 250n + 50n^2 \qquad n \ge 1$$

(A) Write an expression for the average cost per year, $\overline{C}(n)$, for n years.

(B) Graph the average cost function found in part (A).

(C) When is the average cost per year minimum? (This time is frequently referred to as the **replacement time** for this piece of equipment.)

78. *Construction costs.* The management of a manufacturing plant wishes to add a fenced-in rectangular storage yard of 20,000 square feet, using the plant building as one side of the yard (see the figure). If x is the distance (in feet) from the building to the fence parallel to the building, show that the length of the fence required for the yard is given by

$$L(x) = 2x + \frac{20{,}000}{x} \qquad x > 0$$

Storage yard

x

Figure for 78

(A) Graph L.

(B) What are the dimensions of the rectangle requiring the least amount of fencing?

79. *Average and marginal costs.* The total daily cost (in dollars) of producing x park benches is given by

$$C(x) = 1{,}000 + 5x + 0.1x^2$$

(A) Sketch the graphs of the average cost function and the marginal cost function on the same set of coordinate axes. Include any oblique asymptotes.

(B) Find the minimum average cost.

80. *Average and marginal costs.* The total daily cost (in dollars) of producing x picnic tables is given by

$$C(x) = 500 + 2x + 0.2x^2$$

(A) Sketch the graphs of the average cost function and the marginal cost function on the same set of coordinate axes. Include any oblique asymptotes.

(B) Find the minimum average cost.

81. *Minimizing average costs.* The data in the table give the total daily costs y (in dollars) of producing x pepperoni pizzas at various production levels.

Number of Pizzas x	Total Costs y
50	395
100	475
150	640
200	910
250	1,140
300	1,450

(A) Enter the data into a graphing calculator and find a quadratic regression equation for the total costs.

(B) Use the regression equation from part (A) to find the minimum average cost (to the nearest cent) and the corresponding production level (to the nearest integer).

82. *Minimizing average costs.* The data in the table give the total daily costs y (in dollars) of producing x deluxe pizzas at various production levels.

Number of Pizzas x	Total Costs y
50	595
100	755
150	1,110
200	1,380
250	1,875
300	2,410

(A) Enter the data into a graphing calculator and find a quadratic regression equation for the total costs.

(B) Use the regression equation from part (A) to find the minimum average cost (to the nearest cent) and the corresponding production level (to the nearest integer).

83. *Medicine.* A drug is injected into the bloodstream of a patient through her right arm. The concentration of the drug in the bloodstream of the left arm t hours after the injection is given by

$$C(t) = \frac{0.14t}{t^2 + 1}$$

Graph C.

84. *Physiology.* In a study on the speed of muscle contraction in frogs under various loads, researchers W. O. Fems and J. Marsh found that the speed of contraction decreases with increasing loads. More precisely, they found that the relationship between speed of contraction, S (in centimeters per second), and load w (in grams) is given approximately by

$$S(w) = \frac{26 + 0.06w}{w} \qquad w \ge 5$$

Graph S.

85. *Psychology: retention.* An experiment on retention is conducted in a psychology class. Each student in the class is given one day to memorize the same list of 30 special characters. The lists are turned in at the end of the day, and for each succeeding day for 30 days, each student is asked to turn in a list of as many of the symbols as can be recalled. Averages are taken, and it is found that

$$N(t) = \frac{5t + 20}{t} \qquad t \ge 1$$

provides a good approximation of the average number $N(t)$ of symbols retained after t days. Graph N.

Section 5-5 ABSOLUTE MAXIMA AND MINIMA

- Absolute Maxima and Minima
- Second Derivative and Extrema

We are now ready to consider one of the most important applications of the derivative: the use of derivatives to find the absolute maximum or minimum value of a function. An economist may be interested in the price or production level of a commodity that will bring a maximum profit; a doctor may be interested in the time it takes for a drug to reach its maximum concentration in the bloodstream after an injection; and a city planner might be interested in the location of heavy industry in a city in order to produce minimum pollution in residential and business areas. In this section, we develop the procedures needed to find the absolute maximum and absolute minimum values of a function.

■ Absolute Maxima and Minima

Recall that $f(c)$ is a local maximum value if $f(x) \le f(c)$ for x near c and a local minimum value if $f(x) \ge f(c)$ for x near c. Now we are interested in finding the largest and the smallest values of $f(x)$ throughout its domain.

DEFINITION   Absolute Maxima and Minima

If $f(c) \geq f(x)$ for all x in the domain of f, then $f(c)$ is called the **absolute maximum value** of f.

If $f(c) \leq f(x)$ for all x in the domain of f, then $f(x)$ is called the **absolute minimum value** of f.

Figure 1 illustrates some typical examples.

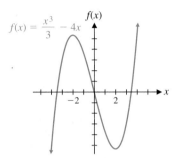

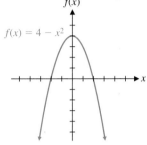

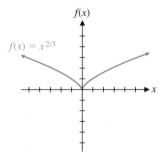

(A) No absolute maximum or minimum
One local maximum at $x = -2$
One local minimum at $x = 2$

(B) Absolute maximum at $x = 0$
No absolute minimum

(C) Absolute minimum at $x = 0$
No absolute maximum

FIGURE 1

INSIGHT

If $f(c)$ is the absolute maximum value of a function f, then $f(c)$ is obviously a "value" of f. It is common practice to omit "value" and to refer to $f(c)$ as the **absolute maximum** of f. In either usage, note that c is a value of x in the domain of f where the absolute maximum occurs. It is incorrect to refer to c as the absolute maximum. Collectively, the absolute maximum and the absolute minimum are referred to as **absolute extrema.**

Explore & Discuss 1   Functions f, g, and h, along with their graphs, are shown in Figure 2.

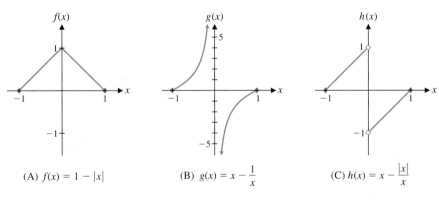

(A) $f(x) = 1 - |x|$

(B) $g(x) = x - \dfrac{1}{x}$

(C) $h(x) = x - \dfrac{|x|}{x}$

FIGURE 2

(A) Which of these functions are continuous on $[-1, 1]$?

(B) Find the absolute maximum and the absolute minimum of each function on $[-1, 1]$, if they exist, and the corresponding values of x that produce these absolute extrema.

(C) Suppose that a function p is continuous on $[-1, 1]$ and satisfies $p(-1) = 0$ and $p(1) = 0$. Sketch a possible graph for p. Does the function you graphed have an absolute maximum? An absolute minimum? Can you modify your sketch so that p does not have an absolute maximum or an absolute minimum on $[-1, 1]$?

In many practical problems, the domain of a function is restricted because of practical or physical considerations. If the domain is restricted to some closed interval, as is often the case, then Theorem 1 applies.

THEOREM 1 EXTREME VALUE THEOREM

A function f that is continuous on a closed interval $[a, b]$ has both an absolute maximum value and an absolute minimum value on that interval.

It is important to understand that the absolute maximum and minimum values depend on both the function f and the interval $[a, b]$. Figure 3 illustrates four cases.

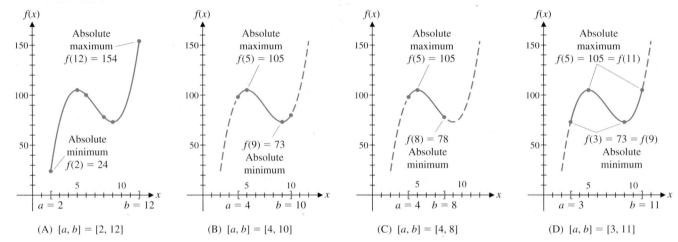

(A) $[a, b] = [2, 12]$ (B) $[a, b] = [4, 10]$ (C) $[a, b] = [4, 8]$ (D) $[a, b] = [3, 11]$

FIGURE 3 Absolute extrema for $f(x) = x^3 - 21x^2 + 135x - 170$ for various closed intervals

In all four cases illustrated in Figure 3, the absolute maximum value and absolute minimum value occur at a critical value or an endpoint. This property is generalized in Theorem 2. Note that both the absolute maximum value and the absolute minimum value are unique, but each can occur at more than one point in the interval (Fig. 3D).

THEOREM 2 LOCATING ABSOLUTE EXTREMA

Absolute extrema (if they exist) must always occur at critical values or at endpoints.

Thus, to find the absolute maximum or minimum value of a continuous function on a closed interval, we simply identify the endpoints and the critical values in the interval, evaluate the function at each, and then choose the largest and smallest values out of this group.

PROCEDURE **Finding Absolute Extrema on a Closed Interval**

Step 1. Check to make certain that f is continuous over $[a, b]$.

Step 2. Find the critical values in the interval (a, b).

Step 3. Evaluate f at the endpoints a and b and at the critical values found in step 2.

Step 4. The absolute maximum $f(x)$ on $[a, b]$ is the largest of the values found in step 3.

Step 5. The absolute minimum $f(x)$ on $[a, b]$ is the smallest of the values found in step 3.

| EXAMPLE 1 | **Finding Absolute Extrema** Find the absolute maximum and absolute minimum values of |

$$f(x) = x^3 + 3x^2 - 9x - 7$$

on each of the following intervals:

(A) $[-6, 4]$ (B) $[-4, 2]$ (C) $[-2, 2]$

| SOLUTION | (A) The function is continuous for all values of x. |

$$f'(x) = 3x^2 + 6x - 9 = 3(x - 1)(x + 3)$$

Thus, $x = -3$ and $x = 1$ are critical values in the interval $(-6, 4)$. Evaluate f at the endpoints and critical values $(-6, -3, 1,$ and $4)$, and choose the maximum and minimum from these:

$$f(-6) = -61 \quad \textit{Absolute minimum}$$
$$f(-3) = 20$$
$$f(1) = -12$$
$$f(4) = 69 \quad \textit{Absolute maximum}$$

(B) Interval: $[-4, 2]$

x	$f(x)$	
-4	13	
-3	20	*Absolute maximum*
1	-12	*Absolute minimum*
2	-5	

(C) Interval: $[-2, 2]$

x	$f(x)$	
-2	15	*Absolute maximum*
1	-12	*Absolute minimum*
2	-5	

The critical value $x = -3$ is not included in this table, because it is not in the interval $[-2, 2]$. ▬

| MATCHED PROBLEM 1 | Find the absolute maximum and absolute minimum values of |

$$f(x) = x^3 - 12x$$

on each of the following intervals:

(A) $[-5, 5]$ (B) $[-3, 3]$ (C) $[-3, 1]$

Now, suppose that we want to find the absolute maximum or minimum value of a function that is continuous on an interval that is not closed. Since Theorem 1 no longer applies, we cannot be certain that the absolute maximum or minimum value exists. Figure 4 illustrates several ways that functions can fail to have absolute extrema.

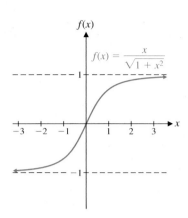

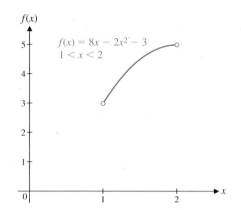

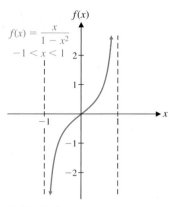

(A) No absolute extrema on $(-\infty, \infty)$:
$-1 < f(x) < 1$ for all x
$[f(x) \neq 1$ or -1 for any $x]$

(B) No absolute extrema on $(1, 2)$:
$3 < f(x) < 5$ for $x \in (1, 2)$
$[f(x) \neq 3$ or 5 for any $x \in (1, 2)]$

(C) No absolute extrema on $(-1, 1)$:
Graph has vertical asymptotes
at $x = -1$ and $x = 1$

FIGURE 4 Functions with no absolute extrema

In general, the best procedure to follow in searching for absolute extrema on an interval that is not of the form $[a, b]$ is to sketch a graph of the function. However, many applications can be solved with a new tool that does not require any graphing.

■ Second Derivative and Extrema

The second derivative can be used to classify the local extrema of a function. Suppose that f is a function satisfying $f'(c) = 0$ and $f''(c) > 0$. First, note that if $f''(c) > 0$, it follows from the properties of limits* that $f''(x) > 0$ in some interval (m, n) containing c. Thus, the graph of f must be concave upward in this interval. But this implies that $f'(x)$ is increasing in the interval. Since $f'(c) = 0$, $f'(x)$ must change from negative to positive at $x = c$, and $f(c)$ is a local minimum (see Fig. 5). Reasoning in the same fashion, we conclude that if $f'(c) = 0$ and $f''(c) < 0$, then $f(c)$ is a local maximum. Of course, it is possible that both $f'(c) = 0$ and $f''(c) = 0$. In this case, the second derivative cannot be used to determine the shape of the graph around $x = c$; $f(c)$ may be a local minimum, a local maximum, or neither.

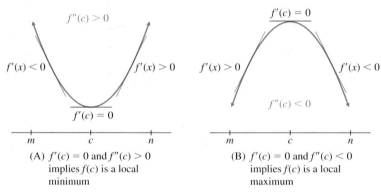

(A) $f'(c) = 0$ and $f''(c) > 0$
implies $f(c)$ is a local
minimum

(B) $f'(c) = 0$ and $f''(c) < 0$
implies $f(c)$ is a local
maximum

FIGURE 5 Second derivative and local extrema

The sign of the second derivative thus provides a simple test for identifying local maxima and minima. This test is most useful when we do not want to draw the graph of the function. If we are interested in drawing the graph and have already constructed the sign chart for $f'(x)$, the first-derivative test can be used to identify the local extrema.

---

* Actually, we are assuming that $f''(x)$ is continuous in an interval containing c. It is very unlikely that we will encounter a function for which $f''(c)$ exists but $f''(x)$ is not continuous in an interval containing c.

RESULT

Second-Derivative Test

Let c be a critical value of $f(x)$.

$f'(c)$	$f''(c)$	Graph of f is:	$f(c)$	Example
0	+	Concave upward	Local minimum	$\smile$
0	−	Concave downward	Local maximum	$\frown$
0	0	?	Test does not apply	

EXAMPLE 2

Testing Local Extrema Find the local maxima and minima for each function. Use the second-derivative test when it applies.

(A) $f(x) = x^3 - 6x^2 + 9x + 1$

(B) $f(x) = xe^{-0.2x}$

(C) $f(x) = \frac{1}{6}x^6 - 4x^5 + 25x^4$

SOLUTION

(A) Take first and second derivatives and find critical values:

$$f(x) = x^3 - 6x^2 + 9x + 1$$
$$f'(x) = 3x^2 - 12x + 9 = 3(x - 1)(x - 3)$$
$$f''(x) = 6x - 12 = 6(x - 2)$$

Critical values are $x = 1$ and $x = 3$.

$$f''(1) = -6 < 0 \quad \text{\textit{f} has a local maximum at x = 1.}$$
$$f''(3) = 6 > 0 \quad \text{\textit{f} has a local minimum at x = 3.}$$

(B)
$$f(x) = xe^{-0.2x}$$
$$f'(x) = e^{-0.2x} + xe^{-0.2x}(-0.2)$$
$$= e^{-0.2x}(1 - 0.2x)$$
$$f''(x) = e^{-0.2x}(-0.2)(1 - 0.2x) + e^{-0.2x}(-0.2)$$
$$= e^{-0.2x}(0.04x - 0.4)$$

Critical value: $x = 1/0.2 = 5$

$$f''(5) = e^{-1}(-0.2) < 0$$

f has a local maximum at $x = 5$.

(C)
$$f(x) = \frac{1}{6}x^6 - 4x^5 + 25x^4$$
$$f'(x) = x^5 - 20x^4 + 100x^3 = x^3(x - 10)^2$$
$$f''(x) = 5x^4 - 80x^3 + 300x^2$$

Critical values are $x = 0$ and $x = 10$.

$$f''(0) = 0 \quad \text{\textit{The second-derivative test fails at both critical values, so}}$$
$$f''(10) = 0 \quad \text{\textit{the first-derivative test must be used.}}$$

Sign chart for $f'(x) = x^3(x - 10)^2$ (partition numbers are 0 and 10):

	$(-\infty, 0)$	$(0, 10)$	$(10, \infty)$
$f'(x)$	− − − − 0	+ + + + 0	+ + + +
		0	10
$f(x)$	Decreasing	Increasing	Increasing

Test Numbers	
x	$f'(x)$
−1	−121 (−)
1	81 (+)
11	1,331 (+)

From the chart, we see that $f(x)$ has a local minimum at $x = 0$ and does not have a local extremum at $x = 10$. ▬

MATCHED PROBLEM 2 | Find the local maxima and minima for each function. Use the second-derivative test when it applies.

(A) $f(x) = x^3 - 9x^2 + 24x - 10$

(B) $f(x) = e^x - 5x$

(C) $f(x) = 10x^6 - 24x^5 + 15x^4$

INSIGHT

The second-derivative test does not apply if $f''(c) = 0$ or if $f''(c)$ is not defined. As Example 2C illustrates, if $f''(c) = 0$, then $f(c)$ may or may not be a local extremum. Some other method, such as the first-derivative test, must be used when $f''(c) = 0$ or $f''(c)$ does not exist.

The solution of many optimization problems involves searching for an absolute extremum. If the function in question has only one critical value, then the second-derivative test not only classifies the local extremum, but also guarantees that the local extremum is, in fact, the absolute extremum.

THEOREM 3 **SECOND-DERIVATIVE TEST FOR ABSOLUTE EXTREMUM**

Let f be continuous on an interval I with only one critical value c in I.

If $f'(c) = 0$ and $f''(c) > 0$, then $f(c)$ is the absolute minimum of f on I.

If $f'(c) = 0$ and $f''(c) < 0$, then $f(c)$ is the absolute maximum of f on I.

Since the second-derivative test cannot be applied when $f''(c) = 0$ or $f''(c)$ does not exist, Theorem 3 makes no mention of these cases.

EXAMPLE 3 | **Finding an Absolute Extremum on an Open Interval** Find the absolute minimum value of each function on $(0, \infty)$.

(A) $f(x) = x + \dfrac{4}{x}$

(B) $f(x) = (\ln x)^2 - 3 \ln x$

SOLUTION (A) $f(x) = x + \dfrac{4}{x}$

$$f'(x) = 1 - \frac{4}{x^2} = \frac{x^2 - 4}{x^2} = \frac{(x-2)(x+2)}{x^2} \quad \text{Critical values are x = -2 and x = 2.}$$

$$f''(x) = \frac{8}{x^3}$$

The only critical value in the interval $(0, \infty)$ is $x = 2$. Since $f'(2) = 1 > 0$, $f(2) = 4$ is the absolute minimum value of f on $(0, \infty)$.

(B)　　$f(x) = (\ln x)^2 - 3 \ln x$

$$f'(x) = (2 \ln x)\frac{1}{x} - \frac{3}{x} = \frac{2 \ln x - 3}{x} \qquad \text{Critical value is } x = e^{3/2}.$$

$$f''(x) = \frac{x\frac{2}{x} - (2 \ln x - 3)}{x^2} = \frac{5 - 2 \ln x}{x^2}$$

The only critical value in the interval $(0, \infty)$ is $x = e^{3/2}$. Since $f''(e^{3/2}) = 2/e^3 > 0$, $f(e^{3/2}) = -2.25$ is the absolute minimum value of f on $(0, \infty)$.　　■

MATCHED PROBLEM 3　Find the absolute maximum value of each function on $(0, \infty)$.

(A) $f(x) = 12 - x - \dfrac{5}{x}$　　(B) $f(x) = 5 \ln x - x$

Answers to Matched Problems

1. (A) Absolute maximum: $f(5) = 65$; absolute minimum: $f(-5) = -65$
   (B) Absolute maximum: $f(-2) = 16$; absolute minimum: $f(2) = -16$
   (C) Absolute maximum: $f(-2) = 16$; absolute minimum: $f(1) = -11$
2. (A) $f(2)$ is a local maximum; $f(4)$ is a local minimum.
   (B) $f(\ln 5) = 5 - 5 \ln 5$ is a local minimum.
   (C) $f(0)$ is a local minimum; there is no local extremum at $x = 1$.
3. (A) $f(\sqrt{5}) = 12 - 2\sqrt{5}$　　(B) $f(5) = 5 \ln 5 - 5$

Exercise 5-5

A　*Problems 1–10 refer to the graph of $y = f(x)$ shown here. Find the absolute minimum and the absolute maximum over the indicated interval.*

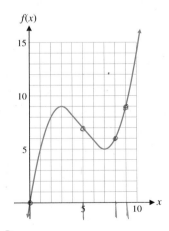

1. $[0, 10]$　　2. $[2, 8]$　　3. $[0, 8]$　　4. $[2, 10]$
5. $[1, 10]$　　6. $[0, 9]$　　7. $[1, 9]$　　8. $[0, 2]$
9. $[2, 5]$　　10. $[5, 8]$

In Problems 11–26, find the absolute maximum and minimum, if either exists, for each function.

11. $f(x) = x^2 - 2x + 3$　　12. $f(x) = x^2 + 4x - 3$
13. $f(x) = -x^2 - 6x + 9$　　14. $f(x) = -x^2 + 2x + 4$
15. $f(x) = x^3 + x$　　16. $f(x) = -x^3 - 2x$
17. $f(x) = 8x^3 - 2x^4$　　18. $f(x) = x^4 - 4x^3$

19. $f(x) = x + \dfrac{16}{x}$　　20. $f(x) = x + \dfrac{25}{x}$

21. $f(x) = \dfrac{x^2}{x^2 + 1}$　　22. $f(x) = \dfrac{1}{x^2 + 1}$

23. $f(x) = \dfrac{2x}{x^2 + 1}$　　24. $f(x) = \dfrac{-8x}{x^2 + 4}$

25. $f(x) = \dfrac{x^2 - 1}{x^2 + 1}$　　26. $f(x) = \dfrac{9 - x^2}{x^2 + 4}$

B　*In Problems 27–50, find the indicated extremum of each function on the given interval.*

27. Absolute minimum value on $[0, \infty)$ for
$$f(x) = 2x^2 - 8x + 6$$

28. Absolute maximum value on $[0, \infty)$ for
$$f(x) = 6x - x^2 + 4$$

29. Absolute maximum value on $[0, \infty)$ for
$$f(x) = 3x^2 - x^3$$

30. Absolute minimum value on $[0, \infty)$ for
$$f(x) = x^3 - 6x^2$$

31. Absolute minimum value on $[0, \infty)$ for
$$f(x) = (x + 4)(x - 2)^2$$

32. Absolute minimum value on $[0, \infty)$ for
$$f(x) = (2 - x)(x + 1)^2$$

33. Absolute maximum value on $(0, \infty)$ for
$$f(x) = 2x^4 - 8x^3$$

34. Absolute maximum value on $(0, \infty)$ for
$$f(x) = 4x^3 - 8x^4$$

35. Absolute maximum value on $(0, \infty)$ for
$$f(x) = 20 - 3x - \frac{12}{x}$$

36. Absolute minimum value on $(0, \infty)$ for
$$f(x) = 4 + x + \frac{9}{x}$$

37. Absolute maximum value on $(0, \infty)$ for
$$f(x) = 10 + 2x + \frac{64}{x^2}$$

38. Absolute maximum value on $(0, \infty)$ for
$$f(x) = 20 - 4x - \frac{250}{x^2}$$

39. Absolute minimum value on $(0, \infty)$ for
$$f(x) = x + \frac{1}{x} + \frac{30}{x^3}$$

40. Absolute minimum value on $(0, \infty)$ for
$$f(x) = 2x + \frac{5}{x} + \frac{4}{x^3}$$

41. Absolute minimum value on $(0, \infty)$ for
$$f(x) = \frac{e^x}{x^2}$$

42. Absolute maximum value on $(0, \infty)$ for
$$f(x) = \frac{x^4}{e^x}$$

43. Absolute maximum value on $(0, \infty)$ for
$$f(x) = \frac{x^3}{e^x}$$

44. Absolute minimum value on $(0, \infty)$ for
$$f(x) = \frac{e^x}{x}$$

45. Absolute maximum value on $(0, \infty)$ for
$$f(x) = 5x - 2x \ln x$$

46. Absolute minimum value on $(0, \infty)$ for
$$f(x) = 4x \ln x - 7x$$

47. Absolute maximum value on $(0, \infty)$ for
$$f(x) = x^2(3 - \ln x)$$

48. Absolute minimum value on $(0, \infty)$ for
$$f(x) = x^3(\ln x - 2)$$

49. Absolute maximum value on $(0, \infty)$ for
$$f(x) = \ln(xe^{-x})$$

50. Absolute maximum value on $(0, \infty)$ for
$$f(x) = \ln(x^2 e^{-x})$$

In Problems 51–56, find the absolute maximum and minimum, if either exists, for each function on the indicated intervals.

51. $f(x) = x^3 - 6x^2 + 9x - 6$
  (A) $[-1, 5]$ (B) $[-1, 3]$ (C) $[2, 5]$

52. $f(x) = 2x^3 - 3x^2 - 12x + 24$
  (A) $[-3, 4]$ (B) $[-2, 3]$ (C) $[-2, 1]$

53. $f(x) = (x - 1)(x - 5)^3 + 1$
  (A) $[0, 3]$ (B) $[1, 7]$ (C) $[3, 6]$

54. $f(x) = x^4 - 8x^2 + 16$
  (A) $[-1, 3]$ (B) $[0, 2]$ (C) $[-3, 4]$

55. $f(x) = x^4 - 4x^3 + 5$
  (A) $[-1, 2]$ (B) $[0, 4]$ (C) $[-1, 1]$

56. $f(x) = x^4 - 18x^2 + 32$
  (A) $[-4, 4]$ (B) $[-1, 1]$ (C) $[1, 3]$

In Problems 57–64, describe the graph of f at the given point relative to the existence of a local maximum or minimum with one of the following phrases: "Local maximum," "Local minimum," "Neither," or "Unable to determine from the given information." Assume that $f(x)$ is continuous on $(-\infty, \infty)$.

57. $(2, f(2))$ if $f'(2) = 0$ and $f''(2) > 0$

58. $(4, f(4))$ if $f'(4) = 1$ and $f''(4) < 0$

59. $(-3, f(-3))$ if $f'(-3) = 0$ and $f''(-3) = 0$

60. $(-1, f(-1))$ if $f'(-1) = 0$ and $f''(-1) < 0$

61. $(6, f(6))$ if $f'(6) = 1$ and $f''(6)$ does not exist

62. $(5, f(5))$ if $f'(5) = 0$ and $f''(5)$ does not exist

63. $(-2, f(-2))$ if $f'(-2) = 0$ and $f''(-2) < 0$

64. $(1, f(1))$ if $f'(1) = 0$ and $f''(1) > 0$

Section 5-6 OPTIMIZATION

- Area and Perimeter
- Maximizing Revenue and Profit
- Inventory Control

Now we want to use the calculus tools we have developed to solve **optimization problems**—problems that involve finding the absolute maximum value or the absolute minimum value of a function. As you work through this section, note that the statement of the problem does not usually include the function that is to be optimized. Often, it is your responsibility to find the function and then to find its absolute extremum.

■ Area and Perimeter

The techniques used to solve optimization problems are best illustrated through examples.

EXAMPLE 1 **Maximizing Area** A homeowner has $320 to spend on building a fence around a rectangular garden. Three sides of the fence will be constructed with wire fencing at a cost of $2 per linear foot. The fourth side is to be constructed with wood fencing at a cost of $6 per linear foot. Find the dimensions and the area of the largest garden that can be enclosed with $320 worth of fencing.

SOLUTION To begin, we draw a figure (Fig. 1), introduce variables, and look for relationships among the variables.

Since we don't know the dimensions of the garden, the lengths of fencing are represented by the variables x and y. The costs of the fencing materials are fixed and are thus represented by constants.

Now we look for relationships among the variables. The area of the garden is

$$A = xy$$

while the cost of the fencing is

$$C = 2y + 2x + 2y + 6x$$
$$= 8x + 4y$$

FIGURE 1

The problem states that the homeowner has $320 to spend on fencing. We make the assumption that enclosing the largest area will use all of the money available for fencing. The problem has now been reduced to

Maximize $A = xy$ subject to $8x + 4y = 320$

Before we can use calculus techniques to find the maximum area A, we must express A as a function of a single variable. We use the cost equation to eliminate one of the variables in the expression for the area (we choose to eliminate y—either will work).

$$8x + 4y = 320$$
$$4y = 320 - 8x$$
$$y = 80 - 2x$$
$$A = xy = x(80 - 2x) = 80x - 2x^2$$

Now we consider the permissible values of x. Because x is one of the dimensions of a rectangle, x must satisfy

$$x \geq 0 \quad \text{Length is always nonnegative.}$$

And because $y = 80 - 2x$ is also a dimension of a rectangle, y must satisfy

$$y = 80 - 2x \geq 0 \quad \text{Width is always nonnegative.}$$
$$80 \geq 2x$$
$$40 \geq x \quad \text{or} \quad x \leq 40$$

We summarize the preceding discussion by stating the following model for this optimization problem:

Maximize $A(x) = 80x - 2x^2$ for $0 \leq x \leq 40$

Next, we find any critical values of A:

$$A'(x) = 80 - 4x = 0$$
$$80 = 4x$$
$$x = \frac{80}{4} = 20 \quad \text{Critical value}$$

TABLE 1

x	$A(x)$
0	0
20	800
40	0

Since $A(x)$ is continuous on $[0, 40]$, the absolute maximum value of A, if it exists, must occur at a critical value or an endpoint. Evaluating A at these values (Table 1), we see that the maximum area is 800 when

$$x = 20 \quad \text{and} \quad y = 80 - 2(20) = 40$$

Finally, we must answer the questions posed in the problem. The dimensions of the garden with the maximum area of 800 square feet are 20 feet by 40 feet, with one 20-foot side with wood fencing.
◼

MATCHED PROBLEM 1   Repeat Example 1 if the wood fencing costs $8 per linear foot and all other information remains the same.

_____

We summarize the steps we followed in the solution to Example 1 in the following box:

> **PROCEDURE** **Strategy for Solving Optimization Problems**
>
> **Step 1.** Introduce variables, look for relationships among the variables, and construct a mathematical model of the form
>
> Maximize (or minimize) $f(x)$ on the interval I
>
> **Step 2.** Find the critical values of $f(x)$.
>
> **Step 3.** Use the procedures developed in Section 5-5 to find the absolute maximum (or minimum) value of $f(x)$ on the interval I and the value(s) of x where this occurs.
>
> **Step 4.** Use the solution to the mathematical model to answer all the questions asked in the problem.

EXAMPLE 2   **Minimizing Perimeter**   Refer to Example 1. The homeowner judges that an area of 800 square feet for the garden is too small and decides to increase the area to 1,250 square feet. What is the minimum cost of building a fence that will enclose a garden with area 1,250 square feet? What are the dimensions of this garden? Assume that the cost of fencing remains unchanged.

SOLUTION   Refer to Figure 1 in the solution for Example 1. This time we want to minimize the cost of the fencing that will enclose 1,250 square feet. The problem can be expressed as

$$\text{Minimize} \quad C = 8x + 4y \quad \text{subject to} \quad xy = 1,250$$

Since x and y represent distances, we know that $x \geq 0$ and $y \geq 0$. But neither variable can equal 0, because their product must be 1,250.

$$xy = 1,250 \qquad \textit{Solve the area equation for y.}$$

$$y = \frac{1,250}{x}$$

$$C(x) = 8x + 4\frac{1,250}{x} \qquad \textit{Substitute for y in the cost equation.}$$

$$= 8x + \frac{5,000}{x} \qquad x > 0$$

The model for this problem is

$$\text{Minimize } C(x) = 8x + \frac{5,000}{x} \qquad \text{for } x > 0$$

$$= 8x + 5,000x^{-1}$$

$$C'(x) = 8 - 5{,}000x^{-2}$$

$$= 8 - \frac{5{,}000}{x^2} = 0$$

$$8 = \frac{5{,}000}{x^2}$$

$$x^2 = \frac{5{,}000}{8} = 625$$

$$x = \sqrt{625} = 25 \qquad \text{\textit{The negative square root is discarded,}}$$
$$\text{\textit{since x > 0.}}$$

We use the second derivative to determine the behavior at $x = 25$.

$$\boxed{C'(x) = 8 - 5{,}000x^{-2}}$$

$$C''(x) = 0 + 10{,}000x^{-3} = \frac{10{,}000}{x^3}$$

$$C''(25) = \frac{10{,}000}{25^3} = 0.64 > 0$$

The second-derivative test shows that $C(x)$ has a local minimum at $x = 25$, and since $x = 25$ is the only critical value of $x > 0$, $C(25)$ must be the absolute minimum value of $C(x)$ for $x > 0$. When $x = 25$, the cost is

$$C(25) = 8(25) + \frac{5{,}000}{25} = 200 + 200 = \$400$$

and

$$y = \frac{1{,}250}{25} = 50$$

The minimal cost for enclosing a 1,250-square-foot garden is $400, and the dimensions are 25 feet by 50 feet, with one 25-foot side with wood fencing. ▬

MATCHED PROBLEM 2   Repeat Example 2 if the homeowner wants to enclose an 1,800-square-foot garden and all other data remain unchanged.

INSIGHT

The restrictions on the variables in the solutions of Examples 1 and 2 are typical of problems involving areas or perimeters (or the cost of the perimeter):

$$8x + 4y = 320 \qquad \text{\textit{Cost of fencing (Example 1)}}$$
$$xy = 1{,}250 \qquad \text{\textit{Area of garden (Example 2)}}$$

The equation in Example 1 restricts the values of x to

$$0 \le x \le 40 \qquad \text{or} \qquad [0, 40]$$

The endpoints are included in the interval for our convenience (a closed interval is easier to work with than an open one). The area function is defined at each endpoint, so it does no harm to include them.

The equation in Example 2 restricts the values of x to

$$x > 0 \qquad \text{or} \qquad (0, \infty)$$

Neither endpoint can be included in this interval. We cannot include 0 because the area is not defined when $x = 0$, and we can never include ∞ as an endpoint. Remember, ∞ is not a number, but a symbol which denotes that the interval is unbounded.

■ Maximizing Revenue and Profit

EXAMPLE 3

Maximizing Revenue An office supply company sells x mechanical pencils per year at $\$p$ per pencil. The price–demand equation for these pencils is $p = 10 - 0.001x$. What price should the company charge for the pencils to maximize revenue? What is the maximum revenue?

SOLUTION

$$\text{Revenue} = \text{price} \times \text{demand}$$
$$R(x) = (10 - 0.001x)x$$
$$= 10x - 0.001x^2$$

Both price and demand must be nonnegative, so

$$x \geq 0 \quad \text{and} \quad p = 10 - 0.001x \geq 0$$
$$10 \geq 0.001x$$
$$10{,}000 \geq x$$

The mathematical model for this problem is

$$\text{Maximize} \quad R(x) = 10x - 0.001x^2 \qquad 0 \leq x \leq 10{,}000$$
$$R'(x) = 10 - 0.002x$$
$$10 - 0.002x = 0$$
$$10 = 0.002x$$
$$x = \frac{10}{0.002} = 5{,}000 \quad \text{\textit{Critical value}}$$

Use the second-derivative test for absolute extrema:

$$R''(x) = -0.002 < 0 \quad \text{for all } x$$
$$\text{Max} \quad R(x) = R(5{,}000) = \$25{,}000$$

When the demand is $x = 5{,}000$, the price is

$$10 - 0.001(5{,}000) = \$5 \quad \textit{p = 10 - 0.001x}$$

The company will realize a maximum revenue of $\$25{,}000$ when the price of a pencil is $\$5$. ■

MATCHED PROBLEM 3

An office supply company sells x heavy-duty paper shredders per year at $\$p$ per shredder. The price–demand equation for these shredders is

$$p = 300 - \frac{x}{30}$$

What price should the company charge for the shredders to maximize revenue? What is the maximum revenue?

EXAMPLE 4

Maximizing Profit The total annual cost of manufacturing x mechanical pencils for the office supply company in Example 3 is

$$C(x) = 5{,}000 + 2x$$

What is the company's maximum profit? What should the company charge for each pencil, and how many pencils should be produced?

SOLUTION Using the revenue model in Example 3, we have

$$\text{Profit} = \text{Revenue} - \text{Cost}$$
$$P(x) = R(x) - C(x)$$
$$= 10x - 0.001x^2 - 5{,}000 - 2x$$
$$= 8x - 0.001x^2 - 5{,}000$$

The mathematical model for profit is

Maximize $P(x) = 8x - 0.001x^2 - 5{,}000$ $0 \le x \le 10{,}000$

(The restrictions on x come from the revenue model in Example 3.)

$$P'(x) = 8 - 0.002x = 0$$
$$8 = 0.002x$$
$$x = \frac{8}{0.002} = 4{,}000 \quad \text{Critical value}$$
$$P''(x) = -0.002 < 0 \quad \text{for all } x$$

Since $x = 4{,}000$ is the only critical value and $P''(x) < 0$,

$$\text{Max } P(x) = P(4{,}000) = \$11{,}000$$

Using the price–demand equation from Example 3 with $x = 4{,}000$, we find that

$$p = 10 - 0.001(4{,}000) = \$6 \quad p = 10 - 0.001x$$

Thus, a maximum profit of \$11,000 is realized when 4,000 pencils are manufactured annually and sold for \$6 each.

The results in Examples 3 and 4 are illustrated in Figure 2.

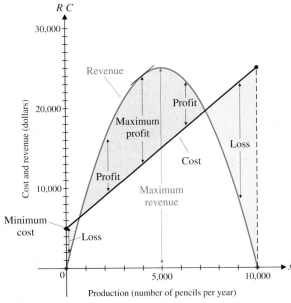

FIGURE 2

INSIGHT

In Figure 2, notice that the maximum revenue and the maximum profit occur at different production levels. The profit is maximum when

$$P'(x) = R'(x) - C'(x) = 0$$

that is, when the marginal revenue is equal to the marginal cost. Notice that the slopes of the revenue function and the cost function are the same at this production level.

MATCHED PROBLEM 4

The annual cost of manufacturing x paper shredders for the office supply company in Matched Problem 3 is $C(x) = 90{,}000 + 30x$. What is the company's maximum profit? What should it charge for each shredder, and how many shredders should it produce?

EXAMPLE 5

Maximizing Profit The government decides to tax the company in Example 4 $2 for each pencil produced. Taking into account this additional cost, how many pencils should the company manufacture each week to maximize its weekly profit? What is the maximum weekly profit? How much should the company charge for the pencils to realize the maximum weekly profit?

SOLUTION The tax of $2 per unit changes the company's cost equation:

$$C(x) = \text{original cost} + \text{tax}$$
$$= 5{,}000 + 2x + 2x$$
$$= 5{,}000 + 4x$$

The new profit function is

$$P(x) = R(x) - C(x)$$
$$= 10x - 0.001x^2 - 5{,}000 - 4x$$
$$= 6x - 0.001x^2 - 5{,}000$$

Thus, we must solve the following equation:

$$\text{Maximize} \quad P(x) = 6x - 0.001x^2 - 5{,}000 \qquad 0 \le x \le 10{,}000$$
$$P'(x) = 6 - 0.002x$$
$$6 - 0.002x = 0$$
$$x = 3{,}000 \quad \text{\small Critical value}$$
$$P''(x) = -0.002 < 0 \quad \text{for all } x$$
$$\text{Max } P(x) = P(3{,}000) = \$4{,}000$$

Using the price–demand equation (Example 3) with $x = 3{,}000$, we find that

$$p = 10 - 0.001(3{,}000) = \$7 \quad \text{\small $p = 10 - 0.001x$}$$

Thus, the company's maximum profit is $4,000 when 3,000 pencils are produced and sold weekly at a price of $7.

Even though the tax caused the company's cost to increase by $2 per pencil, the price that the company should charge to maximize its profit increases by only $1. The company must absorb the other $1, with a resulting decrease of $7,000 in maximum profit. ■

MATCHED PROBLEM 5

The government decides to tax the office supply company in Matched Problem 4 $20 for each shredder produced. Taking into account this additional cost, how many shredders should the company manufacture each week to maximize its weekly profit? What is the maximum weekly profit? How much should the company charge for the shredders to realize the maximum weekly profit?

EXAMPLE 6

Maximizing Revenue When a management training company prices its seminar on management techniques at $400 per person, 1,000 people will attend the seminar. The company estimates that for each $5 reduction in the price, an additional 20 people will attend the seminar. How much should the company charge for the seminar in order to maximize its revenue? What is the maximum revenue?

SOLUTION Let x represent the number of $5 price reductions. Then

$$400 - 5x = \text{price per customer}$$
$$1{,}000 + 20x = \text{number of customers}$$
$$\text{Revenue} = (\text{price per customer})(\text{number of customers})$$
$$R(x) = (400 - 5x) \times (1{,}000 + 20x)$$

Since price cannot be negative, we have

$$400 - 5x \geq 0$$
$$400 \geq 5x$$
$$80 \geq x \quad \text{or} \quad x \leq 80$$

A negative value of x would result in a price increase. Since the problem is stated in terms of price reductions, we must restrict x so that $x \geq 0$. Putting all this together, we have the following model:

$$\text{Maximize} \quad R(x) = (400 - 5x)(1{,}000 + 20x) \quad \text{for } 0 \leq x \leq 80$$
$$R(x) = 400{,}000 + 3{,}000x - 100x^2$$
$$R'(x) = 3{,}000 - 200x = 0$$
$$3{,}000 = 200x$$
$$x = 15 \quad \textit{Critical value}$$

Since $R(x)$ is continuous on the interval $[0, 80]$, we can determine the behavior of the graph by constructing a table. Table 2 shows that $R(15) = \$422{,}500$ is the absolute maximum revenue. The price of attending the seminar at $x = 15$ is $400 - 5(15) = \$325$. The company should charge $\$325$ for the seminar in order to receive a maximum revenue of $\$422{,}500$. ▬

TABLE 2

x	$R(x)$
0	400,000
15	422,500
80	0

MATCHED PROBLEM 6   A walnut grower estimates from past records that if 20 trees are planted per acre, each tree will average 60 pounds of nuts per year. If, for each additional tree planted per acre, the average yield per tree drops 2 pounds, how many trees should be planted to maximize the yield per acre? What is the maximum yield?

Explore & Discuss **1**

In Example 6, letting x be the number of $\$5$ price reductions produced a simple and direct solution to the problem. However, this is not the most obvious choice for a variable. Suppose that we proceed as follows:

Let x be the new price and let y be the attendance at that price level. Then the total revenue is given by xy.

(A) Find y when $x = 400$ and when $x = 395$. Find the equation of the line through these two points.

(B) Use the equation from part (A) to express the revenue in terms of either x or y, and use the expression you came up with to solve Example 6.

(C) Compare this method of solution to the one used in Example 6 with respect to ease of comprehension and ease of computation.

EXAMPLE 7   **Maximizing Revenue**   After additional analysis, the management training company in Example 6 decides that its estimate of attendance was too high. Its new estimate is that only 10 additional people will attend the seminar for each $\$5$ decrease in price. All other information remains the same. How much should the company charge for the seminar now in order to maximize revenue? What is the new maximum revenue?

SOLUTION   Under the new assumption, the model becomes

$$\text{Maximize} \quad R(x) = (400 - 5x)(1{,}000 + 10x) \quad 0 \leq x \leq 80$$
$$= 400{,}000 - 1{,}000x - 50x^2$$
$$R'(x) = -1{,}000 - 100x = 0$$
$$-1{,}000 = 100x$$
$$x = -10 \quad \textit{Critical value}$$

TABLE 3

x	$R(x)$
0	400,000
80	0

Note that $x = -10$ is not in the interval $[0, 80]$. Since $R(x)$ is continuous on $[0, 80]$, we can use a table to find the absolute maximum revenue. Table 3 shows that the maximum revenue is $R(0) = \$400{,}000$. The company should leave the price at $\$400$. Any $\$5$ decreases in price will lower the revenue. ▬

MATCHED PROBLEM 7   After further analysis, the walnut grower in Matched Problem 6 determines that each additional tree planted will reduce the average yield by 4 pounds. All other information remains the same. How many additional trees per acre should the grower plant now in order to maximize the yield? What is the new maximum yield?

INSIGHT

The solution in Example 7 is called an **endpoint solution,** because the optimal value occurs at the endpoint of an interval rather than at a critical value in the interior of the interval. It is always important to verify that the optimal value has been found.

■ Inventory Control

EXAMPLE 8   **Inventory Control** A recording company anticipates that there will be a demand for 20,000 copies of a certain compact disk (CD) during the next year. It costs the company $0.50 to store a CD for one year. Each time it must make additional CDs, it costs $200 to set up the equipment. How many CDs should the company make during each production run to minimize its total storage and setup costs?

SOLUTION   This type of problem is called an **inventory control problem.** One of the basic assumptions made in such problems is that the demand is uniform. For example, if there are 250 working days in a year, the daily demand would be 20,000 ÷ 250 = 80 CDs. The company could decide to produce all 20,000 CDs at the beginning of the year. This would certainly minimize the setup costs, but would result in very large storage costs. At the other extreme, the company could produce 80 CDs each day. This would minimize the storage costs, but would result in very large setup costs. Somewhere between these two extremes is the optimal solution that will minimize the total storage and setup costs. Let

$$x = \text{number of CDs manufactured during each production run}$$
$$y = \text{number of production runs}$$

It is easy to see that the total setup cost for the year is $200y$, but what is the total storage cost? If the demand is uniform, the number of CDs in storage between production runs will decrease from x to 0, and the average number in storage each day is $x/2$. This result is illustrated in Figure 3.

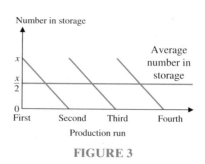

FIGURE 3

Since it costs $0.50 to store a CD for one year, the total storage cost is $0.5(x/2) = 0.25x$ and the total cost is

$$\text{total cost} = \text{setup cost} + \text{storage cost}$$
$$C = 200y + 0.25x$$

In order to write the total cost C as a function of one variable, we must find a relationship between x and y. If the company produces x CDs in each of y production runs, the total number of CDs produced is xy. Thus,

$$xy = 20{,}000$$
$$y = \frac{20{,}000}{x}$$

Certainly, x must be at least 1 and cannot exceed 20,000. Therefore, we must solve the following equation:

$$\text{Minimize} \quad C(x) = 200\left(\frac{20{,}000}{x}\right) + 0.25x \quad 1 \le x \le 20{,}000$$

$$C(x) = \frac{4{,}000{,}000}{x} + 0.25x$$

$$C'(x) = -\frac{4{,}000{,}000}{x^2} + 0.25$$

$$-\frac{4{,}000{,}000}{x^2} + 0.25 = 0$$

$$x^2 = \frac{4{,}000{,}000}{0.25}$$

$$x^2 = 16{,}000{,}000 \quad \textit{−4,000 is not a critical value, since}$$

$$x = 4{,}000 \quad \textit{1 ≤ x ≤ 20,000.}$$

$$C''(x) = \frac{8{,}000{,}000}{x^3} > 0 \quad \text{for } x \in (1, 20{,}000)$$

Thus,

$$\text{Min } C(x) = C(4{,}000) = 2{,}000$$
$$y = \frac{20{,}000}{4{,}000} = 5$$

The company will minimize its total cost by making 4,000 CDs five times during the year. ∎

MATCHED PROBLEM 8   Repeat Example 8 if it costs \$250 to set up a production run and \$0.40 to store a CD for one year.

Answers to Matched Problems

1. The dimensions of the garden with the maximum area of 640 square feet are 16 feet by 40 feet, with one 16-foot side with wood fencing.

2. The minimal cost for enclosing a 1,800-square-foot garden is \$480, and the dimensions are 30 feet by 60 feet, with one 30-foot side with wood fencing.

3. The company will realize a maximum revenue of \$675,000 when the price of a shredder is \$150.

4. A maximum profit of \$456,750 is realized when 4,050 shredders are manufactured annually and sold for \$165 each.

5. A maximum profit of \$378,750 is realized when 3,075 shredders are manufactured annually and sold for \$175 each.

6. The maximum yield is 1,250 pounds per acre when 5 additional trees are planted on each acre.

7. The maximum yield is 1,200 pounds when no additional trees are planted.

8. The company should produce 5,000 CDs four times a year.

Exercise 5-6

Preliminary word problems:

1. How would you divide a 10-inch line so that the product of the two lengths is a maximum?

2. What quantity should be added to 5 and subtracted from 5 to produce the maximum product of the results?

3. Find two numbers whose difference is 30 and whose product is a minimum.

4. Find two positive numbers whose sum is 60 and whose product is a maximum.

5. Find the dimensions of a rectangle with perimeter 100 centimeters that has maximum area. Find the maximum area.

6. Find the dimensions of a rectangle of area 225 square centimeters that has the least perimeter. What is the perimeter?

Problems 7–10 refer to a rectangular area enclosed by a fence that costs $B per foot. Discuss the existence of a solution and the economical implications of each optimization problem.

7. Given a fixed area, minimize the cost of the fencing.

8. Given a fixed area, maximize the cost of the fencing.

9. Given a fixed amount to spend on fencing, maximize the enclosed area.

10. Given a fixed amount to spend on fencing, minimize the enclosed area.

11. *Maximum revenue and profit.* A company manufactures and sells x videophones per week. The weekly price–demand and cost equations are, respectively,

$$p = 500 - 0.5x \quad \text{and} \quad C(x) = 20,000 + 135x$$

(A) What price should the company charge for the phones, and how many phones should be produced to maximize the weekly revenue? What is the maximum weekly revenue?

(B) What is the maximum weekly profit? How much should the company charge for the phones, and how many phones should be produced to realize the maximum weekly profit?

12. *Maximum revenue and profit.* A company manufactures and sells x digital cameras per week. The weekly price–demand and cost equations are, respectively,

$$p = 400 - 0.4x \quad \text{and} \quad C(x) = 2,000 + 160x$$

(A) What price should the company charge for the cameras, and how many cameras should be produced to maximize the weekly revenue? What is the maximum revenue?

(B) What is the maximum weekly profit? How much should the company charge for the cameras, and how many cameras should be produced to realize the maximum weekly profit?

13. *Maximum revenue and profit.* A company manufactures and sells x television sets per month. The monthly cost and price–demand equations are

$$C(x) = 72,000 + 60x$$

$$p = 200 - \frac{x}{30} \quad 0 \le x \le 6,000$$

(A) Find the maximum revenue.

(B) Find the maximum profit, the production level that will realize the maximum profit, and the price the company should charge for each television set.

(C) If the government decides to tax the company $5 for each set it produces, how many sets should the company manufacture each month to maximize its profit? What is the maximum profit? What should the company charge for each set?

14. *Maximum revenue and profit.* Repeat Problem 13 for

$$C(x) = 60,000 + 60x$$

$$p = 200 - \frac{x}{50} \quad 0 \le x \le 10,000$$

15. *Maximum profit.* The following table contains price–demand and total cost data for the production of radial arm saws, where p is the wholesale price (in dollars) of a saw for an annual demand of x saws and C is the total cost (in dollars) of producing x saws:

x	p	C
950	240	130,000
1,200	210	150,000
1,800	160	180,000
2,050	120	190,000

(A) Find a quadratic regression equation for the price–demand data, using x as the independent variable.

(B) Find a linear regression equation for the cost data, using x as the independent variable.

(C) What is the maximum profit? What is the wholesale price per saw that should be charged to realize the maximum profit? Round answers to the nearest dollar.

16. *Maximum profit.* The following table contains price–demand and total cost data for the production of airbrushes, where p is the wholesale price (in dollars) of an airbrush for an annual demand of x airbrushes and C is the total cost (in dollars) of producing x airbrushes:

x	p	C
2,300	98	145,000
3,300	84	170,000
4,500	67	190,000
5,200	51	210,000

(A) Find a quadratic regression equation for the price–demand data, using x as the independent variable.

(B) Find a linear regression equation for the cost data, using x as the independent variable.

(C) What is the maximum profit? What is the wholesale price per airbrush that should be charged to realize the maximum profit? Round answers to the nearest dollar.

17. *Maximum revenue.* A deli sells 640 sandwiches per day at a price of $8 each.

   (A) A market survey shows that for every $0.10 reduction in price, 40 more sandwiches will be sold. How much should the deli charge for a sandwich in order to maximize revenue?

   (B) A different market survey shows that for every $0.20 reduction in the original $8 price, 15 more sandwiches will be sold. Now how much should the deli charge for a sandwich in order to maximize revenue?

18. *Maximum revenue.* A university student center sells 1,600 cups of coffee per day at a price of $2.40.

   (A) A market survey shows that for every $0.05 reduction in price, 50 more cups of coffee will be sold. How much should the student center charge for a cup of coffee in order to maximize revenue?

   (B) A different market survey shows that for every $0.10 reduction in the original $2.40 price, 60 more cups of coffee will be sold. Now how much should the student center charge for a cup of coffee in order to maximize revenue?

19. *Car rental.* A car rental agency rents 200 cars per day at a rate of $30 per day. For each $1 increase in rate, 5 fewer cars are rented. At what rate should the cars be rented to produce the maximum income? What is the maximum income?

20. *Rental income.* A 300-room hotel in Las Vegas is filled to capacity every night at $80 a room. For each $1 increase in rent, 3 fewer rooms are rented. If each rented room costs $10 to service per day, how much should the management charge for each room to maximize gross profit? What is the maximum gross profit?

21. *Agriculture.* A commercial cherry grower estimates from past records that if 30 trees are planted per acre, each tree will yield an average of 50 pounds of cherries per season. If, for each additional tree planted per acre (up to 20), the average yield per tree is reduced by 1 pound, how many trees should be planted per acre to obtain the maximum yield per acre? What is the maximum yield?

22. *Agriculture.* A commercial pear grower must decide on the optimum time to have fruit picked and sold. If the pears are picked now, they will bring 30¢ per pound, with each tree yielding an average of 60 pounds of salable pears. If the average yield per tree increases 6 pounds per tree per week for the next 4 weeks, but the price drops 2¢ per pound per week, when should the pears be picked to realize the maximum return per tree? What is the maximum return?

23. *Manufacturing.* A candy box is to be made out of a piece of cardboard that measures 8 by 12 inches. Squares of equal size will be cut out of each corner, and then the ends and sides will be folded up to form a rectangular box. What size square should be cut from each corner to obtain a maximum volume?

24. *Packaging.* A parcel delivery service will deliver a package only if the length plus girth (distance around) does not exceed 108 inches.

   (A) Find the dimensions of a rectangular box with square ends that satisfies the delivery service's restriction and has maximum volume. What is the maximum volume?

   (B) Find the dimensions (radius and height) of a cylindrical container that meets the delivery service's

Figure for 24

requirement and has maximum volume. What is the maximum volume?

25. *Construction costs.* A fence is to be built to enclose a rectangular area of 800 square feet. The fence along three sides is to be made of material that costs $6 per foot. The material for the fourth side costs $18 per foot. Find the dimensions of the rectangle that will allow the most economical fence to be built.

26. *Construction costs.* The owner of a retail lumber store wants to construct a fence to enclose an outdoor storage area adjacent to the store, using all of the store as part of one side of the area (see the figure). Find the dimensions that will enclose the largest area if

   (A) 240 feet of fencing material are used.

   (B) 400 feet of fencing material are used.

Figure for 26

27. *Inventory control.* A paint manufacturer has a uniform annual demand for 16,000 cans of automobile primer. It costs $4 to store one can of paint for one year and $500 to set up the plant for production of the primer. How many times a year should the company produce this primer in order to minimize the total storage and setup costs?

28. *Inventory control.* A pharmacy has a uniform annual demand for 200 bottles of a certain antibiotic. It costs $10 to store one bottle for one year and $40 to place an order. How many times during the year should the pharmacy order the antibiotic in order to minimize the total storage and reorder costs?

29. *Inventory control.* A publishing company sells 50,000 copies of a certain book each year. It costs the company $1 to store a book for one year. Each time it must print additional copies, it costs the company $1,000 to set up the presses. How many books should the company produce during each printing in order to minimize its total storage and setup costs?

30. *Operational costs.* The cost per hour for fuel to run a train is $v^2/4$ dollars, where v is the speed of the train in miles per hour. (Note that the cost goes up as the square

of the speed.) Other costs, including labor, are $300 per hour. How fast should the train travel on a 360-mile trip to minimize the total cost for the trip?

31. *Construction costs.* A freshwater pipeline is to be run from a source on the edge of a lake to a small resort community on an island 5 miles offshore, as indicated in the figure.

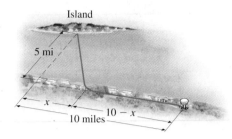

Figure for 31

(A) If it costs 1.4 times as much to lay the pipe in the lake as it does on land, what should x be (in miles) to minimize the total cost of the project?

(B) If it costs only 1.1 times as much to lay the pipe in the lake as it does on land, what should x be to minimize the total cost of the project? [*Note:* Compare with Problem 36.]

32. *Manufacturing costs.* A manufacturer wants to produce cans that will hold 12 ounces (approximately 22 cubic inches) in the form of a right circular cylinder. Find the dimensions (radius of an end and height) of the can that will use the smallest amount of material. Assume that the circular ends are cut out of squares, with the corner portions wasted, and the sides are made from rectangles, with no waste.

33. *Bacteria control.* A lake used for recreational swimming is treated periodically to control harmful bacteria growth. Suppose that t days after a treatment, the concentration of bacteria per cubic centimeter is given by

$$C(t) = 30t^2 - 240t + 500 \qquad 0 \le t \le 8$$

How many days after a treatment will the concentration be minimal? What is the minimum concentration?

34. *Drug concentration.* The concentration $C(t)$, in milligrams per cubic centimeter, of a particular drug in a patient's bloodstream is given by

$$C(t) = \frac{0.16t}{t^2 + 4t + 4}$$

where t is the number of hours after the drug is taken. How many hours after the drug is taken will the concentration be maximum? What is the maximum concentration?

35. *Laboratory management.* A laboratory uses 500 white mice each year for experimental purposes. It costs $4 to feed a mouse for one year. Each time mice are ordered from a supplier, there is a service charge of $10 for processing the order. How many mice should be ordered each time to minimize the total cost of feeding the mice and of placing orders for the mice?

36. *Bird flights.* Some birds tend to avoid flights over large bodies of water during daylight hours. (Scientists speculate that more energy is required to fly over water than land, because air generally rises over land and falls over water during the day.) Suppose that an adult bird with this tendency is taken from its nesting area on the edge

of a large lake to an island 5 miles offshore and is then released (see the figure).

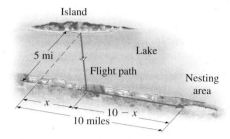

Figure for 36

(A) If it takes 1.4 times as much energy to fly over water as land, how far up the shore (x, in miles) should the bird head to minimize the total energy expended in returning to the nesting area?

(B) If it takes only 1.1 times as much energy to fly over water as land, how far up the shore should the bird head to minimize the total energy expended in returning to the nesting area? [*Note:* Compare with Problem 31.]

37. *Botany.* If it is known from past experiments that the height (in feet) of a certain plant after t months is given approximately by

$$H(t) = 4t^{1/2} - 2t \qquad 0 \le t \le 2$$

how long, on the average, will it take a plant to reach its maximum height? What is the maximum height?

38. *Pollution.* Two heavily industrial areas are located 10 miles apart, as indicated in the figure. If the concentration of particulate matter (in parts per million) decreases as the reciprocal of the square of the distance from the source, and if area A_1 emits eight times the particulate matter as A_2, the concentration of particulate matter at any point between the two areas is given by

$$C(x) = \frac{8k}{x^2} + \frac{k}{(10 - x)^2} \qquad 0.5 \le x \le 9.5, \quad k > 0$$

How far from A_1 will the concentration of particulate matter between the two areas be at a minimum?

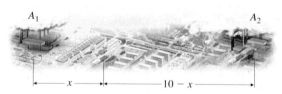

Figure for 38

39. *Politics.* In a newly incorporated city, it is estimated that the voting population (in thousands) will increase according to

$$N(t) = 30 + 12t^2 - t^3 \qquad 0 \le t \le 8$$

where t is time in years. When will the rate of increase be most rapid?

40. *Learning.* A large grocery chain found that, on the average, a checker can recall $P\%$ of a given price list x hours after starting work, as given approximately by

$$P(x) = 96x - 24x^2 \qquad 0 \le x \le 3$$

At what time x does the checker recall a maximum percentage? What is the maximum?

CHAPTER 5 REVIEW

Important Terms, Symbols, and Concepts

5-1 First Derivative and Graphs

Examples

- **Increasing and decreasing properties** of a function can be determined by examining a sign chart for the derivative.

  Ex. 1, p. 276

- A number c is a **partition number** for $f'(x)$ if $f'(c) = 0$ or $f'(x)$ is discontinuous at c. If c is also in the domain of $f(x)$, then c is a **critical value.**

  Ex. 2, p. 277
  Ex. 3, p. 278

- Increasing and decreasing properties and **local extrema** for $f(x)$ can be determined by examining the graph of $f'(x)$.

  Ex. 4, p. 278
  Ex. 5, p. 279
  Ex. 6, p. 280

- **The first-derivative test** is used to locate extrema of a function.

  Ex. 7, p. 283

5-2 Second Derivative and Graphs

- The **second derivative** of a function f can be used to determine the concavity of the graph of f.

  Ex. 1, p. 296

- **Inflection points** on a graph are points where the concavity changes.

  Ex. 2, p. 298
  Ex. 3, p. 298

- The concavity of the graph of $f(x)$ can also be determined by an examination of the graph of $f'(x)$.

  Ex. 4, p. 300

- A **graphing strategy** is used to organize the information obtained from the first and second derivatives.

  Ex. 5, p. 301
  Ex. 6, p. 303

5-3 L'Hôpital's Rule

- Limits at infinity and infinite limits involving powers of $x - c$, e^x, and $\ln x$ are reviewed.

  Ex. 1, p. 312
  Ex. 2, p. 313

- The first version of **L'Hôpital's rule** applies to limits involving the indeterminate form $0/0$ as $x \to c$.

  Ex. 3, p. 314
  Ex. 4, p. 315

- You must always check that L'Hôpital's rule applies.

  Ex. 5, p. 315

- L'Hôpital's rule can be applied more than once.

  Ex. 6, p. 316

- L'Hôpital's rule applies to one-sided limits.

  Ex. 7, p. 317

- L'Hôpital's rule applies to limits at infinity.

  Ex. 8, p. 317

- L'Hôpital's rule applies to limits involving the indeterminate form ∞/∞.

  Ex. 9, p. 318
  Ex. 10, p. 319

5-4 Curve-Sketching Techniques

- The graphing strategy first used in Section 5-2 is expanded to include horizontal and vertical asymptotes.

  Ex. 1, p. 321
  Ex. 2, p. 323
  Ex. 3, p. 324
  Ex. 4, p. 326

- If $f(x) = n(x)/d(x)$ is a rational function with the degree of $n(x)$ 1 more than the degree of $d(x)$, then the graph of $f(x)$ has an **oblique asymptote** of the form $y = mx + b$.

  Ex. 5, p. 327

5-5 Absolute Maxima and Minima

- The steps involved in finding the **absolute maximum and absolute minimum** values of a continuous function on a closed interval are listed in a procedure.

  Ex. 1, .p. 337

- The **second-derivative test for local extrema** can be used to test critical values, but it does not work in all cases.

  Ex. 2, p. 339

- If a function is continuous on an interval I and has only one critical value in I, **the second-derivative test for absolute extrema** can be used to find the absolute extrema, but it not does work in all cases.

  Ex. 3, p. 340

5-6 Optimization

- The methods used to solve **optimization problem** are summarized and illustrated by examples.

  Ex. 1, p. 343
  Ex. 2, p. 344
  Ex. 3, p. 346
  Ex. 4, p. 346
  Ex. 5, p. 348
  Ex. 6, p. 348
  Ex. 7, p. 349
  Ex. 8, p. 350

REVIEW EXERCISE

Work through all the problems in this chapter review, and check your answers in the back of the book. Answers to all review problems are there, along with section numbers in italics to indicate where each type of problem is discussed. Where weaknesses show up, review appropriate sections in the text.

A *Problems 1–8 refer to the graph of $y = f(x)$ that follows. Identify the points or intervals on the x axis that produce the indicated behavior.*

    1. $f(x)$ is increasing.   **2.** $f'(x) < 0$

    3. The graph of f is concave downward.

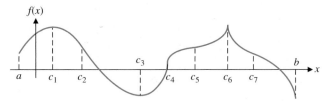

    4. Local minima   **5.** Absolute maxima

    6. $f'(x)$ appears to be 0.   **7.** $f'(x)$ does not exist.

    8. Inflection points

In Problems 9 and 10, use the given information to sketch the graph of f. Assume that f is continuous on its domain and that all intercepts are included in the information given.

9. Domain: All real x

x	-3	-2	-1	0	2	3
$f(x)$	0	3	2	0	-3	0

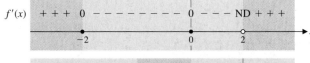

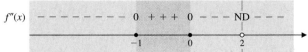

10. Domain: All real x;

    $f(-2) = 1, f(0) = 0, f(2) = 1$;

    $f'(0) = 0; f'(x) < 0$ on $(-\infty, 0)$;

    $f'(x) > 0$ on $(0, \infty)$;

    $f''(-2) = 0, f''(2) = 0$;

    $f''(x) < 0$ on $(-\infty, -2)$ and $(2, \infty)$;

    $f''(x) > 0$ on $(-2, 2)$;

    $\lim\limits_{x \to -\infty} f(x) = 2; \lim\limits_{x \to \infty} f(x) = 2$

11. Find $f''(x)$ for $f(x) = x^4 + 5x^3$.

12. Find y'' for $y = 3x + \dfrac{4}{x}$.

In Problems 13–22, summarize all the pertinent information obtained by applying the final version of the graphing strategy (Section 5-4) to f, and sketch the graph of f.

13. $f(x) = x^3 - 18x^2 + 81x$

14. $f(x) = (x + 4)(x - 2)^2$

15. $f(x) = 8x^3 - 2x^4$ **16.** $f(x) = (x - 1)^3(x + 3)$

17. $f(x) = \dfrac{3x}{x + 2}$ **18.** $f(x) = \dfrac{x^2}{x^2 + 27}$

19. $f(x) = \dfrac{x}{(x + 2)^2}$ **20.** $f(x) = \dfrac{x^3}{x^2 + 3}$

21. $f(x) = 5 - 5e^{-x}$ **22.** $f(x) = x^3 \ln x$

Find each limit in Problems 23–32.

23. $\lim\limits_{x \to 0} \dfrac{e^{3x} - 1}{x}$ **24.** $\lim\limits_{x \to 2} \dfrac{x^2 - 5x + 6}{x^2 + x - 6}$

25. $\lim\limits_{x \to 0^-} \dfrac{\ln(1 + x)}{x^2}$ **26.** $\lim\limits_{x \to 0} \dfrac{\ln(1 + x)}{1 + x}$

27. $\lim\limits_{x \to \infty} \dfrac{e^{4x}}{x^2}$ **28.** $\lim\limits_{x \to 0} \dfrac{e^x + e^{-x} - 2}{x^2}$

29. $\lim\limits_{x \to 0^+} \dfrac{\sqrt{1 + x} - 1}{\sqrt{x}}$ **30.** $\lim\limits_{x \to \infty} \dfrac{\ln x}{x^5}$

31. $\lim\limits_{x \to \infty} \dfrac{\ln(1 + 6x)}{\ln(1 + 3x)}$ **32.** $\lim\limits_{x \to 0} \dfrac{\ln(1 + 6x)}{\ln(1 + 3x)}$

33. Use the graph of $y = f'(x)$ shown here to discuss the graph of $y = f(x)$. Organize your conclusions in a table (see Example 4, Section 5-2). Sketch a possible graph of $y = f(x)$.

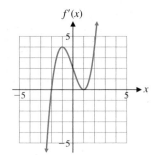

Figure for 33 and 34

34. Refer to the proceding graph of $y = f'(x)$. Which of the following could be the graph of $y = f''(x)$?

(A) 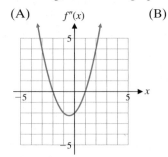   (B)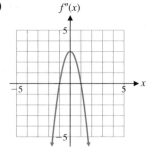

(C)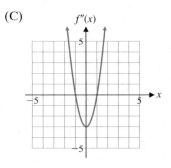

35. Use the second-derivative test to find any local extrema for

$$f(x) = x^3 - 6x^2 - 15x + 12$$

36. Find the absolute maximum and absolute minimum, if either exists, for

$$y = f(x) = x^3 - 12x + 12 \qquad -3 \le x \le 5$$

37. Find the absolute minimum, if it exists, for

$$y = f(x) = x^2 + \frac{16}{x^2} \qquad x > 0$$

38. Find the absolute maximum value, if it exists, for

$$f(x) = 11x - 2x \ln x \qquad x > 0$$

39. Find the absolute maximum value, if it exists, for

$$f(x) = 10xe^{-2x} \qquad x > 0$$

40. Let $y = f(x)$ be a polynomial function with local minima at $x = a$ and $x = b$, $a < b$. Must f have at least one local maximum between a and b? Justify your answer.

41. The derivative of $f(x) = x^{-1}$ is $f'(x) = -x^{-2}$. Since $f'(x) < 0$ for $x \ne 0$, is it correct to say that $f(x)$ is decreasing for all x except $x = 0$? Explain.

42. Discuss the difference between a partition number for $f'(x)$ and a critical value of $f(x)$, and illustrate with examples.

C

43. Find the absolute maximum for $f'(x)$ if

$$f(x) = 6x^2 - x^3 + 8$$

Graph f and f' on the same coordinate system for $0 \le x \le 4$.

44. Find two positive numbers whose product is 400 and whose sum is a minimum. What is the minimum sum?

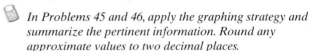 *In Problems 45 and 46, apply the graphing strategy and summarize the pertinent information. Round any approximate values to two decimal places.*

45. $f(x) = x^4 + x^3 - 4x^2 - 3x + 4$

46. $f(x) = 0.25x^4 - 5x^3 + 31x^2 - 70x$

47. Find the absolute maximum value, if it exists, for

$$f(x) = 3x - x^2 + e^{-x} \qquad x > 0$$

48. Find the absolute maximum value, if it exists, for

$$f(x) = \frac{\ln x}{e^x} \qquad x > 0$$

APPLICATIONS

49. *Price analysis.* The graph in the figure approximates the rate of change of the price of tomatoes over a 60-month period, where $p(t)$ is the price of a pound of tomatoes and t is time (in months).

(A) Write a brief verbal description of the graph of $y = p(t)$, including a discussion of local extrema and inflection points.

(B) Sketch a possible graph of $y = p(t)$.

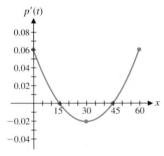

Figure for 49

50. *Maximum revenue and profit.* A company manufactures and sells x electric stoves per month. The monthly cost and price–demand equations are, respectively,

$$C(x) = 350x + 50,000$$

$$p = 500 - 0.025x \qquad 0 \le x \le 20,000$$

(A) Find the maximum revenue.

(B) How many stoves should the company manufacture each month to maximize its profit? What is the maximum monthly profit? How much should the company charge for each stove?

(C) If the government decides to tax the company $20 for each stove it produces, how many stoves should the

company manufacture each month to maximize its profit? What is the maximum monthly profit? How much should the company charge for each stove?

51. *Construction.* A fence is to be built to enclose a rectangular area. The fence along three sides is to be made of material that costs $5 per foot. The material for the fourth side costs $15 per foot.

(A) If the area is 5,000 square feet, find the dimensions of the rectangle that will allow the most economical fence to be built.

(B) If $3,000 is available for the fencing, find the dimensions of the rectangle that will enclose the most area.

52. *Rental income.* A 200-room hotel in Fresno is filled to capacity every night at a rate of $40 per room. For each $1 increase in the nightly rate, 4 fewer rooms are rented. If each rented room costs $8 a day to service, how much should the management charge per room in order to maximize gross profit? What is the maximum gross profit?

53. *Inventory control.* A computer store sells 7,200 boxes of floppy disks annually. It costs the store $0.20 to store a box of disks for one year. Each time it reorders disks, the store must pay a $5.00 service charge for processing the order. How many times during the year should the store order disks to minimize the total storage and reorder costs?

54. *Average cost.* The total cost of producing x garbage disposals per day is given by

$$C(x) = 4,000 + 10x + 0.1x^2$$

Find the minimum average cost. Graph the average cost and the marginal cost functions on the same coordinate system. Include any oblique asymptotes.

55. *Average cost.* The cost of producing x window fans is given by

$$C(x) = 200 + 50x - 50 \ln x \qquad x \geq 1$$

Find the minimum average cost.

56. *Marginal analysis.* The price–demand equation for washing machines at an appliance store is

$$p(x) = 1,000e^{-0.02x}$$

where x is the monthly demand and p is the price in dollars. Find the production level and price per unit that produces the maximum revenue. What is the maximum revenue?

57. *Maximum revenue.* Graph the revenue function from Problem 56 for $0 \leq x \leq 100$.

 58. *Maximum profit.* Refer to Problem 56. If the washing machines cost the store $220 each, find the price (to the nearest cent) that maximizes the profit. What is the maximum profit (to the nearest dollar)?

 59. *Maximum profit.* The data in the table show the daily demand x for cream puffs at a state fair at various price levels p. If it costs $1 to make a cream puff, use logarithmic regression ($p = a + b \ln x$) to find the price (to the nearest cent) that maximizes profit.

Demand	Price per Cream Puff ($)
x	p
3,125	1.99
3,879	1.89
5,263	1.79
5,792	1.69
6,748	1.59
8,120	1.49

60. *Construction costs.* The ceiling supports in a new discount department store are 12 feet apart. Lights are to be hung from these supports by chains in the shape of a "Y." If the lights are 10 feet below the ceiling, what is the shortest length of chain that can be used to support these lights?

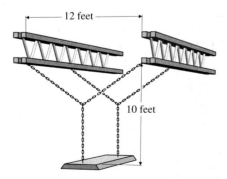

 61. *Average cost.* The data in the table give the total daily cost y (in dollars) of producing x dozen chocolate chip cookies at various production levels.

Dozens of Cookies	Total Cost
x	y
50	119
100	187
150	248
200	382
250	505
300	695

(A) Enter the data into a graphing calculator and find a quadratic regression equation for the total cost.

(B) Use the regression equation from part (A) to find the minimum average cost (to the nearest cent) and the corresponding production level (to the nearest integer).

62. *Advertising—point of diminishing returns.* A company estimates that it will sell $N(x)$ units of a product after spending x thousand on advertising, as given by

$$N(x) = -0.25x^4 + 11x^3 - 108x^2 + 3,000$$
$$9 \leq x \leq 24$$

When is the rate of change of sales increasing and when is it decreasing? What is the point of diminishing returns and the maximum rate of change of sales? Graph N and N' on the same coordinate system.

 63. *Advertising.* A chain of appliance stores uses television ads to promote the sales of refrigerators. Analyzing past records produced the data in the following table, where x is the number of ads placed monthly and y is the number of refrigerators sold that month:

Number of Ads	Number of Refrigerators
x	y
10	271
20	427
25	526
30	629
45	887
48	917

(A) Enter the data into a graphing calculator, set the calculator to display two decimal places, and find a cubic regression equation for the number of refrigerators sold monthly as a function of the number of ads.

(B) How many ads should be placed each month to maximize the rate of change of sales with respect to the number of ads, and how many refrigerators can be expected to be sold with that number of ads? Round answers to the nearest integer.

64. *Bacteria control.* If t days after a treatment the bacteria count per cubic centimeter in a body of water is given by

$$C(t) = 20t^2 - 120t + 800 \qquad 0 \leq t \leq 9$$

in how many days will the count be a minimum?

65. *Politics.* In a new suburb, it is estimated that the number of registered voters will grow according to

$$N = 10 + 6t^2 - t^3 \qquad 0 \leq t \leq 5$$

where t is time in years and N is in thousands. When will the rate of increase be maximum?

Integration

CHAPTER

6

6-1   Antiderivatives and
       Indefinite Integrals

6-2   Integration by Substitution

6-3   Differential Equations;
       Growth and Decay

6-4   The Definite Integral

6-5   The Fundamental Theorem
       of Calculus

Chapter 6 Review

Review Exercise

INTRODUCTION

The preceding three chapters dealt with differential calculus. We now begin the development of the second main part of calculus, called *integral calculus.* Two types of integrals are introduced, the *indefinite integral* and the *definite integral,* each quite different from the other. But through the remarkable *fundamental theorem of calculus,* we show that the two integral forms are intimately related, not only to each other, but also to differentiation.

Section 6-1 ANTIDERIVATIVES AND INDEFINITE INTEGRALS

- Antiderivatives
- Indefinite Integrals: Formulas and Properties
- Applications

Many operations in mathematics have reverses—compare addition and subtraction, multiplication and division, and powers and roots. We now know how to find the derivatives of many functions. The reverse operation, *antidifferentiation* (the reconstruction of a function from its derivative), will receive our attention in this and the next two sections. A function F is an **antiderivative** of a function f if $F'(x) = f(x)$. We develop special antiderivative formulas in much the same way as we developed derivative formulas.

■ Antiderivatives

Explore & Discuss **1**

(A) Find three antiderivatives of $2x$.

(B) How many antiderivatives of $2x$ exist, and how are they related to each other?

(C) What notation would you use to represent all antiderivatives of $2x$?

The function $F(x) = \dfrac{x^3}{3}$ is an antiderivative of the function $f(x) = x^2$ because

$$\frac{d}{dx}\left(\frac{x^3}{3}\right) = x^2$$

However, $F(x)$ is not the only antiderivative of x^2. Note also that

$$\frac{d}{dx}\left(\frac{x^3}{3} + 2\right) = x^2 \qquad \frac{d}{dx}\left(\frac{x^3}{3} - \pi\right) = x^2 \qquad \frac{d}{dx}\left(\frac{x^3}{3} + \sqrt{5}\right) = x^2$$

Therefore,

$$\frac{x^3}{3} + 2 \qquad \frac{x^3}{3} - \pi \qquad \frac{x^3}{3} + \sqrt{5}$$

are also antiderivatives of x^2, because each has x^2 as a derivative. In fact, it appears that

$$\frac{x^3}{3} + C \qquad \text{for any real number } C$$

is an antiderivative of x^2, because

$$\frac{d}{dx}\left(\frac{x^3}{3} + C\right) = x^2$$

Antidifferentiation of a given function, then, leads not to a unique function, but to an entire family of functions.

Does the expression

$$\frac{x^3}{3} + C \qquad \text{with } C \text{ any real number}$$

include all antiderivatives of x^2? Theorem 1 (which we state without proof) indicates that the answer is yes.

THEOREM 1 ON ANTIDERIVATIVES

If the derivatives of two functions are equal on an open interval (a, b), then the functions differ by at most a constant. Symbolically, if F and G are differentiable functions on the interval (a, b) and $F'(x) = G'(x)$ for all x in (a, b), then $F(x) = G(x) + k$ for some constant k.

INSIGHT

Suppose that $F(x)$ is an antiderivative of $f(x)$. If $G(x)$ is any other antiderivative of $f(x)$, then, by Theorem 1, the graph of $G(x)$ is a vertical translation of the graph of $F(x)$ (see Section 2-2).

EXAMPLE 1   **A Family of Antiderivatives**   Note that

$$\frac{d}{dx}\left(\frac{x^2}{2}\right) = x$$

(A) Find all antiderivatives of $f(x) = x$.

(B) Graph the antiderivative of $f(x) = x$ that passes through the point $(0, 0)$; through the point $(0, 1)$; through the point $(0, 2)$.

(C) How are the graphs of the three antiderivatives in part (B) related?

SOLUTION   (A) By Theorem 1, any antiderivative of $f(x)$ has the form

$$F(x) = \frac{x^2}{2} + k$$

where k is a real number.

(B) Because $F(0) = (0^2/2) + k = k$, the functions

$$F_0(x) = \frac{x^2}{2}, \quad F_1(x) = \frac{x^2}{2} + 1, \quad \text{and} \quad F_2(x) = \frac{x^2}{2} + 2$$

pass through the points $(0, 0)$, $(0, 1)$, and $(0, 2)$, respectively (see Fig. 1).

(C) The graphs of the three antiderivatives are vertical translations of each other.

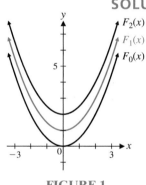

FIGURE 1

MATCHED PROBLEM 1   Note that

$$\frac{d}{dx}(x^3) = 3x^2$$

(A) Find all antiderivatives of $f(x) = 3x^2$.

(B) Graph the antiderivative of $f(x) = 3x^2$ that passes through the point $(0, 0)$; through the point $(0, 1)$; through the point $(0, 2)$.

(C) How are the graphs of the three antiderivatives in part (B) related?

■ Indefinite Integrals: Formulas and Properties

Theorem 1 states that if the derivatives of two functions are equal, then the functions differ by at most a constant. We use the symbol

$$\int f(x)\, dx$$

called the **indefinite integral,** to represent the family of all antiderivatives of $f(x)$, and we write

$$\int f(x)\,dx = F(x) + C \qquad \text{if} \qquad F'(x) = f(x)$$

The symbol $\int$ is called an **integral sign,** and the function $f(x)$ is called the **integrand.** The symbol dx indicates that the antidifferentiation is performed with respect to the variable x. (We will have more to say about the symbols $\int$ and dx later in the chapter.) The arbitrary constant C is called the **constant of integration.** Referring to the preceding discussion, we can write

$$\int x^2\,dx = \frac{x^3}{3} + C \qquad \text{since} \qquad \frac{d}{dx}\left(\frac{x^3}{3} + C\right) = x^2$$

Of course, variables other than x can be used in indefinite integrals. For example,

$$\int t^2\,dt = \frac{t^3}{3} + C \qquad \text{since} \qquad \frac{d}{dt}\left(\frac{t^3}{3} + C\right) = t^2$$

or

$$\int u^2\,du = \frac{u^3}{3} + C \qquad \text{since} \qquad \frac{d}{du}\left(\frac{u^3}{3} + C\right) = u^2$$

The fact that indefinite integration and differentiation are reverse operations, except for the addition of the constant of integration, can be expressed symbolically as

$$\frac{d}{dx}\left[\int f(x)\,dx\right] = f(x) \quad \text{The derivative of the indefinite integral of f(x) is f(x).}$$

and

$$\int F'(x)\,dx = F(x) + C \quad \text{The indefinite integral of the derivative of F(x) is F(x) + C.}$$

We can develop formulas for the indefinite integrals of certain basic functions from the formulas for derivatives that were established in Chapters 3 and 4.

FORMULAS   <u>**Indefinite Integrals of Basic Functions**</u>

For C a constant,

1. $\displaystyle\int x^n\,dx = \frac{x^{n+1}}{n+1} + C, \qquad n \neq -1$

2. $\displaystyle\int e^x\,dx = e^x + C$

3. $\displaystyle\int \frac{1}{x}\,dx = \ln|x| + C, \qquad x \neq 0$

Each formula can be justified by showing that the derivative of the right-hand side is the integrand of the left-hand side (see Problems 101–104 in Exercise 6-1). Note that formula 1 does not give the antiderivative of x^{-1} (because $x^{n+1}/(n+1)$ is undefined when $n = -1$), but formula 3 does.

Explore & Discuss **2**

Formulas 1, 2, and 3 do *not* provide a formula for the indefinite integral of the function $\ln x$. Show, however, that if $x > 0$, then

$$\int \ln x\,dx = x \ln x - x + C$$

by differentiating the right-hand side. [We will discuss a technique for deriving such indefinite integral formulas in Section 7-3.]

We can obtain properties of the indefinite integral from properties of the derivative that were established in Chapter 3.

PROPERTIES   **Of Indefinite Integrals**

For k a constant,

4. $\displaystyle\int kf(x)\,dx = k\int f(x)\,dx$

5. $\displaystyle\int [f(x) \pm g(x)]\,dx = \int f(x)\,dx \pm \int g(x)\,dx$

Property 4 states that

The indefinite integral of a constant times a function is the constant times the indefinite integral of the function.

Property 5 states that

The indefinite integral of the sum of two functions is the sum of the indefinite integrals, and the indefinite integral of the difference of two functions is the difference of the indefinite integrals.

To establish property 4, let F be a function such that $F'(x) = f(x)$. Then

$$k\int f(x)\,dx = k\int F'(x)\,dx = k[F(x) + C_1] = kF(x) + kC_1$$

and since $[kF(x)]' = kF'(x) = kf(x)$, we have

$$\int kf(x)\,dx = \int kF'(x)\,dx = kF(x) + C_2$$

But $kF(x) + kC_1$ and $kF(x) + C_2$ describe the same set of functions, because C_1 and C_2 are arbitrary real numbers. Thus, property 4 is established. Property 5 can be established in a similar manner (see Problems 105–106 in Exercise 6-1).

 CAUTION: It is important to remember that property 4 states that **a constant factor can be moved across an integral sign. A variable factor cannot be moved across an integral sign:**

CONSTANT FACTOR   *VARIABLE FACTOR*

$$\int 5x^{1/2}\,dx = 5\int x^{1/2}\,dx \qquad \int xx^{1/2}\,dx \neq x\int x^{1/2}\,dx$$

Indefinite integral formulas and properties can be used together to find indefinite integrals for many frequently encountered functions. For example, formula 1, gives, for $n = 0$,

$$\int dx = x + C$$

Therefore, by property 4,

$$\int k\,dx = k(x + C) = kx + kC$$

Because kC is a constant, it is customary to replace it by a single symbol that denotes an arbitrary constant (usually C), so we write

$$\int k\,dx = kx + C$$

In words,

The indefinite integral of a constant function with value k is $kx + C$.

Similarly, using property 5 and then formulas 2 and 3 yields

$$\int \left(e^x + \frac{1}{x}\right) dx = \int e^x \, dx + \int \frac{1}{x} \, dx$$
$$= e^x + C_1 + \ln|x| + C_2$$

Because $C_1 + C_2$ is a constant, it is usually replaced by the symbol C, and we write

$$\int \left(e^x + \frac{1}{x}\right) dx = e^x + \ln|x| + C$$

EXAMPLE 2   **Using Indefinite Integral Properties and Formulas**

(A) $\int 5 \, dx = 5x + C$

(B) $\int 9e^x \, dx = 9 \int e^x \, dx = 9e^x + C$

(C) $\int 5t^7 \, dt = 5 \int t^7 \, dt = 5\frac{t^8}{8} + C = \frac{5}{8}t^8 + C$

(D) $\int (4x^3 + 2x - 1) \, dx = \int 4x^3 \, dx + \int 2x \, dx - \int dx$ Property 4 can be extended to the sum and
$$= 4 \int x^3 \, dx + 2 \int x \, dx - \int dx$$ difference of an arbitrary
$$= \frac{4x^4}{4} + \frac{2x^2}{2} - x + C$$ number of functions.
$$= x^4 + x^2 - x + C$$

(E) $\int \left(2e^x + \frac{3}{x}\right) dx = 2 \int e^x \, dx + 3 \int \frac{1}{x} \, dx$
$$= 2e^x + 3 \ln|x| + C$$

To check any of the results in Example 2, we differentiate the final result to obtain the integrand in the original indefinite integral. When you evaluate an indefinite integral, do not forget to include the arbitrary constant C.

MATCHED PROBLEM 2   Find each indefinite integral:

(A) $\int 2 \, dx$

(B) $\int 16e^t \, dt$

(C) $\int 3x^4 \, dx$

(D) $\int (2x^5 - 3x^2 + 1) \, dx$

(E) $\int \left(\frac{5}{x} - 4e^x\right) dx$

EXAMPLE 3   **Using Indefinite Integral Properties and Formulas**

(A) $\displaystyle\int \frac{4}{x^3}\,dx = \int 4x^{-3}\,dx = \frac{4x^{-3+1}}{-3+1} + C = -2x^{-2} + C$

(B) $\displaystyle\int 5\sqrt[3]{u^2}\,du = 5\int u^{2/3}\,du = 5\frac{u^{(2/3)+1}}{\frac{2}{3}+1} + C$

$\displaystyle\qquad\qquad\qquad = 5\frac{u^{5/3}}{\frac{5}{3}} + C = 3u^{5/3} + C$

(C) $\displaystyle\int \frac{x^3 - 3}{x^2}\,dx = \int \left(\frac{x^3}{x^2} - \frac{3}{x^2}\right)dx$

$\displaystyle\qquad\qquad\qquad = \int (x - 3x^{-2})\,dx$

$\displaystyle\qquad\qquad\qquad = \int x\,dx - 3\int x^{-2}\,dx$

$\displaystyle\qquad\qquad\qquad = \frac{x^{1+1}}{1+1} - 3\frac{x^{-2+1}}{-2+1} + C$

$\displaystyle\qquad\qquad\qquad = \tfrac{1}{2}x^2 + 3x^{-1} + C$

(D) $\displaystyle\int \left(\frac{2}{\sqrt[3]{x}} - 6\sqrt{x}\right)dx = \int (2x^{-1/3} - 6x^{1/2})\,dx$

$\displaystyle\qquad\qquad\qquad = 2\int x^{-1/3}\,dx - 6\int x^{1/2}\,dx$

$\displaystyle\qquad\qquad\qquad = 2\frac{x^{(-1/3)+1}}{-\frac{1}{3}+1} - 6\frac{x^{(1/2)+1}}{\frac{1}{2}+1} + C$

$\displaystyle\qquad\qquad\qquad = 2\frac{x^{2/3}}{\frac{2}{3}} - 6\frac{x^{3/2}}{\frac{3}{2}} + C$

$\displaystyle\qquad\qquad\qquad = 3x^{2/3} - 4x^{3/2} + C$

(E) $\displaystyle\int x(x^2 + 2)\,dx = \int (x^3 + 2x)\,dx = \frac{x^4}{4} + x^2 + C$

MATCHED PROBLEM 3   Find each indefinite integral:

(A) $\displaystyle\int \left(2x^{2/3} - \frac{3}{x^4}\right)dx$

(B) $\displaystyle\int 4\sqrt[5]{w^3}\,dw$

(C) $\displaystyle\int \frac{x^4 - 8x^3}{x^2}\,dx$

(D) $\displaystyle\int \left(8\sqrt[3]{x} - \frac{6}{\sqrt{x}}\right)dx$

(E) $\displaystyle\int (x^2 - 2)(x + 3)\,dx$

⚠ CAUTION

1. Note from Example 3(E) that

$$\int x(x^2 + 2)\, dx \neq \frac{x^2}{2}\left(\frac{x^3}{3} + 2x\right) + C$$

In general, the **indefinite integral of a product is not the product of the indefinite integrals.** (This is to be expected, because the derivative of a product is not the product of the derivatives.)

2.
$$\int e^x\, dx \neq \frac{e^{x+1}}{x+1} + C$$

The power rule applies only to power functions of the form x^n, where the exponent n is a real constant not equal to -1 and the base x is the variable. The function e^x is an exponential function with variable exponent x and constant base e. The correct form is

$$\int e^x\, dx = e^x + C$$

3. Not all elementary functions have elementary antiderivatives. It is impossible, for example, to give a formula for the antiderivative of $f(x) = e^{x^2}$ in terms of elementary functions. Nevertheless, finding such a formula, when it exists, can markedly simplify the solution of certain problems.

APPLICATIONS

Let us now consider some applications of the indefinite integral to see why we are interested in finding antiderivatives of functions.

EXAMPLE 4   **Curves**   Find the equation of the curve that passes through $(2, 5)$ if the slope of the curve is given by $dy/dx = 2x$ at any point x.

SOLUTION   We are interested in finding a function $y = f(x)$ such that

$$\frac{dy}{dx} = 2x \tag{1}$$

and

$$y = 5 \quad \text{when} \quad x = 2 \tag{2}$$

If $dy/dx = 2x$, then

$$y = \int 2x\, dx$$

$$= x^2 + C \tag{3}$$

Since $y = 5$ when $x = 2$, we determine the *particular value of C* so that

$$5 = 2^2 + C$$

Thus, $C = 1$, and

$$y = x^2 + 1$$

is the *particular antiderivative* out of all those possible from equation (3) that satisfies both equations (1) and (2) (see Fig. 2). ∎

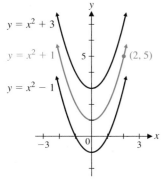

$y = x^2 + 3$

$y = x^2 + 1$

$y = x^2 - 1$

$(2, 5)$

FIGURE 2 $y = x^2 + C$

MATCHED PROBLEM 4   Find the equation of the curve that passes through $(2, 6)$ if the slope of the curve is given by $dy/dx = 3x^2$ at any point x.

In certain situations, it is easier to determine the rate at which something happens than how much of it has happened in a given length of time (for example, population growth rates, business growth rates, the rate of healing of a wound, rates of learning or forgetting). If a rate function (derivative) is given and we know the value of the dependent variable for a given value of the independent variable, then—if the rate function is not too complicated—we can often find the original function by integration.

EXAMPLE 5   **Cost Function**   If the marginal cost of producing x units of a commodity is given by

$$C'(x) = 0.3x^2 + 2x$$

and the fixed cost is $2,000, find the cost function $C(x)$ and the cost of producing 20 units.

SOLUTION   Recall that marginal cost is the derivative of the cost function and that fixed cost is cost at a zero production level. Thus, the mathematical problem is to find $C(x)$, given

$$C'(x) = 0.3x^2 + 2x \qquad C(0) = 2,000$$

We now find the indefinite integral of $0.3x^2 + 2x$ and determine the arbitrary integration constant, using $C(0) = 2,000$:

$$C'(x) = 0.3x^2 + 2x$$

$$C(x) = \int (0.3x^2 + 2x)\,dx$$

$$= 0.1x^3 + x^2 + K \qquad \text{\textit{Since C represents the cost, we use K for the constant of integration.}}$$

But

$$C(0) = (0.1)0^3 + 0^2 + K = 2,000$$

Thus, $K = 2,000$, and the particular cost function is

$$C(x) = 0.1x^3 + x^2 + 2,000$$

We now find $C(20)$, the cost of producing 20 units:

$$C(20) = (0.1)20^3 + 20^2 + 2,000$$
$$= \$3,200$$

See Figure 3 for a geometric representation.   ▬

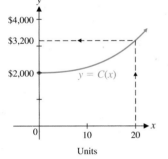

FIGURE 3

MATCHED PROBLEM 5   Find the revenue function $R(x)$ when the marginal revenue is

$$R'(x) = 400 - 0.4x$$

and no revenue results at a zero production level. What is the revenue at a production level of 1,000 units?

EXAMPLE 6   **Advertising**   An FM radio station is launching an aggressive advertising campaign in order to increase the number of daily listeners. The station currently has 27,000 daily listeners, and management expects the number of daily listeners, $S(t)$, to grow at the rate of

$$S'(t) = 60t^{1/2}$$

listeners per day, where t is the number of days since the campaign began. How long should the campaign last if the station wants the number of daily listeners to grow to 41,000?

SOLUTION We must solve the equation $S(t) = 41,000$ for t, given that

$$S'(t) = 60t^{1/2} \qquad \text{and} \qquad S(0) = 27,000$$

First, we use integration to find $S(t)$:

$$S(t) = 60t^{1/2}\, dt$$

$$= 60\frac{t^{3/2}}{\frac{3}{2}} + C$$

$$= 40t^{3/2} + C$$

Since

$$S(0) = 40(0)^{3/2} + C = 27,000$$

we have $C = 27,000$, and

$$S(t) = 40t^{3/2} + 27,000$$

Now we solve the equation $S(t) = 41,000$ for t:

$$40t^{3/2} + 27,000 = 41,000$$
$$40t^{3/2} = 14,000$$
$$t^{3/2} = 350$$
$$t = 350^{2/3} \qquad \text{Use a calculator.}$$
$$= 49.664\ 419\ldots$$

Thus, the advertising campaign should last approximately 50 days. ▬

MATCHED PROBLEM 6 The current monthly circulation of the magazine *Computing News* is 640,000 copies. Due to competition from a new magazine in the same field, the monthly circulation of *Computing News*, $C(t)$, is expected to decrease at the rate of

$$C'(t) = -6,000t^{1/3}$$

copies per month, where t is the time in months since the new magazine began publication. How long will it take for the circulation of *Computing News* to decrease to 460,000 copies per month?

Answers to Matched Problems

1. (A) $x^3 + C$

(B)

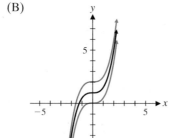

(C) The graphs are vertical translations of each other.

2. (A) $2x + C$ (B) $16e^t + C$ (C) $\frac{3}{5}x^5 + C$

(D) $\frac{1}{3}x^6 - x^3 + x + C$ (E) $5\ln|x| - 4e^x + C$

3. (A) $\frac{6}{5}x^{5/3} + x^{-3} + C$ (B) $\frac{5}{2}w^{8/5} + C$ (C) $\frac{1}{3}x^3 - 4x^2 + C$

(D) $6x^{4/3} - 12x^{1/2} + C$ (E) $\frac{1}{4}x^4 + x^3 - x^2 - 6x + C$

4. $y = x^3 - 2$

5. $R(x) = 400x - 0.2x^2; R(1,000) = \$200,000$

6. $t = (40)^{3/4} \approx 16$ mo

Exercise 6-1

A *In Problems 1–22, find each indefinite integral. (Check by differentiating.)*

1. $\int x^2 \, dx$

2. $\int x^5 \, dx$

3. $\int x^7 \, dx$

4. $\int x^{-4} \, dx$

5. $\int 2 \, dx$

6. $\int -10 \, dx$

7. $\int 5u^{-2} \, du$

8. $\int 7u^3 \, du$

9. $\int e^2 \, dt$

10. $\int \sqrt{2} \, dt$

11. $\int z^{49} \, dz$

12. $\int z^{99} \, dz$

13. $\int 2\pi x \, dx$

14. $\int ex \, dx$

15. $\int 8u^{-1} \, du$

16. $\int \dfrac{9}{u} \, du$

17. $\int 15x^{1/2} \, dx$

18. $\int 8t^{2/3} \, dt$

19. $\int 7t^{-4/3} \, dt$

20. $\int 2x^{-5/2} \, dx$

21. $\int (x - \sqrt{x}) \, dx$

22. $\int (3\sqrt{x} - x^{-1}) \, dx$

In Problems 23–32, find all the antiderivatives for each derivative.

23. $\dfrac{dy}{dx} = 200x^4$

24. $\dfrac{dx}{dt} = 42t^5$

25. $\dfrac{dP}{dx} = 24 - 6x$

26. $\dfrac{dy}{dx} = 3x^2 - 4x^3$

27. $\dfrac{dy}{du} = 2u^5 - 3u^2 - 1$

28. $\dfrac{dA}{dt} = 3 - 12t^3 - 9t^5$

29. $\dfrac{dy}{dx} = e^x + 3$

30. $\dfrac{dy}{dx} = x - e^x$

31. $\dfrac{dx}{dt} = 5t^{-1} + 1$

32. $\dfrac{du}{dv} = \dfrac{4}{v} + \dfrac{v}{4}$

B *In Problems 33–42, discuss the validity of each statement. If the statement is always true, explain why. If not, give a counterexample.*

33. The constant function $f(x) = \pi$ is an antiderivative of the constant function $k(x) = 0$.

34. The constant function $k(x) = 0$ is an antiderivative of the constant function $f(x) = \pi$.

35. If n is an integer, then $x^{n+1}/(n + 1)$ is an antiderivative of x^n.

36. The constant function $k(x) = 0$ is an antiderivative of itself.

37. The function $h(x) = 5e^x$ is an antiderivative of itself.

38. The constant function $g(x) = 5e^\pi$ is an antiderivative of itself.

39. If f and g are both antiderivatives of h, then the function $f + g$ is an antiderivative of h.

40. If f and g are both antiderivatives of h, then $f - g$ is an antiderivative of the constant function $k(x) = 0$.

41. $\dfrac{d}{dx}\left(\int x^2 \, dx \right) = x^2 + C$

42. $\int \dfrac{d}{dx} (x^4 + x^2) \, dx = x^4 + x^2 + C$

In Problems 43–46, could the three graphs be antiderivatives of the same function? Explain.

43.

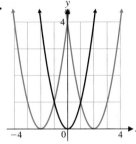

44.

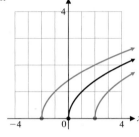

45.

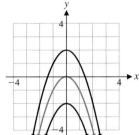

46.

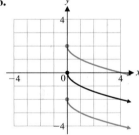

In Problems 47–72, find each indefinite integral. (Check by differentiation.)

47. $\int 5x(1 - x) \, dx$

48. $\int x^2(1 + x^3) \, dx$

49. $\int (2 + x^2)(3 + x^2) \, dx$

50. $\int (1 + x)(1 - x) \, dx$

51. $\int \dfrac{du}{\sqrt{u}}$

52. $\int \dfrac{dt}{\sqrt[3]{t}}$

53. $\int \dfrac{dx}{4x^3}$

54. $\int \dfrac{6 \, dm}{m^2}$

55. $\int \dfrac{4 + u}{u} \, du$

56. $\int \dfrac{1 - y^2}{3y} \, dy$

57. $\int (5e^z + 4) \, dz$

58. $\int \dfrac{e^t - t}{2} \, dt$

59. $\int \left(3x^2 - \dfrac{2}{x^2} \right) dx$

60. $\int \left(4x^3 + \dfrac{2}{x^3} \right) dx$

61. $\int \left(10x^4 - \dfrac{8}{x^5} - 2 \right) dx$

62. $\int \left(\dfrac{6}{x^4} - \dfrac{2}{x^3} + 1 \right) dx$

63. $\int \left(3\sqrt{x} + \dfrac{2}{\sqrt{x}}\right) dx$ **64.** $\int \left(\dfrac{2}{\sqrt[3]{x}} - \sqrt[3]{x^2}\right) dx$

65. $\int \left(\sqrt[3]{x^2} - \dfrac{4}{x^3}\right) dx$ **66.** $\int \left(\dfrac{12}{x^5} - \dfrac{1}{\sqrt[3]{x^2}}\right) dx$

67. $\int \dfrac{e^x - 3x}{4} dx$ **68.** $\int \dfrac{e^x - 3x^2}{2} dx$

69. $\int \dfrac{12 + 5z - 3z^3}{z^4} dz$ **70.** $\int \dfrac{(1 + z^2)^2}{z} dz$

71. $\int \left(\dfrac{6x^2}{5} - \dfrac{2}{3x}\right) dx$ **72.** $\int \left(\dfrac{2}{3x^2} - \dfrac{5}{4x^3}\right) dx$

In Problems 73–82, find the particular antiderivative of each derivative that satisfies the given condition.

73. $\dfrac{dy}{dx} = 2x - 3;\ y(0) = 5$

74. $\dfrac{dy}{dx} = 5 - 4x;\ y(0) = 20$

75. $C'(x) = 6x^2 - 4x;\ C(0) = 3{,}000$

76. $R'(x) = 600 - 0.6x;\ R(0) = 0$

77. $\dfrac{dx}{dt} = \dfrac{20}{\sqrt{t}};\ x(1) = 40$

78. $\dfrac{dR}{dt} = \dfrac{100}{t^2};\ R(1) = 400$

79. $\dfrac{dy}{dx} = 2x^{-2} + 3x^{-1} - 1;\ y(1) = 0$

80. $\dfrac{dy}{dx} = 3x^{-1} + x^{-2};\ y(1) = 1$

81. $\dfrac{dx}{dt} = 4e^t - 2;\ x(0) = 1$

82. $\dfrac{dy}{dt} = 5e^t - 4;\ y(0) = -1$

83. Find the equation of the curve that passes through $(2, 3)$ if its slope is given by

$$\dfrac{dy}{dx} = 4x - 3$$

for each x.

84. Find the equation of the curve that passes through $(1, 3)$ if its slope is given by

$$\dfrac{dy}{dx} = 12x^2 - 12x$$

for each x.

C In Problems 85–90, find each indefinite integral.

85. $\int \dfrac{2x^4 - x}{x^3} dx$ **86.** $\int \dfrac{x^{-1} - x^4}{x^2} dx$

87. $\int \dfrac{x^5 - 2x}{x^4} dx$ **88.** $\int \dfrac{1 - 3x^4}{x^2} dx$

89. $\int \dfrac{x^2 e^x - 2x}{x^2} dx$ **90.** $\int \dfrac{1 - xe^x}{x} dx$

For each derivative in Problems 91–96, find an antiderivative that satisfies the given condition.

91. $\dfrac{dM}{dt} = \dfrac{t^2 - 1}{t^2};\ M(4) = 5$

92. $\dfrac{dR}{dx} = \dfrac{1 - x^4}{x^3};\ R(1) = 4$

93. $\dfrac{dy}{dx} = \dfrac{5x + 2}{\sqrt[3]{x}};\ y(1) = 0$

94. $\dfrac{dx}{dt} = \dfrac{\sqrt{t^3} - t}{\sqrt{t^3}};\ x(9) = 4$

95. $p'(x) = -\dfrac{10}{x^2};\ p(1) = 20$

96. $p'(x) = \dfrac{10}{x^3};\ p(1) = 15$

In Problems 97–100, find the derivative or indefinite integral as indicated.

97. $\dfrac{d}{dx}\left(\int x^3\,dx\right)$ **98.** $\dfrac{d}{dt}\left(\int \dfrac{\ln t}{t}\,dt\right)$

99. $\int \dfrac{d}{dx}(x^4 + 3x^2 + 1)\,dx$

100. $\int \dfrac{d}{du}(e^{u^2})\,du$

101. Use differentiation to justify the formula

$$\int x^n\,dx = \dfrac{x^{n+1}}{n + 1} + C$$

provided that $n \neq -1$.

102. Use differentiation to justify the formula

$$\int e^x\,dx = e^x + C$$

103. Assuming that $x > 0$, use differentiation to justify the formula

$$\int \dfrac{1}{x}\,dx = \ln|x| + C$$

104. Assuming that $x < 0$, use differentiation to justify the formula

$$\int \dfrac{1}{x}\,dx = \ln|x| + C$$

105. Show that the indefinite integral of the sum of two functions is the sum of the indefinite integrals.
[Hint: Assume that $\int f(x)\,dx = F(x) + C_1$ and $\int g(x)\,dx = G(x) + C_2$. Using differentiation, show that $F(x) + C_1 + G(x) + C_2$ is the indefinite integral of the function $s(x) = f(x) + g(x)$.]

106. Show that the indefinite integral of the difference of two functions is the difference of the indefinite integrals.

Applications

107. *Cost function.* The marginal average cost of producing x digital sports watches is given by

$$\overline{C}'(x) = -\frac{1,000}{x^2} \qquad \overline{C}(100) = 25$$

where $\overline{C}(x)$ is the average cost in dollars. Find the average cost function and the cost function. What are the fixed costs?

108. *Renewable energy.* In 2002, U.S. consumption of renewable energy was 5.89 quadrillion Btu (that is, 5.89×10^{15} Btu). Since the 1960s, consumption has been growing at a rate given by

$$f'(t) = 0.004t + 0.062$$

where t is years after 1960. Find $f(t)$ and estimate U.S. consumption of renewable energy in 2020.

109. *Manufacturing costs.* The graph of the marginal cost function from the manufacture of x thousand watches per month [where cost $C(x)$ is in thousands of dollars per month] is given in the figure.

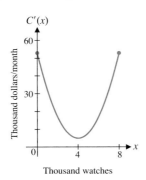

Figure for 109

(A) Using the graph shown, describe the shape of the graph of the cost function $C(x)$ as x increases from 0 to 8,000 watches per month.

(B) Given the equation of the marginal cost function,

$$C'(x) = 3x^2 - 24x + 53$$

find the cost function if monthly fixed costs at 0 output are $30,000. What is the cost of manufacturing 4,000 watches per month? 8,000 watches per month?

(C) Graph the cost function for $0 \le x \le 8$. [Check the shape of the graph relative to the analysis in part (A).]

(D) Why do you think that the graph of the cost function is steeper at both ends than in the middle?

110. *Revenue.* The graph of the marginal revenue function from the sale of x digital sports watches is given in the figure.

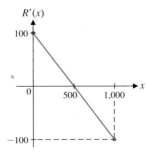

Figure for 110

(A) Using the graph shown, describe the shape of the graph of the revenue function $R(x)$ as x increases from 0 to 1,000.

(B) Find the equation of the marginal revenue function (the linear function shown in the figure).

(C) Find the equation of the revenue function that satisfies $R(0) = 0$. Graph the revenue function over the interval $[0, 1,000]$. [Check the shape of the graph relative to the analysis in part (A).]

(D) Find the price–demand equation and determine the price when the demand is 700 units.

111. *Sales analysis.* Monthly sales of a particular personal computer are expected to decline at the rate of

$$S'(t) = -25t^{2/3}$$

computers per month, where t is time in months and $S(t)$ is the number of computers sold each month. The company plans to stop manufacturing this computer when monthly sales reach 800 computers. If monthly sales now ($t = 0$) are 2,000 computers, find $S(t)$. How long will the company continue to manufacture this computer?

112. *Sales analysis.* The rate of change of the monthly sales of a new home video game cartridge is given by

$$S'(t) = 500t^{1/4} \qquad S(0) = 0$$

where t is the number of months since the game was released and $S(t)$ is the number of cartridges sold each month. Find $S(t)$. When will monthly sales reach 20,000 cartridges?

113. *Sales analysis.* Repeat Problem 111 if $S'(t) = -25t^{2/3} - 70$ and all other information remains the same. Use a graphing calculator to approximate the solution of the equation $S(t) = 800$ to two decimal places.

114. *Sales analysis.* Repeat Problem 112 if $S'(t) = 500t^{1/4} + 300$ and all other information remains the same. Use a graphing calculator to approximate the solution of the equation $S(t) = 20,000$ to two decimal places.

115. *Labor costs and learning.* A defense contractor is starting production on a new missile control system. On the basis of data collected during the assembly of the first 16 control systems, the production manager obtained the following function describing the rate of labor use:

$$g(x) = 2,400x^{-1/2}$$

Here, $g(x)$ is the number of labor-hours required to assemble the xth unit of the control system. For example, after assembly of 16 units, the rate of assembly is 600 labor-hours per unit, and after assembly of 25 units, the rate of assembly is 480 labor-hours per unit. The more units assembled, the more efficient is the process because of learning. If 19,200 labor-hours are required to assemble of the first 16 units, how many labor-hours, $L(x)$, will be required to assemble the first x units? The first 25 units?

116. *Labor costs and learning.* If the rate of labor use in Problem 115 is

$$g(x) = 2,000x^{-1/3}$$

and if the first 8 control units require 12,000 labor-hours, how many labor-hours, $L(x)$, will be required for the first x control units? The first 27 control units?

117. *Weight–height.* For an average person, the rate of change of weight W (in pounds) with respect to height h (in inches) is given approximately by

$$\frac{dW}{dh} = 0.0015h^2$$

Find $W(h)$ if $W(60) = 108$ pounds. Also, find the weight of a person who is 5 feet, 10 inches, tall.

118. *Wound healing.* If the area A of a healing wound changes at a rate given approximately by

$$\frac{dA}{dt} = -4t^{-3} \qquad 1 \le t \le 10$$

where t is time in days and $A(1) = 2$ square centimeters, what will the area of the wound be in 10 days?

119. *Urban growth.* The rate of growth of the population $N(t)$ of a newly incorporated city t years after its incorporation is estimated to be

$$\frac{dN}{dt} = 400 + 600\sqrt{t} \qquad 0 \le t \le 9$$

If the population was 5,000 at the time of incorporation, find the population 9 years later.

120. *Learning.* A beginning high school language class was chosen for an experiment in learning. Using a list of 50 words, the experiment involved measuring the rate of vocabulary memorization at different times during a continuous 5-hour study session. It was found that the average rate of learning for the entire class was inversely proportional to the time spent studying and was given approximately by

$$V'(t) = \frac{15}{t} \qquad 1 \le t \le 5$$

If the average number of words memorized after 1 hour of study was 15 words, what was the average number of words learned after t hours of study for $1 \le t \le 5$? After 4 hours of study? (Round answer to the nearest whole number.)

Section 6-2 INTEGRATION BY SUBSTITUTION

- Reversing the Chain Rule
- Integration by Substitution
- Additional Substitution Techniques
- Application

Many of the indefinite integral formulas introduced in the preceding section are based on corresponding derivative formulas studied earlier. We now consider indefinite integral formulas and procedures based on the chain rule for differentiation.

■ Reversing the Chain Rule

Recall the chain rule:

$$\frac{d}{dx}f[g(x)] = f'[g(x)]g'(x)$$

The expression on the right is formed from the expression on the left by taking the derivative of the outside function f and multiplying it by the derivative of the inside function g. If we recognize an integrand as a chain-rule form $f'[g(x)]g'(x)$, we can easily find an antiderivative and its indefinite integral:

$$\int f'[g(x)]g'(x)\,dx = f[g(x)] + C \qquad (1)$$

Explore & Discuss **1**

(A) Which of the following has e^{x^3-1} as an antiderivative?

$$x^2 e^{x^3-1} \qquad 3x^2 e^{x^3-1} \qquad 3x e^{x^3-1}$$

(B) Which of the following would have e^{x^3-1} as an antiderivative if it were multiplied by a constant factor? A variable factor?

$$3x e^{x^3-1} \qquad x^2 e^{x^3-1}$$

We are interested in finding the indefinite integral

$$\int 3x^2 e^{x^3-1}\, dx \tag{2}$$

The integrand appears to be the chain-rule form $e^{g(x)} g'(x)$, which is the derivative of $e^{g(x)}$. Since

$$\frac{d}{dx} e^{x^3-1} = 3x^2 e^{x^3-1}$$

it follows that

$$\int 3x^2 e^{x^3-1}\, dx = e^{x^3-1} + C \tag{3}$$

How does the following indefinite integral differ from integral (2)?

$$\int x^2 e^{x^3-1}\, dx \tag{4}$$

It is missing the constant factor 3. That is, $x^2 e^{x^3-1}$ is within a constant factor of being the derivative of e^{x^3-1}. But because a constant factor can be moved across the integral sign, this causes us little trouble in finding the indefinite integral of $x^2 e^{x^3-1}$. We introduce the constant factor 3 and at the same time multiply by $\frac{1}{3}$ and move the $\frac{1}{3}$ factor outside the integral sign. This is equivalent to multiplying the integrand in integral (4) by 1:

$$\int x^2 e^{x^3-1}\, dx = \int \frac{3}{3} x^2 e^{x^3-1}\, dx$$

$$= \frac{1}{3} \int 3x^2 e^{x^3-1}\, dx = \frac{1}{3} e^{x^3-1} + C \tag{5}$$

The derivative of the rightmost side of equation (5) is the integrand of the indefinite integral (4). You should check this.

How does the following indefinite integral differ from integral (2)?

$$\int 3x e^{x^3-1}\, dx \tag{6}$$

It is missing a variable factor x. This is more serious. As tempting as it might be, we *cannot* adjust integral (6) by introducing the variable factor x and moving $1/x$ outside the integral sign, as we did with the constant 3 in equation (5). If we could move $1/x$ across the integral sign, what would stop us from moving the entire integrand across the integral sign? Then, indefinite integration would become a trivial exercise and would not give us the results we want: antiderivatives of the integrand.

Summary of the Preceding Integral Forms

$$\int 3x^2 e^{x^3-1}\, dx = e^{x^3-1} + C \qquad \text{Integrand is a chain-rule form.}$$

$$\int x^2 e^{x^3-1}\, dx = \frac{1}{3} e^{x^3-1} + C \qquad \text{Integrand can be adjusted to a chain-rule form.}$$

$$\int 3x e^{x^3-1}\, dx = ? \qquad \text{Integrand cannot be adjusted to be a chain-rule form.}$$

⚠ CAUTION   A constant factor can be moved across an integral sign, but a variable factor cannot.

There is nothing wrong with educated guessing when you are looking for an antiderivative of a given function, and you are encouraged to do so. You have only to check the result by differentiation. If you are right, you go on your way; if you are wrong, you simply try another approach.

In Section 4-4, we saw that the chain rule extends the derivative formulas for x^n, e^x, and $\ln x$ to derivative formulas for $[f(x)]^n$, $e^{f(x)}$, and $\ln[f(x)]$. The chain rule can also be used to extend the indefinite integral formulas discussed in Section 6-1. Some general formulas are summarized in the following box:

FORMULAS   **General Indefinite Integral Formulas**

1. $\displaystyle\int [f(x)]^n f'(x)\, dx = \frac{[f(x)]^{n+1}}{n+1} + C, n \neq -1$

2. $\displaystyle\int e^{f(x)} f'(x)\, dx = e^{f(x)} + C$

3. $\displaystyle\int \frac{1}{f(x)} f'(x)\, dx = \ln|f(x)| + C$

Each formula can be verified by using the chain rule to show that the derivative of the function on the right is the integrand on the left. For example,

$$\frac{d}{dx}[e^{f(x)} + C] = e^{f(x)} f'(x)$$

verifies formula 2.

EXAMPLE 1   **Reversing the Chain Rule**

(A) $\displaystyle\int (3x + 4)^{10}(3)\, dx = \frac{(3x + 4)^{11}}{11} + C$ Formula 1 with $f(x) = 3x + 4$ and $f'(x) = 3$

Check:

$$\frac{d}{dx}\frac{(3x + 4)^{11}}{11} = 11\frac{(3x + 4)^{10}}{11}\frac{d}{dx}(3x + 4) = (3x + 4)^{10}(3)$$

(B) $\displaystyle\int e^{x^2}(2x)\, dx = e^{x^2} + C$ Formula 2 with $f(x) = x^2$ and $f'(x) = 2x$

Check:

$$\frac{d}{dx}e^{x^2} = e^{x^2}\frac{d}{dx}x^2 = e^{x^2}(2x)$$

(C) $\displaystyle\int \frac{1}{1 + x^3}3x^2\, dx = \ln|1 + x^3| + C$ Formula 3 with $f(x) = 1 + x^3$ and $f'(x) = 3x^2$

Check:

$$\frac{d}{dx}\ln|1 + x^3| = \frac{1}{1 + x^3}\frac{d}{dx}(1 + x^3) = \frac{1}{1 + x^3}3x^2$$ ▬

MATCHED PROBLEM 1   Find each indefinite integral.

(A) $\displaystyle\int (2x^3 - 3)^{20}(6x^2)\, dx$ (B) $\displaystyle\int e^{5x}(5)\, dx$

(C) $\displaystyle\int \frac{1}{4 + x^2}2x\, dx$

■ Integration by Substitution

The key step in using formulas 1, 2, and 3 is recognizing the form of the integrand. Some people find it difficult to identify $f(x)$ and $f'(x)$ in these formulas and prefer to use a *substitution* to simplify the integrand. The *method of substitution,* which we now discuss, becomes increasingly useful as one progresses in studies of integration.

We start by introducing the idea of the *differential.* We represented the derivative by the symbol dy/dx taken as a whole. We now define dy and dx as two separate quantities with the property that their ratio is still equal to $f'(x)$:

DEFINITION   Differentials

If $y = f(x)$ defines a differentiable function, then

1. The **differential dx** of the independent variable x is an arbitrary real number.

2. The **differential dy** of the dependent variable y is defined as the product of $f'(x)$ and dx—that is, as

$$dy = f'(x)\,dx$$

Differentials involve mathematical subtleties that are treated carefully in advanced mathematics courses. Here, we are interested in them mainly as a bookkeeping device to aid in the process of finding indefinite integrals. We can always check the results by differentiating.

EXAMPLE 2   Differentials

(A) If $y = f(x) = x^2$, then

$$dy \boxed{= f'(x)\,dx} = 2x\,dx$$

(B) If $u = g(x) = e^{3x}$, then

$$du \boxed{= g'(x)\,dx} = 3e^{3x}\,dx$$

(C) If $w = h(t) = \ln(4 + 5t)$, then

$$dw \boxed{= h'(t)\,dt} = \frac{5}{4 + 5t}\,dt$$

MATCHED PROBLEM 2   (A) Find dy for $y = f(x) = x^3$.

(B) Find du for $u = h(x) = \ln(2 + x^2)$.

(C) Find dv for $v = g(t) = e^{-5t}$.

The **method of substitution** is developed through Examples 3–6.

EXAMPLE 3   **Using Substitution** Find $\int (x^2 + 2x + 5)^5 (2x + 2)\,dx$.

SOLUTION   If

$$u = x^2 + 2x + 5$$

then the differential of u is

$$du = (2x + 2)\,dx$$

Notice that du is one of the factors in the integrand. Substitute u for $x^2 + 2x + 5$ and du for $(2x + 2)\,dx$ to obtain

$$\int (x^2 + 2x + 5)^5(2x + 2)\,dx = \int u^5\,du$$

$$= \frac{u^6}{6} + C$$

$$= \frac{1}{6}(x^2 + 2x + 5)^6 + C \quad \substack{\text{Since}\\ u = x^2 + 2x + 5}$$

Check:

$$\frac{d}{dx}\frac{1}{6}(x^2 + 2x + 5)^6 = \frac{1}{6}(6)(x^2 + 2x + 5)^5\frac{d}{dx}(x^2 + 2x + 5)$$

$$= (x^2 + 2x + 5)^5(2x + 2)$$

MATCHED PROBLEM 3 Find $\int (x^2 - 3x + 7)^4(2x - 3)\,dx$ by substitution.

The substitution method is also called the **change-of-variable method,** since u replaces the variable x in the process. Substituting $u = f(x)$ and $du = f'(x)\,dx$ in formulas 1, 2, and 3 produces the general indefinite integral formulas in the following box:

FORMULAS **General Indefinite Integral Formulas**

4. $\displaystyle\int u^n\,du = \frac{u^{n+1}}{n+1} + C, \qquad n \neq -1$

5. $\displaystyle\int e^u\,du = e^u + C$

6. $\displaystyle\int \frac{1}{u}\,du = \ln|u| + C$

These formulas are valid if u is an independent variable or if u is a function of another variable and du is the differential of u with respect to that variable.

The substitution method for evaluating certain indefinite integrals is outlined in the following box:

PROCEDURE **Integration by Substitution**

Step 1. Select a substitution that appears to simplify the integrand. In particular, try to select u so that du is a factor in the integrand.

Step 2. Express the integrand entirely in terms of u and du, completely eliminating the original variable and its differential.

Step 3. Evaluate the new integral if possible.

Step 4. Express the antiderivative found in step 3 in terms of the original variable.

EXAMPLE 4 **Using Substitution** Use a substitution to find the following:

(A) $\displaystyle\int (3x + 4)^6(3)\,dx$ \qquad (B) $\displaystyle\int e^{t^2}(2t)\,dt$

SOLUTION (A) If we let $u = 3x + 4$, then $du = 3\,dx$, and

$$\int (3x + 4)^6 (3)\,dx = \int u^6\,du \qquad \text{Use formula 4.}$$

$$= \frac{u^7}{7} + C$$

$$= \frac{(3x + 4)^7}{7} + C \quad \text{Since } u = 3x + 4$$

Check:

$$\frac{d}{dx}\frac{(3x + 4)^7}{7} = \frac{7(3x + 4)^6}{7}\frac{d}{dx}(3x + 4) = (3x + 4)^6 (3)$$

(B) If we let $u = t^2$, then $du = 2t\,dt$, and

$$e^{t^2}(2t)\,dt = e^u\,du \qquad \text{Use formula 5.}$$

$$= e^u + C$$

$$= e^{t^2} + C \quad \text{Since } u = t^2$$

Check:

$$\frac{d}{dt}e^{t^2} = e^{t^2}\frac{d}{dt}t^2 = e^{t^2}(2t)$$

■

MATCHED PROBLEM 4   Use a substitution to find each indefinite integral.

(A) $\displaystyle\int (2x^3 - 3)^4 (6x^2)\,dx$

(B) $\displaystyle\int e^{5w}(5)\,dw$

■ Additional Substitution Techniques

In order to use the substitution method, **the integrand must be expressed entirely in terms of u and du.** In some cases, the integrand will have to be modified before making a substitution and using one of the integration formulas. Example 5 illustrates this process.

EXAMPLE 5   **Substitution Techniques** Integrate:

(A) $\displaystyle\int \frac{1}{4x + 7}\,dx$ (B) $\displaystyle\int te^{-t^2}\,dt$ (C) $\displaystyle\int 4x^2\sqrt{x^3 + 5}\,dx$

SOLUTION (A) If $u = 4x + 7$, then $du = 4\,dx$. We are missing a factor of 4 in the integrand to match formula 6 exactly. Recalling that a constant factor can be moved across an integral sign, we proceed as follows:

$$\int \frac{1}{4x + 7}\,dx = \int \frac{1}{4x + 7}\frac{4}{4}\,dx$$

$$= \frac{1}{4}\int \frac{1}{4x + 7}4\,dx \quad \text{Substitute } u = 4x + 7 \text{ and } du = 4\,dx.$$

$$= \frac{1}{4}\int \frac{1}{u}\,du \qquad \text{Use formula 6.}$$

$$= \tfrac{1}{4}\ln|u| + C$$

$$= \tfrac{1}{4}\ln|4x + 7| + C \quad \text{Since } u = 4x + 7$$

Check:

$$\frac{d}{dx}\frac{1}{4}\ln|4x + 7| = \frac{1}{4}\frac{1}{4x + 7}\frac{d}{dx}(4x + 7) = \frac{1}{4}\frac{1}{4x + 7}4 = \frac{1}{4x + 7}$$

(B) If $u = -t^2$, then $du = -2t\, dt$. Proceed as in part (A):

$$\int te^{-t^2}\, dt = \int e^{-t^2}\frac{-2}{-2}t\, dt$$

$$= -\frac{1}{2}\int e^{-t^2}(-2t)\, dt \quad \text{Substitute } u = -t^2 \text{ and } du = -2t\, dt.$$

$$= -\frac{1}{2}\int e^u\, du \quad\quad\quad \text{Use formula 5.}$$

$$= -\frac{1}{2}e^u + C$$

$$= -\frac{1}{2}e^{-t^2} + C \quad\quad\quad \text{Since } u = -t^2$$

Check:

$$\frac{d}{dt}\left(-\frac{1}{2}e^{-t^2}\right) = -\frac{1}{2}e^{-t^2}\frac{d}{dt}(-t^2) = -\frac{1}{2}e^{-t^2}(-2t) = te^{-t^2}$$

(C) $\displaystyle\int 4x^2\sqrt{x^3 + 5}\, dx = 4\int \sqrt{x^3 + 5}(x^2)\, dx$
      Move the 4 across the integral sign and proceed as before.

$$= 4\int \sqrt{x^3 + 5}\,\frac{3}{3}(x^2)\, dx$$

$$= \frac{4}{3}\int \sqrt{x^3 + 5}(3x^2)\, dx \quad \text{Substitute } u = x^3 + 5 \text{ and } du = 3x^2\, dx.$$

$$= \frac{4}{3}\int \sqrt{u}\, du$$

$$= \frac{4}{3}\int u^{1/2}\, du \quad\quad\quad \text{Use formula 4.}$$

$$= \frac{4}{3}\frac{u^{3/2}}{\frac{3}{2}} + C$$

$$= \frac{8}{9}u^{3/2} + C$$

$$= \frac{8}{9}(x^3 + 5)^{3/2} + C \quad\quad \text{Since } u = x^3 + 5$$

Check:

$$\frac{d}{dx}\left[\frac{8}{9}(x^3 + 5)^{3/2}\right] = \frac{4}{3}(x^3 + 5)^{1/2}\frac{d}{dx}(x^3 + 5)$$

$$= \frac{4}{3}(x^3 + 5)^{1/2}(3x^2) = 4x^2\sqrt{x^3 + 5}$$

MATCHED PROBLEM 5   Integrate:

(A) $\displaystyle\int e^{-3x}\, dx$ **(B)** $\displaystyle\int \frac{x}{x^2 - 9}\, dx$

(C) $\displaystyle\int 5t^2(t^3 + 4)^{-2}\, dt$

Even if it is not possible to find a substitution that makes an integrand match one of the integration formulas exactly, a substitution may simplify the integrand sufficiently that other techniques can be used.

EXAMPLE 6   **Substitution Techniques** Find $\int \dfrac{x}{\sqrt{x+2}}\,dx$.

SOLUTION   Proceeding as before, if we let $u = x + 2$, then $du = dx$ and

$$\int \frac{x}{\sqrt{x+2}}\,dx = \int \frac{x}{\sqrt{u}}\,du$$

Notice that this substitution is not yet complete, because we have not expressed the integrand entirely in terms of u and du. As we noted earlier, only a constant factor can be moved across an integral sign, so we cannot move x outside the integral sign (as much as we would like to). Instead, we must return to the original substitution, solve for x in terms of u, and use the resulting equation to complete the substitution:

$$u = x + 2 \quad \text{Solve for x in terms of u.}$$
$$u - 2 = x \qquad \text{Substitute this expression for x.}$$

Thus,

$$\int \frac{x}{\sqrt{x+2}}\,dx = \int \frac{u-2}{\sqrt{u}}\,du \qquad\qquad \text{Simplify the integrand.}$$

$$= \int \frac{u-2}{u^{1/2}}\,du$$

$$= \int (u^{1/2} - 2u^{-1/2})\,du$$

$$\boxed{= \int u^{1/2}\,du - 2\int u^{-1/2}\,du}$$

$$= \frac{u^{3/2}}{\frac{3}{2}} - 2\frac{u^{1/2}}{\frac{1}{2}} + C$$

$$= \tfrac{2}{3}(x+2)^{3/2} - 4(x+2)^{1/2} + C \quad \text{Since } u = x + 2$$

Check:

$$\frac{d}{dx}\left[\tfrac{2}{3}(x+2)^{3/2} - 4(x+2)^{1/2}\right] = (x+2)^{1/2} - 2(x+2)^{-1/2}$$

$$= \frac{x+2}{(x+2)^{1/2}} - \frac{2}{(x+2)^{1/2}}$$

$$= \frac{x}{(x+2)^{1/2}}$$

MATCHED PROBLEM 6   Find $\int x\sqrt{x+1}\,dx$.

INSIGHT

We can find the indefinite integral of some functions in more than one way. For example, we can employ substitution to find

$$\int x(1 + x^2)^2\,dx$$

by letting $u = 1 + x^2$. As a second approach, we can expand the integrand, obtaining

$$\int (x + 2x^3 + x^5)\, dx$$

for which we can easily calculate an antiderivative. In such a case, use the approach that you prefer.

There are also some functions for which substitution is not an effective approach to finding the indefinite integral. For example, substitution is not helpful in finding

$$\int e^{x^2}\, dx \qquad \text{or} \qquad \int \ln x\, dx \qquad \blacksquare$$

APPLICATION

EXAMPLE 7

Price–Demand The market research department of a supermarket chain has determined that, for one store, the marginal price $p'(x)$ at x tubes per week for a certain brand of toothpaste is given by

$$p'(x) = -0.015e^{-0.01x}$$

Find the price–demand equation if the weekly demand is 50 tubes when the price of a tube is \$2.35. Find the weekly demand when the price of a tube is \$1.89.

SOLUTION

$$p(x) = \int -0.015e^{-0.01x}\, dx$$

$$= -0.015 \int e^{-0.01x}\, dx$$

$$= -0.015 \int e^{-0.01x} \frac{-0.01}{-0.01}\, dx$$

$$= \frac{-0.015}{-0.01} \int e^{-0.01x}(-0.01)\, dx \qquad \text{Substitute } u = -0.01x \text{ and } du = -0.01\, dx.$$

$$= 1.5 \int e^u\, du$$

$$= 1.5e^u + C$$

$$= 1.5e^{-0.01x} + C \qquad \text{Since } u = -0.01x$$

We find C by noting that

$$p(50) = 1.5e^{-0.01(50)} + C = \$2.35$$

$$C = \$2.35 - 1.5e^{-0.5} \qquad \text{Use a calculator.}$$

$$= \$2.35 - 0.91$$

$$= \$1.44$$

Thus,

$$p(x) = 1.5e^{-0.01x} + 1.44$$

To find the demand when the price is \$1.89, we solve $p(x) = \$1.89$ for x:

$$1.5e^{-0.01x} + 1.44 = 1.89$$

$$1.5e^{-0.01x} = 0.45$$

$$e^{-0.01x} = 0.3$$

$$-0.01x = \ln 0.3$$

$$x = -100 \ln 0.3 \approx 120 \text{ tubes} \qquad \blacksquare$$

MATCHED PROBLEM 7　The marginal price $p'(x)$ at a supply level of x tubes per week for a certain brand of toothpaste is given by

$$p'(x) = 0.001e^{0.01x}$$

Find the price–supply equation if the supplier is willing to supply 100 tubes per week at a price of \$1.65 each. How many tubes would the supplier be willing to supply at a price of \$1.98 each?

Explore & Discuss **2**　In each of the following examples, explain why $\neq$ is used and then work the problem correctly, using either an appropriate substitution or another method:

1. $\displaystyle\int (x^2 + 3)^2\, dx = \int (x^2 + 3)^2 \frac{2x}{2x}\, dx$

$$\neq \frac{1}{2x} \int (x^2 + 3)^2 (2x)\, dx$$

2. $\displaystyle\int \frac{1}{10x + 3}\, dx = \int \frac{1}{u}\, dx \qquad u = 10x + 3$

$$\neq \ln|u| + C$$

We conclude with two final cautions (the first was stated earlier, but is worth repeating):

⚠ CAUTION

1. A variable cannot be moved across an integral sign!

2. An integral must be expressed entirely in terms of u and du before applying integration formulas 4, 5, and 6.

Answers to Matched Problems　**1.** (A) $\frac{1}{21}(2x^3 - 3)^{21} + C$　　(B) $e^{5x} + C$

    (C) $\ln|4 + x^2| + C$　or　$\ln(4 + x^2) + C$, since $4 + x^2 > 0$

2. (A) $dy = 3x^2\, dx$　　(B) $du = \dfrac{2x}{2 + x^2}\, dx$　　(C) $dv = -5e^{-5t}\, dt$

3. $\frac{1}{5}(x^2 - 3x + 7)^5 + C$

4. (A) $\frac{1}{5}(2x^3 - 3)^5 + C$　　(B) $e^{5w} + C$

5. (A) $-\frac{1}{3}e^{-3x} + C$　　(B) $\frac{1}{2}\ln|x^2 - 9| + C$　　(C) $-\frac{5}{3}(t^3 + 4)^{-1} + C$

6. $\frac{2}{5}(x + 1)^{5/2} - \frac{2}{3}(x + 1)^{3/2} + C$

7. $p(x) = 0.1e^{0.01x} + 1.38$; 179 tubes

Exercise 6-2

A　_In Problems 1–40, find each indefinite integral and check the result by differentiating._

1. $\displaystyle\int (3x + 5)^2 (3)\, dx$

2. $\displaystyle\int (6x - 1)^3 (6)\, dx$

3. $\displaystyle\int (x^2 - 1)^5 (2x)\, dx$

4. $\displaystyle\int (x^6 + 1)^4 (6x^5)\, dx$

5. $\displaystyle\int (5x^3 + 1)^{-3} (15x^2)\, dx$

6. $\displaystyle\int (4x^2 - 3)^{-6} (8x)\, dx$

7. $\displaystyle\int e^{5x} (5)\, dx$

8. $\displaystyle\int e^{x^3} (3x^2)\, dx$

9. $\displaystyle\int \frac{1}{1 + x^2} (2x)\, dx$

10. $\displaystyle\int \frac{1}{5x - 7} (5)\, dx$

11. $\displaystyle\int \sqrt{1 + x^4}\,(4x^3)\, dx$

12. $\displaystyle\int (x^2 + 9)^{-1/2} (2x)\, dx$

B

13. $\displaystyle\int (x + 3)^{10}\, dx$

14. $\displaystyle\int (x - 3)^{-4}\, dx$

15. $\displaystyle\int (6t - 7)^{-2}\, dt$

16. $\displaystyle\int (5t + 1)^3\, dt$

17. $\int (t^2 + 1)^5 \, t \, dt$

18. $\int (t^3 + 4)^{-2} \, t^2 \, dt$

19. $\int x e^{x^2} \, dx$

20. $\int e^{-0.01x} \, dx$

21. $\int \frac{1}{5x + 4} \, dx$

22. $\int \frac{x}{1 + x^2} \, dx$

23. $\int e^{1-t} \, dt$

24. $\int \frac{3}{2 - t} \, dt$

25. $\int \frac{t}{(3t^2 + 1)^4} \, dt$

26. $\int \frac{t^2}{(t^3 - 2)^5} \, dt$

27. $\int \frac{x^2}{(4 - x^3)^2} \, dx$

28. $\int \frac{x}{(5 - 2x^2)^5} \, dx$

29. $\int x \sqrt{x + 4} \, dx$

30. $\int x \sqrt{x - 9} \, dx$

31. $\int \frac{x}{\sqrt{x - 3}} \, dx$

32. $\int \frac{x}{\sqrt{x + 5}} \, dx$

33. $\int x(x - 4)^9 \, dx$

34. $\int x(x + 6)^8 \, dx$

35. $\int e^{2x}(1 + e^{2x})^3 \, dx$

36. $\int e^{-x}(1 - e^{-x})^4 \, dx$

37. $\int \frac{1 + x}{4 + 2x + x^2} \, dx$

38. $\int \frac{x^2 - 1}{x^3 - 3x + 7} \, dx$

39. $\int \frac{x^3 + x}{(x^4 + 2x^2 + 1)^4} \, dx$

40. $\int \frac{x^2 - 1}{(x^3 - 3x + 7)^2} \, dx$

In Problems 41–46, imagine that the indicated "solutions" were given to you by a student whom you are tutoring in this class.

(A) How would you have the student check each solution?

(B) Is the solution right or wrong? If the solution is wrong, explain what is wrong and how it can be corrected.

(C) Show a correct solution for each incorrect solution, and check the result by differentiation.

41. $\int \frac{1}{2x - 3} \, dx = \ln|2x - 3| + C$

42. $\int \frac{x}{x^2 + 5} \, dx = \ln|x^2 + 5| + C$

43. $\int x^3 e^{x^4} \, dx = e^{x^4} + C$

44. $\int e^{4x-5} \, dx = e^{4x-5} + C$

45. $\int 2(x^2 - 2)^2 \, dx = \frac{(x^2 - 2)^2}{3x} + C$

46. $\int (-10x)(x^2 - 3)^{-4} \, dx = (x^2 - 3)^{-5} + C$

C *In Problems 47–58, find each indefinite integral and check the result by differentiating.*

47. $\int x \sqrt{3x^2 + 7} \, dx$

48. $\int x^2 \sqrt{2x^3 + 1} \, dx$

49. $\int x(x^3 + 2)^2 \, dx$

50. $\int x(x^2 + 2)^2 \, dx$

51. $\int x^2(x^3 + 2)^2 \, dx$

52. $\int (x^2 + 2)^2 \, dx$

53. $\int \frac{x^3}{\sqrt{2x^4 + 3}} \, dx$

54. $\int \frac{x^2}{\sqrt{4x^3 - 1}} \, dx$

55. $\int \frac{(\ln x)^3}{x} \, dx$

56. $\int \frac{e^x}{1 + e^x} \, dx$

57. $\int \frac{1}{x^2} e^{-1/x} \, dx$

58. $\int \frac{1}{x \ln x} \, dx$

In Problems 59–64, find the antiderivative of each derivative.

59. $\frac{dx}{dt} = 7t^2(t^3 + 5)^6$

60. $\frac{dm}{dn} = 10n(n^2 - 8)^7$

61. $\frac{dy}{dt} = \frac{3t}{\sqrt{t^2 - 4}}$

62. $\frac{dy}{dx} = \frac{5x^2}{(x^3 - 7)^4}$

63. $\frac{dp}{dx} = \frac{e^x + e^{-x}}{(e^x - e^{-x})^2}$

64. $\frac{dm}{dt} = \frac{\ln(t - 5)}{t - 5}$

Use substitution techniques to derive the integration formulas in Problems 65 and 66. Then check your work by differentiation.

65. $\int e^{au} \, du = \frac{1}{a} e^{au} + C, \quad a \neq 0$

66. $\int \frac{1}{au + b} \, du = \frac{1}{a} \ln|au + b| + C, \quad a \neq 0$

Applications

67. *Price–demand equation.* The marginal price for a weekly demand of x bottles of baby shampoo in a drugstore is given by

$$p'(x) = \frac{-6,000}{(3x + 50)^2}$$

Find the price–demand equation if the weekly demand is 150 when the price of a bottle of shampoo is $4. What is the weekly demand when the price is $2.50?

68. *Price–supply equation.* The marginal price at a supply level of x bottles of baby shampoo per week is given by

$$p'(x) = \frac{300}{(3x + 25)^2}$$

Find the price–supply equation if the distributor of the shampoo is willing to supply 75 bottles a week at a price of $1.60 per bottle. How many bottles would the supplier be willing to supply at a price of $1.75 per bottle?

69. *Cost function.* The weekly marginal cost of producing x pairs of tennis shoes is given by

$$C'(x) = 12 + \frac{500}{x + 1}$$

where $C(x)$ is cost in dollars. If the fixed costs are $2,000 per week, find the cost function. What is the average cost per pair of shoes if 1,000 pairs of shoes are produced each week?

70. *Revenue function.* The weekly marginal revenue from the sale of x pairs of tennis shoes is given by

$$R'(x) = 40 - 0.02x + \frac{200}{x + 1} \qquad R(0) = 0$$

where $R(x)$ is revenue in dollars. Find the revenue function. Find the revenue from the sale of 1,000 pairs of shoes.

71. *Marketing.* An automobile company is ready to introduce a new line of cars through a national sales campaign. After test marketing the line in a carefully selected city, the marketing research department estimates that sales (in millions of dollars) will increase at the monthly rate of

$$S'(t) = 10 - 10e^{-0.1t} \qquad 0 \le t \le 24$$

t months after the campaign has started.

(A) What will be the total sales $S(t)$ t months after the beginning of the national campaign if we assume no sales at the beginning of the campaign?

(B) What are the estimated total sales for the first 12 months of the campaign?

(C) When will the estimated total sales reach $100 million? Use a graphing calculator to approximate the answer to two decimal places.

72. *Marketing.* Repeat Problem 71 if the monthly rate of increase in sales is found to be approximated by

$$S'(t) = 20 - 20e^{-0.05t} \qquad 0 \le t \le 24$$

73. *Oil production.* Using data from the first 3 years of production, as well as data from geological studies, the management of an oil company estimates that oil will be pumped from a field producing at a rate given by

$$R(t) = \frac{100}{t + 1} + 5 \qquad 0 \le t \le 20$$

where $R(t)$ is the rate of production (in thousands of barrels per year) t years after pumping begins. How many barrels of oil $Q(t)$ will the field produce the first t years if $Q(0) = 0$? How many barrels will be produced the first 9 years?

74. *Oil production.* Assume that the rate in Problem 73 is found to be

$$R(t) = \frac{120t}{t^2 + 1} + 3 \qquad 0 \le t \le 20$$

(A) When is the rate of production greatest?

(B) How many barrels of oil $Q(t)$ will the field produce the first t years if $Q(0) = 0$? How many barrels will be produced the first 5 years?

(C) How long (to the nearest tenth of a year) will it take to produce a total of a quarter of a million barrels of oil?

75. *Biology.* A yeast culture is growing at the rate of $W'(t) = 0.2e^{0.1t}$ grams per hour. If the starting culture weighs 2 grams, what will be the weight of the culture, $W(t)$, after t hours? After 8 hours?

76. *Medicine.* The rate of healing for a skin wound (in square centimeters per day) is approximated by $A'(t) = -0.9e^{-0.1t}$. If the initial wound has an area of 9 square centimeters, what will its area $A(t)$ be after t days? After 5 days?

77. *Pollution.* A contaminated lake is treated with a bactericide. The rate of increase in harmful bacteria t days after the treatment is given by

$$\frac{dN}{dt} = -\frac{2,000t}{1 + t^2} \qquad 0 \le t \le 10$$

where $N(t)$ is the number of bacteria per milliliter of water. (Since dN/dt is negative, the count of harmful bacteria is decreasing.)

(A) Find the minimum value of dN/dt.

(B) If the initial count was 5,000 bacteria per milliliter, find $N(t)$ and then find the bacteria count after 10 days.

(C) When (to two decimal places) is the bacteria count 1,000 bacteria per milliliter?

78. *Pollution.* An oil tanker aground on a reef is losing oil and producing an oil slick that is radiating outward at a rate given approximately by

$$\frac{dR}{dt} = \frac{60}{\sqrt{t + 9}} \qquad t \ge 0$$

where R is the radius (in feet) of the circular slick after t minutes. Find the radius of the slick after 16 minutes if the radius is 0 when $t = 0$.

79. *Learning.* In a particular business college, it was found that an average student enrolled in an advanced typing class progressed at a rate of $N'(t) = 6e^{-0.1t}$ words per minute per week t weeks after enrolling in a 15-week course. If, at the beginning of the course, a student could type 40 words per minute, how many words per minute $N(t)$ would the student be expected to type t weeks into the course? After completing the course?

80. *Learning.* In the same business college as in the previous exercise, it was also found that an average student enrolled in a beginning shorthand class progressed at a rate of $N'(t) = 12e^{-0.06t}$ words per minute per week t weeks after enrolling in a 15-week course. If, at the beginning of the course, a student could take dictation in shorthand at zero words per minute, how many words per minute $N(t)$ would the student be expected to handle t weeks into the course? After completing the course?

81. *College enrollment.* The projected rate of increase in enrollment in a new college is estimated by

$$\frac{dE}{dt} = 5,000(t + 1)^{-3/2} \qquad t \ge 0$$

where $E(t)$ is the projected enrollment in t years. If enrollment is 2,000 now ($t = 0$), find the projected enrollment 15 years from now.

Section 6-3 DIFFERENTIAL EQUATIONS; GROWTH AND DECAY

- Differential Equations and Slope Fields
- Continuous Compound Interest Revisited
- Exponential Growth Law
- Population Growth, Radioactive Decay, and Learning
- Comparison of Exponential Growth Phenomena

In the preceding section, we considered equations of the form

$$\frac{dy}{dx} = 6x^2 - 4x \qquad p'(x) = -400e^{-0.04x}$$

These are examples of *differential equations*. In general, an equation is a **differential equation** if it involves an unknown function and one or more of its derivatives. Other examples of differential equations are

$$\frac{dy}{dx} = ky \qquad y'' - xy' + x^2 = 5 \qquad \frac{dy}{dx} = 2xy$$

The first and third equations are called **first-order** equations, because each involves a first derivative, but no higher derivative. The second equation is called a **second-order** equation, because it involves a second derivative and no higher derivatives. Finding solutions of different types of differential equations (functions that satisfy the equation) is the subject matter of entire books and courses on this topic. Here, we consider only a few very special, but very important, first-order equations that have immediate and significant applications.

We start by looking at some first-order equations geometrically, in terms of *slope fields*. We then consider again continuous compound interest as modeled by a first-order differential equation. From this treatment, we will be able to generalize our approach to a wide variety of other types of growth phenomena.

■ Differential Equations and Slope Fields

We introduce the concept of *slope field* through an example. Suppose that we are given the first-order differential equation

$$\frac{dy}{dx} = 0.2y \tag{1}$$

A function f is a solution of equation (1) if $y = f(x)$ satisfies equation (1) for all values of x in the domain of f. Geometrically interpreted, equation (1) gives us the slope of a solution curve that passes through the point (x, y). For example, if $y = f(x)$ is a solution of equation (1) that passes through the point $(0, 2)$, then the slope of f at $(0, 2)$ is given by

$$\frac{dy}{dx} = 0.2(2) = 0.4$$

We indicate this relationship by drawing a short segment of the tangent line at the point $(0, 2)$, as shown in Figure 1A. The procedure is repeated for points $(-3, 1)$ and $(2, 3)$, as also shown in Figure 1A. Assuming that the graph of f passes through all three points, we sketch an approximate graph of f in Figure 1B.

If we continue the process of drawing tangent line segments at each point in the grid in Figure 1—a task easily handled by computers, but not by hand—we obtain a *slope field*. A slope field for differential equation (1), drawn by a computer, is shown

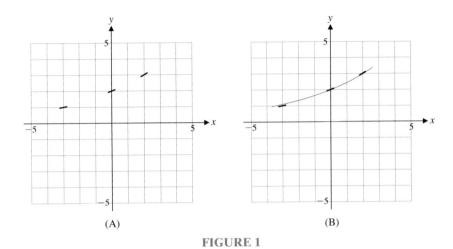

(A)   (B)

FIGURE 1

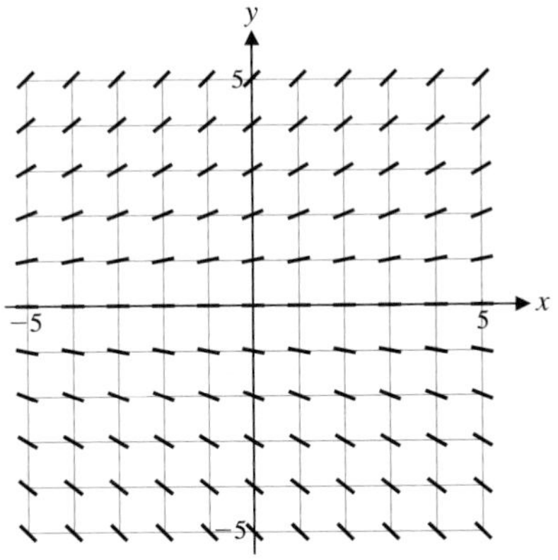

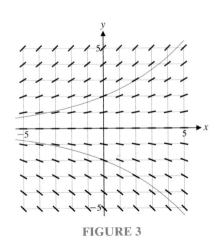

FIGURE 2

in Figure 2. In general, a **slope field** for a first-order differential equation is obtained by drawing tangent line segments determined by the equation at each point in a grid. In a more advanced treatment of the subject, one can find out a lot about the shape and behavior of solution curves of first-order differential equations by looking at slope fields. Our interests are more modest here.

Explore & Discuss **1**

(A) In Figure 1A (or a copy), draw tangent line segments for a solution curve of differential equation (1) that passes through $(-3, -1)$, $(0, -2)$, and $(2, -3)$.

(B) In Figure 1B (or a copy), sketch an approximate graph of the solution curve that passes through the three points given in Part (A). (Repeat the tangent line segments first.)

(C) Make a conjecture as to what type of function, of all the elementary functions discussed in the first two chapters, appears to be a solution of differential equation (1).

In Explore & Discuss 1, if you guessed that all solutions of equation (1) are exponential functions, you are to be congratulated. We now show that

$$y = Ce^{0.2x} \tag{2}$$

is a solution of equation (1) for any real number C. To do this, we substitute $y = Ce^{0.2x}$ into equation (1) to see if the left side is equal to the right side for all real x:

$$\frac{dy}{dx} = 0.2y$$

Left side: $\quad \dfrac{dy}{dx} = \dfrac{d}{dx}(Ce^{0.2x}) = 0.2Ce^{0.2x}$

Right side: $\quad 0.2y = 0.2Ce^{0.2x}$

Thus, equation (2) is a solution of equation (1) for C any real number. [Later in this section, we show how to find equation (2) directly from equation (1).] Which values of C will produce solution curves that pass through $(0, 2)$ and $(0, -2)$, respectively? Substituting the coordinates of each point into equation (2) and solving for C (a task left to the reader), we obtain

$$y = 2e^{0.2x} \qquad \text{and} \qquad y = -2e^{0.2x} \tag{3}$$

FIGURE 3

as can easily be checked. The graphs of equations (3) are shown in Figure 3 and confirm the results shown in Figure 1B.

Explore & Discuss **2**

(A) In Figure 3 (or a copy), sketch in an approximate solution curve that passes through $(0, 3)$ and one that passes through $(0, -3)$.

(B) Use a graphing calculator to graph $y = Ce^{0.2x}$ for $C = -4, -3, -2, 2, 3, 4$, all in the same viewing window. Notice how these solution curves follow the flow of the tangent line segments in the slope field in Figure 2.

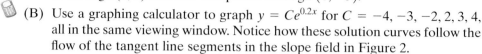

INSIGHT

For a complicated first-order differential equation, say,

$$\frac{dy}{dx} = \frac{3 + \sqrt{xy}}{x^2 - 5y^4}$$

it may be impossible to find a formula analogous to (2) for its solutions. Nevertheless, it is routine to evaluate the right-hand side at each point in a grid. The resulting slope field provides a graphical representation of the solutions of the differential equation.

As indicated, drawing slope fields by hand is not a task for human beings: A 20-by-20 grid would require drawing 400 tangent line segments! Repetitive tasks of this type are what computers are for. A few problems in Exercise 6-3 involve interpreting slope fields, not drawing them.

■ Continuous Compound Interest Revisited

Let P be the initial amount of money deposited in an account, and let A be the amount in the account at any time t. Instead of assuming that the money in the account earns a particular rate of interest, suppose we say that the rate of growth of the amount of money in the account at any time t is proportional to the amount present at that time. Since dA/dt is the rate of growth of A with respect to t, we have

$$\frac{dA}{dt} = rA \qquad A(0) = P \qquad A, P > 0 \tag{4}$$

where r is an appropriate constant. We would like to find a function $A = A(t)$ that satisfies these conditions. Multiplying both sides of equation (4) by $1/A$, we obtain

$$\frac{1}{A}\frac{dA}{dt} = r$$

Now we integrate each side with respect to t:

$$\int \frac{1}{A}\frac{dA}{dt}\,dt = \int r\,dt \qquad \frac{dA}{dt}dt = A'(t)dt = dA$$

$$\int \frac{1}{A}\,dA = \int r\,dt$$

$$\ln|A| = rt + C \qquad |A| = A, \text{ since } A > 0$$

$$\ln A = rt + C$$

We convert this last equation into the equivalent exponential form

$$A = e^{rt+C} \qquad \text{Definition of logarithmic function:}$$
$$\qquad\qquad y = \ln x \text{ if and only if } x = e^y$$

$$= e^C e^{rt} \qquad \text{Property of exponents: } b^m b^n = b^{m+n}$$

Since $A(0) = P$, we evaluate $A(t) = e^C e^{rt}$ at $t = 0$ and set the result equal to P:

$$A(0) = e^C e^0 = e^C = P$$

Hence, $e^C = P$, and we can rewrite $A = e^C e^{rt}$ in the form

$$A = Pe^{rt}$$

This is the same continuous compound interest formula obtained in Section 5-1, where the principal P is invested at an annual nominal rate r compounded continuously for t years.

■ Exponential Growth Law

In general, if the rate of change of a quantity Q with respect to time is proportional to the amount of Q present and $Q(0) = Q_0$, then, proceeding in exactly the same way as we just did, we obtain the following theorem:

THEOREM 1 EXPONENTIAL GROWTH LAW

If $\dfrac{dQ}{dt} = rQ$ and $Q(0) = Q_0$,   then $Q = Q_0 e^{rt}$,

where

Q_0 = amount of Q at $t = 0$

r = relative growth rate (expressed as a decimal)

t = time

Q = quantity at time t

The constant r in the exponential growth law is called the **relative growth rate.** If the relative growth rate is $r = 0.02$, then the quantity Q is growing at a rate $dQ/dt = 0.02Q$ (that is, 2% of the quantity Q per unit of time t). Note the distinction between the relative growth rate r and the rate of growth dQ/dt of the quantity Q. If $r < 0$, then $dQ/dt < 0$ and Q is decreasing. This type of growth is called **exponential decay.**

Once we know that the rate of growth of something is proportional to the amount present, we know that that thing has exponential growth and we can use Theorem 1 without having to solve the differential equation each time. The exponential growth law applies not only to money invested at interest compounded continuously, but also to many other types of problems—population growth, radioactive decay, the depletion of a natural resource, and so on.

■ Population Growth, Radioactive Decay, and Learning

The world population passed 1 billion in 1804, 2 billion in 1927, 3 billion in 1960, 4 billion in 1974, 5 billion in 1987, and 6 billion in 1999, as illustrated in Figure 4. **Population growth** over certain periods often can be approximated by the exponential growth law of Theorem 1.

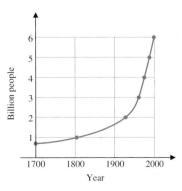

FIGURE 4 World population growth

EXAMPLE 1 **Population Growth** India had a population of about 1.0 billion in 2000 ($t = 0$). Let P represent the population (in billions) t years after 2000, and assume a growth rate of 1.3% compounded continuously.

(A) Find an equation that represents India's population growth after 2000, assuming that the 1.3% growth rate continues.

(B) What is the estimated population (to the nearest tenth of a billion) of India in the year 2030?

(C) Graph the equation found in part (A) from 2000 to 2030.

SOLUTION (A) The exponential growth law applies, and we have

$$\frac{dP}{dt} = 0.013P \qquad P(0) = 1.0$$

Thus,

$$P = 1.0e^{0.013t} \tag{5}$$

(B) Using equation (5), we can estimate the population in India in 2030 ($t = 30$):

$$P = 1.0e^{0.013(30)} = 1.5 \text{ billion people}$$

(C) The graph is shown in Figure 5.

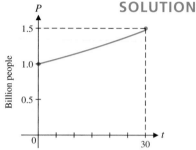

FIGURE 5 Population of India

MATCHED PROBLEM 1 Assuming the same continuous compound growth rate as in Example 1, what will India's population be (to the nearest tenth of a billion) in the year 2015?

EXAMPLE 2 **Population Growth** If the exponential growth law applies to Canada's population growth, at what continuous compound growth rate will the population double over the next 100 years?

SOLUTION The problem is to find r, given that $P = 2P_0$ and $t = 100$:

$$P = P_0 e^{rt}$$

$$2P_0 = P_0 e^{100r}$$

$$2 = e^{100r} \qquad \text{Take the natural logarithm of both sides}$$

$$100r = \ln 2 \qquad \text{and reverse the equation.}$$

$$r = \frac{\ln 2}{100}$$

$$\approx 0.0069 \quad \text{or} \quad 0.69\%$$

MATCHED PROBLEM 2 If the exponential growth law applies to population growth in Nigeria, find the doubling time (to the nearest year) of the population if it grows at 2.1% per year compounded continuously.

We now turn to another type of exponential growth: **radioactive decay.** In 1946, Willard Libby (who later received a Nobel Prize in chemistry) found that as long as a plant or animal is alive, radioactive carbon-14 is maintained at a constant level in its tissues. Once the plant or animal is dead, however, the radioactive carbon-14 diminishes by radioactive decay at a rate proportional to the amount present. Thus,

$$\frac{dQ}{dt} = rQ \qquad Q(0) = Q_0$$

and we have another example of the exponential growth law. The continuous compound rate of decay for radioactive carbon-14 has been found to be 0.000 123 8; thus, $r = -0.000\ 123\ 8$, since decay implies a negative continuous compound growth rate.

EXAMPLE 3

Archaeology A piece of human bone was found at an archaeological site in Africa. If 10% of the original amount of radioactive carbon-14 was present, estimate the age of the bone (to the nearest 100 years).

SOLUTION

Using the exponential growth law for

$$\frac{dQ}{dt} = -0.000\,123\,8Q \qquad Q(0) = Q_0$$

we find that

$$Q = Q_0 e^{-0.0001238t}$$

and our problem is to find t so that $Q = 0.1Q_0$ (since the amount of carbon-14 present now is 10% of the amount Q_0 present at the death of the person). Thus,

$$0.1Q_0 = Q_0 e^{-0.0001238t}$$
$$0.1 = e^{-0.0001238t}$$
$$\ln 0.1 = \ln e^{-0.0001238t}$$
$$t = \frac{\ln 0.1}{-0.000\,123\,8} \approx 18{,}600 \text{ years}$$

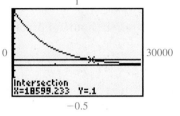

FIGURE 6
$y_1 = e^{-0.0001238x}; \; y_2 = 0.1$

See Figure 6 for a graphical solution to Example 3.

MATCHED PROBLEM 3

Estimate the age of the bone in Example 3 (to the nearest 100 years) if 50% of the original amount of carbon-14 is present.

In learning certain skills, such as typing and swimming, a mathematical model often used is one which assumes that there is a maximum skill attainable—say, M—and the rate of improvement is proportional to the difference between what has been achieved, y, and the maximum attainable, M. Mathematically,

$$\frac{dy}{dt} = k(M - y) \qquad y(0) = 0$$

We solve this type of problem with the same technique that was used to obtain the exponential growth law. First, multiply both sides of the first equation by $1/(M - y)$ to get

$$\frac{1}{M - y}\frac{dy}{dt} = k$$

and then integrate each side with respect to t:

$$\int \frac{1}{M - y}\frac{dy}{dt}\,dt = \int k\,dt$$

$$-\int \frac{1}{M - y}\left(-\frac{dy}{dt}\right) dt = \int k\,dt \qquad \text{Substitute } u = M - y \text{ and}$$

$$-\int \frac{1}{u}\,du = \int k\,dt \qquad du = -dy = -\frac{dy}{dt}\,dt.$$

$$-\ln|u| = kt + C \qquad \text{Substitute } M - y = u.$$

$$-\ln(M - y) = kt + C \qquad \text{Absolute value signs are not required.}$$

$$\ln(M - y) = -kt - C \qquad \text{(Why?)}$$

Change this last equation to an equivalent exponential form:

$$M - y = e^{-kt - C}$$
$$M - y = e^{-C}e^{-kt}$$
$$y = M - e^{-C}e^{-kt}$$

Now, $y(0) = 0$; hence,

$$y(0) = M - e^{-C}e^0 = 0$$

Solving for e^{-C}, we obtain

$$e^{-C} = M$$

and our final solution is

$$y = M - Me^{-kt} = M(1 - e^{-kt})$$

EXAMPLE 4 **Learning** For a particular person who is learning to swim, it is found that the distance y (in feet) the person is able to swim in 1 minute after t hours of practice is given approximately by

$$y = 50(1 - e^{-0.04t})$$

What is the rate of improvement (to two decimal places) after 10 hours of practice?

SOLUTION

$$y = 50 - 50e^{-0.04t}$$
$$y'(t) = 2e^{-0.04t}$$
$$y'(10) = 2e^{-0.04(10)} \approx 1.34 \text{ feet per hour of practice}$$

MATCHED PROBLEM 4 In Example 4, what is the rate of improvement (to two decimal places) after 50 hours of practice?

■ Comparison of Exponential Growth Phenomena

The graphs and equations given in Table 1 compare several widely used growth models. These models are divided basically into two groups: unlimited growth and

TABLE 1 Exponential Growth

Description	Model	Solution	Graph	Uses
Unlimited growth: Rate of growth is proportional to the amount present	$\dfrac{dy}{dt} = ky$ $k, t > 0$ $y(0) = c$	$y = ce^{kt}$		• Short-term population growth (people, bacteria, etc.) • Growth of money at continuous compound interest • Price–supply curves
Exponential decay: Rate of growth is proportional to the amount present	$\dfrac{dy}{dt} = -ky$ $k, t > 0$ $y(0) = c$	$y = ce^{-kt}$		• Depletion of natural resources • Radioactive decay • Absorption of light in water • Price–demand curves • Atmospheric pressure (t is altitude)
Limited growth: Rate of growth is proportional to the difference between the amount present and a fixed limit	$\dfrac{dy}{dt} = k(M - y)$ $k, t > 0$ $y(0) = 0$	$y = M(1 - e^{-kt})$		• Sales fads (for example, skateboards) • Depreciation of equipment • Company growth • Learning
Logistic growth: Rate of growth is proportional to the amount present and to the difference between the amount present and a fixed limit	$\dfrac{dy}{dt} = ky(M - y)$ $k, t > 0$ $y(0) = \dfrac{M}{1 + c}$	$y = \dfrac{M}{1 + ce^{-kMt}}$		• Long-term population growth • Epidemics • Sales of new products • Spread of a rumor • Company growth

limited growth. Following each equation and graph is a short (and necessarily incomplete) list of areas in which the models are used. This list only touches on a subject that has been extensively developed and that you are likely to encounter in greater depth in the future.

Exercise 6-3

A *In Problems 1–12, find the general or particular solution, as indicated, for each differential equation.*

1. $\dfrac{dy}{dx} = 6x$ **2.** $\dfrac{dy}{dx} = 3x^{-2}$

3. $\dfrac{dy}{dx} = \dfrac{7}{x}$ **4.** $\dfrac{dy}{dx} = e^{0.1x}$

5. $\dfrac{dy}{dx} = e^{0.02x}$ **6.** $\dfrac{dy}{dx} = 8x^{-1}$

7. $\dfrac{dy}{dx} = x^2 - x;\ y(0) = 0$

8. $\dfrac{dy}{dx} = \sqrt{x};\ y(0) = 0$

9. $\dfrac{dy}{dx} = -2xe^{-x^2};\ y(0) = 3$

10. $\dfrac{dy}{dx} = e^{x-3};\ y(3) = -5$

11. $\dfrac{dy}{dx} = \dfrac{2}{1 + x};\ y(0) = 5$

12. $\dfrac{dy}{dx} = \dfrac{1}{4(3 - x)};\ y(0) = 1$

B *Problems 13–18 refer to the following slope fields:*

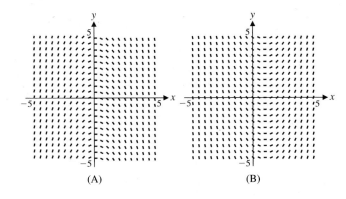

(A)   (B)

13. Which slope field is associated with the differential equation $dy/dx = x - 1$? Briefly justify your answer.

14. Which slope field is associated with the differential equation $dy/dx = -x$? Briefly justify your answer.

15. Solve the differential equation $dy/dx = x - 1$, and find the particular solution that passes through $(0, -2)$.

16. Solve the differential equation $dy/dx = -x$, and find the particular solution that passes through $(0, 3)$.

17. Graph the particular solution found in Problem 15 in the appropriate figure A or B (or a copy).

18. Graph the particular solution found in Problem 16 in the appropriate figure A or B (or a copy).

In Problems 19–26, find the general or particular solution, as indicated, for each differential equation.

19. $\dfrac{dy}{dt} = 2y$ **20.** $\dfrac{dy}{dt} = -3y$

21. $\dfrac{dy}{dx} = -0.5y;\ y(0) = 100$

22. $\dfrac{dy}{dx} = 0.1y;\ y(0) = -2.5$

23. $\dfrac{dx}{dt} = -5x$ **24.** $\dfrac{dx}{dt} = 4t$

25. $\dfrac{dx}{dt} = -5t$ **26.** $\dfrac{dx}{dt} = 4x$

C *Problems 27–34 refer to the following slope fields:*

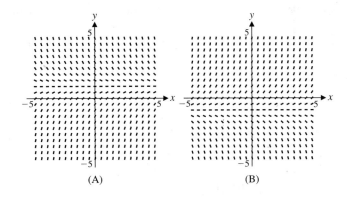

(A)   (B)

27. Which slope field is associated with the differential equation $dy/dx = 1 - y$? Briefly justify your answer.

28. Which slope field is associated with the differential equation $dy/dx = y + 1$? Briefly justify your answer.

29. Show that $y = 1 - Ce^{-x}$ is a solution of the differential equation $dy/dx = 1 - y$ for any real number C. Find the particular solution that passes through $(0, 0)$.

30. Show that $y = Ce^x - 1$ is a solution of the differential equation $dy/dx = y + 1$ for any real number C. Find the particular solution that passes through $(0, 0)$.

31. Graph the particular solution found in Problem 29 in the appropriate figure A or B (or a copy).

32. Graph the particular solution found in Problem 30 in the appropriate figure A or B (or a copy).

33. Use a graphing calculator to graph $y = 1 - Ce^{-x}$ for $C = -2, -1, 1,$ and 2, for $-5 \leq x \leq 5, -5 \leq y \leq 5,$ all in the same viewing window. Observe how the solution curves go with the flow of the tangent line segments in the corresponding slope field shown in figure A or figure B.

34. Use a graphing calculator to graph $y = Ce^x - 1$ for $C = -2, -1, 1,$ and 2, for $-5 \leq x \leq 5, -5 \leq y \leq 5,$ all in the same viewing window. Observe how the solution curves go with the flow of the tangent line segments in the corresponding slope field shown in figure A or figure B.

35. Show that $y = \sqrt{C - x^2}$ is a solution of the differential equation $dy/dx = -x/y$ for any positive real number C. Find the particular solution that passes through $(3, 4)$.

36. Show that $y = \sqrt{x^2 + C}$ is a solution of the differential equation $dy/dx = x/y$ for any real number C. Find the particular solution that passes through $(-6, 7)$.

37. Show that $y = Cx$ is a solution of the differential equation $dy/dx = y/x$ for any real number C. Find the particular solution that passes through $(-8, 24)$.

38. Show that $y = C/x$ is a solution of the differential equation $dy/dx = -y/x$ for any real number C. Find the particular solution that passes through $(2, 5)$.

39. Show that $y = 1/(1 + ce^{-t})$ is a solution of the differential equation $dy/dt = y(1 - y)$ for any real number c. Find the particular solution that passes through $(0, -1)$.

40. Show that $y = 2/(1 + ce^{-6t})$ is a solution of the differential equation $dy/dt = 3y(2 - y)$ for any real number c. Find the particular solution that passes through $(0, 1)$.

In Problems 41–48, use a graphing calculator to graph the given examples of the various cases in Table 1.

41. Unlimited growth:

$$y = 1,000e^{0.08t}$$
$$0 \leq t \leq 15$$
$$0 \leq y \leq 3,500$$

42. Unlimited growth:

$$y = 5,250e^{0.12t}$$
$$0 \leq t \leq 10$$
$$0 \leq y \leq 20,000$$

43. Exponential decay:

$$p = 100e^{-0.05x}$$
$$0 \leq x \leq 30$$
$$0 \leq p \leq 100$$

44. Exponential decay:

$$p = 1,000e^{-0.08x}$$
$$0 \leq x \leq 40$$
$$0 \leq p \leq 1,000$$

45. Limited growth:

$$N = 100(1 - e^{-0.05t})$$
$$0 \leq t \leq 100$$
$$0 \leq N \leq 100$$

46. Limited growth:

$$N = 1,000(1 - e^{-0.07t})$$
$$0 \leq t \leq 70$$
$$0 \leq N \leq 1,000$$

47. Logistic growth:

$$N = \frac{1,000}{1 + 999e^{-0.4t}}$$
$$0 \leq t \leq 40$$
$$0 \leq N \leq 1,000$$

48. Logistic growth:

$$N = \frac{400}{1 + 99e^{-0.4t}}$$
$$0 \leq t \leq 30$$
$$0 \leq N \leq 400$$

49. Show that the rate of logistic growth, $dy/dt = ky(M - y)$, has its maximum value when $y = M/2$.

50. Find the value of t for which the logistic function

$$y = \frac{M}{1 + ce^{-kMt}}$$

is equal to $M/2$.

51. Let $Q(t)$ denote the population of the world at time t. In 1967, the world population was 3.5 billion and increasing at 2.0% per year; in 1999, it was 6.0 billion and increasing at 1.3% per year. In which year, 1967 or 1999, was dQ/dt (the rate of growth of Q with respect to t) greater? Explain.

52. Refer to Problem 51. Explain why the world population function $Q(t)$ does not satisfy an exponential growth law.

Applications

53. *Continuous compound interest.* Find the amount A in an account after t years if

$$\frac{dA}{dt} = 0.08A \quad \text{and} \quad A(0) = 1,000$$

54. *Continuous compound interest.* Find the amount A in an account after t years if

$$\frac{dA}{dt} = 0.12A \quad \text{and} \quad A(0) = 5,250$$

55. *Continuous compound interest.* Find the amount A in an account after t years if

$$\frac{dA}{dt} = rA \quad A(0) = 8,000 \quad A(2) = 9,020$$

56. *Continuous compound interest.* Find the amount A in an account after t years if

$$\frac{dA}{dt} = rA \quad A(0) = 5,000 \quad A(5) = 7,460$$

57. *Price–demand.* The marginal price dp/dx at x units of demand per week is proportional to the price p. There is no weekly demand at a price of $100 per unit $[p(0) = 100]$, and there is a weekly demand of 5 units at a price of $77.88 per unit $[p(5) = 77.88]$.

   (A) Find the price–demand equation.

   (B) At a demand of 10 units per week, what is the price?

   (C) Graph the price–demand equation for $0 \leq x \leq 25$.

58. *Price–supply.* The marginal price dp/dx at x units of supply per day is proportional to the price p. There is no supply at a price of $10 per unit $[p(0) = 10]$, and there is a daily supply of 50 units at a price of $12.84 per unit $[p(50) = 12.84]$.

   (A) Find the price–supply equation.

   (B) At a supply of 100 units per day, what is the price?

   (C) Graph the price–supply equation for $0 \leq x \leq 250$.

59. *Advertising.* A company is trying to expose a new product to as many people as possible through television advertising. Suppose that the rate of exposure to new people is proportional to the number of those who have not seen the product out of L possible viewers. No one is aware of the product at the start of the campaign, and after 10 days 40% of L are aware of the product. Mathematically,

$$\frac{dN}{dt} = k(L - N) \quad N(0) = 0 \quad N(10) = 0.4L$$

   (A) Solve the differential equation.

   (B) What percent of L will have been exposed after 5 days of the campaign?

   (C) How many days will it take to expose 80% of L?

   (D) Graph the solution found in part (A) for $0 \leq t \leq 90$.

60. *Advertising.* Suppose that the differential equation for Problem 59 is

$$\frac{dN}{dt} = k(L - N) \quad N(0) = 0 \quad N(10) = 0.1L$$

   (A) Interpret $N(10) = 0.1L$ verbally.

   (B) Solve the differential equation.

   (C) How many days will it take to expose 50% of L?

   (D) Graph the solution found in part (B) for $0 \leq t \leq 300$.

61. *Biology.* For relatively clear bodies of water, the intensity of light is reduced according to

$$\frac{dI}{dx} = -kI \quad I(0) = I_0$$

where I is the intensity of light at x feet below the surface. For the Sargasso Sea off the West Indies, $k = 0.00942$. Find I in terms of x, and find the depth at which the light is reduced to half of that at the surface.

62. *Blood pressure.* It can be shown under certain assumptions that blood pressure P in the largest artery in the human body (the aorta) changes between beats with respect to time t according to

$$\frac{dP}{dt} = -aP \quad P(0) = P_0$$

where a is a constant. Find $P = P(t)$ that satisfies both conditions.

63. *Drug concentration.* A single injection of a drug is administered to a patient. The amount Q in the body then decreases at a rate proportional to the amount present. For a particular drug, the rate is 4% per hour. Thus,

$$\frac{dQ}{dt} = -0.04Q \quad Q(0) = Q_0$$

where t is time in hours.

   (A) If the initial injection is 3 milliliters $[Q(0) = 3]$, find $Q = Q(t)$ satisfying both conditions.

   (B) How many milliliters (to two decimal places) are in the body after 10 hours?

   (C) How many hours (to two decimal places) will it take for only 1 milliliter of the drug to be left in the body?

   (D) Graph the solution found in part (A).

64. *Simple epidemic.* A community of 1,000 people is assumed to be homogeneously mixed. One person who has just returned from another community has influenza. Assume that the home community has not had influenza shots and all are susceptible. One mathematical model for an influenza epidemic assumes that influenza tends to spread at a rate in direct proportion to the number N who have the disease and to the number it—in this case, $1,000 - N$—who have not yet contracted the disease. Mathematically,

$$\frac{dN}{dt} = kN(1,000 - N) \quad N(0) = 1$$

where N is the number of people who have contracted influenza after t days. For $k = 0.0004$, it can be shown that $N(t)$ is given by

$$N(t) = \frac{1,000}{1 + 999e^{-0.4t}}$$

   (A) How many people have contracted influenza after 10 days? After 20 days?

   (B) How many days will it take until half the community has contracted influenza?

   (C) Find $\lim_{t \to \infty} N(t)$.

   (D) Graph $N = N(t)$ for $0 \leq t \leq 30$.

65. *Nuclear accident.* One of the dangerous radioactive isotopes detected after the Chernobyl nuclear accident in 1986 was cesium-137. If 93.3% of the cesium-137 emitted during the accident is still present 3 years later, find the continuous compound rate of decay of this isotope.

66. *Insecticides.* Many countries have banned the use of the insecticide DDT because of its long-term adverse effects. Five years after a particular country stopped using DDT, the amount of DDT in the ecosystem had declined to 75% of the amount present at the time of the ban. Find the continuous compound rate of decay of DDT.

67. *Archaeology.* A skull from an ancient tomb was discovered and was found to have 5% of the original amount of radioactive carbon-14 present. Estimate the age of the skull. (See Example 3.)

68. *Learning.* For a particular person learning to type, it was found that the number N of words per minute the person was able to type after t hours of practice was given approximately by

$$N = 100(1 - e^{-0.02t})$$

[See Table 1 (limited growth) for a characteristic graph.] What is the rate of improvement after 10 hours of practice? After 40 hours of practice?

69. *Small-group analysis.* In a study on small-group dynamics, sociologists Stephan and Mischler found that when the members of a discussion group of 10 were ranked according to the number of times each participated, the number $N(k)$ of times the kth-ranked person participated was given approximately by

$$N(k) = N_1 e^{-0.11(k-1)} \qquad 1 \le k \le 10$$

where N_1 is the number of times the first-ranked person participated in the discussion. If, in a particular discussion group of 10 people, $N_1 = 180$, estimate how many times the sixth-ranked person participated. How about the 10th-ranked person?

70. *Perception.* One of the oldest laws in mathematical psychology is the Weber–Fechner law (discovered in the middle of the nineteenth century). The law concerns a person's sensed perception of various strengths of stimulation involving weights, sound, light, shock, taste, and so on. One form of the law states that the rate of change of sensed sensation S with respect to stimulus R is inversely proportional to the strength of the stimulus R. Thus,

$$\frac{dS}{dR} = \frac{k}{R}$$

where k is a constant. If we let R_0 be the threshold level at which the stimulus R can be detected (the least

amount of sound, light, weight, and so on, that can be detected), it is appropriate to write

$$S(R_0) = 0$$

Find a function S in terms of R that satisfies the preceding conditions.

71. *Rumor propagation.* A group of 400 parents, relatives, and friends are waiting anxiously at Kennedy Airport for a student charter flight to return after a year in Europe. It is stormy and the plane is late. A particular parent thought he had heard that the plane's radio had gone out, and he related this news to some friends, who in turn passed it on to others, and so on. Sociologists have studied rumor propagation and have found that a rumor tends to spread at a rate in direct proportion to the number x who have heard it and to the number $P - x$ who have not, where P is the total population. Here, the total population is the group waiting at the airport, so $P = 400$ and

$$\frac{dx}{dt} = 0.001x(400 - x) \qquad x(0) = 1$$

where t is time (in minutes). From these conditions, it can be shown that

$$x(t) = \frac{400}{1 + 399e^{-0.4t}}$$

[See Table 1 (logistic growth) for a characteristic graph.]

(A) How many people have heard the rumor after 5 minutes? after 20 minutes?

(B) Find $\lim_{t \to \infty} x(t)$.

(C) Graph $x = x(t)$ for $0 \le t \le 30$.

72. *Rumor propagation.* In Problem 71, how long (to the nearest minute) will it take for half of the group of 400 to have heard the rumor?

Section 6-4 THE DEFINITE INTEGRAL

- Approximating Areas by Left and Right Sums
- The Definite Integral as a Limit of Sums
- Properties of the Definite Integral

The first three sections of this chapter focused on the *indefinite integral*. In this section, we introduce the *definite integral*. The definite integral is used to compute areas, probabilities, average values of functions, future values of continuous income streams, and many other quantities. Initially, the concept of the definite integral may appear to be unrelated to the notion of the indefinite integral. There is, however, a close connection between the two integrals. The fundamental theorem of calculus, discussed in Section 6-5, makes that connection precise.

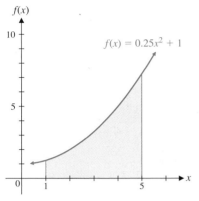

FIGURE 1 What is the shaded area?

■ Approximating Areas by Left and Right Sums

How do we find the shaded area in Figure 1? That is, how do we find the area bounded by the graph of $f(x) = 0.25x^2 + 1$, the x axis, and the vertical lines $x = 1$ and $x = 5$? [This cumbersome description is usually shortened to "the area under the graph of $f(x) = 0.25x^2 + 1$ from $x = 1$ to $x = 5$."] Our standard geometric area formulas do not apply directly, but the formula for the area of a rectangle can be used indirectly. To see how, we look at a method of approximating the area under the graph by using rectangles. This method will give us any accuracy desired, which is quite different from finding the area exactly. Our first area approximation is made by dividing the interval $[1, 5]$ on the x axis into four equal parts, each of length

$$\Delta x = \frac{5 - 1}{4} = 1$$

(It is customary to denote the length of the subintervals by Δx, which is read "delta x," since Δ is the Greek capital letter delta.) We then place a rectangle on each subinterval, with the height of the rectangle determined by the function evaluated at the left endpoint of the subinterval (see Fig. 2).

Summing the areas of the rectangles in Figure 2, we obtain a **left sum** of four rectangles, denoted by L_4, as follows:

$$L_4 = f(1) \cdot 1 + f(2) \cdot 1 + f(3) \cdot 1 + f(4) \cdot 1$$
$$= 1.25 + 2.00 + 3.25 + 5 = 11.5$$

From Figure 3, since $f(x)$ is increasing, it is clear that the left sum L_4 underestimates the area, and we can write

$$11.5 = L_4 < \text{Area}$$

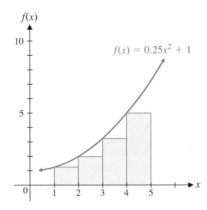

FIGURE 2 Left rectangles

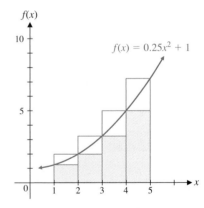

FIGURE 3 Left and right rectangles

Explore & Discuss **1**

If $f(x)$ were decreasing over the interval $[1, 5]$, would the left sum L_4 over- or underestimate the actual area under the curve? Explain.

Now suppose that we use the right endpoint of each subinterval to obtain the height of the rectangle placed on top of it. Superimposing this result on top of Figure 2, we obtain Figure 3.

Summing the areas of the higher rectangles in Figure 3, we obtain the **right sum** of the four rectangles, denoted by R_4, as follows (compare R_4 with L_4 and note that R_4 can be obtained from L_4 by deleting one rectangular area and adding one more):

$$R_4 = f(2) \cdot 1 + f(3) \cdot 1 + f(4) \cdot 1 + f(5) \cdot 1$$
$$= 2.00 + 3.25 + 5.00 + 7.25 = 17.5$$

From Figure 3, since $f(x)$ is increasing, it is clear that the right sum R_4 overestimates the area, and we conclude that the actual area is between 11.5 and 17.5. That is,

$$11.5 = L_4 < \text{Area} < R_4 = 17.5$$

Explore & Discuss 2

If $f(x)$ in Figure 3 were decreasing over the interval [1, 5], would the right sum R_4 over- or underestimate the actual area under the curve? Explain.

The first approximation of the area under the curve in (1) is fairly coarse, but the method outlined can be continued with increasingly accurate results by dividing the interval [1, 5] into more and more subintervals of equal horizontal length. Of course, this is not a job for hand calculation, but a job that computers are designed to do.* Figure 4 shows left- and right-rectangle approximations for 16 equal subdivisions. For this case,

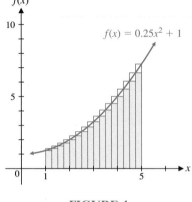

$f(x)$

$f(x) = 0.25x^2 + 1$

FIGURE 4

$$\Delta x = \frac{5 - 1}{16} = 0.25$$
$$L_{16} = f(1) \cdot \Delta x + f(1.25) \cdot \Delta x + \cdots + f(4.75) \cdot \Delta x$$
$$= 13.59$$
$$R_{16} = f(1.25) \cdot \Delta x + f(1.50) \cdot \Delta x + \cdots + f(5) \cdot \Delta x$$
$$= 15.09$$

Thus, we now know that the area under the curve is between 13.59 and 15.09. That is,

$$13.59 = L_{16} < \text{Area} < R_{16} = 15.09$$

For 100 equal subdivisions, computer calculations give us

$$14.214 = L_{100} < \text{Area} < R_{100} = 14.454$$

The **error in an approximation** is the absolute value of the difference between the approximation and the actual value. In general, neither the actual value nor the error in an approximation is known. However, it is often possible to calculate an **error bound**—a positive number such that the error is guaranteed to be less than or equal to that number.

The error in the approximation of the area under the graph of f from $x = 1$ to $x = 5$ by the left sum L_{16} (or the right sum R_{16}) is less than the sum of the areas of the small rectangles in Figure 4. By stacking those rectangles (see Fig. 5), we see that

$$\text{Error} = |\text{Area} - L_{16}| < |f(5) - f(1)| \cdot \Delta x = 1.5$$

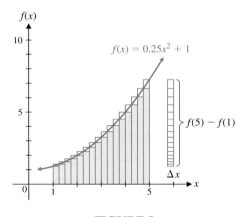

FIGURE 5

---

* The computer software that accompanies this book will perform these calculations (see the preface).

Therefore, 1.5 is an error bound for the approximation of the area under f by L_{16}. We can apply the same stacking argument to any positive function that is increasing on $[a, b]$ or decreasing on $[a, b]$, to obtain the error bound in Theorem 1.

THEOREM 1   **ERROR BOUNDS FOR APPROXIMATIONS OF AREA BY LEFT OR RIGHT SUMS**

If $f(x) > 0$ and is either increasing on $[a, b]$ or decreasing on $[a, b]$, then

$$|f(b) - f(a)| \cdot \frac{b - a}{n}$$

is an error bound for the approximation of the area between the graph of f and the x axis, from $x = a$ to $x = b$, by L_n or R_n.

Because the error bound of Theorem 1 approaches 0 as $n \to \infty$, it can be shown that left and right sums, for certain functions, approach the same limit as $n \to \infty$.

THEOREM 2   **LIMITS OF LEFT AND RIGHT SUMS**

If $f(x) > 0$ and is either increasing on $[a, b]$ or decreasing on $[a, b]$, then its left and right sums approach the same real number as $n \to \infty$.

The number approached as $n \to \infty$ by the left and right sums in Theorem 2 is the area between the graph of f and the x axis from $x = a$ to $x = b$.

EXAMPLE 1   **Approximating Areas**   Given the function $f(x) = 9 - 0.25x^2$, we are interested in approximating the area under $y = f(x)$ from $x = 2$ to $x = 5$.

(A) Graph the function over the interval $[0, 6]$; then draw left and right rectangles for the interval $[2, 5]$ with $n = 6$.

(B) Calculate L_6, R_6, and error bounds for each.

(C) How large should n be chosen for the approximation of the area by L_n or R_n to be within 0.05 of the true value?

SOLUTION   (A) $\Delta x = 0.5$:

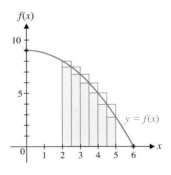

(B) $L_6 = f(2) \cdot \Delta x + f(2.5) \cdot \Delta x + f(3) \cdot \Delta x + f(3.5) \cdot \Delta x + f(4) \cdot \Delta x$
$\quad + f(4.5) \cdot \Delta x = 18.53$

$R_6 = f(2.5) \cdot \Delta x + f(3) \cdot \Delta x + f(3.5) \cdot \Delta x + f(4) \cdot \Delta x$
$\quad + f(4.5) \cdot \Delta x + f(5) \cdot \Delta x = 15.91$

The error bound for L_6 and R_6 is

$$\text{error} \le |f(5) - f(2)|\frac{5 - 2}{6} = |2.75 - 8|(0.5) = 2.625$$

(C) For L_n and R_n, find n such that error ≤ 0.05:

$$|f(b) - f(a)|\frac{b - a}{n} \le 0.05$$

$$|2.75 - 8|\frac{3}{n} \le 0.05$$

$$|-5.25|\frac{3}{n} \le 0.05$$

$$15.75 \le 0.05n$$

$$n \ge \frac{15.75}{0.05} = 315$$

MATCHED PROBLEM 1 Given the function $f(x) = 8 - 0.5x^2$, we are interested in approximating the area under $y = f(x)$ from $x = 1$ to $x = 3$.

(A) Graph the function over the interval $[0, 4]$; then draw left and right rectangles for the interval $[1, 3]$ with $n = 4$.

(B) Calculate L_4, R_4, and error bounds for each.

(C) How large should n be chosen for the approximation of the area by L_n or R_n to be within 0.5 of the true value?

INSIGHT

Note from Example 1(C) that a relatively large value of n ($n = 315$) is required to approximate the area by L_n or R_n to within 0.05. In other words, 315 rectangles must be used, and 315 terms must be summed, to guarantee that the error does not exceed 0.05. We can obtain a more efficient approximation of the area (in the sense that fewer terms must be summed to achieve a given accuracy) by replacing rectangles by trapezoids. The resulting **trapezoidal rule,** and other methods for approximating areas, are discussed in Group Activity 1 at the website for this book.

■ The Definite Integral as a Limit of Sums

Left and right sums are special cases of more general sums, called *Riemann sums* [named after the German mathematician Georg Riemann (1826–1866)], that are used to approximate areas by means of rectangles.

Let f be a function defined on the interval $[a, b]$. We partition $[a, b]$ into n subintervals of equal length $\Delta x = (b - a)/n$ with endpoints

$$a = x_0 < x_1 < x_2 < \cdots < x_n = b$$

Then, using **summation notation** (see Appendix B-1), we have

Left sum: $\quad L_n = f(x_0)\Delta x + f(x_1)\Delta x + \cdots + f(x_{n-1})\Delta x = \sum_{k=1}^{n} f(x_{k-1})\Delta x$

Right sum: $\quad R_n = f(x_1)\Delta x + f(x_2)\Delta x + \cdots + f(x_n)\Delta x = \sum_{k=1}^{n} f(x_k)\Delta x$

Riemann sum: $\quad S_n = f(c_1)\Delta x + f(c_2)\Delta x + \cdots + f(c_n)\Delta x = \sum_{k=1}^{n} f(c_k)\Delta x$

In a **Riemann sum,*** each c_k is required to belong to the subinterval $[x_{k-1}, x_k]$. Left and right sums are the special cases of Riemann sums in which c_k is the left endpoint or right endpoint, respectively, of the subinterval. If $f(x) > 0$, then each term of a Riemann sum S_n represents the area of a rectangle having height $f(c_k)$ and width Δx (see Fig. 6). If $f(x)$ has both positive and negative values, then some terms of S_n represent areas of rectangles, and others represent the negatives of areas of rectangles, depending on the sign of $f(c_k)$ (see Fig. 7).

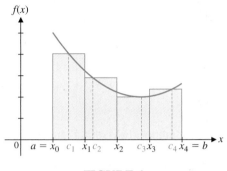

FIGURE 6   FIGURE 7

EXAMPLE 2   **Riemann Sums**   Consider the function $f(x) = 15 - x^2$ on $[1, 5]$. Partition the interval $[1, 5]$ into four subintervals of equal length. For each subinterval $[x_{k-1}, x_k]$, let c_k be the midpoint. Calculate the corresponding Riemann sum S_4. (Riemann sums for which the c_k are the midpoints of the subintervals are called **midpoint sums.**)

SOLUTION
$$\Delta x = \frac{5 - 1}{4} = 1$$

$$S_4 = f(c_1) \cdot \Delta x + f(c_2) \cdot \Delta x + f(c_3) \cdot \Delta x + f(c_4) \cdot \Delta x$$

$$= f(1.5) \cdot 1 + f(2.5) \cdot 1 + f(3.5) \cdot 1 + f(4.5) \cdot 1$$

$$= 12.75 + 8.75 + 2.75 - 5.25 = 19$$

MATCHED PROBLEM 2   Consider the function $f(x) = x^2 - 2x - 10$ on $[2, 8]$. Partition the interval $[2, 8]$ into three subintervals of equal length. For each subinterval $[x_{k-1}, x_k]$, let c_k be the midpoint. Calculate the corresponding Riemann sum S_3.

By analyzing properties of a continuous function on a closed interval, it can be shown that the conclusion of Theorem 2 is valid if f is continuous. Moreover, not just left and right sums, but Riemann sums, have the same limit as $n \to \infty$.

THEOREM 3   **LIMIT OF RIEMANN SUMS**

If f is a continuous function on $[a, b]$, then the Riemann sums for f on $[a, b]$ approach a real number limit I as $n \to \infty$.†

---

* The term *Riemann sum* is often applied to more general sums in which the subintervals $[x_{k-1}, x_k]$ are not required to have the same length. Such sums are not considered in this book.

† The precise meaning of this limit statement is as follows: For each $e > 0$, there exists some $d > 0$ such that $|S_n - I| < e$ whenever S_n is a Riemann sum for f on $[a, b]$ for which $\Delta x < d$.

DEFINITION **Definite Integral**

Let f be a continuous function on $[a, b]$. The limit I of Riemann sums for f on $[a, b]$, guaranteed to exist by Theorem 2, is called the **definite integral** of f from a to b and is denoted

$$\int_a^b f(x)\,dx$$

The **integrand** is $f(x)$, the **lower limit of integration** is a, and the **upper limit of integration** is b.

Because area is a positive quantity, the definite integral has the following geometric interpretation:

$$\int_a^b f(x)\,dx$$

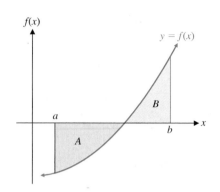

FIGURE 8

$$\int_a^b f(x)\,dx = -A + B$$

represents the cumulative sum of the signed areas between the graph of f and the x axis from $x = a$ to $x = b$, where the areas above the x axis are counted positively and the areas below the x axis are counted negatively (see Fig. 8, where A and B are the actual areas of the indicated regions).

EXAMPLE 3 **Definite Integrals** Calculate the definite integrals by referring to Figure 9 with the indicated areas.

(A) $\int_a^b f(x)\,dx$

(B) $\int_a^c f(x)\,dx$

(C) $\int_b^c f(x)\,dx$

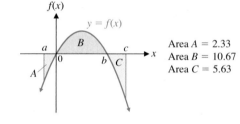

Area $A = 2.33$
Area $B = 10.67$
Area $C = 5.63$

FIGURE 9

SOLUTION (A) $\int_a^b f(x)\,dx = -2.33 + 10.67 = 8.34$

(B) $\int_a^c f(x)\,dx = -2.33 + 10.67 - 5.63 = 2.71$

(C) $\int_b^c f(x)\,dx = -5.63$

MATCHED PROBLEM 3 Referring to the figure for Example 3, calculate the definite integrals.

(A) $\int_a^0 f(x)\,dx$ (B) $\int_0^c f(x)\,dx$ (C) $\int_0^b f(x)\,dx$

Properties of the Definite Integral

Because the definite integral is defined as the limit of Riemann sums, many properties of sums are also properties of the definite integral. Note that Properties 3 and 4 parallel properties given in Section 6-1 for indefinite integrals.

PROPERTIES   Properties of Definite Integrals

1. $\displaystyle\int_a^a f(x)\,dx = 0$

2. $\displaystyle\int_a^b f(x)\,dx = -\int_b^a f(x)\,dx$

3. $\displaystyle\int_a^b kf(x)\,dx = k\int_a^b f(x)\,dx,\quad k \text{ a constant}$

4. $\displaystyle\int_a^b [f(x) \pm g(x)]\,dx = \int_a^b f(x)\,dx \pm \int_a^b g(x)\,dx$

5. $\displaystyle\int_a^b f(x)\,dx = \int_a^c f(x)\,dx + \int_c^b f(x)\,dx$

Example 4 illustrates how properties of definite integrals can be used.

EXAMPLE 4   **Using Properties of the Definite Integral** If

$$\int_0^2 x\,dx = 2 \qquad \int_0^2 x^2\,dx = \frac{8}{3} \qquad \int_2^3 x^2\,dx = \frac{19}{3}$$

then

(A) $\displaystyle\int_0^2 12x^2\,dx = 12\int_0^2 x^2\,dx = 12\left(\frac{8}{3}\right) = 32$

(B) $\displaystyle\int_0^2 (2x - 6x^2)\,dx = 2\int_0^2 x\,dx - 6\int_0^2 x^2\,dx = 2(2) - 6\left(\frac{8}{3}\right) = -12$

(C) $\displaystyle\int_3^2 x^2\,dx = -\int_2^3 x^2\,dx = -\frac{19}{3}$

(D) $\displaystyle\int_5^5 3x^2\,dx = 0$

(E) $\displaystyle\int_0^3 3x^2\,dx = 3\int_0^2 x^2\,dx + 3\int_2^3 x^2\,dx = 3\left(\frac{8}{3}\right) + 3\left(\frac{19}{3}\right) = 27$

MATCHED PROBLEM 4   Using the same integral values given in Example 4, find

(A) $\displaystyle\int_2^3 6x^2\,dx$ (B) $\displaystyle\int_0^2 (9x^2 - 4x)\,dx$ (C) $\displaystyle\int_2^0 3x\,dx$

(D) $\displaystyle\int_{-2}^{-2} 3x\,dx$ (E) $\displaystyle\int_0^3 12x^2\,dx$

Answers to Matched Problems   **1.** (A) $\Delta x = 0.5$:

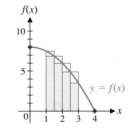

2. $S_3 = 46$

3. (A) -2.33 (B) 5.04 (C) 10.67

4. (A) 38 (B) 16 (C) -6
     (D) 0 (E) 108

(B) $L_4 = 12.625$, $R_4 = 10.625$; error for L_4 and $R_4 = 2$

(C) $n > 16$ for L_n and R_n

Exercise 6-4

A *Problems 1–10 involve estimating the area under the curves in Figures A–D from x = 1 to x = 4. For each figure, divide the interval [1, 4] into three equal subintervals.*

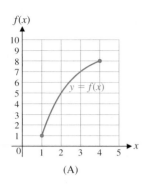

(A)

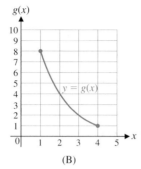

(B)

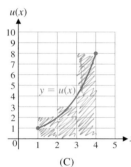

(C)

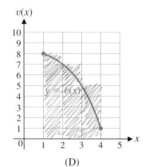

(D)

1. Draw in left and right rectangles for Figures A and B.

2. Draw in left and right rectangles for Figures C and D.

3. Using the results of Problem 1, compute L_3 and R_3 for Figure A and for Figure B.

4. Using the results of Problem 2, compute L_3 and R_3 for Figure C and for Figure D.

5. Replace the question marks with L_3 and R_3 as appropriate. Explain your choice.

$$? \leq \int_1^4 f(x)\,dx \leq ? \qquad ? \leq \int_1^4 g(x)\,dx \leq ?$$

6. Replace the question marks with L_3 and R_3 as appropriate. Explain your choice.

$$? \leq \int_1^4 u(x)\,dx \leq ? \qquad ? \leq \int_1^4 v(x)\,dx \leq ?$$

7. Compute error bounds for L_3 and R_3 found in Problem 3 for both figures.

8. Compute error bounds for L_3 and R_3 found in Problem 4 for both figures.

In Problems 9–12, calculate the indicated Riemann sum S_n for the function $f(x) = 25 - 3x^2$.

9. Partition $[-2, 8]$ into five subintervals of equal length, and for each subinterval $[x_{k-1}, x_k]$, let $c_k = (x_{k-1} + x_k)/2$.

10. Partition $[0, 12]$ into four subintervals of equal length, and for each subinterval $[x_{k-1}, x_k]$, let $c_k = (x_{k-1} + 2x_k)/3$.

11. Partition $[0, 12]$ into four subintervals of equal length, and for each subinterval $[x_{k-1}, x_k]$, let $c_k = (2x_{k-1} + x_k)/3$.

12. Partition $[-5, 5]$ into five subintervals of equal length, and for each subinterval $[x_{k-1}, x_k]$, let $c_k = (x_{k-1} + x_k)/2$.

In Problems 13–16, calculate the indicated Riemann sum S_n for the function $f(x) = x^2 - 5x - 6$.

13. Partition $[0, 3]$ into three subintervals of equal length, and let $c_1 = 0.7$, $c_2 = 1.8$, and $c_3 = 2.4$.

14. Partition $[0, 3]$ into three subintervals of equal length, and let $c_1 = 0.2$, $c_2 = 1.5$, and $c_3 = 2.8$.

15. Partition $[1, 7]$ into six subintervals of equal length, and let $c_1 = 1$, $c_2 = 3$, $c_3 = 3$, $c_4 = 5$, $c_5 = 5$, and $c_6 = 7$.

16. Partition $[1, 7]$ into six subintervals of equal length, and let $c_1 = 2$, $c_2 = 2$, $c_3 = 4$, $c_4 = 4$, $c_5 = 6$, and $c_6 = 6$.

In Problems 17–28, calculate the definite integral by referring to the figure with the indicated areas.

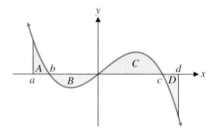

Area $A = 1.408$
Area $B = 2.475$
Area $C = 5.333$
Area $D = 1.792$

17. $\displaystyle\int_b^0 f(x)\,dx$ 18. $\displaystyle\int_0^c f(x)\,dx$

19. $\displaystyle\int_a^c f(x)\,dx$ 20. $\displaystyle\int_b^d f(x)\,dx$

21. $\displaystyle\int_a^d f(x)\,dx$ 22. $\displaystyle\int_0^d f(x)\,dx$

23. $\displaystyle\int_c^0 f(x)\,dx$ 24. $\displaystyle\int_d^a f(x)\,dx$

25. $\displaystyle\int_0^a f(x)\,dx$ 26. $\displaystyle\int_c^a f(x)\,dx$

27. $\displaystyle\int_d^b f(x)\,dx$ 28. $\displaystyle\int_c^b f(x)\,dx$

In Problems 29–40, calculate the definite integral, given that

$$\int_1^4 x\,dx = 7.5 \qquad \int_1^4 x^2\,dx = 21 \qquad \int_4^5 x^2\,dx = \frac{61}{3}$$

29. $\displaystyle\int_1^4 2x\,dx$ 30. $\displaystyle\int_1^4 3x^2\,dx$

31. $\displaystyle\int_1^4 (5x + x^2)\,dx$ 32. $\displaystyle\int_1^4 (7x - 2x^2)\,dx$

33. $\displaystyle\int_1^4 (x^2 - 10x)\,dx$ 34. $\displaystyle\int_1^4 (4x^2 - 9x)\,dx$

35. $\int_1^5 6x^2 \, dx$

36. $\int_1^5 -4x^2 \, dx$

37. $\int_4^4 (7x - 2)^2 \, dx$

38. $\int_5^5 (10 - 7x + x^2) \, dx$

39. $\int_5^4 9x^2 \, dx$

40. $\int_4^1 x(1 - x) \, dx$

x	600	700	800	900	1,000
h(x)	322	388	453	489	500

48. Refer to Problem 47. Use R_{10} to estimate the combined area of both parcels, and calculate an error bound for this estimate. How many subdivisions of the baseline would be required so that the error incurred in using R_n would not exceed 1,000 square feet?

B   *In Problems 41–46, discuss the validity of each statement. If the statement is always true, explain why. If it is not always true, give a counterexample.*

41. If $\int_a^b f(x) \, dx = 0$, then $f(x) = 0$ for all x in $[a, b]$.

42. If $f(x) = 0$ for all x in $[a, b]$, then $\int_a^b f(x) \, dx = 0$.

43. If $f(x) = 2x$ on $[0, 10]$, then there is a positive integer n for which the left sum L_n equals the exact area under the graph of f from $x = 0$ to $x = 10$.

44. If $f(x) = 2x$ on $[0, 10]$ and n is a positive integer, then there is some Riemann sum S_n that equals the exact area under the graph of f from $x = 0$ to $x = 10$.

45. If the area under the graph of f on $[a, b]$ is equal to both the left sum L_n and the right sum R_n for some positive integer n, then f is constant on $[a, b]$.

46. If f is a decreasing function on $[a, b]$, then the area under the graph of f is greater than the left sum L_n and less than the right sum R_n, for any positive integer n.

Problems 47 and 48 refer to the following figure showing two parcels of land along a river:

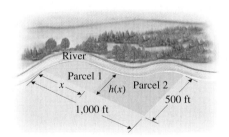

47. You are interested in purchasing both parcels of land shown in the figure and wish to make a quick check on their combined area. There is no equation for the river frontage, so you use the average of the left and right sums of rectangles covering the area. The 1,000-foot baseline is divided into 10 equal parts. At the end of each subinterval, a measurement is made from the baseline to the river, and the results are tabulated. Let x be the distance from the left end of the baseline and let $h(x)$ be the distance from the baseline to the river at x. Use L_{10} to estimate the combined area of both parcels, and calculate an error bound for this estimate. How many subdivisions of the baseline would be required so that the error incurred in using L_n would not exceed 2,500 square feet?

x	0	100	200	300	400	500
h(x)	0	183	235	245	260	286

C   *Problems 49 and 50 refer to the following figure:*

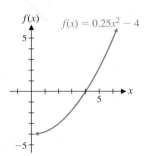

49. Use L_6 and R_6 to approximate $\int_2^5 (0.25x^2 - 4) \, dx$. Compute error bounds for each. (Round answers to two decimal places.) Describe in geometric terms what the definite integral over the interval $[2, 5]$ represents.

50. Use L_5 and R_5 to approximate $\int_1^6 (0.25x^2 - 4) \, dx$. Compute error bounds for each. (Round answers to two decimal places.) Describe in geometric terms what the definite integral over the interval $[1, 6]$ represents.

For Problems 51–54, use a graphing calculator to determine the intervals on which each function is increasing or decreasing.

51. $f(x) = e^{-x^2}$

52. $f(x) = \dfrac{3}{1 + 2e^{-x}}$

53. $f(x) = x^4 - 2x^2 + 3$

54. $f(x) = e^{x^2}$

In Problems 55–58, the left sum L_n or the right sum R_n is used to approximate the definite integral to the indicated accuracy. How large must n be chosen in each case? (Each function is increasing over the indicated interval.)

55. $\int_1^3 \ln x \, dx = R_n \pm 0.1$

56. $\int_0^{10} \ln(x^2 + 1) \, dx = L_n \pm 0.5$

57. $\int_1^3 x^x \, dx = L_n \pm 0.5$

58. $\int_1^4 x^x \, dx = R_n \pm 0.5$

Applications

59. *Employee training.* A company producing computer components has established that, on the average, a new employee can assemble $N(t)$ components per day after t days of on-the-job training, as indicated in the following table (a new employee's productivity increases continuously with time on the job):

t	0	20	40	60	80	100	120
$N(t)$	10	51	68	76	81	84	86

Use left and right sums to estimate the area under the graph of $N(t)$ from $t = 0$ to $t = 60$. Use three subintervals of equal length for each. Calculate an error bound for each estimate.

60. *Employee training.* For a new employee in Problem 59, use left and right sums to estimate the area under the graph of $N(t)$ from $t = 20$ to $t = 100$. Use four equal subintervals for each. Replace the question marks with the values of L_4 or R_4 as appropriate:

$$? \leq \int_{20}^{100} N(t)\, dt \leq ?$$

61. *Medicine.* The rate of healing, $A'(t)$ (in square centimeters per day), for a certain type of abrasive skin wound is given approximately by the following table:

t	0	1	2	3	4	5
$A'(t)$	0.90	0.81	0.74	0.67	0.60	0.55

t	6	7	8	9	10
$A'(t)$	0.49	0.45	0.40	0.36	0.33

(A) Use left and right sums over five equal subintervals to approximate the area under the graph of $A'(t)$ from $t = 0$ to $t = 5$.

(B) Replace the question marks with values of L_5 and R_5 as appropriate:

$$? \leq \int_0^5 A'(t)\, dt \leq ?$$

62. *Medicine.* Refer to Problem 61. Use left and right sums over five equal subintervals to approximate the area under the graph of $A'(t)$ from $t = 5$ to $t = 10$. Calculate an error bound for this estimate.

63. *Learning.* During a special study on learning, a psychologist found that, on the average, the rate of learning a list of special symbols in a code $N'(x)$ after x days of practice was given approximately by the following table values:

x	0	2	4	6	8	10	12
$N'(x)$	29	26	23	21	19	17	15

Use left and right sums over three equal subintervals to approximate the area under the graph of $N'(x)$ from $x = 6$ to $x = 12$. Calculate an error bound for this estimate.

64. *Learning.* For the data in Problem 63, use left and right sums over three equal subintervals to approximate the area under the graph of $N'(x)$ from $x = 0$ to $x = 6$. Replace the question marks with values of L_3 and R_3 as appropriate:

$$? \leq \int_0^6 N'(x)\, dx \leq ?$$

Section 6-5 THE FUNDAMENTAL THEOREM OF CALCULUS

- Introduction to the Fundamental Theorem
- Evaluating Definite Integrals
- Recognizing a Definite Integral: Average Value

The definite integral of a function f on an interval $[a, b]$ is a number, the area (if $f(x) > 0$) between the graph of f and the x axis from $x = a$ to $x = b$. The indefinite integral of a function is a family of antiderivatives. In this section, we explain the connection between these two integrals, a connection that is made precise by the fundamental theorem of calculus.

■ Introduction to the Fundamental Theorem

Suppose that the daily cost function for a small manufacturing firm is given (in dollars) by

$$C(x) = 180x + 200 \qquad 0 \leq x \leq 20$$

Then the marginal cost function is given (in dollars per unit) by

$$C'(x) = 180$$

What is the change in cost as production is increased from $x = 5$ units to $x = 10$ units? That change is equal to

$$C(10) - C(5) = (180 \cdot 10 + 200) - (180 \cdot 5 + 200)$$
$$= 180(10 - 5)$$
$$= \$900$$

Notice that $180(10 - 5)$ is equal to the area between the graph of $C'(x)$ and the x axis from $x = 5$ to $x = 10$. Therefore,

$$C(10) - C(5) = \int_{5}^{10} 180 \, dx$$

In other words, the change in cost from $x = 5$ to $x = 10$ is equal to the area between the marginal cost function and the x axis from $x = 5$ to $x = 10$ (see Fig. 1).

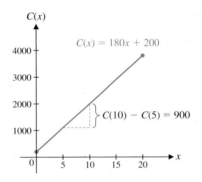

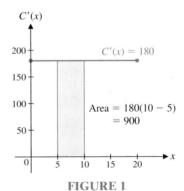

FIGURE 1

INSIGHT

Consider the formula for the slope of a line:

$$m = \frac{y_2 - y_1}{x_2 - x_1}$$

Multiplying both sides of this equation by $x_2 - x_1$ gives

$$y_2 - y_1 = m(x_2 - x_1)$$

The right-hand side, $m(x_2 - x_1)$, is equal to the area of a rectangle of height m and width $x_2 - x_1$. So the change in y coordinates is equal to the area under the constant function with value m from $x = x_1$ to $x = x_2$.

EXAMPLE 1

Change in Cost vs Area under Marginal Cost The daily cost function for a company (in dollars) is given by

$$C(x) = -5x^2 + 210x + 400 \qquad 0 \le x \le 20$$

(A) Graph $C(x)$ for $0 \le x \le 20$, calculate the change in cost from $x = 5$ to $x = 10$, and indicate that change in cost on the graph.

(B) Graph the marginal cost function $C'(x)$ for $0 \le x \le 20$, and use geometric formulas (see Appendix C) to calculate the area between $C'(x)$ and the x axis from $x = 5$ to $x = 10$.

(C) Compare the results of the calculations in parts (A) and (B).

SOLUTION

(A) $C(10) - C(5) = 2,000 - 1,325 = 675$, and this change in cost is indicated in Figure 2A.

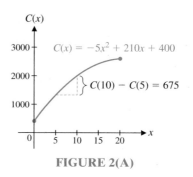

FIGURE 2(A)

(B) $C'(x) = -10x + 210$, so the area between $C'(x)$ and the x axis from $x = 5$ to $x = 10$ (see Fig. 2B) is the area of a trapezoid (geometric formulas are given in Appendix C):

$$\text{Area} = \frac{C(5) + C(10)}{2}(10 - 5) = \frac{160 + 110}{2}(5) = 675$$

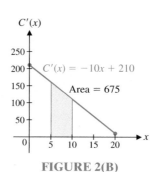

FIGURE 2(B)

(C) The change in cost from $x = 5$ to $x = 10$ is equal to the area between the marginal cost function and the x axis from $x = 5$ to $x = 10$. ∎

MATCHED PROBLEM 1

Repeat Example 1 for the daily cost function

$$C(x) = -7.5x^2 + 305x + 625$$

The connection illustrated in Example 1, between the change in a function from $x = a$ to $x = b$ and the area under the derivative of the function, provides the link between antiderivatives (or indefinite integrals) and the definite integral. This link is known as the fundamental theorem of calculus. (See Problems 59 and 60 in Exercise 6-5 for an outline of its proof.)

THEOREM 1   FUNDAMENTAL THEOREM OF CALCULUS

If f is a continuous function on $[a, b]$, and F is any antiderivative of f, then

$$\int_a^b f(x)\, dx = F(b) - F(a)$$

INSIGHT

Because a definite integral is the limit of Riemann sums, we expect that it would be difficult to calculate definite integrals exactly. The fundamental theorem, however, gives us an easy method for evaluating definite integrals, *provided that we can find an antiderivative $F(x)$ of $f(x)$*: Simply calculate the difference $F(b) - F(a)$. But what if we are unable to find an antiderivative of $f(x)$? In that case, we must resort to left sums, right sums, or other approximation methods to approximate the definite integral. However, it is often useful to remember that such an approximation is also an estimate of the change $F(b) - F(a)$.

■ Evaluating Definite Integrals

By the fundamental theorem, we can evaluate $\int_a^b f(x)\, dx$ easily and exactly whenever we can find an antiderivative $F(x)$ of $f(x)$. We simply calculate the difference $F(b) - F(a)$.

Now you know why we studied techniques of indefinite integration before this section—so that we would have methods of finding antiderivatives of large classes of elementary functions for use with the fundamental theorem. It is important to remember that

Any antiderivative of $f(x)$ can be used in the fundamental theorem. One generally chooses the simplest antiderivative by letting $C = 0$, since any other value of C will drop out in computing the difference $F(b) - F(a)$.

In evaluating definite integrals by the fundamental theorem, it is convenient to use the notation $F(x)\big|_a^b$, which represents the change in $F(x)$ from $x = a$ to $x = b$, as an intermediate step in the calculation. The technique is illustrated in the examples that follow.

EXAMPLE 2   **Evaluating Definite Integrals**   Evaluate $\int_1^2 \left(2x + 3e^x - \dfrac{4}{x}\right) dx$.

SOLUTION
$$\int_1^2 \left(2x + 3e^x - \frac{4}{x}\right) dx = 2\int_1^2 x\, dx + 3\int_1^2 e^x\, dx - 4\int_1^2 \frac{1}{x}\, dx$$

$$= 2\frac{x^2}{2}\bigg|_1^2 + 3e^x\bigg|_1^2 - 4\ln|x|\big|_1^2$$

$$= (2^2 - 1^2) + (3e^2 - 3e^1) - (4\ln 2 - 4\ln 1)$$

$$= 3 + 3e^2 - 3e - 4\ln 2 \approx 14.24$$

▬

MATCHED PROBLEM 2   Evaluate $\int_1^3 \left(4x - 2e^x + \dfrac{5}{x}\right) dx$.

The evaluation of a definite integral is a two-step process: First, find an antiderivative; then find the change in that antiderivative. If *substitution techniques* are required to find the antiderivative, there are two different ways to proceed. The next example illustrates both methods.

EXAMPLE 3   **Definite Integrals and Substitution Techniques**   Evaluate

$$\int_0^5 \frac{x}{x^2 + 10}\, dx$$

SOLUTION   We will solve this problem by using substitution in two different ways.

Method 1. Use substitution in an indefinite integral to find an antiderivative as a function of x; then evaluate the definite integral:

$$\int \frac{x}{x^2 + 10} dx = \frac{1}{2} \int \frac{1}{x^2 + 10} 2x \, dx \quad \text{Substitute } u = x^2 + 10 \text{ and} \\ du = 2x \, dx.$$

$$= \frac{1}{2} \int \frac{1}{u} du$$

$$= \tfrac{1}{2} \ln|u| + C$$

$$= \tfrac{1}{2} \ln(x^2 + 10) + C \quad \text{Since } u = x^2 + 10 > 0$$

We choose $C = 0$ and use the antiderivative $\frac{1}{2}\ln(x^2 + 10)$ to evaluate the definite integral:

$$\int_0^5 \frac{x}{x^2 + 10} dx = \frac{1}{2}\ln(x^2 + 10) \Big|_0^5$$

$$= \tfrac{1}{2}\ln 35 - \tfrac{1}{2}\ln 10 \approx 0.626$$

Method 2. Substitute directly into the definite integral, changing both the variable of integration and the limits of integration: In the definite integral

$$\int_0^5 \frac{x}{x^2 + 10} dx$$

the upper limit is $x = 5$ and the lower limit is $x = 0$. When we make the substitution $u = x^2 + 10$ in this definite integral, we must change the limits of integration to the corresponding values of u:

$$x = 5 \quad \text{implies} \quad u = 5^2 + 10 = 35 \quad \text{New upper limit}$$
$$x = 0 \quad \text{implies} \quad u = 0^2 + 10 = 10 \quad \text{New lower limit}$$

Thus, we have

$$\int_0^5 \frac{x}{x^2 + 10} dx = \frac{1}{2}\int_0^5 \frac{1}{x^2 + 10} 2x \, dx$$

$$= \frac{1}{2}\int_{10}^{35} \frac{1}{u} du$$

$$= \frac{1}{2}\left(\ln|u| \big|_{10}^{35}\right)$$

$$= \tfrac{1}{2}(\ln 35 - \ln 10) \approx 0.626 \quad \blacksquare$$

MATCHED PROBLEM 3   Use both methods described in Example 3 to evaluate $\displaystyle\int_0^1 \frac{1}{2x + 4} dx$.

EXAMPLE 4   **Definite Integrals and Substitution**   Use method 2 described in Example 3 to evaluate

$$\int_{-4}^1 \sqrt{5 - t} \, dt$$

SOLUTION   If $u = 5 - t$, then $du = -dt$, and

$$t = 1 \quad \text{implies} \quad u = 5 - 1 = 4 \quad \text{New upper limit}$$
$$t = -4 \quad \text{implies} \quad u = 5 - (-4) = 9 \quad \text{New lower limit}$$

Notice that the lower limit for u is larger than the upper limit. Be careful not to reverse these two values when substituting into the definite integral:

$$\int_{-4}^{1} \sqrt{5 - t}\, dt = -\int_{-4}^{1} \sqrt{5 - t}\,(-dt)$$

$$= -\int_{9}^{4} \sqrt{u}\, du$$

$$= -\int_{9}^{4} u^{1/2}\, du$$

$$= -\left(\frac{u^{3/2}}{\frac{3}{2}}\Big|_{9}^{4}\right)$$

$$= -\left[\tfrac{2}{3}(4)^{3/2} - \tfrac{2}{3}(9)^{3/2}\right]$$

$$= -\left[\tfrac{16}{3} - \tfrac{54}{3}\right] = \tfrac{38}{3} \approx 12.667$$

MATCHED PROBLEM 4   Use method 2 described in Example 3 to evaluate $\displaystyle\int_{2}^{5} \frac{1}{\sqrt{6 - t}}\, dt$.

Explore & Discuss **1**   Explain why $\neq$ is used in each problem, and finish each correctly:

1. $\displaystyle\int_{0}^{2} e^x\, dx = e^x\big|_{0}^{2} \neq e^2$

2. $\displaystyle\int_{2}^{5} \frac{1}{2x + 3}\, dx \neq \frac{1}{2}\int_{2}^{5} \frac{1}{u}\, du \qquad u = 2x + 3, \qquad du = 2\, dx$

EXAMPLE 5   **Change in Profit**   A company manufactures x television sets per month. The monthly marginal profit (in dollars) is given by

$$P'(x) = 165 - 0.1x \qquad 0 \le x \le 4{,}000$$

The company is currently manufacturing 1,500 sets per month, but is planning to increase production. Find the change in the monthly profit if monthly production is increased to 1,600 sets.

SOLUTION

$$P(1{,}600) - P(1{,}500) = \int_{1{,}500}^{1{,}600} (165 - 0.1x)\, dx$$

$$= (165x - 0.05x^2)\big|_{1{,}500}^{1{,}600}$$

$$= \left[165(1{,}600) - 0.05(1{,}600)^2\right]$$
$$\quad - \left[165(1{,}500) - 0.05(1{,}500)^2\right]$$

$$= 136{,}000 - 135{,}000$$

$$= 1{,}000$$

Thus, increasing monthly production from 1,500 units to 1,600 units will increase the monthly profit by $1,000.

MATCHED PROBLEM 5   Repeat Example 5 if

$$P'(x) = 300 - 0.2x \qquad 0 \le x \le 3{,}000$$

and monthly production is increased from 1,400 sets to 1,500 sets.

EXAMPLE 6 **Useful Life** An amusement company maintains records for each video game it installs in an arcade. Suppose that $C(t)$ and $R(t)$ represent the total accumulated costs and revenues (in thousands of dollars), respectively, t years after a particular game has been installed. Suppose also that

$$C'(t) = 2 \qquad R'(t) = 9e^{-0.5t}$$

The value of t for which $C'(t) = R'(t)$ is called the **useful life** of the game.

(A) Find the useful life of the game, to the nearest year.

(B) Find the total profit accumulated during the useful life of the game.

SOLUTION (A) $R'(t) = C'(t)$

$$9e^{-0.5t} = 2$$

$$e^{-0.5t} = \tfrac{2}{9} \qquad \text{Convert to equivalent logarithmic form.}$$

$$-0.5t = \ln\tfrac{2}{9}$$

$$t = -2\ln\tfrac{2}{9} \approx 3 \text{ years}$$

Thus, the game has a useful life of 3 years. This is illustrated graphically in Figure 3.

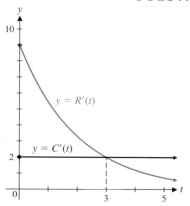

FIGURE 3 Useful life

(B) The total profit accumulated during the useful life of the game is

$$P(3) - P(0) = \int_0^3 P'(t)\,dt$$

$$= \int_0^3 [R'(t) - C'(t)]\,dt$$

$$= \int_0^3 (9e^{-0.5t} - 2)\,dt$$

$$= \left(\frac{9}{-0.5}e^{-0.5t} - 2t\right)\Big|_0^3 \qquad \text{Recall:} \int e^{ax}\,dx = \frac{1}{a}e^{ax} + C$$

$$= (-18e^{-0.5t} - 2t)\big|_0^3$$

$$= (-18e^{-1.5} - 6) - (-18e^0 - 0)$$

$$= 12 - 18e^{-1.5} \approx 7.984 \quad \text{or} \quad \$7,984$$

MATCHED PROBLEM 6 Repeat Example 6 if $C'(t) = 1$ and $R'(t) = 7.5e^{-0.5t}$.

EXAMPLE 7 **Numerical Integration on a Graphing Calculator** Evaluate $\int_{-1}^{2} e^{-x^2}\,dx$ to three decimal places.

SOLUTION The integrand e^{-x^2} does not have an elementary antiderivative, so we are unable to use the fundamental theorem to evaluate the definite integral. Instead, we use a numerical integration routine that has been preprogrammed into a graphing calculator. (Consult your user's manual for specific details.) Such a routine is an approximation algorithm, more powerful than the left-sum and right-sum methods discussed in Section 6-4. From Figure 4,

```
fnInt(e^(-X²),X,
-1,2)
      1.628905524
```

FIGURE 4

$$\int_{-1}^{2} e^{-x^2}\,dx = 1.629$$

MATCHED PROBLEM 7	Evaluate $\displaystyle\int_{1.5}^{4.3} \frac{x}{\ln x}\,dx$ to three decimal places.

■ Recognizing a Definite Integral: Average Value

Recall that the derivative of a function f was defined in Section 3-3 by

$$f'(x) = \lim_{h \to 0} \frac{f(x + h) - f(x)}{h}$$

This form is generally not easy to compute directly, but is easy to recognize in certain practical problems (slope, instantaneous velocity, rates of change, and so on). Once we know that we are dealing with a derivative, we proceed to try to compute the derivative with the use of derivative formulas and rules.

Similarly, evaluating a definite integral with the use of the definition

$$\int_a^b f(x)\,dx = \lim_{n \to \infty} [f(c_1)\Delta x_1 + f(c_2)\Delta x_2 + \cdots + f(c_n)\Delta x_n] \qquad (1)$$

is generally not easy, but the form on the right occurs naturally in many practical problems. We can use the fundamental theorem to evaluate the definite integral (once it is recognized) if an antiderivative can be found; otherwise, we will approximate it with a rectangle sum. We will now illustrate these points by finding the *average value* of a continuous function.

Suppose that the temperature F (in degrees Fahrenheit) in the middle of a small shallow lake from 8 AM ($t = 0$) to 6 PM ($t = 10$) during the month of May is given approximately as shown in Figure 5.

How can we compute the average temperature from 8 AM to 6 PM? We know that the average of a finite number of values $a_1, a_2, \ldots, a_n$ is given by

$$\text{average} = \frac{a_1 + a_2 + \cdots + a_n}{n}$$

But how can we handle a continuous function with infinitely many values? It would seem reasonable to divide the time interval $[0, 10]$ into n equal subintervals, compute the temperature at a point in each subinterval, and then use the average of the temperatures as an approximation of the average value of the continuous function $F = F(t)$ over $[0, 10]$. We would expect the approximations to improve as n increases. In fact, we would be inclined to define the limit of the average of n values as $n \to \infty$ as the *average value of F over* $[0, 10]$ if the limit exists. This is exactly what we will do:

$$\binom{\text{average temperature}}{\text{for } n \text{ values}} = \frac{1}{n}[F(t_1) + F(t_2) + \cdots + F(t_n)] \qquad (2)$$

Here, t_k is a point in the kth subinterval. We will call the limit of equation (2) as $n \to \infty$ the *average temperature over the time interval* $[0, 10]$.

Form (2) resembles form (1), but we are missing the Δt_k. We take care of this by multiplying equation (2) by $(b - a)/(b - a)$, which will change the form of equation (2) without changing its value:

$$\frac{b - a}{b - a} \cdot \frac{1}{n}[F(t_1) + F(t_2) + \cdots + F(t_n)] = \frac{1}{b - a} \cdot \frac{b - a}{n}[F(t_1) + F(t_2) + \cdots + F(t_n)]$$

$$= \frac{1}{b - a}\left[F(t_1)\frac{b - a}{n} + F(t_2)\frac{b - a}{n} + \cdots + F(t_n)\frac{b - a}{n} \right]$$

$$= \frac{1}{b - a}[F(t_1)\Delta t + F(t_2)\Delta t + \cdots + F(t_n)\Delta t]$$

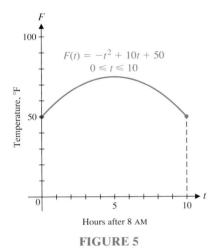

$F(t) = -t^2 + 10t + 50$
$0 \le t \le 10$

Temperature, °F

Hours after 8 AM

FIGURE 5

Thus,

$$\begin{pmatrix} \text{average temperature} \\ \text{over } [a, b] = [0, 10] \end{pmatrix} = \lim_{n \to \infty} \left\{ \frac{1}{b - a} [F(t_1)\Delta t + F(t_2)\Delta t + \cdots + F(t_n)\Delta t] \right\}$$

$$= \frac{1}{b - a} \left\{ \lim_{n \to \infty} [F(t_1)\Delta t + F(t_2)\Delta t + \cdots + F(t_n)\Delta t] \right\}$$

Now, the limit inside the braces is of form (1)—that is, a definite integral. Thus,

$$\begin{pmatrix} \text{average temperature} \\ \text{over } [a, b] = [0, 10] \end{pmatrix} = \frac{1}{b - a} \int_a^b F(t)\, dt$$

$$= \frac{1}{10 - 0} \int_0^{10} (-t^2 + 10t + 50)\, dt \quad \text{We now}$$

$$= \frac{1}{10} \left(-\frac{t^3}{3} + 5t^2 + 50t \right) \Big|_0^{10} \quad \begin{array}{l} \text{use the} \\ \text{fundamental} \\ \text{theorem to} \\ \text{evaluate the} \end{array}$$

$$= \frac{200}{3} \approx 67°\text{F} \quad \begin{array}{l} \text{definite} \\ \text{integral.} \end{array}$$

In general, proceeding as before for an arbitrary continuous function f over an interval $[a, b]$, we obtain the following general formula:

DEFINITION **Average Value of a Continuous Function f over $[a, b]$**

$$\frac{1}{b - a} \int_a^b f(x)\, dx$$

Explore & Discuss **2**   In Figure 6, the rectangle shown has the same area as the area under the graph of $y = f(x)$ from $x = a$ to $x = b$. Explain how the average value of $f(x)$ over the interval $[a, b]$ is related to the height of the rectangle.

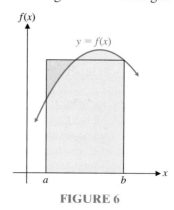

FIGURE 6

EXAMPLE 8   **Average Value of a Function** Find the average value of $f(x) = x - 3x^2$ over the interval $[-1, 2]$.

SOLUTION
$$\frac{1}{b - a} \int_a^b f(x)\, dx = \frac{1}{2 - (-1)} \int_{-1}^2 (x - 3x^2)\, dx$$

$$= \frac{1}{3} \left(\frac{x^2}{2} - x^3 \right) \Big|_{-1}^2 = -\frac{5}{2}$$

MATCHED PROBLEM 8   Find the average value of $g(t) = 6t^2 - 2t$ over the interval $[-2, 3]$.

EXAMPLE 9   **Average Price** Given the demand function

$$p = D(x) = 100e^{-0.05x}$$

find the average price (in dollars) over the demand interval $[40, 60]$.

SOLUTION   Average price $= \dfrac{1}{b-a}\displaystyle\int_a^b D(x)\,dx$

$$= \dfrac{1}{60-40}\int_{40}^{60} 100e^{-0.05x}\,dx$$

$$= \dfrac{100}{20}\int_{40}^{60} e^{-0.05x}\,dx \qquad \text{Use } \int e^{ax}\,dx = \dfrac{1}{a}e^{ax},\, a \neq 0.$$

$$= -\dfrac{5}{0.05}e^{-0.05x}\Big|_{40}^{60}$$

$$= 100(e^{-2} - e^{-3}) \approx \$8.55$$

MATCHED PROBLEM 9   Given the supply equation

$$p = S(x) = 10e^{0.05x}$$

find the average price (in dollars) over the supply interval $[20, 30]$.

Answers to Matched Problems

1. (A) $C(x)$ (B) $C'(x)$

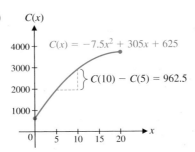

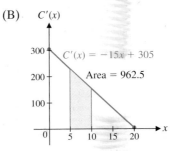

(C) The change in cost from $x = 5$ to $x = 10$ is equal to the area between the marginal cost function and the x axis from $x = 5$ to $x = 10$.

2. $16 + 2e - 2e^3 + 5\ln 3 \approx -13.241$ **3.** $\frac{1}{2}(\ln 6 - \ln 4) \approx 0.203$

4. 2   **5.** \$1,000

6. (A) $-2\ln\frac{2}{15} \approx 4$ yr

   (B) $11 - 15e^{-2} \approx 8.970$ or \$8,970

7. 8.017   **8.** 13   **9.** \$35.27

Exercise 6-5

A *In Problems 1–4,*

   (A) Calculate the change in $F(x)$ from $x = 10$ to $x = 15$.
   (B) Graph $F'(x)$ and use geometric formulas (see Appendix C) to calculate the area between the graph of $F'(x)$ and the x axis from $x = 10$ to $x = 15$.
   (C) Verify that your answers to (A) and (B) are equal, as is guaranteed by the fundamental theorem of calculus.

1. $F(x) = 3x^2 + 160$ **2.** $F(x) = 9x + 120$

3. $F(x) = -x^2 + 42x + 240$

4. $F(x) = x^2 + 30x + 210$

Evaluate the integrals in Problems 5–22.

5. $\displaystyle\int_2^3 2x\,dx$ **6.** $\displaystyle\int_1^2 3x^2\,dx$

7. $\displaystyle\int_3^4 5\,dx$ **8.** $\displaystyle\int_{12}^{20} dx$

9. $\displaystyle\int_1^3 (2x - 3)\,dx$ **10.** $\displaystyle\int_1^3 (6x + 5)\,dx$

11. $\displaystyle\int_{-3}^4 (4 - x^2)\,dx$ **12.** $\displaystyle\int_{-1}^2 (x^2 - 4x)\,dx$

13. $\int_0^1 24x^{11}\,dx$

14. $\int_0^2 30x^5\,dx$

15. $\int_0^1 e^{2x}\,dx$

16. $\int_{-1}^1 e^{5x}\,dx$

17. $\int_1^{3.5} 2x^{-1}\,dx$

18. $\int_1^2 \dfrac{dx}{x}$

19. $\int_1^2 \dfrac{2}{x^3}\,dx$

20. $\int_2^5 3x^{-2}\,dx$

21. $\int_1^4 6x^{-1/2}\,dx$

22. $\int_0^4 9x^{1/2}\,dx$

B *Evaluate the integrals in Problems 23–40.*

23. $\int_1^2 (2x^{-2} - 3)\,dx$

24. $\int_1^2 (5 - 16x^{-3})\,dx$

25. $\int_1^4 3\sqrt{x}\,dx$

26. $\int_4^{25} \dfrac{2}{\sqrt{x}}\,dx$

27. $\int_2^3 12(x^2 - 4)^5 x\,dx$

28. $\int_0^1 32(x^2 + 1)^7 x\,dx$

29. $\int_3^9 \dfrac{1}{x - 1}\,dx$

30. $\int_2^8 \dfrac{1}{x + 1}\,dx$

31. $\int_{-5}^{10} e^{-0.05x}\,dx$

32. $\int_{-10}^{25} e^{-0.01x}\,dx$

33. $\int_1^e \dfrac{\ln t}{t}\,dt$

34. $\int_e^{e^2} \dfrac{(\ln t)^2}{t}\,dt$

35. $\int_0^2 x\sqrt{4 - x^2}\,dx$

36. $\int_0^4 x\sqrt{4 - x}\,dx$

37. $\int_0^1 xe^{-x^2}\,dx$

38. $\int_0^1 xe^{x^2}\,dx$

39. $\int_{-2}^{-1} \dfrac{x^2 + 1}{x}\,dx$

40. $\int_{-2}^{-1} \dfrac{x}{x^2 + 1}\,dx$

In Problems 41–48,

(A) *Find the average value of each function over the indicated interval.*

 (B) *Use a graphing calculator to graph the function and its average value over the indicated interval in the same viewing window.*

41. $f(x) = 500 - 50x;\ [0, 10]$

42. $g(x) = 2x + 7;\ [0, 5]$

43. $f(t) = 3t^2 - 2t;\ [-1, 2]$

44. $g(t) = 4t - 3t^2;\ [-2, 2]$

45. $f(x) = \sqrt[3]{x};\ [1, 8]$

46. $g(x) = \sqrt{x + 1};\ [3, 8]$

47. $f(x) = 4e^{-0.2x};\ [0, 10]$

48. $f(x) = 64e^{0.08x};\ [0, 10]$

C *Evaluate the integrals in Problems 49–54.*

49. $\int_2^3 x\sqrt{2x^2 - 3}\,dx$

50. $\int_0^1 x\sqrt{3x^2 + 2}\,dx$

51. $\int_0^1 \dfrac{x - 1}{x^2 - 2x + 3}\,dx$

52. $\int_1^2 \dfrac{x + 1}{2x^2 + 4x + 4}\,dx$

53. $\int_{-1}^1 \dfrac{e^{-x} - e^x}{(e^{-x} + e^x)^2}\,dx$

54. $\int_6^7 \dfrac{\ln(t - 5)}{t - 5}\,dt$

 Use a numerical integration routine to evaluate each definite integral in Problems 55–58 (to three decimal places).

55. $\int_{1.7}^{3.5} x \ln x\,dx$

56. $\int_{-1}^1 e^{x^2}\,dx$

57. $\int_{-2}^2 \dfrac{1}{1 + x^2}\,dx$

58. $\int_0^3 \sqrt{9 - x^2}\,dx$

59. The **mean value theorem** states that if $F(x)$ is a differentiable function on the interval $[a, b]$, then there exists some number c between a and b such that

$$F'(c) = \dfrac{F(b) - F(a)}{b - a}$$

Explain why the mean value theorem implies that if a car averages 60 miles per hour in some 10-minute interval, then the car's instantaneous velocity is 60 miles per hour at least once in that interval.

60. The fundamental theorem of calculus can be proved by showing that, for every positive integer n, there is a Riemann sum for f on $[a, b]$ that is equal to $F(b) - F(a)$. By the mean value theorem (see Problem 59), within each subinterval $[x_{k-1}, x_k]$ that belongs to a partition of $[a, b]$, there is some c_k such that

$$f(c_k) = F'(c_k) = \dfrac{F(x_k) - F(x_{k-1})}{x_k - x_{k-1}}$$

Multiplying by the denominator $x_k - x_{k-1}$, we get

$$f(c_k)(x_k - x_{k-1}) = F(x_k) - F(x_{k-1})$$

Show that the Riemann sum

$$S_n = \sum_{k=1}^n f(c_k)(x_k - x_{k-1})$$

is equal to $F(b) - F(a)$.

Applications

61. Cost. A company manufactures mountain bikes. The research department produced the marginal cost function

$$C'(x) = 500 - \frac{x}{3} \qquad 0 \le x \le 900$$

where $C'(x)$ is in dollars and x is the number of bikes produced per month. Compute the increase in cost going from a production level of 300 bikes per month to 900 bikes per month. Set up a definite integral and evaluate it.

62. Cost. Referring to Problem 61, compute the increase in cost going from a production level of 0 bikes per month to 600 bikes per month. Set up a definite integral and evaluate it.

63. Salvage value. A new piece of industrial equipment will depreciate in value, rapidly at first and then less rapidly as time goes on. Suppose that the rate (in dollars per year) at which the book value of a new milling machine changes is given approximately by

$$V'(t) = f(t) = 500(t - 12) \qquad 0 \le t \le 10$$

where $V(t)$ is the value of the machine after t years. What is the total loss in value of the machine in the first 5 years? In the second 5 years? Set up appropriate integrals and solve.

64. Maintenance costs. Maintenance costs for an apartment house generally increase as the building gets older. From past records, a managerial service determines that the rate of increase in maintenance costs (in dollars per year) for a particular apartment complex is given approximately by

$$M'(x) = f(x) = 90x^2 + 5,000$$

where x is the age of the apartment complex in years and $M(x)$ is the total (accumulated) cost of maintenance for x years. Write a definite integral that will give the total maintenance costs from the end of the second year to the end of the seventh year after the apartment complex was built, and evaluate the integral.

65. Employee training. A company producing computer components has established that, on the average, a new employee can assemble $N(t)$ components per day after t days of on-the-job training, as indicated in the following table (a new employee's productivity usually increases with time on the job, up to a leveling-off point):

t	0	20	40	60	80	100	120
$N(t)$	10	51	68	76	81	84	85

(A) Find a quadratic regression equation for the data, and graph it and the data set in the same viewing window.

(B) Use the regression equation and a numerical integration routine on a graphing calculator to approximate the number of units assembled by a new employee during the first 100 days on the job.

66. Employee training. Refer to Problem 65.

(A) Find a cubic regression equation for the data, and graph it and the data set in the same viewing window.

(B) Use the regression equation and a numerical integration routine on a graphing calculator to approximate the number of units assembled by a new employee during the second 60 days on the job.

67. Useful life. The total accumulated costs $C(t)$ and revenues $R(t)$ (in thousands of dollars), respectively, for a coin-operated photocopying machine satisfy

$$C'(t) = \tfrac{1}{11}t \qquad \text{and} \qquad R'(t) = 5te^{-t^2}$$

where t is time in years. Find the useful life of the machine, to the nearest year. What is the total profit accumulated during the useful life of the machine?

68. Useful life. The total accumulated costs $C(t)$ and revenues $R(t)$ (in thousands of dollars), respectively, for a coal mine satisfy

$$C'(t) = 3 \qquad \text{and} \qquad R'(t) = 15e^{-0.1t}$$

where t is the number of years the mine has been in operation. Find the useful life of the mine, to the nearest year. What is the total profit accumulated during the useful life of the mine?

69. Average cost. The total cost (in dollars) of manufacturing x auto body frames is $C(x) = 60,000 + 300x$.

(A) Find the average cost per unit if 500 frames are produced. [*Hint:* Recall that $\overline{C}(x)$ is the average cost per unit.]

(B) Find the average value of the cost function over the interval [0, 500].

(C) Discuss the difference between parts (A) and (B).

70. Average cost. The total cost (in dollars) of printing x dictionaries is $C(x) = 20,000 + 10x$.

(A) Find the average cost per unit if 1,000 dictionaries are produced.

(B) Find the average value of the cost function over the interval [0, 1,000].

(C) Discuss the difference between parts (A) and (B).

71. Cost. The marginal cost at various levels of output per month for a company that manufactures watches is shown in the following table, with the output x given in thousands of units per month and the total cost $C(x)$ given in thousands of dollars per month:

x	0	1	2	3	4	5	6	7	8
$C'(x)$	58	30	18	9	5	7	17	33	51

(A) Find a quadratic regression equation for the data, and graph it and the data set in the same viewing window.

(B) Use the regression equation and a numerical integration routine on a graphing calculator to approximate (to the nearest dollar) the increased cost in going from a production level of 2 thousand watches per month to 8 thousand watches per month.

 72. *Cost.* Refer to Problem 71.

(A) Find a cubic regression equation for the data, and graph it and the data set in the same viewing window.

(B) Use the regression equation and a numerical integration routine on a graphing calculator to approximate (to the nearest dollar) the increased cost in going from a production level of 1 thousand watches per month to 7 thousand watches per month.

73. *Supply function.* Given the supply function

$$p = S(x) = 10(e^{0.02x} - 1)$$

find the average price (in dollars) over the supply interval $[20, 30]$.

74. *Demand function.* Given the demand function

$$p = D(x) = \frac{1,000}{x}$$

find the average price (in dollars) over the demand interval $[400, 600]$.

75. *Labor costs and learning.* A defense contractor is starting production on a new missile control system. On the basis of data collected during assembly of the first 16 control systems, the production manager obtained the following function for the rate of labor use:

$$g(x) = 2,400x^{-1/2}$$

Here, $g(x)$ is the number of labor-hours required to assemble the xth unit of a control system. Approximately how many labor-hours will be required to assemble the 17th through the 25th control units? [*Hint*: Let $a = 16$ and $b = 25$.]

76. *Labor costs and learning.* If the rate of labor use in Problem 75 is

$$g(x) = 2,000x^{-1/3}$$

approximately how many labor-hours will be required to assemble the 9th through the 27th control units? [*Hint*: Let $a = 8$ and $b = 27$.]

77. *Inventory.* A store orders 600 units of a product every 3 months. If the product is steadily depleted to 0 by the end of each 3 months, the inventory on hand, I, at any time t during the year is illustrated as shown in the following figure:

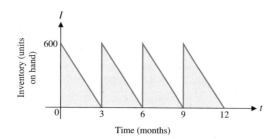

Figure for 77

(A) Write an inventory function (assume that it is continuous) for the first 3 months. [The graph is a straight line joining $(0, 600)$ and $(3, 0)$.]

(B) What is the average number of units on hand for a 3-month period?

78. Repeat Problem 77 with an order of 1,200 units every 4 months.

79. *Oil production.* Using data from the first 3 years of production, as well as geological studies, the management of an oil company estimates that oil will be pumped from a producing field at a rate given by

$$R(t) = \frac{100}{t + 1} + 5 \qquad 0 \le t \le 20$$

where $R(t)$ is the rate of production (in thousands of barrels per year) t years after pumping begins. Approximately how many barrels of oil will the field produce during the first 10 years of production? From the end of the 10th year to the end of the 20th year of production?

80. *Oil production.* In Problem 79, if the rate is found to be

$$R(t) = \frac{120t}{t^2 + 1} + 3 \qquad 0 \le t \le 20$$

approximately how many barrels of oil will the field produce during the first 5 years of production? The second 5 years of production?

81. *Biology.* A yeast culture weighing 2 grams is removed from a refrigerator unit and is expected to grow at the rate of $W'(t) = 0.2e^{0.1t}$ grams per hour at a higher controlled temperature. How much will the weight of the culture increase during the first 8 hours of growth? How much will the weight of the culture increase from the end of the 8th hour to the end of the 16th hour of growth?

82. *Medicine.* The rate of healing of a skin wound (in square centimeters per day) is given approximately by $A'(t) = -0.9e^{-0.1t}$. The initial wound has an area of 9 square centimeters. How much will the area change during the first 5 days? The second 5 days?

83. *Temperature.* If the temperature $C(t)$ in an aquarium is made to change according to

$$C(t) = t^3 - 2t + 10 \qquad 0 \le t \le 2$$

(in degrees Celsius) over a 2-hour period, what is the average temperature over this period?

84. *Medicine.* A drug is injected into the bloodstream of a patient through her right arm. The concentration of the drug in the bloodstream of the left arm t hours after the injection is given by

$$C(t) = \frac{0.14t}{t^2 + 1}$$

What is the average concentration of the drug in the bloodstream of the left arm during the first hour after the injection? During the first 2 hours after the injection?

85. *Politics.* Public awareness of a congressional candidate before and after a successful campaign was approximated by

$$P(t) = \frac{8.4t}{t^2 + 49} + 0.1 \qquad 0 \le t \le 24$$

where t is time in months after the campaign started and $P(t)$ is the fraction of people in the congressional district who could recall the candidate's name. What is the average fraction of people who could recall the candidate's name during the first 7 months after the campaign began? During the first 2 years after the campaign began?

86. *Population composition.* Because of various factors (such as expansion and then contraction of the birthrate, family flights from urban areas, and so on), the number of children in a large city was found to increase and then decrease rather drastically. If the number of children over a 6-year period was found to be given approximately by

$$N(t) = -\tfrac{1}{4}t^2 + t + 4 \qquad 0 \le t \le 6$$

what was the average number of children in the city over the 6-year period? [Assume that $N = N(t)$ is continuous.]

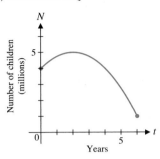

Figure for 86

CHAPTER 6 REVIEW

Important Terms, Symbols, and Concepts

6-1 Antiderivatives and Indefinite Integrals

Examples
Ex. 1, p. 361

- A function F is an **antiderivative** of a function f if $F'(x) = f(x)$.
- If F and G are both antiderivatives of f, then F and G differ by a constant; that is, $F(x) = G(x) + k$ for some constant k.
- We use the symbol $\int f(x)\, dx$, called an **indefinite integral,** to represent the family of all antiderivatives of f, and we write

$$\int f(x)\, dx = F(x) + C$$

The symbol $\int$ is called an **integral sign,** $f(x)$ is the **integrand,** and C is the **constant of integration.**

- Indefinite integrals of basic functions are given by the formulas on page 362.

Ex. 2, p. 364
Ex. 3, p. 365

- Properties of indefinite integrals are given on page 363; in particular, a constant factor can be moved across an integral sign. However, a variable factor *cannot* be moved across an integral sign.

Ex. 4, p. 366
Ex. 5, p. 367
Ex. 6, p. 367

6-2 Integration by Substitution

- The **method of substitution** (also called the **change-of-variable method**) is a technique for finding indefinite integrals. It is based on the following formula, which is obtained by reversing the chain rule:

Ex. 1, p. 374

$$\int f'[g(x)]g'(x)\, dx = f[g(x)] + C$$

- This formula implies the general indefinite integral formulas on page 374.
- In using the method of substitution, it is helpful to employ differentials as a bookkeeping device:

Ex. 2, p. 375
Ex. 3, p. 375
Ex. 4, p. 376

  1. The **differential dx** of the independent variable x is an arbitrary real number.

Ex. 5, p. 377

  2. The **differential dy** of the dependent variable y is defined by $dy = f'(x)\, dx$.

Ex. 6, p. 379

- Guidelines for using the substitution method are given by the procedure on page 376.

Ex. 7, p. 380

6-3 Differential Equations; Growth and Decay

- An equation is a **differential equation** if it involves an unknown function and one or more of its derivatives.

- The equation

$$\frac{dy}{dx} = 3x(1 + xy^2)$$

is a **first-order** differential equation, because it involves the first derivative of the unknown function y, but no second or higher order derivative.

- A **slope field** can be constructed for the preceding differential equation by drawing a tangent line segment with slope $3x(1 + xy^2)$ at each point (x, y) of a grid. The slope field gives a graphical representation of the functions that are solutions of the differential equation.

- The differential equation

$$\frac{dQ}{dt} = rQ$$

(in words, the rate at which the unknown function Q increases is proportional to Q) is called the **exponential growth law.** The constant r is called the **relative growth rate.** The solutions of the exponential growth law are the functions

$$Q(t) = Q_0 e^{rt}$$

where Q_0 denotes $Q(0)$, the amount present at time $t = 0$. These functions can be used to solve problems in population growth, continuous compound interest, radioactive decay, blood pressure, and light absorption.

Ex. 1, p. 388
Ex. 2, p. 388
Ex. 3, p. 389

- Table 1 on page 390 gives the solutions of other first-order differential equations that can be used to model the limited or logistic growth of epidemics, sales, and corporations.

Ex. 4, p. 390

6-4 The Definite Integral

- If the function f is positive on $[a, b]$, then the area between the graph of f and the x axis from $x = a$ to $x = b$ can be approximated by partitioning $[a, b]$ into n subintervals $[x_{k-1}, x_k]$ of equal length $\Delta x = (b - a)/n$ and summing the areas of n rectangles. This can be accomplished by **left sums, right sums,** or, more generally, **Riemann sums:**

Ex. 1, p. 397
Ex. 2, p. 399

Left sum: $\qquad L_n = \sum_{k=1}^{n} f(x_{k-1})\Delta x$

Right sum: $\qquad R_n = \sum_{k=1}^{n} f(x_k)\Delta x$

Riemann sum: $\qquad S_n = \sum_{k=1}^{n} f(c_k)\Delta x$

In a Riemann sum, each c_k is required to belong to the subinterval $[x_{k-1}, x_k]$. Left sums and right sums are the special cases of Riemann sums in which c_k is the left endpoint and right endpoint, respectively, of the subinterval.

- The **error in an approximation** is the absolute value of the difference between the approximation and the actual value. An **error bound** is a positive number such that the error is guaranteed to be less than or equal to that number.

- Theorem 1 on page 397 gives error bounds for the approximation of the area between the graph of a positive function f and the x axis from $x = a$ to $x = b$, by left sums or right sums, if f is either increasing or decreasing.

- If $f(x) > 0$ and is either increasing on $[a, b]$ or decreasing on $[a, b]$, then the left and right sums of $f(x)$ approach the same real number as $n \to \infty$ (Theorem 2, page 397).

- If f is a continuous function on $[a, b]$, then the Riemann sums for f on $[a, b]$ approach a real-number limit I as $n \to \infty$ (Theorem 3, page 399).

- Let f be a continuous function on $[a, b]$. Then the limit I of Riemann sums for f on $[a, b]$, guaranteed to exist by Theorem 3, is called the **definite integral** of f from a to b and is denoted

$$\int_a^b f(x)\, dx$$

The **integrand** is $f(x)$, the **lower limit of integration** is a, and the **upper limit of integration** is b.

- Geometrically, the definite integral

Ex. 3, p. 400

$$\int_a^b f(x)\, dx$$

represents the cumulative sum of the signed areas between the graph of f and the x axis from $x = a$ to $x = b$.

- Properties of the definite integral are given on page 401.

Ex. 4, p. 401

6-5 The Fundamental Theorem of Calculus

- If f is a continuous function on $[a, b]$ and F is any antiderivative of f, then

Ex. 1, p. 406
Ex. 2, p. 407

$$\int_a^b f(x)\, dx = F(b) - F(a)$$

(fundamental theorem of calculus, page 407).

- The fundamental theorem gives an easy and exact method for evaluating definite integrals, provided that we can find an antiderivative $F(x)$ of $f(x)$. In practice, we first find an antiderivative $F(x)$ (when possible), using techniques for computing indefinite integrals. Then we calculate the difference $F(b) - F(a)$. If it is impossible to find an antiderivative, we must resort to left or right sums, or other approximation methods, to evaluate the definite integral. Graphing calculators have a built-in numerical approximation routine, more powerful than left- or right-sum methods, for this purpose.

Ex. 3, p. 407
Ex. 4, p. 408
Ex. 5, p. 409
Ex. 6, p. 410
Ex. 7, p. 410

- If f is a continuous function on $[a, b]$, then the **average value** of f over $[a, b]$ is defined to be

Ex. 8, p. 412

$$\frac{1}{b - a}\int_a^b f(x)\, dx$$

REVIEW EXERCISE

Work through all the problems in this chapter review and check your answers in the back of the book. Answers to all review problems are there, along with section numbers in italics to indicate where each type of problem is discussed. Where weaknesses show up, review appropriate sections of the text.

A   *Find each integral in Problems 1–6.*

1. $\displaystyle\int (6x + 3)\, dx$

2. $\displaystyle\int_{10}^{20} 5\, dx$

3. $\displaystyle\int_0^9 (4 - t^2)\, dt$

4. $\displaystyle\int (1 - t^2)^3 t\, dt$

5. $\displaystyle\int \frac{1 + u^4}{u}\, du$

6. $\displaystyle\int_0^1 xe^{-2x^2}\, dx$

In Problems 7 and 8, find the derivative or indefinite integral as indicated.

7. $\dfrac{d}{dx}\left(\displaystyle\int e^{-x^2}\, dx\right)$

8. $\displaystyle\int \dfrac{d}{dx}\left(\sqrt{4 + 5x}\right) dx$

9. Find a function $y = f(x)$ that satisfies both conditions:

$$\frac{dy}{dx} = 3x^2 - 2 \qquad f(0) = 4$$

10. Find all antiderivatives of

(A) $\dfrac{dy}{dx} = 8x^3 - 4x - 1$

(B) $\dfrac{dx}{dt} = e^t - 4t^{-1}$

11. Approximate $\int_1^5 (x^2 + 1)\, dx$, using a right sum with $n = 2$. Calculate an error bound for this approximation.

12. Evaluate the integral in Problem 11, using the fundamental theorem of calculus, and calculate the actual error $|I - R_2|$ produced by using R_2.

13. Use the following table of values and a left sum with $n = 4$ to approximate $\int_1^{17} f(x)\, dx$:

x	1	5	9	13	17
$f(x)$	1.2	3.4	2.6	0.5	0.1

14. Find the average value of $f(x) = 6x^2 + 2x$ over the interval $[-1, 2]$.

15. Describe a rectangle that would have the same area as the area under the graph of $f(x) = 6x^2 + 2x$ from $x = -1$ to $x = 2$ (see Problem 14).

In Problems 16 and 17, calculate the indicated Riemann sum S_n for the function $f(x) = 100 - x^2$.

16. Partition $[3, 11]$ into four subintervals of equal length, and for each subinterval $[x_{i-1}, x_i]$, let $c_i = (x_{i-1} + x_i)/2$.

17. Partition $[-5, 5]$ into five subintervals of equal length and let $c_1 = -4$, $c_2 = -1$, $c_3 = 1$, $c_4 = 2$, and $c_5 = 5$.

B *Use the graph and actual areas of the indicated regions in the figure to evaluate the integrals in Problems 18–25:*

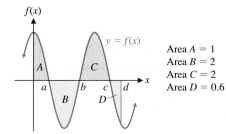

Area $A = 1$
Area $B = 2$
Area $C = 2$
Area $D = 0.6$

18. $\int_a^b 5f(x)\, dx$

19. $\int_b^c \dfrac{f(x)}{5}\, dx$

20. $\int_b^d f(x)\, dx$

21. $\int_a^c f(x)\, dx$

22. $\int_0^d f(x)\, dx$

23. $\int_b^a f(x)\, dx$

24. $\int_c^b f(x)\, dx$

25. $\int_d^0 f(x)\, dx$

Problems 26–31 refer to the slope field shown in the figure:

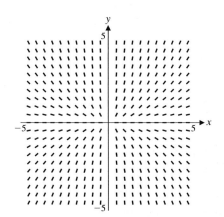

26. (A) For $dy/dx = (2y)/x$, what is the slope of a solution curve at $(2, 1)$? At $(-2, -1)$?

(B) For $dy/dx = (2x)/y$, what is the slope of a solution curve at $(2, 1)$? At $(-2, -1)$?

27. Is the slope field shown in the figure for $dy/dx = (2x)/y$ or for $dy/dx = (2y)/x$? Explain.

28. Show that $y = Cx^2$ is a solution of $dy/dx = (2y)/x$ for any real number C.

29. Referring to Problem 28, find the particular solution of $dy/dx = (2y)/x$ that passes through $(2, 1)$. Through $(-2, -1)$.

30. Graph the two particular solutions found in Problem 29 in the slope field shown (or a copy).

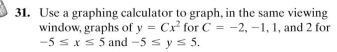 **31.** Use a graphing calculator to graph, in the same viewing window, graphs of $y = Cx^2$ for $C = -2, -1, 1$, and 2 for $-5 \le x \le 5$ and $-5 \le y \le 5$.

Find each integral in Problems 32–42.

32. $\int_{-1}^{1} \sqrt{1 + x}\, dx$

33. $\int_{-1}^{0} x^2(x^3 + 2)^{-2}\, dx$

34. $\int 5e^{-t}\, dt$

35. $\int_{1}^{e} \dfrac{1 + t^2}{t}\, dt$

36. $\int x e^{3x^2}\, dx$

37. $\int_{-3}^{1} \dfrac{1}{\sqrt{2 - x}}\, dx$

38. $\int_{0}^{3} \dfrac{x}{1 + x^2}\, dx$

39. $\int_{0}^{3} \dfrac{x}{(1 + x^2)^2}\, dx$

40. $\int x^3(2x^4 + 5)^5\, dx$

41. $\int \dfrac{e^{-x}}{e^{-x} + 3}\, dx$

42. $\int \dfrac{e^x}{(e^x + 2)^2}\, dx$

43. Find a function $y = f(x)$ that satisfies both conditions:
$$\frac{dy}{dx} = 3x^{-1} - x^{-2} \qquad f(1) = 5$$

44. Find the equation of the curve that passes through $(2, 10)$ if its slope is given by
$$\frac{dy}{dx} = 6x + 1$$
for each x.

45. (A) Find the average value of $f(x) = 3\sqrt{x}$ over the interval $[1, 9]$.

(B) Graph $f(x) = 3\sqrt{x}$ and its average over the interval $[1, 9]$ in the same coordinate system.

C *Find each integral in Problems 46–50.*

46. $\displaystyle\int \frac{(\ln x)^2}{x}\,dx$

47. $\displaystyle\int x(x^3 - 1)^2\,dx$

48. $\displaystyle\int \frac{x}{\sqrt{6 - x}}\,dx$

49. $\displaystyle\int_0^7 x\sqrt{16 - x}\,dx$

50. $\displaystyle\int_{-1}^1 x(x + 1)^4\,dx$

51. Find a function $y = f(x)$ that satisfies both conditions:

$$\frac{dy}{dx} = 9x^2 e^{x^3} \qquad f(0) = 2$$

52. Solve the differential equation

$$\frac{dN}{dt} = 0.06N \qquad N(0) = 800 \qquad N > 0$$

 Graph Problems 53–56 on a graphing calculator, and identify each curve as unlimited growth, exponential decay, limited growth, or logistic growth:

53. $N = 50(1 - e^{-0.07t}); 0 \le t \le 80, 0 \le N \le 60$

54. $p = 500e^{-0.03x}; 0 \le x \le 100, 0 \le p \le 500$

55. $A = 200e^{0.08t}; 0 \le t \le 20, 0 \le A \le 1{,}000$

56. $N = \dfrac{100}{1 + 9e^{-0.3t}}; 0 \le t \le 25, 0 \le N \le 100$

 Use a numerical integration routine to evaluate each definite integral in Problems 57–59 (to three decimal places).

57. $\displaystyle\int_{-0.5}^{0.6} \frac{1}{\sqrt{1 - x^2}}\,dx$

58. $\displaystyle\int_{-2}^3 x^2 e^x\,dx$

59. $\displaystyle\int_{0.5}^{2.5} \frac{\ln x}{x^2}\,dx$

APPLICATIONS

60. *Cost.* A company manufactures downhill skis. The research department produced the marginal cost graph shown in the accompanying figure, where $C'(x)$ is in dollars and x is the number of pairs of skis produced per week. Estimate the increase in cost going from a production level of 200 to 600 pairs of skis per week. Use left and right sums over two equal subintervals. Replace the question marks with the values of L_2 and R_2 as appropriate:

$$? \le \int_{200}^{600} C'(x)\,dx \le ?$$

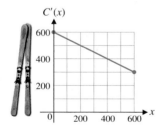

Figure for 60

61. *Cost.* Assuming that the marginal cost function in Problem 60 is linear, find its equation and write a definite integral that represents the increase in costs going from a production level of 200 to 600 pairs of skis per week. Evaluate the definite integral.

62. *Profit and production.* The weekly marginal profit for an output of x units is given approximately by

$$P'(x) = 150 - \frac{x}{10} \qquad 0 \le x \le 40$$

What is the total change in profit for a change in production from 10 units per week to 40 units? Set up a definite integral and evaluate it.

63. *Profit function.* If the marginal profit for producing x units per day is given by

$$P'(x) = 100 - 0.02x \qquad P(0) = 0$$

where $P(x)$ is the profit in dollars, find the profit function P and the profit on 10 units of production per day.

64. *Resource depletion.* An oil well starts out producing oil at the rate of 60,000 barrels of oil per year, but the production rate is expected to decrease by 4,000 barrels per year. Thus, if $P(t)$ is the total production (in thousands of barrels) in t years, then

$$P'(t) = f(t) = 60 - 4t \qquad 0 \le t \le 15$$

Write a definite integral that will give the total production after 15 years of operation, and evaluate the integral.

65. *Inventory.* Suppose that the inventory of a certain item t months after the first of the year is given approximately by

$$I(t) = 10 + 36t - 3t^2 \qquad 0 \le t \le 12$$

What is the average inventory for the second quarter of the year?

66. *Price–supply.* Given the price–supply function

$$p = S(x) = 8(e^{0.05x} - 1)$$

find the average price (in dollars) over the supply interval $[40, 50]$.

67. *Useful life.* The total accumulated costs $C(t)$ and revenues $R(t)$ (in thousands of dollars), respectively, for a coal mine satisfy

$$C'(t) = 3 \quad \text{and} \quad R'(t) = 20e^{-0.1t}$$

where t is the number of years the mine has been in operation. Find the useful life of the mine, to the nearest year. What is the total profit accumulated during the useful life of the mine?

68. *Marketing.* The market research department for an automobile company estimates that sales (in millions of dollars) of a new automobile will increase at the monthly rate of

$$S'(t) = 4e^{-0.08t} \qquad 0 \le t \le 24$$

t months after the introduction of the automobile. What will be the total sales $S(t)$ *t* months after the automobile is introduced if we assume that there were 0 sales at the time the automobile entered the marketplace? What are the estimated total sales during the first 12 months after the introduction of the automobile? How long will it take for the total sales to reach $40 million?

69. *Wound healing.* The area of a small healing surface wound changes at a rate given approximately by

$$\frac{dA}{dt} = -5t^{-2} \qquad 1 \le t \le 5$$

where *t* is time in days and $A(1) = 5$ square centimeters. What will the area of the wound be in 5 days?

70. *Pollution.* An environmental protection agency estimates that the rate of seepage of toxic chemicals from a waste dump (in gallons per year) is given by

$$R(t) = \frac{1,000}{(1 + t)^2}$$

where *t* is the time in years since the discovery of the seepage. Find the total amount of toxic chemicals that seep from the dump during the first 4 years after the seepage is discovered.

71. *Population.* The population of Mexico was 100 million in 2000 and was growing at a rate of 1.5% per year, compounded continuously.

(A) Assuming that the population continues to grow at this rate, estimate the population of Mexico in the year 2025.

(B) At the growth rate indicated, how long will it take the population of Mexico to double?

72. *Archaeology.* The continuous compound rate of decay for carbon-14 is $r = -0.000\,123\,8$. A piece of animal bone found at an archaeological site contains 4% of the original amount of carbon-14. Estimate the age of the bone.

73. *Learning.* In a particular business college, it was found that an average student enrolled in a typing class progressed at a rate of $N'(t) = 7e^{-0.1t}$ words per minute *t* weeks after enrolling in a 15-week course. If a student could type 25 words per minute at the beginning of the course, how many words per minute $N(t)$ would the student be expected to type *t* weeks into the course? After completing the course?

Additional Integration Topics

CHAPTER 7

7-1 Area between Curves

7-2 Applications in Business and Economics

7-3 Integration by Parts

7-4 Integration Using Tables

Chapter 7 Review

Review Exercise

INTRODUCTION

This chapter contains additional topics on integration. Since they are essentially independent of one another, they may be taken up in any order, and certain sections may be omitted if desired.

Section 7-1 AREA BETWEEN CURVES

- Area between Two Curves
- Application: Income Distribution

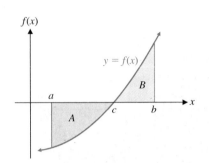

$f(x)$

$y = f(x)$

B

a c b x

A

FIGURE 1

$$\int_a^b f(x)\, dx = -A + B$$

In Chapter 6, we found that the definite integral $\int_a^b f(x)\, dx$ represents the sum of the signed areas between the graph of $y = f(x)$ and the x axis from $x = a$ to $x = b$, where the areas above the x axis are counted positively and the areas below the x axis are counted negatively (see Fig. 1). In this section, we are interested in using the definite integral to find the actual area between a curve and the x axis or the actual area between two curves. These areas are always nonnegative quantities—**area measure is never negative.**

■ Area between Two Curves

Consider the area bounded by $y = f(x)$ and $y = g(x)$, where $f(x) \geq g(x) \geq 0$, for $a \leq x \leq b$, as indicated in Figure 2.

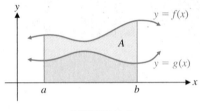

y

$y = f(x)$

A

$y = g(x)$

a b x

FIGURE 2

$$\begin{pmatrix} \text{Area } A \text{ between} \\ f(x) \text{ and } g(x) \end{pmatrix} = \begin{pmatrix} \text{area} \\ \text{under } f(x) \end{pmatrix} - \begin{pmatrix} \text{area} \\ \text{under } g(x) \end{pmatrix} \quad \begin{array}{l}\text{Areas are from } x = a \text{ to} \\ x = b \text{ above the } x \text{ axis.}\end{array}$$

$$= \int_a^b f(x)\, dx - \int_a^b g(x)\, dx \qquad \begin{array}{l}\text{Use definite integral} \\ \text{property 4 (Section 6-4).}\end{array}$$

$$= \int_a^b [f(x) - g(x)]\, dx$$

It can be shown that the preceding result does not require $f(x)$ or $g(x)$ to remain positive over the interval $[a, b]$. A more general result is stated in the following box:

THEOREM 1   AREA BETWEEN TWO CURVES

If f and g are continuous and $f(x) \geq g(x)$ over the interval $[a, b]$, then the area bounded by $y = f(x)$ and $y = g(x)$ for $a \leq x \leq b$ is given exactly by

$$A = \int_a^b [f(x) - g(x)]\, dx$$

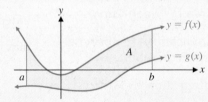

y

$y = f(x)$

A

$y = g(x)$

a b x

INSIGHT

Theorem 1 requires the graph of f to be *above* (or equal to) the graph of g throughout $[a, b]$, but f and g can be either positive, negative, or 0. In Section 6-4, we considered the special cases of Theorem 1 in which (1) f is positive and g is the zero function on $[a, b]$; and (2) f is the zero function and g is negative on $[a, b]$:

Special case 1. If f is continuous and positive over $[a, b]$, then the area bounded by the graph of f and the x axis for $a \leq x \leq b$ is given exactly by

$$\int_a^b f(x)\, dx$$

Special case 2. If g is continous and negative over $[a, b]$, then the area bounded by the graph of g and the x axis for $a \leq x \leq b$ is given exactly by

$$\int_a^b -g(x)\, dx$$

Explore & Discuss **1**   A Riemann sum for the integral representing the area between the graphs of $y = f(x)$ and $y = g(x)$ in Theorem 1 has the form

$$\sum_{k=1}^{n} [f(c_k) - g(c_k)] \Delta x_k$$

Each term in this sum represents the area of a rectangle with height $f(c_k) - g(c_k)$ and width Δx. Discuss the relationship between these rectangles and the area between the graphs of $y = f(x)$ and $y = g(x)$.

EXAMPLE 1   **Area between a Curve and the *x* Axis** Find the area bounded by $f(x) = 6x - x^2$ and $y = 0$ for $1 \le x \le 4$.

SOLUTION   We sketch a graph of the region first (Fig. 3). (The solution of every area problem should begin with a sketch.) Since $f(x) \ge 0$ on $[1, 4]$,

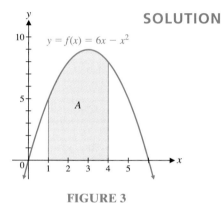

$y = f(x) = 6x - x^2$

A

FIGURE 3

$$A = \int_1^4 (6x - x^2)\, dx = \left(3x^2 - \frac{x^3}{3} \right) \Big|_1^4$$

$$= \left[3(4)^2 - \frac{(4)^3}{3} \right] - \left[3(1)^2 - \frac{(1)^3}{3} \right]$$

$$= 48 - \frac{64}{3} - 3 + \frac{1}{3}$$

$$= 48 - 21 - 3$$

$$= 24$$

MATCHED PROBLEM 1   Find the area bounded by $f(x) = x^2 + 1$ and $y = 0$ for $-1 \le x \le 3$.

EXAMPLE 2   **Area between a Curve and the *x* Axis** Find the area between the graph of $f(x) = x^2 - 2x$ and the *x* axis over the indicated intervals:

(A) $[1, 2]$
(B) $[-1, 1]$

SOLUTION   We begin by sketching the graph of *f*, as shown in Figure 4.

(A) From the graph, we see that $f(x) \le 0$ for $1 \le x \le 2$, so we integrate $-f(x)$:

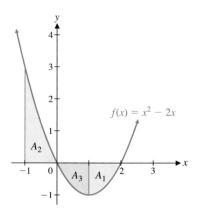

$f(x) = x^2 - 2x$

A_2 A_3 A_1

FIGURE 4

$$A_1 = \int_1^2 [-f(x)]\, dx$$

$$= \int_1^2 (2x - x^2)\, dx$$

$$= \left(x^2 - \frac{x^3}{3} \right) \Big|_1^2$$

$$= \left[(2)^2 - \frac{(2)^3}{3} \right] - \left[(1)^2 - \frac{(1)^3}{3} \right]$$

$$= 4 - \frac{8}{3} - 1 + \frac{1}{3} = \frac{2}{3} \approx 0.667$$

(B) Since the graph shows that $f(x) \geq 0$ on $[-1, 0]$ and $f(x) \leq 0$ on $[0, 1]$, the computation of this area will require two integrals:

$$A = A_2 + A_3$$

$$= \int_{-1}^{0} f(x)\, dx + \int_{0}^{1} [-f(x)]\, dx$$

$$= \int_{-1}^{0} (x^2 - 2x)\, dx + \int_{0}^{1} (2x - x^2)\, dx$$

$$= \left(\frac{x^3}{3} - x^2 \right) \Big|_{-1}^{0} + \left(x^2 - \frac{x^3}{3} \right) \Big|_{0}^{1}$$

$$= \boxed{\frac{4}{3} + \frac{2}{3}} = 2$$

MATCHED PROBLEM 2 Find the area between the graph of $f(x) = x^2 - 9$ and the x axis over the indicated intervals:

(A) $[0, 2]$

(B) $[2, 4]$

EXAMPLE 3 **Area between Two Curves** Find the area bounded by the graphs of $f(x) = \frac{1}{2}x + 3$, $g(x) = -x^2 + 1$, $x = -2$, and $x = 1$.

SOLUTION We first sketch the area (Fig. 5) and then set up and evaluate an appropriate definite integral. We observe from the graph that $f(x) \geq g(x)$ for $-2 \leq x \leq 1$, so

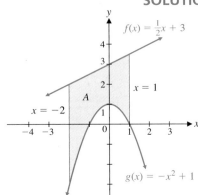

FIGURE 5

$$A = \int_{-2}^{1} [f(x) - g(x)]\, dx = \int_{-2}^{1} \left[\left(\frac{x}{2} + 3 \right) - (-x^2 + 1) \right] dx$$

$$= \int_{-2}^{1} \left(x^2 + \frac{x}{2} + 2 \right) dx$$

$$= \left(\frac{x^3}{3} + \frac{x^2}{4} + 2x \right) \Big|_{-2}^{1}$$

$$= \left(\frac{1}{3} + \frac{1}{4} + 2 \right) - \left(\frac{-8}{3} + \frac{4}{4} - 4 \right) = \frac{33}{4} = 8.25$$

MATCHED PROBLEM 3 Find the area bounded by $f(x) = x^2 - 1$, $g(x) = -\frac{1}{2}x - 3$, $x = -1$, and $x = 2$.

EXAMPLE 4 **Area between Two Curves** Find the area bounded by $f(x) = 5 - x^2$ and $g(x) = 2 - 2x$.

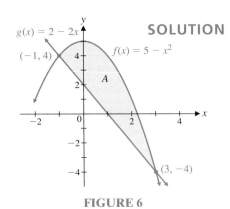

FIGURE 6

SOLUTION First, graph f and g on the same coordinate system, as shown in Figure 6. Since the statement of the problem does not include any limits on the values of x, we must determine the appropriate values from the graph. The graph of f is a parabola and the graph of g is a line, as shown. The area bounded by these two graphs extends from the intersection point on the left to the intersection point on the right. To find these intersection points, we solve the equation $f(x) = g(x)$ for x:

$$f(x) = g(x)$$
$$5 - x^2 = 2 - 2x$$
$$x^2 - 2x - 3 = 0$$
$$x = -1, 3$$

You should check these values in the original equations. (Note that the area between the graphs for $x < -1$ is unbounded on the left, and the area between the graphs for $x > 3$ is unbounded on the right.) Figure 6 shows that $f(x) \geq g(x)$ over the interval $[-1, 3]$, so we have

$$A = \int_{-1}^{3} [f(x) - g(x)] \, dx = \int_{-1}^{3} [5 - x^2 - (2 - 2x)] \, dx$$

$$= \int_{-1}^{3} (3 + 2x - x^2) \, dx$$

$$= \left(3x + x^2 - \frac{x^3}{3} \right) \Big|_{-1}^{3}$$

$$= \left[3(3) + (3)^2 - \frac{(3)^3}{3} \right] - \left[3(-1) + (-1)^2 - \frac{(-1)^3}{3} \right] = \frac{32}{3} \approx 10.667$$

MATCHED PROBLEM 4   Find the area bounded by $f(x) = 6 - x^2$ and $g(x) = x$.

EXAMPLE 5   **Area between Two Curves** Find the area bounded by $f(x) = x^2 - x$ and $g(x) = 2x$ for $-2 \leq x \leq 3$.

SOLUTION   The graphs of f and g are shown in Figure 7. Examining the graph, we see that $f(x) \geq g(x)$ on the interval $[-2, 0]$, but $g(x) \geq f(x)$ on the interval $[0, 3]$. Thus, two integrals are required to compute this area:

$$A_1 = \int_{-2}^{0} [f(x) - g(x)] \, dx \quad f(x) \geq g(x) \text{ on } [-2, 0]$$

$$= \int_{-2}^{0} [x^2 - x - 2x] \, dx$$

$$= \int_{-2}^{0} (x^2 - 3x) \, dx$$

$$= \left(\frac{x^3}{3} - \frac{3}{2} x^2 \right) \Big|_{-2}^{0}$$

$$= (0) - \left[\frac{(-2)^3}{3} - \frac{3}{2}(-2)^2 \right] = \frac{26}{3} \approx 8.667$$

$$A_2 = \int_{0}^{3} [g(x) - f(x)] \, dx \quad g(x) \geq f(x) \text{ on } [0, 3]$$

$$= \int_{0}^{3} [2x - (x^2 - x)] \, dx$$

$$= \int_{0}^{3} (3x - x^2) \, dx$$

$$= \left(\frac{3}{2} x^2 - \frac{x^3}{3} \right) \Big|_{0}^{3}$$

$$= \left[\frac{3}{2}(3)^2 - \frac{(3)^3}{3} \right] - (0) = \frac{9}{2} = 4.5$$

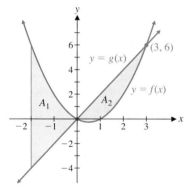

FIGURE 7

The total area between the two graphs is

$$A = A_1 + A_2 = \frac{26}{3} + \frac{9}{2} = \frac{79}{6} \approx 13.167$$

MATCHED PROBLEM 5 | Find the area bounded by $f(x) = 2x^2$ and $g(x) = 4 - 2x$ for $-2 \leq x \leq 2$.

EXAMPLE 6 **Computing Areas with a Numerical Integration Routine** Find the area (to three decimal places) bounded by $f(x) = e^{-x^2}$ and $g(x) = x^2 - 1$.

SOLUTION First, we use a graphing calculator to graph the functions f and g and find their intersection points (see Fig. 8A). We see that the graph of f is bell shaped and the graph of g is a parabola, and we note that $f(x) \geq g(x)$ on the interval $[-1.131, 1.131]$. Now we compute the area A by a numerical integration routine (see Fig. 8B):

$$A = \int_{-1.131}^{1.131} \left[e^{-x^2} - (x^2 - 1) \right] dx = 2.876$$

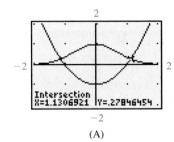

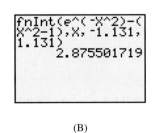

(A)   (B)

FIGURE 8

MATCHED PROBLEM 6 | Find the area (to three decimal places) bounded by the graphs of $f(x) = x^2 \ln x$ and $g(x) = 3x - 3$.

■ Application: Income Distribution

The U.S. Census Bureau compiles and analyzes a great deal of data having to do with the distribution of income among families in the United States. For 2003, the Bureau reported that the lowest 20% of families received 4% of all family income and the top 20% received 47%. Table 1 and Figure 9 give a detailed picture of the distribution of family income in 2003.

The graph of $y = f(x)$ in Figure 9 is called a **Lorenz curve** and is generally found with the use of *regression analysis,* a technique of fitting a particular elementary function to a data set over a given interval. The variable **x represents the cumulative percentage of families at or below a given income level,** and **y represents the cumulative percentage of total family income received.** For example, data point

TABLE 1 Family Income Distribution in the United States, 2003		
Income Level	***x***	***y***
Under $24,000	0.20	0.04
Under $42,000	0.40	0.14
Under $65,000	0.60	0.30
Under $98,000	0.80	0.53

Source: U.S. Census Bureau

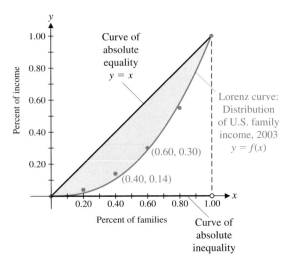

FIGURE 9 Lorenz chart

(0.40, 0.14) in Table 1 indicates that the bottom 40% of families (those with incomes under $42,000) received 14% of the total income for all families in 2003, data point (0.60, 0.30) indicates that the bottom 60% of families receive 30% of the total income for all families that year, and so on.

Absolute equality of income would occur if the area between the Lorenz curve and $y = x$ were 0. In this case, the Lorenz curve would be $y = x$ and all families would receive equal shares of the total income. That is, 5% of the families would receive 5% of the income, 20% of the families would receive 20% of the income, 65% of the families would receive 65% of the income, and so on. The maximum possible area between a Lorenz curve and $y = x$ is $\frac{1}{2}$, the area of the triangle below $y = x$. In this case, we would have **absolute inequality:** All the income would be in the hands of one family and the rest would have none. In actuality, Lorenz curves lie between these two extremes. But as the shaded area increases, the greater is the inequality of income distribution.

We use a single number, the **Gini index** [named after the Italian sociologist Corrado Gini (1884–1965)], to measure income concentration. The Gini index is the ratio of two areas: the area between $y = x$ and the Lorenz curve, and the area between $y = x$ and the x axis, from $x = 0$ to $x = 1$. The first area equals $\int_0^1 [x - f(x)]\, dx$ and the second (triangular) area equals $\frac{1}{2}$, giving the following definition:

DEFINITION **Gini Index of Income Concentration**

If $y = f(x)$ is the equation of a Lorenz curve, then

$$\text{Gini index} = 2 \int_0^1 [x - f(x)]\, dx$$

The Gini index is always a number between 0 and 1:

A measure of 0 indicates absolute equality—all people share equally in the income. A measure of 1 indicates absolute inequality—one person has all the income and the rest have none.

The closer the index is to 0, the closer the income is to being equally distributed. The closer the index is to 1, the closer the income is to being concentrated in a few hands. The Gini index of income concentration is used to compare income distributions at various points in time, between different groups of people, before and after taxes are paid, between different countries, and so on.

EXAMPLE 7 **Distribution of Income** The Lorenz curve for the distribution of income in a certain country in 2005 is given by $f(x) = x^{2.6}$. Economists predict that the Lorenz curve for the country in the year 2020 will be given by $g(x) = x^{1.8}$. Find the Gini index of income concentration for each curve, and interpret the results.

SOLUTION The Lorenz curves are shown in Figure 10.

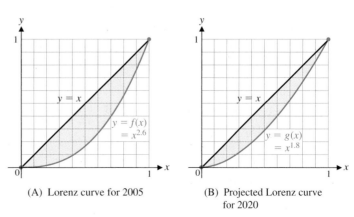

(A) Lorenz curve for 2005

(B) Projected Lorenz curve for 2020

FIGURE 10

The Gini index in 2005 is (see Fig. 10A)

$$2 \int_0^1 [x - f(x)]\, dx = 2 \int_0^1 [x - x^{2.6}]\, dx = 2\left(\frac{1}{2}x^2 - \frac{1}{3.6}x^{3.6}\right)\Big|_0^1$$

$$= 2\left(\frac{1}{2} - \frac{1}{3.6}\right) \approx 0.444$$

The projected Gini index in 2020 is (see Fig. 10B)

$$2 \int_0^1 [x - g(x)]\, dx = 2 \int_0^1 [x - x^{1.8}]\, dx = 2\left(\frac{1}{2}x^2 - \frac{1}{2.8}x^{2.8}\right)\Big|_0^1$$

$$= 2\left(\frac{1}{2} - \frac{1}{2.8}\right) \approx 0.286$$

If this projection is correct, the Gini index will decrease, and income will be more equally distributed in the year 2020 than in 2005. ▬

MATCHED PROBLEM 7   Repeat Example 7 if the projected Lorenz curve in the year 2020 is given by $g(x) = x^{3.8}$.

Explore & Discuss **2**   Do you agree or disagree with each of the following statements (explain your answers by reference to the data in Table 2):

(A) In countries with a low Gini index, there is little incentive for individuals to strive for success, and therefore productivity is low.

(B) In countries with a high Gini index, it is almost impossible to rise out of poverty, and therefore productivity is low.

TABLE 2		
Country	Gini Index	Per Capita Gross Domestic Product
Brazil	0.60	$6,500
China	0.40	3,600
France	0.33	24,400
Germany	0.30	23,400
Japan	0.25	24,900
Jordan	0.36	3,500
Mexico	0.54	9,100
Russia	0.50	7,700
Sweden	0.25	22,200
United States	0.41	36,200

Source: The World Bank

Answers to Matched Problems

1. $A = \int_{-1}^{3} (x^2 + 1)\, dx = \frac{40}{3} \approx 13.333$

2. (A) $A = \int_0^2 (9 - x^2)\, dx = \frac{46}{3} \approx 15.333$

   (B) $A = \int_2^3 (9 - x^2)\, dx + \int_3^4 (x^2 - 9)\, dx = 6$

3. $A = \int_{-1}^{2}\left[(x^2 - 1) - \left(-\frac{x}{2} - 3\right)\right] dx = \frac{39}{4} = 9.75$

4. $A = \int_{-3}^{2} [(6 - x^2) - x] \, dx = \frac{125}{6} \approx 20.833$

5. $A = \int_{-2}^{1} [(4 - 2x) - 2x^2] \, dx + \int_{1}^{2} [2x^2 - (4 - 2x)] \, dx = \frac{38}{3} \approx 12.667$

6. 0.443

7. Gini index of income concentration ≈ 0.583; income will be less equally distributed in 2020.

Exercise 7-1

A Problems 1–6 refer to Figures A–D. Set up definite integrals in Problems 1–4 that represent the indicated shaded area.

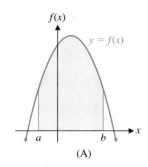

(A)

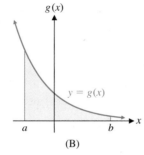

(B)

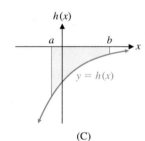

(C)

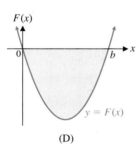

(D)

1. Shaded area in Figure B

2. Shaded area in Figure A

3. Shaded area in Figure C

4. Shaded area in Figure D

5. Explain why $\int_{a}^{b} h(x) \, dx$ does not represent the area between the graph of $y = h(x)$ and the x axis from $x = a$ to $x = b$ in Figure C.

6. Explain why $\int_{a}^{b} [-h(x)] \, dx$ represents the area between the graph of $y = h(x)$ and the x axis from $x = a$ to $x = b$ in Figure C.

In Problems 7–24, find the area bounded by the graphs of the indicated equations over the given interval. Compute answers to three decimal places.

7. $y = x + 4$; $y = 0, 0 \le x \le 4$

8. $y = -x + 10$; $y = 0, -2 \le x \le 2$

9. $y = -2x - 5$; $y = 0, -1 \le x \le 2$

10. $y = 3x - 14$; $y = 0, 0 \le x \le 4$

11. $y = x^2 - 20$; $y = 0, -3 \le x \le 0$

12. $y = x^2 + 2$; $y = 0, 0 \le x \le 3$

13. $y = -x^2 + 10$; $y = 0, -3 \le x \le 3$

14. $y = -2x^2$; $y = 0, -6 \le x \le 0$

15. $y = x^3 + 1$; $y = 0, 0 \le x \le 2$

16. $y = -x^3 + 3$; $y = 0, -2 \le x \le 1$

17. $y = x(1 - x)$; $y = 0, -1 \le x \le 0$

18. $y = -x(3 - x)$; $y = 0, 1 \le x \le 2$

19. $y = -e^x$; $y = 0, -1 \le x \le 1$

20. $y = e^x$; $y = 0, 0 \le x \le 1$

21. $y = \dfrac{1}{x}$; $y = 0, 1 \le x \le e$

22. $y = -\dfrac{1}{x}$; $y = 0, -1 \le x \le -\dfrac{1}{e}$

23. $y = -e^{-2x}$; $y = 0, 0 \le x \le 1$

24. $y = -e^{3x}$; $y = 0, -1 \le x \le 0$

B Problems 25–34 refer to Figures A and B. Set up definite integrals in Problems 25–32 that represent the indicated shaded areas over the given intervals.

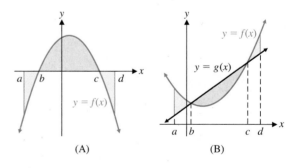

(A)   (B)

25. Over interval $[a, b]$ in Figure A

26. Over interval $[c, d]$ in Figure A

27. Over interval $[b, d]$ in Figure A

28. Over interval $[a, c]$ in Figure A

29. Over interval $[c, d]$ in Figure B

30. Over interval $[a, b]$ in Figure B

31. Over interval $[a, c]$ in Figure B

32. Over interval $[b, d]$ in Figure B

33. Referring to Figure B, explain how you would use definite integrals and the functions f and g to find the area bounded by the two functions from $x = a$ to $x = d$.

34. Referring to Figure A, explain how you would use definite integrals to find the area between the graph of $y = f(x)$ and the x axis from $x = a$ to $x = d$.

In Problems 35–50, find the area bounded by the graphs of the indicated equations over the given intervals (when stated). Compute answers to three decimal places.

35. $y = -x; y = 0, -2 \le x \le 1$

36. $y = -x + 1; y = 0, -1 \le x \le 2$

37. $y = x^2 - 4; y = 0, 0 \le x \le 3$

38. $y = 4 - x^2; y = 0, 0 \le x \le 4$

39. $y = x^2 - 3x; y = 0, -2 \le x \le 2$

40. $y = -x^2 - 2x; y = 0, -2 \le x \le 1$

41. $y = -2x + 8; y = 12; -1 \le x \le 2$

42. $y = 2x + 6; y = 3; -1 \le x \le 2$

43. $y = 3x^2; y = 12$

44. $y = x^2; y = 9$

45. $y = 4 - x^2; y = -5$

46. $y = x^2 - 1; y = 3$

47. $y = x^2 + 1; y = 2x - 2; -1 \le x \le 2$

48. $y = x^2 - 1; y = x - 2; -2 \le x \le 1$

49. $y = e^{0.5x}; y = -\dfrac{1}{x}; 1 \le x \le 2$

50. $y = \dfrac{1}{x}; y = -e^x; 0.5 \le x \le 1$

In Problems 51–56, set up a definite integral that represents the area bounded by the graphs of the indicated equations over the given interval. Find the areas to three decimal places. [Hint: A circle of radius r, with center at the origin, has equation $x^2 + y^2 = r^2$ and area πr^2].

51. $y = \sqrt{9 - x^2}; y = 0, -3 \le x \le 3$

52. $y = \sqrt{25 - x^2}; y = 0, -5 \le x \le 5$

53. $y = -\sqrt{16 - x^2}; y = 0, 0 \le x \le 4$

54. $y = -\sqrt{36 - x^2}; y = 0, -6 \le x \le 0$

55. $y = -\sqrt{4 - x^2}; y = \sqrt{4 - x^2}, -2 \le x \le 2$

56. $y = -\sqrt{100 - x^2}; y = \sqrt{100 - x^2}, -10 \le x \le 10$

 In Problems 57–60, use a graphing calculator to graph the equations and find relevant intersection points. Then find the area bounded by the curves. Compute answers to three decimal places.

57. $y = 3 - 5x - 2x^2; y = 2x^2 + 3x - 2$

58. $y = 3 - 2x^2; y = 2x^2 - 4x$

59. $y = -0.5x + 2.25; y = \dfrac{1}{x}$

60. $y = x - 4.25; y = -\dfrac{1}{x}$

C *In Problems 61–68, find the area bounded by the graphs of the indicated equations over the given intervals (when stated). Compute answers to three decimal places.*

61. $y = e^x; y = e^{-x}; 0 \le x \le 4$

62. $y = e^x; y = -e^{-x}; 1 \le x \le 2$

63. $y = x^3; y = 4x$

64. $y = x^3 + 1; y = x + 1$

65. $y = x^3 - 3x^2 - 9x + 12; y = x + 12$

66. $y = x^3 - 6x^2 + 9x; y = x$

67. $y = x^4 - 4x^2 + 1; y = x^2 - 3$

68. $y = x^4 - 6x^2; y = 4x^2 - 9$

 In Problems 69–74, use a graphing calculator to graph the equations and find relevant intersection points. Then find the area bounded by the curves. Compute answers to three decimal places.

69. $y = x^3 - x^2 + 2; y = -x^3 + 8x - 2$

70. $y = 2x^3 + 2x^2 - x; y = -2x^3 - 2x^2 + 2x$

71. $y = e^{-x}; y = 3 - 2x$

72. $y = 2 - (x + 1)^2; y = e^{x+1}$

73. $y = e^x; y = 5x - x^3$

74. $y = 2 - e^x; y = x^3 + 3x^2$

 In Problems 75–78, use a numerical integration routine on a graphing calculator to find the area bounded by the graphs of the indicated equations over the given interval (when stated). Compute answers to three decimal places.

75. $y = e^{-x}; y = \sqrt{\ln x}; 2 \le x \le 5$

76. $y = x^2 + 3x + 1; y = e^{e^x}; -3 \le x \le 0$

77. $y = e^{x^2}; y = x + 2$

78. $y = \ln(\ln x); y = 0.01x$

Applications

In the applications that follow, it is helpful to sketch graphs to get a clearer understanding of each problem and to interpret results. A graphing calculator will prove useful if you have one, but it is not necessary.

79. *Oil production.* Using data from the first 3 years of production, as well as geological studies, the management of

an oil company estimates that oil will be pumped from a producing field at a rate given by

$$R(t) = \frac{100}{t + 10} + 10 \qquad 0 \le t \le 15$$

where $R(t)$ is the rate of production (in thousands of barrels per year) t years after pumping begins. Find the area between the graph of R and the t axis over the interval $[5, 10]$, and interpret the results.

80. *Oil production.* In Problem 79, if the rate is found to be

$$R(t) = \frac{100t}{t^2 + 25} + 4 \qquad 0 \le t \le 25$$

find the area between the graph of R and the t axis over the interval $[5, 15]$ and interpret the results.

81. *Useful life.* An amusement company maintains records for each video game it installs in an arcade. Suppose that $C(t)$ and $R(t)$ represent the total accumulated costs and revenues (in thousands of dollars), respectively, t years after a particular game has been installed. If

$$C'(t) = 2 \qquad \text{and} \qquad R'(t) = 9e^{-0.3t}$$

find the area between the graphs of C' and R' over the interval on the t axis from 0 to the useful life of the game and interpret the results.

82. *Useful life.* Repeat Problem 81 if

$$C'(t) = 2t \qquad \text{and} \qquad R'(t) = 5te^{-0.1t^2}$$

83. *Income distribution.* As part of a study of the effects of World War II on the economy of the United States, an economist used data from the U.S. Census Bureau to produce the following Lorenz curves for the distribution of income in the United States in 1935 and in 1947:

$$f(x) = x^{2.4} \quad \text{Lorenz curve for 1935}$$
$$g(x) = x^{1.6} \quad \text{Lorenz curve for 1947}$$

Find the Gini index of income concentration for each Lorenz curve, and interpret the results.

84. *Income distribution.* Using data from the U.S. Census Bureau, an economist produced the following Lorenz curves for the distribution of income in the United States in 1962 and in 1972:

$$f(x) = \tfrac{3}{10}x + \tfrac{7}{10}x^2 \quad \text{Lorenz curve for 1962}$$
$$g(x) = \tfrac{1}{2}x + \tfrac{1}{2}x^2 \quad \text{Lorenz curve for 1972}$$

Find the Gini index of income concentration for each Lorenz curve, and interpret the results.

85. *Distribution of wealth.* Lorenz curves also can be used to provide a relative measure of the distribution of the total assets of a country. Using data in a report by the U.S. Congressional Joint Economic Committee, an economist produced the following Lorenz curves for the distribution of total assets in the United States in 1963 and in 1983:

$$f(x) = x^{10} \quad \text{Lorenz curve for 1963}$$
$$g(x) = x^{12} \quad \text{Lorenz curve for 1983}$$

Find the Gini index of income concentration for each Lorenz curve, and interpret the results.

86. *Income distribution.* The government of a small country is planning sweeping changes in the tax structure in order to provide a more equitable distribution of income. The Lorenz curves for the current income distribution and for the projected income distribution after enactment of the tax changes are as follows:

$$f(x) = x^{2.3} \qquad \text{Current Lorenz curve}$$
$$g(x) = 0.4x + 0.6x^2 \quad \text{Projected Lorenz curve after changes in tax laws}$$

Find the Gini index of income concentration for each Lorenz curve. Will the proposed changes provide a more equitable income distribution? Explain.

87. *Distribution of wealth.* The data in the table describe the distribution of wealth in a country:

x	0	0.20	0.40	0.60	0.80	1
y	0	0.12	0.31	0.54	0.78	1

(A) Use quadratic regression to find the equation of a Lorenz curve for the data.

(B) Use the regression equation and a numerical integration routine to approximate the Gini index of income concentration.

88. *Distribution of wealth.* Refer to Problem 87.

(A) Use cubic regression to find the equation of a Lorenz curve for the data.

(B) Use the cubic regression equation you found in Part (A) and a numerical integration routine to approximate the Gini index of income concentration.

89. *Biology.* A yeast culture is growing at a rate of $W'(t) = 0.3e^{0.1t}$ grams per hour. Find the area between the graph of W' and the t axis over the interval $[0, 10]$, and interpret the results.

90. *Natural resource depletion.* The instantaneous rate of change of the demand for lumber in the United States since 1970 ($t = 0$), in billions of cubic feet per year, is estimated to be given by

$$Q'(t) = 12 + 0.006t^2 \qquad 0 \le t \le 50$$

Find the area between the graph of Q' and the t axis over the interval $[15, 20]$, and interpret the results.

91. *Learning.* A beginning high school language class was chosen for an experiment on learning. Using a list of 50 words, the experiment involved measuring the rate of vocabulary memorization at different times during a continuous 5-hour study session. It was found that the average rate of learning for the entire class was inversely proportional to the time spent studying and was given approximately by

$$V'(t) = \frac{15}{t} \qquad 1 \le t \le 5$$

Find the area between the graph of V' and the t axis over the interval $[2, 4]$, and interpret the results.

92. *Learning.* Repeat Problem 91 if $V'(t) = 13/t^{1/2}$ and the interval is changed to $[1, 4]$.

Section 7-2 APPLICATIONS IN BUSINESS AND ECONOMICS

- Probability Density Functions
- Continuous Income Stream
- Future Value of a Continuous Income Stream
- Consumers' and Producers' Surplus

This section contains a number of important applications of the definite integral to business and economics. Included are three independent topics: probability density functions, continuous income streams, and consumers' and producers' surplus. Any of the three may be covered as time and interests dictate, and in any order.

■ Probability Density Functions

We now take a brief look at the use of the definite integral to determine probabilities. Our approach will be intuitive and informal. A more formal treatment of the subject requires the use of the special "improper" integral form $\int_{-\infty}^{\infty} f(x)\,dx$, which we have not defined or discussed at this point.

Suppose that an experiment is designed in such a way that any real number x on the interval $[c, d]$ is a possible outcome. For example, x may represent an IQ score, the height of a person in inches, or the life of a lightbulb in hours. Technically, we refer to x as a *continuous random variable*.

In certain situations, it is possible to find a function f with x as an independent variable such that the function f can be used to determine the probability that the outcome x of an experiment will be in the interval $[c, d]$. Such a function, called a **probability density function,** must satisfy the following three conditions (see Fig. 1):

1. $f(x) \geq 0$ for all real x.
2. The area under the graph of $f(x)$ over the interval $(-\infty, \infty)$ is exactly 1.
3. If $[c, d]$ is a subinterval of $(-\infty, \infty)$, then

$$\text{Probability } (c \leq x \leq d) = \int_{c}^{d} f(x)\,dx$$

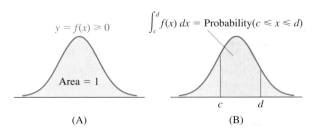

FIGURE 1 Probability density function

Explore & Discuss **1**   Let $f(x) = \begin{cases} \frac{1}{4}e^{-x/4} & \text{if } x \geq 0 \\ 0 & \text{otherwise} \end{cases}$

(A) Explain why $f(x) \geq 0$ over the interval $(-\infty, \infty)$.

(B) Find $\int_{0}^{10} f(x)\,dx$, $\int_{0}^{20} f(x)\,dx$, and $\int_{0}^{30} f(x)\,dx$.

(C) On the basis of part (B), what would you conjecture the area under the graph of $f(x)$ over the interval $(-\infty, \infty)$ to be equal to?

EXAMPLE 1

Duration of Telephone Calls Suppose that the length of telephone calls (in minutes) in a public telephone booth is a continuous random variable with probability density function shown in Figure 2:

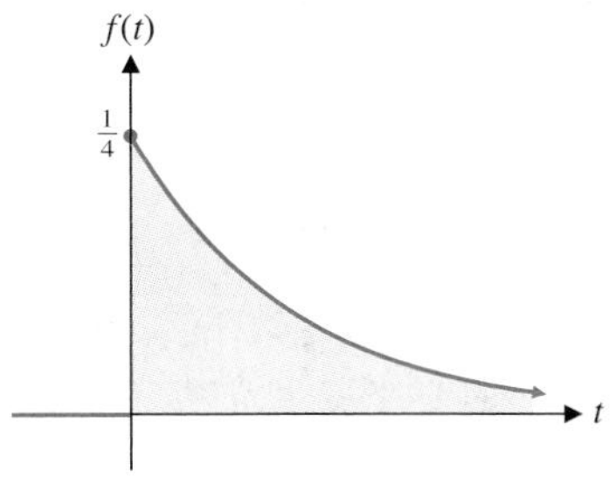

$$f(t) = \begin{cases} \frac{1}{4}e^{-t/4} & \text{if } t \geq 0 \\ 0 & \text{otherwise} \end{cases}$$

(A) Determine the probability that a call selected at random will last between 2 and 3 minutes.

(B) Find b (to two decimal places) so that the probability of a call selected at random lasting between 2 and b minutes is .5.

$f(t)$

FIGURE 2

SOLUTION

(A) Probability $(2 \leq t \leq 3) = \int_2^3 \frac{1}{4}e^{-t/4}\,dt$

$$= (-e^{-t/4})\big|_2^3$$

$$= -e^{-3/4} + e^{-1/2} \approx .13$$

(B) We want to find b such that Probability $(2 \leq t \leq b) = .5$.

$$\int_2^b \frac{1}{4}e^{-t/4}\,dt = .5$$

$$-e^{-b/4} + e^{-1/2} = .5 \qquad \text{Solve for } b.$$

$$e^{-b/4} = e^{-.5} - .5$$

$$-\frac{b}{4} = \ln(e^{-.5} - .5)$$

$$b = 8.96 \text{ minutes}$$

Thus, the probability of a call selected at random lasting from 2 to 8.96 minutes is .5.

MATCHED PROBLEM 1

(A) In Example 1, find the probability that a call selected at random will last 4 minutes or less.

(B) Find b (to two decimal places) so that the probability of a call selected at random lasting b minutes or less is .9

INSIGHT

The probability that a phone call in Example 1 lasts exactly 2 minutes (not 1.999 minutes, not 1.999 999 minutes) is given by

$$\text{Probability } (2 \leq t \leq 2) = \int_2^2 \frac{1}{4}e^{-t/4}\,dt \qquad \text{Use Property 1, Section 6-4}$$

$$= 0$$

In fact, for any *continuous* random variable x with probability density function $f(x)$, the probability that x is exactly equal to a constant c is equal to 0:

$$\text{Probability } (c \leq x \leq c) = \int_c^c f(x)\,dx \qquad \text{Use Property 1, Section 6-4}$$

$$= 0$$

In this respect, a *continuous* random variable differs from a *discrete* random variable. If x, for example, is the discrete random variable that represents the number of dots that appear on the top face when a fair die is rolled, then

$$\text{Probability } (2 \leq x \leq 2) = \tfrac{1}{6}$$

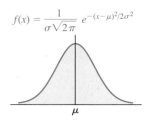

$$f(x) = \frac{1}{\sigma\sqrt{2\pi}} e^{-(x-\mu)^2/2\sigma^2}$$

FIGURE 3 Normal curve

One of the most important probability density functions, the **normal probability density function,** is defined as follows and graphed in Figure 3:

$$f(x) = \frac{1}{\sigma\sqrt{2\pi}} e^{-(x-\mu)^2/2\sigma^2} \quad \begin{array}{l} \mu \text{ is the mean.} \\ \sigma \text{ is the standard deviation.} \end{array}$$

It can be shown (but not easily) that the area under the normal curve in Figure 3 over the interval $(-\infty, \infty)$ is exactly 1. Since $\int e^{-x^2}\, dx$ is nonintegrable in terms of elementary functions (that is, the antiderivative cannot be expressed as a finite combination of simple functions), probabilities such as

$$\text{Probability } (c \le x \le d) = \frac{1}{\sigma\sqrt{2\pi}} \int_c^d e^{-(x-\mu)^2/2\sigma^2}\, dx$$

are generally determined by making an appropriate substitution in the integrand and then using a table of areas under the standard normal curve (that is, the normal curve with $\mu = 0$ and $\sigma = 1$). Such tables are readily available in most mathematical handbooks. A table can be constructed by using a rectangle rule, as discussed in Section 6-5; however, computers that employ refined techniques are generally used for this purpose. Some calculators have the capability of computing the areas of normal curves directly.

■ Continuous Income Stream

We start with a simple example having an obvious solution and generalize the concept to examples having less obvious solutions.

Suppose an aunt has established a trust that pays you $2,000 a year for 10 years. What is the total amount you will receive from the trust by the end of the 10th year? Since there are 10 payments of $2,000 each, you will receive

$$10 \times \$2{,}000 = \$20{,}000$$

We now look at the same problem from a different point of view, a point of view that will be useful in more complex problems. Let us assume that the income stream is continuous at a rate of $2,000 per year. In Figure 4, the area under the graph of $f(t) = 2{,}000$ from 0 to t represents the income accumulated t years after the start. For example, for $t = \frac{1}{4}$ year, the income would be $\frac{1}{4}(2{,}000) = \$500$; for $t = \frac{1}{2}$ year, the income would be $\frac{1}{2}(2{,}000) = \$1{,}000$; for $t = 1$ year, the income would be $1(2{,}000) = \$2{,}000$; for $t = 5.3$ years, the income would be $5.3(2{,}000) = \$10{,}600$; and for $t = 10$ years, the income would be $10(2{,}000) = \$20{,}000$. The total income over a 10-year period—that is, the area under the graph of $f(t) = 2{,}000$ from 0 to 10—is also given by the definite integral

$$\int_0^{10} 2{,}000\, dt = 2{,}000t\Big|_0^{10} = 2{,}000(10) - 2{,}000(0) = \$20{,}000$$

We now apply the idea of a continuous income stream to a less obvious problem.

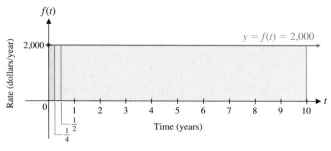

FIGURE 4 Continuous income stream

EXAMPLE 2 **Continuous Income Stream** The rate of change of the income produced by a vending machine located at an airport is given by

$$f(t) = 5,000e^{0.04t}$$

where t is time in years since the installation of the machine. Find the total income produced by the machine during the first 5 years of operation.

SOLUTION The area under the graph of the rate-of-change function from 0 to 5 represents the total change in income over the first 5 years (Fig. 5), and hence is given by a definite integral:

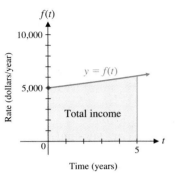

$$\text{Total income} = \int_0^5 5,000e^{0.04t}\, dt$$

$$= 125,000e^{0.04t}\Big|_0^5$$

$$= 125,000e^{0.04(5)} - 125,000e^{0.04(0)}$$

$$= 152,675 - 125,000$$

$$= \$27,675 \quad \textit{Rounded to the nearest dollar}$$

FIGURE 5 Continuous income stream

Thus, the vending machine produces a total income of $27,675 during the first 5 years of operation.

MATCHED PROBLEM 2 Referring to Example 2, find the total income produced (to the nearest dollar) during the second 5 years of operation.

In reality, income from a vending machine is not usually received as a single payment at the end of each year, even though the rate is given as a yearly rate. Income is usually collected on a daily or weekly basis. In problems of this type, it is convenient to assume that income is actually received in a **continuous stream;** that is, we assume that the rate at which income is received is a continuous function of time. The rate of change is called the **rate of flow** of the continuous income stream. In general, we have the following definition:

DEFINITION **Total Income for a Continuous Income Stream**

If $f(t)$ is the rate of flow of a continuous income stream, the **total income** produced during the period from $t = a$ to $t = b$ is

$$\text{Total income} = \int_a^b f(t)\, dt$$

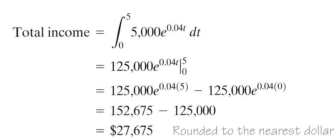

■ Future Value of a Continuous Income Stream

In Section 4-1, we discussed the continuous compound interest formula

$$A = Pe^{rt}$$

where P is the principal (or present value), A is the amount (or future value), r is the annual rate of continuous compounding (expressed as a decimal), and t is time in years. For example, if money is worth 12% compounded continuously, then the future value of a $10,000 investment in 5 years is (to the nearest dollar)

$$A = 10{,}000e^{0.12(5)} = \$18{,}221$$

Now we want to apply the future value concept to the income produced by a continuous income stream. Suppose that $f(t)$ is the rate of flow of a continuous income stream and the income produced by this continuous income stream is invested as soon as it is received at a rate r, compounded continuously. We already know how to find the total income produced after T years, but how can we find the total of the income produced and the interest earned by this income? Since the income is received in a continuous flow, we cannot just use the formula $A = Pe^{rt}$. This formula is valid only for a single deposit P, not for a continuous flow of income. Instead, we use a Riemann sum approach that will allow us to apply the formula $A = Pe^{rt}$ repeatedly. To begin, we divide the time interval $[0, T]$ into n equal subintervals of length Δt and choose an arbitrary point c_k in each subinterval, as illustrated in Figure 6.

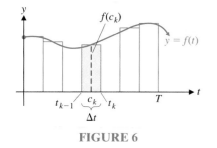

FIGURE 6

The total income produced during the period from $t = t_{k-1}$ to $t = t_k$ is equal to the area under the graph of $f(t)$ over this subinterval and is approximately equal to $f(c_k) \Delta t$, the area of the shaded rectangle in Figure 6. The income received during this period will earn interest for approximately $T - c_k$ years. Thus, from the future-value formula $A = Pe^{rt}$ with $P = f(c_k) \Delta t$ and $t = T - c_k$, the future-value of the income produced during the period from $t = t_{k-1}$ to $t = t_k$ is approximately equal to

$$f(c_k) \Delta t\, e^{(T-c_k)r}$$

The total of these approximate future values over n subintervals is then

$$f(c_1) \Delta t\, e^{(T-c_1)r} + f(c_2) \Delta t\, e^{(T-c_2)r} + \cdots + f(c_n) \Delta t\, e^{(T-c_n)r} = \sum_{k=1}^{n} f(c_k) e^{r(T-c_k)} \Delta t$$

This equation has the form of a Riemann sum, the limit of which is a definite integral. (See the definition of the definite integral in Section 6-4.) Thus, the *future value FV* of the income produced by the continuous income stream is given by

$$FV = \int_0^T f(t) e^{r(T-t)}\, dt$$

Since r and T are constants, we also can write

$$FV = \int_0^T f(t) e^{rT} e^{-rt}\, dt = e^{rT} \int_0^T f(t) e^{-rt}\, dt \qquad (1)$$

This last form is preferable, since the integral is usually easier to evaluate than the first form.

DEFINITION **Future Value of a Continuous Income Stream**

If $f(t)$ is the rate of flow of a continuous income stream, $0 \le t \le T$, and if the income is continuously invested at a rate r, compounded continuously, then the **future value FV** at the end of T years is given by

$$FV = \int_0^T f(t) e^{r(T-t)}\, dt = e^{rT} \int_0^T f(t) e^{-rt}\, dt$$

The future value of a continuous income stream is the total value of all money produced by the continuous income stream (income and interest) at the end of T years.

We return to the trust set up for you by your aunt. Suppose that the $2,000 per year you receive from the trust is invested as soon as it is received at 8%, compounded continuously. We consider the trust income a continuous income stream with a flow rate of $2,000 per year. What is its future value (to the nearest dollar) by the end of the 10th year? Using the definite integral for future value from the box, we have

$$FV = e^{rT} \int_0^T f(t)e^{-rt} dt$$

$$FV = e^{0.08(10)} \int_0^{10} 2,000e^{-0.08t} dt \qquad r = 0.08, T = 10, f(t) = 2,000$$

$$= 2,000e^{0.8} \int_0^{10} e^{-0.08t} dt$$

$$= 2,000e^{0.8} \left[\frac{e^{-0.08t}}{-0.08} \right] \Big|_0^{10}$$

$$= 2,000e^{0.8}[-12.5e^{-0.8} + 12.5] = \$30,639$$

Thus, at the end of 10 years you will have received $30,639, including interest. How much is interest? Since you received $20,000 in income from the trust, the interest is the difference between the future value and income. Thus,

$$\$30,639 - \$20,000 = \$10,639$$

is the interest earned by the income received from the trust over the 10-year period.

Explore & Discuss **2**   Suppose that a trust fund is set up to pay you $2,000 per year for 20 years, with the continuous income stream invested at 8%, compounded continuously. When will its future value be equal to $50,000?

We now apply the same analysis to Example 2, the slightly more involved vending machine problem.

EXAMPLE 3   **Future Value of a Continuous Income Stream**   Using the continuous income rate of flow for the vending machine in Example 2, namely,

$$f(t) = 5,000e^{0.04t}$$

find the future value of this income stream at 12%, compounded continuously for 5 years, and find the total interest earned. Compute answers to the nearest dollar.

SOLUTION   Using the formula

$$FV = e^{rT} \int_0^T f(t)e^{-rt} dt$$

with $r = 0.12, T = 5$, and $f(t) = 5,000e^{0.04t}$, we have

$$FV = e^{0.12(5)} \int_0^5 5,000e^{0.04t}e^{-0.12t} dt$$

$$= 5,000e^{0.6} \int_0^5 e^{-0.08t} dt$$

$$= 5,000e^{0.6} \left(\frac{e^{-0.08t}}{-0.08} \right) \Big|_0^5$$

$$= 5,000e^{0.6}(-12.5e^{-0.4} + 12.5)$$

$$= \$37,545 \qquad \text{Rounded to the nearest dollar}$$

Thus, the future value of the income stream at 12% compounded continuously at the end of 5 years is $37,545.

In Example 2, we saw that the total income produced by this vending machine over a 5-year period was $27,675. The difference between future value and income is interest. Thus,

$$\$37{,}545 - \$27{,}675 = \$9{,}870$$

is the interest earned by the income produced by the vending machine during the 5-year period.

MATCHED PROBLEM 3 Repeat Example 3 if the interest rate is 9%, compounded continuously.

■ Consumers' and Producers' Surplus

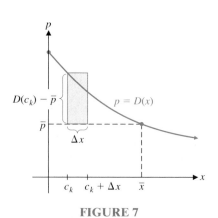

FIGURE 7

Let $p = D(x)$ be the price–demand equation for a product, where x is the number of units of the product that consumers will purchase at a price of $\$p$ per unit. Suppose that $\overline{p}$, is the current price and $\overline{x}$ is the number of units that can be sold at that price. Then the price–demand curve in Figure 7 shows that if the price is higher than $\overline{p}$, the demand x is less than $\overline{x}$, but some consumers are still willing to pay the higher price. Consumers who are willing to pay more than $\overline{p}$, but who are still able to buy the product at $\overline{p}$, have saved money. We want to determine the total amount saved by all the consumers who are willing to pay a price higher than $\overline{p}$ for the product.

To do this, consider the interval $[c_k, c_k + \Delta x]$, where $c_k + \Delta x < \overline{x}$. If the price remained constant over that interval, the savings on each unit would be the difference between $D(c_k)$, the price consumers are willing to pay, and $\overline{p}$, the price they actually pay. Since Δx represents the number of units purchased by consumers over the interval, the total savings to consumers over this interval is approximately equal to

$$[D(c_k) - \overline{p}] \Delta x \quad \text{(savings per unit)} \times \text{(number of units)}$$

which is the area of the shaded rectangle shown in Figure 7. If we divide the interval $[0, \overline{x}]$ into n equal subintervals, the total savings to consumers is approximately equal to

$$[D(c_1) - \overline{p}] \Delta x + [D(c_2) - \overline{p}] \Delta x + \cdots + [D(c_n) - \overline{p}] \Delta x = \sum_{k=1}^{n} [D(c_k) - \overline{p}] \Delta x$$

which we recognize as a Riemann sum for the integral

$$\int_{0}^{\overline{x}} [D(x) - \overline{p}] \, dx$$

Thus, we define the *consumers' surplus* to be this integral.

DEFINITION Consumers' Surplus

If $(\overline{x}, \overline{p})$ is a point on the graph of the price–demand equation $p = D(x)$ for a particular product, then the **consumers' surplus CS** at a price level of $\overline{p}$ is

$$CS = \int_{0}^{\overline{x}} [D(x) - \overline{p}] \, dx$$

which is the area between $p = \overline{p}$ and $p = D(x)$ from $x = 0$ to $x = \overline{x}$, as shown in the margin.

The consumers' surplus represents the total savings to consumers who are willing to pay more than $\overline{p}$ for the product, but are still able to buy the product for $\overline{p}$.

EXAMPLE 4 **Consumers' Surplus** Find the consumers' surplus at a price level of $8 for the price–demand equation

$$p = D(x) = 20 - 0.05x$$

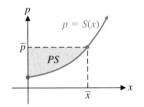

FIGURE 8

SOLUTION　**Step 1.** Find $\overline{x}$, the demand when the price is $\overline{p} = 8$:

$$\overline{p} = 20 - 0.05\overline{x}$$
$$8 = 20 - 0.05\overline{x}$$
$$0.05\overline{x} = 12$$
$$\overline{x} = 240$$

Step 2. Sketch a graph, as shown in Figure 8.

Step 3. Find the consumers' surplus (the shaded area in the graph):

$$CS = \int_0^{\overline{x}} [D(x) - \overline{p}]\, dx$$

$$= \int_0^{240} (20 - 0.05x - 8)\, dx$$

$$= \int_0^{240} (12 - 0.05x)\, dx$$

$$= (12x - 0.025x^2)\big|_0^{240}$$

$$= 2{,}880 - 1{,}440 = \$1{,}440$$

Thus, the total savings to consumers who are willing to pay a higher price for the product is $1,440.

MATCHED PROBLEM 4　Repeat Example 4 for a price level of $4.

If $p = S(x)$ is the price–supply equation for a product, $\overline{p}$ is the current price, and $\overline{x}$ is the current supply, some suppliers are still willing to supply some units at a lower price than $\overline{p}$. The additional money that these suppliers gain from the higher price is called the *producers' surplus* and can be expressed in terms of a definite integral (proceeding as we did for the consumers' surplus).

DEFINITION　**Producers' Surplus**

If $(\overline{x}, \overline{p})$ is a point on the graph of the price–supply equation $p = S(x)$, then the **producers' surplus PS** at a price level of $\overline{p}$ is

$$PS = \int_0^{\overline{x}} [\overline{p} - S(x)]\, dx$$

which is the area between $p = \overline{p}$ and $p = S(x)$ from $x = 0$ to $x = \overline{x}$, as shown in the margin.

The producers' surplus represents the total gain to producers who are willing to supply units at a lower price than $\overline{p}$, but are still able to supply units at $\overline{p}$.

EXAMPLE 5　**Producers' Surplus**　Find the producers' surplus at a price level of $20 for the price–supply equation

$$p = S(x) = 2 + 0.0002x^2$$

SOLUTION　**Step 1.** Find $\overline{x}$, the supply when the price is $\overline{p} = 20$:

$$\overline{p} = 2 + 0.0002\overline{x}^2$$
$$20 = 2 + 0.0002\overline{x}^2$$
$$0.0002\overline{x}^2 = 18$$
$$\overline{x}^2 = 90{,}000$$
$$\overline{x} = 300 \qquad \text{There is only one solution, since } \overline{x} \geq 0.$$

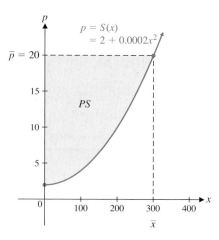

FIGURE 9

Step 2. Sketch a graph, as shown in Figure 9.

Step 3. Find the producers' surplus (the shaded area in the graph):

$$PS = \int_0^{\bar{x}} [\bar{p} - S(x)] \, dx = \int_0^{300} [20 - (2 + 0.0002x^2)] \, dx$$

$$= \int_0^{300} (18 - 0.0002x^2) \, dx = \left(18x - 0.0002\frac{x^3}{3}\right)\Big|_0^{300}$$

$$= 5{,}400 - 1{,}800 = \$3{,}600$$

Thus, the total gain to producers who are willing to supply units at a lower price is $3,600.

| MATCHED PROBLEM 5 | Repeat Example 5 for a price level of $4. |

In a free competitive market, the price of a product is determined by the relationship between supply and demand. If $p = D(x)$ and $p = S(x)$ are the price–demand and price–supply equations, respectively, for a product and if $(\bar{x}, \bar{p})$ is the point of intersection of these equations, $\bar{p}$ is called the **equilibrium price** and $\bar{x}$ is called the **equilibrium quantity.** If the price stabilizes at the equilibrium price $\bar{p}$, this is the price level that will determine both the consumers' surplus and the producers' surplus.

EXAMPLE 6 **Equilibrium Price and Consumers' and Producers' Surplus** Find the equilibrium price, and then find the consumers' surplus and producers' surplus at the equilibrium price level, if

$$p = D(x) = 20 - 0.05x \quad \text{and} \quad p = S(x) = 2 + 0.0002x^2$$

SOLUTION **Step 1.** Find the equilibrium point. Set $D(x)$ equal to $S(x)$ and solve:

$$D(x) = S(x)$$
$$20 - 0.05x = 2 + 0.0002x^2$$
$$0.0002x^2 + 0.05x - 18 = 0$$
$$x^2 + 250x - 90{,}000 = 0$$
$$x = 200, \; -450$$

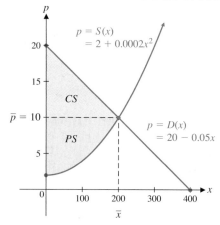

FIGURE 10

Since x cannot be negative, the only solution is $x = 200$. The equilibrium price can be determined by using $D(x)$ or $S(x)$. We will use both to check our work:

$$\bar{p} = D(200) \qquad\qquad \bar{p} = S(200)$$
$$= 20 - 0.05(200) = 10 \qquad = 2 + 0.0002(200)^2 = 10$$

Thus, the equilibrium price is $\bar{p} = 10$, and the equilibrium quantity is $\bar{x} = 200$.

Step 2. Sketch a graph, as shown in Figure 10.

Step 3. Find the consumers' surplus:

$$CS = \int_0^{\bar{x}} [D(x) - \bar{p}] \, dx = \int_0^{200} (20 - 0.05x - 10) \, dx$$

$$= \int_0^{200} (10 - 0.05x) \, dx$$

$$= (10x - 0.025x^2)\big|_0^{200}$$

$$= 2{,}000 - 1{,}000 = \$1{,}000$$

Step 4. Find the producers' surplus:

$$PS = \int_0^{\bar{x}} [\bar{p} - S(x)] \, dx$$

$$= \int_0^{200} [10 - (2 + 0.0002x^2)] \, dx$$

$$= \int_0^{200} (8 - 0.0002x^2) \, dx$$

$$= \left(8x - 0.0002\frac{x^3}{3} \right) \Big|_0^{200}$$

$$= 1{,}600 - \tfrac{1{,}600}{3} \approx \$1{,}067 \quad \textit{Rounded to the nearest dollar}$$

A graphing calculator offers an alternative approach to finding the equilibrium point for Example 6 (Fig. 11A). A numerical integration routine can then be used to find the consumers' and producers' surplus (Fig. 11B).

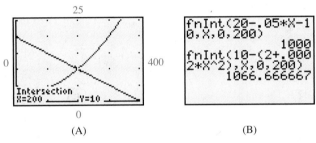

(A)   (B)

FIGURE 11

MATCHED PROBLEM 6   Repeat Example 6 for

$$p = D(x) = 25 - 0.001x^2 \quad \text{and} \quad p = S(x) = 5 + 0.1x$$

Answers to Matched Problems   **1.** (A) .63   (B) 9.21 min   **2.** \$33,803   **3.** $FV = \$34{,}691$; interest $= \$7{,}016$

4. \$2,560   **5.** \$133   **6.** $\bar{p} = 15$; $CS = \$667$; $PS = \$500$

Exercise 7-2

A *In Problems 1–10, evaluate each definite integral to two decimal places.*

1. $\displaystyle\int_0^1 e^{-2t} \, dt$

2. $\displaystyle\int_1^3 e^{-t} \, dt$

3. $\displaystyle\int_0^2 e^{4(2-t)} \, dt$

4. $\displaystyle\int_0^1 e^{3(1-t)} \, dt$

5. $\displaystyle\int_0^8 e^{0.06(8-t)} \, dt$

6. $\displaystyle\int_1^{10} e^{0.07(10-t)} \, dt$

7. $\displaystyle\int_0^{20} e^{0.08t} e^{0.12(20-t)} \, dt$

8. $\displaystyle\int_0^{15} e^{0.05t} e^{0.06(15-t)} \, dt$

9. $\displaystyle\int_0^{30} 500 e^{0.02t} e^{0.09(30-t)} \, dt$

10. $\displaystyle\int_0^{25} 900 e^{0.03t} e^{0.04(25-t)} \, dt$

B *In Problems 11 and 12, explain which of (A), (B), and (C) are equal before evaluating the expressions. Then evaluate each expression to two decimal places.*

11. (A) $\int_0^8 e^{0.07(8-t)}\, dt$ (B) $\int_0^8 (e^{0.56} - e^{0.07t})\, dt$

   (C) $e^{0.56} \int_0^8 e^{-0.07t}\, dt$

12. (A) $\int_0^{10} 2{,}000 e^{0.05t} e^{0.12(10-t)}\, dt$

   (B) $2{,}000 e^{1.2} \int_0^{10} e^{-0.07t}\, dt$

   (C) $2{,}000 e^{0.05} \int_0^{10} e^{0.12(10-t)}\, dt$

Applications

Unless stated to the contrary, compute all monetary answers to the nearest dollar.

13. The life expectancy (in years) of a certain brand of clock radio is a continuous random variable with probability density function

$$f(x) = \begin{cases} 2/(x+2)^2 & \text{if } x \geq 0 \\ 0 & \text{otherwise} \end{cases}$$

   (A) Find the probability that a randomly selected clock radio lasts at most 6 years.
   (B) Find the probability that a randomly selected clock radio lasts from 6 to 12 years.
   (C) Graph $y = f(x)$ for $[0, 12]$ and show the shaded region for part (A).

14. The shelf life (in years) of a certain brand of flashlight battery is a continuous random variable with probability density function

$$f(x) = \begin{cases} 1/(x+1)^2 & \text{if } x \geq 0 \\ 0 & \text{otherwise} \end{cases}$$

   (A) Find the probability that a randomly selected battery has a shelf life of 3 years or less.
   (B) Find the probability that a randomly selected battery has a shelf life of from 3 to 9 years.
   (C) Graph $y = f(x)$ for $[0, 10]$ and show the shaded region for part (A).

15. In Problem 13, find d so that the probability of a randomly selected clock radio lasting d years or less is .8.

16. In Problem 14, find d so that the probability of a randomly selected battery lasting d years or less is .5.

17. A manufacturer guarantees a product for 1 year. The time to failure of the product after it is sold is given by the probability density function

$$f(t) = \begin{cases} .01 e^{-.01t} & \text{if } t \geq 0 \\ 0 & \text{otherwise} \end{cases}$$

   where t is time in months. What is the probability that a buyer chosen at random will have a product failure

   (A) During the warranty period?
   (B) During the second year after purchase?

18. In a certain city, the daily use of water (in hundreds of gallons) per household is a continuous random variable with probability density function

$$f(x) = \begin{cases} .15 e^{-.15x} & \text{if } x \geq 0 \\ 0 & \text{otherwise} \end{cases}$$

Find the probability that a household chosen at random will use

   (A) At most 400 gallons of water per day
   (B) Between 300 and 600 gallons of water per day

19. In Problem 17, what is the probability that the product will last at least 1 year? [*Hint:* Recall that the total area under the probability density function curve is 1.]

20. In Problem 18, what is the probability that a household will use more than 400 gallons of water per day? [See the hint in Problem 19.]

21. Find the total income produced by a continuous income stream in the first 5 years if the rate of flow is $f(t) = 2{,}500$.

22. Find the total income produced by a continuous income stream in the first 10 years if the rate of flow is $f(t) = 3{,}000$.

23. Interpret the results of Problem 21 with both a graph and a verbal description of the graph.

24. Interpret the results of Problem 22 with both a graph and a verbal description of the graph.

25. Find the total income produced by a continuous income stream in the first 3 years if the rate of flow is $f(t) = 400 e^{0.05t}$.

26. Find the total income produced by a continuous income stream in the first 2 years if the rate of flow is $f(t) = 600 e^{0.06t}$.

27. Interpret the results of Problem 25 with both a graph and a verbal description of the graph.

28. Interpret the results of Problem 26 with both a graph and a verbal description of the graph.

29. Starting at age 25, you deposit $2,000 a year into an IRA account for retirement. Treat the yearly deposits into the account as a continuous income stream. If money in the account earns 5%, compounded continuously, how much will be in the account 40 years later, when you retire at age 65? How much of the final amount is interest?

30. Suppose in Problem 29 that you start the IRA deposits at age 30, but the account earns 6%, compounded continuously. Treat the yearly deposits into the account as a continuous income stream. How much will be in the account 35 years later, when you retire at age 65? How much of the final amount is interest?

31. Find the future value at 6.25% interest, compounded continuously for 4 years, of the continuous income stream with rate of flow $f(t) = 1{,}650 e^{-0.02t}$.

32. Find the future value, at 5.75% interest, compounded continuously for 6 years, of the continuous income stream with rate of flow $f(t) = 2,000e^{0.06t}$.

33. Compute the interest earned in Problem 31.

34. Compute the interest earned in Problem 32.

35. An investor is presented with a choice of two investments: an established clothing store and a new computer store. Each choice requires the same initial investment and each produces a continuous income stream of 10%, compounded continuously. The rate of flow of income from the clothing store is $f(t) = 12,000$, and the rate of flow of income from the computer store is expected to be $g(t) = 10,000e^{0.05t}$. Compare the future values of these investments to determine which is the better choice over the next 5 years.

36. Refer to Problem 35. Which investment is the better choice over the next 10 years?

37. An investor has $10,000 to invest in either a bond that matures in 5 years or a business that will produce a continuous stream of income over the next 5 years with rate of flow $f(t) = 2,000$. If both the bond and the continuous income stream earn 8%, compounded continuously, which is the better investment?

38. Refer to Problem 37. Which is the better investment if the rate of the income from the business is $f(t) = 3,000$?

39. A business is planning to purchase a piece of equipment that will produce a continuous stream of income for 8 years with rate of flow $f(t) = 9,000$. If the continuous income stream earns 6.95%, compounded continuously, what single deposit into an account earning the same interest rate will produce the same future value as the continuous income stream? (This deposit is called the **present value** of the continuous income stream.)

40. Refer to Problem 39. Find the present value of a continuous income stream at 7.65%, compounded continuously for 12 years, if the rate of flow is $f(t) = 1,000e^{0.03t}$.

41. Find the future value at a rate r, compounded continuously for T years, of a continuous income stream with rate of flow $f(t) = k$, where k is a constant.

42. Find the future value at a rate r, compounded continuously for T years, of a continuous income stream with rate of flow $f(t) = ke^{ct}$, where c and k are constants, $c \neq r$.

43. Find the consumers' surplus at a price level of $\bar{p} = \$150$ for the price-demand equation

$$p = D(x) = 400 - 0.05x$$

44. Find the consumers' surplus at a price level of $\bar{p} = \$120$ for the price-demand equation

$$p = D(x) = 200 - 0.02x$$

45. Interpret the results of Problem 43 with both a graph and a verbal description of the graph.

46. Interpret the results of Problem 44 with both a graph and a verbal description of the graph.

47. Find the producers' surplus at a price level of $\bar{p} = \$67$ for the price-supply equation

$$p = S(x) = 10 + 0.1x + 0.0003x^2$$

48. Find the producers' surplus at a price level of $\bar{p} = \$55$ for the price-supply equation

$$p = S(x) = 15 + 0.1x + 0.003x^2$$

49. Interpret the results of Problem 47 with both a graph and a verbal description of the graph.

50. Interpret the results of Problem 48 with both a graph and a verbal description of the graph.

In Problems 51–58, find the consumers' surplus and the producers' surplus at the equilibrium price level for the given price–demand and price–supply equations. Include a graph that identifies the consumers' surplus and the producers' surplus. Round all values to the nearest integer.

51. $p = D(x) = 50 - 0.1x; p = S(x) = 11 + 0.05x$

52. $p = D(x) = 25 - 0.004x^2; p = S(x) = 5 + 0.004x^2$

53. $p = D(x) = 80e^{-0.001x}; p = S(x) = 30e^{0.001x}$

54. $p = D(x) = 185e^{-0.005x}; p = S(x) = 25e^{0.005x}$

55. $p = D(x) = 80 - 0.04x; p = S(x) = 30e^{0.001x}$

56. $p = D(x) = 190 - 0.2x; p = S(x) = 25e^{0.005x}$

57. $p = D(x) = 80e^{-0.001x}; p = S(x) = 15 + 0.0001x^2$

58. $p = D(x) = 185e^{-0.005x}; p = S(x) = 20 + 0.002x^2$

59. The following tables give price–demand and price–supply data for the sale of soybeans at a grain market, where x is the number of bushels of soybeans (in thousands of bushels) and p is the price per bushel (in dollars):

Price–Demand		Price–Supply	
x	$p = D(x)$	x	$p = S(x)$
0	6.70	0	6.43
10	6.59	10	6.45
20	6.52	20	6.48
30	6.47	30	6.53
40	6.45	40	6.62

Use quadratic regression to model the price–demand data and linear regression to model the price–supply data.

(A) Find the equilibrium quantity (to three decimal places) and equilibrium price (to the nearest cent).

(B) Use a numerical integration routine to find the consumers' surplus and producers' surplus at the equilibrium price level.

60. Repeat Problem 59, using quadratic regression to model both sets of data.

Section 7-3 INTEGRATION BY PARTS

In Section 6-1, we said that we would return later to the indefinite integral

$$\int \ln x \, dx$$

since none of the integration techniques considered up to that time could be used to find an antiderivative for ln x. We now develop a very useful technique, called *integration by parts,* that will enable us to find not only the preceding integral, but also many others, including integrals such as

$$\int x \ln x \, dx \quad \text{and} \quad \int xe^x \, dx$$

The technique of integration by parts is also used to derive many integration formulas that are tabulated in mathematical handbooks. Some of these handbook formulas are discussed in the next section.

The method of integration by parts is based on the product formula for derivatives. If f and g are differentiable functions, then

$$\frac{d}{dx}[f(x)g(x)] = f(x)g'(x) + g(x)f'(x)$$

which can be written in the equivalent form

$$f(x)g'(x) = \frac{d}{dx}[f(x)g(x)] - g(x)f'(x)$$

Integrating both sides, we obtain

$$\int f(x)g'(x) \, dx = \int \frac{d}{dx}[f(x)g(x)] \, dx - \int g(x)f'(x) \, dx$$

The first integral to the right of the equals sign is $f(x)g(x) + C$. (Why?) We will leave out the constant of integration for now, since we can add it after integrating the second integral to the right of the equals sign. So we have

$$\int f(x)g'(x) \, dx = f(x)g(x) - \int g(x)f'(x) \, dx$$

This equation can be transformed into a more convenient form by letting $u = f(x)$ and $v = g(x)$; then $du = f'(x) \, dx$ and $dv = g'(x) \, dx$. Making these substitutions, we obtain the **integration-by-parts formula:**

Integration-by-Parts Formula

$$\int u \, dv = uv - \int v \, du$$

This formula can be very useful when the integral on the left is difficult or impossible to integrate with standard formulas. If u and dv are chosen with care—this is the crucial part of the process—then the integral on the right side may be easier to integrate than the one on the left. The formula provides us with another tool that is helpful in many, but not all, cases. We are able to easily check the results by differentiating to get the original integrand, a good habit to develop. Several examples will demonstrate the use of the formula.

EXAMPLE 1 **Integration by Parts** Find $\int xe^x \, dx$, using integration by parts, and check the result.

SOLUTION First, write the integration-by-parts formula:

$$\int u \, dv = uv - \int v \, du \tag{1}$$

Now try to identify u and dv in $\int xe^x\,dx$ so that $\int v\,du$ on the right side of (1) is easier to integrate than $\int u\,dv = \int xe^x\,dx$ on the left side. There are essentially two reasonable choices in selecting u and dv in $\int xe^x\,dx$:

$$
\underbrace{\int \overset{u}{x}\,\overbrace{e^x\,dx}^{dv}}_{\text{Choice 1}} \qquad \underbrace{\int \overset{u}{e^x}\,\overbrace{x\,dx}^{dv}}_{\text{Choice 2}}
$$

We pursue choice 1 and leave choice 2 for you to explore (see Explore & Discuss 1 following this example).

From choice 1, $u = x$ and $dv = e^x\,dx$. Looking at formula (1), we need du and v to complete the right side. It is convenient to proceed with the following arrangement: Let

$$
u = x \qquad dv = e^x\,dx
$$

Then

$$
du = dx \qquad \int dv = \int e^x\,dx
$$

$$
v = e^x
$$

Any constant may be added to v, but we will always choose 0 for simplicity. The general arbitrary constant of integration will be added at the end of the process.

Substituting these results into formula (1), we obtain

$$
\int u\,dv = uv - \int v\,du
$$

$$
\int xe^x\,dx = xe^x - \int e^x\,dx \qquad \text{The right integral is easy to integrate.}
$$

$$
= xe^x - e^x + C \qquad \text{Now add the arbitrary constant C.}
$$

Check:

$$
\frac{d}{dx}(xe^x - e^x + C) = xe^x + e^x - e^x = xe^x \qquad \blacksquare
$$

Explore & Discuss **1**   Pursue choice 2 in Example 1, using the integration-by-parts formula, and explain why this choice does not work out.

| MATCHED PROBLEM 1 | Find $\int xe^{2x}\,dx$.

EXAMPLE 2   **Integration by Parts**   Find $\int x\ln x\,dx$.

SOLUTION   As before, we have essentially two choices in choosing u and dv:

$$
\underbrace{\int \overset{u}{x}\,\overbrace{\ln x\,dx}^{dv}}_{\text{Choice 1}} \qquad \underbrace{\int \overset{u}{\ln x}\,\overbrace{x\,dx}^{dv}}_{\text{Choice 2}}
$$

Choice 1 is rejected, since we do not yet know how to find an antiderivative of $\ln x$. So we move to choice 2 and choose $u = \ln x$ and $dv = x\,dx$; then we proceed as in Example 1. Let

$$
u = \ln x \qquad dv = x\,dx
$$

Then

$$
du = \frac{1}{x}dx \qquad \int dv = \int x\,dx
$$

$$
v = \frac{x^2}{2}
$$

Substitute these results into the integration-by-parts formula:

$$\int u \, dv = uv - \int v \, du$$

$$\int x \ln x \, dx = (\ln x)\left(\frac{x^2}{2}\right) - \int \left(\frac{x^2}{2}\right)\left(\frac{1}{x}\right) dx$$

$$= \frac{x^2}{2} \ln x - \int \frac{x}{2} dx \qquad \textit{An easy integral to evaluate}$$

$$= \frac{x^2}{2} \ln x - \frac{x^2}{4} + C$$

Check:

$$\frac{d}{dx}\left(\frac{x^2}{2}\ln x - \frac{x^2}{4} + C\right) = x \ln x + \left(\frac{x^2}{2} \cdot \frac{1}{x}\right) - \frac{x}{2} = x \ln x$$

MATCHED PROBLEM 2   Find $\int x \ln 2x \, dx$.

INSIGHT

As you should have discovered in Explore & Discuss 1, some choices for u and dv will lead to integrals that are more complicated than the original integral. This does not mean that there is an error in either the calculations or the integration-by-parts formula. It simply means that the particular choice of u and dv does not change the problem into one we can solve. When this happens, we must look for a different choice of u and dv. In some problems, it is possible that no choice will work.

Guidelines for selecting u and dv for integration by parts are summarized in the following box:

SUMMARY   **INTEGRATION BY PARTS: SELECTION OF u AND dv**

For $\int u \, dv = uv - \int v \, du$,

1. The product $u \, dv$ must equal the original integrand.
2. It must be possible to integrate dv (preferably by using standard formulas or simple substitutions).
3. The new integral $\int v \, du$ should not be more complicated than the original integral $\int u \, dv$.
4. For integrals involving $x^p e^{ax}$, try

$$u = x^p \qquad \text{and} \qquad dv = e^{ax} \, dx$$

5. For integrals involving $x^p (\ln x)^q$, try

$$u = (\ln x)^q \qquad \text{and} \qquad dv = x^p \, dx$$

In some cases, repeated use of the integration-by-parts formula will lead to the evaluation of the original integral. The next example provides an illustration of such a case.

EXAMPLE 3   **Repeated Use of Integration by Parts**   Find $\int x^2 e^{-x} \, dx$.

SOLUTION   Following suggestion 4 in the box, we choose

$$u = x^2 \qquad dv = e^{-x} \, dx$$

Then

$$du = 2x \, dx \qquad v = -e^{-x}$$

and

$$x^2 e^{-x}\, dx = x^2(-e^{-x}) - (-e^{-x})2x\, dx$$
$$= -x^2 e^{-x} + 2x e^{-x}\, dx \tag{2}$$

The new integral is not one we can evaluate by standard formulas, but it is simpler than the original integral. Applying the integration-by-parts formula to it will produce an even simpler integral. For the integral $\int x e^{-x}\, dx$, we choose

$$u = x \qquad dv = e^{-x}\, dx$$

Then

$$du = dx \qquad v = -e^{-x}$$

and

$$\int x e^{-x}\, dx = x(-e^{-x}) - \int (-e^{-x})\, dx$$

$$= -x e^{-x} + \int e^{-x}\, dx$$

$$= -x e^{-x} - e^{-x} \qquad \textit{Choose 0 for the constant.} \tag{3}$$

Substituting equation (3) into equation (2), we have

$$\int x^2 e^{-x}\, dx = -x^2 e^{-x} + 2(-x e^{-x} - e^{-x}) + C \qquad \textit{Add an arbitrary constant here.}$$

$$= -x^2 e^{-x} - 2x e^{-x} - 2e^{-x} + C$$

Check:

$$\frac{d}{dx}(-x^2 e^{-x} - 2x e^{-x} - 2e^{-x} + C) = x^2 e^{-x} - 2x e^{-x} + 2x e^{-x} - 2e^{-x} + 2e^{-x}$$

$$= x^2 e^{-x}$$　▄

MATCHED PROBLEM 3　Find $\int x^2 e^{2x}\, dx$.

EXAMPLE 4　**Using Integration by Parts**　Find $\int_1^e \ln x\, dx$ and interpret the result geometrically.

SOLUTION　First, we will find $\int \ln x\, dx$. Then we will return to the definite integral. Following suggestion 5 in the box (with $p = 0$), we choose

$$u = \ln x \qquad dv = dx$$

Then

$$du = \frac{1}{x}\, dx \qquad v = x$$

Hence,

$$\int \ln x\, dx = (\ln x)(x) - \int (x)\frac{1}{x}\, dx$$

$$= x \ln x - x + C$$

Note that this is the important result we mentioned at the beginning of this section. Now we have

$$\int_1^e \ln x\, dx = (x \ln x - x)\Big|_1^e$$

$$= (e \ln e - e) - (1 \ln 1 - 1)$$

$$= (e - e) - (0 - 1)$$

$$= 1$$

The integral represents the area under the curve $y = \ln x$ from $x = 1$ to $x = e$, as shown in Figure 1.　▄

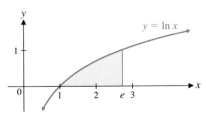

FIGURE 1

MATCHED PROBLEM 4   Find $\int_1^2 \ln 3x\, dx$.

Explore & Discuss 2   Try using the integration-by-parts formula for $\int e^{x^2}\, dx$, and explain why it does not solve the problem.

Answers to Matched Problems   **1.** $\dfrac{x}{2}e^{2x} - \dfrac{1}{4}e^{2x} + C$ **2.** $\dfrac{x^2}{2}\ln 2x - \dfrac{x^2}{4} + C$

3. $\dfrac{x^2}{2}e^{2x} - \dfrac{x}{2}e^{2x} + \dfrac{1}{4}e^{2x} + C$ **4.** $2\ln 6 - \ln 3 - 1 \approx 1.4849$

Exercise 7-3

A   *In Problems 1–4, integrate by parts. Assume that $x > 0$ whenever the natural logarithm function is involved.*

1. $\displaystyle\int xe^{3x}\, dx$ **2.** $\displaystyle\int xe^{4x}\, dx$

3. $\displaystyle\int x^2 \ln x\, dx$ **4.** $\displaystyle\int x^3 \ln x\, dx$

B

5. If you want to use integration by parts to find $\int (x+1)^5(x+2)\, dx$, which is the better choice for u: $u = (x+1)^5$ or $u = x+2$? Explain your choice, and then integrate.

6. If you want to use integration by parts to find $\int (5x-7)(x-1)^4\, dx$, which is the better choice for u: $u = 5x-7$ or $u = (x-1)^4$? Explain your choice, and then integrate.

Problems 7–24 are mixed—some require integration by parts, and others can be solved with techniques we have considered earlier. Integrate as indicated, assuming $x > 0$ whenever the natural logarithm function is involved.

7. $\displaystyle\int xe^{-x}\, dx$ **8.** $\displaystyle\int (x-1)e^{-x}\, dx$

9. $\displaystyle\int xe^{x^2}\, dx$ **10.** $\displaystyle\int xe^{-x^2}\, dx$

11. $\displaystyle\int_0^1 (x-3)e^x\, dx$ **12.** $\displaystyle\int_0^1 (x+1)e^x\, dx$

13. $\displaystyle\int_1^3 \ln 2x\, dx$ **14.** $\displaystyle\int_1^2 \ln\!\left(\dfrac{x}{2}\right) dx$

15. $\displaystyle\int \dfrac{2x}{x^2+1}\, dx$ **16.** $\displaystyle\int \dfrac{x^2}{x^3+5}\, dx$

17. $\displaystyle\int \dfrac{\ln x}{x}\, dx$ **18.** $\displaystyle\int \dfrac{e^x}{e^x+1}\, dx$

19. $\displaystyle\int \sqrt{x}\, \ln x\, dx$ **20.** $\displaystyle\int \dfrac{\ln x}{\sqrt{x}}\, dx$

21. $\displaystyle\int (x-1)^6(x+3)\, dx$ **22.** $\displaystyle\int (x+2)^9(x-4)\, dx$

23. $\displaystyle\int (x+1)^2(x-1)^2\, dx$ **24.** $\displaystyle\int x^5(x+1)^2\, dx$

In Problems 25–28, illustrate each integral graphically and describe what the integral represents in terms of areas.

25. Problem 11   **26.** Problem 12

27. Problem 13   **28.** Problem 14

C

Problems 29–50 are mixed—some may require use of the integration-by-parts formula along with techniques we have considered earlier; others may require repeated use of the integration-by-parts formula. Assume that $g(x) > 0$ whenever $\ln g(x)$ is involved.

29. $\displaystyle\int x^2 e^x\, dx$ **30.** $\displaystyle\int x^3 e^x\, dx$

31. $\displaystyle\int xe^{ax}\, dx,\ a \neq 0$ **32.** $\displaystyle\int \ln(ax)\, dx,\ a > 0$

33. $\displaystyle\int_1^e \dfrac{\ln x}{x^2}\, dx$ **34.** $\displaystyle\int_1^2 x^3 e^{x^2}\, dx$

35. $\displaystyle\int_0^2 \ln(x+4)\, dx$ **36.** $\displaystyle\int_0^2 \ln(4-x)\, dx$

37. $\displaystyle\int xe^{x-2}\, dx$ **38.** $\displaystyle\int xe^{x+1}\, dx$

39. $\displaystyle\int x\ln(1+x^2)\, dx$ **40.** $\displaystyle\int x\ln(1+x)\, dx$

41. $\displaystyle\int e^x \ln(1+e^x)\, dx$ **42.** $\displaystyle\int \dfrac{\ln(1+\sqrt{x})}{\sqrt{x}}\, dx$

43. $\displaystyle\int (\ln x)^2\, dx$ **44.** $\displaystyle\int x(\ln x)^2\, dx$

45. $\displaystyle\int (\ln x)^3\, dx$ **46.** $\displaystyle\int x(\ln x)^3\, dx$

47. $\displaystyle\int_1^e \ln(x^2)\, dx$ **48.** $\displaystyle\int_1^e \ln(x^4)\, dx$

49. $\displaystyle\int_0^1 \ln(e^{x^2})\, dx$ **50.** $\displaystyle\int_1^2 \ln(xe^x)\, dx$

In Problems 51–54, use a graphing calculator to graph each equation over the indicated interval, and find the area between the curve and the x axis over that interval. Find answers to two decimal places.

51. $y = x - 2 - \ln x; 1 \le x \le 4$

52. $y = 6 - x^2 - \ln x; 1 \le x \le 4$

53. $y = 5 - xe^x; 0 \le x \le 3$

54. $y = xe^x + x - 6; 0 \le x \le 3$

Applications

55. *Profit.* If the marginal profit (in millions of dollars per year) is given by

$$P'(t) = 2t - te^{-t}$$

use an appropriate definite integral to find the total profit (to the nearest million dollars) earned over the first 5 years of operation.

56. *Production.* An oil field is estimated to produce oil at a rate of $R(t)$ thousand barrels per month t months from now, as given by

$$R(t) = 10te^{-0.1t}$$

Use an appropriate definite integral to find the total production (to the nearest thousand barrels) in the first year of operation.

57. *Profit.* Interpret the results of Problem 55 with both a graph and a verbal description of the graph.

58. *Production.* Interpret the results of Problem 56 with both a graph and a verbal description of the graph.

59. *Continuous income stream.* Find the future value at 8%, compounded continuously, for 5 years of a continuous income stream with a rate of flow of

$$f(t) = 1,000 - 200t$$

60. *Continuous income stream.* Find the interest earned at 10%, compounded continuously, for 4 years for a continuous income stream with a rate of flow of

$$f(t) = 1,000 - 250t$$

61. *Income distribution.* Find the Gini index of income concentration for the Lorenz curve with equation

$$y = xe^{x-1}$$

62. *Income distribution.* Find the Gini index of income concentration for the Lorenz curve with equation

$$y = x^2 e^{x-1}$$

63. *Income distribution.* Interpret the results of Problem 61 with both a graph and a verbal description of the graph.

64. *Income distribution.* Interpret the results of Problem 62 with both a graph and a verbal description of the graph.

 65. *Sales analysis.* Monthly sales of a particular personal computer are expected to decline at the rate of

$$S'(t) = -4te^{0.1t}$$

computers per month, where t is time in months and $S(t)$ is the number of computers sold each month. The company plans to stop manufacturing this computer when monthly sales reach 800 computers. If monthly sales now ($t = 0$) are 2,000 computers, find $S(t)$. How long, to the nearest month, will the company continue to manufacture the computer?

66. *Sales analysis.* The rate of change of the monthly sales of a new home video game cartridge is given by

$$S'(t) = 350 \ln(t + 1) \qquad S(0) = 0$$

where t is the number of months since the game was released and $S(t)$ is the number of cartridges sold each month. Find $S(t)$. When, to the nearest month, will monthly sales reach 15,000 cartridges?

67. *Consumers' surplus.* Find the consumers' surplus (to the nearest dollar) at a price level of $\bar{p} = \$2.089$ for the price–demand equation

$$p = D(x) = 9 - \ln(x + 4)$$

Use $\bar{x}$ computed to the nearest higher unit.

68. *Producers' surplus.* Find the producers' surplus (to the nearest dollar) at a price level of $\bar{p} = \$26$ for the price–supply equation

$$p = S(x) = 5 \ln(x + 1)$$

Use $\bar{x}$ computed to the nearest higher unit.

69. *Consumers' surplus.* Interpret the results of Problem 67 with both a graph and a verbal description of the graph.

70. *Producers' surplus.* Interpret the results of Problem 68 with both a graph and a verbal description of the graph.

71. *Pollution.* The concentration of particulate matter (in parts per million) t hours after a factory ceases operation for the day is given by

$$C(t) = \frac{20 \ln(t + 1)}{(t + 1)^2}$$

Find the average concentration for the period from $t = 0$ to $t = 5$.

72. *Medicine.* After a person takes a pill, the drug contained in the pill is assimilated into the bloodstream. The rate of assimilation t minutes after taking the pill is

$$R(t) = te^{-0.2t}$$

Find the total amount of the drug that is assimilated into the bloodstream during the first 10 minutes after the pill is taken.

73. *Learning.* In a particular business college, it was found that an average student enrolled in an advanced typing class progressed at a rate of

$$N'(t) = (t + 6)e^{-0.25t}$$

words per minute per week t weeks after enrolling in a 15-week course. If a student could type 40 words per minute at the beginning of the course, how many words per minute $N(t)$ would the student be expected to type t weeks into the course? How long, to the nearest week, should it take the student to achieve the 70-word-per-minute level? How many words per minute should the student be able to type by the end of the course?

 74. *Learning.* In the same business college as in Problem 73, it was also found that an average student enrolled in a beginning shorthand class progressed at a rate of

$$N'(t) = (t + 10)e^{-0.1t}$$

words per minute per week *t* weeks after enrolling in a 15-week course. If a student had no knowledge of shorthand (that is, if the student could take dictation in shorthand at 0 words per minute) at the beginning of the course, how many words per minute $N(t)$ would the

student be expected to handle *t* weeks into the course? How long, to the nearest week, should it take the student to achieve 90 words per minute? How many words per minute should the student be able to handle by the end of the course?

 75. *Politics.* The number of voters (in thousands) in a certain city is given by

$$N(t) = 20 + 4t - 5te^{-0.1t}$$

where *t* is time in years. Find the average number of voters during the period from $t = 0$ to $t = 5$.

Section 7-4 INTEGRATION USING TABLES

- Using a Table of Integrals
- Substitution and Integral Tables
- Reduction Formulas
- Application

A **table of integrals** is a list of integration formulas used to evaluate integrals. Table II of Appendix C contains a short list of integral formulas illustrating the types found in more extensive tables. Some of these formulas can be derived with the integration techniques discussed earlier, while others require techniques we have not considered. However, it is possible to verify each formula by differentiating the right side.

■ Using a Table of Integrals

The formulas in Table II (and in larger integral tables) are organized by categories, such as "Integrals Involving $a + bu$," "Integrals Involving $\sqrt{u^2 - a^2}$," and so on. The variable *u* is the variable of integration. All other symbols represent constants. To use a table to evaluate an integral, you must first find the category that most closely agrees with the form of the integrand and then find a formula in that category that you can make to match the integrand exactly by assigning values to the constants in the formula. The examples that follow illustrate this process.

EXAMPLE 1	**Integration Using Tables** Use Table II to find

$$\int \frac{x}{(5 + 2x)(4 - 3x)}\,dx$$

SOLUTION Since the integrand

$$f(x) = \frac{x}{(5 + 2x)(4 - 3x)}$$

is a rational function involving terms of the form $a + bu$ and $c + du$, we examine formulas 15 to 20 in Table II to see if any of the integrands in these formulas can be made to match $f(x)$ exactly. Comparing the integrand in formula 16 with $f(x)$, we see that this integrand will match $f(x)$ if we let $a = 5, b = 2, c = 4$, and $d = -3$. Letting $u = x$ and substituting for a, b, c, and d in formula 16, we have

$$\int \frac{u}{(a + bu)(c + du)}\,du = \frac{1}{ad - bc}\left(\frac{a}{b}\ln|a + bu| - \frac{c}{d}\ln|c + du|\right) \quad \text{Formula 16}$$

$$\int \frac{x}{\underset{a\quad\; b\quad\; c\quad\; d}{(5 + 2x)(4 - 3x)}}\,dx = \frac{1}{5\cdot(-3) - 2\cdot 4}\left(\frac{5}{2}\ln|5 + 2x| - \frac{4}{-3}\ln|4 - 3x|\right) + C$$

$$\qquad\qquad\qquad\qquad\qquad\qquad\quad a\cdot d - b\cdot c = 5\cdot(-3) - 2\cdot 4 = -23$$

$$= -\tfrac{5}{46}\ln|5 + 2x| - \tfrac{4}{69}\ln|4 - 3x| + C$$

Notice that the constant of integration, C, is not included in any of the formulas in Table II. However, you must still include C in all antiderivatives.

MATCHED PROBLEM 1   Use Table II to find $\displaystyle\int \frac{1}{(5 + 3x)^2(1 + x)}\,dx$.

EXAMPLE 2   **Integration Using Tables**   Evaluate $\displaystyle\int_3^4 \frac{1}{x\sqrt{25 - x^2}}\,dx$.

SOLUTION   First, we use Table II to find

$$\int \frac{1}{x\sqrt{25 - x^2}}\,dx$$

Since the integrand involves the expression $\sqrt{25 - x^2}$, we examine formulas 29 to 31 and select formula 29 with $a^2 = 25$ and $a = 5$:

$$\int \frac{1}{u\sqrt{a^2 - u^2}}\,du = -\frac{1}{a}\ln\left|\frac{a + \sqrt{a^2 - u^2}}{u}\right| \qquad \text{Formula 29}$$

$$\int \frac{1}{x\sqrt{25 - x^2}}\,dx = -\frac{1}{5}\ln\left|\frac{5 + \sqrt{25 - x^2}}{x}\right| + C$$

Thus,

$$\int_3^4 \frac{1}{x\sqrt{25 - x^2}}\,dx = -\frac{1}{5}\ln\left|\frac{5 + \sqrt{25 - x^2}}{x}\right|\Bigg|_3^4$$

$$= -\frac{1}{5}\ln\left|\frac{5 + 3}{4}\right| + \frac{1}{5}\ln\left|\frac{5 + 4}{3}\right|$$

$$= -\tfrac{1}{5}\ln 2 + \tfrac{1}{5}\ln 3 = \tfrac{1}{5}\ln 1.5 \approx 0.0811$$

MATCHED PROBLEM 2   Evaluate $\displaystyle\int_6^8 \frac{1}{x^2\sqrt{100 - x^2}}\,dx$.

■ Substitution and Integral Tables

As Examples 1 and 2 illustrate, if the integral we want to evaluate can be made to match one in the table exactly, evaluating the indefinite integral consists simply of substituting the correct values of the constants into the formula. But what happens if we cannot match an integral with one of the formulas in the table? In many cases, a substitution will change the given integral into one that corresponds to a table entry. The examples that follow illustrate several frequently used substitutions.

EXAMPLE 3   **Integration Using Substitution and Tables**   Find $\displaystyle\int \frac{x^2}{\sqrt{16x^2 - 25}}\,dx$.

SOLUTION   In order to relate this integral to one of the formulas involving $\sqrt{u^2 - a^2}$ (formulas 40 to 45), we observe that if $u = 4x$, then

$$u^2 = 16x^2 \qquad \text{and} \qquad \sqrt{16x^2 - 25} = \sqrt{u^2 - 25}$$

Thus, we will use the substitution $u = 4x$ to change this integral into one that appears in the table:

$$\int \frac{x^2}{\sqrt{16x^2 - 25}} dx = \frac{1}{4} \int \frac{\frac{1}{16}u^2}{\sqrt{u^2 - 25}} du \qquad \text{Substitution:} \\ u = 4x, \, du = 4 \, dx, \, x = \frac{1}{4}u$$

$$= \frac{1}{64} \int \frac{u^2}{\sqrt{u^2 - 25}} du$$

This last integral can be evaluated with the aid of formula 44 with $a = 5$:

$$\int \frac{u^2}{\sqrt{u^2 - a^2}} du = \frac{1}{2}(u\sqrt{u^2 - a^2} + a^2 \ln|u + \sqrt{u^2 - a^2}|) \qquad \text{Formula 44}$$

$$\int \frac{x^2}{\sqrt{16x^2 - 25}} dx = \frac{1}{64} \int \frac{u^2}{\sqrt{u^2 - 25}} du \qquad \text{Use formula 44 with } a = 5.$$

$$= \frac{1}{128}(u\sqrt{u^2 - 25} + 25 \ln|u + \sqrt{u^2 - 25}|) + C \qquad \text{Substitute } u = 4x.$$

$$= \frac{1}{128}(4x\sqrt{16x^2 - 25} + 25 \ln|4x + \sqrt{16x^2 - 25}|) + C$$

MATCHED PROBLEM 3 Find $\int \sqrt{9x^2 - 16} \, dx$.

EXAMPLE 4 **Integration Using Substitution and Tables** Find $\displaystyle\int \frac{x}{\sqrt{x^4 + 1}} dx$.

SOLUTION None of the formulas in the table involve fourth powers; however, if we let $u = x^2$, then

$$\sqrt{x^4 + 1} = \sqrt{u^2 + 1}$$

and this form does appear in formulas 32 to 39. Thus, we substitute $u = x^2$:

$$\int \frac{1}{\sqrt{x^4 + 1}} x \, dx = \frac{1}{2} \int \frac{1}{\sqrt{u^2 + 1}} du \qquad \text{Substitution:} \\ u = x^2, \, du = 2x \, dx$$

We recognize the last integral as formula 36 with $a = 1$:

$$\int \frac{1}{\sqrt{u^2 + a^2}} du = \ln|u + \sqrt{u^2 + a^2}| \qquad \text{Formula 36}$$

$$\int \frac{x}{\sqrt{x^4 + 1}} dx = \frac{1}{2} \int \frac{1}{\sqrt{u^2 + 1}} du \qquad \text{Use formula 36 with } a = 1.$$

$$= \frac{1}{2} \ln|u + \sqrt{u^2 + 1}| + C \qquad \text{Substitute } u = x^2.$$

$$= \frac{1}{2} \ln|x^2 + \sqrt{x^4 + 1}| + C$$

MATCHED PROBLEM 4 Find $\int x\sqrt{x^4 + 1} \, dx$.

■ Reduction Formulas

EXAMPLE 5 **Using Reduction Formulas** Use Table II to find $\int x^2 e^{3x} \, dx$.

SOLUTION Since the integrand involves the function e^{3x}, we examine formulas 46–48 and conclude that formula 47 can be used for this problem. Letting $u = x, n = 2$, and $a = 3$ in formula 47, we have

$$\int u^n e^{au} \, du = \frac{u^n e^{au}}{a} - \frac{n}{a} \int u^{n-1} e^{au} \, du \qquad \text{Formula 47}$$

$$\int x^2 e^{3x} \, dx = \frac{x^2 e^{3x}}{3} - \frac{2}{3} \int x e^{3x} \, dx$$

Notice that the expression on the right still contains an integral, but the exponent of x has been reduced by 1. Formulas of this type are called **reduction formulas** and are designed to be applied repeatedly until an integral that can be evaluated is obtained. Applying formula 47 to $\int xe^{3x}\,dx$ with $n = 1$, we have

$$\int x^2 e^{3x}\,dx = \frac{x^2 e^{3x}}{3} - \frac{2}{3}\left(\frac{xe^{3x}}{3} - \frac{1}{3}\int e^{3x}\,dx\right)$$

$$= \frac{x^2 e^{3x}}{3} - \frac{2xe^{3x}}{9} + \frac{2}{9}\int e^{3x}\,dx$$

This last expression contains an integral that is easy to evaluate:

$$\int e^{3x}\,dx = \tfrac{1}{3}e^{3x}$$

After making a final substitution and adding a constant of integration, we have

$$\int x^2 e^{3x}\,dx = \frac{x^2 e^{3x}}{3} - \frac{2xe^{3x}}{9} + \frac{2}{27}e^{3x} + C$$ ▬

MATCHED PROBLEM 5 Use Table II to find $\int(\ln x)^2\,dx$.

APPLICATION

EXAMPLE 6 **Producers' Surplus** Find the producers' surplus at a price level of $20 for the price–supply equation

$$p = S(x) = \frac{5x}{500 - x}$$

SOLUTION Step 1. Find $\bar{x}$, the supply when the price is $\bar{p} = 20$:

$$\bar{p} = \frac{5\bar{x}}{500 - \bar{x}}$$

$$20 = \frac{5\bar{x}}{500 - \bar{x}}$$

$$10{,}000 - 20\bar{x} = 5\bar{x}$$

$$10{,}000 = 25\bar{x}$$

$$\bar{x} = 400$$

Step 2. Sketch a graph, as shown in Figure 1.

Step 3. Find the producers' surplus (the shaded area of the graph):

$$PS = \int_0^{\bar{x}} [\bar{p} - S(x)]\,dx$$

$$= \int_0^{400}\left(20 - \frac{5x}{500 - x}\right)dx$$

$$= \int_0^{400}\frac{10{,}000 - 25x}{500 - x}\,dx$$

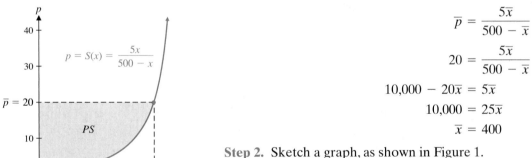

FIGURE 1

Use formula 20 with $a = 10{,}000$, $b = -25$, $c = 500$, and $d = -1$:

$$\int\frac{a + bu}{c + du}\,du = \frac{bu}{d} + \frac{ad - bc}{d^2}\ln|c + du| \qquad \text{Formula 20}$$

$$PS = (25x + 2{,}500\ln|500 - x|)\Big|_0^{400}$$

$$= 10{,}000 + 2{,}500\ln|100| - 2{,}500\ln|500|$$

$$\approx \$5{,}976$$ ▬

MATCHED PROBLEM 6   Find the consumers' surplus at a price level of \$10 for the price–demand equation

$$p = D(x) = \frac{20x - 8{,}000}{x - 500}$$

Explore & Discuss **1**   Use algebraic manipulation, including algebraic long division, on the integrand in Example 6 to show that

$$\frac{5\overline{x}}{500 - \overline{x}} = \frac{-5\overline{x}}{\overline{x} - 500} = -5 - \frac{2{,}500}{\overline{x} - 500}$$

Now use this result to find the indefinite integral in Example 6 without resorting to table formulas.

Answers to Matched Problems
1. $\dfrac{1}{2}\left(\dfrac{1}{5 + 3x}\right) + \dfrac{1}{4}\ln\left|\dfrac{1 + x}{5 + 3x}\right| + C$

2. $\dfrac{7}{1{,}200} \approx 0.0058$

3. $\dfrac{1}{6}(3x\sqrt{9x^2 - 16} - 16\ln|3x + \sqrt{9x^2 - 16}|) + C$

4. $\dfrac{1}{4}(x^2\sqrt{x^4 + 1} + \ln|x^2 + \sqrt{x^4 + 1}|) + C$

5. $x(\ln x)^2 - 2x\ln x + 2x + C$

6. $3{,}000 + 2{,}000\ln 200 - 2{,}000\ln 500 \approx \$1{,}167$

Exercise 7-4

A   *Use Table II to find each indefinite integral in Problems 1–14.*

1. $\displaystyle\int \frac{1}{x(1 + x)}\,dx$

2. $\displaystyle\int \frac{1}{x^2(1 + x)}\,dx$

3. $\displaystyle\int \frac{1}{(3 + x)^2(5 + 2x)}\,dx$

4. $\displaystyle\int \frac{x}{(5 + 2x)^2(2 + x)}\,dx$

5. $\displaystyle\int \frac{x}{\sqrt{16 + x}}\,dx$

6. $\displaystyle\int \frac{1}{x\sqrt{16 + x}}\,dx$

7. $\displaystyle\int \frac{1}{x\sqrt{1 - x^2}}\,dx$

8. $\displaystyle\int \frac{\sqrt{9 - x^2}}{x}\,dx$

9. $\displaystyle\int \frac{1}{x\sqrt{x^2 + 4}}\,dx$

10. $\displaystyle\int \frac{1}{x^2\sqrt{x^2 - 16}}\,dx$

11. $\displaystyle\int x^2 \ln x\,dx$

12. $\displaystyle\int x^3 \ln x\,dx$

13. $\displaystyle\int \frac{1}{1 + e^x}\,dx$

14. $\displaystyle\int \frac{1}{5 + 2e^{3x}}\,dx$

Evaluate each definite integral in Problems 15–20. Use Table II to find the antiderivative.

15. $\displaystyle\int_1^3 \frac{x^2}{3 + x}\,dx$

16. $\displaystyle\int_2^6 \frac{x}{(6 + x)^2}\,dx$

17. $\displaystyle\int_0^7 \frac{1}{(3 + x)(1 + x)}\,dx$

18. $\displaystyle\int_0^7 \frac{x}{(3 + x)(1 + x)}\,dx$

19. $\displaystyle\int_0^4 \frac{1}{\sqrt{x^2 + 9}}\,dx$

20. $\displaystyle\int_4^5 \sqrt{x^2 - 16}\,dx$

B   *In Problems 21–32, use substitution techniques and Table II to find each indefinite integral.*

21. $\displaystyle\int \frac{\sqrt{4x^2 + 1}}{x^2}\,dx$

22. $\displaystyle\int x^2\sqrt{9x^2 - 1}\,dx$

23. $\displaystyle\int \frac{x}{\sqrt{x^4 - 16}}\,dx$

24. $\displaystyle\int x\sqrt{x^4 - 16}\,dx$

25. $\displaystyle\int x^2\sqrt{x^6 + 4}\,dx$

26. $\displaystyle\int \frac{x^2}{\sqrt{x^6 + 4}}\,dx$

27. $\displaystyle\int \frac{1}{x^3\sqrt{4 - x^4}}\,dx$

28. $\displaystyle\int \frac{\sqrt{x^4 + 4}}{x}\,dx$

29. $\displaystyle\int \frac{e^x}{(2 + e^x)(3 + 4e^x)}\,dx$

30. $\displaystyle\int \frac{e^x}{(4 + e^x)^2(2 + e^x)}\,dx$

31. $\displaystyle\int \frac{\ln x}{x\sqrt{4 + \ln x}}\,dx$

32. $\displaystyle\int \frac{1}{x \ln x\sqrt{4 + \ln x}}\,dx$

C   *In Problems 33–38, use Table II to find each indefinite integral.*

33. $\displaystyle\int x^2 e^{5x}\,dx$

34. $\displaystyle\int x^2 e^{-4x}\,dx$

35. $\displaystyle\int x^3 e^{-x}\,dx$

36. $\displaystyle\int x^3 e^{2x}\,dx$

37. $\displaystyle\int (\ln x)^3\,dx$

38. $\displaystyle\int (\ln x)^4\,dx$

Problems 39–46 are mixed—some require the use of Table II, and others can be solved with techniques we considered earlier.

39. $\displaystyle\int_3^5 x\sqrt{x^2 - 9}\,dx$

40. $\displaystyle\int_3^5 x^2\sqrt{x^2 - 9}\,dx$

41. $\displaystyle\int_2^4 \frac{1}{x^2 - 1}\,dx$

42. $\displaystyle\int_2^4 \frac{x}{(x^2 - 1)^2}\,dx$

43. $\displaystyle\int \frac{\ln x}{x^2}\,dx$

44. $\displaystyle\int \frac{(\ln x)^2}{x}\,dx$

45. $\displaystyle\int \frac{x}{\sqrt{x^2 - 1}}\,dx$

46. $\displaystyle\int \frac{x^2}{\sqrt{x^2 - 1}}\,dx$

 In Problems 47–50, find the area bounded by the graphs of $y = f(x)$ and $y = g(x)$ to two decimal places. Use a graphing calculator to approximate intersection points to two decimal places.

47. $f(x) = \dfrac{10}{\sqrt{x^2 + 1}};\ g(x) = x^2 + 3x$

48. $f(x) = \sqrt{1 + x^2};\ g(x) = 5x - x^2$

49. $f(x) = x\sqrt{4 + x};\ g(x) = 1 + x$

50. $f(x) = \dfrac{x}{\sqrt{x + 4}};\ g(x) = x - 2$

Applications

Use Table II to evaluate all integrals involved in any solutions of Problems 51–74.

51. *Consumers' surplus.* Find the consumers' surplus at a price level of $\bar{p} = \$15$ for the price–demand equation

$$p = D(x) = \frac{7{,}500 - 30x}{300 - x}$$

52. *Producers' surplus.* Find the producers' surplus at a price level of $\bar{p} = \$20$ for the price–supply equation

$$p = S(x) = \frac{10x}{300 - x}$$

53. *Consumers' surplus.* Graph the price–demand equation and the price-level equation $\bar{p} = 15$ of Problem 51 in the same coordinate system. What region represents the consumers' surplus?

54. *Producers' surplus.* Graph the price–supply equation and the price-level equation $\bar{p} = 20$ of Problem 52 in the same coordinate system. What region represents the producers' surplus?

55. *Cost.* A company manufactures downhill skis. It has fixed costs of $25,000 and a marginal cost given by

$$C'(x) = \frac{250 + 10x}{1 + 0.05x}$$

where $C(x)$ is the total cost at an output of x pairs of skis. Find the cost function $C(x)$ and determine the production level (to the nearest unit) that produces a cost of $150,000. What is the cost (to the nearest dollar) for a production level of 850 pairs of skis?

56. *Cost.* A company manufactures a portable CD player. It has fixed costs of $11,000 per week and a marginal cost given by

$$C'(x) = \frac{65 + 20x}{1 + 0.4x}$$

where $C(x)$ is the total cost per week at an output of x players per week. Find the cost function $C(x)$ and determine the production level (to the nearest unit) that produces a cost of $52,000 per week. What is the cost (to the nearest dollar) for a production level of 700 players per week?

57. *Continuous income stream.* Find the future value at 10%, compounded continuously, for 10 years for the continuous income stream with rate of flow $f(t) = 50t^2$.

58. *Continuous income stream.* Find the interest earned at 8%, compounded continuously, for 5 years for the continuous income stream with rate of flow $f(t) = 200t$.

59. *Income distribution.* Find the Gini index of income concentration for the Lorenz curve with equation

$$y = \tfrac{1}{2}x\sqrt{1 + 3x}$$

60. *Income distribution.* Find the Gini index of income concentration for the Lorenz curve with equation

$$y = \tfrac{1}{2}x^2\sqrt{1 + 3x}$$

61. *Income distribution.* Graph $y = x$ and the Lorenz curve of Problem 59 over the interval $[0, 1]$. Discuss the effect of the area bounded by $y = x$ and the Lorenz curve getting smaller relative to the equitable distribution of income.

62. *Income distribution.* Graph $y = x$ and the Lorenz curve of Problem 60 over the interval $[0, 1]$. Discuss the effect of the area bounded by $y = x$ and the Lorenz curve getting larger relative to the equitable distribution of income.

63. *Marketing.* After test marketing a new high-fiber cereal, the market research department of a major food producer estimates that monthly sales (in millions of dollars) will grow at the monthly rate of

$$S'(t) = \frac{t^2}{(1 + t)^2}$$

t months after the cereal is introduced. If we assume 0 sales at the time the cereal is introduced, find the total sales $S(t)$ t months after the cereal is introduced. Find the total sales during the first 2 years the cereal is on the market.

64. *Average price.* At a discount department store, the price–demand equation for premium motor oil is given by

$$p = D(x) = \frac{50}{\sqrt{100 + 6x}}$$

where x is the number of cans of oil that can be sold at a price of p. Find the average price over the demand interval $[50, 250]$.

65. *Marketing.* For the cereal of Problem 63, show the sales over the first 2 years geometrically, and verbally describe the geometric representation.

66. *Price–demand.* For the motor oil of Problem 64, graph the price–demand equation and the line representing the average price in the same coordinate system over the interval $[50, 250]$. Describe how the areas under the two curves over the interval $[50, 250]$ are related.

 67. *Profit.* The marginal profit for a small car agency that sells x cars per week is given by

$$P'(x) = x\sqrt{2 + 3x}$$

where $P(x)$ is the profit in dollars. The agency's profit on the sale of only 1 car per week is $-$2,000. Find the profit function and the number of cars that must be sold (to the nearest unit) to produce a profit of $13,000 per week. How much weekly profit (to the nearest dollar) will the agency have if 80 cars are sold per week?

 68. *Revenue.* The marginal revenue for a company that manufactures and sells x solar-powered calculators per week is given by

$$R'(x) = \frac{x}{\sqrt{1 + 2x}} \qquad R(0) = 0$$

where $R(x)$ is the revenue in dollars. Find the revenue function and the number of calculators that must be sold (to the nearest unit) to produce $10,000 in revenue per week. How much weekly revenue (to the nearest dollar) will the company have if 1,000 calculators are sold per week?

69. *Pollution.* An oil tanker aground on a reef is losing oil and producing an oil slick that is radiating outward at a rate given approximately by

$$\frac{dR}{dt} = \frac{100}{\sqrt{t^2 + 9}} \qquad t \geq 0$$

where R is the radius (in feet) of the circular slick after t minutes. Find the radius of the slick after 4 minutes if the radius is 0 when $t = 0$.

70. Pollution. The concentration of particulate matter (in parts per million) during a 24-hour period is given approximately by

$$C(t) = t\sqrt{24 - t} \qquad 0 \le t \le 24$$

where t is time in hours. Find the average concentration during the period from $t = 0$ to $t = 24$.

71. Learning. A person learns N items at a rate given approximately by

$$N'(t) = \frac{60}{\sqrt{t^2 + 25}} \qquad t \ge 0$$

where t is the number of hours of continuous study. Determine the total number of items learned in the first 12 hours of continuous study.

72. Politics. The number of voters (in thousands) in a metropolitan area is given approximately by

$$f(t) = \frac{500}{2 + 3e^{-t}} \qquad t \ge 0$$

where t is time in years. Find the average number of voters during the period from $t = 0$ to $t = 10$.

73. Learning. Interpret Problem 71 geometrically. Verbally describe the geometric interpretation.

74. Politics. For the voters of Problem 72, graph $y = f(t)$ and the line representing the average number of voters over the interval $[0, 10]$ in the same coordinate system. Describe how the areas under the two curves over the interval $[0, 10]$ are related.

CHAPTER 7 REVIEW

Important Terms, Symbols, and Concepts

7-1 Area between Curves

Examples

- If f and g are continuous and $f(x) \ge g(x)$ over the interval $[a, b]$, then the area bounded by $y = f(x)$ and $y = g(x)$ for $a \le x \le b$ is given exactly by

$$A = \int_a^b [f(x) - g(x)] \, dx$$

Ex. 1, p. 425
Ex. 2, p. 425
Ex. 3, p. 426
Ex. 4, p. 426
Ex. 5, p. 427
Ex. 6, p. 428

- A graphical representation of the distribution of income among a population can be obtained by plotting data points (x, y), where **x represents the cumulative percentage of families at or below a given income level** and **y represents the cumulative percentage of total family income received.** Regression analysis can be used to find a particular function $y = f(x)$, called a **Lorenz curve,** that best fits the data.

- A single number, the **Gini index,** measures income concentration:

Ex. 7, p. 429

$$\text{Gini index} = 2 \int_0^1 [x - f(x)] \, dx$$

A Gini index of 0 indicates **absolute equality:** All families share equally in the income. A Gini index of 1 indicates **absolute inequality:** One family has all of the income and the rest have none.

7-2 Applications in Business and Economics

- ***Probability Density Functions.*** If any real number x in an interval is a possible outcome of an experiment, then x is said to be a **continuous random variable.** The probability distribution of a continuous random variable is described by a **probability density function** f that satisfies the following conditions:

Ex. 1, p. 435

1. $f(x) \ge 0$ for all real x.
2. The area under the graph of $f(x)$ over the interval $(-\infty, \infty)$ is exactly 1.
3. If $[c, d]$ is a subinterval of $(-\infty, \infty)$, then

$$\text{Probability} \ \ (c \le x \le d) = \int_c^d f(x) \, dx$$

- ***Continuous Income Stream*** If the rate at which income is received—its **rate of flow**—is a continuous function $f(t)$ of time, then the income is said to be a **continuous income stream.** The **total income** produced by a continuous income stream from $t = a$ to $t = b$ is

Ex. 2, p. 437

$$\text{Total income} = \int_a^b f(t) \, dt$$

The **future value** of a continuous income stream that is invested at rate r, compounded continuously, for $0 \le t \le T$, is

$$FV = \int_0^T f(t)e^{r(T-t)}\, dt$$

Ex. 3, p. 439

- **Consumers' and Producers' Surplus**   If $(\bar{x}, \bar{p})$ is a point on the graph of a price–demand equation $p = D(x)$, then the **consumers' surplus** at a price level of $\bar{p}$ is

$$CS = \int_0^{\bar{x}} [D(x) - \bar{p}]\, dx$$

Ex. 4, p. 440

The consumers' surplus represents the total savings to consumers who are willing to pay more than $\bar{p}$, but are still able to buy the product for $\bar{p}$.

Similarly, for a point $(\bar{x}, \bar{p})$ on the graph of a price–supply equation $p = S(x)$, the **producers' surplus** at a price level of $\bar{p}$ is

$$PS = \int_0^{\bar{x}} [\bar{p} - S(x)]\, dx$$

Ex. 5, p. 441

The producers' surplus represents the total gain to producers who are willing to supply units at a lower price $\bar{p}$, but are still able to supply units at $\bar{p}$.

If $(\bar{x}, \bar{p})$ is the intersection point of a price–demand equation $p = D(x)$ and a price–supply equation $p = S(x)$, then $\bar{p}$ is called the **equilibrium price** and $\bar{x}$ is called the **equilibrium quantity.**

Ex. 6, p. 442

7-3 Integration by Parts

- Some indefinite integrals, but not all, can be found by means of the **integration-by-parts formula**

$$\int u\, dv = uv - \int v\, du$$

Ex. 1, p. 446
Ex. 2, p. 447
Ex. 3, p. 448
Ex. 4, p. 449

Select u and dv with the help of the guidelines on page 448.

7-4 Integration Using Tables

- A **table of integrals** is a list of integration formulas that can be used to find indefinite or definite integrals of frequently encountered functions. Such a list appears in Table II of Appendix C.

Ex. 1, p. 452
Ex. 2, p. 453
Ex. 3, p. 453
Ex. 4, p. 454
Ex. 5, p. 454
Ex. 6, p. 455

REVIEW EXERCISE

Work through all the problems in this chapter review and check your answers in the back of the book. Answers to all review problems are there, along with section numbers in italics to indicate where each type of problem is discussed. Where weaknesses show up, review appropriate sections of the text.

Compute all numerical answers to three decimal places unless directed otherwise.

A   In Problems 1–3, set up definite integrals that represent the shaded areas in the figure over the indicated intervals.

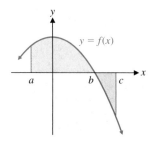

1. Interval $[a, b]$
2. Interval $[b, c]$
3. Interval $[a, c]$
4. Sketch a graph of the area between the graphs of $y = \ln x$ and $y = 0$ over the interval $[0.5, e]$, and find the area.

In Problems 5–10, evaluate each integral.

5. $\displaystyle\int xe^{4x}\, dx$

6. $\displaystyle\int x \ln x\, dx$

7. $\displaystyle\int \frac{\ln x}{x}\, dx$

8. $\displaystyle\int \frac{x}{1 + x^2}\, dx$

9. $\int \dfrac{1}{x(1+x)^2}\,dx$

10. $\int \dfrac{1}{x^2\sqrt{1+x}}\,dx$

In Problems 11–16, find the area bounded by the graphs of the indicated equations over the given interval.

11. $y = 5 - 2x - 6x^2;\ y = 0,\ 1 \le x \le 2$

12. $y = 5x + 7;\ y = 12,\ -3 \le x \le 1$

13. $y = -x + 2;\ y = x^2 + 3,\ -1 \le x \le 4$

14. $y = \dfrac{1}{x};\ y = -e^{-x},\ 1 \le x \le 2$

15. $y = x;\ y = -x^3,\ -2 \le x \le 2$

16. $y = x^2;\ y = -x^4;\ -2 \le x \le 2$

B *In Problems 17–20, set up definite integrals that represent the shaded areas in the figure over the indicated intervals.*

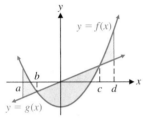

17. Interval $[a, b]$

18. Interval $[b, c]$

19. Interval $[b, d]$

20. Interval $[a, d]$

21. Sketch a graph of the area bounded by the graphs of $y = x^2 - 6x + 9$ and $y = 9 - x$, and find the area.

In Problems 22–27, evaluate each integral.

22. $\int_0^1 xe^x\,dx$

23. $\int_0^3 \dfrac{x^2}{\sqrt{x^2 + 16}}\,dx$

24. $\int \sqrt{9x^2 - 49}\,dx$

25. $\int te^{-0.5t}\,dt$

26. $\int x^2 \ln x\,dx$

27. $\int \dfrac{1}{1 + 2e^x}\,dx$

28. Sketch a graph of the area bounded by the indicated graphs, and find the area. In part (B), approximate intersection points and area to two decimal places.

   (A) $y = x^3 - 6x^2 + 9x;\ y = x$

 (B) $y = x^3 - 6x^2 + 9x;\ y = x + 1$

C *In Problems 29–36, evaluate each integral.*

29. $\int \dfrac{(\ln x)^2}{x}\,dx$

30. $\int x(\ln x)^2\,dx$

31. $\int \dfrac{x}{\sqrt{x^2 - 36}}\,dx$

32. $\int \dfrac{x}{\sqrt{x^4 - 36}}\,dx$

33. $\int_0^4 x\ln(10 - x)\,dx$

34. $\int (\ln x)^2\,dx$

35. $\int xe^{-2x^2}\,dx$

36. $\int x^2 e^{-2x}\,dx$

37. Use a numerical integration routine on a graphing calculator to find the area in the first quadrant that is below the graph of

$$y = \dfrac{6}{2 + 5e^{-x}}$$

and above the graph of $y = 0.2x + 1.6$.

APPLICATIONS

38. *Product warranty.* A manufacturer warrants a product for parts and labor for 1 year and for parts only for a second year. The time to a failure of the product after it is sold is given by the probability density function

$$f(t) = \begin{cases} 0.21e^{-0.21t} & \text{if } t \ge 0 \\ 0 & \text{otherwise} \end{cases}$$

What is the probability that a buyer chosen at random will have a product failure

   (A) During the first year of warranty?

   (B) During the second year of warranty?

39. *Product warranty.* Graph the probability density function for Problem 38 over the interval $[0, 3]$, interpret part (B) of Problem 38 geometrically, and verbally describe the geometric representation.

40. *Revenue function.* The weekly marginal revenue from the sale of x hair dryers is given by

$$R'(x) = 65 - 6\ln(x + 1) \qquad R(0) = 0$$

where $R(x)$ is the revenue in dollars. Find the revenue function and the production level (to the nearest unit) for a revenue of \$20,000 per week. What is the weekly revenue (to the nearest dollar) at a production level of 1,000 hair dryers per week?

41. *Continuous income stream.* The rate of flow (in dollars per year) of a continuous income stream for a 5-year period is given by

$$f(t) = 2{,}500e^{0.05t} \qquad 0 \le t \le 5$$

(A) Graph $y = f(t)$ over $[0, 5]$, and shade in the area that represents the total income received from the end of the first year to the end of the fourth year.

(B) Find the total income received, to the nearest dollar, from the end of the first year to the end of the fourth year.

42. *Future value of a continuous income stream.* The continuous income stream in Problem 41 is invested as it is received at 15%, compounded continuously.

(A) Find the future value (to the nearest dollar) at the end of the 5-year period.

(B) Find the interest earned (to the nearest dollar) during this 5-year period.

43. *Income distribution.* An economist produced the following Lorenz curves for the current income distribution and the projected income distribution 10 years from now in a certain country:

$$f(x) = 0.1x + 0.9x^2 \quad \text{Current Lorenz curve}$$
$$g(x) = x^{1.5} \quad\qquad \text{Projected Lorenz curve}$$

(A) Graph $y = x$ and the current Lorenz curve on one set of coordinate axes for $[0, 1]$, and graph $y = x$ and the projected Lorenz curve on another set of coordinate axes over the same interval.

(B) Looking at the areas bounded by the Lorenz curves and $y = x$, can you say that the income will be more or less equitably distributed 10 years from now?

(C) Compute the Gini index of income concentration (to one decimal place) for the current and projected curves. Now what can you say about the distribution of income 10 years from now? Is it more equitable or less?

44. *Consumers' and producers' surplus.* Find the consumers' surplus and the producers' surplus at the equilibrium price level for each pair of price–demand and price–supply equations. Include a graph that identifies the consumers' surplus and the producers' surplus. Round all values to the nearest integer.

(A) $p = D(x) = 70 - 0.2x;$
$p = S(x) = 13 + 0.0012x^2$

(B) $p = D(x) = 70 - 0.2x; p = S(x) = 13e^{0.006x}$

45. *Producers' surplus.* The following table gives price–supply data for the sale of hogs at a livestock market, where x is the number of pounds (in thousands) and p is the price per pound (in cents):

Price–Supply	
x	$p = S(x)$
0	43.50
10	46.74
20	50.05
30	54.72
40	59.18

(A) Using quadratic regression to model the data, find the demand at a price of 52.50 cents per pound.

(B) Use a numerical integration routine to find the producers' surplus (to the nearest dollar) at a price level of 52.50 cents per pound.

46. *Drug assimilation.* The rate at which the body eliminates a certain drug (in milliliters per hour) is given by

$$R(t) = \frac{60t}{(t + 1)^2(t + 2)}$$

where t is the number of hours since the drug was administered. How much of the drug is eliminated during the first hour after it was administered? During the fourth hour?

47. With the aid of a graphing calculator, illustrate Problem 46 geometrically.

48. *Medicine.* For a particular doctor, the length of time (in hours) spent with a patient per office visit has the probability density function

$$f(t) = \begin{cases} \dfrac{\frac{4}{3}}{(t + 1)^2} & \text{if } 0 \le t \le 3 \\ 0 & \text{otherwise} \end{cases}$$

(A) What is the probability that this doctor will spend less than 1 hour with a randomly selected patient?

(B) What is the probability that this doctor will spend more than 1 hour with a randomly selected patient?

49. *Medicine.* Illustrate part (B) in Problem 48 geometrically. Describe the geometric interpretation verbally.

50. *Politics.* The rate of change of the voting population of a city with respect to time t (in years) is estimated to be

$$N'(t) = \frac{100t}{(1 + t^2)^2}$$

where $N(t)$ is in thousands. If $N(0)$ is the current voting population, how much will this population increase during the next 3 years?

51. *Psychology.* Rats were trained to go through a maze by rewarding them with a food pellet upon successful completion of the run. After the seventh successful run, it was found that the probability density function for length of time (in minutes) until success on the eighth trial was given by

$$f(t) = \begin{cases} .5e^{-.5t} & \text{if } t \ge 0 \\ 0 & \text{otherwise} \end{cases}$$

What is the probability that a rat selected at random after seven successful runs will take 2 or more minutes to complete the eighth run successfully? [Recall that the area under a probability density function curve from $-\infty$ to ∞ is 1.]

Multivariable Calculus

CHAPTER

8

8-1 Functions of Several Variables

8-2 Partial Derivatives

8-3 Maxima and Minima

8-4 Maxima and Minima Using Lagrange Multipliers

8-5 Method of Least Squares

8-6 Double Integrals over Rectangular Regions

8-7 Double Integrals over More General Regions

Chapter 8 Review

Review Exercise

Section 8-1 FUNCTIONS OF SEVERAL VARIABLES

- Functions of Two or More Independent Variables
- Examples of Functions of Several Variables
- Three-Dimensional Coordinate Systems

■ Functions of Two or More Independent Variables

In Section 2-1, we introduced the concept of a function with one independent variable. Now we broaden the concept to include functions with more than one independent variable. We start with an example.

A small manufacturing company produces a standard type of surfboard and no other products. If fixed costs are $500 per week and variable costs are $70 per board produced, the weekly cost function is given by

$$C(x) = 500 + 70x \tag{1}$$

where x is the number of boards produced per week. The cost function is a function of a single independent variable x. For each value of x from the domain of C, there exists exactly one value of $C(x)$ in the range of C.

Now, suppose that the company decides to add a high-performance competition board to its line. If the fixed costs for the competition board are $200 per week and the variable costs are $100 per board, the cost function (1) must be modified to

$$C(x, y) = 700 + 70x + 100y \tag{2}$$

where $C(x, y)$ is the cost for a weekly output of x standard boards and y competition boards. Equation (2) is an example of a function with two independent variables x and y. Of course, as the company expands its product line even further, its weekly cost function must be modified to include more and more independent variables, one for each new product produced.

In general, an equation of the form

$$z = f(x, y)$$

describes a **function of two independent variables** if, for each permissible ordered pair (x, y), there is one and only one value of z determined by $f(x, y)$. The variables x and y are **independent variables,** and the variable z is a **dependent variable.** The set of all ordered pairs of permissible values of x and y is the **domain** of the function, and the set of all corresponding values $f(x, y)$ is the **range** of the function. Unless otherwise stated, we will assume that the domain of a function specified by an equation of the form $z = f(x, y)$ is the set of all ordered pairs of real numbers (x, y) such that $f(x, y)$ is also a real number. It should be noted, however, that certain conditions in practical problems often lead to further restrictions on the domain of a function.

We can similarly define functions of three independent variables, $w = f(x, y, z)$; of four independent variables, $u = f(w, x, y, z)$; and so on. In this chapter, we concern ourselves primarily with functions of two independent variables.

EXAMPLE 1 **Evaluating a Function of Two Independent Variables** For the cost function $C(x, y) = 700 + 70x + 100y$ described earlier, find $C(10, 5)$.

SOLUTION
$$C(10, 5) = 700 + 70(10) + 100(5)$$
$$= \$1,900$$

MATCHED PROBLEM 1 Find $C(20, 10)$ for the cost function in Example 1.

EXAMPLE 2 **Evaluating a Function of Three Independent Variables** For the function $f(x, y, z) = 2x^2 - 3xy + 3z + 1$, find $f(3, 0, -1)$.

SOLUTION
$$f(3, 0, -1) = 2(3)^2 - 3(3)(0) + 3(-1) + 1$$
$$= 18 - 0 - 3 + 1 = 16$$

MATCHED PROBLEM 2 Find $f(-2, 2, 3)$ for f in Example 2.

EXAMPLE 3 **Revenue, Cost, and Profit Functions** Suppose the surfboard company discussed at the beginning of this section has determined that the demand equations for the two types of boards it produces are given by

$$p = 210 - 4x + y$$
$$q = 300 + x - 12y$$

where p is the price of the standard board, q is the price of the competition board, x is the weekly demand for standard boards, and y is the weekly demand for competition boards.

(A) Find the weekly revenue function $R(x, y)$, and evaluate $R(20, 10)$.

(B) If the weekly cost function is

$$C(x, y) = 700 + 70x + 100y$$

find the weekly profit function $P(x, y)$ and evaluate $P(20, 10)$.

SOLUTION   (A)

$$\text{Revenue} = \left(\begin{array}{c}\text{demand for}\\\text{standard}\\\text{boards}\end{array}\right) \times \left(\begin{array}{c}\text{price of a}\\\text{standard}\\\text{board}\end{array}\right) + \left(\begin{array}{c}\text{demand for}\\\text{competition}\\\text{boards}\end{array}\right) \times \left(\begin{array}{c}\text{price of a}\\\text{competition}\\\text{board}\end{array}\right)$$

$$\begin{aligned}
R(x, y) &= xp + yq \\
&= x(210 - 4x + y) + y(300 + x - 12y) \\
&= 210x + 300y - 4x^2 + 2xy - 12y^2 \\
R(20, 10) &= 210(20) + 300(10) - 4(20)^2 + 2(20)(10) - 12(10)^2 \\
&= \$4,800
\end{aligned}$$

(B) $\text{Profit} = \text{revenue} - \text{cost}$

$$\begin{aligned}
P(x, y) &= R(x, y) - C(x, y) \\
&= 210x + 300y - 4x^2 + 2xy - 12y^2 - 700 - 70x - 100y \\
&= 140x + 200y - 4x^2 + 2xy - 12y^2 - 700 \\
P(20, 10) &= 140(20) + 200(10) - 4(20)^2 + 2(20)(10) - 12(10)^2 - 700 \\
&= \$1,700
\end{aligned}$$

MATCHED PROBLEM 3   Repeat Example 3 if the demand and cost equations are given by

$$\begin{aligned}
p &= 220 - 6x + y \\
q &= 300 + 3x - 10y \\
C(x, y) &= 40x + 80y + 1,000
\end{aligned}$$

▪ Examples of Functions of Several Variables

A number of concepts we have already considered can be thought of in terms of functions of two or more variables. We list a few of these as follows:

Area of a rectangle $A(x, y) = xy$

Volume of a box $V(x, y, z) = xyz$

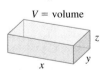

Volume of a right circular cylinder $V(r, h) = \pi r^2 h$

Simple interest	$A(P, r, t) = P(1 + rt)$	A = amount
		P = principal
		r = annual rate
		t = time in years

Compound interest	$A(P, r, t, n) = P\left(1 + \dfrac{r}{n}\right)^{nt}$	A = amount
		P = principal
		r = annual rate
		t = time in years
		n = number of compound periods per year

IQ	$Q(M, C) = \dfrac{M}{C}(100)$	Q = IQ = intelligence quotient
		M = MA = mental age
		C = CA
		= chronological age

Resistance for blood flow in a vessel (Poiseuille's law)	$R(L, r) = k\dfrac{L}{r^4}$	R = resistance
		L = length of vessel
		r = radius of vessel
		k = constant

EXAMPLE 4 **Package Design** A company uses a box with a square base and an open top for one of its products (see figure). If x is the length (in inches) of each side of the base and y is the height (in inches), find the total amount of material $M(x, y)$ required to construct one of these boxes, and evaluate $M(5, 10)$.

SOLUTION

$$\text{Area of base} = x^2$$
$$\text{Area of one side} = xy$$
$$\text{Total material} = (\text{area of base}) + 4(\text{area of one side})$$
$$M(x, y) = x^2 + 4xy$$
$$M(5, 10) = (5)^2 + 4(5)(10)$$
$$= 225 \text{ square inches}$$

MATCHED PROBLEM 4 For the box in Example 4, find the volume $V(x, y)$ and evaluate $V(5, 10)$.

The next example concerns the **Cobb–Douglas production function**

$$f(x, y) = kx^m y^n$$

where k, m, and n are positive constants with $m + n = 1$. Economists use this function to describe the number of units $f(x, y)$ produced from the utilization of x units of labor and y units of capital (for equipment such as tools, machinery, buildings, and so on). Cobb–Douglas production functions are also used to describe the productivity of a single industry, of a group of industries producing the same product, or even of an entire country.

EXAMPLE 5

Productivity   The productivity of a steel-manufacturing company is given approximately by the function

$$f(x, y) = 10x^{0.2}y^{0.8}$$

with the utilization of x units of labor and y units of capital. If the company uses 3,000 units of labor and 1,000 units of capital, how many units of steel will be produced?

SOLUTION   The number of units of steel produced is given by

$$f(3,000, 1,000) = 10(3,000)^{0.2}(1,000)^{0.8} \quad \textit{Use a calculator.}$$
$$\approx 12,457 \text{ units}$$

MATCHED PROBLEM 5   Refer to Example 5. Find the steel production if the company uses 1,000 units of labor and 2,000 units of capital.

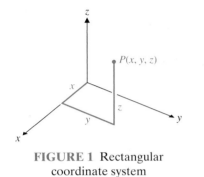

FIGURE 1 Rectangular coordinate system

■ Three-Dimensional Coordinate Systems

We now take a brief look at some graphs of functions of two independent variables. Since functions of the form $z = f(x, y)$ involve two independent variables x and y, and one dependent variable z, we need a *three-dimensional coordinate system* for their graphs. A **three-dimensional coordinate system** is formed by three mutually perpendicular number lines intersecting at their origins (see Fig. 1). In such a system, every ordered **triplet of numbers** (x, y, z) can be associated with a unique point, and conversely.

EXAMPLE 6

Three-Dimensional Coordinates   Locate $(-3, 5, 2)$ in a rectangular coordinate system.

SOLUTION

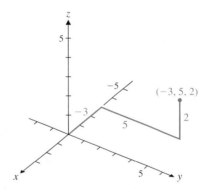

MATCHED PROBLEM 6   Find the coordinates of the corners A, C, G, and D of the rectangular box shown in the figure.

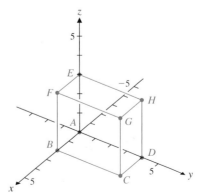

Explore & Discuss **1**

Imagine that you are facing the front of a classroom whose rectangular walls meet at right angles. Suppose that the point of intersection of the floor, front wall, and left-side wall is the origin of a three-dimensional coordinate system in which every point in the room has nonnegative coordinates. Then the plane $z = 0$ (or, equivalently, the xy plane) can be described as "the floor," and the plane $z = 2$ can be described as "the plane parallel to, but 2 units above, the floor." Give similar descriptions of the following planes:

(A) $x = 0$ (B) $x = 3$ (C) $y = 0$
(D) $y = 4$ (E) $x = -1$

What does the graph of $z = x^2 + y^2$ look like? If we let $x = 0$ and graph $z = 0^2 + y^2 = y^2$ in the yz plane, we obtain a parabola; if we let $y = 0$ and graph $z = x^2 + 0^2 = x^2$ in the xz plane, we obtain another parabola. It can be shown that the graph of $z = x^2 + y^2$ is either one of these parabolas rotated around the z axis (see Fig. 2). This cup-shaped figure is a *surface* and is called a **paraboloid.**

In general, the graph of any function of the form $z = f(x, y)$ is called a **surface.** The graph of such a function is the graph of all ordered triplets of numbers (x, y, z) that satisfy the equation that is the function. Graphing functions of two independent variables is often a very difficult task, and the general process will not be dealt with in this book. We present only a few simple graphs to suggest extensions of earlier geometric interpretations of the derivative and local maxima and minima to functions of two variables. Note that $z = f(x, y) = x^2 + y^2$ appears (see Fig. 2) to have a local minimum at $(x, y) = (0, 0)$. Figure 3 shows a local maximum at $(x, y) = (0, 0)$.

Figure 4 shows a point at $(x, y) = (0, 0)$, called a **saddle point,** that is neither a local minimum nor a local maximum. Note that if $x = 0$ the saddle point is a local minimum, and if $y = 0$ the saddle point is a local maximum. More will be said about local maxima and minima in Section 8-3.

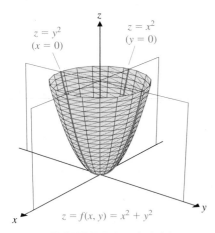

FIGURE 2 Paraboloid

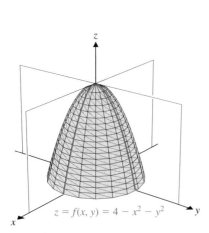

FIGURE 3 Local maximum:
$f(0, 0) = 4$

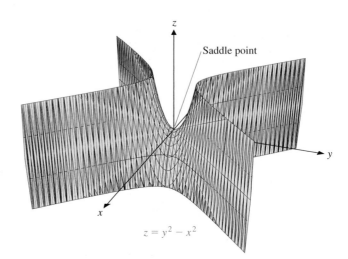

FIGURE 4 Saddle point at $(0, 0, 0)$

Explore & Discuss **2**

(A) Let $f(x, y) = x^2 + y^2$. The cross section of the surface $z = f(x, y)$ produced by cutting it with the plane $x = 2$ is a parabola, the graph of $z = f(2, y) = 4 + y^2$ (see Fig. 5). Explain why each cross section of $z = f(x, y)$ produced by cutting it with a plane parallel to the yz plane is a parabola that opens upward. Explain why each cross section of $z = f(x, y)$ produced by cutting it with a plane parallel to the xz plane is a parabola that opens upward.

(B) Let $g(x, y) = y^2 - x^2$. Explain why each cross section of $z = g(x, y)$ produced by cutting it with a plane parallel to the yz plane is a parabola that

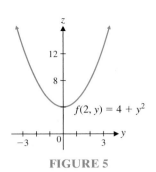

FIGURE 5

opens upward (see Fig. 4). Explain why each cross section of $z = f(x, y)$ produced by cutting it with a plane parallel to the xz plane is a parabola that opens downward.

Some graphing calculators are designed to draw graphs (like those of Figs. 2, 3, and 4) of functions of two independent variables. Others, such as the graphing calculator used for the displays in this book, are designed to draw graphs of functions of just a single independent variable. When using the latter type of calculator, we can graph cross sections produced by cutting surfaces with planes parallel to the xz plane or yz plane to gain insight into the graph of a function of two independent variables.

EXAMPLE 7   Graphing Cross Sections

(A) Describe the cross sections of $f(x, y) = 2x^2 + y^2$ in the planes $y = 0$, $y = 1$, $y = 2$, $y = 3$, and $y = 4$.

(B) Describe the cross sections of $f(x, y) = 2x^2 + y^2$ in the planes $x = 0$, $x = 1$, $x = 2$, $x = 3$, and $x = 4$.

SOLUTION   (A) The cross section of $f(x, y) = 2x^2 + y^2$ produced by cutting it with the plane $y = 0$ is the graph of the function $f(x, 0) = 2x^2$ in this plane. We can examine the shape of this cross section by graphing $y_1 = 2x^2$ on a graphing calculator (Fig. 6). Similarly, the graphs of $y_2 = f(x, 1) = 2x^2 + 1$, $y_3 = f(x, 2) = 2x^2 + 4$, $y_4 = f(x, 3) = 2x^2 + 9$, and $y_5 = f(x, 4) = 2x^2 + 16$ show the shapes of the other four cross sections (see Fig. 6). Each of these is a parabola that opens upward. Note the correspondence between the graphs in Figure 6 and the actual cross sections of $f(x, y) = 2x^2 + y^2$ shown in Figure 7.

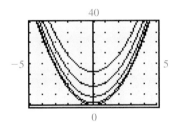

FIGURE 6

$y_1 = 2x^2$ $y_4 = 2x^2 + 9$
$y_2 = 2x^2 + 1$ $y_5 = 2x^2 + 16$
$y_3 = 2x^2 + 4$

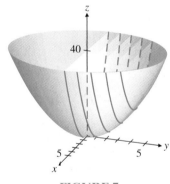

FIGURE 7

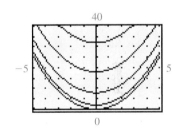

FIGURE 8

$y_1 = x^2$ $y_4 = 18 + x^2$
$y_2 = 2 + x^2$ $y_5 = 32 + x^2$
$y_3 = 8 + x^2$

(B) The five cross sections are represented by the graphs of the functions $f(0, y) = y^2$, $f(1, y) = 2 + y^2$, $f(2, y) = 8 + y^2$, $f(3, y) = 18 + y^2$, and $f(4, y) = 32 + y^2$. These five functions are graphed in Figure 8. (Note that changing the name of the independent variable from y to x for graphing purposes does not affect the graph displayed.) Each of the five cross sections is a parabola that opens upward.

MATCHED PROBLEM 7   (A) Describe the cross sections of $g(x, y) = y^2 - x^2$ in the planes $y = 0$, $y = 1$, $y = 2$, $y = 3$, and $y = 4$.

(B) Describe the cross sections of $g(x, y) = y^2 - x^2$ in the planes $x = 0$, $x = 1$, $x = 2$, $x = 3$, and $x = 4$.

INSIGHT

The graph of the *equation*

$$x^2 + y^2 + z^2 = 4 \qquad (3)$$

is the graph of all ordered triplets of numbers (x, y, z) that satisfy the equation. The Pythagorean theorem can be used to show that the distance from the point (x, y, z) to the origin $(0, 0, 0)$ is equal to

$$\sqrt{x^2 + y^2 + z^2}$$

Therefore, the graph of (3) consists of all points that are a distance 2 from the origin—that is, all points on the sphere of radius 2 and with center at the origin. Recall that a circle in the plane is *not* the graph of a function $y = f(x)$, because it fails the vertical-line test (Section 2-1). Similarly, a sphere is *not* the graph of a *function* $z = f(x, y)$ of two variables.

Answers to Matched Problems   **1.** $3,100   **2.** 30

3. (A) $R(x, y) = 220x + 300y - 6x^2 + 4xy - 10y^2$; $R(20, 10) = \$4,800$
    (B) $P(x, y) = 180x + 220y - 6x^2 + 4xy - 10y^2 - 1,000$; $P(20, 10) = \$2,200$

4. $V(x, y) = x^2y$; $V(5, 10) = 250$ in.^{3} **5.** 17,411 units

6. $A(0, 0, 0)$; $C(2, 4, 0)$; $G(2, 4, 3)$; $D(0, 4, 0)$

7. (A) Each cross section is a parabola that opens downward.
    (B) Each cross section is a parabola that opens upward.

Exercise 8-1

A In Problems 1–10, find the indicated values of the functions

$$f(x, y) = 2x + 7y - 5 \quad \text{and} \quad g(x, y) = \frac{88}{x^2 + 3y}$$

1. $f(4, -1)$ **2.** $f(0, 10)$

3. $f(8, 0)$ **4.** $f(5, 6)$

5. $g(1, 7)$ **6.** $g(-2, 0)$

7. $g(3, -3)$ **8.** $g(0, 0)$

9. $3f(-2, 2) + 5g(-2, 2)$

10. $2f(10, -4) - 7g(10, -4)$

In Problems 11–14, find the indicated values of

$$f(x, y, z) = 2x - 3y^2 + 5z^3 - 1$$

11. $f(0, 0, 0)$ **12.** $f(0, 0, 2)$

13. $f(6, -5, 0)$ **14.** $f(-10, 4, -3)$

B In Problems 15–24, find the indicated value of the given function.

15. $V(2, 4)$ for $V(r, h) = \pi r^2 h$

16. $S(4, 2)$ for $S(x, y) = 5x^2y^3$

17. $R(1, 2)$ for
    $R(x, y) = -5x^2 + 6xy - 4y^2 + 200x + 300y$

18. $P(2, 2)$ for
    $P(x, y) = -x^2 + 2xy - 2y^2 - 4x + 12y + 5$

19. $R(6, 0.5)$ for $R(L, r) = 0.002\dfrac{L}{r^4}$

20. $L(2,000, 50)$ for $L(w, v) = (1.25 \times 10^{-5})wv^2$

21. $A(100, 0.06, 3)$ for $A(P, r, t) = P + Prt$

22. $A(10, 0.04, 3, 2)$ for $A(P, r, t, n) = P\left(1 + \dfrac{r}{n}\right)^{tn}$

23. $P(0.05, 12)$ for $P(r, T) = \displaystyle\int_0^T 4,000e^{-rt}\, dt$

24. $F(0.07, 10)$ for $F(r, T) = \displaystyle\int_0^T 4,000e^{r(T-t)}\, dt$

In Problems 25–30, find the indicated function f of a single variable.

25. $f(x) = G(x, 0)$ for $G(x, y) = x^2 + 3xy + y^2 - 7$

26. $f(y) = H(0, y)$ for $H(x, y) = x^2 - 5xy - y^2 + 2$

27. $f(y) = K(4, y)$ for $K(x, y) = 10xy + 3x - 2y + 8$

28. $f(x) = L(x, -2)$ for $L(x, y) = 25 - x + 5y - 6xy$

29. $f(y) = M(y, y)$ for $M(x, y) = x^2y - 3xy^2 + 5$

30. $f(x) = N(x, 2x)$ for $N(x, y) = 3xy + x^2 - y^2 + 1$

31. Let $F(x, y) = 2x + 3y - 6$. Find all values of y such that $F(0, y) = 0$.

32. Let $F(x, y) = 5x - 4y + 12$. Find all values of x such that $F(x, 0) = 0$.

33. Let $F(x, y) = 2xy + 3x - 4y - 1$. Find all values of x such that $F(x, x) = 0$.

34. Let $F(x, y) = xy + 2x^2 + y^2 - 25$. Find all values of y such that $F(y, y) = 0$.

 35. Let $F(x, y) = x^2 + e^x y - y^2$. Find all values of x such that $F(x, 2) = 0$.

 36. Let $G(a, b, c) = a^3 + b^3 + c^3 - (ab + ac + bc) - 6$. Find all values of b such that $G(2, b, 1) = 0$.

C

37. For the function $f(x, y) = x^2 + 2y^2$, find

$$\frac{f(x + h, y) - f(x, y)}{h}$$

38. For the function $f(x, y) = x^2 + 2y^2$, find

$$\frac{f(x, y + k) - f(x, y)}{k}$$

39. For the function $f(x, y) = 2xy^2$, find

$$\frac{f(x + h, y) - f(x, y)}{h}$$

40. For the function $f(x, y) = 2xy^2$, find

$$\frac{f(x, y + k) - f(x, y)}{k}$$

41. Find the coordinates of E and F in the figure for Matched Problem 6 in the text.

42. Find the coordinates of B and H in the figure for Matched Problem 6 in the text.

 In Problems 43–48, use a graphing calculator as necessary to explore the graphs of the indicated cross sections.

43. Let $f(x, y) = x^2$.

(A) Explain why the cross sections of the surface $z = f(x, y)$ produced by cutting it with planes parallel to $y = 0$ are parabolas.

(B) Describe the cross sections of the surface in the planes $x = 0$, $x = 1$, and $x = 2$.

(C) Describe the surface $z = f(x, y)$.

44. Let $f(x, y) = \sqrt{4 - y^2}$.

(A) Explain why the cross sections of the surface $z = f(x, y)$ produced by cutting it with planes parallel to $x = 0$ are semicircles of radius 2.

(B) Describe the cross sections of the surface in the planes $y = 0$, $y = 2$, and $y = 3$.

(C) Describe the surface $z = f(x, y)$.

45. Let $f(x, y) = \sqrt{36 - x^2 - y^2}$.

(A) Describe the cross sections of the surface $z = f(x, y)$ produced by cutting it with the planes $y = 1$, $y = 2$, $y = 3$, $y = 4$, and $y = 5$.

(B) Describe the cross sections of the surface in the planes $x = 0$, $x = 1$, $x = 2$, $x = 3$, $x = 4$, and $x = 5$.

(C) Describe the surface $z = f(x, y)$.

46. Let $f(x, y) = 100 + 10x + 25y - x^2 - 5y^2$.

(A) Describe the cross sections of the surface $z = f(x, y)$ produced by cutting it with the planes $y = 0$, $y = 1$, $y = 2$, and $y = 3$.

(B) Describe the cross sections of the surface in the planes $x = 0$, $x = 1$, $x = 2$, and $x = 3$.

(C) Describe the surface $z = f(x, y)$.

47. Let $f(x, y) = e^{-(x^2 + y^2)}$.

(A) Explain why $f(a, b) = f(c, d)$ whenever (a, b) and (c, d) are points on the same circle centered at the origin in the xy plane.

(B) Describe the cross sections of the surface $z = f(x, y)$ produced by cutting it with the planes $x = 0$, $y = 0$, and $x = y$.

(C) Describe the surface $z = f(x, y)$.

48. Let $f(x, y) = 4 - \sqrt{x^2 + y^2}$.

(A) Explain why $f(a, b) = f(c, d)$ whenever (a, b) and (c, d) are points on the same circle with center at the origin in the xy plane.

(B) Describe the cross sections of the surface $z = f(x, y)$ produced by cutting it with the planes $x = 0$, $y = 0$, and $x = y$.

(C) Describe the surface $z = f(x, y)$.

Applications

49. *Cost function.* A small manufacturing company produces two models of a surfboard: a standard model and a competition model. If the standard model is produced at a variable cost of \$70 each and the competition model at a variable cost of \$100 each, and if the total fixed costs per month are \$2,000, then the monthly cost function is given by

$$C(x, y) = 2,000 + 70x + 100y$$

where x and y are the numbers of standard and competition models produced per month, respectively. Find $C(20, 10)$, $C(50, 5)$, and $C(30, 30)$.

50. *Advertising and sales.* A company spends $\$x$ thousand per week on newspaper advertising and $\$y$ thousand per week on television advertising. Its weekly sales are found to be given by

$$S(x, y) = 5x^2y^3$$

Find $S(3, 2)$ and $S(2, 3)$.

51. *Revenue function.* A supermarket sells two brands of coffee: brand A at $\$p$ per pound and brand B at $\$q$ per pound. The daily demand equations for brands A and B are, respectively,

$$x = 200 - 5p + 4q$$
$$y = 300 + 2p - 4q$$

(both in pounds). Find the daily revenue function $R(p, q)$. Evaluate $R(2, 3)$ and $R(3, 2)$.

52. *Revenue, cost, and profit functions.* A company manufactures 10- and 3-speed bicycles. The weekly demand and cost equations are

$$p = 230 - 9x + y$$
$$q = 130 + x - 4y$$
$$C(x, y) = 200 + 80x + 30y$$

where $\$p$ is the price of a 10-speed bicycle, $\$q$ is the price of a 3-speed bicycle, x is the weekly demand for 10-speed bicycles, y is the weekly demand for 3-speed bicycles, and $C(x, y)$ is the cost function. Find the weekly revenue function $R(x, y)$ and the weekly profit function $P(x, y)$. Evaluate $R(10, 15)$ and $P(10, 15)$.

53. *Productivity.* The Cobb–Douglas production function for a petroleum company is given by

$$f(x, y) = 20x^{0.4}y^{0.6}$$

where x is the utilization of labor and y is the utilization of capital. If the company uses 1,250 units of labor and 1,700 units of capital, how many units of petroleum will be produced?

54. *Productivity.* The petroleum company in Problem 53 is taken over by another company that decides to double both the units of labor and the units of capital utilized in the production of petroleum. Use the Cobb–Douglas production function given in Problem 53 to find the amount of petroleum that will be produced by this increased utilization of labor and capital. What is the effect on productivity of doubling both the units of labor and the units of capital?

55. *Future value.* At the end of each year, $2,000 is invested into an IRA earning 9% compounded annually.

(A) How much will be in the account at the end of 30 years? Use the annuity formula

$$F(P, i, n) = P\frac{(1 + i)^n - 1}{i}$$

where

$$P = \text{periodic payment}$$
$$i = \text{rate per period}$$
$$n = \text{number of payments (periods)}$$
$$F = \text{FV} = \text{future value}$$

 (B) Use graphical approximation methods to determine the rate of interest that would produce $500,000 in the account at the end of 30 years.

56. *Package design.* The packaging department in a company has been asked to design a rectangular box with no top and a partition down the middle (see the figure). Let x, y, and z be the dimensions of the box (in inches).

FIGURE FOR 56

(A) Find the total amount of material $M(x, y, z)$ used in constructing one of these boxes, and evaluate $M(10, 12, 6)$.

 (B) Suppose that the box is to have a square base and a volume of 720 cubic inches. Use graphical approximation methods to determine the dimensions that require the least amount of material.

57. *Marine biology.* In using scuba-diving gear, a marine biologist estimates the time of a dive according to the equation

$$T(V, x) = \frac{33V}{x + 33}$$

where

$$T = \text{time of dive in minutes}$$
$$V = \text{volume of air, at sea level pressure,}$$
$$\text{compressed into tanks}$$
$$x = \text{depth of dive in feet}$$

Find $T(70, 47)$ and $T(60, 27)$.

58. *Blood flow.* Poiseuille's law states that the resistance R for blood flowing in a blood vessel varies directly as the length L of the vessel and inversely as the fourth power of its radius r. Stated as an equation,

$$R(L, r) = k\frac{L}{r^4} \quad k \text{ a constant}$$

Find $R(8, 1)$ and $R(4, 0.2)$.

59. *Physical anthropology.* Anthropologists, in their study of race and human genetic groupings, often use an index called the *cephalic index.* The cephalic index C varies directly as the width W of the head and inversely as the length L of the head (both viewed from the top). In terms of an equation,

$$C(W, L) = 100\frac{W}{L}$$

where

$$W = \text{width in inches}$$
$$L = \text{length in inches}$$

Find $C(6, 8)$ and $C(8.1, 9)$

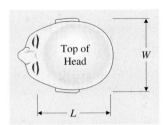

60. *Safety research.* Under ideal conditions, if a person driving a car slams on the brakes and skids to a stop, the length of the skid marks (in feet) is given by the formula

$$L(w, v) = kwv^2$$

where

$$k = \text{constant}$$
$$w = \text{weight of car in pounds}$$
$$v = \text{speed of car in miles per hour}$$

For $k = 0.000\ 013\ 3$, find $L(2,000, 40)$ and $L(3,000, 60)$.

61. *Psychology.* The intelligence quotient (IQ) is defined to be the ratio of mental age (MA), as determined by certain tests, to chronological age (CA), multiplied by 100. Stated as an equation,

$$Q(M, C) = \frac{M}{C} \cdot 100$$

where

$$Q = \text{IQ} \qquad M = \text{MA} \qquad C = \text{CA}$$

Find $Q(12, 10)$ and $Q(10, 12)$.

Section 8-2   PARTIAL DERIVATIVES

- Partial Derivatives
- Second-Order Partial Derivatives

■ Partial Derivatives

We know how to differentiate many kinds of functions of one independent variable and how to interpret the derivatives that result. What about functions with two or more independent variables? Let us return to the surfboard example considered at the beginning of the chapter.

For the company producing only the standard board, the cost function was

$$C(x) = 500 + 70x$$

Differentiating with respect to x, we obtain the marginal cost function

$$C'(x) = 70$$

Since the marginal cost is constant, $70 is the change in cost for a 1-unit increase in production at any output level.

For the company producing two types of boards—a standard model and a competition model—the cost function was

$$C(x, y) = 700 + 70x + 100y$$

Now suppose that we differentiate with respect to x, holding y fixed, and denote the resulting function by $C_x(x, y)$; or suppose we differentiate with respect to y, holding x fixed, and denote the resulting function by $C_y(x, y)$. Differentiating in this way, we obtain

$$C_x(x, y) = 70 \qquad C_y(x, y) = 100$$

Each of these functions is called a **partial derivative,** and, in this example, each represents marginal cost. The first is the change in cost due to a 1-unit increase in production of the standard board with the production of the competition model held fixed. The second is the change in cost due to a 1-unit increase in production of the competition board with the production of the standard board held fixed.

In general, if $z = f(x, y)$, then the **partial derivative of f with respect to x,** denoted $\partial z/\partial x$, f_x, or $f_x(x, y)$, is defined by

$$\frac{\partial z}{\partial x} = \lim_{h \to 0} \frac{f(x + h, y) - f(x, y)}{h}$$

provided that the limit exists. We recognize this formula as the ordinary derivative of f with respect to x, holding y constant. Thus, we are able to continue to use all the derivative rules and properties discussed in Chapters 3 to 5 and apply them to partial derivatives.

Similarly, the **partial derivative of f with respect to y,** denoted $\partial z/\partial y$, f_y, or $f_y(x, y)$, is defined by

$$\frac{\partial z}{\partial y} = \lim_{k \to 0} \frac{f(x, y + k) - f(x, y)}{k}$$

which is the ordinary derivative with respect to y, holding x constant.

Parallel definitions and interpretations hold for functions with three or more independent variables.

EXAMPLE 1 **Partial Derivatives** For $z = f(x, y) = 2x^2 - 3x^2y + 5y + 1$, find

(A) $\partial z/\partial x$ (B) $f_x(2, 3)$

SOLUTION (A) $z = 2x^2 - 3x^2y + 5y + 1$

Differentiating with respect to x, holding y constant (that is, treating y as a constant), we obtain

$$\frac{\partial z}{\partial x} = 4x - 6xy$$

(B) $f(x, y) = 2x^2 - 3x^2y + 5y + 1$

First, differentiate with respect to x. From part (A), we have

$$f_x(x, y) = 4x - 6xy$$

Then evaluate this equation at $(2, 3)$:

$$f_x(2, 3) = 4(2) - 6(2)(3) = -28$$

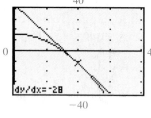

 In part 1(B), an alternative approach would be to substitute $y = 3$ into $f(x, y)$ and graph the function $f(x, 3) = -7x^2 + 16$, which represents the cross section of the surface $z = f(x, y)$ produced by cutting it with the plane $y = 3$. Then determine the slope of the tangent line when $x = 2$. Again, we conclude that $f_x(2, 3) = -28$ (see Fig. 1).

FIGURE 1 $y_1 = -7x^2 + 16$

MATCHED PROBLEM 1 For f in Example 1, find

(A) $\partial z/\partial y$ (B) $f_y(2, 3)$

EXAMPLE 2 **Partial Derivatives Using the Chain Rule** For $z = f(x, y) = e^{x^2+y^2}$, find

(A) $\partial z/\partial x$ (B) $f_y(2, 1)$

SOLUTION (A) Using the chain rule [thinking of $z = e^u, u = u(x)$; y is held constant], we obtain

$$\frac{\partial z}{\partial x} = e^{x^2+y^2}\frac{\partial(x^2 + y^2)}{\partial x}$$
$$= 2xe^{x^2+y^2}$$

(B)
$$f_y(x, y) = e^{x^2+y^2}\frac{\partial(x^2 + y^2)}{\partial y} = 2ye^{x^2+y^2}$$
$$f_y(2, 1) = 2(1)e^{(2)^2+(1)^2}$$
$$= 2e^5$$

MATCHED PROBLEM 2	For $z = f(x, y) = (x^2 + 2xy)^5$, find

(A) $\partial z/\partial y$ (B) $f_x(1, 0)$

EXAMPLE 3   **Profit** The profit function for the surfboard company in Example 3 in Section 8-1 was

$$P(x, y) = 140x + 200y - 4x^2 + 2xy - 12y^2 - 700$$

Find $P_x(15, 10)$ and $P_x(30, 10)$, and interpret the results.

SOLUTION
$$P_x(x, y) = 140 - 8x + 2y$$
$$P_x(15, 10) = 140 - 8(15) + 2(10) = 40$$
$$P_x(30, 10) = 140 - 8(30) + 2(10) = -80$$

At a production level of 15 standard and 10 competition boards per week, increasing the production of standard boards by 1 unit and holding the production of competition boards fixed at 10 will increase profit by approximately $40. At a production level of 30 standard and 10 competition boards per week, increasing the production of standard boards by 1 unit and holding the production of competition boards fixed at 10 will decrease profit by approximately $80.

MATCHED PROBLEM 3	For the profit function in Example 3, find $P_y(25, 10)$ and $P_y(25, 15)$, and interpret the results.

Explore & Discuss **1**   Let $P(x, y)$ be the profit function of Example 3 and Matched Problem 3.

(A) Assume that the production of competition boards remains fixed at 10. Which production level of standard boards will yield the maximum profit? Calculate that maximum profit.

(B) Assume that the production of standard boards remains fixed at 25. Which production level of competition boards will yield the maximum profit? Calculate that maximum profit.

EXAMPLE 4   **Productivity** The productivity of a major computer manufacturer is given approximately by the Cobb–Douglas production function

$$f(x, y) = 15x^{0.4}y^{0.6}$$

with the utilization of x units of labor and y units of capital. The partial derivative $f_x(x, y)$ represents the rate of change of productivity with respect to labor and is called the **marginal productivity of labor.** The partial derivative $f_y(x, y)$ represents the rate of change of productivity with respect to capital and is called the **marginal productivity of capital.** If the company is currently utilizing 4,000 units of labor and 2,500 units of capital, find the marginal productivity of labor and the marginal productivity of capital. For the greatest increase in productivity, should the management of the company encourage increased use of labor or increased use of capital?

SOLUTION
$$f_x(x, y) = 6x^{-0.6}y^{0.6}$$

$$f_x(4,000, 2,500) = 6(4,000)^{-0.6}(2,500)^{0.6}$$
$$\approx 4.53 \qquad \text{Marginal productivity of labor}$$

$$f_y(x, y) = 9x^{0.4}y^{-0.4}$$

$$f_y(4,000, 2,500) = 9(4,000)^{0.4}(2,500)^{-0.4}$$
$$\approx 10.86 \qquad \text{Marginal productivity of capital}$$

At the current level of utilization of 4,000 units of labor and 2,500 units of capital, each 1-unit increase in labor utilization (keeping capital utilization fixed at 2,500 units) will increase production by approximately 4.53 units, and each 1-unit increase in capital utilization (keeping labor utilization fixed at 4,000 units) will increase production by approximately 10.86 units. Thus, the management of the company should encourage increased use of capital. ▬

MATCHED PROBLEM 4 | The productivity of an airplane-manufacturing company is given approximately by the Cobb–Douglas production function

$$f(x, y) = 40x^{0.3}y^{0.7}$$

(A) Find $f_x(x, y)$ and $f_y(x, y)$.

(B) If the company is currently using 1,500 units of labor and 4,500 units of capital, find the marginal productivity of labor and the marginal productivity of capital.

(C) For the greatest increase in productivity, should the management of the company encourage increased use of labor or increased use of capital?

Partial derivatives have simple geometric interpretations, as indicated in Figure 2. If we hold x fixed at, say, $x = a$, then $f_y(a, y)$ is the slope of the curve obtained by intersecting the surface $z = f(x, y)$ with the plane $x = a$. A similar interpretation is given to $f_x(x, b)$.

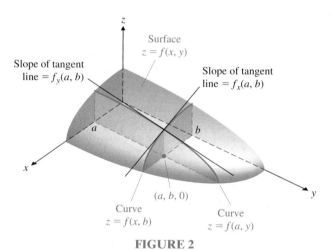

FIGURE 2

Explore & Discuss **2**   Let $f(x, y) = x^2y - 2xy^2 + 3$ and consider the surface $z = f(x, y)$.

(A) The cross section of the surface produced by cutting it with the plane $x = 2$ is the graph of $f(2, y) = 4y - 4y^2 + 3$, which is a parabola. Use an ordinary derivative to find the slope of the tangent line to this parabola when $y = \frac{1}{2}$.

(B) Use partial derivatives to confirm your answer to part (A).

(C) Explain why the cross section of the surface produced by cutting it with the plane $y = 1$ is also a parabola. Use an ordinary derivative to find the slope of the tangent line to this parabola when $x = 3$.

(D) Use partial derivatives to confirm your answer to part (C).

■ Second-Order Partial Derivatives

The function

$$z = f(x, y) = x^4y^7$$

has two **first-order partial derivatives:**

$$\frac{\partial z}{\partial x} = f_x = f_x(x, y) = 4x^3 y^7 \qquad \text{and} \qquad \frac{\partial z}{\partial y} = f_y = f_y(x, y) = 7x^4 y^6$$

Each of these partial derivatives, in turn, has two partial derivatives called **second-order partial derivatives** of $z = f(x, y)$. Generalizing the various notations we have for first-order partial derivatives, we write the four second-order partial derivatives of $z = f(x, y) = x^4 y^7$ as

Equivalent notations

$$f_{xx} = f_{xx}(x, y) = \frac{\partial^2 z}{\partial x^2} = \frac{\partial}{\partial x}\left(\frac{\partial z}{\partial x}\right) = \frac{\partial}{\partial x}(4x^3 y^7) = 12x^2 y^7$$

$$f_{xy} = f_{xy}(x, y) = \frac{\partial^2 z}{\partial y\, \partial x} = \frac{\partial}{\partial y}\left(\frac{\partial z}{\partial x}\right) = \frac{\partial}{\partial y}(4x^3 y^7) = 28x^3 y^6$$

$$f_{yx} = f_{yx}(x, y) = \frac{\partial^2 z}{\partial x\, \partial y} = \frac{\partial}{\partial x}\left(\frac{\partial z}{\partial y}\right) = \frac{\partial}{\partial x}(7x^4 y^6) = 28x^3 y^6$$

$$f_{yy} = f_{yy}(x, y) = \frac{\partial^2 z}{\partial y^2} = \frac{\partial}{\partial y}\left(\frac{\partial z}{\partial y}\right) = \frac{\partial}{\partial y}(7x^4 y^6) = 42x^4 y^5$$

In the mixed partial derivative $\partial^2 z/\partial y\, \partial x = f_{xy}$, we started with $z = f(x, y)$ and first differentiated with respect to x (holding y constant). Then we differentiated with respect to y (holding x constant). In the other mixed partial derivative, $\partial^2 z/\partial x\, \partial y = f_{yx}$, the order of differentiation was reversed; however, the final result was the same—that is, $f_{xy} = f_{yx}$. Although it is possible to find functions for which $f_{xy} \neq f_{yx}$, such functions rarely occur in applications involving partial derivatives. Thus, for all the functions in this book, we will assume that $f_{xy} = f_{yx}$.

In general, we have the following definitions:

DEFINITION   **Second-Order Partial Derivatives**

If $z = f(x, y)$, then

$$f_{xx} = f_{xx}(x, y) = \frac{\partial^2 z}{\partial x^2} = \frac{\partial}{\partial x}\left(\frac{\partial z}{\partial x}\right)$$

$$f_{xy} = f_{xy}(x, y) = \frac{\partial^2 z}{\partial y\, \partial x} = \frac{\partial}{\partial y}\left(\frac{\partial z}{\partial x}\right)$$

$$f_{yx} = f_{yx}(x, y) = \frac{\partial^2 z}{\partial x\, \partial y} = \frac{\partial}{\partial x}\left(\frac{\partial z}{\partial y}\right)$$

$$f_{yy} = f_{yy}(x, y) = \frac{\partial^2 z}{\partial y^2} = \frac{\partial}{\partial y}\left(\frac{\partial z}{\partial y}\right)$$

EXAMPLE 5   **Second-Order Partial Derivatives** For $z = f(x, y) = 3x^2 - 2xy^3 + 1$, find

(A) $\dfrac{\partial^2 z}{\partial x\, \partial y}, \dfrac{\partial^2 z}{\partial y\, \partial x}$ (B) $\dfrac{\partial^2 z}{\partial x^2}$ (C) $f_{yx}(2, 1)$

SOLUTION   (A) First differentiate with respect to y and then with respect to x:

$$\frac{\partial z}{\partial y} = -6xy^2 \qquad \frac{\partial^2 z}{\partial x\, \partial y} = \frac{\partial}{\partial x}\left(\frac{\partial z}{\partial y}\right) = \frac{\partial}{\partial x}(-6xy^2) = -6y^2$$

Now differentiate with respect to x and then with respect to y:

$$\frac{\partial z}{\partial x} = 6x - 2y^3 \qquad \frac{\partial^2 z}{\partial y\, \partial x} = \frac{\partial}{\partial y}\left(\frac{\partial z}{\partial x}\right) = \frac{\partial}{\partial y}(6x - 2y^3) = -6y^2$$

(B) Differentiate with respect to x twice:

$$\frac{\partial z}{\partial x} = 6x - 2y^3 \qquad \frac{\partial^2 z}{\partial x^2} = \frac{\partial}{\partial x}\left(\frac{\partial z}{\partial x}\right) = 6$$

(C) First find $f_{yx}(x, y)$; then evaluate the resulting equation at $(2, 1)$. Again, remember that f_{yx} signifies differentiation first with respect to y and then with respect to x. Thus,

$$f_y(x, y) = -6xy^2 \qquad f_{yx}(x, y) = -6y^2$$

and

$$f_{yx}(2, 1) = -6(1)^2 = -6$$

▬

MATCHED PROBLEM 5   For $z = f(x, y) = x^3y - 2y^4 + 3$, find

(A) $\dfrac{\partial^2 z}{\partial y\, \partial x}$ (B) $\dfrac{\partial^2 z}{\partial y^2}$

(C) $f_{xy}(2, 3)$ (D) $f_{yx}(2, 3)$

INSIGHT

Although the mixed second-order partial derivatives f_{xy} and f_{yx} are equal for all functions considered in this book, it is a good idea to compute both of them, as in Example 5(A), as a check on your work. By contrast, the other two second-order partial derivatives, f_{xx} and f_{yy}, are generally not equal to each other. For example, for the function

$$f(x, y) = 3x^2 - 2xy^3 + 1$$

of Example 5,

$$f_{xx} = 6 \quad \text{and} \quad f_{yy} = -12xy$$

Answers to Matched Problems

1. (A) $\partial z/\partial y = -3x^2 + 5$ (B) $f_y(2, 3) = -7$
2. (A) $10x(x^2 + 2xy)^4$ (B) 10
3. $P_y(25, 10) = 10$: At a production level of $x = 25$ and $y = 10$, increasing y by 1 unit and holding x fixed at 25 will increase profit by approximately \$10; $P_y(25, 15) = -110$: At a production level of $x = 25$ and $y = 15$, increasing y by 1 unit and holding x fixed at 25 will decrease profit by approximately \$110
4. (A) $f_x(x, y) = 12x^{-0.7}y^{0.7}$; $f_y(x, y) = 28x^{0.3}y^{-0.3}$
   (B) Marginal productivity of labor ≈ 25.89; marginal productivity of capital ≈ 20.14
   (C) Labor
5. (A) $3x^2$ (B) $-24y^2$ (C) 12 (D) 12

Exercise 8-2

A   *In Problems 1–16, find the indicated first-order partial derivative for each function $z = f(x, y)$.*

1. $f_x(x, y)$ if $f(x, y) = 4x - 3y + 6$
2. $f_x(x, y)$ if $f(x, y) = 7x + 8y - 2$
3. $f_y(x, y)$ if $f(x, y) = 5x + 6y - 8$
4. $f_y(x, y)$ if $f(x, y) = -2x + 9y - 1$
5. $f_y(x, y)$ if $f(x, y) = x^2 - 3xy + 2y^2$
6. $f_y(x, y)$ if $f(x, y) = 3x^2 + 2xy - 7y^2$
7. $f_x(x, y)$ if $f(x, y) = -4x^2 + 5xy + 6y^2$
8. $f_x(x, y)$ if $f(x, y) = 6x^2 - 8xy + 3y^2$

9. $\dfrac{\partial z}{\partial x}$ if $z = x^3 + 4x^2y + 2y^3$

10. $\dfrac{\partial z}{\partial x}$ if $z = 3x^3 + 5xy^2 - y^3$

11. $\dfrac{\partial z}{\partial x}$ if $z = x^3 - y^3 - x + y$

12. $\dfrac{\partial z}{\partial y}$ if $z = 4x^2y - 5xy^2$

13. $\dfrac{\partial z}{\partial y}$ if $z = (5x + 2y)^{10}$

14. $\dfrac{\partial z}{\partial x}$ if $z = (2x - 3y)^8$

15. $\dfrac{\partial z}{\partial x}$ if $z = (x^2 - 4xy + y^2)^5$

16. $\dfrac{\partial z}{\partial y}$ if $z = (2x^2 - 3xy + 5y^2)^7$

In Problems 17–24, find the indicated value.

17. $f_x(1, 3)$ if $f(x, y) = 5x^3y - 4xy^2$

18. $f_x(4, 1)$ if $f(x, y) = x^2y^2 - 5xy^3$

19. $f_y(1, 0)$ if $f(x, y) = 3xe^y$

20. $f_y(2, 4)$ if $f(x, y) = x^4 \ln y$

21. $f_y(2, 1)$ if $f(x, y) = e^{x^2} - 4y$

22. $f_y(3, 3)$ if $f(x, y) = e^{3x} - y^2$

23. $f_x(1, -1)$ if $f(x, y) = \dfrac{2xy}{1 + x^2y^2}$

24. $f_x(-1, 2)$ if $f(x, y) = \dfrac{x^2 - y^2}{1 + x^2}$

B *In Problems 25–40, find the indicated second-order partial derivative for each function $f(x, y)$.*

25. $f_{xx}(x, y)$ if $f(x, y) = 6x - 5y + 3$

26. $f_{yx}(x, y)$ if $f(x, y) = -2x + y + 8$

27. $f_{xy}(x, y)$ if $f(x, y) = 4x^2 + 6y^2 - 10$

28. $f_{yy}(x, y)$ if $f(x, y) = x^2 + 9y^2 - 4$

29. $f_{yy}(x, y)$ if $f(x, y) = -3x^3 + 7x^2y + y^3$

30. $f_{xy}(x, y)$ if $f(x, y) = 5x^2y = 4xy^2$

31. $f_{yx}(x, y)$ if $f(x, y) = 2x^6y^4 + 3x^3y^7$

32. $f_{xx}(x, y)$ if $f(x, y) = -4x^3y^5 + 9x^6y^2$

33. $f_{xy}(x, y)$ if $f(x, y) = e^{xy^2}$

34. $f_{yx}(x, y)$ if $f(x, y) = e^{3x+2y}$

35. $f_{yy}(x, y)$ if $f(x, y) = \dfrac{\ln x}{y}$

36. $f_{xx}(x, y)$ if $f(x, y) = \dfrac{3 \ln x}{y^2}$

37. $f_{xx}(x, y)$ if $f(x, y) = (2x + y)^5$

38. $f_{yx}(x, y)$ if $f(x, y) = (3x - 8y)^6$

39. $f_{xy}(x, y)$ if $f(x, y) = (x^2 + y^4)^{10}$

40. $f_{yy}(x, y)$ if $f(x, y) = (1 + 2xy^2)^8$

In Problems 41–48, find the indicated function or value if
$C(x, y) = 3x^2 + 10xy - 8y^2 + 4x - 15y - 120.$

41. $C_x(x, y)$ **42.** $C_y(x, y)$

43. $C_x(3, -2)$ **44.** $C_y(3, -2)$

45. $C_{xx}(x, y)$ **46.** $C_{yy}(x, y)$

47. $C_{xy}(x, y)$ **48.** $C_{yx}(x, y)$

49. $C_{xx}(3, -2)$ **50.** $C_{yy}(3, -2)$

In Problems 51–62, find the indicated function or value if $S(x, y) = x^3 \ln y + 4y^2e^x$.

51. $S_y(x, y)$ **52.** $S_x(x, y)$

53. $S_y(-1, 1)$ **54.** $S_x(-1, 1)$

55. $S_{yx}(x, y)$ **56.** $S_{xy}(x, y)$

57. $S_{yy}(x, y)$ **58.** $S_{xx}(x, y)$

59. $S_{yx}(-1, 1)$ **60.** $S_{xy}(-1, 1)$

61. $S_{yy}(-1, 1)$ **62.** $S_{xx}(-1, 1)$

In Problems 63–72, find $f_x(x, y)$ and $f_y(x, y)$ for each function f.

63. $f(x, y) = (x^2 - y^3)^3$ **64.** $f(x, y) = \sqrt{2x - y^2}$

65. $f(x, y) = (3x^2y - 1)^4$ **66.** $f(x, y) = (3 + 2xy^2)^3$

67. $f(x, y) = \ln(x^2 + y^2)$ **68.** $f(x, y) = \ln(2x - 3y)$

69. $f(x, y) = y^2e^{xy^2}$ **70.** $f(x, y) = x^3e^{x^2y}$

71. $f(x, y) = \dfrac{x^2 - y^2}{x^2 + y^2}$ **72.** $f(x, y) = \dfrac{2x^2y}{x^2 + y^2}$

73. (A) Let $f(x, y) = y^3 + 4y^2 - 5y + 3$. Show that $\partial f/\partial x = 0$.

   (B) Explain why there are an infinite number of functions $g(x, y)$ such that $\partial g/\partial x = 0$.

74. (A) Find an example of a function $f(x, y)$ such that $\partial f/\partial x = 3$ and $\partial f/\partial y = 2$.

   (B) How many such functions are there? Explain.

In Problems 75–80, find $f_{xx}(x, y), f_{xy}(x, y), f_{yx}(x, y),$ and $f_{yy}(x, y)$ for each function f.

75. $f(x, y) = x^2y^2 + x^3 + y$

76. $f(x, y) = x^3y^3 + x + y^2$

77. $f(x, y) = \dfrac{x}{y} - \dfrac{y}{x}$ **78.** $f(x, y) = \dfrac{x^2}{y} - \dfrac{y^2}{x}$

79. $f(x, y) = xe^{xy}$ **80.** $f(x, y) = x \ln(xy)$

C

81. For
$$P(x, y) = -x^2 + 2xy - 2y^2 - 4x + 12y - 5$$
find all values of x and y such that
$$P_x(x, y) = 0 \quad \text{and} \quad P_y(x, y) = 0$$
simultaneously.

82. For
$$C(x, y) = 2x^2 + 2xy + 3y^2 - 16x - 18y + 54$$
find all values of x and y such that
$$C_x(x, y) = 0 \quad \text{and} \quad C_y(x, y) = 0$$
simultaneously.

83. For

$$F(x, y) = x^3 - 2x^2y^2 - 2x - 4y + 10$$

find all values of x and y such that

$$F_x(x, y) = 0 \quad \text{and} \quad F_y(x, y) = 0$$

simultaneously.

 84. For

$$G(x, y) = x^2 \ln y - 3x - 2y + 1$$

find all values of x and y such that

$$G_x(x, y) = 0 \quad \text{and} \quad G_y(x, y) = 0$$

simultaneously.

85. Let $f(x, y) = 3x^2 + y^2 - 4x - 6y + 2$.

(A) Find the minimum value of $f(x, y)$ when $y = 1$.

(B) Explain why the answer to part (A) is not the minimum value of the function $f(x, y)$.

86. Let $f(x, y) = 5 - 2x + 4y - 3x^2 - y^2$.

(A) Find the maximum value of $f(x, y)$ when $x = 2$.

(B) Explain why the answer to part (A) is not the maximum value of the function $f(x, y)$.

 87. Let $f(x, y) = 4 - x^4y + 3xy^2 + y^5$.

(A) Use graphical approximation methods to find c (to three decimal places) such that $f(c, 2)$ is the maximum value of $f(x, y)$ when $y = 2$.

(B) Find $f_x(c, 2)$ and $f_y(c, 2)$.

 88. Let $f(x, y) = e^x + 2e^y + 3xy^2 + 1$.

(A) Use graphical approximation methods to find d (to three decimal places) such that $f(1, d)$ is the minimum value of $f(x, y)$ when $x = 1$.

(B) Find $f_x(1, d)$ and $f_y(1, d)$.

In Problems 89 and 90, show that the function f satisfies
$$f_{xx}(x, y) + f_{yy}(x, y) = 0.$$

89. $f(x, y) = \ln(x^2 + y^2)$

90. $f(x, y) = x^3 - 3xy^2$

91. For $f(x, y) = x^2 + 2y^2$, find

(A) $\lim\limits_{h \to 0} \dfrac{f(x + h, y) - f(x, y)}{h}$

(B) $\lim\limits_{k \to 0} \dfrac{f(x, y + k) - f(x, y)}{k}$

92. For $f(x, y) = 2xy^2$, find

(A) $\lim\limits_{h \to 0} \dfrac{f(x + h, y) - f(x, y)}{h}$

(B) $\lim\limits_{k \to 0} \dfrac{f(x, y + k) - f(x, y)}{k}$

Applications

93. *Profit function.* A firm produces two types of calculators each week, x of type A and y of type B. The weekly revenue and cost functions (in dollars) are

$$R(x, y) = 80x + 90y + 0.04xy - 0.05x^2 - 0.05y^2$$

$$C(x, y) = 8x + 6y + 20{,}000$$

Find $P_x(1{,}200, 1{,}800)$ and $P_y(1{,}200, 1{,}800)$, and interpret the results.

94. *Advertising and sales.* A company spends $\$x$ per week on newspaper advertising and $\$y$ per week on television advertising. Its weekly sales were found to be given by

$$S(x, y) = 10x^{0.4}y^{0.8}$$

Find $S_x(3{,}000, 2{,}000)$ and $S_y(3{,}000, 2{,}000)$, and interpret the results.

95. *Demand equations.* A supermarket sells two brands of coffee: brand A at $\$p$ per pound and brand B at $\$q$ per pound. The daily demands x and y (in pounds) for brands A and B, respectively, are given by

$$x = 200 - 5p + 4q$$
$$y = 300 + 2p - 4q$$

Find $\partial x / \partial p$ and $\partial y / \partial p$, and interpret the results.

96. *Revenue and profit functions.* A company manufactures 10- and 3-speed bicycles. The weekly demand and cost functions are

$$p = 230 - 9x + y$$
$$q = 130 + x - 4y$$
$$C(x, y) = 200 + 80x + 30y$$

where $\$p$ is the price of a 10-speed bicycle, $\$q$ is the price of a 3-speed bicycle, x is the weekly demand for 10-speed bicycles, y is the weekly demand for 3-speed bicycles, and $C(x, y)$ is the cost function. Find $R_x(10, 5)$ and $P_x(10, 5)$, and interpret the results.

97. *Productivity.* The productivity of a certain third-world country is given approximately by the function

$$f(x, y) = 10x^{0.75}y^{0.25}$$

with the utilization of x units of labor and y units of capital.

(A) Find $f_x(x, y)$ and $f_y(x, y)$.

(B) If the country is now using 600 units of labor and 100 units of capital, find the marginal productivity of labor and the marginal productivity of capital.

(C) For the greatest increase in the country's productivity, should the government encourage increased use of labor or increased use of capital?

98. *Productivity.* The productivity of an automobile-manufacturing company is given approximately by the function

$$f(x, y) = 50\sqrt{xy} = 50x^{0.5}y^{0.5}$$

with the utilization of x units of labor and y units of capital.

(A) Find $f_x(x, y)$ and $f_y(x, y)$.

(B) If the company is now using 250 units of labor and 125 units of capital, find the marginal productivity of labor and the marginal productivity of capital.

(C) For the greatest increase in the company's productivity, should the management encourage increased use of labor or increased use of capital?

Problems 99–102 refer to the following: If a decrease in demand for one product results in an increase in demand for another product, the two products are said to be **competitive,** *or* **substitute, products.** *(Real whipping cream and imitation whipping cream are examples of competitive, or substitute, products.) If a decrease in demand for one product results in a decrease in demand for another product, the two products are said to be* **complementary products.** *(Fishing boats and outboard motors are examples of complementary products.) Partial derivatives can be used to test whether two products are competitive, complementary, or neither. We start with demand functions for two products such that the demand for either depends on the prices for both:*

$$x = f(p, q) \quad \text{Demand function for product } A$$
$$y = g(p, q) \quad \text{Demand function for product } B$$

The variables x and y represent the number of units demanded of products A and B, respectively, at a price p for 1 unit of product A and a price q for 1 unit of product B. Normally, if the price of A increases while the price of B is held constant, then the demand for A will decrease; that is, $f_p(p, q) < 0$. Then, if A and B are competitive products, the demand for B will increase; that is, $g_p(p, q) > 0$. Similarly, if the price of B increases while the price of A is held constant, the demand for B will decrease; that is, $g_q(p, q) < 0$. Then, if A and B are competitive products, the demand for A will increase; that is, $f_q(p, q) > 0$. Reasoning similarly for complementary products, we arrive at the following test:

Test for Competitive and Complementary Products

Partial Derivatives	Products A and B
$f_q(p, q) > 0$ and $g_p(p, q) > 0$	Competitive (substitute)
$f_q(p, q) < 0$ and $g_p(p, q) < 0$	Complementary
$f_q(p, q) \geq 0$ and $g_p(p, q) \leq 0$	Neither
$f_q(p, q) \leq 0$ and $g_p(p, q) \geq 0$	Neither

Use this test in Problems 99–102 to determine whether the indicated products are competitive, complementary, or neither.

99. *Product demand.* The weekly demand equations for the sale of butter and margarine in a supermarket are

$$x = f(p, q) = 8,000 - 0.09p^2 + 0.08q^2 \quad \text{Butter}$$
$$y = g(p, q) = 15,000 + 0.04p^2 - 0.3q^2 \quad \text{Margarine}$$

100. *Product demand.* The daily demand equations for the sale of brand A coffee and brand B coffee in a supermarket are

$$x = f(p, q) = 200 - 5p + 4q \quad \text{Brand A coffee}$$
$$y = g(p, q) = 300 + 2p - 4q \quad \text{Brand B coffee}$$

101. *Product demand.* The monthly demand equations for the sale of skis and ski boots in a sporting goods store are

$$x = f(p, q) = 800 - 0.004p^2 - 0.003q^2 \quad \text{Skis}$$
$$y = g(p, q) = 600 - 0.003p^2 - 0.002q^2 \quad \text{Ski boots}$$

102. *Product demand.* The monthly demand equations for the sale of tennis rackets and tennis balls in a sporting goods store are

$$x = f(p, q) = 500 - 0.5p - q^2 \quad \text{Tennis rackets}$$
$$y = g(p, q) = 10,000 - 8p - 100q^2 \quad \text{Tennis balls (cans)}$$

103. *Medicine.* The following empirical formula relates the surface area A (in square inches) of an average human body to its weight w (in pounds) and its height h (in inches):

$$A = f(w, h) = 15.64w^{0.425}h^{0.725}$$

Knowing the surface area of a human body is useful, for example, in studies pertaining to hypothermia (heat loss due to exposure).

(A) Find $f_w(w, h)$ and $f_h(w, h)$.

(B) For a 65-pound child who is 57 inches tall, find $f_w(65, 57)$ and $f_h(65, 57)$, and interpret the results.

104. *Blood flow.* Poiseuille's law states that the resistance R for blood flowing in a blood vessel varies directly as the length L of the vessel and inversely as the fourth power of its radius r. Stated as an equation,

$$R(L, r) = k\frac{L}{r^4} \quad k \text{ a constant}$$

Find $R_L(4, 0.2)$ and $R_r(4, 0.2)$, and interpret the results.

105. *Physical anthropology.* Anthropologists, in their study of race and human genetic groupings, often use the cephalic index C, which varies directly as the width W of the head and inversely as the length L of the head (both viewed from the top). In terms of an equation,

$$C(W, L) = 100\frac{W}{L}$$

where

$$W = \text{width in inches} \qquad L = \text{length in inches}$$

Find $C_W(6, 8)$ and $C_L(6, 8)$, and interpret the results.

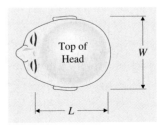

106. *Safety research.* Under ideal conditions, if a person driving a car slams on the brakes and skids to a stop, the length of the skid marks (in feet) is given by the formula

$$L(w, v) = kwv^2$$

where

k = constant
w = weight of car in pounds
v = speed of car in miles per hour

For $k = 0.000\ 013\ 3$, find $L_w(2,500, 60)$ and $L_v(2,500, 60)$, and interpret the results.

Section 8-3 MAXIMA AND MINIMA

We are now ready to undertake a brief, but useful, analysis of local maxima and minima for functions of the type $z = f(x, y)$. Basically, we are going to extend the second-derivative test developed for functions of a single independent variable. To start, we assume that all second-order partial derivatives exist for the function f in some circular region in the xy plane. This guarantees that the surface $z = f(x, y)$ has no sharp points, breaks, or ruptures. In other words, we are dealing only with smooth surfaces with no edges (like the edge of a box), breaks (like an earthquake fault), or sharp points (like the bottom point of a golf tee). (See Figure 1.)

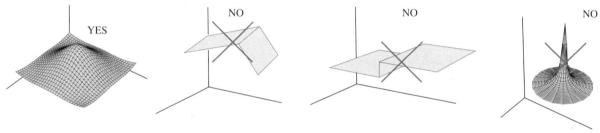

FIGURE 1

In addition, we will not concern ourselves with boundary points or absolute maxima–minima theory. Despite these restrictions, the procedure we are now going to describe will help us solve a large number of useful problems.

What does it mean for $f(a, b)$ to be a local maximum or a local minimum? We say that **$f(a, b)$ is a local maximum** if there exists a circular region in the domain of f with (a, b) as the center, such that

$$f(a, b) \geq f(x, y)$$

for all (x, y) in the region. Similarly, we say that **$f(a, b)$ is a local minimum** if there exists a circular region in the domain of f with (a, b) as the center, such that

$$f(a, b) \leq f(x, y)$$

for all (x, y) in the region. Figure 2A illustrates a local maximum, Figure 2B a local minimum, and Figure 2C a **saddle point,** which is neither a local maximum nor a local minimum.

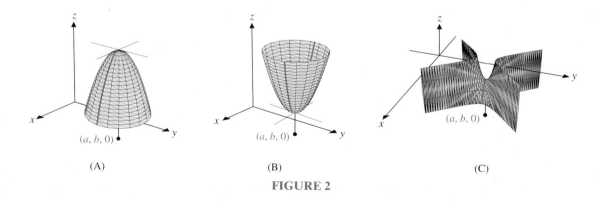

(A) (B) (C)

FIGURE 2

What happens to $f_x(a, b)$ and $f_y(a, b)$ if $f(a, b)$ is a local minimum or a local maximum and the partial derivatives of f exist in a circular region containing (a, b)? Figure 2 suggests that $f_x(a, b) = 0$ and $f_y(a, b) = 0$, since the tangent lines to the given curves are horizontal. Theorem 1 indicates that our intuitive reasoning is correct.

THEOREM 1   LOCAL EXTREMA AND PARTIAL DERIVATIVES

Let $f(a, b)$ be a local extremum (a local maximum or a local minimum) for the function f. If both f_x and f_y exist at (a, b), then

$$f_x(a, b) = 0 \qquad \text{and} \qquad f_y(a, b) = 0 \tag{1}$$

The converse of this theorem is false. That is, if $f_x(a, b) = 0$ and $f_y(a, b) = 0$, then $f(a, b)$ may or may not be a local extremum; for example, the point $(a, b, f(a, b))$ may be a saddle point (see Fig. 2C).

Theorem 1 gives us *necessary* (but not *sufficient*) conditions for $f(a, b)$ to be a local extremum. We thus find all points (a, b) such that $f_x(a, b) = 0$ and $f_y(a, b) = 0$ and test these further to determine whether $f(a, b)$ is a local extremum or a saddle point. Points (a, b) such that conditions (1) hold are called **critical points.**

Explore & Discuss 1

(A) Let $f(x, y) = y^2 + 1$. Explain why $f(x, y)$ has a local minimum at every point on the x axis. Verify that every point on the x axis is a critical point. Explain why the graph of $z = f(x, y)$ could be described as a trough.

(B) Let $g(x, y) = x^3$. Show that every point on the y axis is a critical point. Explain why no point on the y axis is a local extremum. Explain why the graph of $z = g(x, y)$ could be described as a slide.

The next theorem, using second-derivative tests, gives us *sufficient* conditions for a critical point to produce a local extremum or a saddle point. (As was the case with Theorem 1, we state this theorem without proof.)

THEOREM 2   SECOND-DERIVATIVE TEST FOR LOCAL EXTREMA

If

1. $z = f(x, y)$

2. $f_x(a, b) = 0$ and $f_y(a, b) = 0$ [(a, b) is a critical point]

3. All second-order partial derivatives of f exist in some circular region containing (a, b) as center.

4. $A = f_{xx}(a, b)$, $B = f_{xy}(a, b)$, $C = f_{yy}(a, b)$

Then

Case 1. If $AC - B^2 > 0$ and $A < 0$, then $f(a, b)$ is a local maximum.

Case 2. If $AC - B^2 > 0$ and $A > 0$, then $f(a, b)$ is a local minimum.

Case 3. If $AC - B^2 < 0$, then f has a saddle point at (a, b).

Case 4. If $AC - B^2 = 0$, the test fails.

INSIGHT

The condition $A = f_{xx}(a, b) < 0$ in case 1 of Theorem 2 is analogous to the condition $f''(c) < 0$ in the second-derivative test for a function of one variable (Section 5-5), which implies that the function is concave downward and therefore has a local maximum. Similarly, the condition $A = f_{xx}(a, b) > 0$ in case 2 is analogous to the condition $f''(c) > 0$ in the earlier second-derivative test, which implies that the function is concave upward and therefore has a local minimum.

To illustrate the use of Theorem 2, we will first find the local extremum for a very simple function whose solution is almost obvious: $z = f(x, y) = x^2 + y^2 + 2$. From

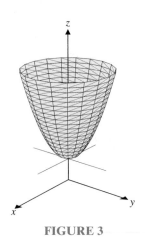

FIGURE 3

the function f itself and its graph (Fig. 3), it is clear that a local minimum is found at $(0, 0)$. Let us see how Theorem 2 confirms this observation.

Step 1. Find critical points: Find (x, y) such that $f_x(x, y) = 0$ and $f_y(x, y) = 0$ simultaneously:

$$f_x(x, y) = 2x = 0 \qquad f_y(x, y) = 2y = 0$$
$$x = 0 \qquad\qquad y = 0$$

The only critical point is $(a, b) = (0, 0)$.

Step 2. Compute $A = f_{xx}(0, 0)$, $B = f_{xy}(0, 0)$, and $C = f_{yy}(0, 0)$:

$$f_{xx}(x, y) = 2; \qquad \text{thus,} \qquad A = f_{xx}(0, 0) = 2$$
$$f_{xy}(x, y) = 0; \qquad \text{thus,} \qquad B = f_{xy}(0, 0) = 0$$
$$f_{yy}(x, y) = 2; \qquad \text{thus,} \qquad C = f_{yy}(0, 0) = 2$$

Step 3. Evaluate $AC - B^2$ and try to classify the critical point $(0, 0)$ by using Theorem 2:

$$AC - B^2 = (2)(2) - (0)^2 = 4 > 0 \qquad \text{and} \qquad A = 2 > 0$$

Therefore, case 2 in Theorem 2 holds. That is, $f(0, 0) = 2$ is a local minimum.

We will now use Theorem 2 in the examples that follow to analyze extrema without the aid of graphs.

EXAMPLE 1 **Finding Local Extrema** Use Theorem 2 to find local extrema of

$$f(x, y) = -x^2 - y^2 + 6x + 8y - 21$$

SOLUTION **Step 1.** Find critical points: Find (x, y) such that $f_x(x, y) = 0$ and $f_y(x, y) = 0$ simultaneously:

$$f_x(x, y) = -2x + 6 = 0 \qquad f_y(x, y) = -2y + 8 = 0$$
$$x = 3 \qquad\qquad y = 4$$

The only critical point is $(a, b) = (3, 4)$.

Step 2. Compute $A = f_{xx}(3, 4)$, $B = f_{xy}(3, 4)$, and $C = f_{yy}(3, 4)$:

$$f_{xx}(x, y) = -2; \qquad \text{thus,} \qquad A = f_{xx}(3, 4) = -2$$
$$f_{xy}(x, y) = 0; \qquad \text{thus,} \qquad B = f_{xy}(3, 4) = 0$$
$$f_{yy}(x, y) = -2; \qquad \text{thus,} \qquad C = f_{yy}(3, 4) = -2$$

Step 3. Evaluate $AC - B^2$ and try to classify the critical point $(3, 4)$ by using Theorem 2:

$$AC - B^2 = (-2)(-2) - (0)^2 = 4 > 0 \qquad \text{and} \qquad A = -2 < 0$$

Therefore, case 1 in Theorem 2 holds. That is, $f(3, 4) = 4$ is a local maximum. ∎

MATCHED PROBLEM 1 Use Theorem 2 to find local extrema of

$$f(x, y) = x^2 + y^2 - 10x - 2y + 36$$

EXAMPLE 2 **Finding Local Extrema: Multiple Critical Points** Use Theorem 2 to find local extrema of

$$f(x, y) = x^3 + y^3 - 6xy$$

SOLUTION   **Step 1.**   Find critical points of $f(x, y) = x^3 + y^3 - 6xy$:

$$f_x(x, y) = 3x^2 - 6y = 0 \qquad \textit{Solve for y.}$$
$$6y = 3x^2$$
$$y = \tfrac{1}{2}x^2 \qquad\qquad\qquad (2)$$

$$f_y(x, y) = 3y^2 - 6x = 0$$
$$3y^2 = 6x \qquad \textit{Use equation (2) to eliminate y.}$$
$$3(\tfrac{1}{2}x^2)^2 = 6x$$
$$\tfrac{3}{4}x^4 = 6x \qquad \textit{Solve for x.}$$
$$3x^4 - 24x = 0$$
$$3x(x^3 - 8) = 0$$

$$x = 0 \quad \text{or} \quad x = 2$$
$$y = 0 \qquad\qquad y = \tfrac{1}{2}(2)^2 = 2$$

The critical points are $(0, 0)$ and $(2, 2)$.

Since there are two critical points, steps 2 and 3 must be performed twice.

Test (0, 0)   **Step 2.**   Compute $A = f_{xx}(0, 0)$, $B = f_{xy}(0, 0)$, and $C = f_{yy}(0, 0)$:

$$f_{xx}(x, y) = 6x; \qquad \text{thus,} \qquad A = f_{xx}(0, 0) = 0$$
$$f_{xy}(x, y) = -6; \qquad \text{thus,} \qquad B = f_{xy}(0, 0) = -6$$
$$f_{yy}(x, y) = 6y; \qquad \text{thus,} \qquad C = f_{yy}(0, 0) = 0$$

Step 3.   Evaluate $AC - B^2$ and try to classify the critical point $(0, 0)$ by using Theorem 2:

$$AC - B^2 = (0)(0) - (-6)^2 = -36 < 0$$

Therefore, case 3 in Theorem 2 applies. That is, f has a saddle point at $(0, 0)$.

Now we will consider the second critical point, $(2, 2)$:

Test (2, 2)   **Step 2.**   Compute $A = f_{xx}(2, 2)$, $B = f_{xy}(2, 2)$, and $C = f_{yy}(2, 2)$:

$$f_{xx}(x, y) = 6x; \qquad \text{thus,} \qquad A = f_{xx}(2, 2) = 12$$
$$f_{xy}(x, y) = -6; \qquad \text{thus,} \qquad B = f_{xy}(2, 2) = -6$$
$$f_{yy}(x, y) = 6y; \qquad \text{thus,} \qquad C = f_{yy}(2, 2) = 12$$

Step 3.   Evaluate $AC - B^2$ and try to classify the critical point $(2, 2)$ by using Theorem 2:

$$AC - B^2 = (12)(12) - (-6)^2 = 108 > 0 \qquad \text{and} \qquad A = 12 > 0$$

Thus, case 2 in Theorem 2 applies, and $f(2, 2) = -8$ is a local minimum. ▬

Our conclusions in Example 2 may be confirmed geometrically by graphing cross sections of the function f. The cross sections of f in the planes $y = 0$, $x = 0$, $y = x$, and $y = -x$ [each of these planes contains $(0, 0)$] are represented by the graphs of the functions $f(x, 0) = x^3$, $f(0, y) = y^3$, $f(x, x) = 2x^3 - 6x^2$, and $f(x, -x) = 6x^2$, respectively, as shown in Figure 4A (note that the first two functions have the same graph). The cross sections of f in the planes $y = 2$, $x = 2$, $y = x$, and $y = 4 - x$ [each of these planes contains $(2, 2)$] are represented by the graphs of the functions $f(x, 2) = x^3 - 12x + 8$, $f(2, y) = y^3 - 12y + 8$, $f(x, x) = 2x^3 - 6x^2$, and $f(x, 4 - x) = x^3 + (4 - x)^3 + 6x^2 - 24x$, respectively, as shown in Figure 4B (the first two functions have the same graph). Figure 4B illustrates the fact that since f has a local minimum at $(2, 2)$, each of the cross sections of f through $(2, 2)$ has a local minimum of -8 at $(2, 2)$. Figure 4A, by contrast, indicates that some cross sections of f through $(0, 0)$ have a local minimum, some a local maximum, and some neither one, at $(0, 0)$.

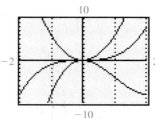
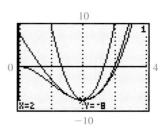

(A) $y_1 = x^3$
$y_2 = 2x^3 - 6x^2$
$y_3 = 6x^2$

(B) $y_1 = x^3 - 12x + 8$
$y_2 = 2x^3 - 6x^2$
$y_3 = x^3 + (4 - x)^3 + 6x^2 - 24x$

FIGURE 4

MATCHED PROBLEM 2   Use Theorem 2 to find local extrema for $f(x, y) = x^3 + y^2 - 6xy$.

Explore & Discuss **2**   Let $f(x, y) = x^4 + y^2 + 3$, $g(x, y) = 10 - x^2 - y^4$, and $h(x, y) = x^3 + y^2$.

    (A) Show that each function has only $(0, 0)$ as a critical point.

    (B) Explain why $f(x, y)$ has a minimum at $(0, 0)$, $g(x, y)$ has a maximum at $(0, 0)$, and $h(x, y)$ has neither a minimum nor a maximum at $(0, 0)$.

    (C) Are the results of part (B) consequences of Theorem 2? Explain.

EXAMPLE 3   **Profit** Suppose that the surfboard company discussed earlier has developed the yearly profit equation

$$P(x, y) = -22x^2 + 22xy - 11y^2 + 110x - 44y - 23$$

where x is the number (in thousands) of standard surfboards produced per year, y is the number (in thousands) of competition surfboards produced per year, and P is profit (in thousands of dollars). How many of each type of board should be produced per year to realize a maximum profit? What is the maximum profit?

SOLUTION   **Step 1.** Find critical points:

$$P_x(x, y) = -44x + 22y + 110 = 0$$
$$P_y(x, y) = 22x - 22y - 44 = 0$$

Solving this system, we obtain $(3, 1)$ as the only critical point.

Step 2. Compute $A = P_{xx}(3, 1)$, $B = P_{xy}(3, 1)$, and $C = P_{yy}(3, 1)$:

$$P_{xx}(x, y) = -44; \quad \text{thus,} \quad A = P_{xx}(3, 1) = -44$$
$$P_{xy}(x, y) = 22; \quad \text{thus,} \quad B = P_{xy}(3, 1) = 22$$
$$P_{yy}(x, y) = -22; \quad \text{thus,} \quad C = P_{yy}(3, 1) = -22$$

Step 3. Evaluate $AC - B^2$ and try to classify the critical point $(3, 1)$ by using Theorem 2:

$$AC - B^2 = (-44)(-22) - 22^2 = 484 > 0 \quad \text{and} \quad A = -44 < 0$$

Therefore, case 1 in Theorem 2 applies. That is, $P(3, 1) = 120$ is a local maximum. A maximum profit of $120,000 is obtained by producing and selling 3,000 standard boards and 1,000 competition boards per year.

MATCHED PROBLEM 3   Repeat Example 3 with

$$P(x, y) = -66x^2 + 132xy - 99y^2 + 132x - 66y - 19$$

EXAMPLE 4

Package Design The packaging department in a company has been asked to design a rectangular box with no top and a partition down the middle. The box must have a volume of 48 cubic inches. Find the dimensions that will minimize the amount of material used to construct the box.

SOLUTION Refer to Figure 5. The amount of material used in constructing this box is

$$\begin{array}{ccc} & \text{Front,} & \text{Sides,} \\ \text{Base} & \text{back} & \text{partition} \end{array}$$

$$M = xy + 2xz + 3yz \tag{3}$$

The volume of the box is

$$V = xyz = 48 \tag{4}$$

Since Theorem 2 applies only to functions with two independent variables, we must use equation (4) to eliminate one of the variables in equation (3):

$$M = xy + 2xz + 3yz \qquad \text{Substitute } z = 48/xy.$$
$$= xy + 2x\left(\frac{48}{xy}\right) + 3y\left(\frac{48}{xy}\right)$$
$$= xy + \frac{96}{y} + \frac{144}{x}$$

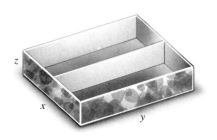

FIGURE 5

Thus, we must find the minimum value of

$$M(x, y) = xy + \frac{96}{y} + \frac{144}{x} \qquad x > 0 \qquad \text{and} \qquad y > 0$$

Step 1. Find critical points:

$$M_x(x, y) = y - \frac{144}{x^2} = 0$$
$$y = \frac{144}{x^2} \tag{5}$$
$$M_y(x, y) = x - \frac{96}{y^2} = 0$$
$$x = \frac{96}{y^2} \qquad \text{Solve for } y^2.$$
$$y^2 = \frac{96}{x} \qquad \text{Use equation (5) to eliminate } y \text{ and solve for } x.$$
$$\left(\frac{144}{x^2}\right)^2 = \frac{96}{x}$$
$$\frac{20{,}736}{x^4} = \frac{96}{x} \qquad \text{Multiply both sides by } x^4/96 \text{ (recall that } x > 0\text{).}$$
$$x^3 = \frac{20{,}736}{96} = 216$$
$$x = 6 \qquad \text{Use equation (5) to find } y.$$
$$y = \frac{144}{36} = 4$$

Thus, $(6, 4)$ is the only critical point.

Step 2. Compute $A = M_{xx}(6, 4)$, $B = M_{xy}(6, 4)$, and $C = M_{yy}(6, 4)$:

$$M_{xx}(x, y) = \frac{288}{x^3}; \qquad \text{thus,} \qquad A = M_{xx}(6, 4) = \frac{288}{216} = \frac{4}{3}$$
$$M_{xy}(x, y) = 1; \qquad \text{thus,} \qquad B = M_{xy}(6, 4) = 1$$
$$M_{yy}(x, y) = \frac{192}{y^3}; \qquad \text{thus,} \qquad C = M_{yy}(6, 4) = \frac{192}{64} = 3$$

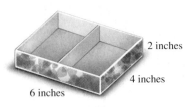

2 inches

4 inches

6 inches

FIGURE 6

Step 3. Evaluate $AC - B^2$ and try to classify the critical point $(6, 4)$ by using Theorem 2:

$$AC - B^2 = \left(\tfrac{4}{3}\right)(3) - (1)^2 = 3 > 0 \qquad \text{and} \qquad A = \tfrac{4}{3} > 0$$

Therefore, case 2 in Theorem 2 applies, and $M(x, y)$ has a local minimum at $(6, 4)$. If $x = 6$ and $y = 4$, then

$$z = \frac{48}{xy} = \frac{48}{(6)(4)} = 2$$

Thus, the dimensions that will require the minimum amount of material are 6 inches by 4 inches by 2 inches (see Fig. 6). ▬

MATCHED PROBLEM 4   If the box in Example 4 must have a volume of 384 cubic inches, find the dimensions that will require the least amount of material.

Answers to Matched Problems   **1.** $f(5, 1) = 10$ is a local minimum

2. f has a saddle point at $(0, 0)$; $f(6, 18) = -108$ is a local minimum

3. Local maximum for $x = 2$ and $y = 1$; $P(2, 1) = 80$; a maximum profit of \$80,000 is obtained by producing and selling 2,000 standard boards and 1,000 competition boards

4. 12 in. by 8 in. by 4 in.

Exercise 8-3

A In Problems 1–4, find $f_x(x, y)$ and $f_y(x, y)$, and explain, using Theorem 1, why $f(x, y)$ has no local extrema.

1. $f(x, y) = 4x + 5y - 6$

2. $f(x, y) = 10 - 2x - 3y + x^2$

3. $f(x, y) = 3.7 - 1.2x + 6.8y + 0.2y^3 + x^4$

4. $f(x, y) = x^3 - y^2 + 7x + 3y + 1$

Use Theorem 2 to find local extrema in Problems 5–24.

5. $f(x, y) = 6 - x^2 - 4x - y^2$

6. $f(x, y) = 3 - x^2 - y^2 + 6y$

7. $f(x, y) = x^2 + y^2 + 2x - 6y + 14$

8. $f(x, y) = x^2 + y^2 - 4x + 6y + 23$

B

9. $f(x, y) = xy + 2x - 3y - 2$

10. $f(x, y) = x^2 - y^2 + 2x + 6y - 4$

11. $f(x, y) = -3x^2 + 2xy - 2y^2 + 14x + 2y + 10$

12. $f(x, y) = -x^2 + xy - 2y^2 + x + 10y - 5$

13. $f(x, y) = 2x^2 - 2xy + 3y^2 - 4x - 8y + 20$

14. $f(x, y) = 2x^2 - xy + y^2 - x - 5y + 8$

C

15. $f(x, y) = e^{xy}$

16. $f(x, y) = x^2y - xy^2$

17. $f(x, y) = x^3 + y^3 - 3xy$

18. $f(x, y) = 2y^3 - 6xy - x^2$

19. $f(x, y) = 2x^4 + y^2 - 12xy$

20. $f(x, y) = 16xy - x^4 - 2y^2$

21. $f(x, y) = x^3 - 3xy^2 + 6y^2$

22. $f(x, y) = 2x^2 - 2x^2y + 6y^3$

23. $f(x, y) = y^3 + 2x^2y^2 - 3x - 2y + 8$

24. $f(x, y) = x \ln y + x^2 - 4x - 5y + 3$

25. Explain why $f(x, y) = x^2$ has an infinite number of local extrema.

26. (A) Find the local extrema of the functions $f(x, y) = x + y$, $g(x, y) = x^2 + y^2$, and $h(x, y) = x^3 + y^3$.

(B) Discuss the local extrema of the function $k(x, y) = x^n + y^n$, where n is a positive integer.

27. (A) Show that $(0, 0)$ is a critical point of the function $f(x, y) = x^4e^y + x^2y^4 + 1$, but that the second-derivative test for local extrema fails.

(B) Use cross sections, as in Example 2, to decide whether f has a local maximum, a local minimum, or a saddle point at $(0, 0)$.

28. (A) Show that $(0, 0)$ is a critical point of the function $g(x, y) = e^{xy^2} + x^2y^3 + 2$, but that the second-derivative test for local extrema fails.

(B) Use cross sections, as in Example 2, to decide whether g has a local maximum, a local minimum, or a saddle point at $(0, 0)$.

Applications

29. *Product mix for maximum profit.* A firm produces two types of calculators per year: x thousand of type A and y thousand of type B. If the revenue and cost equations for the year are (in millions of dollars)

$$R(x, y) = 2x + 3y$$
$$C(x, y) = x^2 - 2xy + 2y^2 + 6x - 9y + 5$$

determine how many of each type of calculator should be produced per year to maximize profit. What is the maximum profit?

30. *Automation–labor mix for minimum cost.* The annual labor and automated equipment cost (in millions of dollars) for a company's production of television sets is given by

$$C(x, y) = 2x^2 + 2xy + 3y^2 - 16x - 18y + 54$$

where x is the amount spent per year on labor and y is the amount spent per year on automated equipment (both in millions of dollars). Determine how much should be spent on each per year to minimize this cost. What is the minimum cost?

31. *Maximizing profit.* A department store sells two brands of inexpensive calculators. The store pays \$6 for each brand A calculator and \$8 for each brand B calculator. The research department has estimated the weekly demand equations for these two competitive products to be

$x = 116 - 30p + 20q$ *Demand equation for brand A*

$y = 144 + 16p - 24q$ *Demand equation for brand B*

where p is the selling price for brand A and q is the selling price for brand B.

(A) Determine the demands x and y when $p = \$10$ and $q = \$12$; when $p = \$11$ and $q = \$11$.

(B) How should the store price each calculator to maximize weekly profits? What is the maximum weekly profit? [*Hint:* $C = 6x + 8y, R = px + qy$, and $P = R - C$.]

32. *Maximizing profit.* A store sells two brands of color print film. The store pays \$2 for each roll of brand A film and \$3 for each roll of brand B film. A consulting firm has estimated the daily demand equations for these two competitive products to be

$x = 75 - 40p + 25q$ *Demand equation for brand A*

$y = 80 + 20p - 30q$ *Demand equation for brand B*

where p is the selling price for brand A and q is the selling price for brand B.

(A) Determine the demands x and y when $p = \$4$ and $q = \$5$; when $p = \$4$ and $q = \$4$.

(B) How should the store price each brand of film to maximize daily profits? What is the maximum daily profit? [*Hint:* $C = 2x + 3y, R = px + qy$, and $P = R - C$.]

33. *Minimizing cost.* A satellite television reception station is to be located at $P(x, y)$ so that the sum of the squares of the distances from P to the three towns A, B, and C is a minimum (see the figure). Find the coordinates of P, the location that will minimize the cost of providing satellite cable television for all three towns.

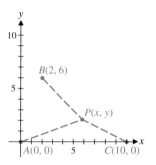

Figure for 33

34. *Minimizing cost.* Repeat Problem 33, replacing the coordinates of B with $B(6, 9)$ and the coordinates of C with $C(9, 0)$.

35. *Minimum material.* A rectangular box with no top and two parallel partitions (see the figure) is to be made to hold a volume of 64 cubic inches. Find the dimensions that will require the least amount of material.

Figure for 35

36. *Minimum material.* A rectangular box with no top and two intersecting partitions (see the figure) is to be made to hold a volume of 72 cubic inches. What should the dimensions of the box be in order to use the least amount of material in its construction?

Figure for 36

37. *Maximum volume.* A mailing service states that a rectangular package shall have the sum of its length and girth not to exceed 120 inches (see the figure). What are the dimensions of the largest (in volume) mailing carton that can be constructed to meet these restrictions?

Figure for 37

38. *Maximum shipping volume.* A shipping box is to be reinforced with steel bands in all three directions, as indicated in the figure. A total of 150 inches of steel tape is to be used, with 6 inches of waste because of a 2-inch overlap in each direction. Find the dimensions of the box with maximum volume that can be taped as indicated.

Figure for 38

Section 8-4 MAXIMA AND MINIMA USING LAGRANGE MULTIPLIERS

- Functions of Two Independent Variables
- Functions of Three Independent Variables

■ Functions of Two Independent Variables

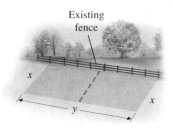

FIGURE 1

We now consider a particularly powerful method of solving a certain class of maxima–minima problems. The method is due to Joseph Louis Lagrange (1736–1813), an eminent eighteenth-century French mathematician, and it is called the **method of Lagrange multipliers.** We introduce the method through an example; then we formalize the discussion in the form of a theorem.

A rancher wants to construct two feeding pens of the same size along an existing fence (see Fig. 1). If the rancher has 720 feet of fencing materials available, how long should x and y be in order to obtain the maximum total area? What is the maximum area?

The total area is given by

$$f(x, y) = xy$$

which can be made as large as we like, provided that there are no restrictions on x and y. But there are restrictions on x and y, since we have only 720 feet of fencing. That is, x and y must be chosen so that

$$3x + y = 720$$

This restriction on x and y, called a **constraint,** leads to the following maxima–minima problem:

$$\text{Maximize} \quad f(x, y) = xy \tag{1}$$

$$\text{subject to} \quad 3x + y = 720, \quad \text{or} \quad 3x + y - 720 = 0 \tag{2}$$

This problem is a special case of a general class of problems of the form

$$\text{Maximize (or minimize)} \quad z = f(x, y) \tag{3}$$

$$\text{subject to} \quad g(x, y) = 0 \tag{4}$$

Of course, we could try to solve equation (4) for y in terms of x, or for x in terms of y, then substitute the result into equation (3), and use methods developed in Section 5-5 for functions of a single variable. But what if equation (4) is more complicated than equation (2), and solving for one variable in terms of the other is either very difficult or impossible? In the method of Lagrange multipliers, we will work with $g(x, y)$ directly and avoid having to solve equation (4) for one variable in terms of the other. In addition, the method generalizes to functions of arbitrarily many variables subject to one or more constraints.

Now to the method: We form a new function F, using functions f and g in equations (3) and (4), as follows:

$$F(x, y, \lambda) = f(x, y) + \lambda g(x, y) \tag{5}$$

Here, λ (the Greek lowercase letter lambda) is called a **Lagrange multiplier.** Theorem 1 gives the basis for the method.

THEOREM 1 **METHOD OF LAGRANGE MULTIPLIERS FOR FUNCTIONS OF TWO VARIABLES**

Any local maxima or minima of the function $z = f(x, y)$ subject to the constraint $g(x, y) = 0$ will be among those points (x_0, y_0) for which (x_0, y_0, λ_0) is a solution of the system

$$F_x(x, y, \lambda) = 0$$
$$F_y(x, y, \lambda) = 0$$
$$F_\lambda(x, y, \lambda) = 0$$

where $F(x, y, \lambda) = f(x, y) + \lambda g(x, y)$, provided that all the partial derivatives exist.

We now use the method of Lagrange multipliers to solve the fence problem.

Step 1. Formulate the problem in the form of equations (3) and (4):

$$\text{Maximize} \quad f(x, y) = xy$$
$$\text{subject to} \quad g(x, y) = 3x + y - 720 = 0$$

Step 2. Form the function F, introducing the Lagrange multiplier λ:

$$F(x, y, \lambda) = f(x, y) + \lambda g(x, y)$$
$$= xy + \lambda(3x + y - 720)$$

Step 3. Solve the system $F_x = 0, F_y = 0, F_\lambda = 0$ (the solutions are called **critical points** of F):

$$F_x = y + 3\lambda = 0$$
$$F_y = x + \lambda = 0$$
$$F_\lambda = 3x + y - 720 = 0$$

From the first two equations, we see that

$$y = -3\lambda$$
$$x = -\lambda$$

Substitute these values for x and y into the third equation and solve for λ:

$$-3\lambda - 3\lambda = 720$$
$$-6\lambda = 720$$
$$\lambda = -120$$

Thus,

$$y = -3(-120) = 360 \text{ feet}$$
$$x = -(-120) = 120 \text{ feet}$$

and $(x_0, y_0, \lambda_0) = (120, 360, -120)$ is the only critical point of F.

Step 4. According to Theorem 1, if the function $f(x, y)$, subject to the constraint $g(x, y) = 0$, has a local maximum or minimum, that maximum or minimum must occur at $x = 120$, $y = 360$. Although it is possible to develop a test similar to Theorem 2 in Section 8-3 to determine the nature of this local extremum, we will not do so. [Note that Theorem 2 cannot be applied to $f(x, y)$ at $(120, 360)$, since this point is not a critical point of the unconstrained function $f(x, y)$.] We simply assume that the maximum value of $f(x, y)$ must occur for $x = 120$, $y = 360$. Thus,

$$\text{Max } f(x, y) = f(120, 360)$$
$$= (120)(360) = 43{,}200 \text{ square feet}$$

The key steps in applying the method of Lagrange multipliers are as follows:

PROCEDURE

Method of Lagrange Multipliers: Key Steps

Step 1. Write the problem in the form

$$\text{Maximize (or minimize)} \quad z = f(x, y)$$
$$\text{subject to} \quad g(x, y) = 0$$

Step 2. Form the function F:

$$F(x, y, \lambda) = f(x, y) + \lambda g(x, y)$$

Step 3. Find the critical points of F; that is, solve the system

$$F_x(x, y, \lambda) = 0$$
$$F_y(x, y, \lambda) = 0$$
$$F_\lambda(x, y, \lambda) = 0$$

Step 4. If (x_0, y_0, λ_0) is the only critical point of F, we assume that (x_0, y_0) will always produce the solution to the problems we consider. If F has more than one critical point, we evaluate $z = f(x, y)$ at (x_0, y_0) for each critical point (x_0, y_0, λ_0) of F. For the problems we consider, we assume that the largest of these values is the maximum value of $f(x, y)$, subject to the constraint $g(x, y) = 0$, and the smallest is the minimum value of $f(x, y)$, subject to the constraint $g(x, y) = 0$.

EXAMPLE 1 **Minimization Subject to a Constraint** Minimize $f(x, y) = x^2 + y^2$ subject to $x + y = 10$.

SOLUTION **Step 1.**
$$\text{Minimize} \quad f(x, y) = x^2 + y^2$$
$$\text{subject to} \quad g(x, y) = x + y - 10 = 0$$

Step 2. $F(x, y, \lambda) = x^2 + y^2 + \lambda(x + y - 10)$

Step 3.
$$F_x = 2x + \lambda = 0$$
$$F_y = 2y + \lambda = 0$$
$$F_\lambda = x + y - 10 = 0$$

From the first two equations, $x = -\lambda/2$ and $y = -\lambda/2$. Substituting these values into the third equation, we obtain

$$-\frac{\lambda}{2} - \frac{\lambda}{2} = 10$$
$$-\lambda = 10$$
$$\lambda = -10$$

The only critical point is $(x_0, y_0, \lambda_0) = (5, 5, -10)$.

Step 4. Since $(5, 5, -10)$ is the only critical point of F, we conclude that (see step 4 in the box)

$$\text{Min } f(x, y) = f(5, 5) = (5)^2 + (5)^2 = 50$$

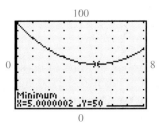

Since $g(x, y)$ in Example 1 has a relatively simple form, an alternative to the method of Lagrange multipliers is to solve $g(x, y) = 0$ for y and then substitute into $f(x, y)$ to obtain the function $h(x) = f(x, 10 - x) = x^2 + (10 - x)^2$ in the single variable x. Then we minimize h (see Fig. 2). From Figure 2, we conclude that Min $f(x, y) = f(5, 5) = 50$. This technique depends on being able to solve the constraint for one of the two variables and thus is not always available as an alternative to the method of Lagrange multipliers.

FIGURE 2 $h(x) = x^2 + (10 - x)^2$

| MATCHED PROBLEM 1 | Maximize $f(x, y) = 25 - x^2 - y^2$ subject to $x + y = 4$.

Figures 3 and 4 illustrate the results obtained in Example 1 and Matched Problem 1, respectively.

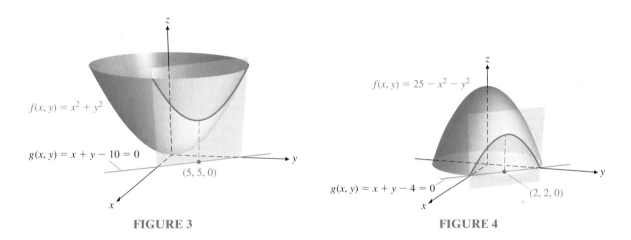

FIGURE 3   **FIGURE 4**

Explore & Discuss **1**   Consider the problem of minimizing $f(x, y) = 3x^2 + 5y^2$ subject to the constraint $g(x, y) = 2x + 3y - 6 = 0$.

(A) Compute the value of $f(x, y)$ when x and y are integers, $0 \le x \le 3$, $0 \le y \le 2$. Record your answers in the empty boxes next to the points (x, y) in Figure 5.

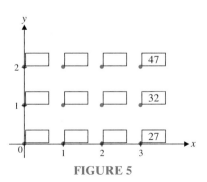

FIGURE 5

(B) Graph the constraint $g(x, y) = 0$.

(C) Estimate the minimum value of f on the basis of your graph and the computations from part (A).

(D) Use the method of Lagrange multipliers to solve the minimization problem.

EXAMPLE 2 **Productivity** The Cobb–Douglas production function for a new product is given by

$$N(x, y) = 16x^{0.25}y^{0.75}$$

where x is the number of units of labor and y is the number of units of capital required to produce $N(x, y)$ units of the product. Each unit of labor costs \$50 and each unit of capital costs \$100. If \$500,000 has been budgeted for the production of this product, how should that amount be allocated between labor and capital in order to maximize production? What is the maximum number of units that can be produced?

SOLUTION The total cost of using x units of labor and y units of capital is $50x + 100y$. Thus, the constraint imposed by the \$500,000 budget is

$$50x + 100y = 500,000$$

Step 1. Maximize $N(x, y) = 16x^{0.25}y^{0.75}$

subject to $g(x, y) = 50x + 100y - 500,000 = 0$

Step 2. $F(x, y, \lambda) = 16x^{0.25}y^{0.75} + \lambda(50x + 100y - 500,000)$

Step 3. $F_x = 4x^{-0.75}y^{0.75} + 50\lambda = 0$

$F_y = 12x^{0.25}y^{-0.25} + 100\lambda = 0$

$F_\lambda = 50x + 100y - 500,000 = 0$

From the first two equations,

$$\lambda = -\tfrac{2}{25}x^{-0.75}y^{0.75} \quad \text{and} \quad \lambda = -\tfrac{3}{25}x^{0.25}y^{-0.25}$$

Thus,

$$-\tfrac{2}{25}x^{-0.75}y^{0.75} = -\tfrac{3}{25}x^{0.25}y^{-0.25} \qquad \text{Multiply both sides by } x^{0.75}\,y^{0.25}.$$

$$-\tfrac{2}{25}y = -\tfrac{3}{25}x \qquad \text{(We can assume that } x \neq 0 \text{ and } y \neq 0.)$$

$$y = \tfrac{3}{2}x$$

Now substitute for y in the third equation and solve for x:

$$50x + 100\left(\tfrac{3}{2}x\right) - 500,000 = 0$$

$$200x = 500,000$$

$$x = 2,500$$

Thus,

$$y = \tfrac{3}{2}(2,500) = 3,750$$

and

$$\lambda = -\tfrac{2}{25}(2,500)^{-0.75}(3,750)^{0.75} \approx -0.1084$$

The only critical point of F is $(2,500, 3,750, -0.1084)$.

Step 4. Since F has only one critical point, we conclude that maximum productivity occurs when 2,500 units of labor and 3,750 units of capital are used (see step 4 in the method of Lagrange multipliers). Thus,

$$\text{Max } N(x, y) = N(2,500, 3,750)$$

$$= 16(2,500)^{0.25}(3,750)^{0.75}$$

$$\approx 54,216 \text{ units} \qquad \blacksquare$$

The negative of the value of the Lagrange multiplier found in step 3 is called the **marginal productivity of money** and gives the approximate increase in production for each additional dollar spent on production. In Example 2, increasing the production budget from \$500,000 to \$600,000 would result in an approximate increase in production of

$$0.1084(100,000) = 10,840 \text{ units}$$

Note that simplifying the constraint equation

$$50x + 100y - 500,000 = 0$$

to

$$x + 2y - 10,000 = 0$$

before forming the function $F(x, y, \lambda)$ would make it difficult to interpret $-\lambda$ correctly. Thus, **in marginal productivity problems, the constraint equation should not be simplified.**

| MATCHED PROBLEM 2 | The Cobb–Douglas production function for a new product is given by

$$N(x, y) = 20x^{0.5}y^{0.5}$$

where x is the number of units of labor and y is the number of units of capital required to produce $N(x, y)$ units of the product. Each unit of labor costs $40 and each unit of capital costs $120.

(A) If $300,000 has been budgeted for the production of this product, how should that amount be allocated in order to maximize production? What is the maximum production?

(B) Find the marginal productivity of money in this case, and estimate the increase in production if an additional $40,000 is budgeted for production.

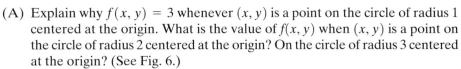

Explore & Discuss 2

Consider the problem of maximizing $f(x, y) = 4 - x^2 - y^2$ subject to the constraint $g(x, y) = y - x^2 + 1 = 0$.

(A) Explain why $f(x, y) = 3$ whenever (x, y) is a point on the circle of radius 1 centered at the origin. What is the value of $f(x, y)$ when (x, y) is a point on the circle of radius 2 centered at the origin? On the circle of radius 3 centered at the origin? (See Fig. 6.)

(B) Explain why some points on the parabola $y - x^2 + 1 = 0$ lie inside the circle $x^2 + y^2 = 1$.

(C) In light of part (B), would you guess that the maximum value of $f(x, y)$ subject to the constraint is greater than 3? Explain.

(D) Use Lagrange multipliers to solve the maximization problem.

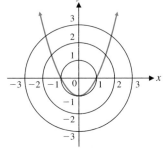

FIGURE 6

■ Functions of Three Independent Variables

We have indicated that the method of Lagrange multipliers can be extended to functions with arbitrarily many independent variables with one or more constraints. We now state a theorem for functions with three independent variables and one constraint, and we consider an example that will demonstrate the advantage of the method of Lagrange multipliers over the method used in Section 8-3.

THEOREM 2   **METHOD OF LAGRANGE MULTIPLIERS FOR FUNCTIONS OF THREE VARIABLES**

Any local maxima or minima of the function $w = f(x, y, z)$ subject to the constraint $g(x, y, z) = 0$ will be among the set of points (x_0, y_0, z_0) for which $(x_0, y_0, z_0, \lambda_0)$ is a solution of the system

$$F_x(x, y, z, \lambda) = 0$$
$$F_y(x, y, z, \lambda) = 0$$
$$F_z(x, y, z, \lambda) = 0$$
$$F_\lambda(x, y, z, \lambda) = 0$$

where $F(x, y, z, \lambda) = f(x, y, z) + \lambda g(x, y, z)$, provided that all the partial derivatives exist.

EXAMPLE 3

Package Design A rectangular box with an open top and one partition is to be constructed from 162 square inches of cardboard (Fig. 7). Find the dimensions that will result in a box with the largest possible volume.

SOLUTION

We must maximize

$$V(x, y, z) = xyz$$

subject to the constraint that the amount of material used is 162 square inches. Thus, x, y, and z must satisfy

$$xy + 2xz + 3yz = 162$$

Step 1. Maximize $V(x, y, z) = xyz$

subject to $g(x, y, z) = xy + 2xz + 3yz - 162 = 0$

Step 2. $F(x, y, z, \lambda) = xyz + \lambda(xy + 2xz + 3yz - 162)$

Step 3.
$$F_x = yz + \lambda(y + 2z) = 0$$
$$F_y = xz + \lambda(x + 3z) = 0$$
$$F_z = xy + \lambda(2x + 3y) = 0$$
$$F_\lambda = xy + 2xz + 3yz - 162 = 0$$

From the first two equations, we can write

$$\lambda = \frac{-yz}{y + 2z} \qquad \lambda = \frac{-xz}{x + 3z}$$

Eliminating λ, we have

$$\frac{-yz}{y + 2z} = \frac{-xz}{x + 3z}$$
$$-xyz - 3yz^2 = -xyz - 2xz^2$$
$$3yz^2 = 2xz^2 \qquad \text{We can assume that } z \neq 0.$$
$$3y = 2x$$
$$x = \tfrac{3}{2}y$$

From the second and third equations,

$$\lambda = \frac{-xz}{x + 3z} \qquad \lambda = \frac{-xy}{2x + 3y}$$

Eliminating λ, we have

$$\frac{-xz}{x + 3z} = \frac{-xy}{2x + 3y}$$
$$-2x^2z - 3xyz = -x^2y - 3xyz$$
$$2x^2z = x^2y \qquad \text{We can assume that } x \neq 0.$$
$$2z = y$$
$$z = \tfrac{1}{2}y$$

Substituting $x = \tfrac{3}{2}y$ and $z = \tfrac{1}{2}y$ into the fourth equation, we have

$$\left(\tfrac{3}{2}y\right)y + 2\left(\tfrac{3}{2}y\right)\left(\tfrac{1}{2}y\right) + 3y\left(\tfrac{1}{2}y\right) - 162 = 0$$
$$\tfrac{3}{2}y^2 + \tfrac{3}{2}y^2 + \tfrac{3}{2}y^2 = 162$$
$$y^2 = 36 \qquad \text{We can assume that } y > 0.$$
$$y = 6$$
$$x = \tfrac{3}{2}(6) = 9 \quad \text{Using } x = \tfrac{3}{2}y$$
$$z = \tfrac{1}{2}(6) = 3 \quad \text{Using } z = \tfrac{1}{2}y$$

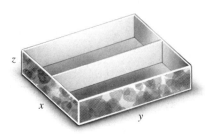

FIGURE 7

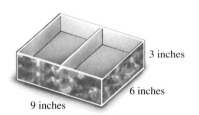

FIGURE 8

9 inches
6 inches
3 inches

and finally,

$$\lambda = \frac{-(6)(3)}{6 + 2(3)} = -\frac{3}{2} \quad \text{Using } \lambda = \frac{-yz}{y + 2z}$$

Thus, the only critical point of F with x, y, and z all positive is $(9, 6, 3, -\frac{3}{2})$.

Step 4. The box with maximum volume has dimensions 9 inches by 6 inches by 3 inches (see Fig. 8).

MATCHED PROBLEM 3

A box of the same type as described in Example 3 is to be constructed from 288 square inches of cardboard. Find the dimensions that will result in a box with the largest possible volume.

INSIGHT

An alternative to the method of Lagrange multipliers would be to solve Example 3 by means of Theorem 2 (the second-derivative test for local extrema) of Section 8-3. That approach would involve solving the material constraint for one of the variables, say, z:

$$z = \frac{162 - xy}{2x + 3y}$$

Then we would eliminate z in the volume function to obtain a function of two variables:

$$V(x, y) = xy \frac{162 - xy}{2x + 3y}$$

The method of Lagrange multipliers allows us to avoid the formidable tasks of calculating the partial derivatives of V and finding the critical points of V in order to apply Theorem 2.

Answers to Matched Problems

1. Max $f(x, y) = f(2, 2) = 17$ (see Fig. 4)

2. (A) 3,750 units of labor and 1,250 units of capital;
   Max $N(x, y) = N(3,750, 1,250) \approx 43,301$ units

   (B) Marginal productivity of money ≈ 0.1443; increase in production $\approx 5,774$ units

3. 12 in. by 8 in. by 4 in.

Exercise 8-4

A *Use the method of Lagrange multipliers in Problems 1–4.*

1. Maximize $f(x, y) = 2xy$
   subject to $x + y = 6$

2. Minimize $f(x, y) = 6xy$
   subject to $y - x = 6$

3. Minimize $f(x, y) = x^2 + y^2$
   subject to $3x + 4y = 25$

4. Maximize $f(x, y) = 25 - x^2 - y^2$
   subject to $2x + y = 10$

B *In Problems 5 and 6, use Theorem 1 to explain why no maxima or minima exist.*

5. Minimize $f(x, y) = 4y - 3x$
   subject to $2x + 5y = 3$

6. Maximize $f(x, y) = 6x + 5y + 24$
   subject to $3x + 2y = 4$

Use the method of Lagrange multipliers in Problems 7–16.

7. Find the maximum and minimum of $f(x, y) = 2xy$ subject to $x^2 + y^2 = 18$.

8. Find the maximum and minimum of $f(x, y) = x^2 - y^2$ subject to $x^2 + y^2 = 25$.

9. Maximize the product of two numbers if their sum must be 10.

10. Minimize the product of two numbers if their difference must be 10.

C

11. Minimize $f(x, y, z) = x^2 + y^2 + z^2$
   subject to $2x - y + 3z = -28$

12. Maximize $f(x, y, z) = xyz$
   subject to $2x + y + 2z = 120$

13. Maximize and minimize $f(x, y, z) = x + y + z$
   subject to $x^2 + y^2 + z^2 = 12$

14. Maximize and minimize $f(x, y, z) = 2x + 4y + 4z$
subject to $x^2 + y^2 + z^2 = 9$

 15. Maximize $f(x, y) = y + xy^2$
subject to $x + y^2 = 1$

 16. Maximize and minimize $f(x, y) = x + e^y$
subject to $x^2 + y^2 = 1$

In Problems 17 and 18, use Theorem 1 to explain why no maxima or minima exist.

17. Maximize $f(x, y) = e^x + 3e^y$
subject to $x - 2y = 6$

18. Minimize $f(x, y) = x^3 + 2y^3$
subject to $6x - 2y = 1$

19. Consider the problem of maximizing $f(x, y)$ subject to $g(x, y) = 0$, where $g(x, y) = y - 5$. Explain how the maximization problem can be solved without using the method of Lagrange multipliers.

20. Consider the problem of minimizing $f(x, y)$ subject to $g(x, y) = 0$, where $g(x, y) = 4x - y + 3$. Explain how the minimization problem can be solved without using the method of Lagrange multipliers.

 21. Consider the problem of maximizing $f(x, y) = e^{-(x^2+y^2)}$ subject to the constraint $g(x, y) = x^2 + y - 1 = 0$.

(A) Solve the constraint equation for y, and then substitute into $f(x, y)$ to obtain a function $h(x)$ of the single variable x. Graph h and solve the original maximization problem by maximizing h (round answers to three decimal places).

(B) Confirm your answer by the method of Lagrange multipliers.

 22. Consider the problem of minimizing

$$f(x, y) = x^2 + 2y^2$$

subject to the constraint $g(x, y) = ye^{x^2} - 1 = 0$.

(A) Solve the constraint equation for y, and then substitute into $f(x, y)$ to obtain a function $h(x)$ of the single variable x. Graph h and solve the original minimization problem by minimizing h (round answers to three decimal places).

(B) Confirm your answer by the method of Lagrange multipliers.

Applications

23. *Budgeting for least cost.* A manufacturing company produces two models of a television set per week, x units of model A and y units of model B at a cost (in dollars) of

$$C(x, y) = 6x^2 + 12y^2$$

If it is necessary (because of shipping considerations) that

$$x + y = 90$$

how many of each type of set should be manufactured per week to minimize cost? What is the minimum cost?

24. *Budgeting for maximum production.* A manufacturing firm has budgeted $60,000 per month for labor and materials. If $x thousand is spent on labor and $y thousand is spent on materials, and if the monthly output (in units) is given by

$$N(x, y) = 4xy - 8x$$

how should the $60,000 be allocated to labor and materials in order to maximize N? What is the maximum N?

25. *Productivity.* A consulting firm for a manufacturing company arrived at the following Cobb–Douglas production function for a particular product:

$$N(x, y) = 50x^{0.8}y^{0.2}$$

In this equation, x is the number of units of labor and y is the number of units of capital required to produce $N(x, y)$ units of the product. Each unit of labor costs $40 and each unit of capital costs $80.

(A) If $400,000 is budgeted for production of the product, determine how that amount should be allocated to maximize production, and find the maximum production.

(B) Find the marginal productivity of money in this case, and estimate the increase in production if an additional $50,000 is budgeted for the production of the product.

26. *Productivity.* The research department of a manufacturing company arrived at the following Cobb–Douglas production function for a particular product:

$$N(x, y) = 10x^{0.6}y^{0.4}$$

In this equation, x is the number of units of labor and y is the number of units of capital required to produce $N(x, y)$ units of the product. Each unit of labor costs $30 and each unit of capital costs $60.

(A) If $300,000 is budgeted for production of the product, determine how that amount should be allocated to maximize production, and find the maximum production.

(B) Find the marginal productivity of money in this case, and estimate the increase in production if an additional $80,000 is budgeted for the production of the product.

27. *Maximum volume.* A rectangular box with no top and two intersecting partitions is to be constructed from 192 square inches of cardboard (see the figure). Find the dimensions that will maximize the volume.

Figure for 27

28. *Maximum volume.* A mailing service states that a rectangular package shall have the sum of its length and girth not to exceed 120 inches (see the figure). What are the dimensions of the largest (in volume) mailing carton that can be constructed to meet these restrictions?

Figure for 28

29. *Agriculture.* Three pens of the same size are to be built along an existing fence (see the figure). If 400 feet of fencing is available, what length should x and y be to produce the maximum total area? What is the maximum area?

Figure for 29

30. *Diet and minimum cost.* A group of guinea pigs is to receive 25,600 calories per week. Two available foods produce $200xy$ calories for a mixture of x kilograms of type M food and y kilograms of type N food. If type M costs $1 per kilogram and type N costs $2 per kilogram, how much of each type of food should be used to minimize weekly food costs? What is the minimum cost?

Note: $x \geq 0, y \geq 0$

Section 8-5 METHOD OF LEAST SQUARES

- Least Squares Approximation
- Applications

Least Squares Approximation

Regression analysis is the process of fitting an elementary function to a set of data points by the **method of least squares.** The mechanics of using regression techniques were introduced in Chapter 1. Now, using the optimization techniques of Section 8-3, we are able to develop and explain the mathematical foundation of the method of least squares. We begin with **linear regression,** the process of finding the equation of the line that is the "best" approximation to a set of data points.

A manufacturer wants to approximate the cost function for a product. The value of the cost function has been determined for certain levels of production, as listed in Table 1. Although these points do not all lie on a line (see Fig. 1), they are very close to being linear. The manufacturer would like to approximate the cost function by a linear function—that is, determine values a and b so that the line

$$y = ax + b$$

is, in some sense, the "best" approximation to the cost function.

What do we mean by "best"? Since the line $y = ax + b$ will not go through all four points, it is reasonable to examine the differences between the y coordinates of the points listed in the table and the y coordinates of the corresponding points on the line. Each of these differences is called the **residual** at that point (see Fig. 2). For example, at $x = 2$, the point from Table 1 is $(2, 4)$ and the point on the line is $(2, 2a + b)$, so the residual is

$$4 - (2a + b) = 4 - 2a - b$$

All the residuals are listed in Table 2.

TABLE 1

Number of Units	Cost
x (hundreds)	y (thousand $)
2	4
5	6
6	7
9	8

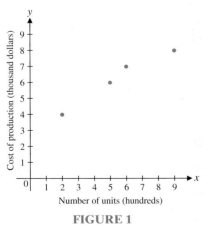

FIGURE 1

TABLE 2			
x	y	$ax + b$	Residual
2	4	$2a + b$	$4 - 2a - b$
5	6	$5a + b$	$6 - 5a - b$
6	7	$6a + b$	$7 - 6a - b$
9	8	$9a + b$	$8 - 9a - b$

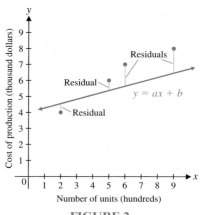

FIGURE 2

Our criterion for the "best" approximation is the following: Determine the values of a and b that *minimize the sum of the squares* of the residuals. The resulting line is called the **least squares line,** or the **regression line.** To this end, we minimize

$$F(a, b) = (4 - 2a - b)^2 + (6 - 5a - b)^2 + (7 - 6a - b)^2 + (8 - 9a - b)^2$$

Step 1. Find critical points:

$$\begin{aligned}
F_a(a, b) &= 2(4 - 2a - b)(-2) + 2(6 - 5a - b)(-5) \\
&\quad + 2(7 - 6a - b)(-6) + 2(8 - 9a - b)(-9) \\
&= -304 + 292a + 44b = 0
\end{aligned}$$

$$\begin{aligned}
F_b(a, b) &= 2(4 - 2a - b)(-1) + 2(6 - 5a - b)(-1) \\
&\quad + 2(7 - 6a - b)(-1) + 2(8 - 9a - b)(-1) \\
&= -50 + 44a + 8b = 0
\end{aligned}$$

After dividing each equation by 2, we solve the system

$$146a + 22b = 152$$
$$22a + 4b = 25$$

obtaining $(a, b) = (0.58, 3.06)$ as the only critical point.

Step 2. Compute $A = F_{aa}(a, b)$, $B = F_{ab}(a, b)$, and $C = F_{bb}(a, b)$:

$$\begin{aligned}
F_{aa}(a, b) &= 292; &\text{thus,}&& A &= F_{aa}(0.58, 3.06) = 292 \\
F_{ab}(a, b) &= 44; &\text{thus,}&& B &= F_{ab}(0.58, 3.06) = 44 \\
F_{bb}(a, b) &= 8; &\text{thus,}&& C &= F_{bb}(0.58, 3.06) = 8
\end{aligned}$$

Step 3. Evaluate $AC - B^2$ and try to classify the critical point (a, b) by using Theorem 2 in Section 8-3:

$$AC - B^2 = (292)(8) - (44)^2 = 400 > 0 \qquad \text{and} \qquad A = 292 > 0$$

Therefore, case 2 in Theorem 2 applies, and $F(a, b)$ has a local minimum at the critical point $(0.58, 3.06)$.

Thus, the least squares line for the given data is

$$y = 0.58x + 3.06 \quad \text{Least squares line}$$

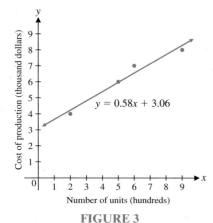

FIGURE 3

The sum of the squares of the residuals is minimized for this choice of a and b (see Fig. 3).

This linear function can now be used by the manufacturer to estimate any of the quantities normally associated with the cost function—such as costs, marginal costs, average costs, and so on. For example, the cost of producing 2,000 units is approximately

$$y = (0.58)(20) + 3.06 = 14.66, \qquad \text{or} \qquad \$14,660$$

The marginal cost function is

$$\frac{dy}{dx} = 0.58$$

The average cost function is

$$\bar{y} = \frac{0.58x + 3.06}{x}$$

In general, if we are given a set of n points $(x_1, y_1), (x_2, y_2), \ldots, (x_n, y_n)$, we want to determine the line $y = ax + b$ for which the sum of the squares of the residuals is minimized. Using summation notation, we find that the sum of the squares of the residuals is given by

$$F(a, b) = \sum_{k=1}^{n} (y_k - ax_k - b)^2$$

Note that in this expression the variables are a and b, and the x_k and y_k are all known values. To minimize $F(a, b)$, we thus compute the partial derivatives with respect to a and b and set them equal to 0:

$$F_b(a, b) = \sum_{k=1}^{n} 2(y_k - ax_k - b)(-x_k) = 0$$

$$F_b(a, b) = \sum_{k=1}^{n} 2(y_k - ax_k - b)(-1) = 0$$

Dividing each equation by 2 and simplifying, we see that the coefficients a and b of the least squares line $y = ax + b$ must satisfy the following system of *normal equations:*

$$\left(\sum_{k=1}^{n} x_k^2 \right) a + \left(\sum_{k=1}^{n} x_k \right) b = \sum_{k=1}^{n} x_k y_k$$

$$\left(\sum_{k=1}^{n} x_k \right) a + nb = \sum_{k=1}^{n} y_k$$

Solving this system for a and b produces the formulas given in Theorem 1.

THEOREM 1 LEAST SQUARES APPROXIMATION

For a set of n points $(x_1, y_1), (x_2, y_2), \ldots, (x_n, y_n)$, the coefficients of the least squares line $y = ax + b$ are the solutions of the system of **normal equations**

$$\left(\sum_{k=1}^{n} x_k^2 \right) a + \left(\sum_{k=1}^{n} x_k \right) b = \sum_{k=1}^{n} x_k y_k \tag{1}$$

$$\left(\sum_{k=1}^{n} x_k \right) a + nb = \sum_{k=1}^{n} y_k$$

and are given by the formulas

$$a = \frac{n\left(\sum_{k=1}^{n} x_k y_k \right) - \left(\sum_{k=1}^{n} x_k \right)\left(\sum_{k=1}^{n} y_k \right)}{n\left(\sum_{k=1}^{n} x_k^2 \right) - \left(\sum_{k=1}^{n} x_k \right)^2} \tag{2}$$

$$b = \frac{\sum_{k=1}^{n} y_k - a\left(\sum_{k=1}^{n} x_k \right)}{n} \tag{3}$$

Now we return to the data in Table 1 and tabulate the sums required for the normal equations and their solution in Table 3.

TABLE 3

	x_k	y_k	$x_k y_k$	x_k^2
	2	4	8	4
	5	6	30	25
	6	7	42	36
	9	8	72	81
Totals	22	25	152	146

The normal equations (1) are then

$$146a + 22b = 152$$
$$22a + 4b = 25$$

The solution of the normal equations given by equations (2) and (3) is

$$a = \frac{4(152) - (22)(25)}{4(146) - (22)^2} = 0.58$$

$$b = \frac{25 - 0.58(22)}{4} = 3.06$$

Compare these results with step 1 on page 500. Note that Table 3 provides a convenient format for the computation of step 1.

Many graphing calculators have a linear regression feature that solves the system of normal equations obtained by setting the partial derivatives of the sum of squares of the residuals equal to 0. Therefore, in practice, we simply enter the given data points and use the linear regression feature to determine the line $y = ax + b$ that best fits the data (see Fig. 4). There is no need to compute partial derivatives or even to tabulate sums (as in Table 3).

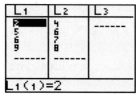

(A)

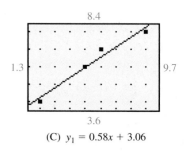

(B) (C) $y_1 = 0.58x + 3.06$

FIGURE 4

***Explore & Discuss* 1**

(A) Plot the four points $(0, 0)$, $(0, 1)$, $(10, 0)$, and $(10, 1)$. Which line would you guess "best" fits these four points? Use formulas (2) and (3) to test your conjecture.

(B) Plot the four points $(0, 0)$, $(0, 10)$, $(1, 0)$ and $(1, 10)$. Which line would you guess "best" fits these four points? Use formulas (2) and (3) to test your conjecture.

(C) If either of your conjectures was wrong, explain how your reasoning was mistaken.

INSIGHT

Formula (2) for a is undefined if the denominator equals 0. When can this happen? Suppose $n = 3$. Then

$$n\left(\sum_{k=1}^{n} x_k^2\right) - \left(\sum_{k=1}^{n} x_k\right)^2$$

$$= 3\left(x_1^2 + x_2^2 + x_3^2\right) - (x_1 + x_2 + x_3)^2$$
$$= 3\left(x_1^2 + x_2^2 + x_3^2\right) - \left(x_1^2 + x_2^2 + x_3^2 + 2x_1x_2 + 2x_1x_3 + 2x_2x_3\right)$$
$$= 2\left(x_1^2 + x_2^2 + x_3^2\right) - (2x_1x_2 + 2x_1x_3 + 2x_2x_3)$$
$$= \left(x_1^2 + x_2^2\right) + \left(x_1^2 + x_3^2\right) + \left(x_2^2 + x_3^2\right) - (2x_1x_2 + 2x_1x_3 + 2x_2x_3)$$
$$= \left(x_1^2 - 2x_1x_2 + x_2^2\right) + \left(x_1^2 - 2x_1x_3 + x_3^2\right) + \left(x_2^2 - 2x_2x_3 + x_3^2\right)$$
$$= (x_1 - x_2)^2 + (x_1 - x_3)^2 + (x_2 - x_3)^2$$

and the last expression is equal to 0 if and only if $x_1 = x_2 = x_3$ (i.e., if and only if the three points all lie on the same vertical line). A similar algebraic manipulation works for any integer $n > 1$, showing that, in formula (2) for a, the denominator equals 0 if and only if all n points lie on the same vertical line.

The method of least squares can also be applied to find the quadratic equation $y = ax^2 + bx + c$ that best fits a set of data points. In this case, the sum of the squares of the residuals is a function of three variables:

$$F(a, b, c) = \sum_{k=1}^{n} \left(y_k - ax_k^2 - bx_k - c\right)^2$$

There are now three partial derivatives to compute and set equal to 0:

$$F_a(a, b, c) = \sum_{k=1}^{n} 2\left(y_k - ax_k^2 - bx_k - c\right)\left(-x_k^2\right) = 0$$

$$F_b(a, b, c) = \sum_{k=1}^{n} 2\left(y_k - ax_k^2 - bx_k - c\right)(-x_k) = 0$$

$$F_c(a, b, c) = \sum_{k=1}^{n} 2\left(y_k - ax_k^2 - bx_k - c\right)(-1) = 0$$

The resulting set of three linear equations in the three variables a, b, and c is called the *set of normal equations for quadratic regression.*

A quadratic regression feature on a calculator is designed to solve such normal equations after the given set of points has been entered. Figure 5 illustrates the computation for the data of Table 1.

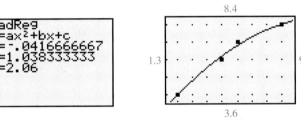

(A)

(B)

(C) $y_1 = -0.0417x^2 + 1.0383x + 2.06$

FIGURE 5

Explore & Discuss **2**   (A) Use the graphs in Figures 4 and 5 to predict which technique, linear regression or quadratic regression, yields the smaller sum of squares of the residuals for the data of Table 1. Explain.

(B) Confirm your prediction by computing the sum of squares of the residuals in each case.

The method of least squares can also be applied to other regression equations—for example, cubic, quartic, logarithmic, exponential, and power regression models. Details are explored in some of the exercises at the end of this section.

APPLICATIONS

EXAMPLE 1   **Educational Testing**   Table 4 lists the midterm and final examination scores of 10 students in a calculus course.

TABLE 4			
Midterm	**Final**	**Midterm**	**Final**
49	61	78	77
53	47	83	81
67	72	85	79
71	76	91	93
74	68	99	99

(A) Use formulas (1), (2), and (3) to find the normal equations and the least squares line for the data given in Table 4.

(B) Use the linear regression feature on a graphing calculator to find and graph the least squares line.

(C) Use the least squares line to predict the final examination score of a student who scored 95 on the midterm examination.

SOLUTION (A) Table 5 shows a convenient way to compute all the sums in the formulas for a and b.

TABLE 5			
x_k	y_k	$x_k y_k$	x_k^2
49	61	2,989	2,401
53	47	2,491	2,809
67	72	4,824	4,489
71	76	5,396	5,041
74	68	5,032	5,476
78	77	6,006	6,084
83	81	6,723	6,889
85	79	6,715	7,225
91	93	8,463	8,281
99	99	9,801	9,801
Totals 750	753	58,440	58,496

From the last line in Table 5, we have

$$\sum_{k=1}^{10} x_k = 750 \qquad \sum_{k=1}^{10} y_k = 753 \qquad \sum_{k=1}^{10} x_k y_k = 58,440 \qquad \sum_{k=1}^{10} x_k^2 = 58,496$$

and the normal equations are

$$58,496a + 750b = 58,440$$
$$750a + 10b = 753$$

Using formulas (2) and (3), we obtain

$$a = \frac{10(58,440) - (750)(753)}{10(58,496) - (750)^2} = \frac{19,650}{22,460} \approx 0.875$$

$$b = \frac{753 - 0.875(750)}{10} = 9.675$$

The least squares line is given (approximately) by

$$y = 0.875x + 9.675$$

(B) We enter the data and use the linear regression feature, as shown in Figure 6. [The discrepancy between values of a and b in the preceding calculations and those in Figure 6B is due to rounding in part (A).]

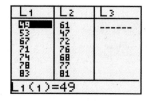

(A)

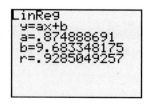

(B)

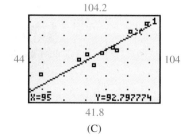

(C)

FIGURE 6

(C) If $x = 95$, then $y = 0.875(95) + 9.675 \approx 92.8$ is the predicted score on the final exam. This is also indicated in Figure 6C. If we assume that the exam score must be an integer, we would predict a score of 93.

MATCHED PROBLEM 1

Repeat Example 1 for the scores listed in Table 6.

TABLE 6			
Midterm	**Final**	**Midterm**	**Final**
54	50	84	80
60	66	88	95
75	80	89	85
76	68	97	94
78	71	99	86

EXAMPLE 2

Energy Consumption The use of fuel oil for home heating in the United States has declined steadily for several decades. Table 7 lists the percentage of occupied housing units in the United States that were heated by fuel oil for various years between 1960 and 2000. Use the data in the table and linear regression to predict the percentage of occupied housing units in the United States that will be heated by fuel oil in the year 2010.

TABLE 7	Occupied Housing Units Heated by Fuel Oil		
Year	**Percent**	**Year**	**Percent**
1960	32.4	1990	13.2
1970	26.0	2000	9.0
1980	18.1		

SOLUTION

We enter the data, with $x = 0$ representing 1960, $x = 10$ representing 1970, and so on, and use linear regression as shown in Figure 7.

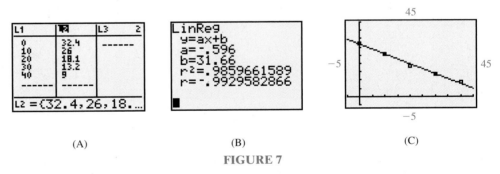

(A)   (B)   (C)

FIGURE 7

Figure 7 indicates that the least squares line is $y = -0.596x + 31.66$. To estimate the percentage of occupied housing units heated by fuel oil in the year 2010 (corresponding to $x = 50$), we substitute $x = 50$ in the equation of the least squares line: $-0.596(50) + 31.66 = 1.86$. Thus, the estimated percentage in 2010 is 1.86%.

MATCHED PROBLEM 2

In 1950, coal was still a major source of fuel for home energy consumption, and the percentage of occupied housing units heated by fuel oil was only 22.1%. Add the data for 1950 to the data for Example 2, and compute the new least squares line and the new estimate for the percentage of occupied housing units heated by fuel oil in the year 2010. Discuss the discrepancy between the two estimates. (As in Example 2, let $x = 0$ represent 1960.)

Explore & Discuss 3

The data in Table 8 give the amount A (in milligrams) of the radioactive isotope gallium-67, used in the diagnosis of malignant tumors, at time t (in hours).

TABLE 8	
t (hours)	*A* (milligrams)
0	50.0
5	46.4
10	43.1
15	39.9
20	37.1

(A) Since the data describe radioactive decay, we would expect the relationship between A and t to be exponential; that is, $A = ae^{bt}$ for some constants a and b. Show that if $A = ae^{bt}$, then the relationship between $\ln A$ and t is linear.

(B) Compute $\ln A$ for each data point and find the line that "best" fits the data $(t, \ln A)$.

(C) Determine the "best" values of a and b.

(D) Confirm your results by using the exponential regression feature on a graphing calculator.

Answers to Matched Problems

1. (A) $y = 0.85x + 9.47$

(B)

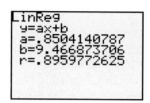

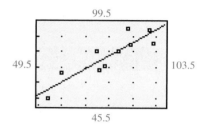

(C) 90.3

2. $y = -0.37x + 25.75; 7.03\%$

Exercise 8-5

A *In Problems 1–6, find the least squares line. Graph the data and the least squares line.*

1.

x	*y*
1	1
2	3
3	4
4	3

2.

x	*y*
1	−2
2	−1
3	3
4	5

3.

x	*y*
1	8
2	5
3	4
4	0

4.

x	*y*
1	20
2	14
3	11
4	3

5.

x	*y*
1	3
2	4
3	5
4	6

6.

x	*y*
1	2
2	3
3	3
4	2

B *In Problems 7–14, find the least squares line and use it to estimate y for the indicated value of x. Round answers to two decimal places.*

7.

x	*y*
1	3
2	1
2	2
3	0

Estimate *y* when *x* = 2.5.

8.

x	*y*
1	0
3	1
3	6
3	4

Estimate *y* when *x* = 3.

9.

x	*y*
0	10
5	22
10	31
15	46
20	51

Estimate *y* when *x* = 25.

10.

x	*y*
−5	60
0	50
5	30
10	20
15	15

Estimate *y* when *x* = 20.

11.

x	*y*
−1	14
1	12
3	8
5	6
7	5

Estimate *y* when *x* = 2.

12.

x	*y*
2	−4
6	0
10	8
14	12
18	14

Estimate *y* when *x* = 15.

13.

x	*y*	*x*	*y*
0.5	25	9.5	12
2	22	11	11
3.5	21	12.5	8
5	21	14	5
6.5	18	15.5	1

Estimate *y* when *x* = 8.

14.

x	*y*	*x*	*y*
0	−15	12	11
2	−9	14	13
4	−7	16	19
6	−7	18	25
8	−1	20	33

Estimate *y* when *x* = 10.

15. To find the coefficients of the parabola

$$y = ax^2 + bx + c$$

that is the "best" fit to the points $(1, 2)$, $(2, 1)$, $(3, 1)$, and $(4, 3)$, minimize the sum of the squares of the residuals

$$
\begin{aligned}
F(a, b, c) = & (a + b + c - 2)^2 \\
& + (4a + 2b + c - 1)^2 \\
& + (9a + 3b + c - 1)^2 \\
& + (16a + 4b + c - 3)^2
\end{aligned}
$$

by solving the system of normal equations

$$F_a(a, b, c) = 0 \qquad F_b(a, b, c) = 0 \qquad F_c(a, b, c) = 0$$

for a, b, and c. Graph the points and the parabola.

16. Repeat Problem 15 for the points $(-1, -2)$, $(0, 1)$, $(1, 2)$, and $(2, 0)$.

Problems 17 and 18 refer to the system of normal equations and the formulas for a and b given in the text.

17. Verify formulas (2) and (3) by solving the system of normal equations (1) for a and b.

18. If

$$\bar{x} = \frac{1}{n}\sum_{k=1}^{n} x_k \qquad \text{and} \qquad \bar{y} = \frac{1}{n}\sum_{k=1}^{n} y_k$$

are the averages of the x and y coordinates, respectively, show that the point $(\bar{x}, \bar{y})$ satisfies the equation of the least squares line, $y = ax + b$.

19. (A) Suppose that $n = 5$ and that the x coordinates of the data points (x_1, y_1), $(x_2, y_2), \ldots, (x_n, y_n)$ are

$-2, -1, 0, 1, 2$. Show that system (1) in the text implies that

$$a = \frac{\sum x_k y_k}{\sum x_k^2}$$

and that b is equal to the average of the values of y_k.

(B) Show that the conclusion of part (A) holds whenever the average of the x coordinates of the data points is 0.

20. (A) Give an example of a set of six data points such that half of the points lie above the least squares line and half lie below.

(B) Give an example of a set of six data points such that just one of the points lies above the least squares line and five lie below.

21. (A) Find the linear and quadratic functions that best fit the data points $(0, 1.3)$, $(1, 0.6)$, $(2, 1.5)$, $(3, 3.6)$, and $(4, 7.4)$. (Round coefficients to two decimal places.)

(B) Which of the two functions best fits the data? Explain.

22. (A) Find the linear, quadratic, and logarithmic functions that best fit the data points $(1, 3.2)$, $(2, 4.2)$, $(3, 4.7)$, $(4, 5.0)$, and $(5, 5.3)$. (Round coefficients to two decimal places.)

(B) Which of the three functions best fits the data? Explain.

23. Describe the normal equations for cubic regression. How many equations are there? What are the variables? What techniques could be used to solve the equations?

24. Describe the normal equations for quartic regression. How many equations are there? What are the variables? What techniques could be used to solve the equations?

Applications

25. *Motor vehicle production.* Data for motor vehicle production in China for the years 1997 to 2004 are given in the table.

Motor Vehicle Production in China			
Year	**Thousands**	**Year**	**Thousands**
1997	1,578	2001	2,332
1998	1,628	2002	3,251
1999	1,805	2003	4,444
2000	2,009	2004	5,071

(A) Find the least squares line for the data, using $x = 0$ for 1990.

(B) Use the least squares line to estimate the annual production of motor vehicles in China in 2010.

26. *Beef production.* Data for U.S. production of beef for the years 1990 to 2004 are given in the table.

U.S. Beef Production			
Year	**Million Pounds**	**Year**	**Million Pounds**
1990	22,743	1998	25,760
1992	23,086	2000	26,888
1994	24,386	2002	27,192
1996	25,525	2004	24,650

(A) Find the least squares line for the data, using $x = 0$ for 1990.

(B) Use the least squares line to estimate the annual production of beef in the United States in 2016.

27. *Maximizing profit.* The market research department for a drugstore chain chose two summer resort areas to test market a new sunscreen lotion packaged in 4-ounce plastic bottles. After a summer of varying the selling price and recording the monthly demand, the research department arrived at the following demand table, where y is the number of bottles purchased per month (in thousands) at x dollars per bottle:

x	y
5.0	2.0
5.5	1.8
6.0	1.4
6.5	1.2
7.0	1.1

(A) Use the method of least squares to find a demand equation.

(B) If each bottle of sunscreen costs the drugstore chain $4, how should the sunscreen be priced to achieve a maximum monthly profit? [*Hint:* Use the result of part (A), with $C = 4y$, $R = xy$, and $P = R - C$.]

28. *Maximizing profit.* A market research consultant for a supermarket chain chose a large city to test market a new brand of mixed nuts packaged in 8-ounce cans. After a year of varying the selling price and recording the monthly demand, the consultant arrived at the following demand table, where y is the number of cans purchased per month (in thousands) at x dollars per can:

x	y
4.0	4.2
4.5	3.5
5.0	2.7
5.5	1.5
6.0	0.7

(A) Use the method of least squares to find a demand equation.

(B) If each can of nuts costs the supermarket chain $3, how should the nuts be priced to achieve a maximum monthly profit?

29. *Medicine.* If a person dives into cold water, a neural reflex response automatically shuts off blood circulation to the skin and muscles and reduces the pulse rate. A medical research team conducted an experiment using a group of 10 two-year-olds. A child's face was placed momentarily in cold water, and the corresponding reduction in pulse rate was recorded. The data for the average reduction in heart rate for each temperature are summarized in the following table:

Water Temperature (°F)	Pulse Rate Reduction
50	15
55	13
60	10
65	6
70	2

(A) If T is water temperature (in degrees Fahrenheit) and P is pulse rate reduction (in beats per minute), use the method of least squares to find a linear equation relating T and P.

(B) Use the equation found in part (A) to find P when $T = 57$.

30. *Biology.* In biology, there is an approximate rule, called the *bioclimatic rule for temperate climates,* that has been known for a couple of hundred years. This rule states that in spring and early summer, periodic phenomena such as the blossoming of flowers, the appearance of insects, and the ripening of fruit usually come about 4 days later for each 500 feet of altitude. Stated as a formula, the rule becomes

$$d = 8h \qquad 0 \le h \le 4$$

where d is the change in days and h is the altitude (in thousands of feet). To test this rule, an experiment was set up to record the difference in blossoming time of the same type of apple tree at different altitudes. A summary of the results is given in the following table:

h	d
0	0
1	7
2	18
3	28
4	33

(A) Use the method of least squares to find a linear equation relating h and d. Does the bioclimatic rule $d = 8h$ appear to be approximately correct?

(B) How much longer will it take this type of apple tree to blossom at 3.5 thousand feet than at sea level? [Use the linear equation found in part (A).]

 31. *Global warming.* Average global temperatures from 1885 to 2005 are given in the table.

Average Global Temperatures			
Year	°F	Year	°F
1885	56.65	1955	57.06
1895	56.64	1965	57.05
1905	56.52	1975	57.04
1915	56.57	1985	57.36
1925	56.74	1995	57.64
1935	57.00	2005	58.59
1945	57.13		

(A) Find the least squares line for the data, using $x = 0$ for 1885.

(B) Use the least squares line to estimate the average global temperature in 2085.

 32. *Air pollution.* Data for emissions of sulfur dioxide in the United States from selected years from 1972 to 2002 are given in the table.

Emissions of Sulfur Dioxide in the United States			
Year	Million Short Tons	Year	Million Short Tons
1972	30.4	1992	22.1
1977	28.6	1997	18.8
1982	23.2	2002	15.4
1987	22.2		

(A) Find the least squares line for the data, using $x = 0$ for 1972.

(B) Use the least squares line to estimate the emissions of sulfur dioxide in the United States in 2015.

33. *Political science.* The association between economic class and party affiliation did not start with Roosevelt's New Deal; it goes back to the time of Andrew Jackson (1767–1845). Paul Lazarsfeld of Columbia University published an article in the November 1950 issue of *Scientific American* in which he discusses statistical investigations of the relationships between economic class and party affiliation. The data in the accompanying table are taken from this article.

(A) If A represents the average assessed value per person in a given ward in 1836 and D represents the percentage of people in that ward voting Democratic in 1836, use the method of least squares to find a linear equation relating A and D.

Political Affiliations, 1836		
Ward	**Average Assessed Value per Person (Hundred $)**	**Democratic Votes (%)**
12	1.7	51
3	2.1	49
1	2.3	53
5	2.4	36
2	3.6	65
11	3.7	35
10	4.7	29
4	6.2	40
6	7.1	34
9	7.4	29
8	8.7	20
7	11.9	23

(B) If the average assessed value per person in a ward had been $300, what is the predicted percentage of people in that ward who would have voted Democratic?

34. *Education.* The table lists the high school grade-point averages (GPAs) of 10 students, along with their grade-point averages after one semester of college.

High School GPA	College GPA	High School GPA	College GPA
2.0	1.5	3.0	2.3
2.2	1.5	3.1	2.5
2.4	1.6	3.3	2.9
2.7	1.8	3.4	3.2
2.9	2.1	3.7	3.5

(A) Find the least squares line for the data.

(B) Estimate the college GPA for a student with a high school GPA of 3.5.

(C) Estimate the high school GPA necessary for a college GPA of 2.7.

 35. *Olympic Games.* The table gives the winning heights in the pole vault in the Olympic Games from 1896 to 2004.

Olympic Pole Vault Winning Height			
Year	**Height (ft)**	**Year**	**Height (ft)**
1896	10.81	1956	14.96
1900	10.82	1960	15.43
1904	11.50	1964	16.73
1906	11.60	1968	17.71
1908	12.17	1972	18.04
1912	12.96	1976	18.04
1920	13.46	1980	18.96
1924	12.96	1984	18.85
1928	13.78	1988	19.35
1932	14.16	1992	19.02
1936	14.27	1996	19.42
1948	14.10	2000	19.35
1952	14.93	2004	19.52

(A) Use a graphing calculator to find the least squares line for the data, letting $x = 0$ for 1896.

(B) Estimate the winning height in the pole vault in the Olympic Games of 2020.

Section 8-6 DOUBLE INTEGRALS OVER RECTANGULAR REGIONS

- Introduction
- Definition of the Double Integral
- Average Value over Rectangular Regions
- Volume and Double Integrals

Introduction

We have generalized the concept of differentiation to functions with two or more independent variables. How can we do the same with integration, and how can we interpret the results? Let us first look at the operation of antidifferentiation. We can antidifferentiate a function of two or more variables with respect to one of the variables by treating all the other variables as though they were constants. Thus, this operation is the reverse operation of partial differentiation, just as ordinary antidifferentiation is the reverse operation of ordinary differentiation. We write $\int f(x, y)\, dx$ to indicate that we are to antidifferentiate $f(x, y)$ with respect to x, holding y fixed; we write $\int f(x, y)\, dy$ to indicate that we are to antidifferentiate $f(x, y)$ with respect to y, holding x fixed.

EXAMPLE 1

Partial Antidifferentiation Evaluate:

(A) $\displaystyle\int (6xy^2 + 3x^2)\, dy$ (B) $\displaystyle\int (6xy^2 + 3x^2)\, dx$

SOLUTION   (A) Treating x as a constant and using the properties of antidifferentiation from Section 6-1, we have

$$\int (6xy^2 + 3x^2)\, dy = \int 6xy^2\, dy + \int 3x^2\, dy$$

> The dy tells us that we are looking for the antiderivative of $6xy^2 + 3x^2$ with respect to y only, holding x constant.

$$= 6x \int y^2\, dy + 3x^2 \int dy$$

$$= 6x\left(\frac{y^3}{3}\right) + 3x^2(y) + C(x)$$

$$= 2xy^3 + 3x^2y + C(x)$$

Notice that the constant of integration actually can be *any function of x alone*, since, for any such function,

$$\frac{\partial}{\partial y} C(x) = 0$$

Check:

We can verify that our answer is correct by using partial differentiation:

$$\frac{\partial}{\partial y}[2xy^3 + 3x^2y + C(x)] = 6xy^2 + 3x^2 + 0$$

$$= 6xy^2 + 3x^2$$

(B) Now we treat y as a constant:

$$\int (6xy^2 + 3x^2)\, dx = \int 6xy^2\, dx + \int 3x^2\, dx$$

$$= 6y^2 \int x\, dx + 3 \int x^2\, dx$$

$$= 6y^2\left(\frac{x^2}{2}\right) + 3\left(\frac{x^3}{3}\right) + E(y)$$

$$= 3x^2y^2 + x^3 + E(y)$$

This time, the antiderivative contains an arbitrary function $E(y)$ of y alone.

Check:

$$\frac{\partial}{\partial x}[3x^2y^2 + x^3 + E(y)] = 6xy^2 + 3x^2 + 0$$

$$= 6xy^2 + 3x^2$$

MATCHED PROBLEM 1 Evaluate: (A) $\displaystyle\int (4xy + 12x^2y^3)\, dy$ (B) $\displaystyle\int (4xy + 12x^2y^3)\, dx$

Now that we have extended the concept of antidifferentiation to functions with two variables, we also can evaluate definite integrals of the form

$$\int_a^b f(x, y)\, dx \qquad \text{or} \qquad \int_c^d f(x, y)\, dy$$

EXAMPLE 2   **Evaluating a Partial Antiderivative** Evaluate, substituting the limits of integration in y if dy is used and in x if dx is used:

(A) $\displaystyle\int_0^2 (6xy^2 + 3x^2)\, dy$ (B) $\displaystyle\int_0^1 (6xy^2 + 3x^2)\, dx$

SOLUTION (A) From Example 1A, we know that

$$\int (6xy^2 + 3x^2)\, dy = 2xy^3 + 3x^2y + C(x)$$

According to properties of the definite integral for a function of one variable, we can use any antiderivative to evaluate the definite integral. Thus, choosing $C(x) = 0$, we have

$$\int_0^2 (6xy^2 + 3x^2)\, dy = (2xy^3 + 3x^2y)\big|_{y=0}^{y=2}$$

$$= [2x(2)^3 + 3x^2(2)] - [2x(0)^3 + 3x^2(0)]$$

$$= 16x + 6x^2$$

(B) From Example 1B, we know that

$$\int (6xy^2 + 3x^2)\, dx = 3x^2y^2 + x^3 + E(y)$$

Thus, choosing $E(y) = 0$, we have

$$\int_0^1 (6xy^2 + 3x^2)\, dx = (3x^2y^2 + x^3)\big|_{x=0}^{x=1}$$

$$= [3y^2(1)^2 + (1)^3] - [3y^2(0)^2 + (0)^3]$$

$$= 3y^2 + 1 \qquad\blacksquare$$

MATCHED PROBLEM 2   Evaluate:

(A) $\displaystyle\int_0^1 (4xy + 12x^2y^3)\, dy$ (B) $\displaystyle\int_0^3 (4xy + 12x^2y^3)\, dx$

Notice that integrating and evaluating a definite integral with integrand $f(x, y)$ with respect to y produces a function of x alone (or a constant). Likewise, integrating and evaluating a definite integral with integrand $f(x, y)$ with respect to x produces a function of y alone (or a constant). Each of these results, involving at most one variable, can now be used as an integrand in a second definite integral.

EXAMPLE 3   **Evaluating Iterated Integrals** Evaluate:

(A) $\displaystyle\int_0^1 \left[\int_0^2 (6xy^2 + 3x^2)\, dy \right] dx$ (B) $\displaystyle\int_0^2 \left[\int_0^1 (6xy^2 + 3x^2)\, dx \right] dy$

SOLUTION (A) Example 2A showed that

$$\int_0^2 (6xy^2 + 3x^2)\, dy = 16x + 6x^2$$

Thus,

$$\int_0^1 \left[\int_0^2 (6xy^2 + 3x^2)\, dy \right] dx = \int_0^1 (16x + 6x^2)\, dx$$
$$= (8x^2 + 2x^3)\big|_{x=0}^{x=1}$$
$$= [8(1)^2 + 2(1)^3] - [8(0)^2 + 2(0)^3] = 10$$

(B) Example 2B showed that

$$\int_0^1 (6xy^2 + 3x^2)\, dx = 3y^2 + 1$$

Thus,

$$\int_0^2 \left[\int_0^1 (6xy^2 + 3x^2)\, dx \right] dy = \int_0^2 (3y^2 + 1)\, dy$$
$$= (y^3 + y)\big|_{y=0}^{y=2}$$
$$= [(2)^3 + 2] - [(0)^3 + 0] = 10$$

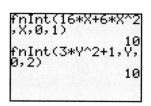

FIGURE 1

A numerical integration routine can be used as an alternative to the fundamental theorem of calculus to evaluate the last integrals in Examples 3A and 3B, $\int_0^1 (16x + 6x^2)\, dx$ and $\int_0^2 (3y^2 + 1)\, dy$, since the integrand in each case is a function of a single variable (see Fig. 1).

MATCHED PROBLEM 3   Evaluate:

(A) $\displaystyle\int_0^3 \left[\int_0^1 (4xy + 12x^2y^3)\, dy \right] dx$

(B) $\displaystyle\int_0^1 \left[\int_0^3 (4xy + 12x^2y^3)\, dx \right] dy$

■ Definition of the Double Integral

Notice that the answers in Examples 3A and 3B are identical. This is not an accident. In fact, it is this property that enables us to define the *double integral*, as follows:

DEFINITION   **Double Integral**

The **double integral** of a function $f(x, y)$ over a rectangle

$$R = \{(x, y) | a \le x \le b, \quad c \le y \le d\}$$

is

$$\iint\limits_R f(x, y)\, dA = \int_a^b \left[\int_c^d f(x, y)\, dy \right] dx$$
$$= \int_c^d \left[\int_a^b f(x, y)\, dx \right] dy$$

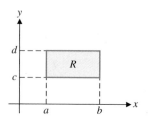

In the double integral $\iint_R f(x, y)\,dA$, $f(x, y)$ is called the **integrand** and R is called the **region of integration.** The expression dA indicates that this is an integral over a two-dimensional region. The integrals

$$\int_a^b \left[\int_c^d f(x, y)\,dy\right] dx \quad \text{and} \quad \int_c^d \left[\int_a^b f(x, y)\,dx\right] dy$$

are referred to as **iterated integrals** (the brackets are often omitted), and the order in which dx and dy are written indicates the order of integration. This is not the most general definition of the double integral over a rectangular region; however, it is equivalent to the general definition for all the functions we will consider.

EXAMPLE 4 **Evaluating a Double Integral** Evaluate

$$\iint_R (x + y)\,dA \quad \text{over} \quad R = \{(x, y) | 1 \le x \le 3, \quad -1 \le y \le 2\}$$

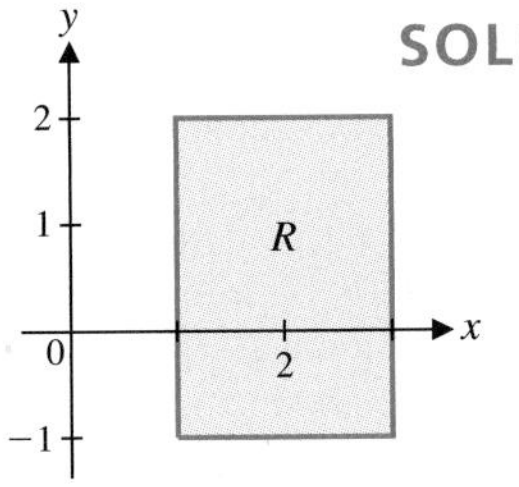

FIGURE 2

SOLUTION Region R is illustrated in Figure 2. We can choose either order of iteration. As a check, we will evaluate the integral both ways:

$$\begin{aligned}
\iint_R (x + y)\,dA &= \int_1^3 \int_{-1}^2 (x + y)\,dy\,dx \\
&= \int_1^3 \left[\left(xy + \frac{y^2}{2}\right)\Big|_{y=-1}^{y=2}\right] dx \\
&= \int_1^3 \left[(2x + 2) - \left(-x + \tfrac{1}{2}\right)\right] dx \\
&= \int_1^3 \left(3x + \tfrac{3}{2}\right) dx \\
&= \left(\tfrac{3}{2}x^2 + \tfrac{3}{2}x\right)\Big|_{x=1}^{x=3} \\
&= \left(\tfrac{27}{2} + \tfrac{9}{2}\right) - \left(\tfrac{3}{2} + \tfrac{3}{2}\right) = 18 - 3 = 15
\end{aligned}$$

$$\begin{aligned}
\iint_R (x + y)\,dA &= \int_{-1}^2 \int_1^3 (x + y)\,dx\,dy \\
&= \int_{-1}^2 \left[\left(\frac{x^2}{2} + xy\right)\Big|_{x=1}^{x=3}\right] dy \\
&= \int_{-1}^2 \left[\left(\tfrac{9}{2} + 3y\right) - \left(\tfrac{1}{2} + y\right)\right] dy \\
&= \int_{-1}^2 (4 + 2y)\,dy \\
&= (4y + y^2)\Big|_{y=-1}^{y=2} \\
&= (8 + 4) - (-4 + 1) = 12 - (-3) = 15
\end{aligned}$$

MATCHED PROBLEM 4 Evaluate

$$\iint_R (2x - y)\,dA \quad \text{over} \quad R = \{(x, y) | -1 \le x \le 5, \quad 2 \le y \le 4\}$$

both ways.

EXAMPLE 5 **Double Integral of an Exponential Function** Evaluate

$$\iint_R 2xe^{x^2+y} \, dA \quad \text{over} \quad R = \{(x, y)|0 \le x \le 1, \ -1 \le y \le 1\}$$

SOLUTION Region R is illustrated in Figure 3.

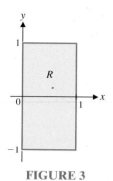

FIGURE 3

$$\iint_R 2xe^{x^2+y} \, dA = \int_{-1}^{1} \int_{0}^{1} 2xe^{x^2+y} \, dx \, dy$$

$$= \int_{-1}^{1} \left[(e^{x^2+y}) \Big|_{x=0}^{x=1} \right] dy$$

$$= \int_{-1}^{1} (e^{1+y} - e^y) \, dy$$

$$= (e^{1+y} - e^y)|_{y=-1}^{y=1}$$

$$= (e^2 - e) - (e^0 - e^{-1})$$

$$= e^2 - e - 1 + e^{-1}$$

MATCHED PROBLEM 5 Evaluate $\quad \iint_R \dfrac{x}{y^2} e^{x/y} \, dA \quad$ over $\quad R = \{(x, y)|0 \le x \le 1, \ 1 \le y \le 2\}.$

Average Value over Rectangular Regions

In Section 6-5, the average value of a function $f(x)$ over an interval $[a, b]$ was defined as

$$\frac{1}{b - a} \int_a^b f(x) \, dx$$

This definition is easily extended to functions of two variables over rectangular regions as follows (notice that the denominator $(b - a)(d - c)$ is simply the area of the rectangle R):

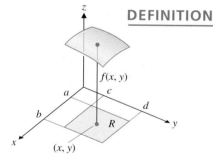

DEFINITION **Average Value over Rectangular Regions**

The **average value** of the function $f(x, y)$ over the rectangle

$$R = \{(x, y)|a \le x \le b, \ c \le y \le d\}$$

is

$$\frac{1}{(b - a)(d - c)} \iint_R f(x, y) \, dA$$

EXAMPLE 6 **Average Value** Find the average value of $f(x, y) = 4 - \frac{1}{2}x - \frac{1}{2}y$ over the rectangle $R = \{(x, y)|0 \le x \le 2, \ 0 \le y \le 2\}.$

SOLUTION Region R is illustrated in Figure 4. We have

$$\frac{1}{(b - a)(d - c)} \iint_R f(x, y) \, dA = \frac{1}{(2 - 0)(2 - 0)} \iint_R \left(4 - \frac{1}{2}x - \frac{1}{2}y \right) dA$$

$$= \frac{1}{4} \int_0^2 \int_0^2 \left(4 - \frac{1}{2}x - \frac{1}{2}y \right) dy \, dx$$

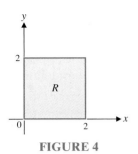

FIGURE 4

$$= \frac{1}{4} \int_0^2 \left[\left(4y - \frac{1}{2}xy - \frac{1}{4}y^2 \right) \Big|_{y=0}^{y=2} \right] dx$$

$$= \frac{1}{4} \int_0^2 (7 - x) \, dx$$

$$= \frac{1}{4} \left(7x - \frac{1}{2}x^2 \right) \Big|_{x=0}^{x=2}$$

$$= \frac{1}{4}(12) = 3$$

Figure 5 illustrates the surface $z = f(x, y)$, and our calculations show that 3 is the average of the z values over the region R.

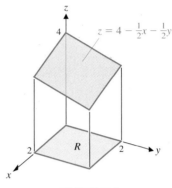

FIGURE 5

MATCHED PROBLEM 6 Find the average value of $f(x, y) = x + 2y$ over the rectangle

$$R = \{(x, y) | 0 \leq x \leq 2, \quad 0 \leq y \leq 1\}$$

Explore & Discuss **1**

(A) Which of the functions $f(x, y) = 4 - x^2 - y^2$ and $g(x, y) = 4 - x - y$ would you guess has the greater average value over the rectangle $R = \{(x, y) | 0 \leq x \leq 1, \quad 0 \leq y \leq 1\}$? Explain.

(B) Use double integrals to check the correctness of your guess in part (A).

■ Volume and Double Integrals

One application of the definite integral of a function with one variable is the calculation of areas, so it is not surprising that the definite integral of a function of two variables can be used to calculate volumes of solids.

THEOREM 1 **VOLUME UNDER A SURFACE**

If $f(x, y) \geq 0$ over a rectangle $R = \{(x, y) | a \leq x \leq b, \quad c \leq y \leq d\}$, then the volume of the solid formed by graphing f over the rectangle R is given by

$$V = \iint_R f(x, y) \, dA$$

A proof of Theorem 1 is left to a more advanced text.

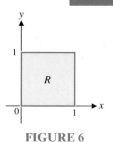

FIGURE 6

EXAMPLE 7

Volume Find the volume of the solid under the graph of $f(x, y) = 1 + x^2 + y^2$ over the rectangle $R = \{(x, y)|0 \leq x \leq 1, \ 0 \leq y \leq 1\}$.

SOLUTION Figure 6 shows the region R, and Figure 7 illustrates the volume under consideration. We have

$$V = \iint\limits_{R} (1 + x^2 + y^2) \, dA$$

$$= \int_0^1 \int_0^1 (1 + x^2 + y^2) \, dx \, dy$$

$$= \int_0^1 \left[\left(x + \tfrac{1}{3}x^3 + xy^2 \right) \Big|_{x=0}^{x=1} \right] dy$$

$$= \int_0^1 \left(\tfrac{4}{3} + y^2 \right) dy$$

$$= \left(\tfrac{4}{3}y + \tfrac{1}{3}y^3 \right)\Big|_{y=0}^{y=1} = \tfrac{5}{3} \text{ cubic units}$$

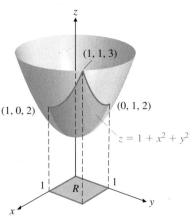

FIGURE 7

MATCHED PROBLEM 7

Find the volume of the solid under the graph of $f(x, y) = 1 + x + y$ over the rectangle $R = \{(x, y)|0 \leq x \leq 1, \ 0 \leq y \leq 2\}$.

Explore & Discuss **2**

Consider the solid under the graph of $f(x, y) = 4 - y^2$ and above the rectangle $R = \{(x, y)|0 \leq x \leq 3, \ 0 \leq y \leq 2\}$.

(A) Explain why each cross section of the solid produced by cutting it with a plane parallel to the yz plane has the same area, and compute that area.

(B) Compute the areas of the cross sections of the solid produced by cutting it with the planes $y = 0$, $y = \tfrac{1}{2}$, and $y = 1$.

(C) Compute the volume of the solid in two different ways.

INSIGHT

Double integrals can be defined over regions that are more general than rectangles. For example, let $R > 0$. Then the function $f(x, y) = \sqrt{R^2 - (x^2 + y^2)}$ can be integrated over the circular region $C = \{(x, y)|x^2 + y^2 \leq R^2\}$. In fact, it can be shown that

$$\iint\limits_{C} \sqrt{R^2 - (x^2 + y^2)} \, dx \, dy = \frac{2\pi R^3}{3}$$

Because $x^2 + y^2 + z^2 = R^2$ is the equation of a sphere of radius R centered at the origin, the double integral over C represents the volume of the upper hemisphere. Therefore, the volume of a sphere of radius R is given by

$$V = \frac{4\pi R^3}{3} \quad \text{Volume of sphere of radius R}$$

Double integrals can also be used to obtain volume formulas for other geometric figures (see Table 1, Appendix C).

Answers to Matched Problems   **1.** (A) $2xy^2 + 3x^2y^4 + C(x)$ (B) $2x^2y + 4x^3y^3 + E(y)$

2. (A) $2x + 3x^2$ (B) $18y + 108y^3$ **3.** (A) 36   (B) 36

4. 12   **5.** $e - 2e^{1/2} + 1$ **6.** 2   **7.** 5 cubic units

Exercise 8-6

A *In Problems 1–8, find each antiderivative. Then use the antiderivative to evaluate the definite integral.*

1. (A) $\int 12x^2y^3 \, dy$ (B) $\int_0^1 12x^2y^3 \, dy$

2. (A) $\int 12x^2y^3 \, dx$ (B) $\int_{-1}^2 12x^2y^3 \, dx$

3. (A) $\int (4x + 6y + 5) \, dx$

(B) $\int_{-2}^3 (4x + 6y + 5) \, dx$

4. (A) $\int (4x + 6y + 5) \, dy$

(B) $\int_1^4 (4x + 6y + 5) \, dy$

5. (A) $\int \frac{x}{\sqrt{y + x^2}} \, dx$ (B) $\int_0^2 \frac{x}{\sqrt{y + x^2}} \, dx$

6. (A) $\int \frac{x}{\sqrt{y + x^2}} \, dy$ (B) $\int_1^5 \frac{x}{\sqrt{y + x^2}} \, dy$

7. (A) $\int \frac{\ln x}{xy} \, dy$ (B) $\int_1^{e^2} \frac{\ln x}{xy} \, dy$

8. (A) $\int \frac{\ln x}{xy} \, dx$ (B) $\int_1^e \frac{\ln x}{xy} \, dx$

B *In Problems 9–16, evaluate each iterated integral. (See the indicated problem for the evaluation of the inner integral.)*

9. $\int_{-1}^2 \int_0^1 12x^2y^3 \, dy \, dx$

(See Problem 1.)

10. $\int_0^1 \int_{-1}^2 12x^2y^3 \, dx \, dy$

(See Problem 2.)

11. $\int_1^4 \int_{-2}^3 (4x + 6y + 5) \, dx \, dy$

(See Problem 3.)

12. $\int_{-2}^3 \int_1^4 (4x + 6y + 5) \, dy \, dx$

(See Problem 4.)

13. $\int_1^5 \int_0^2 \frac{x}{\sqrt{y + x^2}} \, dx \, dy$

(See Problem 5.)

14. $\int_0^2 \int_1^5 \frac{x}{\sqrt{y + x^2}} \, dy \, dx$

(See Problem 6.)

15. $\int_1^e \int_1^{e^2} \frac{\ln x}{xy} \, dy \, dx$

(See Problem 7.)

16. $\int_1^{e^2} \int_1^e \frac{\ln x}{xy} \, dx \, dy$

(See Problem 8.)

Use both orders of iteration to evaluate each double integral in Problems 17–20.

17. $\iint\limits_R xy \, dA; R = \{(x, y) | 0 \le x \le 2, \ 0 \le y \le 4\}$

18. $\iint\limits_R \sqrt{xy} \, dA; R = \{(x, y) | 1 \le x \le 4, \ 1 \le y \le 9\}$

19. $\iint\limits_R (x + y)^5 \, dA;$

$R = \{(x, y) | -1 \le x \le 1, \ 1 \le y \le 2\}$

20. $\iint\limits_R xe^y \, dA; R = \{(x, y) | -2 \le x \le 3, \ 0 \le y \le 2\}$

In Problems 21–24, find the average value of each function over the given rectangle.

21. $f(x, y) = (x + y)^2$;
$R = \{(x, y)|1 \le x \le 5, \quad -1 \le y \le 1\}$

22. $f(x, y) = x^2 + y^2$;
$R = \{(x, y)|-1 \le x \le 2, \quad 1 \le y \le 4\}$

23. $f(x, y) = x/y$; $R = \{(x, y)|1 \le x \le 4, \quad 2 \le y \le 7\}$

24. $f(x, y) = x^2 y^3$;
$R = \{(x, y)|-1 \le x \le 1, \quad 0 \le y \le 2\}$

In Problems 25–28, find the volume of the solid under the graph of each function over the given rectangle.

25. $f(x, y) = 2 - x^2 - y^2$;
$R = \{(x, y)|0 \le x \le 1, \quad 0 \le y \le 1\}$

26. $f(x, y) = 5 - x$;
$R = \{(x, y)|0 \le x \le 5, \quad 0 \le y \le 5\}$

27. $f(x, y) = 4 - y^2$;
$R = \{(x, y)|0 \le x \le 2, \quad 0 \le y \le 2\}$

28. $f(x, y) = e^{-x-y}$;
$R = \{(x, y)|0 \le x \le 1, \quad 0 \le y \le 1\}$

C *Evaluate each double integral in Problems 29–32. Select the order of integration carefully; each problem is easy to do one way and difficult the other.*

29. $\iint\limits_R xe^{xy} \, dA$; $R = \{(x, y)|0 \le x \le 1, \quad 1 \le y \le 2\}$

30. $\iint\limits_R xye^{x^2 y} \, dA$; $R = \{(x, y)|0 \le x \le 1, \quad 1 \le y \le 2\}$

31. $\iint\limits_R \dfrac{2y + 3xy^2}{1 + x^2} \, dA$;
$R = \{(x, y)|0 \le x \le 1, \quad -1 \le y \le 1\}$

32. $\iint\limits_R \dfrac{2x + 2y}{1 + 4y + y^2} \, dA$;
$R = \{(x, y)|1 \le x \le 3, \quad 0 \le y \le 1\}$

33. Show that $\int_0^2 \int_0^2 (1 - y) \, dx \, dy = 0$. Does the double integral represent the volume of a solid? Explain.

34. (A) Find the average values of the functions
$f(x, y) = x + y, g(x, y) = x^2 + y^2$, and
$h(x, y) = x^3 + y^3$ over the rectangle
$$R = \{(x, y)|0 \le x \le 1, \quad 0 \le y \le 1\}$$

(B) Does the average value of $k(x, y) = x^n + y^n$ over the rectangle
$$R_1 = \{(x, y)|0 \le x \le 1, \quad 0 \le y \le 1\}$$
increase or decrease as n increases? Explain.

(C) Does the average value of $k(x, y) = x^n + y^n$ over the rectangle
$$R_2 = \{(x, y)|0 \le x \le 2, \quad 0 \le y \le 2\}$$
increase or decrease as n increases? Explain.

 35. Let $f(x, y) = x^3 + y^2 - e^{-x} - 1$.

(A) Find the average value of $f(x, y)$ over the rectangle
$R = \{(x, y)|-2 \le x \le 2, \quad -2 \le y \le 2\}$.

(B) Graph the set of all points (x, y) in R for which $f(x, y) = 0$.

(C) For which points (x, y) in R is $f(x, y)$ greater than 0? Less than 0? Explain.

 36. Find the dimensions of the square S centered at the origin for which the average value of $f(x, y) = x^2 e^y$ over S is equal to 100.

Applications

37. *Multiplier principle.* Suppose that Congress enacts a one-time-only 10% tax rebate that is expected to infuse y billion, $5 \le y \le 7$, into the economy. If every person and every corporation is expected to spend a proportion $x, 0.6 \le x \le 0.8$, of each dollar received, then, by the **multiplier principle** in economics, the total amount of spending S (in billions of dollars) generated by this tax rebate is given by

$$S(x, y) = \frac{y}{1 - x}$$

What is the average total amount of spending for the indicated ranges of the values of x and y? Set up a double integral and evaluate it.

38. *Multiplier principle.* Repeat Problem 37 if $6 \le y \le 10$ and $0.7 \le x \le 0.9$.

39. *Cobb–Douglas production function.* If an industry invests x thousand labor-hours, $10 \le x \le 20$, and y million, $1 \le y \le 2$, in the production of N thousand units of a certain item, then N is given by
$$N(x, y) = x^{0.75} y^{0.25}$$

What is the average number of units produced for the indicated ranges of x and y? Set up a double integral and evaluate it.

40. *Cobb–Douglas production function.* Repeat Problem 39 for
$$N(x, y) = x^{0.5} y^{0.5}$$
where $10 \le x \le 30$ and $1 \le y \le 3$.

41. *Population distribution.* In order to study the population distribution of a certain species of insect, a biologist has constructed an artificial habitat in the shape of a rectangle 16 feet long and 12 feet wide. The only food available to the insects in this habitat is located at its center. The biologist has determined that the concentration C of insects per square foot at a point d units from the food supply (see the figure) is given approximately by
$$C = 10 - \tfrac{1}{10} d^2$$

What is the average concentration of insects throughout the habitat? Express C as a function of x and y, set up a double integral, and evaluate it.

Figure for 41

42. *Population distribution.* Repeat Problem 41 for a square habitat that measures 12 feet on each side, where the insect concentration is given by

$$C = 8 - \frac{1}{10}d^2$$

43. *Pollution.* A heavy industrial plant located in the center of a small town emits particulate matter into the atmosphere. Suppose that the concentration of particulate matter (in parts per million) at a point d miles from the plant (see the figure) is given by

$$C = 100 - 15d^2$$

If the boundaries of the town form a rectangle 4 miles long and 2 miles wide, what is the average concentration of particulate matter throughout the city? Express C as a function of x and y, set up a double integral, and evaluate it.

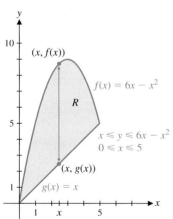

Figure for 43

44. *Pollution.* Repeat Problem 43 if the boundaries of the town form a rectangle 8 miles long and 4 miles wide and the concentration of particulate matter is given by

$$C = 100 - 3d^2$$

45. *Safety research.* Under ideal conditions, if a person driving a car slams on the brakes and skids to a stop, the length of the skid marks (in feet) is given by the formula

$$L = 0.000\,013\,3xy^2$$

where x is the weight of the car (in pounds) and y is the speed of the car (in miles per hour). What is the average length of the skid marks for cars weighing between 2,000 and 3,000 pounds and traveling at speeds between 50 and 60 miles per hour? Set up a double integral and evaluate it.

46. *Safety research.* Repeat Problem 45 for cars weighing between 2,000 and 2,500 pounds and traveling at speeds between 40 and 50 miles per hour.

47. *Psychology.* The intelligence quotient Q for a person with mental age x and chronological age y is given by

$$Q(x, y) = 100\frac{x}{y}$$

In a group of sixth-graders, the mental age varies between 8 and 16 years and the chronological age varies between 10 and 12 years. What is the average intelligence quotient for this group? Set up a double integral and evaluate it.

48. *Psychology.* Repeat Problem 47 for a group with mental ages between 6 and 14 years and chronological ages between 8 and 10 years.

Section 8-7 DOUBLE INTEGRALS OVER MORE GENERAL REGIONS

- Regular Regions
- Double Integrals over Regular Regions
- Reversing the Order of Integration
- Volume and Double Integrals

In this section, we extend the concept of double integration discussed in Section 8-6 to nonrectangular regions. We begin with an example and some new terminology.

■ Regular Regions

Let R be the region graphed in Figure 1. We can describe R with the following inequalities:

$$R = \{(x, y) | x \le y \le 6x - x^2, \quad 0 \le x \le 5\}$$

The region R can be viewed as a union of vertical line segments. For each x in the interval $[0, 5]$, the line segment from the point $(x, g(x))$ to the point $(x, f(x))$ lies in the region R. Any region that can be covered by vertical line segments in this manner is called a *regular x region*.

FIGURE 1

(In Figure 1: $(x, f(x))$; $f(x) = 6x - x^2$; R; $x \le y \le 6x - x^2$; $0 \le x \le 5$; $(x, g(x))$; $g(x) = x$)

Now consider the region S in Figure 2. It can be described with the following inequalities:

$$S = \{(x, y) \mid y^2 \le x \le y + 2, \quad -1 \le y \le 2\}$$

The region S can be viewed as a union of horizontal line segments going from the graph of $h(y) = y^2$ to the graph of $k(y) = y + 2$ on the interval $[-1, 2]$. Regions that can be described in this manner are called *regular y regions*.

In general, *regular regions* are defined as follows:

DEFINITION **Regular Regions**

A region R in the xy plane is a **regular x region** if there exist functions $f(x)$ and $g(x)$ and numbers a and b such that

$$R = \{(x, y) \mid g(x) \le y \le f(x), \quad a \le x \le b\}$$

A region R is a **regular y region** if there exist functions $h(y)$ and $k(y)$ and numbers c and d such that

$$R = \{(x, y) \mid h(y) \le x \le k(y), \quad c \le y \le d\}$$

See Figure 3 for a geometric interpretation.

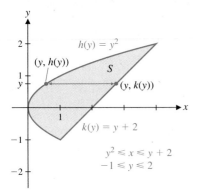

FIGURE 2

INSIGHT

If, for some region R, there is a horizontal line that has a nonempty intersection I with R, and if I is neither a closed interval nor a point, then R is *not* a regular y region. Similarly, if, for some region R, there is a vertical line that has a nonempty intersection I with R, and if I is neither a closed interval nor a point, then R is *not* a regular x region (see Fig. 3).

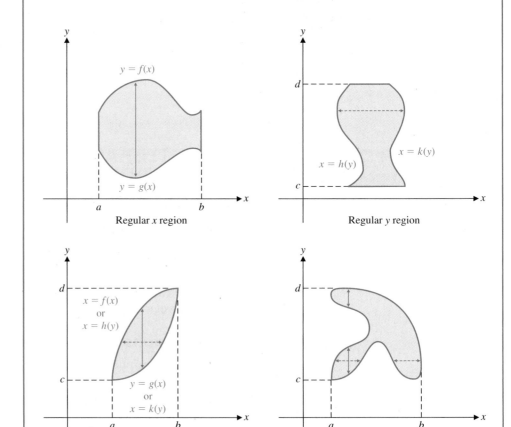

FIGURE 3

EXAMPLE 1

Describing a Regular _x_ Region The region R is bounded by the graphs of $y = 4 - x^2$ and $y = x - 2$, $x \geq 0$, and the y axis. Graph R and use set notation and double inequalities to describe R as a regular x region.

SOLUTION

As the solid line in the following figure indicates, R can be covered by vertical line segments that go from the graph of $y = x - 2$ to the graph of $y = 4 - x^2$. Thus, R is a regular x region. In terms of set notation and double inequalities, we can write

$$R = \{(x, y) \mid x - 2 \leq y \leq 4 - x^2, \ 0 \leq x \leq 2\}$$

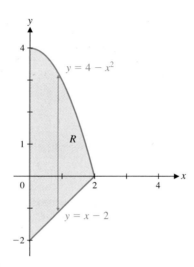

INSIGHT

The region R of Example 1 is also a regular y region, since $R = \{(x, y) \mid 0 \leq x \leq k(y), -2 \leq y \leq 4\}$, where

$$k(y) = \begin{cases} 2 + y & \text{if } -2 \leq y \leq 0 \\ \sqrt{4 - y} & \text{if } 0 \leq y \leq 4 \end{cases}$$

But because $k(y)$ is piecewise defined, this description is more complicated than the description of R in Example 1 as a regular x region.

MATCHED PROBLEM 1

Describe the region R bounded by the graphs of $x = 6 - y$ and $x = y^2$, $y \geq 0$, and the x axis as a regular y region.

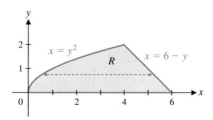

EXAMPLE 2

Describing Regular Regions The region R is bounded by the graphs of $x + y^2 = 9$ and $x + 3y = 9$. Graph R and describe R as a regular x region, a regular y region, both, or neither. Represent R in set notation and with double inequalities.

SOLUTION

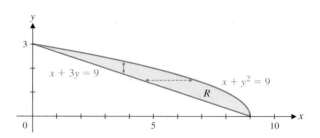

Region R can be covered by vertical line segments that go from the graph of $x + 3y = 9$ to the graph of $x + y^2 = 9$. Thus, R is a regular x region. In order to describe R with inequalities, we must solve each equation for y in terms of x:

$$x + 3y = 9 \qquad\qquad x + y^2 = 9$$
$$3y = 9 - x \qquad\qquad y^2 = 9 - x \qquad \text{We use the positive square}$$
$$y = 3 - \tfrac{1}{3}x \qquad\qquad y = \sqrt{9 - x} \qquad \text{root, since the graph is in the first quadrant.}$$

Thus,

$$R = \{(x, y)|3 - \tfrac{1}{3}x \le y \le \sqrt{9 - x}, \ 0 \le x \le 9\}$$

Since region R also can be covered by horizontal line segments (see the dashed line in the preceding figure) that go from the graph of $x + 3y = 9$ to the graph of $x + y^2 = 9$, it is a regular y region. Now we must solve each equation for x in terms of y:

$$x + 3y = 9 \qquad\qquad x + y^2 = 9$$
$$x = 9 - 3y \qquad\qquad x = 9 - y^2$$

Thus,

$$R = \{(x, y)|9 - 3y \le x \le 9 - y^2, \ 0 \le y \le 3\}$$

MATCHED PROBLEM 2   Repeat Example 2 for the region bounded by the graphs of $2y - x = 4$ and $y^2 - x = 4$, as shown in the following figure:

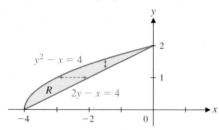

Explore & Discuss **1**

A E I O U

Consider the vowels A, E, I, O, U, written in block letters as shown in the margin, to be regions of the plane. One of the vowels is a regular x region, but not a regular y region; one is a regular y region, but not a regular x region; one is both; two are neither. Explain.

■ Double Integrals over Regular Regions

Now we want to extend the definition of double integration to include regular x regions and regular y regions. Notice in the box that the order of integration now depends on the nature of the region R. If R is a regular x region, we integrate with respect to y first, while if R is a regular y region, we integrate with respect to x first.

It is also important to note that the variable limits of integration (when present) are always on the inner integral, and the constant limits of integration are always on the outer integral.

DEFINITION Double Integration over Regular Regions

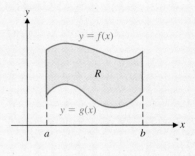

Regular x Region

If $R = \{(x, y) | g(x) \leq y \leq f(x), \quad a \leq x \leq b\}$, then

$$\iint\limits_{R} F(x, y)\, dA = \int_{a}^{b} \left[\int_{g(x)}^{f(x)} F(x, y)\, dy \right] dx$$

Regular y Region

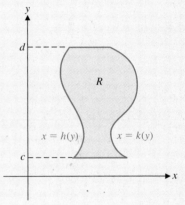

If $R = \{(x, y) | h(y) \leq x \leq k(y), \quad c \leq y \leq d\}$, then

$$\iint\limits_{R} F(x, y)\, dA = \int_{c}^{d} \left[\int_{h(y)}^{k(y)} F(x, y)\, dx \right] dy$$

EXAMPLE 3

Evaluating a Double Integral Evaluate $\iint\limits_{R} 2xy\, dA$, where R is the region bounded by the graphs of $y = -x$ and $y = x^2$, $x \geq 0$, and the graph of $x = 1$.

SOLUTION   From the graph, we can see that R is a regular x region described by

$$R = \{(x, y) | -x \leq y \leq x^2, \quad 0 \leq x \leq 1\}$$

Thus,

$$\iint\limits_{R} 2xy\, dA = \int_{0}^{1} \left[\int_{-x}^{x^2} 2xy\, dy \right] dx$$

$$= \int_{0}^{1} \left[xy^2 \Big|_{y=-x}^{y=x^2} \right] dx$$

$$= \int_{0}^{1} \left[x(x^2)^2 - x(-x)^2 \right] dx$$

$$= \int_{0}^{1} (x^5 - x^3)\, dx$$

$$= \left(\frac{x^6}{6} - \frac{x^4}{4} \right) \Big|_{x=0}^{x=1}$$

$$= \left(\tfrac{1}{6} - \tfrac{1}{4} \right) - (0 - 0) = -\tfrac{1}{12}$$

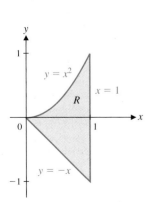

MATCHED PROBLEM 3   Evaluate $\iint\limits_{R} 3xy^2\, dA$, where R is the region in Example 3.

EXAMPLE 4 **Evaluating a Double Integral** Evaluate $\iint\limits_{R} (2x + y)\, dA$, where R is the region bounded by the graphs of $y = \sqrt{x}$, $x + y = 2$, and $y = 0$.

SOLUTION From the graph, we can see that R is a regular y region. After solving each equation for x, we can write

$$R = \{(x, y)\,|\, y^2 \le x \le 2 - y, \ \ 0 \le y \le 1\}$$

Thus,

$$\iint\limits_{R} (2x + y)\, dA = \int_0^1 \left[\int_{y^2}^{2-y} (2x + y)\, dx \right] dy$$

$$= \int_0^1 \left[(x^2 + yx) \Big|_{x=y^2}^{x=2-y} \right] dy$$

$$= \int_0^1 \{[(2 - y)^2 + y(2 - y)] - [(y^2)^2 + y(y^2)]\}\, dy$$

$$= \int_0^1 (4 - 2y - y^3 - y^4)\, dy$$

$$= \left(4y - y^2 - \tfrac{1}{4}y^4 - \tfrac{1}{5}y^5 \right) \Big|_{y=0}^{y=1}$$

$$= \left(4 - 1 - \tfrac{1}{4} - \tfrac{1}{5} \right) - 0 = \tfrac{51}{20}$$

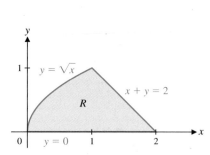

MATCHED PROBLEM 4 Evaluate $\iint\limits_{R} (y - 4x)\, dA$, where R is the region in Example 4.

EXAMPLE 5 **Evaluating a Double Integral** The region R is bounded by the graphs of $y = \sqrt{x}$ and $y = \tfrac{1}{2}x$. Evaluate $\iint\limits_{R} 4xy^3\, dA$ two different ways.

SOLUTION Region R is both a regular x region and a regular y region:

$$R = \{(x, y)\,|\,\tfrac{1}{2}x \le y \le \sqrt{x}, \ \ 0 \le x \le 4\} \quad \text{Regular } x \text{ region}$$
$$R = \{(x, y)\,|\, y^2 \le x \le 2y, \ \ 0 \le y \le 2\} \quad \text{Regular } y \text{ region}$$

Using the first representation (a regular x region), we obtain

$$\iint\limits_{R} 4xy^3\, dA = \int_0^4 \left[\int_{x/2}^{\sqrt{x}} 4xy^3\, dy \right] dx$$

$$= \int_0^4 \left[xy^4 \Big|_{y=x/2}^{y=\sqrt{x}} \right] dx$$

$$= \int_0^4 [x(\sqrt{x})^4 - x(\tfrac{1}{2}x)^4]\, dx$$

$$= \int_0^4 \left(x^3 - \tfrac{1}{16}x^5 \right) dx$$

$$= \left(\tfrac{1}{4}x^4 - \tfrac{1}{96}x^6 \right) \Big|_{x=0}^{x=4}$$

$$= \left(64 - \tfrac{128}{3} \right) - 0 = \tfrac{64}{3}$$

Using the second representation (a regular y region), we obtain

$$\iint\limits_R 4xy^3 \, dA = \int_0^2 \left[\int_{y^2}^{2y} 4xy^3 \, dx \right] dy$$

$$= \int_0^2 \left[2x^2y^3 \Big|_{x=y^2}^{x=2y} \right] dy$$

$$= \int_0^2 \left[2(2y)^2y^3 - 2(y^2)^2y^3 \right] dy$$

$$= \int_0^2 (8y^5 - 2y^7) \, dy$$

$$= \left(\tfrac{4}{3}y^6 - \tfrac{1}{4}y^8 \right) \Big|_{y=0}^{y=2}$$

$$= \left(\tfrac{256}{3} - 64 \right) - 0 = \tfrac{64}{3}$$

MATCHED PROBLEM 5   The region R is bounded by the graphs of $y = x$ and $y = \tfrac{1}{2}x^2$. Evaluate $\iint\limits_R 4x^3y \, dA$ two different ways.

■ Reversing the Order of Integration

Example 5 shows that

$$\iint\limits_R 4xy^3 \, dA = \int_0^4 \left[\int_{x/2}^{\sqrt{x}} 4xy^3 \, dy \right] dx = \int_0^2 \left[\int_{y^2}^{2y} 4xy^3 \, dx \right] dy$$

In general, if R is both a regular x region and a regular y region, the two iterated integrals are equal. In rectangular regions, reversing the order of integration in an iterated integral was a simple matter. As Example 5 illustrates, the process is more complicated in nonrectangular regions. The next example illustrates how to start with an iterated integral and reverse the order of integration. Since we are interested in the reversal process and not in the value of either integral, the integrand will not be specified.

EXAMPLE 6   **Reversing the Order of Integration**   Reverse the order of integration in

$$\int_1^3 \left[\int_0^{x-1} f(x, y) \, dy \right] dx$$

SOLUTION   The order of integration indicates that the region of integration is a regular x region:

$$R = \{(x, y) \mid 0 \le y \le x - 1, \ 1 \le x \le 3\}$$

Graph region R to determine whether it is also a regular y region. The graph shows that R is also a regular y region, and we can write

$$R = \{(x, y) \mid y + 1 \le x \le 3, \ 0 \le y \le 2\}$$

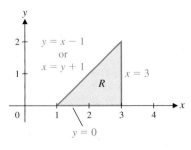

Thus,

$$\int_1^3 \left[\int_0^{x-1} f(x, y) \, dy \right] dx = \int_0^2 \left[\int_{y+1}^3 f(x, y) \, dx \right] dy$$

MATCHED PROBLEM 6   Reverse the order of integration in $\int_2^4 \left[\int_0^{4-x} f(x, y) \, dy \right] dx$.

Explore & Discuss **2**   Explain the difficulty in evaluating $\int_0^2 \int_{x^2}^4 xe^{y^2} dy \, dx$ and how it can be overcome by reversing the order of integration.

■ Volume and Double Integrals

In Section 8-6, we used the double integral to calculate the volume of a solid with a rectangular base. In general, if a solid can be described by the graph of a positive function $f(x, y)$ over a regular region R (not necessarily a rectangle), then the double integral of the function f over the region R still represents the volume of the corresponding solid.

EXAMPLE 7   **Volume** The region R is bounded by the graphs of $x + y = 1$, $y = 0$, and $x = 0$. Find the volume of the solid under the graph of $z = 1 - x - y$ over the region R.

SOLUTION

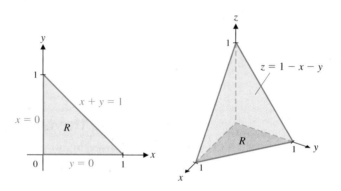

The graph of R indicates that R is both a regular x region and a regular y region. We choose to use the regular x region:

$$R = \{(x, y) \mid 0 \le y \le 1 - x, \ 0 \le x \le 1\}$$

Thus, the volume of the solid is

$$V = \iint\limits_R (1 - x - y) \, dA = \int_0^1 \left[\int_0^{1-x} (1 - x - y) \, dy \right] dx$$

$$= \int_0^1 \left[\left(y - xy - \tfrac{1}{2}y^2 \right) \Big|_{y=0}^{y=1-x} \right] dx$$

$$= \int_0^1 \left[(1 - x) - x(1 - x) - \tfrac{1}{2}(1 - x)^2 \right] dx$$

$$= \int_0^1 \left(\tfrac{1}{2} - x + \tfrac{1}{2}x^2 \right) dx$$

$$= \left(\tfrac{1}{2}x - \tfrac{1}{2}x^2 + \tfrac{1}{6}x^3 \right) \Big|_{x=0}^{x=1}$$

$$= \left(\tfrac{1}{2} - \tfrac{1}{2} + \tfrac{1}{6} \right) - 0 = \tfrac{1}{6}$$

MATCHED PROBLEM 7   The region R is bounded by the graphs of $y + 2x = 2$, $y = 0$, and $x = 0$. Find the volume of the solid under the graph of $z = 2 - 2x - y$ over the region R. [*Hint:* Sketch the region first; the solid does not have to be sketched.]

Answers to Matched Problems

1. $R = \{(x, y)|y^2 \le x \le 6 - y, \; 0 \le y \le 2\}$

2. R is both a regular x region and a regular y region:

$R = \{(x, y) \mid \frac{1}{2}x + 2 \le y \le \sqrt{x + 4}, \; -4 \le x \le 0\}$

$R = \{(x, y)|y^2 - 4 \le x \le 2y - 4, \; 0 \le y \le 2\}$

3. $\frac{13}{40}$ **4.** $-\frac{77}{20}$ **5.** $\frac{16}{3}$

6. $\int_0^2 \int_2^{4-y} f(x, y)\, dx\, dy$

7. $\frac{2}{3}$

Exercise 8-7

A *In Problems 1–6, graph the region R bounded by the graphs of the equations. Use set notation and double inequalities to describe R as a regular x region and as a regular y region in Problems 1 and 2, and as a regular x region or a regular y region, whichever is simpler, in Problems 3–6.*

1. $y = 4 - x^2, y = 0, 0 \le x \le 2$

2. $y = x^2, y = 9, 0 \le x \le 3$

3. $y = x^3, y = 12 - 2x, x = 0$

4. $y = 5 - x, y = 1 + x, y = 0$

5. $y^2 = 2x, y = x - 4$

6. $y = 4 + 3x - x^2, x + y = 4$

Evaluate each integral in Problems 7–10.

7. $\int_0^1 \int_0^x (x + y)\, dy\, dx$

8. $\int_0^2 \int_0^y xy\, dx\, dy$

9. $\int_0^1 \int_{y^3}^{\sqrt{y}} (2x + y)\, dx\, dy$

10. $\int_1^4 \int_x^{x^2} (x^2 + 2y)\, dy\, dx$

B *In Problems 11–14, give a verbal description of the region R and determine whether R is a regular x region, a regular y region, both, or neither.*

11. $R = \{(x, y)||x| \le 2, \; |y| \le 3\}$

12. $R = \{(x, y)|1 \le x^2 + y^2 \le 4\}$

13. $R = \{(x, y)|x^2 + y^2 \ge 1, \; |x| \le 2, \; 0 \le y \le 2\}$

14. $R = \{(x, y)||x| + |y| \le 1\}$

In Problems 15–20, use the description of the region R to evaluate the indicated integral.

15. $\iint_R (x^2 + y^2)\, dA;$

$R = \{(x, y)|0 \le y \le 2x, \; 0 \le x \le 2\}$

16. $\iint_R 2x^2 y\, dA;$

$R = \{(x, y)|0 \le y \le 9 - x^2, \; -3 \le x \le 3\}$

17. $\iint_R (x + y - 2)^3\, dA;$

$R = \{(x, y)|0 \le x \le y + 2, \; 0 \le y \le 1\}$

18. $\iint_R (2x + 3y)\, dA;$

$R = \{(x, y)|y^2 - 4 \le x \le 4 - 2y, \; 0 \le y \le 2\}$

19. $\iint_R e^{x+y}\, dA;$

$R = \{(x, y)|-x \le y \le x, \; 0 \le x \le 2\}$

20. $\iint_R \frac{x}{\sqrt{x^2 + y^2}}\, dA;$

$R = \{(x, y)|0 \le x \le \sqrt{4y - y^2}, \; 0 \le y \le 2\}$

In Problems 21–26, graph the region R bounded by the graphs of the indicated equations. Describe R in set notation with double inequalities, and evaluate the indicated integral.

21. $y = x + 1, \quad y = 0, \quad x = 0, \quad x = 1;$

$\iint_R \sqrt{1 + x + y}\, dA$

22. $y = x^2, \quad y = \sqrt{x}; \quad \iint_R 12xy\, dA$

23. $y = 4x - x^2, \quad y = 0; \quad \iint_R \sqrt{y + x^2}\, dA$

24. $x = 1 + 3y, \quad x = 1 - y, \quad y = 1;$

$\iint_R (x + y + 1)^3\, dA$

25. $y = 1 - \sqrt{x}, \quad y = 1 + \sqrt{x}, \quad x = 4;$

$\iint_R x(y - 1)^2\, dA$

26. $y = \frac{1}{2}x, \quad y = 6 - x, \quad y = 1; \quad \iint_R \frac{1}{x + y}\, dA$

In Problems 27–32, evaluate each integral. Graph the region of integration, reverse the order of integration, and then evaluate the integral with the order reversed.

27. $\int_0^3 \int_0^{3-x} (x + 2y)\, dy\, dx$

28. $\int_0^2 \int_0^y (y - x)^4\, dx\, dy$

29. $\int_0^1 \int_0^{1-x^2} x\sqrt{y}\, dy\, dx$

30. $\int_0^2 \int_{x^3}^{4x} (1 + 2y) \, dy \, dx$

31. $\int_0^4 \int_{x/4}^{\sqrt{x/2}} x \, dy \, dx$

32. $\int_0^4 \int_{y^2/4}^{2\sqrt{y}} (1 + 2xy) \, dx \, dy$

In Problems 33–36, find the volume of the solid under the graph of $f(x, y)$ over the region R bounded by the graphs of the indicated equations. Sketch the region R; the solid does not have to be sketched.

33. $f(x, y) = 4 - x - y$; R is the region bounded by the graphs of $x + y = 4$, $y = 0$, $x = 0$

34. $f(x, y) = (x - y)^2$; R is the region bounded by the graphs of $y = x$, $y = 2$, $x = 0$

35. $f(x, y) = 4$; R is the region bounded by the graphs of $y = 1 - x^2$ and $y = 0$ for $0 \le x \le 1$

36. $f(x, y) = 4xy$; R is the region bounded by the graphs of $y = \sqrt{1 - x^2}$ and $y = 0$ for $0 \le x \le 1$

C In Problems 37–40, reverse the order of integration for each integral. Evaluate the integral with the order reversed. Do not attempt to evaluate the integral in the original form.

37. $\int_0^2 \int_{x^2}^4 \frac{4x}{1 + y^2} \, dy \, dx$

38. $\int_0^1 \int_y^1 \sqrt{1 - x^2} \, dx \, dy$

39. $\int_0^1 \int_{y^2}^1 4ye^{x^2} \, dx \, dy$

40. $\int_0^4 \int_{\sqrt{x}}^2 \sqrt{3x + y^2} \, dy \, dx$

 In Problems 41–46, use a graphing calculator to graph the region R bounded by the graphs of the indicated equations. Use approximation techniques to find intersection points correct to two decimal places. Describe R in set notation with double inequalities, and evaluate the indicated integral correct to two decimal places.

41. $y = 1 + \sqrt{x}$, $\quad y = x^2$, $\quad x = 0$; $\quad \iint_R x \, dA$

42. $y = 1 + \sqrt[3]{x}$, $y = x$, $\quad x = 0$; $\quad \iint_R x \, dA$

43. $y = \sqrt[3]{x}$, $\quad y = 1 - x$, $\quad y = 0$; $\quad \iint_R 24xy \, dA$

44. $y = x^3$, $\quad y = 1 - x$, $\quad y = 0$; $\quad \iint_R 48xy \, dA$

45. $y = e^{-x}$, $\quad y = 3 - x$; $\quad \iint_R 4y \, dA$

46. $y = e^x$, $\quad y = 2 + x$; $\quad \iint_R 8y \, dA$

CHAPTER 8 REVIEW

Important Terms, Symbols, and Concepts

8-1 Functions of Several Variables

Examples

- An equation of the form $z = f(x, y)$ describes a **function of two independent variables** if, for each permissible ordered pair (x, y), there is one and only one value of z determined by $f(x, y)$. The variables x and y are **independent variables,** and z is a **dependent variable.** The set of all ordered pairs of permissible values of x and y is the **domain** of the function, and the set of all corresponding values $f(x, y)$ is the **range.** Functions of more than two independent variables are defined similarly.

Ex. 1, p. 464
Ex. 2, p. 464
Ex. 3, p. 464
Ex. 4, p. 466
Ex. 5, p. 467

- The graph of $z = f(x, y)$ consists of all triples (x, y, z) in a **three-dimensional coordinate system** that satisfy the equation. The graphs of the functions $z = f(x, y) = x^2 + y^2$ and $z = g(x, y) = x^2 - y^2$, for example, are **surfaces;** the first has a local minimum, and the second has a **saddle point,** at $(0, 0)$.

Ex. 6, p. 467
Ex. 7, p. 469

8-2 Partial Derivatives

- If $z = f(x, y)$, then the **partial derivative of f with respect to x,** denoted $\partial z/\partial x$, f_x, or $f_x(x, y)$, is

Ex. 1, p. 474
Ex. 2, p. 474
Ex. 3, p. 475
Ex. 4, p. 475

$$\frac{\partial z}{\partial x} = \lim_{h \to 0} \frac{f(x + h, y) - f(x, y)}{h}$$

Similarly, the **partial derivative of f with respect to y,** denoted $\partial z/\partial y$, f_y, or $f_y(x, y)$, is

$$\frac{\partial z}{\partial y} = \lim_{k \to 0} \frac{f(x, y + k) - f(x, y)}{k}$$

The partial derivatives $\partial z/\partial x$ and $\partial z/\partial y$ are said to be **first-order partial derivatives.**

- There are four **second-order partial derivatives** of $z = f(x, y)$:

Ex. 5, p. 477

$$f_{xx} = f_{xx}(x, y) = \frac{\partial^2 z}{\partial x^2} = \frac{\partial}{\partial x}\left(\frac{\partial z}{\partial x}\right)$$

$$f_{xy} = f_{xy}(x, y) = \frac{\partial^2 z}{\partial y\, \partial x} = \frac{\partial}{\partial y}\left(\frac{\partial z}{\partial x}\right)$$

$$f_{yx} = f_{yx}(x, y) = \frac{\partial^2 z}{\partial x\, \partial y} = \frac{\partial}{\partial x}\left(\frac{\partial z}{\partial y}\right)$$

$$f_{yy} = f_{yy}(x, y) = \frac{\partial^2 z}{\partial y^2} = \frac{\partial}{\partial y}\left(\frac{\partial z}{\partial y}\right)$$

8-3 Maxima and Minima

- If $f(a, b) \geq f(x, y)$ for all (x, y) in a circular region in the domain of f with (a, b) as center, then $f(a, b)$ is a **local maximum.** If $f(a, b) \leq f(x, y)$ for all (x, y) in such a region, then $f(a, b)$ is a **local minimum.**
- If a function $f(x, y)$ has a local maximum or minimum at the point (a, b), and f_x and f_y exist at (a, b), then both first-order partial derivatives equal 0 at (a, b) [Theorem 1, p. 483].
- The second-derivative test for local extrema (Theorem 2, p. 483) gives conditions on the first- and second-order partial derivatives of $f(x, y)$ which guarantee that $f(a, b)$ is a local maximum, local minimum, or saddle point.

Ex. 1, p. 484
Ex. 2, p. 484
Ex. 3, p. 486
Ex. 4, p. 487

8-4 Maxima and Minima Using Lagrange Multipliers

- The method of Lagrange multipliers can be used to find local extrema of a function $z = f(x, y)$ subject to the constraint $g(x, y) = 0$. A procedure that lists the key steps in the method is given on page 492.
- The method of Lagrange multipliers can be extended to functions with arbitrarily many independent variables with one or more constraints (see Theorem 1, p. 491, and Theorem 2, p. 495, for the method when there are two and three independent variables, respectively).

Ex. 1, p. 492
Ex. 2, p. 494
Ex. 3, p. 496

8-5 Method of Least Squares

- **Linear regression** is the process of fitting a line $y = ax + b$ to a set of data points $(x_1, y_1), (x_2, y_2), \dots, (x_n, y_n)$ by using the **method of least squares.**
- We minimize $F(a, b) = \sum_{k=1}^{n}(y_k - ax_k - b)^2$, the **sum of the squares of the residuals,** by computing the first-order partial derivatives of F and setting them equal to 0. Solving for a and b gives the formulas

Ex. 1, p. 503

$$a = \frac{n\left(\sum\limits_{k=1}^{n}x_k y_k\right) - \left(\sum\limits_{k=1}^{n}x_k\right)\left(\sum\limits_{k=1}^{n}y_k\right)}{n\left(\sum\limits_{k=1}^{n}x_k^2\right) - \left(\sum\limits_{k=1}^{n}x_k\right)^2}$$

$$b = \frac{\sum\limits_{k=1}^{n}y_k - a\left(\sum\limits_{k=1}^{n}x_k\right)}{n}$$

- Graphing calculators have built-in routines to calculate linear—as well as quadratic, cubic, quartic, logarithmic, exponential, power, and trigonometric—regression equations.

Ex. 2, p. 505

8-6 Double Integrals over Rectangular Regions

- The **double integral** of a function $f(x, y)$ over a rectangle

$$R = \{(x, y)\,|\,a \leq x \leq b, \quad c \leq y \leq d\}$$

is

Ex. 1, p. 510
Ex. 2, p. 511
Ex. 3, p. 511
Ex. 4, p. 513
Ex. 5, p. 514

$$\iint\limits_{R} f(x, y)\, dA = \int_{a}^{b}\left[\int_{c}^{d} f(x, y)\, dy\right] dx$$

$$= \int_{c}^{d}\left[\int_{a}^{b} f(x, y)\, dx\right] dy$$

- In the double integral $\iint_R f(x, y)\, dA$, $f(x, y)$ is called the **integrand** and R is called the **region of integration.** The expression dA indicates that this is an integral over a two-dimensional region. The integrals

$$\int_a^b \left[\int_c^d f(x, y)\, dy \right] dx \quad \text{and} \quad \int_c^d \left[\int_a^b f(x, y)\, dx \right] dy$$

are referred to as **iterated integrals** (the brackets are often omitted), and the order in which dx and dy are written indicates the order of integration.

- The **average value** of the function $f(x, y)$ over the rectangle   Ex. 6, p. 514

$$R = \{(x, y)\,|\,a \le x \le b, \ c \le y \le d\}$$

is

$$\frac{1}{(b - a)(d - c)} \iint_R f(x, y)\, dA$$

- If $f(x, y) \ge 0$ over a rectangle $R = \{(x, y)\,|\,a \le x \le b, c \le y \le d\}$, then the volume of the solid   Ex. 7, p. 516
formed by graphing f over the rectangle R is given by

$$V = \iint_R f(x, y)\, dA$$

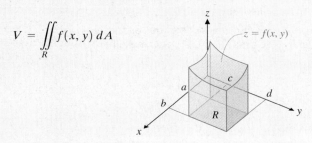

8-7 Double Integrals over More General Regions

- A region R in the xy plane is a **regular x region** if there exist functions $f(x)$ and $g(x)$ and numbers a and   Ex. 1, p. 521
b such that

$$R = \{(x, y)\,|\,g(x) \le y \le f(x), \ a \le x \le b\}$$

- A region R in the xy plane is a **regular y region** if there exist functions $h(y)$ and $k(y)$ and numbers c and   Ex. 2, p. 521
d such that

$$R = \{(x, y)\,|\,h(y) \le x \le k(y), \ c \le y \le d\}$$

- The double integral of a function $F(x, y)$ over a regular x region $R = \{(x, y)\,|\,g(x) \le y \le f(x),$ Ex. 3, p. 523
$a \le x \le b\}$ is   Ex. 4, p. 524
   Ex. 5, p. 524
$$\iint_R F(x, y)\, dA = \int_a^b \left[\int_{g(x)}^{f(x)} F(x, y)\, dy \right] dx$$ Ex. 6, p. 525
   Ex. 7, p. 526

- The double integral of a function $F(x, y)$ over a regular y region $R = \{(x, y)\,|\,h(y) \le x \le k(y),$
$c \le y \le d\}$ is

$$\iint_R F(x, y)\, dA = \int_c^d \left[\int_{h(y)}^{k(y)} F(x, y)\, dx \right] dy$$

REVIEW EXERCISE

*Work through all the problems in this chapter review and
check your answers in the back of the book. Answers to all review problems are there, along with section numbers in italics
to indicate where each type of problem is discussed. Where
weaknesses show up, review appropriate sections of the text.*

A

1. For $f(x, y) = 2,000 + 40x + 70y$, find $f(5, 10)$,
$f_x(x, y)$, and $f_y(x, y)$.

2. For $z = x^3 y^2$, find $\partial^2 z/\partial x^2$ and $\partial^2 z/\partial x\, \partial y$.

3. Evaluate $\int (6xy^2 + 4y)\, dy$.

4. Evaluate $\int (6xy^2 + 4y)\, dx$.

5. Evaluate $\int_0^1 \int_0^1 4xy\, dy\, dx$.

6. For $f(x, y) = 6 + 5x - 2y + 3x^2 + x^3$, find $f_x(x, y)$ and
$f_y(x, y)$, and explain why $f(x, y)$ has no local extrema.

B

7. For $f(x, y) = 3x^2 - 2xy + y^2 - 2x + 3y - 7$, find $f(2, 3)$, $f_y(x, y)$, and $f_y(2, 3)$.

8. For $f(x, y) = -4x^2 + 4xy - 3y^2 + 4x + 10y + 81$, find $[f_{xx}(2, 3)][f_{yy}(2, 3)] - [f_{xy}(2, 3)]^2$

9. If $f(x, y) = x + 3y$ and $g(x, y) = x^2 + y^2 - 10$, find the critical points of $F(x, y, \lambda) = f(x, y) + \lambda g(x, y)$.

10. Use the least squares line for the data in the table to estimate y when $x = 10$.

x	y
2	12
4	10
6	7
8	3

11. For $R = \{(x, y)|-1 \le x \le 1, \ 1 \le y \le 2\}$, evaluate the following in two ways:

$$\iint_R (4x + 6y) \, dA$$

12. For $R = \{(x, y)|\sqrt{y} \le x \le 1, \ 0 \le y \le 1\}$, evaluate

$$\iint_R (6x + y) \, dA$$

C

13. For $f(x, y) = e^{x^2 + 2y}$, find f_x, f_y, and f_{xy}.

14. For $f(x, y) = (x^2 + y^2)^5$, find f_x and f_{xy}.

15. Find all critical points and test for extrema for

$$f(x, y) = x^3 - 12x + y^2 - 6y$$

16. Use Lagrange multipliers to maximize $f(x, y) = xy$ subject to $2x + 3y = 24$.

17. Use Lagrange multipliers to minimize $f(x, y, z) = x^2 + y^2 + z^2$ subject to $2x + y + 2z = 9$.

18. Find the least squares line for the data in the following table.

x	y	x	y
10	50	60	80
20	45	70	85
30	50	80	90
40	55	90	90
50	65	100	110

19. Find the average value of $f(x, y) = x^{2/3}y^{1/3}$ over the rectangle

$$R = \{(x, y)|-8 \le x \le 8, \ 0 \le y \le 27\}$$

20. Find the volume of the solid under the graph of $z = 3x^2 + 3y^2$ over the rectangle

$$R = \{(x, y)|0 \le x \le 1, \ -1 \le y \le 1\}$$

21. Without doing any computation, predict the average value of $f(x, y) = x + y$ over the rectangle $R = \{(x, y)|-10 \le x \le 10, \ -10 \le y \le 10\}$. Then check the correctness of your prediction by evaluating a double integral.

22. (A) Find the dimensions of the square S centered at the origin such that the average value of

$$f(x, y) = \frac{e^x}{y + 10}$$

over S is equal to 5.

(B) Is there a square centered at the origin over which

$$f(x, y) = \frac{e^x}{y + 10}$$

has average value 0.05? Explain.

23. Explain why the function $f(x, y) = 4x^3 - 5y^3$, subject to the constraint $3x + 2y = 7$, has no maxima or minima.

24. Find the volume of the solid under the graph of $F(x, y) = 60x^2y$ over the region R bounded by the graph of $x + y = 1$ and the coordinate axes.

APPLICATIONS

25. *Maximizing profit.* A company produces x units of product A and y units of product B (both in hundreds per month). The monthly profit equation (in thousands of dollars) is found to be

$$P(x, y) = -4x^2 + 4xy - 3y^2 + 4x + 10y + 81$$

(A) Find $P_x(1, 3)$ and interpret the results.

(B) How many of each product should be produced each month to maximize profit? What is the maximum profit?

26. *Minimizing material.* A rectangular box with no top and six compartments (see the figure) is to have a volume of 96 cubic inches. Find the dimensions that will require the least amount of material.

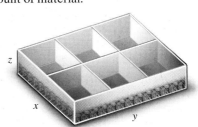

Figure for 26

27. *Profit.* A company's annual profits (in millions of dollars) over a 5-year period are given in the table. Use the least squares line to estimate the profit for the sixth year.

Year	Profit
1	2
2	2.5
3	3.1
4	4.2
5	4.3

28. *Productivity.* The Cobb–Douglas production function for a product is

$$N(x, y) = 10x^{0.8}y^{0.2}$$

where x is the number of units of labor and y is the number of units of capital required to produce N units of the product.

(A) Find the marginal productivity of labor and the marginal productivity of capital at $x = 40$ and $y = 50$. For the greatest increase in productivity, should management encourage increased use of labor or increased use of capital?

(B) If each unit of labor costs $100, each unit of capital costs $50, and $10,000 is budgeted for production of this product, use the method of Lagrange multipliers to determine the allocations of labor and capital that will maximize the number of units produced and find the maximum production. Find the marginal productivity of money and approximate the increase in production that would result from an increase of $2,000 in the amount budgeted for production.

(C) If $50 \le x \le 100$ and $20 \le y \le 40$, find the average number of units produced. Set up a double integral, and evaluate it.

29. *Marine biology.* The function used for timing dives with scuba gear is

$$T(V, x) = \frac{33V}{x + 33}$$

where T is the time of the dive in minutes, V is the volume of air (in cubic feet, at sea-level pressure) compressed into tanks, and x is the depth of the dive in feet. Find $T_x(70, 17)$ and interpret the results.

30. *Pollution.* A heavy industrial plant located in the center of a small town emits particulate matter into the atmosphere. Suppose that the concentration of particulate matter (in parts per million) at a point d miles from the plant is given by

$$C = 100 - 24d^2$$

If the boundaries of the town form a square 4 miles long and 4 miles wide, what is the average concentration of particulate matter throughout the town? Express C as a function of x and y, set up a double integral and evaluate it.

31. *Sociology.* Joseph Cavanaugh, a sociologist, found that the number of long-distance telephone calls, n, between two cities during a given period varied (approximately) jointly as the populations P_1 and P_2 of the two cities and varied inversely as the distance d between the cities. An equation for a period of 1 week is

$$n(P_1, P_2, d) = 0.001\frac{P_1P_2}{d}$$

Find $n(100,000, 50,000, 100)$.

32. *Education.* At the beginning of the semester, students in a foreign language course are given a proficiency exam. The same exam is given at the end of the semester. The results for 5 students are shown in the accompanying table. Use the least squares line to estimate the second exam score of a student who scored 40 on the first exam.

First Exam	Second Exam
30	60
50	75
60	80
70	85
90	90

33. *Population density.* The following table gives the U.S. population per square mile for the years 1900–2000:

U.S. Population Density			
Year	Population (per Square Mile)	Year	Population (per Square Mile)
1900	25.6	1960	50.6
1910	31.0	1970	57.4
1920	35.6	1980	64.0
1930	41.2	1990	70.4
1940	44.2	2000	77.8
1950	50.7		

(A) Find the least squares line for the data, using $x = 0$ for 1900.

(B) Use the least squares line to estimate the population density in the United States in the year 2020.

 (C) Now use quadratic regression and exponential regression to obtain the estimate of part (B).

34. *Life expectancy.* The following table gives life expectancies for males and females in a sample of Central and South American countries:

Life Expectancies for Central and South American Countries			
Males	Females	Males	Females
62.30	67.50	70.15	74.10
68.05	75.05	62.93	66.58
72.40	77.04	68.43	74.88
63.39	67.59	66.68	72.80
55.11	59.43		

(A) Find the least squares line for the data.

(B) Use least squares line to estimate the life expectancy of a female in a Central or South American country in which the life expectancy for males is 60 years.

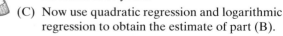 (C) Now use quadratic regression and logarithmic regression to obtain the estimate of part (B).

Trigonometric Functions

CHAPTER 9

9-1 Trigonometric Functions
Review

9-2 Derivatives of
Trigonometric Functions

9-3 Integration of
Trigonometric Functions

Chapter 9 Review

Review Exercise

INTRODUCTION

Thus far, we have restricted our attention to algebraic, logarithmic, and exponential functions. These functions were used to model many real-life situations from business, economics, and the life and social sciences. Now we turn our attention to another important class of functions, called the *trigonometric functions.* These functions are particularly useful in describing periodic phenomena—that is, phenomena which repeat in cycles. Consider the sunrise times for a 2-year period starting January 1, as pictured in the figure. We see that the cycle repeats after 1 year. Business cycles, blood pressure in the aorta, seasonal growth, water waves, and amounts of pollution in the atmosphere are often periodic and can be modeled with similar types of graphs.

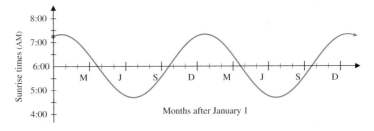

We assume that the reader has had a course in trigonometry. In Section 9-1, we provide a brief review of the topics that are most important for our purposes.

Section 9-1 TRIGONOMETRIC FUNCTIONS REVIEW

- Angles: Degree–Radian Measure
- Trigonometric Functions
- Graphs of the Sine and Cosine Functions
- Four Other Trigonometric Functions

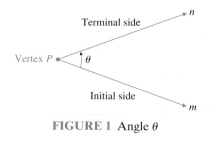

FIGURE 1 Angle θ

■ Angles: Degree–Radian Measure

In a plane, an **angle** is formed by rotating a ray m, called the **initial side** of the angle, around its endpoint until the ray coincides with a ray n, called the **terminal side** of the angle. The common endpoint P of m and n is called the **vertex** (see Fig. 1).

There is no restriction on the amount or direction of rotation. A counterclockwise rotation produces a **positive** angle (Fig. 2A), and a clockwise rotation produces a **negative** angle (Fig. 2B). Two different angles may have the same initial and terminal sides, as shown in Figure 2C. Such angles are said to be **coterminal.**

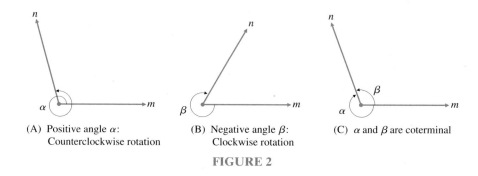

(A) Positive angle α:
Counterclockwise rotation

(B) Negative angle β:
Clockwise rotation

(C) α and β are coterminal

FIGURE 2

There are two widely used measures of angles: the *degree* and the *radian*. When a central angle of a circle is subtended by an arc $\frac{1}{360}$ of the circumference of the circle, the angle is said to have **degree measure 1,** written **1°** (see Fig. 3A). It follows that a central angle subtended by an arc $\frac{1}{4}$ of the circumference has degree measure 90, $\frac{1}{2}$ of the circumference has degree measure 180, and the whole circumference of a circle has degree measure 360.

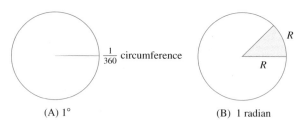

(A) 1°

(B) 1 radian

FIGURE 3 Degree and radian measure

The other measure of angles, which we use extensively in the next two sections, is radian measure. A central angle subtended by an arc of length equal to the radius (R) of the circle is said to have **radian measure 1,** written **1 radian** or **1 rad** (see Fig. 3B). In general, a central angle subtended by an arc of length s has radian measure determined as follows:

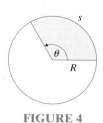

FIGURE 4

$$\theta_{\text{rad}} = \text{radian measure of } \theta = \frac{\text{arc length}}{\text{radius}} = \frac{s}{R}$$

(See Figure 4.) [*Note:* If $R = 1$, then $\theta_{\text{rad}} = s$.]

What is the radian measure of an angle of 180°? A central angle of 180° is subtended by an arc that is $\frac{1}{2}$ of the circumference of a circle. Thus,

$$s = \frac{C}{2} = \frac{2\pi R}{2} = \pi R \qquad \text{and} \qquad \theta_{\text{rad}} = \frac{s}{R} = \frac{\pi R}{R} = \pi \text{ rad}$$

The following proportion can be used to convert degree measure to radian measure, and vice versa:

Degree–Radian Conversion

$$\frac{\theta_{\text{deg}}}{180°} = \frac{\theta_{\text{rad}}}{\pi \text{ rad}}$$

EXAMPLE 1	**From Degrees to Radians** Find the radian measure of 1°.

SOLUTION

$$\frac{1°}{180°} = \frac{\theta_{\text{rad}}}{\pi \text{ rad}}$$

$$\theta_{\text{rad}} = \frac{\pi}{180} \text{rad} \approx 0.0175 \text{ rad}$$

MATCHED PROBLEM 1

Find the degree measure of 1 rad.

A comparison of degree and radian measure for a few important angles is given in the following table:

Radian	0	$\pi/6$	$\pi/4$	$\pi/3$	$\pi/2$	π	2π
Degree	0	30°	45°	60°	90°	180°	360°

An angle in a rectangular coordinate system is said to be in **standard position** if its vertex is at the origin and its initial side is on the positive *x* axis. Figure 5 shows three angles in standard position.

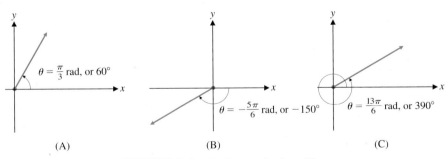

(A)   (B)   (C)

FIGURE 5 Angles in standard position

Trigonometric Functions

Let us locate a unit circle (radius 1) in a coordinate system with center at the origin (Fig. 6). The terminal side of any angle in standard position will pass through this circle at some point *P*. The abscissa of this point *P* is called the **cosine of *θ*** (abbreviated **cos *θ***), and the ordinate of the point is the **sine of *θ*** (abbreviated **sin *θ***). Thus, the set of all ordered pairs of the form $(\theta, \cos \theta)$ and the set of all ordered pairs of the form $(\theta, \sin \theta)$ constitute, respectively, the **cosine** and **sine functions.** The

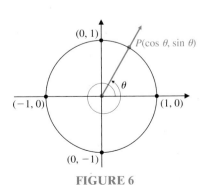

FIGURE 6

domain of these two functions is the set of all angles, positive or negative, with measure either in degrees or radians. The **range** is a subset of the set of real numbers.

It is desirable, and necessary for our work in calculus, to define these two trigonometric functions in terms of real-number domains. This is easily done as follows:

DEFINITION **Sine and Cosine Functions with Real-Number Domains**

For any real number x,

$$\sin x = \sin(x \text{ radians}) \qquad \cos x = \cos(x \text{ radians})$$

EXAMPLE 2 **Evaluating Sine and Cosine Functions** Referring to Figure 6, find

(A) $\cos 90°$ (B) $\sin(-\pi/2 \text{ rad})$ (C) $\cos \pi$

SOLUTION (A) The terminal side of an angle of degree measure 90 passes through $(0, 1)$ on the unit circle. This point has abscissa 0. Thus,

$$\cos 90° = 0$$

(B) The terminal side of an angle of radian measure $-\pi/2$ $(-90°)$ passes through $(0, -1)$ on the unit circle. This point has ordinate -1. Thus,

$$\sin\left(-\frac{\pi}{2} \text{rad}\right) = -1$$

(C) $\cos \pi = \cos(\pi \text{ rad}) = -1$, since the terminal side of an angle of radian measure $\pi (180°)$ passes through $(-1, 0)$ on the unit circle and this point has abscissa -1.

MATCHED PROBLEM 2 Referring to Figure 6, find

(A) $\sin 180°$ (B) $\cos(2\pi \text{ rad})$ (C) $\sin(-\pi)$

Explore & Discuss **1** (A) For the sine and cosine functions with angle domains, discuss the range of each function by referring to Figure 6.

(B) For the sine and cosine functions with real-number domains, discuss the range of each function by referring to Figure 6.

Finding the value of either the sine or the cosine function for any angle or any real number by direct use of the definition is not easy. Calculators with $\boxed{\text{sin}}$ and $\boxed{\text{cos}}$ keys are used. Calculators generally have degree and radian options, so we can use a calculator to evaluate these functions for most of the real numbers in which we might have an interest. The following table includes a few values produced by a calculator in radian mode:

x	1	-7	35.26	-105.9
$\sin x$	0.8415	-0.6570	-0.6461	0.7920
$\cos x$	0.5403	0.7539	-0.7632	0.6105

Explore & Discuss **2**

Many errors in trigonometry can be traced to having the calculator in the wrong mode, radian instead of degree, or vice versa, when performing calculations. The window display from a calculator shown here gives two different values for cos 30. Experiment with your calculator and explain the discrepancy.

```
cos 30
       .1542514499
cos 30
       .8660254038
```

Exact values of the sine and cosine functions can be obtained for multiples of the special angles shown in the triangles in Figure 7, because these triangles can be used to find the coordinate of the intersection of the terminal side of each angle with the unit circle.

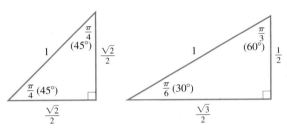

FIGURE 7

EXAMPLE 3   **Finding Exact Values for Special "Angles"** Use Figure 7 to find the exact value of each of the following:

(A) $\cos\dfrac{\pi}{4}$ (B) $\sin\dfrac{\pi}{6}$ (C) $\sin\left(-\dfrac{\pi}{6}\right)$

SOLUTION

(A) $\cos\dfrac{\pi}{4} = \dfrac{\sqrt{2}}{2}$ (B) $\sin\dfrac{\pi}{6} = \dfrac{1}{2}$ (C) $\sin\left(-\dfrac{\pi}{6}\right) = -\dfrac{1}{2}$

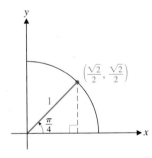
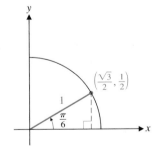
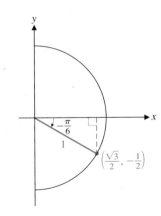

MATCHED PROBLEM 3   Use Figure 7 to find the exact value of each of the following:

(A) $\sin\dfrac{\pi}{4}$ (B) $\cos\dfrac{\pi}{3}$ (C) $\cos\left(-\dfrac{\pi}{3}\right)$

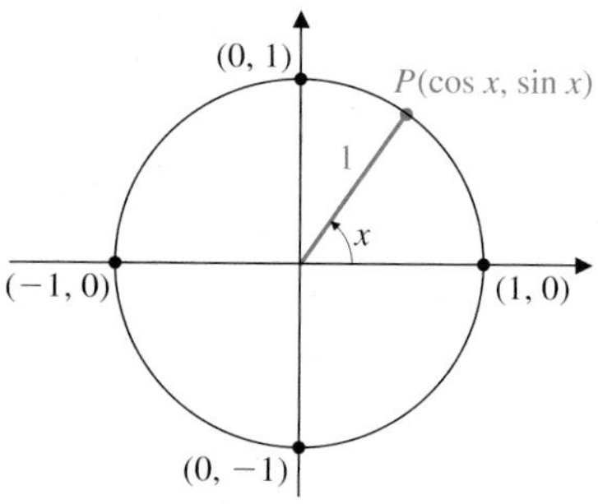

FIGURE 8

Graphs of the Sine and Cosine Functions

To graph $y = \sin x$ or $y = \cos x$ for x a real number, we could use a calculator to produce a table and then plot the ordered pairs from the table in a coordinate system. However, we can speed up the process by returning to basic definitions. Referring to Figure 8, since $\cos x$ and $\sin x$ are the coordinates of a point on the unit circle, we see that

$$-1 \leq \sin x \leq 1 \qquad \text{and} \qquad -1 \leq \cos x \leq 1$$

for all real numbers x. Furthermore, as x increases and P moves around the unit circle in a counterclockwise (positive) direction, both $\sin x$ and $\cos x$ behave in uniform ways, as indicated in the following table:

As *x* Increases from	*y* = sin *x*	*y* = cos *x*
0 to $\pi/2$	Increases from 0 to 1	Decreases from 1 to 0
$\pi/2$ to π	Decreases from 1 to 0	Decreases from 0 to -1
π to $3\pi/2$	Decreases from 0 to -1	Increases from -1 to 0
$3\pi/2$ to 2π	Increases from -1 to 0	Increases from 0 to 1

Note that P has completed one revolution and is back at its starting place. If we let x continue to increase, the second and third columns in the table will be repeated every 2π units. In general, it can be shown that

$$\sin(x + 2\pi) = \sin x \qquad \cos(x + 2\pi) = \cos x$$

for all real numbers x. Functions such that

$$f(x + p) = f(x)$$

for some positive constant p and all real numbers x for which the functions are defined are said to be **periodic.** The smallest such value of p is called the **period** of the function. Thus, both the sine and cosine functions are periodic (a very important property) with period 2π.

Putting all this information together and perhaps adding a few values obtained from a calculator or Figure 7, we obtain the graphs of the sine and cosine functions illustrated in Figure 9. Notice that these curves are continuous. It can be shown that **the sine and cosine functions are continuous for all real numbers.**

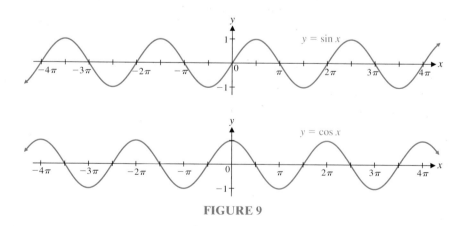

FIGURE 9

Four Other Trigonometric Functions

The sine and cosine functions are only two of six trigonometric functions. They are, however, the most important of the six for many applications. The other four trigonometric functions are the **tangent, cotangent, secant,** and **cosecant.**

DEFINITION Four Other Trigonometric Functions

$$\tan x = \frac{\sin x}{\cos x} \quad \cos x \neq 0 \qquad \sec x = \frac{1}{\cos x} \quad \cos x \neq 0$$

$$\cot x = \frac{\cos x}{\sin x} \quad \sin x \neq 0 \qquad \csc x = \frac{1}{\sin x} \quad \sin x \neq 0$$

INSIGHT

The functions $\sin x$ and $\cos x$ are periodic with period 2π, so

$$\tan(x + 2\pi) = \frac{\sin(x + 2\pi)}{\cos(x + 2\pi)} = \frac{\sin x}{\cos x} = \tan x$$

One might therefore guess that $\tan x$ is periodic with period 2π. However, 2π is not the *smallest* positive constant p such that $\tan(x + p) = \tan x$. Because the points $(\cos x, \sin x)$ and $(\cos(x + \pi), \sin(x + \pi))$ are diametrically opposed on the unit circle,

$$\sin(x + \pi) = -\sin x \qquad \text{and} \quad \cos(x + \pi) = -\cos x$$

Therefore,

$$\tan(x + \pi) = \frac{\sin(x + \pi)}{\cos(x + \pi)} = \frac{-\sin x}{-\cos x} = \tan x$$

It follows that the functions $\tan x$ and $\cot x$ have period π. The other four trigonometric functions—$\sin x$, $\cos x$, $\sec x$, and $\csc x$—all have period 2π.

Answers to Matched Problems   **1.** $180/\pi \approx 57.3°$ **2.** (A) 0 (B) 1 (C) 0
3. (A) $\sqrt{2}/2$ (B) $\frac{1}{2}$ (C) $\frac{1}{2}$

Exercise 9-1

A *Recall that 180° corresponds to π radians. Mentally convert each degree measure given in Problems 1–6 to radian measure in terms of π.*

1. 18°   **2.** 60°   **3.** 90°

4. 135°   **5.** 540°   **6.** 360°

In Problems 7–12, indicate the quadrant in which the terminal side of each angle lies.

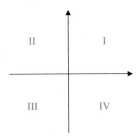

7. 85°   **8.** −105°   **9.** −210°

10. 175°   **11.** 4 rad   **12.** −6 rad

Use Figure 6 to find the exact value of each expression in Problems 13–18.

13. $\sin(-90°)$ **14.** $\cos 180°$ **15.** $\cos\left(\dfrac{\pi}{2}\right)$

16. $\sin\left(\dfrac{-3\pi}{2}\right)$ **17.** $\sin 540°$ **18.** $\cos 360°$

B *Recall that π rad corresponds to 180°. Mentally convert each radian measure given in Problems 19–24 to degree measure.*

19. $2\pi/3$ rad   **20.** $\pi/4$ rad   **21.** 2π rad

22. 3π rad   **23.** $3\pi/4$ rad   **24.** $5\pi/3$ rad

In Problems 25–40, use Figure 6 or Figure 7 (or both) to find the exact value of each expression.

25. $\cos 45°$ **26.** $\sin 30°$

27. $\sin 150°$ **28.** $\cos 135°$

29. $\cos(-60°)$ **30.** $\sin(-45°)$

31. $\sin 180°$ **32.** $\cos 270°$

33. $\sin\dfrac{3\pi}{2}$ **34.** $\cos(-\pi)$

35. $\cos\dfrac{\pi}{6}$

36. $\sin\dfrac{2\pi}{3}$

37. $\cos\left(-\dfrac{3\pi}{4}\right)$

38. $\sin\dfrac{5\pi}{4}$

39. $\sin\left(-\dfrac{7\pi}{6}\right)$

40. $\cos\left(-\dfrac{\pi}{4}\right)$

Use a calculator (set in radian mode) to find the value (to four decimal places) of each expression in Problems 41–46.

41. $\sin 3$ **42.** $\cos 13$ **43.** $\cos 33.74$

44. $\sin 325.9$ **45.** $\sin(-43.06)$ **46.** $\cos(-502.3)$

C   *In Problems 47 and 48, convert to radian measure.*

47. $27°$ **48.** $18°$

In Problems 49 and 50, convert to degree measure.

49. $\pi/12$ rad   **50.** $\pi/60$ rad

In Problems 51–66, use Figure 6 or Figure 7 (or both) to find the exact value of each expression.

51. $\cot 60°$

52. $\sec 45°$

53. $\tan 45°$

54. $\csc 30°$

55. $\sec 120°$

56. $\cot(-150°)$

57. $\csc(-135°)$

58. $\tan 315°$

59. $\cot \pi$

60. $\csc\dfrac{\pi}{2}$

61. $\sec\dfrac{3\pi}{2}$

62. $\tan\left(-\dfrac{3\pi}{4}\right)$

63. $\csc\left(-\dfrac{5\pi}{4}\right)$

64. $\cot 0$

65. $\tan\dfrac{2\pi}{3}$

66. $\sec\dfrac{7\pi}{6}$

67. Refer to Figure 6 and use the Pythagorean theorem to show that
$$(\sin x)^2 + (\cos x)^2 = 1$$
for all x.

68. Use the results of Problem 67 and basic definitions to show that

  (A) $(\tan x)^2 + 1 = (\sec x)^2$

  (B) $1 + (\cot x)^2 = (\csc x)^2$

 In Problems 69–72, use a graphing calculator set in radian mode to graph each trigonometric form.

69. $y = 2\sin\pi x; 0 \le x \le 2, -2 \le y \le 2$

70. $y = -0.5\cos 2x; 0 \le x \le 2\pi, -0.5 \le y \le 0.5$

71. $y = 4 - 4\cos\dfrac{\pi x}{2}; 0 \le x \le 8, 0 \le y \le 8$

72. $y = 6 + 6\sin\dfrac{\pi x}{26}; 0 \le x \le 104, 0 \le y \le 12$

73. Find the domain of the tangent function.

74. Find the domain of the cotangent function.

75. Find the domain of the secant function.

76. Find the domain of the cosecant function.

77. Explain why the range of the cosecant function is $(-\infty, -1] \cup [1, \infty)$.

78. Explain why the range of the secant function is $(-\infty, -1] \cup [1, \infty)$.

79. Explain why the range of the cotangent function is $(-\infty, \infty)$.

80. Explain why the range of the tangent function is $(-\infty, \infty)$.

Applications

81. *Seasonal business cycle.* Suppose that profits on the sale of swimming suits in a department store over a 2-year period are given approximately by
$$P(t) = 5 - 5\cos\dfrac{\pi t}{26} \qquad 0 \le t \le 104$$
where P is profit (in hundreds of dollars) for a week of sales t weeks after January 1. The graph of the profit function is shown in the figure.

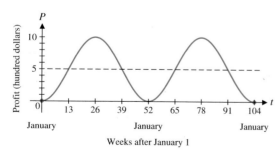

Figure for 81

  (A) Find the exact values of $P(13)$, $P(26)$, $P(39)$, and $P(52)$ without using a calculator.

  (B) Use a calculator to find $P(30)$ and $P(100)$. Interpret the results verbally.

   (C) Use a graphing calculator to confirm the graph shown here for $y = P(t)$.

82. *Seasonal business cycle.* Revenues from sales of a soft drink over a 2-year period are given approximately by
$$R(t) = 4 - 3\cos\dfrac{\pi t}{6} \qquad 0 \le t \le 24$$
where $R(t)$ is revenue (in millions of dollars) for a month of sales t months after February 1. The graph of the revenue function is shown in the figure.

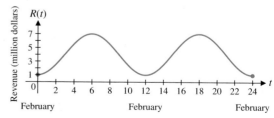

Figure for 82

(A) Find the exact values of $R(0)$, $R(2)$, $R(3)$, and $R(18)$ without using a calculator.

(B) Use a calculator to find $R(5)$ and $R(23)$. Interpret the results verbally.

 (C) Use a graphing calculator to confirm the graph shown here for $y = R(t)$.

83. *Physiology.* A normal seated adult inhales and exhales about 0.8 liter of air every 4 seconds. The volume $V(t)$ of air in the lungs t seconds after exhaling is given approximately by

$$V(t) = 0.45 - 0.35 \cos\frac{\pi t}{2} \qquad 0 \le t \le 8$$

The graph for two complete respirations is shown in the figure.

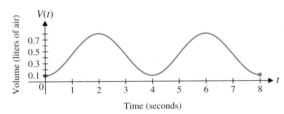

Figure for 83

(A) Find the exact value of $V(0)$, $V(1)$, $V(2)$, $V(3)$, and $V(7)$ without using a calculator.

(B) Use a calculator to find $V(3.5)$ and $V(5.7)$. Interpret the results verbally.

 (C) Use a graphing calculator to confirm the graph shown here for $y = V(t)$.

84. *Pollution.* In a large city, the amount of sulfur dioxide pollutant released into the atmosphere due to the burning of coal and oil for heating purposes varies seasonally. Suppose that the number of tons of pollutant released into the atmosphere during the nth week after January 1 is given approximately by

$$P(n) = 1 + \cos\frac{\pi n}{26} \qquad 0 \le n \le 104$$

The graph of the pollution function is shown in the figure.

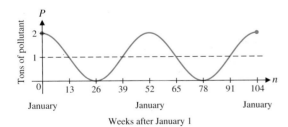

Figure for 84

(A) Find the exact values of $P(0)$, $P(39)$, $P(52)$, and $P(65)$ without using a calculator.

(B) Use a calculator to find $P(10)$ and $P(95)$. Interpret the results verbally.

 (C) Use a graphing calculator to confirm the graph shown here for $y = P(n)$.

85. *Psychology: perception.* An important area of study in psychology is perception. Individuals perceive objects differently in different settings. Consider the well-known illusions shown in Figure A. Lines that appear parallel in one setting may appear to be curved in another (the two vertical lines are actually parallel). Lines of the same length may appear to be of different lengths in two different settings (the two horizontal lines are actually the same length). An interesting experiment in visual perception was conducted by the psychologists Berliner and Berliner (*American Journal of Psychology,* 1952, 65:271–277), who reported that when subjects were presented with a large tilted field of parallel lines and were asked to estimate the position of a horizontal line in the field, most of the subjects were consistently off (Figure B). They found that the difference d in degrees between the estimates and the actual horizontal could be approximated by the equation

$$d = a + b \sin 4\theta$$

where a and b are constants associated with a particular person and θ is the angle of tilt of the visual field (in degrees). Suppose that, for a given person, $a = -2.1$ and $b = -4$. Find d if

(A) $\theta = 30°$ \qquad\qquad (B) $\theta = 10°$

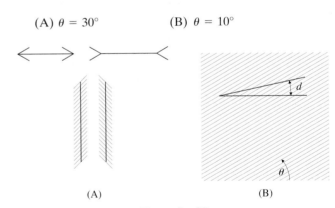

Figure for 85

Section 9-2 DERIVATIVES OF TRIGONOMETRIC FUNCTIONS

- Derivative Formulas
- Application

■ Derivative Formulas

In this section, we discuss derivative formulas for the sine and cosine functions. Once we have these formulas, we automatically have integral formulas for the same functions, which we discuss in the next section.

From the definition of the derivative (Section 3-3),

$$\frac{d}{dx}\sin x = \lim_{h \to 0} \frac{\sin(x + h) - \sin x}{h}$$

On the basis of trigonometric identities and some special trigonometric limits, it can be shown that the limit on the right is cos x. Similarly, it can be shown that

$$\frac{d}{dx}\cos x = -\sin x$$

We now add the following important derivative formulas to our list of derivative formulas:

Derivative of Sine and Cosine

Basic Form

$$\frac{d}{dx}\sin x = \cos x \qquad \frac{d}{dx}\cos x = -\sin x$$

Generalized Form
For $u = u(x)$,

$$\frac{d}{dx}\sin u = \cos u \frac{du}{dx} \qquad \frac{d}{dx}\cos u = -\sin u \frac{du}{dx}$$

INSIGHT

The derivative formula for the function $y = \sin x$ implies that each line tangent to the graph of the function has a slope between -1 and 1. Furthermore, the slope of the line tangent to $y = \sin x$ is equal to 1 if and only if $\cos x = 1$—that is, at $x = 0, \pm 2\pi, \pm 4\pi, \ldots$. Similarly, the derivative formula for $y = \cos x$ implies that each line tangent to the graph of the function has a slope between -1 and 1. The slope of the tangent line is equal to 1 if and only if $-\sin x = 1$—that is, at $x = 3\pi/2, (3\pi/2) \pm 2\pi, (3\pi/2) \pm 4\pi, \ldots$. Note that these observations are consistent with the graphs of $y = \sin x$ and $y = \cos x$ shown in Figure 9, Section 9-1.

EXAMPLE 1   Derivatives Involving Sine and Cosine

(A) $\dfrac{d}{dx}\sin x^2 = (\cos x^2)\dfrac{d}{dx}x^2 = (\cos x^2)2x = 2x \cos x^2$

(B) $\dfrac{d}{dx}\cos(2x - 5) = -\sin(2x - 5)\dfrac{d}{dx}(2x - 5) = -2\sin(2x - 5)$

(C) $\dfrac{d}{dx}(3x^2 - x)\cos x = (3x^2 - x)\dfrac{d}{dx}\cos x + (\cos x)\dfrac{d}{dx}(3x^2 - x)$

$\qquad = -(3x^2 - x)\sin x + (6x - 1)\cos x$

$\qquad = (x - 3x^2)\sin x + (6x - 1)\cos x$

MATCHED PROBLEM 1 Find each of the following derivatives:

(A) $\dfrac{d}{dx}\cos x^3$ (B) $\dfrac{d}{dx}\sin(5-3x)$ (C) $\dfrac{d}{dx}\dfrac{\sin x}{x}$

EXAMPLE 2   **Slope** Find the slope of the graph of $f(x) = \sin x$ at $(\pi/2, 1)$, and sketch in the line tangent to the graph at this point.

SOLUTION Slope at $(\pi/2, 1) = f'(\pi/2) = \cos(\pi/2) = 0$.

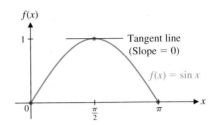

MATCHED PROBLEM 2 Find the slope of the graph of $f(x) = \cos x$ at $(\pi/6, \sqrt{3}/2)$.

Explore & Discuss **1**   From the graph of $y = f'(x)$ shown in Figure 1, describe the shape of the graph of $y = f(x)$ relative to where it is increasing, where it is decreasing, its concavity, and the locations of local maxima and minima. Make a sketch of a possible graph of $y = f(x)$, $0 \le x \le 2\pi$, given that it has x intercepts at $(0, 0)$, $(\pi, 0)$, and $(2\pi, 0)$. Can you identify $f(x)$ and $f'(x)$ in terms of sine or cosine functions?

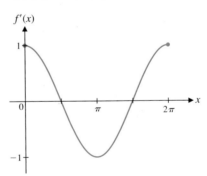

FIGURE 1

EXAMPLE 3   **Derivative of Secant** Find $\dfrac{d}{dx}\sec x$.

SOLUTION
$$\frac{d}{dx}\sec x = \frac{d}{dx}\frac{1}{\cos x} \qquad \sec x = \frac{1}{\cos x}$$
$$= \frac{d}{dx}(\cos x)^{-1}$$
$$= -(\cos x)^{-2}\frac{d}{dx}\cos x$$
$$= -(\cos x)^{-2}(-\sin x)$$
$$= \frac{\sin x}{(\cos x)^2}$$
$$= \left(\frac{\sin x}{\cos x}\right)\left(\frac{1}{\cos x}\right) \qquad \tan x = \frac{\sin x}{\cos x}$$
$$= \tan x \sec x$$

MATCHED PROBLEM 3   Find $\dfrac{d}{dx}\csc x$.

APPLICATION

EXAMPLE 4   **Revenue** Revenues from the sale of ski jackets in a sporting-goods store are given approximately by

$$R(t) = 1.55 + 1.45\cos\frac{\pi t}{26} \qquad 0 \le t \le 104$$

where $R(t)$ is revenue (in thousands of dollars) for a week of sales t weeks after January 1.

(A) What is the rate of change of revenue t weeks after the first of the year?

(B) What is the rate of change of revenue 10 weeks after the first of the year? 26 weeks after the first of the year? 40 weeks after the first of the year?

(C) Find all local maxima and minima for $0 < t < 104$.

(D) Find the absolute maximum and minimum for $0 \le t \le 104$.

 (E) Illustrate the results from parts (A)–(D) by sketching a graph of $y = R(t)$ with the aid of a graphing calculator.

SOLUTION   (A) $R'(t) = -\dfrac{1.45\pi}{26}\sin\dfrac{\pi t}{26} \qquad 0 \le t \le 104$

(B) $R'(10) \approx -\$0.164$ thousand, or $-\$164$ per week

$R'(26) = \$0$ per week

$R'(40) \approx \$0.174$ thousand, or $\$174$ per week

(C) Find the critical points:

$$R'(t) = -\frac{1.45\pi}{26}\sin\frac{\pi t}{26} = 0 \qquad 0 < t < 104$$

$$\sin\frac{\pi t}{26} = 0$$

$$\frac{\pi t}{26} = \pi, 2\pi, 3\pi \qquad \text{Note: } 0 < t < 104 \text{ implies that } 0 < \frac{\pi t}{26} < 4\pi.$$

$$t = 26, 52, 78$$

Use the second-derivative test to get the results shown in Table 1.

$$R''(t) = -\frac{1.45\pi^2}{26^2}\cos\frac{\pi t}{26}$$

(D) Evaluate $R(t)$ at endpoints $t = 0$ and $t = 104$ and at the critical points found in part (C), as listed in Table 2.

TABLE 1

t	$R''(t)$	Graph of R
26	+	Local minimum
52	−	Local maximum
78	+	Local minimum

TABLE 2

t	$R(t)$	
0	$3,000	Absolute maximum
26	$100	Absolute minimum
52	$3,000	Absolute maximum
78	$100	Absolute minimum
104	$3,000	Absolute maximum

(E) The results from parts (A)–(D) can be visualized as shown in the graph of $y = R(t)$ in Figure 2.

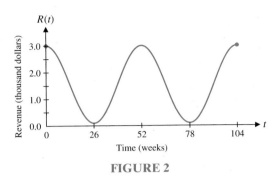

FIGURE 2

MATCHED PROBLEM 4

Suppose that in Example 4 revenues from the sale of ski jackets are given approximately by

$$R(t) = 6.2 + 5.8 \cos \frac{\pi t}{6} \qquad 0 \le t \le 24$$

where $R(t)$ is revenue (in thousands of dollars) for a month of sales t months after January 1.

(A) What is the rate of change of revenue t months after the first of the year?

(B) What is the rate of change of revenue 2 months after the first of the year? 12 months after the first of the year? 23 months after the first of the year?

(C) Find all local maxima and minima for $0 < t < 24$.

(D) Find the absolute maximum and minimum for $0 \le t \le 24$.

(E) Illustrate the results from parts (A)–(D) by sketching a graph of $y = R(t)$ with the aid of a graphing calculator.

Answers to Matched Problems

1. (A) $-3x^2 \sin x^3$ (B) $-3 \cos(5 - 3x)$ (C) $\dfrac{x \cos x - \sin x}{x^2}$

2. $-\frac{1}{2}$ **3.** $-\cot x \csc x$

4. (A) $R'(t) = -\dfrac{5.8\pi}{6} \sin \dfrac{\pi t}{6}, 0 < t < 24$

(B) $R'(2) \approx -\$2.630$ thousand, or $-\$2,630$/month; $R'(12) = \$0$/month;
$R'(23) \approx \$1.518$ thousand, or $\$1,518$/month

(C) Local minima at $t = 6$ and $t = 18$; local maximum at $t = 12$

(D)

	t	$R(t)$	
Endpoint	0	$12,000	Absolute maximum
	6	$400	Absolute minimum
	12	$12,000	Absolute maximum
Endpoint	18	$400	Absolute minimum
	24	$12,000	Absolute maximum

(E)

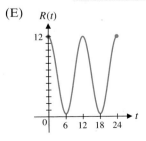

Exercise 9-2

Find the indicated derivatives in Problems 1–14.

A

1. $\dfrac{d}{dt}\cos t$

2. $\dfrac{d}{dw}\sin w$

3. $\dfrac{d}{dx}\sin x^3$

4. $\dfrac{d}{dx}\cos(x^2 - 1)$

B

5. $\dfrac{d}{dt}t \sin t$

6. $\dfrac{d}{du}u \cos u$

7. $\dfrac{d}{dx}\sin x \cos x$

8. $\dfrac{d}{dx}\dfrac{\sin x}{\cos x}$

9. $\dfrac{d}{dx}(\sin x)^5$

10. $\dfrac{d}{dx}(\cos x)^8$

11. $\dfrac{d}{dx}\sqrt{\sin x}$

12. $\dfrac{d}{dx}\sqrt{\cos x}$

13. $\dfrac{d}{dx}\cos\sqrt{x}$

14. $\dfrac{d}{dx}\sin\sqrt{x}$

15. Find the slope of the graph of $f(x) = \sin x$ at $x = \pi/6$.

16. Find the slope of the graph of $f(x) = \cos x$ at $x = \pi/4$.

17. From the graph of $y = f'(x)$ shown here, describe the shape of the graph of $y = f(x)$ relative to where it is increasing, where it is decreasing, its concavity, and the locations of local maxima and minima. Make a sketch of a possible graph of $y = f(x)$, $-\pi \le x \le \pi$, given that it has x intercepts at $(-\pi/2, 0)$ and $(\pi/2, 0)$. Identify $f(x)$ and $f'(x)$ as particular trigonometric functions.

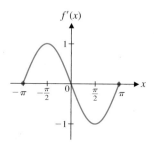

Figure for 17

18. From the graph of $y = f'(x)$ shown here, describe the shape of the graph of $y = f(x)$ relative to where it is increasing, where it is decreasing, its concavity, and the locations of local maxima and minima. Make a sketch of a possible graph of $y = f(x)$, $-\pi \le x \le \pi$, given that it has x intercepts at $(-\pi/2, 0)$ and $(\pi/2, 0)$. Identify $f(x)$ and $f'(x)$ as particular trigonometric functions.

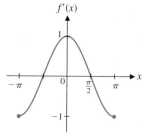

Figure for 18

C *Find the indicated derivatives in Problems 19–22.*

19. $\dfrac{d}{dx}\tan x$

20. $\dfrac{d}{dx}\cot x$

21. $\dfrac{d}{dx}\sin\sqrt{x^2 - 1}$

22. $\dfrac{d}{dx}\cos\sqrt{x^4 - 1}$

In Problems 23 and 24, find $f''(x)$.

23. $f(x) = e^x \sin x$

24. $f(x) = e^x \cos x$

In Problems 25–30, graph each function on a graphing calculator.

25. $y = x \sin \pi x$; $0 \le x \le 9$, $-9 \le y \le 9$

26. $y = -x \cos \pi x$; $0 \le x \le 9$, $-9 \le y \le 9$

27. $y = \dfrac{\cos \pi x}{x}$; $0 \le x \le 8$, $-2 \le y \le 3$

28. $y = \dfrac{\sin \pi x}{0.5x}$; $0 \le x \le 8$, $-2 \le y \le 3$

29. $y = e^{-0.3x} \sin \pi x$; $0 \le x \le 10$, $-1 \le y \le 1$

30. $y = e^{-0.2x} \cos \pi x$; $0 \le x \le 10$, $-1 \le y \le 1$

Applications

31. *Profit.* Suppose that profits on the sale of swimming suits in a department store are given approximately by

$$P(t) = 5 - 5\cos\frac{\pi t}{26} \qquad 0 \le t \le 104$$

where $P(t)$ is profit (in hundreds of dollars) for a week of sales t weeks after January 1.

(A) What is the rate of change of profit t weeks after the first of the year?

(B) What is the rate of change of profit 8 weeks after the first of the year? 26 weeks after the first of the year? 50 weeks after the first of the year?

(C) Find all local maxima and minima for $0 < t < 104$.

(D) Find the absolute maximum and minimum for $0 \le t \le 104$.

(E) Repeat part (C), using a graphing calculator.

32. *Revenue.* Revenues from sales of a soft drink over a 2-year period are given approximately by

$$R(t) = 4 - 3\cos\frac{\pi t}{6} \qquad 0 \le t \le 24$$

where $R(t)$ is revenue (in millions of dollars) for a month of sales t months after February 1.

(A) What is the rate of change of revenue t months after February 1?

(B) What is the rate of change of revenue 1 month after February 1? 6 months after February 1? 11 months after February 1?

(C) Find all local maxima and minima for $0 < t < 24$.

(D) Find the absolute maximum and minimum for $0 \leq t \leq 24$.

 (E) Repeat part (C), using a graphing calculator.

33. *Physiology.* A normal seated adult inhales and exhales about 0.8 liter of air every 4 seconds. The volume of air $V(t)$ in the lungs t seconds after exhaling is given approximately by

$$V(t) = 0.45 - 0.35 \cos \frac{\pi t}{2} \qquad 0 \leq t \leq 8$$

(A) What is the rate of flow of air t seconds after exhaling?

(B) What is the rate of flow of air 3 seconds after exhaling? 4 seconds after exhaling? 5 seconds after exhaling?

(C) Find all local maxima and minima for $0 < t < 8$.

(D) Find the absolute maximum and minimum for $0 \leq t \leq 8$.

 (E) Repeat part (C), using a graphing calculator.

34. *Pollution.* In a large city, the amount of sulfur dioxide pollutant released into the atmosphere due to the burning of coal and oil for heating purposes varies seasonally. Suppose that the number of tons of pollutant released into the atmosphere during the nth week after January 1 is given approximately by

$$P(n) = 1 + \cos \frac{\pi n}{26} \qquad 0 \leq n \leq 104$$

(A) What is the rate of change of pollutant n weeks after the first of the year?

(B) What is the rate of change of pollutant 13 weeks after the first of the year? 26 weeks after the first of the year? 30 weeks after the first of the year?

(C) Find all local maxima and minima for $0 < t < 104$.

(D) Find the absolute maximum and minimum for $0 \leq t \leq 104$.

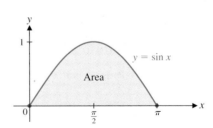 (E) Repeat part (C), using a graphing calculator.

Section 9-3 INTEGRATION OF TRIGONOMETRIC FUNCTIONS

- Integral Formulas
- Application

■ Integral Formulas

Now that we know the derivative formulas

$$\frac{d}{dx} \sin x = \cos x \qquad \text{and} \qquad \frac{d}{dx} \cos x = -\sin x$$

we automatically have the two integral formulas from the definition of the indefinite integral of a function (Section 6-1):

$$\int \cos x \, dx = \sin x + C \qquad \text{and} \qquad \int \sin x \, dx = -\cos x + C$$

EXAMPLE 1 **Area under a Sine Curve** Find the area under the sine curve $y = \sin x$ from 0 to π.

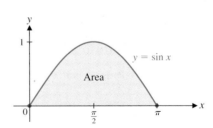

SOLUTION
$$\text{Area} = \int_0^{\pi} \sin x \, dx = -\cos x \big|_0^{\pi}$$
$$= (-\cos \pi) - (-\cos 0)$$
$$= [-(-1)] - [-(1)] = 2$$

| MATCHED PROBLEM 1 | Find the area under the cosine curve $y = \cos x$ from 0 to $\pi/2$.

Explore & Discuss **1**   From the graph of $y = \sin x, 0 \leq x \leq 2\pi$, guess the value for $\int_0^{2\pi} \sin x \, dx$ and explain the reasoning behind your guess. Confirm the guess by direct evaluation of the definite integral.

From the general derivative formulas

$$\frac{d}{dx} \sin u = \cos u \frac{du}{dx} \quad \text{and} \quad \frac{d}{dx} \cos u = -\sin u \frac{du}{dx}$$

we obtain the following general integral formulas:

Indefinite Integrals of Sine and Cosine

For $u = u(x)$,

$$\int \sin u \, du = -\cos u + C \quad \text{and} \quad \int \cos u \, du = \sin u + C$$

| EXAMPLE 2 | **Indefinite Integrals and Trigonometric Functions** Find $\int x \sin x^2 \, dx$.

SOLUTION

$$\int x \sin x^2 \, dx = \frac{1}{2} \int 2x \sin x^2 \, dx$$

$$= \frac{1}{2} \int (\sin x^2) 2x \, dx \quad \text{Let } u = x^2; \text{ then } du = 2x \, dx.$$

$$= \frac{1}{2} \int \sin u \, du$$

$$= -\frac{1}{2} \cos u + C$$

$$= -\frac{1}{2} \cos x^2 + C \quad \text{Since } u = x^2$$

Check:

To check, we differentiate the result to obtain the original integrand:

$$\frac{d}{dx} \left(-\frac{1}{2} \cos x^2 \right) = -\frac{1}{2} \frac{d}{dx} \cos x^2$$

$$= -\frac{1}{2} (-\sin x^2) \frac{d}{dx} x^2$$

$$= -\frac{1}{2} (-\sin x^2)(2x)$$

$$= x \sin x^2$$

| MATCHED PROBLEM 2 | Find $\int \cos 20\pi t \, dt$.

| EXAMPLE 3 | **Indefinite Integrals and Trigonometric Functions**
Find $\int (\sin x)^5 \cos x \, dx$.

SOLUTION This integrand is of the form $\int u^p \, du$, where $u = \sin x$ and $du = \cos x \, dx$. Thus,

$$\int (\sin x)^5 \cos x \, dx = \frac{(\sin x)^6}{6} + C$$

| MATCHED PROBLEM 3 | Find $\int \sqrt{\sin x} \cos x \, dx$.

EXAMPLE 4   **Definite Integrals and Trigonometric Functions**

Evaluate $\int_2^{3.5} \cos x \, dx$.

SOLUTION
$$\int_2^{3.5} \cos x \, dx = \sin x \Big|_2^{3.5}$$

$$= \sin 3.5 - \sin 2 \qquad \text{Use a calculator in radian mode.}$$

$$= -0.3508 - 0.9093$$

$$= -1.2601$$

MATCHED PROBLEM 4   Use a calculator to evaluate $\int_1^{1.5} \sin x \, dx$.

INSIGHT

Recall that $y = \sin x$ is a periodic function with period 2π, and let c be any real number. Then

$$\int_c^{c+2\pi} \cos x \, dx = \sin x \Big|_c^{c+2\pi} = \sin(c + 2\pi) - \sin c = 0$$

In other words, over any interval of the form $[c, c + 2\pi]$, the area that is above the x axis, but below the graph of $y = \cos x$, is equal to the area that is below the x axis, but above the graph of $y = \cos x$ (see Figure 9, Section 9-1). Similarly, for any real number c,

$$\int_c^{c+2\pi} \sin x \, dx = 0$$

APPLICATION

EXAMPLE 5   **Total Revenue**   In Example 4 of Section 9-2, we were given the following revenue equation from the sale of ski jackets:

$$R(t) = 1.55 + 1.45 \cos \frac{\pi t}{26} \qquad 0 \le t \le 104$$

Here, $R(t)$ is revenue (in thousands of dollars) for a week of sales t weeks after January 1.

(A) Find the total revenue taken in over the 2-year period—that is, from $t = 0$ to $t = 104$.

(B) Find the total revenue taken in from $t = 39$ to $t = 65$.

SOLUTION   (A) The area under the graph of the revenue equation for the 2-year period approximates the total revenue taken in for that period:

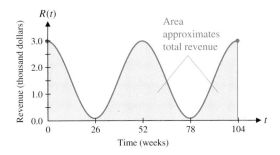

This area (and therefore the total revenue) is given by the following definite integral:

$$\text{Total revenue} \approx \int_0^{104} \left(1.55 + 1.45 \cos \frac{\pi t}{26}\right) dt$$

$$= \left[1.55t + 1.45\left(\frac{26}{\pi}\right) \sin \frac{\pi t}{26}\right]\Bigg|_0^{104}$$

$$= \$161.200 \text{ thousand,} \quad \text{or} \quad \$161,200$$

(B) The total revenue from $t = 39$ to $t = 65$ is approximated by the area under the curve from $t = 39$ to $t = 65$:

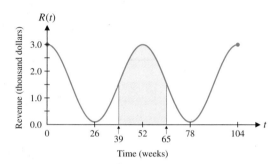

$$\text{Total revenue} \approx \int_{39}^{65} \left(1.55 + 1.45 \cos \frac{\pi t}{26}\right) dt$$

$$= \left[1.55t + 1.45\left(\frac{26}{\pi}\right) \sin \frac{\pi t}{26}\right]\Bigg|_{39}^{65}$$

$$= \$64.301 \text{ thousand,} \quad \text{or} \quad \$64,301$$

MATCHED PROBLEM 5 Suppose that in Example 5 revenues from the sale of ski jackets are given approximately by

$$R(t) = 6.2 + 5.8 \cos \frac{\pi t}{6} \qquad 0 \le t \le 24$$

where $R(t)$ is revenue (in thousands of dollars) for a month of sales t months after January 1.

(A) Find the total revenue taken in over the 2 year-period—that is, from $t = 0$ to $t = 24$.

(B) Find the total revenue taken in from $t = 4$ to $t = 8$.

Answers to Matched Problems   **1.** 1   **2.** $\dfrac{1}{20\pi} \sin 20\pi t + C$ **3.** $\frac{2}{3}(\sin x)^{3/2} + C$ **4.** 0.4696
5. (A) \$148.8 thousand, or \$148,800   (B) \$5.614 thousand, or \$5,614

Exercise 9-3

Find each of the indefinite integrals in Problems 1–10.

5. $\displaystyle\int (\sin x)^{12} \cos x \, dx$ **6.** $\displaystyle\int \sin x \cos x \, dx$

A

B

1. $\displaystyle\int \sin t \, dt$ **2.** $\displaystyle\int \cos w \, dw$ **7.** $\displaystyle\int \sqrt[3]{\cos x} \sin x \, dx$ **8.** $\displaystyle\int \frac{\cos x}{\sqrt{\sin x}} dx$

3. $\displaystyle\int \cos 3x \, dx$ **4.** $\displaystyle\int \sin 2x \, dx$ **9.** $\displaystyle\int x^2 \cos x^3 \, dx$ **10.** $\displaystyle\int (x + 1) \sin(x^2 + 2x) \, dx$

Evaluate each of the definite integrals in Problems 11–14.

11. $\int_0^{\pi/2} \cos x \, dx$ **12.** $\int_0^{\pi/4} \cos x \, dx$

13. $\int_{\pi/2}^{\pi} \sin x \, dx$ **14.** $\int_{\pi/6}^{\pi/3} \sin x \, dx$

15. Find the shaded area under the cosine curve in the figure:

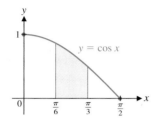

16. Find the shaded area under the sine curve in the figure:

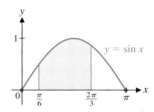

Use a calculator to evaluate the definite integrals in Problems 17–20 after performing the indefinite integration. (Remember that the limits are real numbers, so radian mode must be used on the calculator.)

17. $\int_0^2 \sin x \, dx$ **18.** $\int_0^{0.5} \cos x \, dx$

19. $\int_1^2 \cos x \, dx$ **20.** $\int_1^3 \sin x \, dx$

C *Find each of the indefinite integrals in Problems 21–26.*

21. $\int e^{\sin x} \cos x \, dx$ **22.** $\int e^{\cos x} \sin x \, dx$

23. $\int \frac{\cos x}{\sin x} dx$ **24.** $\int \frac{\sin x}{\cos x} dx$

25. $\int \tan x \, dx$ **26.** $\int \cot x \, dx$

27. Given the definite integral

$$I = \int_0^3 e^{-x} \sin x \, dx$$

(A) Graph the integrand $f(x) = e^{-x} \sin x$ over $[0, 3]$.
(B) Use the left sum L_6 (see Section 6-5) to approximate I.

28. Given the definite integral

$$I = \int_0^3 e^{-x} \cos x \, dx$$

(A) Graph the integrand $f(x) = e^{-x} \cos x$ over $[0, 3]$.
(B) Use the right sum R_6 (see Section 6-5) to approximate I.

Applications

29. *Seasonal business cycle.* Suppose that profits on the sale of swimming suits in a department store are given approximately by

$$P(t) = 5 - 5 \cos \frac{\pi t}{26} \quad 0 \le t \le 104$$

where $P(t)$ is profit (in hundreds of dollars) for a week of sales t weeks after January 1. Use definite integrals to approximate

(A) The total profit earned during the 2-year period
(B) The total profit earned from $t = 13$ to $t = 26$
(C) Illustrate part (B) graphically with an appropriate shaded region representing the total profit earned.

30. *Seasonal business cycle.* Revenues from sales of a soft drink over a 2-year period are given approximately by

$$R(t) = 4 - 3 \cos \frac{\pi t}{6} \quad 0 \le t \le 24$$

where $R(t)$ is revenue (in millions of dollars) for a month of sales t months after February 1. Use definite integrals to approximate

(A) Total revenues taken in over the 2-year period
(B) Total revenues taken in from $t = 8$ to $t = 14$
(C) Illustrate part (B) graphically with an appropriate shaded region representing the total revenues taken in.

31. *Pollution.* In a large city, the amount of sulfur dioxide pollutant released into the atmosphere due to the burning of coal and oil for heating purposes is given approximately by

$$P(n) = 1 + \cos \frac{\pi n}{26} \quad 0 \le n \le 104$$

where $P(n)$ is the amount of sulfur dioxide (in tons) released during the nth week after January 1.

(A) How many tons of pollutants were emitted into the atmosphere over the 2-year period?
(B) How many tons of pollutants were emitted into the atmosphere from $n = 13$ to $n = 52$?
(C) Illustrate part (B) graphically with an appropriate shaded region representing the total tons of pollutants emitted into the atmosphere.

CHAPTER 9 REVIEW

Important Terms, Symbols, and Concepts

9-1 Trigonometric Functions Review

Examples

- In a plane, an **angle** is formed by rotating a ray m, called the **initial side** of the angle, around its endpoint until the ray coincides with a ray n, called the **terminal side** of the angle. The common endpoint of m and n is called the **vertex.**

- A counterclockwise rotation produces a **positive** angle, and a clockwise rotation produces a **negative** angle.

- Two angles with the same initial and terminal sides are said to be **coterminal.**

- An angle of **degree measure 1** is $\frac{1}{360}$ of a complete rotation. In a circle, an angle of **radian measure 1** is the central angle subtended by an arc having the same length as the radius of the circle.

- Degree measure can be converted to radian measure, and vice versa, by the proportion

$$\frac{\theta_{\deg}}{180°} = \frac{\theta_{\text{rad}}}{\pi \text{ rad}}$$

Ex. 1, p. 535

- An angle in a rectangular coordinate system is in **standard position** if its vertex is at the origin and its initial side is on the positive x axis.

- If θ is an angle in standard position, its terminal side intersects the unit circle at a point P. We denote the coordinates of P by $(\cos \theta, \sin \theta)$ [see Fig. 6, p. 535].

- The set of all ordered pairs of the form $(\theta, \sin \theta)$ is the **sine function,** and the set of all ordered pairs of the form $(\theta, \cos \theta)$ is the **cosine function.** For work in calculus, we define these functions for angles measured in radians. (See Fig. 9, p. 538, for the graphs of $y = \sin x$ and $y = \cos x$.)

Ex. 2, p. 536
Ex. 3, p. 537

- A function is **periodic** if $f(x + p) = f(x)$ for some positive constant p and all real numbers x for which $f(x)$ is defined. The smallest such constant p is called the **period.** Both $\sin x$ and $\cos x$ are periodic continuous functions with period 2π.

- Four additional trigonometric functions—the **tangent, cotangent, secant,** and **cosecant** functions—are defined in terms of $\sin x$ and $\cos x$:

$$\tan x = \frac{\sin x}{\cos x} \quad \cos x \neq 0 \qquad \sec x = \frac{1}{\cos x} \quad \cos x \neq 0$$

$$\cot x = \frac{\cos x}{\sin x} \quad \sin x \neq 0 \qquad \csc x = \frac{1}{\sin x} \quad \sin x \neq 0$$

9-2 Derivatives of Trigonometric Functions

- The derivatives of the functions $\sin x$ and $\cos x$ are

$$\frac{d}{dx}\sin x = \cos x \qquad \frac{d}{dx}\cos x = -\sin x$$

For $u = u(x)$,

$$\frac{d}{dx}\sin u = \cos u \frac{du}{dx} \qquad \frac{d}{du}\cos u = -\sin u \frac{du}{dx}$$

Ex. 1, p. 542
Ex. 2, p. 543
Ex. 3, p. 543
Ex. 4, p. 544

9-3 Integration of Trigonometric Functions

- Indefinite integrals of the functions $\sin x$ and $\cos x$ are

$$\int \sin u \, du = -\cos u + C \qquad \int \cos u \, du = \sin u + C$$

Ex. 1, p. 547
Ex. 2, p. 548
Ex. 3, p. 548
Ex. 4, p. 549
Ex. 5, p. 549

REVIEW EXERCISE

Work through all the problems in this chapter review and check your answers in the back of the book. Answers to all review problems are there, along with section numbers in italics to indicate where each type of problem is discussed. Where weaknesses show up, review appropriate sections of the text.

A

1. Convert to radian measure in terms of π:
  (A) 30°   (B) 45°   (C) 60°   (D) 90°

2. Evaluate without using a calculator:
  (A) $\cos \pi$ (B) $\sin 0$ (C) $\sin \dfrac{\pi}{2}$

In Problems 3–6, find each derivative or integral.

3. $\dfrac{d}{dm}\cos m$

4. $\dfrac{d}{du}\sin u$

5. $\dfrac{d}{dx}\sin(x^2 - 2x + 1)$

6. $\displaystyle\int \sin 3t \, dt$

B

7. Convert to degree measure:
  (A) $\pi/6$ (B) $\pi/4$ (C) $\pi/3$ (D) $\pi/2$

8. Evaluate without using a calculator:
  (A) $\sin \dfrac{\pi}{6}$ (B) $\cos \dfrac{\pi}{4}$ (C) $\sin \dfrac{\pi}{3}$

9. Evaluate with the use of a calculator:
  (A) $\cos 33.7$ (B) $\sin(-118.4)$

In Problems 10–16, find each derivative or integral.

10. $\dfrac{d}{dx}(x^2 - 1)\sin x$

11. $\dfrac{d}{dx}(\sin x)^6$

12. $\dfrac{d}{dx}\sqrt[3]{\sin x}$

13. $\displaystyle\int t\cos(t^2 - 1)\,dt$

14. $\displaystyle\int_0^\pi \sin u \, du$ **15.** $\displaystyle\int_0^{\pi/3} \cos x \, dx$ **16.** $\displaystyle\int_1^{2.5} \cos x \, dx$

17. Find the slope of the cosine curve $y = \cos x$ at $x = \pi/4$.

18. Find the area under the sine curve $y = \sin x$ from $x = \pi/4$ to $x = 3\pi/4$.

19. Given the definite integral
$$I = \int_1^5 \frac{\sin x}{x}\,dx$$
  (A) Graph the integrand
$$f(x) = \frac{\sin x}{x}$$
  over $[1, 5]$.
  (B) Use the right sum R_4 to approximate I.

C

20. Convert 15° to radian measure.

21. Evaluate without using a calculator:
  (A) $\sin \dfrac{3\pi}{2}$ (B) $\cos \dfrac{5\pi}{6}$ (C) $\sin\left(\dfrac{-\pi}{6}\right)$

In Problems 22–26, find each derivative or integral.

22. $\dfrac{d}{du}\tan u$

23. $\dfrac{d}{dx}e^{\cos x^2}$

24. $\displaystyle\int e^{\sin x}\cos x \, dx$

25. $\displaystyle\int \tan x \, dx$

26. $\displaystyle\int_2^5 (5 + 2\cos 2x)\,dx$

 In Problems 27–29, graph each function on a graphing calculator set in radian mode.

27. $y = \dfrac{\sin \pi x}{0.2x}; 1 \le x \le 8, -4 \le y \le 4$

28. $y = 0.5x \cos \pi x; 0 \le x \le 8, -5 \le y \le 5$

29. $y = 3 - 2\cos \pi x; 0 \le x \le 6, 0 \le y \le 5$

APPLICATIONS

Problems 30–32 refer to the following: Revenues from sweater sales in a sportswear chain are given approximately by
$$R(t) = 3 + 2\cos\frac{\pi t}{6} \qquad 0 \le t \le 24$$
where $R(t)$ is the revenue (in thousands of dollars) for a month of sales t months after January 1.

30. (A) Find the exact values of $R(0)$, $R(2)$, $R(3)$, and $R(6)$ without using a calculator.
  (B) Use a calculator to find $R(1)$ and $R(22)$. Interpret the results verbally.

31. (A) What is the rate of change of revenue t months after January 1?

(B) What is the rate of change of revenue 3 months after January 1? 10 months after January 1? 18 months after January 1?

(C) Find all local maxima and minima for $0 < t < 24$.

(D) Find the absolute maximum and minimum for $0 \le t \le 24$.

 (E) Repeat part (C), using a graphing calculator.

32. (A) Find the total revenues taken in over the 2-year period.

(B) Find the total revenues taken in from $t = 5$ to $t = 9$.

 (C) Illustrate part (B) graphically with an appropriate shaded region representing the total revenue taken in.

APPENDIX A   Basic Algebra Review

SELF-TEST ON BASIC ALGEBRA
A-1 Algebra and Real Numbers
A-2 Operations on Polynomials
A-3 Factoring Polynomials
A-4 Operations on Rational Expressions
A-5 Integer Exponents and Scientific Notation
A-6 Rational Exponents and Radicals
A-7 Quadratic Equations

INTRODUCTION

Appendix A reviews some important basic algebra concepts usually studied in earlier courses. The material may be studied systematically before beginning the rest of the book or reviewed as needed. The Self-Test on Basic Algebra that precedes Section A-1 may be taken to locate areas of weakness. All the answers to the self-test are in the answer section in the back of the book and are keyed to the sections in Appendix A where the related topics are discussed.

SELF-TEST ON BASIC ALGEBRA

Work through all the problems in this self-test and check your answers in the back of the book. All answers are there and are keyed to relevant sections in Appendix A. Where weaknesses show up, review appropriate sections in the appendix.

1. Replace each question mark with an appropriate expression that will illustrate the use of the indicated real number property:

    (A) Commutative $(\cdot)$: $x(y + z) = ?$
    (B) Associative $(+)$: $2 + (x + y) = ?$
    (C) Distributive: $(2 + 3)x = ?$

Problems 2–6 refer to the following polynomials:
    (A) $3x - 4$
    (B) $x + 2$
    (C) $2 - 3x^2$
    (D) $x^3 + 8$

2. Add all four.

3. Subtract the sum of (A) and (C) from the sum of (B) and (D).

4. Multiply (C) and (D).

5. What is the degree of each polynomial?

6. What is the leading coefficient of each polynomial?

In Problems 7–12, perform the indicated operations and simplify.

7. $5x^2 - 3x[4 - 3(x - 2)]$

8. $(2x + y)(3x - 4y)$

9. $(2a - 3b)^2$

10. $(2x - y)(2x + y) - (2x - y)^2$

11. $(3x^3 - 2y)^2$

12. $(x - 2y)^3$

13. Write in scientific notation:

    (A) 4,065,000,000,000
    (B) 0.0073

14. Write in standard decimal form:

    (A) 2.55×10^8
    (B) 4.06×10^{-4}

15. Indicate true (T) or false (F):

    (A) A natural number is a rational number.
    (B) A number with a repeating decimal expansion is an irrational number.

16. Give an example of an integer that is not a natural number.

Simplify Problems 17–25 and write answers using positive exponents only. All variables represent positive real numbers.

17. $6(xy^3)^5$

18. $\dfrac{9u^8v^6}{3u^4v^8}$

19. $(2 \times 10^5)(3 \times 10^{-3})$

20. $(x^{-3}y^2)^{-2}$

21. $u^{5/3}u^{2/3}$

22. $(9a^4b^{-2})^{1/2}$

23. $\dfrac{5^0}{3^2} + \dfrac{3^{-2}}{2^{-2}}$

24. $(x^{1/2} + y^{1/2})^2$

25. $(3x^{1/2} - y^{1/2})(2x^{1/2} + 3y^{1/2})$

Write Problems 26–31 in completely factored form relative to the integers. If a polynomial cannot be factored further relative to the integers, say so.

26. $12x^2 + 5x - 3$

27. $8x^2 - 18xy + 9y^2$

28. $t^2 - 4t - 6$

29. $6n^3 - 9n^2 - 15n$

30. $(4x - y)^2 - 9x^2$

31. $6x(2x + 1)^2 - 15x^2(2x + 1)$

In Problems 32–37, perform the indicated operations and reduce to lowest terms. Represent all compound fractions as simple fractions reduced to lowest terms.

32. $\dfrac{2}{5b} - \dfrac{4}{3a^3} - \dfrac{1}{6a^2b^2}$

33. $\dfrac{3x}{3x^2 - 12x} + \dfrac{1}{6x}$

34. $\dfrac{x}{x^2 - 16} - \dfrac{x + 4}{x^2 - 4x}$

35. $\dfrac{(x + y)^2 - x^2}{y}$

36. $\dfrac{\dfrac{1}{7 + h} - \dfrac{1}{7}}{h}$

37. $\dfrac{x^{-1} + y^{-1}}{x^{-2} - y^{-2}}$

38. Each statement illustrates the use of one of the following real number properties or definitions. Indicate which one.

Commutative $(+, \cdot)$ Associative $(+, \cdot)$ Distributive

Identity $(+, \cdot)$ Inverse $(+, \cdot)$ Subtraction

Division   Negatives   Zero

(A) $(-7) - (-5) = (-7) + [-(-5)]$

(B) $5u + (3v + 2) = (3v + 2) + 5u$

(C) $(5m - 2)(2m + 3) =$
    $(5m - 2)2m + (5m - 2)3$

(D) $9 \cdot (4y) = (9 \cdot 4)y$

(E) $\dfrac{u}{-(v - w)} = -\dfrac{u}{v - w}$

(F) $(x - y) + 0 = (x - y)$

39. Change to rational exponent form:

$$6\sqrt[5]{x^2} - 7\sqrt[4]{(x - 1)^3}$$

40. Change to radical form: $2x^{1/2} - 3x^{2/3}$

41. Write in the form $ax^p + bx^q$, where a and b are real numbers and p and q are rational numbers:

$$\frac{4\sqrt{x} - 3}{2\sqrt{x}}$$

In Problems 42 and 43, rationalize the denominator.

42. $\dfrac{3x}{\sqrt{3x}}$

43. $\dfrac{x - 5}{\sqrt{x} - \sqrt{5}}$

In Problems 44 and 45, rationalize the numerator.

44. $\dfrac{\sqrt{x} - 5}{x - 5}$

45. $\dfrac{\sqrt{u + h} - \sqrt{u}}{h}$

Solve Problems 46–49 for x.

46. $x^2 = 5x$

47. $3x^2 - 21 = 0$

48. $x^2 - x - 20 = 0$

49. $-6x^2 + 7x - 1 = 0$

Section A-1 ALGEBRA AND REAL NUMBERS

- Set of Real Numbers
- Real Number Line
- Basic Real Number Properties
- Further Properties
- Fraction Properties

The rules for manipulating and reasoning with symbols in algebra depend, in large measure, on properties of the real numbers. In this section we look at some of the important properties of this number system. To make our discussions here and elsewhere in the book clearer and more precise, we occasionally make use of simple *set* concepts and notation.

■ Set of Real Numbers

What number system have you been using most of your life? The *real number system.* Informally, a **real number** is any number that has a decimal representation. Table 1 describes the set of real numbers and some of its important subsets. Figure 1 illustrates how these sets of numbers are related.

The set of integers contains all the natural numbers and something else—their negatives and 0. The set of rational numbers contains all the integers and something else—noninteger ratios of integers. And the set of real numbers contains all the rational numbers and something else—irrational numbers.

TABLE 1 Set of Real Numbers			
Symbol	**Name**	**Description**	**Examples**
N	Natural numbers	Counting numbers (also called positive integers)	$1, 2, 3, \ldots$
Z	Integers	Natural numbers, their negatives, and 0	$\ldots, -2, -1, 0, 1, 2, \ldots$
Q	Rational numbers	Numbers that can be represented as a/b, where a and b are integers and $b \neq 0$; decimal representations are repeating or terminating	$-4, 0, 1, 25, \frac{-3}{5}, \frac{2}{3}, 3.67, -0.33\overline{3}, 5.272\,\overline{727}^*$
I	Irrational numbers	Numbers that can be represented as nonrepeating and nonterminating decimal numbers	$\sqrt{2}, \pi, \sqrt[3]{7}, 1.414\,213\ldots, 2.718\,281\,82\ldots$
R	Real numbers	Rational and irrational numbers	

*The overbar indicates that the number (or block of numbers) repeats indefinitely. The space after every third digit is used to help keep track of the number of decimal places.

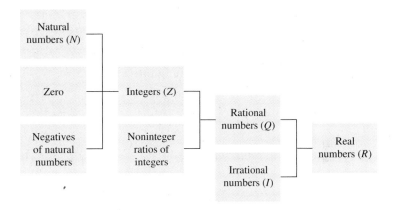

FIGURE 1 Real numbers and important subsets

■ Real Number Line

A one-to-one correspondence exists between the set of real numbers and the set of points on a line. That is, each real number corresponds to exactly one point, and each point corresponds to exactly one real number. A line with a real number associated with each point, and vice versa, as shown in Figure 2, is called a **real number line,** or simply a **real line.** Each number associated with a point is called the **coordinate** of the point.

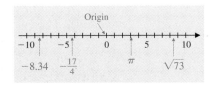

FIGURE 2 Real number line

The point with coordinate 0 is called the **origin.** The arrow on the right end of the line indicates a positive direction. The coordinates of all points to the right of the origin are called **positive real numbers,** and those to the left of the origin are called **negative real numbers.** The real number 0 is neither positive nor negative.

■ Basic Real Number Properties

We now take a look at some of the basic properties of the real number system that enable us to convert algebraic expressions into *equivalent forms*. These assumed properties become operational rules in the algebra of real numbers.

<u>SUMMARY</u>   BASIC PROPERTIES OF THE SET OF REAL NUMBERS

Let a, b, and c be arbitrary elements in the set of real numbers R.

Addition Properties

Associative:	$(a + b) + c = a + (b + c)$
Commutative:	$a + b = b + a$
Identity:	0 is the additive identity; that is, $0 + a = a + 0 = a$ for all a in R, and 0 is the only element in R with this property.
Inverse:	For each a in R, $-a$ is its unique additive inverse; that is, $a + (-a) = (-a) + a = 0$, and $-a$ is the only element in R relative to a with this property.

Multiplication Properties

Associative:	$(ab)c = a(bc)$
Commutative:	$ab = ba$
Identity:	1 is the multiplicative identity; that is, $(1)a = a(1) = a$ for all a in R, and 1 is the only element in R with this property.
Inverse:	For each a in R, $a \neq 0$, $1/a$ is its unique multiplicative inverse; that is, $a(1/a) = (1/a)a = 1$, and $1/a$ is the only element in R relative to a with this property.

Distributive Properties

$$a(b + c) = ab + ac \qquad (a + b)c = ac + bc$$

Do not be intimidated by the names of these properties. Most of the ideas presented here are quite simple. In fact, you have been using many of these properties in arithmetic for a long time.

You are already familiar with the **commutative properties** for addition and multiplication. They indicate that the order in which the addition or multiplication of two numbers is performed does not matter. For example,

$$7 + 2 = 2 + 7 \qquad \text{and} \qquad 3 \cdot 5 = 5 \cdot 3$$

Is there a commutative property relative to subtraction or division? That is, does $a - b = b - a$ or does $a \div b = b \div a$ for all real numbers a and b (division by 0 excluded)? The answer is no, since, for example,

$$8 - 6 \neq 6 - 8 \qquad \text{and} \qquad 10 \div 5 \neq 5 \div 10$$

When computing

$$3 + 2 + 6 \qquad \text{or} \qquad 3 \cdot 2 \cdot 6$$

why do we not need parentheses to indicate which two numbers are to be added or multiplied first? The answer is to be found in the **associative properties.** These properties allow us to write

$$(3 + 2) + 6 = 3 + (2 + 6) \quad \text{and} \quad (3 \cdot 2) \cdot 6 = 3 \cdot (2 \cdot 6)$$

so it does not matter how we group numbers relative to either operation. Is there an associative property for subtraction or division? The answer is no, since, for example,

$$(12 - 6) - 2 \neq 12 - (6 - 2) \quad \text{and} \quad (12 \div 6) \div 2 \neq 12 \div (6 \div 2)$$

Evaluate each side of each equation to see why.

INSIGHT

Why are we concerned with the commutative and associative properties? When manipulating algebraic expressions involving addition or multiplication, these properties permit us to change the order of operation at will and insert or remove parentheses as we please. We don't have the same freedom with subtraction or division.

What number added to a given number will give that number back again? What number times a given number will give that number back again? The answers are 0 and 1, respectively. Because of this, 0 and 1 are called the **identity elements** for the real numbers. Hence, for any real numbers a and b,

$$0 + 5 = 5 \quad \text{and} \quad (a + b) + 0 = a + b$$
$$1 \cdot 4 = 4 \quad \text{and} \quad (a + b) \cdot 1 = a + b$$

We now consider **inverses.** For each real number a, there is a unique real number $-a$ such that $a + (-a) = 0$. The number $-a$ is called the **additive inverse** of a, or the **negative** of a. For example, the additive inverse (or negative) of 7 is -7, since $7 + (-7) = 0$. The additive inverse (or negative) of -7 is $-(-7) = 7$, since $-7 + [-(-7)] = 0$.

INSIGHT

Do not confuse negation with the sign of a number. If a is a real number, $-a$ is the negative of a and may be positive or negative. Specifically, if a is negative, then $-a$ is positive and if a is positive, then $-a$ is negative.

For each nonzero real number a, there is a unique real number $1/a$ such that $a(1/a) = 1$. The number $1/a$ is called the **multiplicative inverse** of a, or the **reciprocal** of a. For example, the multiplicative inverse (or reciprocal) of 4 is $\frac{1}{4}$, since $4(\frac{1}{4}) = 1$. (Also note that 4 is the multiplicative inverse of $\frac{1}{4}$.) The number 0 has no multiplicative inverse.

We now turn to the **distributive properties,** which involve both multiplication and addition. Consider the following two computations:

$$5(3 + 4) = 5 \cdot 7 = 35 \qquad 5 \cdot 3 + 5 \cdot 4 = 15 + 20 = 35$$

Thus,

$$5(3 + 4) = 5 \cdot 3 + 5 \cdot 4$$

and we say that multiplication by 5 *distributes* over the sum $(3 + 4)$. In general, **multiplication distributes over addition** in the real number system. Two more illustrations are

$$9(m + n) = 9m + 9n \qquad (7 + 2)u = 7u + 2u$$

EXAMPLE 1 **Real Number Properties** State the real number property that justifies the indicated statement.

STATEMENT	PROPERTY ILLUSTRATED
(A) $x(y + z) = (y + z)x$	Commutative $(\cdot)$
(B) $5(2y) = (5 \cdot 2)y$	Associative $(\cdot)$
(C) $2 + (y + 7) = 2 + (7 + y)$	Commutative $(+)$
(D) $4z + 6z = (4 + 6)z$	Distributive
(E) If $m + n = 0$, then $n = -m$.	Inverse $(+)$

MATCHED PROBLEM 1 State the real number property that justifies the indicated statement.

(A) $8 + (3 + y) = (8 + 3) + y$
(B) $(x + y) + z = z + (x + y)$
(C) $(a + b)(x + y) = a(x + y) + b(x + y)$
(D) $5xy + 0 = 5xy$
(E) If $xy = 1, x \neq 0$, then $y = 1/x$.

■ Further Properties

Subtraction and *division* can be defined in terms of addition and multiplication, respectively:

DEFINITION **Subtraction and Division**

For all real numbers a and b,

Subtraction: $a - b = a + (-b)$ $7 - (-5) = 7 + [-(-5)]$
$= 7 + 5 = 12$

Division: $a \div b = a\left(\dfrac{1}{b}\right), b \neq 0$ $9 \div 4 = 9\left(\dfrac{1}{4}\right) = \dfrac{9}{4}$

Thus, to subtract b from a, add the negative (the additive inverse) of b to a. To divide a by b, multiply a by the reciprocal (the multiplicative inverse) of b. Note that division by 0 is not defined, since 0 does not have a reciprocal. Thus, **0 can never be used as a divisor!**

The following properties of negatives can be proved using the preceding assumed properties and definitions.

THEOREM 1 **NEGATIVE PROPERTIES**

For all real numbers a and b,

1. $-(-a) = a$
2. $(-a)b = -(ab)$
   $= a(-b) = -ab$
3. $(-a)(-b) = ab$
4. $(-1)a = -a$

5. $\dfrac{-a}{b} = -\dfrac{a}{b} = \dfrac{a}{-b}, b \neq 0$

6. $\dfrac{-a}{-b} = -\dfrac{-a}{b} = -\dfrac{a}{-b} = \dfrac{a}{b}, b \neq 0$

We now state two important properties involving 0.

THEOREM 2 **ZERO PROPERTIES**

For all real numbers a and b,

1. $a \cdot 0 = 0$ $0 \cdot 0 = 0$ $(-35)(0) = 0$

2. $ab = 0$ if and only if $a = 0$ or $b = 0$
   If $(3x + 2)(x - 7) = 0$, then either $3x + 2 = 0$ or $x - 7 = 0$.

Explore & Discuss **1**   In general, a set of numbers is closed under an operation if performing the operation on numbers in the set always produces another number in the set. For example, the real numbers R are closed under addition, multiplication, subtraction, and division, excluding division by 0. Replace each ? in the following tables with T (true) or F (false), and illustrate each false statement with an example. (See Table 1 for the definitions of the sets N, Z, Q, I, and R.)

	Closed under Addition	Closed under Multiplication
N	?	?
Z	?	?
Q	?	?
I	?	?
R	T	T

	Closed under Subtraction	Closed under Division*
N	?	?
Z	?	?
Q	?	?
I	?	?
R	T	T

*Excluding division by 0.

■ Fraction Properties

Recall that the quotient $a \div b(b \neq 0)$ written in the form a/b is called a **fraction.** The quantity a is called the **numerator,** and the quantity b is called the **denominator.**

THEOREM 3   **FRACTION PROPERTIES**

For all real numbers a, b, c, d, and k (division by 0 excluded):

1. $\dfrac{a}{b} = \dfrac{c}{d}$ if and only if $ad = bc$ $\dfrac{4}{6} = \dfrac{6}{9}$ since $4 \cdot 9 = 6 \cdot 6$

2. $\dfrac{ka}{kb} = \dfrac{a}{b}$

$\dfrac{7 \cdot 3}{7 \cdot 5} = \dfrac{3}{5}$

3. $\dfrac{a}{b} \cdot \dfrac{c}{d} = \dfrac{ac}{bd}$

$\dfrac{3}{5} \cdot \dfrac{7}{8} = \dfrac{3 \cdot 7}{5 \cdot 8}$

4. $\dfrac{a}{b} \div \dfrac{c}{d} = \dfrac{a}{b} \cdot \dfrac{d}{c}$

$\dfrac{2}{3} \div \dfrac{5}{7} = \dfrac{2}{3} \cdot \dfrac{7}{5}$

5. $\dfrac{a}{b} + \dfrac{c}{b} = \dfrac{a+c}{b}$

$\dfrac{3}{6} + \dfrac{5}{6} = \dfrac{3+5}{6}$

6. $\dfrac{a}{b} - \dfrac{c}{b} = \dfrac{a-c}{b}$

$\dfrac{7}{8} - \dfrac{3}{8} = \dfrac{7-3}{8}$

7. $\dfrac{a}{b} + \dfrac{c}{d} = \dfrac{ad + bc}{bd}$

$\dfrac{2}{3} + \dfrac{3}{5} = \dfrac{2 \cdot 5 + 3 \cdot 3}{3 \cdot 5}$

Answers to Matched Problems   **1.** (A) Associative $(+)$ (B) Commutative $(+)$ (C) Distributive
(D) Identity $(+)$ (E) Inverse $(\cdot)$

Exercise A-1

All variables represent real numbers.

A *In Problems 1–6, replace each question mark with an appropriate expression that will illustrate the use of the indicated real number property.*

1. Commutative property $(\cdot)$: $uv = ?$

2. Commutative property $(+)$: $x + 7 = ?$

3. Associative property $(+)$: $3 + (7 + y) = ?$

4. Associative property $(\cdot)$: $x(yz) = ?$

5. Identity property $(\cdot)$: $1(u + v) = ?$

6. Identity property $(+)$: $0 + 9m = ?$

In Problems 7–26, indicate true (T) or false (F).

7. $5(8m) = (5 \cdot 8)m$

8. $a + cb = a + bc$

9. $5x + 7x = (5 + 7)x$

10. $uv(w + x) = uvw + uvx$

11. $-2(-a)(2x - y) = 2a(-4x + y)$

12. $8 \div (-5) = 8\left(\frac{1}{-5}\right)$

13. $(x + 3) + 2x = 2x + (x + 3)$

14. $\dfrac{x}{3y} \div \dfrac{5y}{x} = \dfrac{15y^2}{x^2}$

15. $\dfrac{2x}{-(x+3)} = -\dfrac{2x}{x+3}$

16. $-\dfrac{2x}{-(x-3)} = \dfrac{2x}{x-3}$

17. $(-3)\left(\dfrac{1}{-3}\right) = 1$

18. $(-0.5) + (0.5) = 0$

19. $-x^2y^2 = (-1)x^2y^2$

20. $[-(x+2)](-x) = (x+2)x$

21. $\dfrac{a}{b} + \dfrac{c}{d} = \dfrac{a+c}{b+d}$

22. $\dfrac{k}{k+b} = \dfrac{1}{1+b}$

23. $(x+8)(x+6) = (x+8)x + (x+8)6$

24. $u(u-2v) + v(u-2v) = (u+v)(u-2v)$

25. If $(x-2)(2x+3) = 0$, then either $x-2 = 0$ or $2x+3 = 0$.

26. If either $x-2 = 0$ or $2x+3 = 0$, then $(x-2)(2x+3) = 0$.

B

27. If $uv = 1$, does either u or v have to be 1? Explain.

28. If $uv = 0$, does either u or v have to be 0? Explain.

29. Indicate whether the following are true (T) or false (F):

(A) All integers are natural numbers.
(B) All rational numbers are real numbers.
(C) All natural numbers are rational numbers.

30. Indicate whether the following are true (T) or false (F):

(A) All natural numbers are integers.
(B) All real numbers are irrational.
(C) All rational numbers are real numbers.

31. Give an example of a real number that is not a rational number.

32. Give an example of a rational number that is not an integer.

33. Given the sets of numbers N (natural numbers), Z (integers), Q (rational numbers), and R (real numbers), indicate to which set(s) each of the following numbers belongs:

(A) 8   (B) $\sqrt{2}$ (C) -1.414 (D) $\dfrac{-5}{2}$

34. Given the sets of numbers N, Z, Q, and R (see Problem 33), indicate to which set(s) each of the following numbers belongs:

(A) -3 (B) 3.14 (C) π (D) $\dfrac{2}{3}$

35. Indicate true (T) or false (F), and for each false statement find real number replacements for a, b, and c that will provide a counterexample. For all real numbers a, b, and c,

(A) $a(b-c) = ab - c$
(B) $(a-b) - c = a - (b-c)$
(C) $a(bc) = (ab)c$
(D) $(a \div b) \div c = a \div (b \div c)$

36. Indicate true (T) or false (F), and for each false statement find real number replacements for a and b that will provide a counterexample. For all real numbers a and b,

(A) $a + b = b + a$
(B) $a - b = b - a$
(C) $ab = ba$
(D) $a \div b = b \div a$

C

37. If $c = 0.151\,515\ldots$, then $100c = 15.151\,5\ldots$ and

$$100c - c = 15.151\,5\ldots - 0.151\,515\ldots$$
$$99c = 15$$
$$c = \frac{15}{99} = \frac{5}{33}$$

Proceeding similarly, convert the repeating decimal $0.090\,909\ldots$ into a fraction. (All repeating decimals are rational numbers, and all rational numbers have repeating decimal representations.)

38. Repeat Problem 37 for $0.181\,818\ldots$.

Use a calculator to express each number in Problems 39 and 40 as a decimal to the capacity of your calculator. Observe the repeating decimal representation of the rational numbers and the nonrepeating decimal representation of the irrational numbers.

39. (A) $\dfrac{13}{6}$ (B) $\sqrt{21}$ (C) $\dfrac{7}{16}$ (D) $\dfrac{29}{111}$

40. (A) $\dfrac{8}{9}$ (B) $\dfrac{3}{11}$ (C) $\sqrt{5}$ (D) $\dfrac{11}{8}$

Section A-2 OPERATIONS ON POLYNOMIALS

- Natural Number Exponents
- Polynomials
- Combining Like Terms
- Addition and Subtraction
- Multiplication
- Combined Operations

This section covers basic operations on *polynomials*, a mathematical form that is encountered frequently. Our discussion starts with a brief review of natural number exponents. Integer and rational exponents and their properties will be discussed in detail in subsequent sections. (Natural numbers, integers, and rational numbers are important parts of the real number system; see Table 1 and Figure 1 in Appendix A-1.)

■ Natural Number Exponents

We define a **natural number exponent** as follows:

DEFINITION   **Natural Number Exponent**

For n a natural number and b any real number,

$$b^n = b \cdot b \cdot \cdots \cdot b \qquad n \text{ factors of } b$$
$$3^5 = 3 \cdot 3 \cdot 3 \cdot 3 \cdot 3 \qquad 5 \text{ factors of } 3$$

where n is called the **exponent** and b is called the **base.**

Along with this definition, we state the **first property of exponents:**

THEOREM 1   **FIRST PROPERTY OF EXPONENTS**

For any natural numbers m and n, and any real number b:

$$b^m b^n = b^{m+n} \qquad (2t^4)(5t^3) = 2 \cdot 5 t^{4+3} = 10t^7$$

■ Polynomials

Algebraic expressions are formed by using constants and variables and the algebraic operations of addition, subtraction, multiplication, division, raising to powers, and taking roots. Special types of algebraic expressions are called *polynomials*. A **polynomial in one variable** x is constructed by adding or subtracting constants and terms of the form ax^n, where a is a real number and n is a natural number. A **polynomial in two variables** x and y is constructed by adding and subtracting constants and terms of the form $ax^m y^n$, where a is a real number and m and n are natural numbers. Polynomials in three and more variables are defined in a similar manner.

POLYNOMIALS		*NOT POLYNOMIALS*	
8	0	$\dfrac{1}{x}$	$\dfrac{x-y}{x^2+y^2}$
$3x^3 - 6x + 7$	$6x + 3$		
$2x^2 - 7xy - 8y^2$	$9y^3 + 4y^2 - y + 4$	$\sqrt{x^3 - 2x}$	$2x^{-2} - 3x^{-1}$
$2x - 3y + 2$	$u^5 - 3u^3v^2 + 2uv^4 - v^4$		

Polynomial forms are encountered frequently in mathematics. For the efficient study of polynomials it is useful to classify them according to their *degree*. If a term in a polynomial has only one variable as a factor, then the **degree of the term** is the power of the variable. If two or more variables are present in a term as factors, then the **degree of the term** is the sum of the powers of the variables. The **degree of a polynomial** is the degree of the nonzero term with the highest degree in the polynomial. Any nonzero constant is defined to be a **polynomial of degree 0.** The number 0 is also a polynomial but is not assigned a degree.

EXAMPLE 1   **Degree**

(A) The degree of the first term in $5x^3 + \sqrt{3}x - \frac{1}{2}$ is 3, the degree of the second term is 1, the degree of the third term is 0, and the degree of the whole polynomial is 3 (the same as the degree of the term with the highest degree).

(B) The degree of the first term in $8u^3v^2 - \sqrt{7}uv^2$ is 5, the degree of the second term is 3, and the degree of the whole polynomial is 5.   ▬

(A) Given the polynomial $6x^5 + 7x^3 - 2$, what is the degree of the first term? The second term? The third term? The whole polynomial?

(B) Given the polynomial $2u^4v^2 - 5uv^3$, what is the degree of the first term? The second term? The whole polynomial?

In addition to classifying polynomials by degree, we also call a single-term polynomial a **monomial,** a two-term polynomial a **binomial,** and a three-term polynomial a **trinomial.**

■ Combining Like Terms

The concept of *coefficient* plays a central role in the process of combining *like terms*. A constant in a term of a polynomial, including the sign that precedes it, is called the **numerical coefficient,** or simply, the **coefficient,** of the term. If a constant does not appear, or only a + sign appears, the coefficient is understood to be 1. If only a − sign appears, the coefficient is understood to be −1. Thus, given the polynomial

$$5x^4 - x^3 - 3x^2 + x - 7 \quad \boxed{= 5x^4 + (-1)x^3 + (-3)x^2 + 1x + (-7)}$$

the coefficient of the first term is 5, the coefficient of the second term is −1, the coefficient of the third term is −3, the coefficient of the fourth term is 1, and the coefficient of the fifth term is −7.

The following distributive properties are fundamental to the process of combining *like terms*.

THEOREM 2   DISTRIBUTIVE PROPERTIES OF REAL NUMBERS

1. $a(b + c) = (b + c)a = ab + ac$

2. $a(b - c) = (b - c)a = ab - ac$

3. $a(b + c + \cdots + f) = ab + ac + \cdots + af$

Two terms in a polynomial are called **like terms** if they have exactly the same variable factors to the same powers. The numerical coefficients may or may not be the same. Since constant terms involve no variables, all constant terms are like terms. If a polynomial contains two or more like terms, these terms can be combined into a single term by making use of distributive properties. The following example illustrates the reasoning behind the process:

$$3x^2y - 5xy^2 + x^2y - 2x^2y \quad \boxed{\begin{aligned} &= 3x^2y + x^2y - 2x^2y - 5xy^2 \\ &= (3x^2y + 1x^2y - 2x^2y) - 5xy^2 \\ &= (3 + 1 - 2)x^2y - 5xy^2 \end{aligned}} \quad \begin{aligned} &\textit{Note the} \\ &\textit{use of} \\ &\textit{distributive} \\ &\textit{properties.} \end{aligned}$$
$$= 2x^2y - 5xy^2$$

It should be clear that free use is made of the real number properties discussed in Appendix A-1.

INSIGHT

The steps shown in the dashed box are usually done mentally. The same result can be obtained by simply adding the numerical coefficients of like terms.

How can we simplify expressions such as $4(x - 2y) - 3(2x - 7y)$? We clear the expression of parentheses using distributive properties, and combine like terms:

$$4(x - 2y) - 3(2x - 7y) = 4x - 8y - 6x + 21y$$
$$= -2x + 13y$$

EXAMPLE 2   **Removing Parentheses**   Remove parentheses and simplify:

(A) $2(3x^2 - 2x + 5) + (x^2 + 3x - 7)$ $\boxed{= 2(3x^2 - 2x + 5) + 1(x^2 + 3x - 7)}$

$$= 6x^2 - 4x + 10 + x^2 + 3x - 7$$

$$= 7x^2 - x + 3$$

(B) $(x^3 - 2x - 6) - (2x^3 - x^2 + 2x - 3)$

$\boxed{= 1(x^3 - 2x - 6) + (-1)(2x^3 - x^2 + 2x - 3)}$ *Be careful with the sign here*

$$= x^3 - 2x - 6 - 2x^3 + x^2 - 2x + 3$$

$$= -x^3 + x^2 - 4x - 3$$

(C) $[3x^2 - (2x + 1)] - (x^2 - 1) = [3x^2 - 2x - 1] - (x^2 - 1)$ *Remove inner parentheses first.*

$$= 3x^2 - 2x - 1 - x^2 + 1$$

$$= 2x^2 - 2x$$

MATCHED PROBLEM 2   Remove parentheses and simplify:

(A) $3(u^2 - 2v^2) + (u^2 + 5v^2)$

(B) $(m^3 - 3m^2 + m - 1) - (2m^3 - m + 3)$

(C) $(x^3 - 2) - [2x^3 - (3x + 4)]$

■ Addition and Subtraction

Addition and subtraction of polynomials can be thought of in terms of removing parentheses and combining like terms, as illustrated in Example 2. Horizontal and vertical arrangements are illustrated in the next two examples. You should be able to work either way, letting the situation dictate your choice.

EXAMPLE 3   **Adding Polynomials**   Add horizontally and vertically:

$$x^4 - 3x^3 + x^2, \quad -x^3 - 2x^2 + 3x, \quad \text{and} \quad 3x^2 - 4x - 5$$

SOLUTION   Add horizontally:

$$(x^4 - 3x^3 + x^2) + (-x^3 - 2x^2 + 3x) + (3x^2 - 4x - 5)$$

$$= x^4 - 3x^3 + x^2 - x^3 - 2x^2 + 3x + 3x^2 - 4x - 5$$

$$= x^4 - 4x^3 + 2x^2 - x - 5$$

Or vertically, by lining up like terms and adding their coefficients:

$$
\begin{array}{r}
x^4 - 3x^3 + x^2 \phantom{{}- 4x - 5} \\
- x^3 - 2x^2 + 3x \phantom{{}- 5} \\
3x^2 - 4x - 5 \\
\hline
x^4 - 4x^3 + 2x^2 - x - 5
\end{array}
$$

MATCHED PROBLEM 3   Add horizontally and vertically:

$$3x^4 - 2x^3 - 4x^2, \quad x^3 - 2x^2 - 5x, \quad \text{and} \quad x^2 + 7x - 2$$

EXAMPLE 4   **Subtracting Polynomials**   Subtract $4x^2 - 3x + 5$ from $x^2 - 8$, both horizontally and vertically.

SOLUTION $(x^2 - 8) - (4x^2 - 3x + 5)$ or

$$\begin{aligned} &x^2 \qquad\quad - 8 \\ &\underline{-4x^2 + 3x - 5} \leftarrow \text{Change signs and add.} \\ &-3x^2 + 3x - 13 \end{aligned}$$

$$= x^2 - 8 - 4x^2 + 3x - 5$$
$$= -3x^2 + 3x - 13$$

MATCHED PROBLEM 4 Subtract $2x^2 - 5x + 4$ from $5x^2 - 6$, both horizontally and vertically.

■ Multiplication

Multiplication of algebraic expressions involves the extensive use of distributive properties for real numbers, as well as other real number properties.

EXAMPLE 5 **Multiplying Polynomials** Multiply: $(2x - 3)(3x^2 - 2x + 3)$

SOLUTION

$$(2x - 3)(3x^2 - 2x + 3) = 2x(3x^2 - 2x + 3) - 3(3x^2 - 2x + 3)$$
$$= 6x^3 - 4x^2 + 6x - 9x^2 + 6x - 9$$
$$= 6x^3 - 13x^2 + 12x - 9$$

Or, using a vertical arrangement,

$$\begin{aligned} &3x^2 - 2x + 3 \\ &\underline{2x - 3} \\ &6x^3 - 4x^2 + 6x \\ &\underline{\qquad - 9x^2 + 6x - 9} \\ &6x^3 - 13x^2 + 12x - 9 \end{aligned}$$

MATCHED PROBLEM 5 Multiply: $(2x - 3)(2x^2 + 3x - 2)$

Thus, to multiply two polynomials, multiply each term of one by each term of the other, and combine like terms.

Products of binomial factors occur frequently, so it is useful to develop procedures that will enable us to write down their products by inspection. To find the product $(2x - 1)(3x + 2)$, we proceed as follows:

$$(2x - 1)(3x + 2) \quad = 6x^2 + 4x - 3x - 2 \qquad \begin{array}{l}\text{The inner and outer products} \\ \text{are like terms, so combine into} \\ \text{a single term.}\end{array}$$
$$= 6x^2 + x - 2$$

To speed the process, we do the step in the dashed box mentally.

Products of certain binomial factors occur so frequently that it is useful to learn formulas for their products. The following formulas are easily verified by multiplying the factors on the left.

THEOREM 3 **SPECIAL PRODUCTS**

1. $(a - b)(a + b) = a^2 - b^2$
2. $(a + b)^2 = a^2 + 2ab + b^2$
3. $(a - b)^2 = a^2 - 2ab + b^2$

Explore & Discuss **1**   (A) Explain the relationship between special product formula 1 and the areas of the rectangles in the figure.

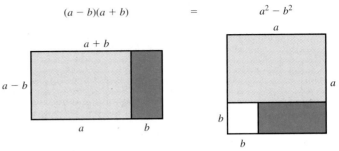

$$(a - b)(a + b) \qquad = \qquad a^2 - b^2$$

(B) Construct similar figures to provide geometric interpretations for special product formulas 2 and 3.

EXAMPLE 6   **Special Products**   Multiply mentally where possible:

(A) $(2x - 3y)(5x + 2y)$ (B) $(3a - 2b)(3a + 2b)$

(C) $(5x - 3)^2$ (D) $(m + 2n)^3$

SOLUTION   (A) $(2x - 3y)(5x + 2y) \boxed{= 10x^2 + 4xy - 15xy - 6y^2}$

$\qquad\qquad\qquad\qquad\qquad = 10x^2 - 11xy - 6y^2$

(B) $(3a - 2b)(3a + 2b) \boxed{= (3a)^2 - (2b)^2}$

$\qquad\qquad\qquad\qquad\quad = 9a^2 - 4b^2$

(C) $(5x - 3)^2 \boxed{= (5x)^2 - 2(5x)(3) + 3^2}$

$\qquad\qquad\quad = 25x^2 - 30x + 9$

(D) $(m + 2n)^3 = (m + 2n)^2(m + 2n)$

$\qquad\qquad\quad = (m^2 + 4mn + 4n^2)(m + 2n)$

$\qquad\qquad\quad = m^2(m + 2n) + 4mn(m + 2n) + 4n^2(m + 2n)$

$\qquad\qquad\quad = m^3 + 2m^2n + 4m^2n + 8mn^2 + 4mn^2 + 8n^3$

$\qquad\qquad\quad = m^3 + 6m^2n + 12mn^2 + 8n^3$

MATCHED PROBLEM 6   Multiply mentally where possible:

(A) $(4u - 3v)(2u + v)$

(B) $(2xy + 3)(2xy - 3)$

(C) $(m + 4n)(m - 4n)$

(D) $(2u - 3v)^2$

(E) $(2x - y)^3$

■ Combined Operations

We complete this section by considering several examples that use all the operations just discussed. Note that in simplifying, we usually remove grouping symbols starting from the inside. That is, we remove parentheses ( ) first, then brackets [ ], and finally braces { }, if present. Also,

DEFINITION   **Order of Operations**

Multiplication and division precede addition and subtraction, and taking powers precedes multiplication and division.

$$2 \cdot 3 + 4 = 6 + 4 = 10, \quad not \quad 2 \cdot 7 = 14$$

$$\frac{10^2}{2} = \frac{100}{2} = 50, \quad not \quad 5^2 = 25$$

EXAMPLE 7 **Combined Operations** Perform the indicated operations and simplify:

$$
\begin{aligned}
\text{(A) } 3x - \{5 - 3[x - x(3 - x)]\} &= 3x - \{5 - 3[x - 3x + x^2]\} \\
&= 3x - \{5 - 3x + 9x - 3x^2\} \\
&= 3x - 5 + 3x - 9x + 3x^2 \\
&= 3x^2 - 3x - 5
\end{aligned}
$$

$$
\begin{aligned}
\text{(B) } (x - 2y)(2x + 3y) - (2x + y)^2 &= 2x^2 - xy - 6y^2 - (4x^2 + 4xy + y^2) \\
&= 2x^2 - xy - 6y^2 - 4x^2 - 4xy - y^2 \\
&= -2x^2 - 5xy - 7y^2
\end{aligned}
$$

MATCHED PROBLEM 7 Perform the indicated operations and simplify:

(A) $2t - \{7 - 2[t - t(4 + t)]\}$

(B) $(u - 3v)^2 - (2u - v)(2u + v)$

Answers to Matched Problems

1. (A) $5, 3, 0, 5$ (B) $6, 4, 6$

2. (A) $4u^2 - v^2$ (B) $-m^3 - 3m^2 + 2m - 4$ (C) $-x^3 + 3x + 2$

3. $3x^4 - x^3 - 5x^2 + 2x - 2$ **4.** $3x^2 + 5x - 10$ **5.** $4x^3 - 13x + 6$

6. (A) $8u^2 - 2uv - 3v^2$ (B) $4x^2y^2 - 9$ (C) $m^2 - 16n^2$

(D) $4u^2 - 12uv + 9v^2$ (E) $8x^3 - 12x^2y + 6xy^2 - y^3$

7. (A) $-2t^2 - 4t - 7$ (B) $-3u^2 - 6uv + 10v^2$

Exercise A-2

A *Problems 1–8 refer to the following polynomials:*

(A) $2x - 3$ (B) $2x^2 - x + 2$ (C) $x^3 + 2x^2 - x + 3$

1. What is the degree of (C)?

2. What is the degree of (A)?

3. Add (B) and (C).

4. Add (A) and (B).

5. Subtract (B) from (C).

6. Subtract (A) from (B).

7. Multiply (B) and (C).

8. Multiply (A) and (C).

In Problems 9–30, perform the indicated operations and simplify.

9. $2(u - 1) - (3u + 2) - 2(2u - 3)$

10. $2(x - 1) + 3(2x - 3) - (4x - 5)$

11. $4a - 2a[5 - 3(a + 2)]$

12. $2y - 3y[4 - 2(y - 1)]$

13. $(a + b)(a - b)$

14. $(m - n)(m + n)$

15. $(3x - 5)(2x + 1)$

16. $(4t - 3)(t - 2)$

17. $(2x - 3y)(x + 2y)$

18. $(3x + 2y)(x - 3y)$

19. $(3y + 2)(3y - 2)$

20. $(2m - 7)(2m + 7)$

21. $-(2x - 3)^2$

22. $-(5 - 3x)^2$

23. $(4m + 3n)(4m - 3n)$

24. $(3x - 2y)(3x + 2y)$

25. $(3u + 4v)^2$

26. $(4x - y)^2$

27. $(a - b)(a^2 + ab + b^2)$

28. $(a + b)(a^2 - ab + b^2)$

29. $[(x - y) + 3z][(x - y) - 3z]$

30. $[a - (2b - c)][a + (2b - c)]$

B *In Problems 31–44, perform the indicated operations and simplify.*

31. $m - \{m - [m - (m - 1)]\}$

32. $2x - 3\{x + 2[x - (x + 5)] + 1\}$

33. $(x^2 - 2xy + y^2)(x^2 + 2xy + y^2)$

34. $(3x - 2y)^2(2x + 5y)$

35. $(5a - 2b)^2 - (2b + 5a)^2$

36. $(2x - 1)^2 - (3x + 2)(3x - 2)$

37. $(m - 2)^2 - (m - 2)(m + 2)$

38. $(x - 3)(x + 3) - (x - 3)^2$

39. $(x - 2y)(2x + y) - (x + 2y)(2x - y)$

40. $(3m + n)(m - 3n) - (m + 3n)(3m - n)$

41. $(u + v)^3$

42. $(x - y)^3$

43. $(x - 2y)^3$

44. $(2m - n)^3$

45. Subtract the sum of the last two polynomials from the sum of the first two: $2x^2 - 4xy + y^2, 3xy - y^2,$ $x^2 - 2xy - y^2, -x^2 + 3xy - 2y^2$

46. Subtract the sum of the first two polynomials from the sum of the last two: $3m^2 - 2m + 5, 4m^2 - m,$ $3m^2 - 3m - 2, m^3 + m^2 + 2$

C *In Problems 47–50, perform the indicated operations and simplify.*

47. $[(2x - 1)^2 - x(3x + 1)]^2$

48. $[5x(3x + 1) - 5(2x - 1)^2]^2$

49. $2\{(x - 3)(x^2 - 2x + 1) - x[3 - x(x - 2)]\}$

50. $-3x\{x[x - x(2 - x)] - (x + 2)(x^2 - 3)\}$

51. If you are given two polynomials, one of degree *m* and the other of degree *n*, where *m* is greater than *n*, what is the degree of their product?

52. What is the degree of the sum of the two polynomials in Problem 51?

53. How does the answer to Problem 51 change if the two polynomials can have the same degree?

54. How does the answer to Problem 52 change if the two polynomials can have the same degree?

55. Show by example that, in general, $(a + b)^2 \neq a^2 + b^2$. Discuss possible conditions on *a* and *b* that would make this a valid equation.

56. Show by example that, in general, $(a - b)^2 \neq a^2 - b^2$. Discuss possible conditions on *a* and *b* that would make this a valid equation.

Applications

57. *Investment.* You have $10,000 to invest, part at 9% and the rest at 12%. If *x* is the amount invested at 9%, write an algebraic expression that represents the total annual income from both investments. Simplify the expression.

58. *Investment.* A person has $100,000 to invest. If $*x* are invested in a money market account yielding 7% and twice that amount in certificates of deposit yielding 9%, and if the rest is invested in high-grade bonds yielding 11%, write an algebraic expression that represents the total annual income from all three investments. Simplify the expression.

59. *Gross receipts.* Four thousand tickets are to be sold for a musical show. If *x* tickets are to be sold for $10 each and three times that number for $30 each, and if the rest are sold for $50 each, write an algebraic expression that represents the gross receipts from ticket sales, assuming all tickets are sold. Simplify the expression.

60. *Gross receipts.* Six thousand tickets are to be sold for a concert, some for $9 each and the rest for $15 each. If *x* is the number of $9 tickets sold, write an algebraic expression that represents the gross receipts from ticket sales, assuming all tickets are sold. Simplify the expression.

61. *Nutrition.* Food mix *A* contains 2% fat, and food mix *B* contains 6% fat. A 10-kilogram diet mix of foods *A* and *B* is formed. If *x* kilograms of food *A* are used, write an algebraic expression that represents the total number of kilograms of fat in the final food mix. Simplify the expression.

62. *Nutrition.* Each ounce of food *M* contains 8 units of calcium, and each ounce of food *N* contains 5 units of calcium. A 160-ounce diet mix is formed using foods *M* and *N*. If *x* is the number of ounces of food *M* used, write an algebraic expression that represents the total number of units of calcium in the diet mix. Simplify the expression.

Section A-3 FACTORING POLYNOMIALS

- Common Factors
- Factoring by Grouping
- Factoring Second-Degree Polynomials
- Special Factoring Formulas
- Combined Factoring Techniques

A polynomial is written in factored form if it is written as the product of two or more polynomials. The following polynomials are written in factored form:

$$4x^2y - 6xy^2 = 2xy(2x - 3y) \qquad 2x^3 - 8x = 2x(x - 2)(x + 2)$$
$$x^2 - x - 6 = (x - 3)(x + 2) \qquad 5m^2 + 20 = 5(m^2 + 4)$$

Factoring Polynomials
Unless stated to the contrary, we will limit our discussion of factoring of polynomials to polynomials with integer coefficients.

A polynomial with integer coefficients is said to be **factored completely** if each factor cannot be expressed as the product of two or more polynomials with integer coefficients, other than itself or 1. All the polynomials above, as we will see by the conclusion of this section, are factored completely.

Writing polynomials in completely factored form is often a difficult task. But accomplishing it can lead to the simplification of certain algebraic expressions and to the solution of certain types of equations and inequalities. The distributive properties for real numbers are central to the factoring process.

■ Common Factors

Generally, a first step in any factoring procedure is to factor out all factors common to all terms.

EXAMPLE 1   **Common Factors**   Factor out all factors common to all terms:

(A) $3x^3y - 6x^2y^2 - 3xy^3$
(B) $3y(2y + 5) + 2(2y + 5)$

SOLUTION   (A) $3x^3y - 6x^2y^2 - 3xy^3 \boxed{= (3xy)x^2 - (3xy)2xy - (3xy)y^2}$
$$= 3xy(x^2 - 2xy - y^2)$$

(B) $3y(2y + 5) + 2(2y + 5) \boxed{= 3y(2y + 5) + 2(2y + 5)}$
$$= (3y + 2)(2y + 5)$$

MATCHED PROBLEM 1   Factor out all factors common to all terms:
(A) $2x^3y - 8x^2y^2 - 6xy^3$ (B) $2x(3x - 2) - 7(3x - 2)$

■ Factoring by Grouping

Occasionally, polynomials can be factored by grouping terms in such a way that we obtain results that look like Example 1B. We can then complete the factoring following the steps used in that example. This process will prove useful in the next subsection, where an efficient method is developed for factoring a second-degree polynomial as the product of two first-degree polynomials, if such factors exist.

EXAMPLE 2   **Factoring by Grouping**   Factor by grouping:

(A) $3x^2 - 3x - x + 1$
(B) $4x^2 - 2xy - 6xy + 3y^2$
(C) $y^2 + xz + xy + yz$

SOLUTION   (A) $3x^2 - 3x - x + 1$

$= (3x^2 - 3x) - (x - 1)$

$= 3x(x - 1) - (x - 1)$

$= (x - 1)(3x - 1)$

Group the first two and the last two terms. Factor out any common factors from each group. The common factor $(x - 1)$ can be taken out, and the factoring is complete.

(B) $4x^2 - 2xy - 6xy + 3y^2 = (4x^2 - 2xy) - (6xy - 3y^2)$

$= 2x(2x - y) - 3y(2x - y)$

$= (2x - y)(2x - 3y)$

(C)   If, as in parts (A) and (B), we group the first two terms and the last two terms of $y^2 + xz + xy + yz$, no common factor can be taken out of each group to complete the factoring. However, if the two middle terms are reversed, we can proceed as before:

$$y^2 + xz + xy + yz = y^2 + xy + xz + yz$$
$$= (y^2 + xy) + (xz + yz)$$
$$= y(y + x) + z(x + y)$$
$$= y(x + y) + z(x + y)$$
$$= (x + y)(y + z)$$

MATCHED PROBLEM 2

Factor by grouping:

(A) $6x^2 + 2x + 9x + 3$

(B) $2u^2 + 6uv - 3uv - 9v^2$

(C) $ac + bd + bc + ad$

■ Factoring Second-Degree Polynomials

We now turn our attention to factoring second-degree polynomials of the form

$$2x^2 - 5x - 3 \quad \text{and} \quad 2x^2 + 3xy - 2y^2$$

into the product of two first-degree polynomials with integer coefficients. Since many second-degree polynomials with integer coefficients cannot be factored in this way, it would be useful to know ahead of time that the factors we are seeking actually exist. The factoring approach we use, involving the *ac test*, determines at the beginning whether first-degree factors with integer coefficients do exist. Then, if they exist, the test provides a simple method for finding them.

THEOREM 1   *ac* **TEST FOR FACTORABILITY**

If in polynomials of the form

$$ax^2 + bx + c \quad \text{or} \quad ax^2 + bxy + cy^2 \qquad (1)$$

the product ac has two integer factors p and q whose sum is the coefficient b of the middle term; that is, if integers p and q exist so that

$$pq = ac \quad \text{and} \quad p + q = b \qquad (2)$$

then the polynomials have first-degree factors with integer coefficients. If no integers p and q exist that satisfy equations (2), then the polynomials in equations (1) will not have first-degree factors with integer coefficients.

If integers p and q exist that satisfy equations (2) in the *ac* test, the factoring always can be completed as follows: Using $b = p + q$, split the middle terms in equations (1) to obtain

$$ax^2 + bx + c = ax^2 + px + qx + c$$
$$ax^2 + bxy + cy^2 = ax^2 + pxy + qxy + cy^2$$

Complete the factoring by grouping the first two terms and the last two terms as in Example 2. This process always works, and it does not matter if the two middle terms on the right are interchanged.

Several examples should make the process clear. After a little practice, you will perform many of the steps mentally and will find the process fast and efficient.

EXAMPLE 3 **Factoring Second-Degree Polynomials** Factor, if possible, using integer coefficients:

(A) $4x^2 - 4x - 3$

(B) $2x^2 - 3x - 4$

(C) $6x^2 - 25xy + 4y^2$

SOLUTION (A) $4x^2 - 4x - 3$

Step 1. Use the *ac* test to test for factorability. Comparing $4x^2 - 4x - 3$ with $ax^2 + bx + c$, we see that $a = 4$, $b = -4$, and $c = -3$. Multiply a and c to obtain

$$ac = (4)(-3) = -12$$

pq
$(1)(-12)$
$(-1)(12)$
$(2)(-6)$
$(-2)(6)$
$(3)(-4)$
$(-3)(4)$

All factor pairs of $-12 = ac$

List all pairs of integers whose product is -12, as shown in the margin. These are called **factor pairs** of -12. Then try to find a factor pair that sums to $b = -4$, the coefficient of the middle term in $4x^2 - 4x - 3$. (In practice, this part of step 1 is often done mentally and can be done rather quickly.) Notice that the factor pair 2 and -6 sums to -4. Thus, by the *ac* test, $4x^2 - 4x - 3$ has first-degree factors with integer coefficients.

Step 2. Split the middle term, using $b = p + q$, and complete the factoring by grouping. Using $-4 = 2 + (-6)$, we split the middle term in $4x^2 - 4x - 3$ and complete the factoring by grouping:

$$4x^2 - 4x - 3 = 4x^2 + 2x - 6x - 3$$
$$= (4x^2 + 2x) - (6x + 3)$$
$$= 2x(2x + 1) - 3(2x + 1)$$
$$= (2x + 1)(2x - 3)$$

The result can be checked by multiplying the two factors to obtain the original polynomial.

(B) $2x^2 - 3x - 4$

Step 1. Use the *ac* test to test for factorability:

$$ac = (2)(-4) = -8$$

pq
$(-1)(8)$
$(1)(-8)$
$(-2)(4)$
$(2)(-4)$

All factor pairs of $-8 = ac$

Does -8 have a factor pair whose sum is -3? None of the factor pairs listed in the margin sums to $-3 = b$, the coefficient of the middle term in $2x^2 - 3x - 4$. According to the *ac* test, we can conclude that $2x^2 - 3x - 4$ does not have first-degree factors with integer coefficients, and we say that the polynomial is **not factorable.**

(C) $6x^2 - 25xy + 4y^2$

Step 1. Use the *ac* test to test for factorability:

$$ac = (6)(4) = 24$$

Mentally checking through the factor pairs of 24, keeping in mind that their sum must be $-25 = b$, we see that if $p = -1$ and $q = -24$, then

$$pq = (-1)(-24) = 24 = ac$$

and

$$p + q = (-1) + (-24) = -25 = b$$

Thus, the polynomial is factorable.

Step 2. Split the middle term, using $b = p + q$, and complete the factoring by grouping. Using $-25 = (-1) + (-24)$, we split the middle term in $6x^2 - 25xy + 4y^2$ and complete the factoring by grouping:

$$6x^2 - 25xy + 4y^2 = 6x^2 - xy - 24xy + 4y^2$$
$$= (6x^2 - xy) - (24xy - 4y^2)$$
$$= x(6x - y) - 4y(6x - y)$$
$$= (6x - y)(x - 4y)$$

The check is left to the reader.

MATCHED PROBLEM 3 Factor, if possible, using integer coefficients:

(A) $2x^2 + 11x - 6$

(B) $4x^2 + 11x - 6$

(C) $6x^2 + 5xy - 4y^2$

■ Special Factoring Formulas

The factoring formulas listed in the following box will enable us to factor certain polynomial forms that occur frequently. These formulas can be established by multiplying the factors on the right.

THEOREM 2 **SPECIAL FACTORING FORMULAS**

Perfect square:	1. $u^2 + 2uv + v^2 = (u + v)^2$
Perfect square:	2. $u^2 - 2uv + v^2 = (u - v)^2$
Difference of squares:	3. $u^2 - v^2 = (u - v)(u + v)$
Difference of cubes:	4. $u^3 - v^3 = (u - v)(u^2 + uv + v^2)$
Sum of cubes:	5. $u^3 + v^3 = (u + v)(u^2 - uv + v^2)$

⚠ CAUTION: Notice that $u^2 + v^2$ is not included in the list of special factoring formulas. In general,

$$u^2 + v^2 \neq (au + bv)(cu + dv)$$

for any choice of real number coefficients a, b, c, and d.

EXAMPLE 4 **Factoring** Factor completely:

(A) $4m^2 - 12mn + 9n^2$ (B) $x^2 - 16y^2$

(C) $z^3 - 1$ (D) $m^3 + n^3$

(E) $a^2 - 4(b + 2)^2$

SOLUTION (A) $4m^2 - 12mn + 9n^2 = (2m - 3n)^2$

(B) $x^2 - 16y^2 \boxed{= x^2 - (4y)^2} = (x - 4y)(x + 4y)$

(C) $z^3 - 1 = (z - 1)(z^2 + z + 1)$ Use the ac test to verify that $z^2 + z + 1$ cannot be factored.

(D) $m^3 + n^3 = (m + n)(m^2 - mn + n^2)$ Use the ac test to verify that $m^2 - mn + n^2$ cannot be factored.

(E) $a^2 - 4(b + 2)^2 = [a - 2(b + 2)][a + 2(b + 2)]$

| MATCHED PROBLEM 4 | Factor completely:

(A) $x^2 + 6xy + 9y^2$ (B) $9x^2 - 4y^2$

(C) $8m^3 - 1$ (D) $x^3 + y^3z^3$

(E) $9(m - 3)^2 - 4n^2$

Explore & Discuss 1

(A) Verify the following factoring formulas for $u^4 - v^4$:

$$u^4 - v^4 = (u - v)(u + v)(u^2 + v^2)$$
$$= (u - v)(u^3 + u^2v + uv^2 + v^3)$$

(B) Discuss the pattern in the following formulas:

$$u^2 - v^2 = (u - v)(u + v)$$
$$u^3 - v^3 = (u - v)(u^2 + uv + v^2)$$
$$u^4 - v^4 = (u - v)(u^3 + u^2v + uv^2 + v^3)$$

(C) Use the pattern you discovered in part (B) to write similar formulas for $u^5 - v^5$ and $u^6 - v^6$. Verify your formulas by multiplication.

■ Combined Factoring Techniques

We complete this section by considering several factoring problems that involve combinations of the preceding techniques.

PROCEDURE | **Factoring Polynomials**

Step 1. Take out any factors common to all terms.

Step 2. Use any of the special formulas listed in Theorem 2 that are applicable.

Step 3. Apply the *ac* test to any remaining second-degree polynomial factors.

Note: It may be necessary to perform some of these steps more than once. Furthermore, the order of applying these steps can vary.

EXAMPLE 5 **Combined Factoring Techniques** Factor completely:

(A) $3x^3 - 48x$

(B) $3u^4 - 3u^3v - 9u^2v^2$

(C) $3m^4 - 24mn^3$

(D) $3x^4 - 5x^2 + 2$

SOLUTION

(A) $3x^3 - 48x = 3x(x^2 - 16) = 3x(x - 4)(x + 4)$

(B) $3u^4 - 3u^3v - 9u^2v^2 = 3u^2(u^2 - uv - 3v^2)$

(C) $3m^4 - 24mn^3 = 3m(m^3 - 8n^3) = 3m(m - 2n)(m^2 + 2mn + 4n^2)$

(D) $3x^4 - 5x^2 + 2 = (3x^2 - 2)(x^2 - 1) = (3x^2 - 2)(x - 1)(x + 1)$

| MATCHED PROBLEM 5 | Factor completely:

(A) $18x^3 - 8x$

(B) $4m^3n - 2m^2n^2 + 2mn^3$

(C) $2t^4 - 16t$

(D) $2y^4 - 5y^2 - 12$

Answers to Matched Problems

1. (A) $2xy(x^2 - 4xy - 3y^2)$ (B) $(2x - 7)(3x - 2)$

2. (A) $(3x + 1)(2x + 3)$ (B) $(u + 3v)(2u - 3v)$ (C) $(a + b)(c + d)$

3. (A) $(2x - 1)(x + 6)$ (B) Not factorable   (C) $(3x + 4y)(2x - y)$

4. (A) $(x + 3y)^2$ (B) $(3x - 2y)(3x + 2y)$

   (C) $(2m - 1)(4m^2 + 2m + 1)$ (D) $(x + yz)(x^2 - xyz + y^2z^2)$

   (E) $[3(m - 3) - 2n][3(m - 3) + 2n]$

5. (A) $2x(3x - 2)(3x + 2)$ (B) $2mn(2m^2 - mn + n^2)$

   (C) $2t(t - 2)(t^2 + 2t + 4)$ (D) $(2y^2 + 3)(y - 2)(y + 2)$

Exercise A-3

A In Problems 1–8, factor out all factors common to all terms.

1. $6m^4 - 9m^3 - 3m^2$

2. $6x^4 - 8x^3 - 2x^2$

3. $8u^3v - 6u^2v^2 + 4uv^3$

4. $10x^3y + 20x^2y^2 - 15xy^3$

5. $7m(2m - 3) + 5(2m - 3)$

6. $5x(x + 1) - 3(x + 1)$

7. $4ab(2c + d) - (2c + d)$

8. $12a(b - 2c) - 15b(b - 2c)$

In Problems 9–18, factor by grouping.

9. $2x^2 - x + 4x - 2$

10. $x^2 - 3x + 2x - 6$

11. $3y^2 - 3y + 2y - 2$

12. $2x^2 - x + 6x - 3$

13. $2x^2 + 8x - x - 4$

14. $6x^2 + 9x - 2x - 3$

15. $wy - wz + xy - xz$

16. $ac + ad + bc + bd$

17. $am - 3bm + 2na - 6bn$

18. $ab + 6 + 2a + 3b$

B In Problems 19–56, factor completely. If a polynomial cannot be factored, say so.

19. $3y^2 - y - 2$

20. $2x^2 + 5x - 3$

21. $u^2 - 2uv - 15v^2$

22. $x^2 - 4xy - 12y^2$

23. $m^2 - 6m - 3$

24. $x^2 + x - 4$

25. $w^2x^2 - y^2$

26. $25m^2 - 16n^2$

27. $9m^2 - 6mn + n^2$

28. $x^2 + 10xy + 25y^2$

29. $y^2 + 16$

30. $u^2 + 81$

31. $4z^2 - 28z + 48$

32. $6x^2 + 48x + 72$

33. $2x^4 - 24x^3 + 40x^2$

34. $2y^3 - 22y^2 + 48y$

35. $4xy^2 - 12xy + 9x$

36. $16x^2y - 8xy + y$

37. $6m^2 - mn - 12n^2$

38. $6s^2 + 7st - 3t^2$

39. $4u^3v - uv^3$

40. $x^3y - 9xy^3$

41. $2x^3 - 2x^2 + 8x$

42. $3m^3 - 6m^2 + 15m$

43. $8x^3 - 27y^3$

44. $5x^3 + 40y^3$

45. $x^4y + 8xy$

46. $8a^3 - 1$

C

47. $(x + 2)^2 - 9y^2$

48. $(a - b)^2 - 4(c - d)^2$

49. $5u^2 + 4uv - 2v^2$

50. $3x^2 - 2xy - 4y^2$

51. $6(x - y)^2 + 23(x - y) - 4$

52. $4(A + B)^2 - 5(A + B) - 6$

53. $y^4 - 3y^2 - 4$

54. $m^4 - n^4$

55. $15y(x - y)^3 + 12x(x - y)^2$

56. $15x^2(3x - 1)^4 + 60x^3(3x - 1)^3$

In Problems 57–60, discuss the validity of each statement. If the statement is true, explain why. If not, give a counterexample.

57. If n is a positive integer greater than 1, then $u^n - v^n$ can be factored.

58. If m and n are positive integers and $m \neq n$, then $u^m - v^n$ is not factorable.

59. If n is a positive integer greater than 1, then $u^n + v^n$ can be factored.

60. If k is a positive integer, then $u^{2k+1} + v^{2k+1}$ can be factored.

Section A-4 OPERATIONS ON RATIONAL EXPRESSIONS

- Reducing to Lowest Terms
- Multiplication and Division
- Addition and Subtraction
- Compound Fractions

We now turn our attention to fractional forms. A quotient of two algebraic expressions (division by 0 excluded) is called a **fractional expression.** If both the numerator and the denominator are polynomials, the fractional expression is called a **rational expression.** Some examples of rational expressions are

$$\frac{1}{x^3 + 2x} \qquad \frac{5}{x} \qquad \frac{x + 7}{3x^2 - 5x + 1} \qquad \frac{x^2 - 2x + 4}{1}$$

In this section we discuss basic operations on rational expressions, including multiplication, division, addition, and subtraction.

Since variables represent real numbers in the rational expressions we will consider, the properties of real number fractions summarized in Appendix A-1 will play a central role in much of the work that we will do.

AGREEMENT

Variable Restriction

Even though not always explicitly stated, we always assume that variables are restricted so that division by 0 is excluded.

For example, given the rational expression

$$\frac{2x + 5}{x(x + 2)(x - 3)}$$

the variable x is understood to be restricted from being 0, -2, or 3, since these values would cause the denominator to be 0.

■ Reducing to Lowest Terms

Central to the process of reducing rational expressions to *lowest terms* is the *fundamental property of fractions,* which we restate here for convenient reference:

THEOREM 1 FUNDAMENTAL PROPERTY OF FRACTIONS

If a, b, and k are real numbers with b, $k \neq 0$, then

$$\frac{ka}{kb} = \frac{a}{b} \qquad \frac{5 \cdot 2}{5 \cdot 7} = \frac{2}{7} \qquad \frac{x(x + 4)}{2(x + 4)} = \frac{x}{2}, \quad x \neq -4$$

Using this property from left to right to eliminate all common factors from the numerator and the denominator of a given fraction is referred to as **reducing a fraction to lowest terms.** We are actually dividing the numerator and denominator by the same nonzero common factor.

Using the property from right to left—that is, multiplying the numerator and denominator by the same nonzero factor—is referred to as **raising a fraction to higher terms.** We will use the property in both directions in the material that follows.

| EXAMPLE 1 | **Reducing to Lowest Terms**   Reduce each rational expression to lowest terms. |

(A) $\dfrac{6x^2 + x - 1}{2x^2 - x - 1} = \dfrac{(2x + 1)(3x - 1)}{(2x + 1)(x - 1)}$ *Factor numerator and denominator completely.*

$= \dfrac{3x - 1}{x - 1}$ *Divide numerator and denominator by the common factor* (2x + 1).

(B) $\dfrac{x^4 - 8x}{3x^3 - 2x^2 - 8x} = \dfrac{x(x - 2)(x^2 + 2x + 4)}{x(x - 2)(3x + 4)}$

$= \dfrac{x^2 + 2x + 4}{3x + 4}$

| MATCHED PROBLEM 1 | Reduce each rational expression to lowest terms. |

(A) $\dfrac{x^2 - 6x + 9}{x^2 - 9}$ (B) $\dfrac{x^3 - 1}{x^2 - 1}$

INSIGHT

Using Theorem 1 to divide the numerator and denominator of a fraction by a common factor is often referred to as **canceling.** This operation can be denoted by drawing a slanted line through each common factor and writing any remaining factors above or below the common factor. Canceling is often incorrectly applied to individual terms in the numerator or denominator, instead of to common factors. For example,

$$\frac{14 - 5}{2} = \frac{9}{2}$$ *Theorem 1 does not apply. There are no common factors in the numerator.*

$$\overset{7}{\frac{14 - 5}{2}} \neq \frac{\cancel{14} - 5}{\underset{1}{\cancel{2}}} = 2$$ *Incorrect use of Theorem 1. To cancel 2 in the denominator, 2 must be a factor of each term in the numerator.*

■ Multiplication and Division

Since we are restricting variable replacements to real numbers, multiplication and division of rational expressions follow the rules for multiplying and dividing real number fractions summarized in Appendix A-1.

| THEOREM 2 | **MULTIPLICATION AND DIVISION** |

If $a, b, c,$ and d are real numbers, then

1. $\dfrac{a}{b} \cdot \dfrac{c}{d} = \dfrac{ac}{bd},\quad b, d \neq 0$ $\qquad \dfrac{3}{5} \cdot \dfrac{x}{x + 5} = \dfrac{3x}{5(x + 5)}$

2. $\dfrac{a}{b} \div \dfrac{c}{d} = \dfrac{a}{b} \cdot \dfrac{d}{c},\quad b, c, d \neq 0$ $\qquad \dfrac{3}{5} \div \dfrac{x}{x + 5} = \dfrac{3}{5} \cdot \dfrac{x + 5}{x}$

Explore & Discuss **1**   Write a verbal description of the process of multiplying two rational expressions. Do the same for the quotient of two rational expressions.

| EXAMPLE 2 | **Multiplication and Division**   Perform the indicated operations and reduce to lowest terms. |

(A) $\dfrac{10x^3y}{3xy + 9y} \cdot \dfrac{x^2 - 9}{4x^2 - 12x}$ Factor numerators and denominators. Then divide any numerator and any denominator with a like common factor.

$$= \dfrac{\overset{5x^2}{\cancel{10x^3y}}}{\underset{3 \cdot 1}{3\cancel{y}\cancel{(x + 3)}}} \cdot \dfrac{\overset{1 \cdot 1}{\cancel{(x - 3)}\cancel{(x + 3)}}}{\underset{2 \cdot 1}{4\cancel{x}\cancel{(x - 3)}}}$$

$$= \dfrac{5x^2}{6}$$

(B) $\dfrac{4 - 2x}{4} \div (x - 2) = \dfrac{\overset{1}{\cancel{2}}(2 - x)}{\underset{2}{\cancel{4}}} \cdot \dfrac{1}{x - 2}$ $x - 2 = \dfrac{x - 2}{1}$

$$= \dfrac{2 - x}{2(x - 2)} = \dfrac{\overset{-1}{\cancel{-(x - 2)}}}{\underset{1}{2\cancel{(x - 2)}}}$$ $b - a = -(a - b)$, a useful change in some problems

$$= -\dfrac{1}{2}$$

MATCHED PROBLEM 2 | Perform the indicated operations and reduce to lowest terms.

(A) $\dfrac{12x^2y^3}{2xy^2 + 6xy} \cdot \dfrac{y^2 + 6y + 9}{3y^3 + 9y^2}$ (B) $(4 - x) \div \dfrac{x^2 - 16}{5}$

■ Addition and Subtraction

Again, because we are restricting variable replacements to real numbers, addition and subtraction of rational expressions follow the rules for adding and subtracting real number fractions.

THEOREM 3   ADDITION AND SUBTRACTION

For a, b, and c real numbers,

1. $\dfrac{a}{b} + \dfrac{c}{b} = \dfrac{a + c}{b}$, $b \neq 0$ $\dfrac{x}{x + 5} + \dfrac{8}{x + 5} = \dfrac{x + 8}{x + 5}$

2. $\dfrac{a}{b} - \dfrac{c}{b} = \dfrac{a - c}{b}$, $b \neq 0$ $\dfrac{x}{3x^2y^2} - \dfrac{x + 7}{3x^2y^2} = \dfrac{x - (x + 7)}{3x^2y^2}$

Thus, we add rational expressions with the same denominators by adding or subtracting their numerators and placing the result over the common denominator. If the denominators are not the same, we raise the fractions to higher terms, using the fundamental property of fractions to obtain common denominators, and then proceed as described.

Even though any common denominator will do, our work will be simplified if the *least common denominator (LCD)* is used. Often, the LCD is obvious, but if it is not, the steps in the next box describe how to find it.

PROCEDURE   **Least Common Denominator**
The least common denominator (LCD) of two or more rational expressions is found as follows:

1. Factor each denominator completely, including integer factors.
2. Identify each different factor from all the denominators.
3. Form a product using each different factor to the highest power that occurs in any one denominator. This product is the LCD.

EXAMPLE 3   **Addition and Subtraction**   Combine into a single fraction and reduce to lowest terms.

(A) $\dfrac{3}{10} + \dfrac{5}{6} - \dfrac{11}{45}$ (B) $\dfrac{4}{9x} - \dfrac{5x}{6y^2} + 1$ (C) $\dfrac{1}{x-1} - \dfrac{1}{x} - \dfrac{2}{x^2-1}$

SOLUTION   (A) To find the LCD, factor each denominator completely:

$$\left.\begin{array}{l} 10 = 2 \cdot 5 \\ 6 = 2 \cdot 3 \\ 45 = 3^2 \cdot 5 \end{array}\right\} \quad LCD = 2 \cdot 3^2 \cdot 5 = 90$$

Now use the fundamental property of fractions to make each denominator 90:

$$\frac{3}{10} + \frac{5}{6} - \frac{11}{45} = \frac{9 \cdot 3}{9 \cdot 10} + \frac{15 \cdot 5}{15 \cdot 6} - \frac{2 \cdot 11}{2 \cdot 45}$$

$$= \boxed{\frac{27}{90} + \frac{75}{90} - \frac{22}{90}}$$

$$= \frac{27 + 75 - 22}{90} = \frac{80}{90} = \frac{8}{9}$$

(B) $\left.\begin{array}{l} 9x = 3^2 x \\ 6y^2 = 2 \cdot 3y^2 \end{array}\right\} \quad LCD = 2 \cdot 3^2 xy^2 = 18xy^2$

$$\frac{4}{9x} - \frac{5x}{6y^2} + 1 = \frac{2y^2 \cdot 4}{2y^2 \cdot 9x} - \frac{3x \cdot 5x}{3x \cdot 6y^2} + \frac{18xy^2}{18xy^2}$$

$$= \frac{8y^2 - 15x^2 + 18xy^2}{18xy^2}$$

(C) $\dfrac{1}{x-1} - \dfrac{1}{x} - \dfrac{2}{x^2-1}$

$$= \frac{1}{x-1} - \frac{1}{x} - \frac{2}{(x-1)(x+1)} \quad LCD = x(x-1)(x+1)$$

$$= \frac{x(x+1) - (x-1)(x+1) - 2x}{x(x-1)(x+1)}$$

$$= \frac{x^2 + x - x^2 + 1 - 2x}{x(x-1)(x+1)}$$

$$= \frac{1 - x}{x(x-1)(x+1)}$$

$$= \frac{\overset{-1}{\cancel{-(x-1)}}}{x\underset{1}{\cancel{(x-1)}}(x+1)} = \frac{-1}{x(x+1)}$$

| MATCHED PROBLEM 3 | Combine into a single fraction and reduce to lowest terms. |

(A) $\dfrac{5}{28} - \dfrac{1}{10} + \dfrac{6}{35}$

(B) $\dfrac{1}{4x^2} - \dfrac{2x + 1}{3x^3} + \dfrac{3}{12x}$

(C) $\dfrac{2}{x^2 - 4x + 4} + \dfrac{1}{x} - \dfrac{1}{x - 2}$

Explore & Discuss **2**　What is the value of $\dfrac{\frac{16}{4}}{2}$?

What is the result of entering $16 \div 4 \div 2$ on a calculator?

What is the difference between $16 \div (4 \div 2)$ and $(16 \div 4) \div 2$?

How could you use fraction bars to distinguish between these two cases when writing $\dfrac{\frac{16}{4}}{2}$?

■ Compound Fractions

A fractional expression with fractions in its numerator, denominator, or both is called a **compound fraction.** It is often necessary to represent a compound fraction as a **simple fraction**—that is (in all cases we will consider), as the quotient of two polynomials. The process does not involve any new concepts. It is a matter of applying old concepts and processes in the correct sequence.

| EXAMPLE 4 | **Simplifying Compound Fractions** Express as a simple fraction reduced to lowest terms: |

(A) $\dfrac{\dfrac{1}{5 + h} - \dfrac{1}{5}}{h}$

(B) $\dfrac{\dfrac{y}{x^2} - \dfrac{x}{y^2}}{\dfrac{y}{x} - \dfrac{x}{y}}$

SOLUTION　We will simplify the expressions in parts (A) and (B) using two different methods—each is suited to the particular type of problem.

(A) We simplify this expression by combining the numerator into a single fraction and using division of rational forms.

$$\dfrac{\dfrac{1}{5 + h} - \dfrac{1}{5}}{h} = \left[\dfrac{1}{5 + h} - \dfrac{1}{5}\right] \div \dfrac{h}{1}$$

$$= \dfrac{5 - 5 - h}{5(5 + h)} \cdot \dfrac{1}{h}$$

$$= \dfrac{-h}{5(5 + h)h} = \dfrac{-1}{5(5 + h)}$$

(B) The method used here makes effective use of the fundamental property of fractions in the form

$$\dfrac{a}{b} = \dfrac{ka}{kb} \qquad b, k \neq 0$$

Multiply the numerator and denominator by the LCD of all fractions in the numerator and denominator—in this case, x^2y^2:

$$\frac{x^2y^2\left(\dfrac{y}{x^2} - \dfrac{x}{y^2}\right)}{x^2y^2\left(\dfrac{y}{x} - \dfrac{x}{y}\right)} = \frac{x^2y^2\dfrac{y}{x^2} - x^2y^2\dfrac{x}{y^2}}{x^2y^2\dfrac{y}{x} - x^2y^2\dfrac{x}{y}} = \frac{y^3 - x^3}{xy^3 - x^3y}$$

$$= \frac{\overset{1}{\cancel{(y - x)}}(y^2 + xy + x^2)}{xy\underset{1}{\cancel{(y - x)}}(y + x)}$$

$$= \frac{y^2 + xy + x^2}{xy(y + x)} \quad \text{or} \quad \frac{x^2 + xy + y^2}{xy(x + y)}$$

MATCHED PROBLEM 4　Express as a simple fraction reduced to lowest terms:

(A) $\dfrac{\dfrac{1}{2 + h} - \dfrac{1}{2}}{h}$

(B) $\dfrac{\dfrac{a}{b} - \dfrac{b}{a}}{\dfrac{a}{b} + 2 + \dfrac{b}{a}}$

Answers to Matched Problems

1.　(A) $\dfrac{x - 3}{x + 3}$　(B) $\dfrac{x^2 + x + 1}{x + 1}$

2.　(A) $2x$　(B) $\dfrac{-5}{x + 4}$

3.　(A) $\dfrac{1}{4}$　(B) $\dfrac{3x^2 - 5x - 4}{12x^3}$　(C) $\dfrac{4}{x(x - 2)^2}$

4.　(A) $\dfrac{-1}{2(2 + h)}$　(B) $\dfrac{a - b}{a + b}$

Exercise A-4

A　*In Problems 1–18, perform the indicated operations and reduce answers to lowest terms.*

1.　$\dfrac{d^5}{3a} \div \left(\dfrac{d^2}{6a^2} \cdot \dfrac{a}{4d^3}\right)$

2.　$\left(\dfrac{d^5}{3a} \div \dfrac{d^2}{6a^2}\right) \cdot \dfrac{a}{4d^3}$

3.　$\dfrac{x^2}{12} + \dfrac{x}{18} - \dfrac{1}{30}$

4.　$\dfrac{2y}{18} - \dfrac{-1}{28} - \dfrac{y}{42}$

5.　$\dfrac{4m - 3}{18m^3} + \dfrac{3}{4m} - \dfrac{2m - 1}{6m^2}$

6.　$\dfrac{3x + 8}{4x^2} - \dfrac{2x - 1}{x^3} - \dfrac{5}{8x}$

7.　$\dfrac{x^2 - 9}{x^2 - 3x} \div (x^2 - x - 12)$

8.　$\dfrac{2x^2 + 7x + 3}{4x^2 - 1} \div (x + 3)$

9.　$\dfrac{2}{x} - \dfrac{1}{x - 3}$

10.　$\dfrac{5}{m - 2} - \dfrac{3}{2m + 1}$

11.　$\dfrac{2}{(x + 1)^2} - \dfrac{5}{x^2 - x - 2}$

12.　$\dfrac{3}{x^2 - 5x + 6} - \dfrac{5}{(x - 2)^2}$

13.　$\dfrac{x + 1}{x - 1} - 1$

14.　$m - 3 - \dfrac{m - 1}{m - 2}$

15.　$\dfrac{3}{a - 1} - \dfrac{2}{1 - a}$

16. $\dfrac{5}{x-3} - \dfrac{2}{3-x}$

17. $\dfrac{2x}{x^2-16} - \dfrac{x-4}{x^2+4x}$

18. $\dfrac{m+2}{m^2-2m} - \dfrac{m}{m^2-4}$

B In Problems 19–30, perform the indicated operations and reduce answers to lowest terms. Represent any compound fractions as simple fractions reduced to lowest terms.

19. $\dfrac{x^2}{x^2+2x+1} + \dfrac{x-1}{3x+3} - \dfrac{1}{6}$

20. $\dfrac{y}{y^2-y-2} - \dfrac{1}{y^2+5y-14} - \dfrac{2}{y^2+8y+7}$

21. $\dfrac{1-\dfrac{x}{y}}{2-\dfrac{y}{x}}$

22. $\dfrac{2}{5-\dfrac{3}{4x+1}}$

23. $\dfrac{c+2}{5c-5} - \dfrac{c-2}{3c-3} + \dfrac{c}{1-c}$

24. $\dfrac{x+7}{ax-bx} + \dfrac{y+9}{by-ay}$

25. $\dfrac{1+\dfrac{3}{x}}{x-\dfrac{9}{x}}$

26. $\dfrac{1-\dfrac{y^2}{x^2}}{1-\dfrac{y}{x}}$

27. $\dfrac{\dfrac{1}{2(x+h)} - \dfrac{1}{2x}}{h}$

28. $\dfrac{\dfrac{1}{x+h} - \dfrac{1}{x}}{h}$

29. $\dfrac{\dfrac{x}{y} - 2 + \dfrac{y}{x}}{\dfrac{x}{y} - \dfrac{y}{x}}$

30. $\dfrac{1 + \dfrac{2}{x} - \dfrac{15}{x^2}}{1 + \dfrac{4}{x} - \dfrac{5}{x^2}}$

In Problems 31–38, imagine that the indicated "solutions" were given to you by a student whom you were tutoring in this class.

(A) Is the solution correct? If the solution is incorrect, explain what is wrong and how it can be corrected.

(B) Show a correct solution for each incorrect solution.

31. $\dfrac{x^2+4x+3}{x+3} = \dfrac{x^2+4x}{x} = x+4$

32. $\dfrac{x^2-3x-4}{x-4} = \dfrac{x^2-3x}{x} = x-3$

33. $\dfrac{(x+h)^2 - x^2}{h} = (x+1)^2 - x^2 = 2x+1$

34. $\dfrac{(x+h)^3 - x^3}{h} = (x+1)^3 - x^3 = 3x^2 + 3x + 1$

35. $\dfrac{x^2-3x}{x^2-2x-3} + x - 3 = \dfrac{x^2-3x+x-3}{x^2-2x-3} = 1$

36. $\dfrac{2}{x-1} - \dfrac{x+3}{x^2-1} = \dfrac{2x+2-x-3}{x^2-1} = \dfrac{1}{x+1}$

37. $\dfrac{2x^2}{x^2-4} - \dfrac{x}{x-2} = \dfrac{2x^2-x^2-2x}{x^2-4} = \dfrac{x}{x+2}$

38. $x + \dfrac{x-2}{x^2-3x+2} = \dfrac{x+x-2}{x^2-3x+2} = \dfrac{2}{x-2}$

C Represent the compound fractions in Problems 39–42 as simple fractions reduced to lowest terms.

39. $\dfrac{\dfrac{1}{3(x+h)^2} - \dfrac{1}{3x^2}}{h}$

40. $\dfrac{\dfrac{1}{(x+h)^2} - \dfrac{1}{x^2}}{h}$

41. $x - \dfrac{2}{1-\dfrac{1}{x}}$

42. $2 - \dfrac{1}{1-\dfrac{2}{a+2}}$

Section A-5 INTEGER EXPONENTS AND SCIENTIFIC NOTATION

■ Integer Exponents
■ Scientific Notation

We now review basic operations on integer exponents and scientific notation and its use.

■ Integer Exponents

DEFINITION	**Integer Exponents**

For n an integer and a a real number:

1. For n a positive integer,

$$a^n = a \cdot a \cdot \cdots \cdot a \qquad n \text{ factors of } a \qquad 5^4 = 5 \cdot 5 \cdot 5 \cdot 5$$

2. For $n = 0$,

$$a^0 = 1 \qquad a \neq 0 \qquad 12^0 = 1$$

0^0 is not defined.

3. For n a negative integer,

$$a^n = \frac{1}{a^{-n}} \qquad a \neq 0 \qquad a^{-3} = \frac{1}{a^{-(-3)}} = \frac{1}{a^3}$$

[If n is negative, then $(-n)$ is positive.]

Note: It can be shown that for *all* integers n,

$$a^{-n} = \frac{1}{a^n} \qquad \text{and} \qquad a^n = \frac{1}{a^{-n}} \qquad a \neq 0 \qquad a^5 = \frac{1}{a^{-5}}, \quad a^{-5} = \frac{1}{a^5}$$

The following integer exponent properties are very useful in manipulating integer exponent forms.

THEOREM 1	**EXPONENT PROPERTIES**

For n and m integers and a and b real numbers,

1. $a^m a^n = a^{m+n}$ $\qquad a^8 a^{-3} = a^{8+(-3)} = a^5$

2. $(a^n)^m = a^{mn}$ $\qquad (a^{-2})^3 = a^{3(-2)} = a^{-6}$

3. $(ab)^m = a^m b^m$ $\qquad (ab)^{-2} = a^{-2}b^{-2}$

4. $\left(\dfrac{a}{b}\right)^m = \dfrac{a^m}{b^m}$ $\quad b \neq 0 \qquad \left(\dfrac{a}{b}\right)^5 = \dfrac{a^5}{b^5}$

5. $\dfrac{a^m}{a^n} = a^{m-n} = \dfrac{1}{a^{n-m}}$ $\quad a \neq 0 \qquad \dfrac{a^{-3}}{a^7} = \dfrac{1}{a^{7-(-3)}} = \dfrac{1}{a^{10}}$

Explore & Discuss **1**   Property 1 in Theorem 1 can be expressed verbally as follows:

To find the product of two exponential forms with the same base, add the exponents and use the same base.

Express the other properties in Theorem 1 verbally. Decide which you find easier to remember—a formula or a verbal description.

Exponent forms are frequently encountered in algebraic applications. You should sharpen your skills in using these forms by reviewing the preceding basic definitions and properties and the examples that follow.

EXAMPLE 1	**Simplifying Exponent Forms** Simplify, and express the answers using positive exponents only.

(A) $(2x^3)(3x^5) \;\boxed{= 2 \cdot 3x^{3+5}} = 6x^8$

(B) $x^5 x^{-9} = x^{-4} = \dfrac{1}{x^4}$

(C) $\dfrac{x^5}{x^7} \boxed{= x^{5-7}} = x^{-2} = \dfrac{1}{x^2}$ or $\dfrac{x^5}{x^7} = \boxed{\dfrac{1}{x^{7-5}}} = \dfrac{1}{x^2}$

(D) $\dfrac{x^{-3}}{y^{-4}} = \dfrac{y^4}{x^3}$

(E) $(u^{-3}v^2)^{-2} \boxed{= (u^{-3})^{-2}(v^2)^{-2}} = u^6 v^{-4} = \dfrac{u^6}{v^4}$

(F) $\left(\dfrac{y^{-5}}{y^{-2}}\right)^{-2} \boxed{= \dfrac{(y^{-5})^{-2}}{(y^{-2})^{-2}}} = \dfrac{y^{10}}{y^4} = y^6$

(G) $\dfrac{4m^{-3}n^{-5}}{6m^{-4}n^3} \boxed{= \dfrac{2m^{-3-(-4)}}{3n^{3-(-5)}}} = \dfrac{2m}{3n^8}$

MATCHED PROBLEM 1 | Simplify, and express the answers using positive exponents only.

(A) $(3y^4)(2y^3)$ (B) $m^2 m^{-6}$ (C) $(u^3 v^{-2})^{-2}$

(D) $\left(\dfrac{y^{-6}}{y^{-2}}\right)^{-1}$ (E) $\dfrac{8x^{-2}y^{-4}}{6x^{-5}y^2}$

INSIGHT

Remember from Section A-2 that taking powers always precedes multiplication and division unless parentheses are used to indicate otherwise. Thus,

$$-5x^3 = -5 \cdot x \cdot x \cdot x$$

while

$$(-5x)^3 = (-5)^3 x^3 = -125 \cdot x \cdot x \cdot x$$

EXAMPLE 2 | **Converting to a Simple Fraction** Write as a simple fraction with positive exponents:

$$\dfrac{1 - x}{x^{-1} - 1}$$

SOLUTION First note that

$$\dfrac{1 - x}{x^{-1} - 1} \neq \dfrac{x(1 - x)}{-1} \qquad \textit{A common error}$$

The original expression is a complex fraction, and we proceed to simplify it as follows:

$$\dfrac{1 - x}{x^{-1} - 1} = \dfrac{1 - x}{\dfrac{1}{x} - 1} \qquad \textit{Multiply numerator and denominator by x to clear internal fractions.}$$

$$= \dfrac{x(1 - x)}{x\left(\dfrac{1}{x} - 1\right)}$$

$$= \dfrac{x(1 - x)}{1 - x} = x$$

MATCHED PROBLEM 2 | Write as a simple fraction with positive exponents: $\dfrac{1 + x^{-1}}{1 - x^{-2}}$

■ Scientific Notation

In the real world, one often encounters very large numbers. For example,

The public debt in the United States in 1998, to the nearest billion dollars, was

$$\$5,526,000,000,000$$

The world population in the year 2025, to the nearest million, is projected to be

$$7,896,000,000$$

Very small numbers are also encountered:

The sound intensity of a normal conversation is

$$0.000\ 000\ 000\ 316 \text{ watt per square centimeter*}$$

It is generally troublesome to write and work with numbers of this type in standard decimal form. The first and last example cannot even be entered into many calculators as they are written. But with exponents defined for all integers, we can now express any finite decimal form as the product of a number between 1 and 10 and an integer power of 10, that is, in the form

$$a \times 10^n \qquad 1 \le a < 10, \quad a \text{ in decimal form}, \quad n \text{ an integer}$$

A number expressed in this form is said to be in **scientific notation.** The following are some examples of numbers in standard decimal notation and in scientific notation:

DECIMAL AND SCIENTIFIC NOTATION

$7 = 7 \times 10^0$	$0.5 = 5 \times 10^{-1}$
$67 = 6.7 \times 10$	$0.45 = 4.5 \times 10^{-1}$
$580 = 5.8 \times 10^2$	$0.0032 = 3.2 \times 10^{-3}$
$43,000 = 4.3 \times 10^4$	$0.000\ 045 = 4.5 \times 10^{-5}$
$73,400,000 = 7.34 \times 10^7$	$0.000\ 000\ 391 = 3.91 \times 10^{-7}$

Note that the power of 10 used corresponds to the number of places we move the decimal to form a number between 1 and 10. The power is positive if the decimal is moved to the left and negative if it is moved to the right. Positive exponents are associated with numbers greater than or equal to 10; negative exponents are associated with positive numbers less than 1; and a zero exponent is associated with a number that is 1 or greater, but less than 10.

EXAMPLE 3

Scientific Notation

(A) Write each number in scientific notation:

$$7,320,000 \qquad \text{and} \qquad 0.000\ 000\ 54$$

(B) Write each number in standard decimal form:

$$4.32 \times 10^6 \qquad \text{and} \qquad 4.32 \times 10^{-5}$$

SOLUTION   (A) $7,320,000 = 7.320\ 000. \times 10^6 = 7.32 \times 10^6$

6 places left
Positive exponent

$0.000\ 000\ 54 = 0.000\ 000\ 5.4 \times 10^{-7} = 5.4 \times 10^{-7}$

7 places right
Negative exponent

---

* We write 0.000 000 000 316 in place of 0.000000000316, because it is then easier to keep track of the number of decimal places. We follow this convention when there are more than five decimal places to the right of the decimal.

(B) $4.32 \times 10^6 = 4,320,000$ $4.32 \times 10^{-5} = \dfrac{4.32}{10^5} = 0.000\ 043\ 2$

6 places right   5 places left

Positive exponent 6   Negative exponent −5

MATCHED PROBLEM 3

(A) Write each number in scientific notation: 47,100; 2,443,000,000; 1.45
(B) Write each number in standard decimal form: 3.07×10^8; 5.98×10^{-6}

Explore & Discuss **2**

Scientific and graphing calculators can calculate in either standard decimal mode or scientific notation mode. If the result of a computation in decimal mode is either too large or too small to be displayed, most calculators will automatically display the answer in scientific notation. Read the manual for your calculator and experiment with some operations on very large and very small numbers in both decimal mode and scientific notation mode. For example, show that

$$\frac{216,700,000,000}{0.000\ 000\ 000\ 000\ 078\ 8} = 2.75 \times 10^{24}$$

Answers to Matched Problems   **1.** (A) $6y^7$ (B) $\dfrac{1}{m^4}$ (C) $\dfrac{v^4}{u^6}$ (D) y^4 (E) $\dfrac{4x^3}{3y^6}$ **2.** $\dfrac{x}{x-1}$

3. (A) 4.7×10^4; 2.443×10^9; 1.45×10^0 (B) $307,000,000$; $0.000\ 005\ 98$

Exercise A-5

A *In Problems 1–14, simplify and express answers using positive exponents only. Variables are restricted to avoid division by 0.*

1. $2x^{-9}$ **2.** $3y^{-5}$ **3.** $\dfrac{3}{2w^{-7}}$

4. $\dfrac{5}{4x^{-9}}$ **5.** $2x^{-8}x^5$ **6.** $3c^{-9}c^4$

7. $\dfrac{w^{-8}}{w^{-3}}$ **8.** $\dfrac{m^{-11}}{m^{-5}}$ **9.** $(2a^{-3})^2$

10. $7d^{-4}d^4$ **11.** $(a^{-3})^2$ **12.** $(5b^{-2})^2$

13. $(2x^4)^{-3}$ **14.** $(a^{-3}b^4)^{-3}$

Write each number in Problems 15–20 in scientific notation.

15. 82,300,000,000   **16.** 5,380,000

17. 0.783   **18.** 0.019

19. 0.000 034   **20.** 0.000 000 007 832

Write each number in Problems 21–28 in standard decimal notation.

21. 4×10^4 **22.** 9×10^6

23. 7×10^{-3} **24.** 2×10^{-5}

25. 6.171×10^7 **26.** 3.044×10^3

27. 8.08×10^{-4} **28.** 1.13×10^{-2}

B *In Problems 29–38, simplify and express answers using positive exponents only.*

29. $(22 + 31)^0$ **30.** $(2x^3y^4)^0$

31. $\dfrac{10^{-3} \cdot 10^4}{10^{-11} \cdot 10^{-2}}$ **32.** $\dfrac{10^{-17} \cdot 10^{-5}}{10^{-3} \cdot 10^{-14}}$

33. $(5x^2y^{-3})^{-2}$ **34.** $(2m^{-3}n^2)^{-3}$

35. $\left(\dfrac{-5}{2x^3}\right)^{-2}$ **36.** $\left(\dfrac{2a}{3b^2}\right)^{-3}$

37. $\dfrac{8x^{-3}y^{-1}}{6x^2y^{-4}}$ **38.** $\dfrac{9m^{-4}n^3}{12m^{-1}n^{-1}}$

In Problems 39–42, write each expression in the form $ax^p + bx^q$ or $ax^p + bx^q + cx^r$, where a, b, and c are real numbers and p, q, and r are integers. For example,

$$\frac{2x^4 - 3x^2 + 1}{2x^3} \boxed{= \frac{2x^4}{2x^3} - \frac{3x^2}{2x^3} + \frac{1}{2x^3}} = x - \frac{3}{2}x^{-1} + \frac{1}{2}x^{-3}$$

39. $\dfrac{7x^5 - x^2}{4x^5}$ **40.** $\dfrac{5x^3 - 2}{3x^2}$

41. $\dfrac{5x^4 - 3x^2 + 8}{2x^2}$ **42.** $\dfrac{2x^3 - 3x^2 + x}{2x^2}$

Write each expression in Problems 43–46 with positive exponents only, and as a single fraction reduced to lowest terms.

43. $\dfrac{3x^2(x-1)^2 - 2x^3(x-1)}{(x-1)^4}$

44. $\dfrac{5x^4(x+3)^2 - 2x^5(x+3)}{(x+3)^4}$

45. $2x^{-2}(x-1) - 2x^{-3}(x-1)^2$

46. $2x(x+3)^{-1} - x^2(x+3)^{-2}$

In Problems 47–50, convert each number to scientific notation and simplify. Express the answer in both scientific notation and in standard decimal form.

47. $\dfrac{9{,}600{,}000{,}000}{(1{,}600{,}000)(0.000\,000\,25)}$

48. $\dfrac{(60{,}000)(0.000\,003)}{(0.0004)(1{,}500{,}000)}$

49. $\dfrac{(1{,}250{,}000)(0.000\,38)}{0.0152}$

50. $\dfrac{(0.000\,000\,82)(230{,}000)}{(625{,}000)(0.0082)}$

51. What is the result of entering 2^{3^2} on a calculator?

52. Refer to Problem 51. What is the difference between $2^{(3^2)}$ and $(2^3)^2$? Which agrees with the value of 2^{3^2} obtained with a calculator?

53. If $n = 0$, then property 1 in Theorem 1 implies that $a^m a^0 = a^{m+0} = a^m$. Explain how this helps motivate the definition of a^0.

54. If $m = -n$, then property 1 in Theorem 1 implies that $a^{-n}a^n = a^0 = 1$. Explain how this helps motivate the definition of a^{-n}.

C *Write the fractions in Problems 55–58 as simple fractions reduced to lowest terms.*

55. $\dfrac{u+v}{u^{-1} + v^{-1}}$

56. $\dfrac{x^{-2} - y^{-2}}{x^{-1} + y^{-1}}$

57. $\dfrac{b^{-2} - c^{-2}}{b^{-3} - c^{-3}}$

58. $\dfrac{xy^{-2} - yx^{-2}}{y^{-1} - x^{-1}}$

Applications

Problems 59 and 60 refer to Table 1.

TABLE 1	Assets of the Five Largest U.S. Commercial Banks, 2006
Bank	**Assets ($)**
Bank of America	1,105,000,000,000
J. P. Morgan	1,093,400,000,000
Citibank	749,300,000,000
Wachovia	496,600,000,000
Wells Fargo	415,800,000,000

59. *Financial assets*

(A) Write Bank of America's assets in scientific notation.

(B) After converting to scientific notation, determine the ratio of the assets of Bank of America to the assets of Citibank. Write the answer in standard decimal form to four decimal places.

(C) Repeat part (B) with the banks reversed.

60. *Financial assets*

(A) Write J. P. Morgan's assets in scientific notation.

(B) After converting to scientific notation, determine the ratio of the assets of J. P. Morgan to the assets of Wells Fargo. Write the answer in standard decimal form to four decimal places.

(C) Repeat part (B) with the banks reversed.

Problems 61 and 62 refer to Table 2.

TABLE 2	U.S. Public Debt, Interest on Debt, and Population		
Year	**Public Debt ($)**	**Interest on Debt ($)**	**Population**
1990	3,233,300,000,000	264,800,000,000	248,765,170
2006	8,528,000,000,000	360,804,000,000	299,449,400

61. *Public debt.* Carry out the following computations using scientific notation, and write final answers in standard decimal form.

(A) What was the per capita debt in 2006 (to the nearest dollar)?

(B) What was the per capita interest paid on the debt in 2006 (to the nearest dollar)?

(C) What was the percentage interest paid on the debt in 2006 (to two decimal places)?

62. *Public debt.* Carry out the following computations using scientific notation, and write final answers in standard decimal form.

(A) What was the per capita debt in 1990 (to the nearest dollar)?

(B) What was the per capita interest paid on the debt in 1990 (to the nearest dollar)?

(C) What was the percentage interest paid on the debt in 1990 (to two decimal places)?

Air pollution. Air quality standards establish maximum amounts of pollutants considered acceptable in the air. The amounts are frequently given in parts per million (ppm). A standard of 30 ppm also can be expressed as follows:

$$30 \text{ ppm} = \frac{30}{1,000,000} = \frac{3 \times 10}{10^6}$$
$$= 3 \times 10^{-5} = 0.000\,03 = 0.003\%$$

In Problems 63 and 64, express the given standard:

(A) In scientific notation

(B) In standard decimal notation

(C) As a percent

63. 9 ppm, the standard for carbon monoxide, when averaged over a period of 8 hours

64. 0.03 ppm, the standard for sulfur oxides, when averaged over a year

65. *Crime.* In 2006, the United States had a violent crime rate of 493.8 per 100,000 people and a population of 299.4 million people. How many violent crimes occurred that year? Compute the answer using scientific notation and convert the answer to standard decimal form (to the nearest thousand).

66. *Population density.* The United States had a 2006 population of 299.4 million people and a land area of 3,539,000 square miles. What was the population density? Compute the answer using scientific notation and convert the answer to standard decimal form (to one decimal place).

Section A-6 RATIONAL EXPONENTS AND RADICALS

- *n*th Roots of Real Numbers
- Rational Exponents and Radicals
- Properties of Radicals

Square roots may now be generalized to *nth roots,* and the meaning of exponent may be generalized to include all rational numbers.

■ *n*th Roots of Real Numbers

Consider a square of side *r* with area 36 square inches. We can write

$$r^2 = 36$$

and conclude that side *r* is a number whose square is 36. We say that *r* is a **square root** of *b* if $r^2 = b$. Similarly, we say that *r* is a **cube root** of *b* if $r^3 = b$. And, in general,

---

DEFINITION *n*th Root

For any natural number *n*,

$$r \text{ is an } \textbf{\textit{n}th root} \text{ of } b \text{ if } r^n = b$$

Thus, 4 is a square root of 16, since $4^2 = 16$, and -2 is a cube root of -8, since $(-2)^3 = -8$. Since $(-4)^2 = 16$, we see that -4 is also a square root of 16. It can be shown that any positive number has two real square roots, two real 4th roots, and, in general, two real *n*th roots if *n* is even. Negative numbers have no real square roots, no real 4th roots, and, in general, no real *n*th roots if *n* is even. The reason is that no real number raised to an even power can be negative. For odd roots the situation is simpler. Every real number has exactly one real cube root, one real 5th root, and, in general, one real *n*th root if *n* is odd.

Additional roots can be considered in the *complex number system.* But in this book we restrict our interest to *real roots of real numbers,* and *root* will always be interpreted to mean "real root."

■ Rational Exponents and Radicals

We now turn to the question of what symbols to use to represent *n*th roots. For *n* a natural number greater than 1, we use

$$b^{1/n} \quad \text{or} \quad \sqrt[n]{b}$$

to represent a **real *n*th root of *b*.** The exponent form is motivated by the fact that $(b^{1/n})^n = b$ if exponent laws are to continue to hold for rational exponents. The other form is called an ***n*th root radical.** In the expression below, the symbol $\sqrt{}$ is called a **radical,** *n* is the **index** of the radical, and *b* is the **radicand:**

Index $\longrightarrow$ $\ulcorner$ Radical
$$\sqrt[n]{b}$$
$\llcorner$ Radicand

When the index is 2, it is usually omitted. That is, when dealing with square roots, we simply use $\sqrt{b}$ rather than $\sqrt[2]{b}$. If there are two real nth roots, both $b^{1/n}$ and $\sqrt[n]{b}$ denote the positive root, called the **principal nth root.**

EXAMPLE 1

Finding nth Roots   Evaluate each of the following:

(A) $4^{1/2}$ and $\sqrt{4}$ (B) $-4^{1/2}$ and $-\sqrt{4}$ (C) $(-4)^{1/2}$ and $\sqrt{-4}$

(D) $8^{1/3}$ and $\sqrt[3]{28}$ (E) $(-8)^{1/3}$ and $\sqrt[3]{-8}$ (F) $-8^{1/3}$ and $-\sqrt[3]{8}$

SOLUTION   (A) $4^{1/2} = \sqrt{4} = 2$ $(\sqrt{4} \neq \pm 2)$ (B) $-4^{1/2} = -\sqrt{4} = -2$

(C) $(-4)^{1/2}$ and $\sqrt{-4}$ are not real numbers

(D) $8^{1/3} = \sqrt[3]{8} = 2$ (E) $(-8)^{1/3} = \sqrt[3]{-8} = -2$

(F) $-8^{1/3} = -\sqrt[3]{8} = -2$

MATCHED PROBLEM 1

Evaluate each of the following:

(A) $16^{1/2}$ (B) $-\sqrt{16}$ (C) $\sqrt[3]{-27}$ (D) $(-9)^{1/2}$ (E) $(\sqrt[4]{81})^3$

 COMMON ERROR: The symbol $\sqrt{4}$ represents the single number 2, not ± 2. Do not confuse $\sqrt{4}$ with the solutions of the equation $x^2 = 4$, which are usually written in the form $x = \pm\sqrt{4} = \pm 2$.

We now define b^r for any rational number $r = m/n$.

DEFINITION   Rational Exponents

If m and n are natural numbers without common prime factors, b is a real number, and b is nonnegative when n is even, then

$$b^{m/n} = \begin{cases} (b^{1/n})^m = (\sqrt[n]{b})^m \\ (b^m)^{1/n} = \sqrt[n]{b^m} \end{cases} \qquad \begin{aligned} & 8^{2/3} = (8^{1/3})^2 = (\sqrt[3]{8})^2 = 2^2 = 4 \\ & 8^{2/3} = (8^2)^{1/3} = \sqrt[3]{8^2} = \sqrt[3]{64} = 4 \end{aligned}$$

and

$$b^{-m/n} = \frac{1}{b^{m/n}} \quad b \neq 0 \qquad 8^{-2/3} = \frac{1}{8^{2/3}} = \frac{1}{4}$$

Note that the two definitions of $b^{m/n}$ are equivalent under the indicated restrictions on m, n, and b.

INSIGHT

All the properties for integer exponents listed in Theorem 1 in Section A-5 also hold for rational exponents, provided that b is nonnegative when n is even. This restriction on b is necessary to avoid nonreal results. For example,

$$(-4)^{3/2} = \sqrt{(-4)^3} = \sqrt{-64} \quad \text{Not a real number}$$

To avoid nonreal results, all variables in the remainder of this discussion represent positive real numbers.

EXAMPLE 2	**From Rational Exponent Form to Radical Form and Vice Versa** Change rational exponent form to radical form.

(A) $x^{1/7} = \sqrt[7]{-8}$

(B) $(3u^2v^3)^{3/5} = \sqrt[5]{(3u^2v^3)^3}$ or $(\sqrt[5]{3u^2v^3})^3$ The first is usually preferred.

(C) $y^{-2/3} = \dfrac{1}{y^{2/3}} = \dfrac{1}{\sqrt[3]{y^2}}$ or $\sqrt[3]{y^{-2}}$ or $\sqrt[3]{\dfrac{1}{y^2}}$

Change radical form to rational exponent form.

(D) $\sqrt[5]{6} = 6^{1/5}$

(E) $-\sqrt[3]{x^2} = -x^{2/3}$

(F) $\sqrt{x^2 + y^2} = (x^2 + y^2)^{1/2}$ Note that $(x^2 + y^2)^{1/2} \neq x + y$. Why?   ▬

MATCHED PROBLEM 2	Convert to radical form.

(A) $u^{1/5}$ (B) $(6x^2y^5)^{2/9}$ (C) $(3xy)^{-3/5}$

Convert to rational exponent form.

(D) $\sqrt[4]{9u}$ (E) $-\sqrt{(2x)^4}$ (F) $\sqrt[3]{x^3 + y^3}$

EXAMPLE 3	**Working with Rational Exponents** Simplify each and express answers using positive exponents only. If rational exponents appear in final answers, convert to radical form.

(A) $(3x^{1/3})(2x^{1/2}) = 6x^{1/3+1/2} = 6x^{5/6} = 6\sqrt[6]{x^5}$

(B) $(-8)^{5/3} = [(-8)^{1/3}]^5 = (-2)^5 = -32$

(C) $(2x^{1/3}y^{-2/3})^3 = 8xy^{-2} = \dfrac{8x}{y^2}$

(D) $\left(\dfrac{4x^{1/3}}{x^{1/2}}\right)^{1/2} = \dfrac{4^{1/2}x^{1/6}}{x^{1/4}} = \dfrac{2}{x^{1/4-1/6}} = \dfrac{2}{x^{1/12}} = \dfrac{2}{\sqrt[12]{x}}$ ▬

MATCHED PROBLEM 3	Simplify each and express answers using positive exponents only. If rational exponents appear in final answers, convert to radical form.

(A) $9^{3/2}$ (B) $(-27)^{4/3}$ (C) $(5y^{1/4})(2y^{1/3})$ (D) $(2x^{-3/4}y^{1/4})^4$

(E) $\left(\dfrac{8x^{1/2}}{x^{2/3}}\right)^{1/3}$

Explore & Discuss **1**   In each of the following, evaluate both radical forms:

$$16^{3/2} = \sqrt{16^3} = (\sqrt{16})^3$$

$$27^{2/3} = \sqrt[3]{27^2} = (\sqrt[3]{27})^2$$

Which radical conversion form is easier to use if you are performing the calculations by hand?

EXAMPLE 4	**Working with Rational Exponents** Multiply, and express answers using positive exponents only.

(A) $3y^{2/3}(2y^{1/3} - y^2)$ (B) $(2u^{1/2} + v^{1/2})(u^{1/2} - 3v^{1/2})$

SOLUTION   (A) $3y^{2/3}(2y^{1/3} - y^2)$ $\boxed{= 6y^{2/3+1/3} - 3y^{2/3+2}}$

$= 6y - 3y^{8/3}$

(B) $(2u^{1/2} + v^{1/2})(u^{1/2} - 3v^{1/2}) = 2u - 5u^{1/2}v^{1/2} - 3v$

MATCHED PROBLEM 4	Multiply, and express answers using positive exponents only.

(A) $2c^{1/4}(5c^3 - c^{3/4})$ (B) $(7x^{1/2} - y^{1/2})(2x^{1/2} + 3y^{1/2})$

EXAMPLE 5	**Working with Rational Exponents** Write the following expression in the form $ax^p + bx^q$, where a and b are real numbers and p and q are rational numbers:

$$\frac{2\sqrt{x} - 3\sqrt[3]{x^2}}{2\ \sqrt[3]{x}}$$

SOLUTION $\dfrac{2\sqrt{x} - 3\sqrt[3]{x^2}}{2\sqrt[3]{x}} = \dfrac{2x^{1/2} - 3x^{2/3}}{2x^{1/3}}$ *Change to rational exponent form.*

$= \dfrac{2x^{1/2}}{2x^{1/3}} - \dfrac{3x^{2/3}}{2x^{1/3}}$ *Separate into two fractions.*

$= x^{1/6} - 1.5x^{1/3}$

MATCHED PROBLEM 5	Write the following expression in the form $ax^p + bx^q$, where a and b are real numbers and p and q are rational numbers:

$$\frac{5\sqrt[3]{x} - 4\sqrt{x}}{2\sqrt{x^3}}$$

■ Properties of Radicals

Changing or simplifying radical expressions is aided by several properties of radicals that follow directly from the properties of exponents considered earlier.

THEOREM 1   **PROPERTIES OF RADICALS**

If c, n, and m are natural numbers greater than or equal to 2, and if x and y are positive real numbers, then

1. $\sqrt[n]{x^n} = x$ $\sqrt[3]{x^3} = x$

2. $\sqrt[n]{xy} = \sqrt[n]{x}\sqrt[n]{y}$ $\sqrt[5]{xy} = \sqrt[5]{x}\sqrt[5]{y}.$

3. $\sqrt[n]{\dfrac{x}{y}} = \dfrac{\sqrt[n]{x}}{\sqrt[n]{y}}$ $\sqrt[4]{\dfrac{x}{y}} = \dfrac{\sqrt[4]{x}}{\sqrt[4]{y}}$

EXAMPLE 6	**Applying Properties of Radicals** Simplify using properties of radicals:

(A) $\sqrt[4]{(3x^4y^3)^4}$ (B) $\sqrt[4]{8}\sqrt[4]{2}$ (C) $\sqrt[3]{\dfrac{xy}{27}}$

SOLUTION   (A) $\sqrt[4]{(3x^4y^3)^4} = 3x^4y^3$ *Property 1*

        (B) $\sqrt[4]{8}\sqrt[4]{2} = \sqrt[4]{16} = \sqrt[4]{2^4} = 2$ *Properties 2 and 1*

        (C) $\sqrt[3]{\dfrac{xy}{27}} = \dfrac{\sqrt[3]{xy}}{\sqrt[3]{27}} = \dfrac{\sqrt[3]{xy}}{3}$ or $\dfrac{1}{3}\sqrt[3]{xy}$ *Properties 3 and 1*

MATCHED PROBLEM 6   Simplify using properties of radicals:

    (A) $\sqrt[7]{(x^3 + y^3)^7}$ (B) $\sqrt[3]{8y^3}$ (C) $\dfrac{\sqrt[3]{16x^4y}}{\sqrt[3]{2xy}}$

Explore & Discuss **2**   Multiply:

$$(\sqrt{a} - \sqrt{b})(\sqrt{a} + \sqrt{b})$$

How can this product be used to simplify radical forms?

    A question arises regarding the best form in which a radical expression should be left. There are many answers, depending on what use we wish to make of the expression. In deriving certain formulas, it is sometimes useful to clear either a denominator or a numerator of radicals. The process is referred to as **rationalizing** the denominator or numerator. Examples 7 and 8 illustrate the rationalizing process.

EXAMPLE 7   **Rationalizing Denominators** Rationalize each denominator:

    (A) $\dfrac{6x}{\sqrt{2x}}$ (B) $\dfrac{6}{\sqrt{7} - \sqrt{5}}$ (C) $\dfrac{x - 4}{\sqrt{x} + 2}$

SOLUTION   (A) $\dfrac{6x}{\sqrt{2x}} = \dfrac{6x}{\sqrt{2x}} \cdot \dfrac{\sqrt{2x}}{\sqrt{2x}} = \dfrac{6x\sqrt{2x}}{2x} = 3\sqrt{2x}$

        (B) $\dfrac{6}{\sqrt{7} - \sqrt{5}} = \dfrac{6}{\sqrt{7} - \sqrt{5}} \cdot \dfrac{\sqrt{7} + \sqrt{5}}{\sqrt{7} + \sqrt{5}}$

               $= \dfrac{6(\sqrt{7} + \sqrt{5})}{2} = 3(\sqrt{7} + \sqrt{5})$

        (C) $\dfrac{x - 4}{\sqrt{x} + 2} = \dfrac{x - 4}{\sqrt{x} + 2} \cdot \dfrac{\sqrt{x} - 2}{\sqrt{x} - 2}$

               $= \dfrac{(x - 4)(\sqrt{x} - 2)}{x - 4} = \sqrt{x} - 2$

MATCHED PROBLEM 7   Rationalize each denominator:

    (A) $\dfrac{12ab^2}{\sqrt{3ab}}$ (B) $\dfrac{9}{\sqrt{6} + \sqrt{3}}$ (C) $\dfrac{x^2 - y^2}{\sqrt{x} - \sqrt{y}}$

EXAMPLE 8   **Rationalizing Numerators** Rationalize each numerator:

    (A) $\dfrac{\sqrt{2}}{2\sqrt{3}}$ (B) $\dfrac{3 + \sqrt{m}}{9 - m}$ (C) $\dfrac{\sqrt{2 + h} - \sqrt{2}}{h}$

SOLUTION   (A) $\dfrac{\sqrt{2}}{2\sqrt{3}} = \dfrac{\sqrt{2}}{2\sqrt{3}} \cdot \dfrac{\sqrt{2}}{\sqrt{2}} = \dfrac{2}{2\sqrt{6}} = \dfrac{1}{\sqrt{6}}$

(B) $\dfrac{3 + \sqrt{m}}{9 - m} = \dfrac{3 + \sqrt{m}}{9 - m} \cdot \dfrac{3 - \sqrt{m}}{3 - \sqrt{m}} = \dfrac{9 - m}{(9 - m)(3 - \sqrt{m})} = \dfrac{1}{3 - \sqrt{m}}$

(C) $\dfrac{\sqrt{2 + h} - \sqrt{2}}{h} = \dfrac{\sqrt{2 + h} - \sqrt{2}}{h} \cdot \dfrac{\sqrt{2 + h} + \sqrt{2}}{\sqrt{2 + h} + \sqrt{2}}$

$= \dfrac{h}{h(\sqrt{2 + h} + \sqrt{2})} = \dfrac{1}{\sqrt{2 + h} + \sqrt{2}}$

MATCHED PROBLEM 8   Rationalize each numerator:

(A) $\dfrac{\sqrt{3}}{3\sqrt{2}}$ (B) $\dfrac{2 - \sqrt{n}}{4 - n}$ (C) $\dfrac{\sqrt{3 + h} - \sqrt{3}}{h}$

Answers to Matched Problems   **1.** (A) 4   (B) −4   (C) −3

(D) Not a real number   (E) 27

2. (A) $\sqrt[5]{u}$ (B) $\sqrt[9]{(6x^2y^5)^2}$ or $\left(\sqrt[9]{6x^2y^5}\right)^2$ (C) $1/\sqrt[5]{(3xy)^3}$

(D) $(9u)^{1/4}$ (E) $-(2x)^{4/7}$ (F) $(x^3 + y^3)^{1/3}$ (not $x + y$)

3. (A) 27   (B) 81   (C) $10y^{7/12} = 10\sqrt[12]{y^7}$ (D) $16y/x^3$

(E) $2/x^{1/18} = 2/\sqrt[18]{x}$

4. (A) $10c^{13/4} - 2c$ (B) $14x + 19x^{1/2}y^{1/2} - 3y$ **5.** $2.5x^{-7/6} - 2x^{-1}$

6. (A) $x^3 + y^3$ (B) $2y$ (C) $2x$

7. (A) $4b\sqrt{3ab}$ (B) $3(\sqrt{6} - \sqrt{3})$ (C) $(x + y)(\sqrt{x} + \sqrt{y})$

8. (A) $\dfrac{1}{\sqrt{6}}$ (B) $\dfrac{1}{2 + \sqrt{n}}$ (C) $\dfrac{1}{\sqrt{3 + h} + \sqrt{3}}$

Exercise A-6

A   Change each expression in Problems 1–6 to radical form. Do not simplify.

1. $6x^{3/5}$ **2.** $7y^{2/5}$ **3.** $(32x^2y^3)^{3/5}$

4. $(7x^2y)^{5/7}$ **5.** $(x^2 + y^2)^{1/2}$ **6.** $x^{1/2} + y^{1/2}$

Change each expression in Problems 7–12 to rational exponent form. Do not simplify.

7. $5\sqrt[4]{x^3}$ **8.** $7m\sqrt[5]{n^2}$ **9.** $\sqrt[5]{(2x^2y)^3}$

10. $\sqrt[7]{(8x^4y)^3}$ **11.** $\sqrt[3]{x} + \sqrt[3]{y}$ **12.** $\sqrt[3]{x^2 + y^3}$

In Problems 13–24, find rational number representations for each, if they exist.

13. $25^{1/2}$ **14.** $64^{1/3}$ **15.** $16^{3/2}$

16. $16^{3/4}$ **17.** $-49^{1/2}$ **18.** $(-49)^{1/2}$

19. $-64^{2/3}$ **20.** $(-64)^{2/3}$ **21.** $\left(\frac{4}{25}\right)^{3/2}$

22. $\left(\frac{8}{27}\right)^{2/3}$ **23.** $9^{-3/2}$ **24.** $8^{-2/3}$

In Problems 25–34, simplify each expression and write answers using positive exponents only. All variables represent positive real numbers.

25. $x^{4/5}x^{-2/5}$ **26.** $y^{-3/7}y^{4/7}$ **27.** $\dfrac{m^{2/3}}{m^{-1/3}}$

28. $\dfrac{x^{1/4}}{x^{3/4}}$ **29.** $(8x^3y^{-6})^{1/3}$ **30.** $(4u^{-2}v^4)^{1/2}$

31. $\left(\dfrac{4x^{-2}}{y^4}\right)^{-1/2}$ **32.** $\left(\dfrac{w^4}{9x^{-2}}\right)^{-1/2}$ **33.** $\dfrac{(8x)^{-1/3}}{12x^{1/4}}$

34. $\dfrac{6a^{3/4}}{15a^{-1/3}}$

Simplify each expression in Problems 35–40 using properties of radicals. All variables represent positive real numbers.

35. $\sqrt[5]{(2x + 3)^5}$ **36.** $\sqrt[3]{(7 + 2y)^3}$

37. $\sqrt{6x}\sqrt{15x^3}\sqrt{30x^7}$ **38.** $\sqrt[5]{16a^4}\sqrt[5]{4a^2}\sqrt[5]{8a^3}$

39. $\dfrac{\sqrt{6x}\sqrt{10}}{\sqrt{15x}}$ **40.** $\dfrac{\sqrt{8}\sqrt{12y}}{\sqrt{6y}}$

B In Problems 41–48, multiply, and express answers using positive exponents only.

41. $3x^{3/4}(4x^{1/4} - 2x^8)$

42. $2m^{1/3}(3m^{2/3} - m^6)$

43. $(3u^{1/2} - v^{1/2})(u^{1/2} - 4v^{1/2})$

44. $(a^{1/2} + 2b^{1/2})(a^{1/2} - 3b^{1/2})$

45. $(6m^{1/2} + n^{-1/2})(6m - n^{-1/2})$

46. $(2x - 3y^{1/3})(2x^{1/3} + 1)$

47. $(3x^{1/2} - y^{1/2})^2$

48. $(x^{1/2} + 2y^{1/2})^2$

Write each expression in Problems 49–54 in the form $ax^p + bx^q$, where a and b are real numbers and p and q are rational numbers.

49. $\dfrac{\sqrt[3]{x^2} + 2}{2\sqrt[3]{x}}$

50. $\dfrac{12\sqrt{x} - 3}{4\sqrt{x}}$

51. $\dfrac{2\sqrt[4]{x^3} + \sqrt[3]{x}}{3x}$

52. $\dfrac{3\sqrt[3]{x^2} + \sqrt{x}}{5x}$

53. $\dfrac{2\sqrt[3]{x} - \sqrt{x}}{4\sqrt{x}}$

54. $\dfrac{x^2 - 4\sqrt{x}}{2\sqrt[3]{x}}$

Rationalize the denominators in Problems 55–60.

55. $\dfrac{12mn^2}{\sqrt{3mn}}$

56. $\dfrac{14x^2}{\sqrt{7x}}$

57. $\dfrac{2(x + 3)}{\sqrt{x} - 2}$

58. $\dfrac{3(x + 1)}{\sqrt{x} + 4}$

59. $\dfrac{7(x - y)^2}{\sqrt{x} - \sqrt{y}}$

60. $\dfrac{3a - 3b}{\sqrt{a} + \sqrt{b}}$

Rationalize the numerators in Problems 61–66.

61. $\dfrac{\sqrt{5xy}}{5x^2y^2}$

62. $\dfrac{\sqrt{3mn}}{3mn}$

63. $\dfrac{\sqrt{x + h} - \sqrt{x}}{h}$

64. $\dfrac{\sqrt{2(a + h)} - \sqrt{2a}}{h}$

65. $\dfrac{\sqrt{t} - \sqrt{x}}{t^2 - x^2}$

66. $\dfrac{\sqrt{x} - \sqrt{y}}{\sqrt{x} + \sqrt{y}}$

Problems 67–70 illustrate common errors involving rational exponents. In each case, find numerical examples that show that the left side is not always equal to the right side.

67. $(x + y)^{1/2} \neq x^{1/2} + y^{1/2}$

68. $(x^3 + y^3)^{1/3} \neq x + y$

69. $(x + y)^{1/3} \neq \dfrac{1}{(x + y)^3}$

70. $(x + y)^{-1/2} \neq \dfrac{1}{(x + y)^2}$

C In Problems 71–82, discuss the validity of each statement. If the statement is true, explain why. If not, give a counterexample.

71. $\sqrt{x^2} = x$ for all real numbers x

72. $\sqrt{x^2} = |x|$ for all real numbers x

73. $\sqrt[3]{x^3} = |x|$ for all real numbers x

74. $\sqrt[3]{x^3} = x$ for all real numbers x

75. If $r < 0$, then r has no cube roots.

76. If $r < 0$, then r has no square roots.

77. If $r > 0$, then r has two square roots.

78. If $r > 0$, then r has three cube roots.

79. The fourth roots of 100 are $\sqrt{10}$ and $-\sqrt{10}$.

80. The square roots of $2\sqrt{6} - 5$ are $\sqrt{3} - \sqrt{2}$ and $\sqrt{2} - \sqrt{3}$.

81. $\sqrt{355 - 60\sqrt{35}} = 5\sqrt{7} - 6\sqrt{5}$

82. $\sqrt[3]{7 - 5\sqrt{2}} = 1 - \sqrt{2}$

In Problems 83–88, simplify by writing each expression as a simple or single fraction reduced to lowest terms and without negative exponents.

83. $-\frac{1}{2}(x - 2)(x + 3)^{-3/2} + (x + 3)^{-1/2}$

84. $2(x - 2)^{-1/2} - \frac{1}{2}(2x + 3)(x - 2)^{-3/2}$

85. $\dfrac{(x - 1)^{1/2} - x\left(\frac{1}{2}\right)(x - 1)^{-1/2}}{x - 1}$

86. $\dfrac{(2x - 1)^{1/2} - (x + 2)\left(\frac{1}{2}\right)(2x - 1)^{-1/2}(2)}{2x - 1}$

87. $\dfrac{(x + 2)^{2/3} - x\left(\frac{2}{3}\right)(x + 2)^{-1/3}}{(x + 2)^{4/3}}$

88. $\dfrac{2(3x - 1)^{1/3} - (2x + 1)\left(\frac{1}{3}\right)(3x - 1)^{-2/3}(3)}{(3x - 1)^{2/3}}$

In Problems 89–94, evaluate using a calculator. (Refer to the instruction book for your calculator to see how exponential forms are evaluated.)

89. $22^{3/2}$

90. $15^{5/4}$

91. $827^{-3/8}$

92. $103^{-3/4}$

93. $37.09^{7/3}$

94. $2.876^{8/5}$

In Problems 95 and 96, evaluate each expression on a calculator and determine which pairs have the same value. Verify these results algebraically.

95. (A) $\sqrt{3} + \sqrt{5}$

(B) $\sqrt{2 + \sqrt{3}} + \sqrt{2 - \sqrt{3}}$

(C) $1 + \sqrt{3}$

(D) $\sqrt[3]{10 + 6\sqrt{3}}$

(E) $\sqrt{8 + \sqrt{60}}$

(F) $\sqrt{6}$

96. (A) $2\sqrt[3]{2} + \sqrt{5}$

(B) $\sqrt{8}$

(C) $\sqrt{3} + \sqrt{7}$

(D) $\sqrt{3 + \sqrt{8}} + \sqrt{3 - \sqrt{8}}$

(E) $\sqrt{10 + \sqrt{84}}$

(F) $1 + \sqrt{5}$

Section A-7   QUADRATIC EQUATIONS

- Solution by Square Root
- Solution by Factoring
- Quadratic Formula
- Quadratic Formula and Factoring
- Application: Supply and Demand

In this section we consider equations involving second-degree polynomials.

DEFINITION   **Quadratic Equation**

A **quadratic equation** in one variable is any equation that can be written in the form

$$ax^2 + bx + c = 0 \qquad a \neq 0 \qquad \text{Standard form}$$

where x is a variable and a, b, and c are constants.

The equations

$$5x^2 - 3x + 7 = 0 \qquad \text{and} \qquad 18 = 32t^2 - 12t$$

are both quadratic equations, since they are either in the standard form or can be transformed into this form.

We restrict our review to finding real solutions to quadratic equations.

■ Solution by Square Root

The easiest type of quadratic equation to solve is the special form where the first-degree term is missing:

$$ax^2 + c = 0 \qquad a \neq 0$$

The method of solution of this special form makes direct use of the square root property:

THEOREM 1   **SQUARE ROOT PROPERTY**

If $a^2 = b$, then $a = \pm\sqrt{b}$.

Explore & Discuss **1**   Determine whether each of the following pairs of equations are equivalent. Explain.

(A) $x^2 = 4$ and $x = 2$
(B) $x^2 = 4$ and $x = -2$
(C) $x = \sqrt{4}$ and $x = 2$
(D) $x = \sqrt{4}$ and $x = -2$
(E) $x = -\sqrt{4}$ and $x = -2$

EXAMPLE 1   **Square Root Method**   Use the square root property to solve each equation.

(A) $x^2 - 7 = 0$ (B) $2x^2 - 10 = 0$ (C) $3x^2 + 27 = 0$
(D) $(x - 8)^2 = 9$

SOLUTION   (A) $x^2 - 7 = 0$

$$x^2 = 7 \qquad \text{What real number squared is 7?}$$
$$x = \pm\sqrt{7} \qquad \text{Short for } \sqrt{7} \text{ and } -\sqrt{7}$$

(B) $2x^2 - 10 = 0$

$\qquad 2x^2 = 10$

$\qquad\quad x^2 = 5 \qquad$ *What real number squared is 5?*

$\qquad\quad\ \, x = \pm\sqrt{5}$

(C) $3x^2 + 27 = 0$

$\qquad 3x^2 = -27$

$\qquad\quad x^2 = -9 \qquad$ *What real number squared is −9?*

No real solution, since no real number squared is negative.

(D) $(x - 8)^2 = 9$

$\qquad x - 8 = \pm\sqrt{9}$

$\qquad x - 8 = \pm 3$

$\qquad\qquad x = 8 \pm 3 = 5 \quad$ or $\quad 11$

∎

MATCHED PROBLEM 1 | Use the square root property to solve each equation.

(A) $x^2 - 6 = 0$

(B) $3x^2 - 12 = 0$

(C) $x^2 + 4 = 0$

(D) $(x + 5)^2 = 1$

■ Solution by Factoring

If the left side of a quadratic equation when written in standard form can be factored, the equation can be solved very quickly. The method of solution by factoring rests on a basic property of real numbers, first mentioned in Section A-1.

I N S I G H T

Theorem 2 in Section A-1 states that if a and b are real numbers, then $ab = 0$ if and only if $a = 0$ or $b = 0$. To see that this property is useful for solving quadratic equations, consider the following:

$$x^2 - 4x + 3 = 0 \qquad\qquad (1)$$
$$(x - 1)(x - 3) = 0$$

$$x - 1 = 0 \quad\text{or}\quad x - 3 = 0$$
$$x = 1 \quad\text{or}\quad x = 3$$

You should check these solutions in equation (1).

If one side of the equation is not 0, then this method cannot be used. For example, consider

$$x^2 - 4x + 3 = 8 \qquad\qquad (2)$$
$$(x - 1)(x - 3) = 8$$
$$x - 1 \neq 8 \quad\text{or}\quad x - 3 \neq 8 \qquad \textit{ab = 8 does not imply}$$
$$\textit{that a = 8 or b = 8.}$$
$$x = 9 \quad\text{or}\quad x = 11$$

Verify that neither $x = 9$ nor $x = 11$ is a solution for equation (2).

EXAMPLE 2 | **Factoring Method** Solve by factoring using integer coefficients, if possible.

(A) $3x^2 - 6x - 24 = 0$ (B) $3y^2 = 2y$ (C) $x^2 - 2x - 1 = 0$

SOLUTION (A) $3x^2 - 6x - 24 = 0$ *Divide both sides by 3, since 3 is a factor of each coefficient.*

$$x^2 - 2x - 8 = 0$$ *Factor the left side, if possible.*
$$(x - 4)(x + 2) = 0$$
$$x - 4 = 0 \quad \text{or} \quad x + 2 = 0$$
$$x = 4 \quad \text{or} \quad x = -2$$

(B) $$3y^2 = 2y$$
$$3y^2 - 2y = 0 \quad \text{\textit{We lose the solution } } y = 0 \text{ \textit{if both sides are divided}}$$
$$y(3y - 2) = 0 \quad \text{\textit{by } y \text{ } (3y^2 = 2y \text{ and } 3y = 2 \text{ are not equivalent}).}$$
$$y = 0 \quad \text{or} \quad 3y - 2 = 0$$
$$3y = 2$$
$$y = \tfrac{2}{3}$$

(C) $x^2 - 2x - 1 = 0$

This equation cannot be factored using integer coefficients. We will solve this type of equation by another method, considered below. ■

MATCHED PROBLEM 2 Solve by factoring using integer coefficients, if possible.

(A) $2x^2 + 4x - 30 = 0$

(B) $2x^2 = 3x$

(C) $2x^2 - 8x + 3 = 0$

Note that an equation such as $x^2 = 25$ can be solved by either the square root or the factoring method, and the results are the same (as they should be). Solve this equation both ways and compare.

Also, note that the factoring method can be extended to higher-degree polynomial equations. Consider the following:

$$x^3 - x = 0$$
$$x(x^2 - 1) = 0$$
$$x(x - 1)(x + 1) = 0$$
$$x = 0 \quad \text{or} \quad x - 1 = 0 \quad \text{or} \quad x + 1 = 0$$
$$\text{Solution: } x = 0, 1, -1$$

Check these solutions in the original equation.

The factoring and square root methods are fast and easy to use when they apply. However, there are quadratic equations that look simple but cannot be solved by either method. For example, as was noted in Example 2C, the polynomial in

$$x^2 - 2x - 1 = 0$$

cannot be factored using integer coefficients. This brings us to the well-known and widely used *quadratic formula*.

■ Quadratic Formula

There is a method called *completing the square* that will work for all quadratic equations. After briefly reviewing this method, we will then use it to develop the famous quadratic formula—a formula that will enable us to solve any quadratic equation quite mechanically.

Explore & Discuss **2** Replace ? in each of the following with a number that makes the equation valid.

(A) $(x + 1)^2 = x^2 + 2x + ?$ \qquad (B) $(x + 2)^2 = x^2 + 4x + ?$

(C) $(x + 3)^2 = x^2 + 6x + ?$ \qquad (D) $(x + 4)^2 = x^2 + 8x + ?$

Replace **?** in each of the following with a number that makes the trinomial a perfect square.

(E) $x^2 + 10x + ?$ (F) $x^2 + 12x + ?$ (G) $x^2 + bx + ?$

The method of **completing the square** is based on the process of transforming a quadratic equation in standard form,

$$ax^2 + bx + c = 0$$

into the form

$$(x + A)^2 = B$$

where A and B are constants. Then, this last equation can be solved easily (if it has a real solution) by the square root method discussed above.

Consider the equation from Example 2C:

$$x^2 - 2x - 1 = 0 \qquad (3)$$

Since the left side does not factor using integer coefficients, we add 1 to each side to remove the constant term from the left side:

$$x^2 - 2x = 1 \qquad (4)$$

Now we try to find a number that we can add to each side to make the left side a square of a first-degree polynomial. Note the following square of a binomial:

$$(x + m)^2 = x^2 + 2mx + m^2$$

We see that the third term on the right is the square of one-half the coefficient of x in the second term on the right. To complete the square in equation (4), we add the square of one-half the coefficient of x, $(-\frac{2}{2})^2 = 1$, to each side. (This rule works only when the coefficient of x^2 is 1, that is, $a = 1$.) Thus,

$$x^2 - 2x + 1 = 1 + 1$$

The left side is the square of $x - 1$, and we write

$$(x - 1)^2 = 2$$

What number squared is 2?

$$x - 1 = \pm\sqrt{2}$$
$$x = 1 \pm \sqrt{2}$$

And equation (3) is solved!

Let us try the method on the general quadratic equation

$$ax^2 + bx + c = 0 \qquad a \neq 0 \qquad (5)$$

and solve it once and for all for x in terms of the coefficients a, b, and c. We start by multiplying both sides of equation (5) by $1/a$ to obtain

$$x^2 + \frac{b}{a}x + \frac{c}{a} = 0$$

Add $-c/a$ to both sides:

$$x^2 + \frac{b}{a}x = -\frac{c}{a}$$

Now we complete the square on the left side by adding the square of one-half the coefficient of x, that is, $(b/2a)^2 = b^2/4a^2$, to each side:

$$x^2 + \frac{b}{a}x + \frac{b^2}{4a^2} = \frac{b^2}{4a^2} - \frac{c}{a}$$

Writing the left side as a square and combining the right side into a single fraction, we obtain

$$\left(x + \frac{b}{2a}\right)^2 = \frac{b^2 - 4ac}{4a^2}$$

Now we solve by the square root method:

$$x + \frac{b}{2a} = \pm\sqrt{\frac{b^2 - 4ac}{4a^2}}$$

$$x = -\frac{b}{2a} \pm \frac{\sqrt{b^2 - 4ac}}{2a} \qquad \text{Since } \pm\sqrt{4a^2} = \pm 2a \text{ for any real number } a$$

When this is written as a single fraction, it becomes the **quadratic formula:**

Quadratic Formula

If $ax^2 + bx + c = 0$, $a \neq 0$, then

$$x = \frac{-b \pm \sqrt{b^2 - 4ac}}{2a}$$

This formula is generally used to solve quadratic equations when the square root or factoring methods do not work. The quantity $b^2 - 4ac$ under the radical is called the **discriminant,** and it gives us the useful information about solutions listed in Table 1.

TABLE 1	
$b^2 - 4ac$	$ax^2 + bx + c = 0$
Positive	Two real solutions
Zero	One real solution
Negative	No real solutions

EXAMPLE 3 **Quadratic Formula Method**   Solve $x^2 - 2x - 1 = 0$ using the quadratic formula.

SOLUTION

$$x^2 - 2x - 1 = 0$$

$$x = \frac{-b \pm \sqrt{b^2 - 4ac}}{2a} \qquad a = 1, b = -2, c = -1$$

$$= \frac{-(-2) \pm \sqrt{(-2)^2 - 4(1)(-1)}}{2(1)}$$

$$= \frac{2 \pm \sqrt{8}}{2} = \frac{2 \pm 2\sqrt{2}}{2} = 1 \pm \sqrt{2} \approx -0.414 \quad \text{or} \quad 2.414$$

CHECK: $x^2 - 2x - 1 = 0$

When $x = 1 + \sqrt{2}$,

$$(1 + \sqrt{2})^2 - 2(1 + \sqrt{2}) - 1 = 1 + 2\sqrt{2} + 2 - 2 - 2\sqrt{2} - 1 = 0$$

When $x = 1 - \sqrt{2}$,

$$(1 - \sqrt{2})^2 - 2(1 - \sqrt{2}) - 1 = 1 - 2\sqrt{2} + 2 - 2 + 2\sqrt{2} - 1 = 0$$

MATCHED PROBLEM 3   Solve $2x^2 - 4x - 3 = 0$ using the quadratic formula.

If we try to solve $x^2 - 6x + 11 = 0$ using the quadratic formula, we obtain

$$x = \frac{6 \pm \sqrt{-8}}{2}$$

which is not a real number. (Why?)

■ Quadratic Formula and Factoring

As in Section A-3, we restrict our interest in factoring to polynomials with integer coefficients. If a polynomial cannot be factored as a product of lower-degree polynomials with integer coefficients, we say that the polynomial is **not factorable in the integers.**

INSIGHT

Suppose you were asked to apply the *ac* test discussed in Section A-3 to the following polynomial:

$$x^2 - 13x - 2,310 \qquad (6)$$

Your first step would be to construct a table listing the 26 factor pairs for 2,310—a tedious process at best. Fortunately, there is a better way. The quadratic formula provides a simple and efficient method of factoring a second-degree polynomial with integer coefficients as the product of two first-degree polynomials with integer coefficients, if such factors exist.

We illustrate the method using polynomial (6), and generalize the process from this experience. We start by solving the corresponding quadratic equation using the quadratic formula:

$$x^2 - 13x - 2,310 = 0$$

$$x = \frac{-(-13) \pm \sqrt{(-13)^2 - 4(1)(-2,310)}}{2}$$

$$x = \frac{-(-13) \pm \sqrt{9,409}}{2}$$

$$= \frac{13 \pm 97}{2} = 55 \quad \text{or} \quad -42$$

Now we write

$$x^2 - 13x - 2,310 = [x - 55][x - (-42)] = (x - 55)(x + 42)$$

Multiplying the two factors on the right produces the second-degree polynomial on the left.

What is behind this procedure? The following two theorems justify and generalize the process:

THEOREM 2   FACTORABILITY THEOREM

A second-degree polynomial, $ax^2 + bx + c$, with integer coefficients can be expressed as the product of two first-degree polynomials with integer coefficients if and only if $\sqrt{b^2 - 4ac}$ is an integer.

THEOREM 3   FACTOR THEOREM

If r_1 and r_2 are solutions to the second-degree equation $ax^2 + bx + c = 0$, then

$$ax^2 + bx + c = a(x - r_1)(x - r_2)$$

EXAMPLE 4

Factoring with the Aid of the Discriminant Factor, if possible, using integer coefficients:

(A) $4x^2 - 65x + 264$ 　　　　　　(B) $2x^2 - 33x - 306$

SOLUTION (A) $4x^2 - 65x + 264$

Step 1. Test for factorability:

$$\sqrt{b^2 - 4ac} = \sqrt{(-65)^2 - 4(4)(264)} = 1$$

Since the result is an integer, the polynomial has first-degree factors with integer coefficients.

Step 2. Factor, using the factor theorem. Find the solutions to the corresponding quadratic equation using the quadratic formula:

$$4x^2 - 65x + 264 = 0 \quad \text{From step 1}$$

$$x = \frac{-(-65) \pm 1}{2 \cdot 4} = \frac{33}{4} \quad \text{or} \quad 8$$

Thus,

$$4x^2 - 65x + 264 = 4\left(x - \frac{33}{4}\right)(x - 8)$$
$$= (4x - 33)(x - 8)$$

(B) $2x^2 - 33x - 306$

Step 1. Test for factorability:

$$\sqrt{b^2 - 4ac} = \sqrt{(-33)^2 - 4(2)(-306)} = \sqrt{3,537}$$

Since $\sqrt{3,537}$ is not an integer, the polynomial is not factorable in the integers.

MATCHED PROBLEM 4

Factor, if possible, using integer coefficients:

(A) $3x^2 - 28x - 464$ 　　　　　　(B) $9x^2 + 320x - 144$

APPLICATION: SUPPLY AND DEMAND

Supply and demand analysis is a very important part of business and economics. In general, producers are willing to supply more of an item as the price of an item increases, and less of an item as the price decreases. Similarly, buyers are willing to buy less of an item as the price increases, and more of an item as the price decreases. Thus, we have a dynamic situation where the price, supply, and demand fluctuate until a price is reached at which the supply is equal to the demand. In economic theory, this point is called the **equilibrium point**—if the price increases from this point, the supply will increase and the demand will decrease; if the price decreases from this point, the supply will decrease and the demand will increase.

EXAMPLE 5

Supply and Demand At a large beach resort in the summer, the weekly supply and demand equations for folding beach chairs are

$$p = \frac{x}{140} + \frac{3}{4} \quad \text{Supply equation}$$

$$p = \frac{5,670}{x} \quad \text{Demand equation}$$

The supply equation indicates that the supplier is willing to sell x units at a price of p dollars per unit. The demand equation indicates that consumers are willing to buy

x units at a price of p dollars per unit. How many units are required for supply to equal demand? At what price will supply equal demand?

SOLUTION   Set the right side of the supply equation equal to the right side of the demand equation and solve for x:

$$\frac{x}{140} + \frac{3}{4} = \frac{5,670}{x} \qquad \text{Multiply by } 140x, \text{ the LCD.}$$

$$x^2 + 105x = 793,800 \qquad \text{Write in standard form.}$$

$$x^2 + 105x - 793,800 = 0 \qquad \text{Use the quadratic formula.}$$

$$x = \frac{-105 \pm \sqrt{105^2 - 4(1)(-793,800)}}{2}$$

$$x = 840 \text{ units}$$

The negative root is discarded, since a negative number of units cannot be produced or sold. Substitute $x = 840$ back into either the supply equation or the demand equation to find the equilibrium price (we use the demand equation):

$$p = \frac{5,670}{x} = \frac{5,670}{840} = \$6.75$$

Thus, at a price of \$6.75 the supplier is willing to supply 840 chairs and consumers are willing to buy 840 chairs during a week.   ▬

MATCHED PROBLEM 5   Repeat Example 5 if near the end of summer the supply and demand equations are

$$p = \frac{x}{80} - \frac{1}{20} \qquad \text{Supply equation}$$

$$p = \frac{1,264}{x} \qquad \text{Demand equation}$$

Answers to Matched Problems   **1.** (A) $\pm\sqrt{6}$ (B) ± 2 (C) No real solution   (D) $-6, -4$

2. (A) $-5, 3$ (B) $0, \frac{3}{2}$ (C) Cannot be factored using integer coefficients

3. $(2 \pm \sqrt{10})/2$

4. (A) Cannot be factored using integer coefficients   (B) $(9x - 4)(x + 36)$

5. 320 chairs at \$3.95 each

Exercise A-7

Find only real soluztions in the problems below. If there are no real solutions, say so.

A *Solve Problems 1–4 by the square root method.*

1. $2x^2 - 22 = 0$ **2.** $3m^2 - 21 = 0$

3. $(3x - 1)^2 = 25$ **4.** $(2x + 1)^2 = 16$

Solve Problems 5–8 by factoring.

5. $2u^2 - 8u - 24 = 0$ **6.** $3x^2 - 18x + 15 = 0$

7. $x^2 = 2x$ **8.** $n^2 = 3n$

Solve Problems 9–12 by using the quadratic formula.

9. $x^2 - 6x - 3 = 0$ **10.** $m^2 + 8m + 3 = 0$

11. $3u^2 + 12u + 6 = 0$ **12.** $2x^2 - 20x - 6 = 0$

B *Solve Problems 13–30 by using any method.*

13. $\frac{2x^2}{3} = 5x$ **14.** $x^2 = -\frac{3}{4}x$

15. $4u^2 - 9 = 0$ **16.** $9y^2 - 25 = 0$

17. $8x^2 + 20x = 12$ **18.** $9x^2 - 6 = 15x$

19. $x^2 = 1 - x$ **20.** $m^2 = 1 - 3m$

21. $2x^2 = 6x - 3$ **22.** $2x^2 = 4x - 1$

23. $y^2 - 4y = -8$ **24.** $x^2 - 2x = -3$

25. $(2x + 3)^2 = 11$ **26.** $(5x - 2)^2 = 7$

27. $\dfrac{3}{p} = p$

28. $x - \dfrac{7}{x} = 0$

29. $2 - \dfrac{2}{m^2} = \dfrac{3}{m}$

30. $2 + \dfrac{5}{u} = \dfrac{3}{u^2}$

In Problems 31–38, factor, if possible, as the product of two first-degree polynomials with integer coefficients. Use the quadratic formula and the factor theorem.

31. $x^2 + 40x - 84$

32. $x^2 - 28x - 128$

33. $x^2 - 32x + 144$

34. $x^2 + 52x + 208$

35. $2x^2 + 15x - 108$

36. $3x^2 - 32x - 140$

37. $4x^2 + 241x - 434$

38. $6x^2 - 427x - 360$

C

39. Solve $A = P(1 + r)^2$ for r in terms of A and P; that is, isolate r on the left side of the equation (with coefficient 1) and end up with an algebraic expression on the right side involving A and P but not r. Write the answer using positive square roots only.

40. Solve $x^2 + 3mx - 3n = 0$ for x in terms of m and n.

41. Consider the quadratic equation

$$x^2 + 4x + c = 0$$

where c is a real number. Discuss the relationship between the values of c and the three types of roots listed in Table 1.

42. Consider the quadratic equation

$$x^2 - 2x + c = 0$$

where c is a real number. Discuss the relationship between the values of c and the three types of roots listed in Table 1.

Applications

43. *Supply and demand.* A company wholesales a certain brand of shampoo in a particular city. Their marketing research department established the following weekly supply and demand equations:

$$p = \dfrac{x}{450} + \dfrac{1}{2} \quad \text{Supply equation}$$

$$p = \dfrac{6{,}300}{x} \quad \text{Demand equation}$$

How many units are required for supply to equal demand? At what price per bottle will supply equal demand?

44. *Supply and demand.* An importer sells a certain brand of automatic camera to outlets in a large metropolitan area. During the summer, the weekly supply and demand equations are

$$p = \dfrac{x}{6} + 9 \quad \text{Supply equation}$$

$$p = \dfrac{24{,}840}{x} \quad \text{Demand equation}$$

How many units are required for supply to equal demand? At what price will supply equal demand?

45. *Interest rate.* If P dollars are invested at $100r$ percent compounded annually, at the end of 2 years it will grow to $A = P(1 + r)^2$. At what interest rate will \$484 grow to \$625 in 2 years? (*Note:* If $A = 625$ and $P = 484$ find r.)

46. *Interest rate.* Using the formula in Problem 45, determine the interest rate that will make \$1,000 grow to \$1,210 in 2 years?

47. *Ecology.* An important element in the erosive force of moving water is its velocity. To measure the velocity v (in feet per second) of a stream, we position a hollow L-shaped tube with one end under the water pointing upstream and the other end pointing straight up a couple of feet out of the water. The water will then be pushed up the tube a certain distance h (in feet) above the surface of the stream. Physicists have shown that $v^2 = 64h$. Approximately how fast is a stream flowing if $h = 1$ foot? If $h = 0.5$ foot?

48. *Safety research.* It is of considerable importance to know the least number of feet d in which a car can be stopped, including reaction time of the driver, at various speeds v (in miles per hour). Safety research has produced the formula $d = 0.044v^2 + 1.1v$. If it took a car 550 feet to stop, estimate the car's speed at the moment the stopping process was started.

B-1 Sequences, Series, and Summation Notation
B-2 Arithmetic and Geometric Sequences
B-3 Binomial Theorem

Section B-1   SEQUENCES, SERIES, AND SUMMATION NOTATION

- Sequences
- Series and Summation Notation

If someone asked you to list all natural numbers that are perfect squares, you might begin by writing

$$1, 4, 9, 16, 25, 36$$

But you would soon realize that it is impossible to actually list all the perfect squares, since there are an infinite number of them. However, you could represent this collection of numbers in several different ways. One common method is to write

$$1, 4, 9, \ldots, n^2, \ldots \qquad n \in N$$

where N is the set of natural numbers. A list of numbers such as this is generally called a *sequence*. Sequences and related topics form the subject matter of this section.

■ Sequences

Consider the function f given by

$$f(n) = 2n + 1 \qquad (1)$$

where the domain of f is the set of natural numbers N. Note that

$$f(1) = 3, \quad f(2) = 5, \quad f(3) = 7, \quad \ldots$$

The function f is an example of a sequence. In general, a **sequence** is a function with domain a set of successive integers. Instead of the standard function notation used in equation (1), sequences are usually defined in terms of a special notation.

The range value $f(n)$ is usually symbolized more compactly with a symbol such as a_n. Thus, in place of equation (1), we write

$$a_n = 2n + 1$$

and the domain is understood to be the set of natural numbers unless something is said to the contrary or the context indicates otherwise. The elements in the range are called **terms of the sequence;** a_1 is the first term, a_2 is the second term, and a_n is the **nth term,** or **general term.**

$$a_1 = 2(1) + 1 = 3 \qquad \text{First term}$$
$$a_2 = 2(2) + 1 = 5 \qquad \text{Second term}$$
$$a_3 = 2(3) + 1 = 7 \qquad \text{Third term}$$
$$\vdots$$
$$a_n = 2n + 1 \qquad \text{General term}$$

The ordered list of elements

$$3, 5, 7, \ldots, 2n + 1, \ldots$$

obtained by writing the terms of the sequence in their natural order with respect to the domain values is often informally referred to as a sequence. A sequence also may be represented in the abbreviated form $\{a_n\}$, where a symbol for the nth term is written within braces. For example, we could refer to the sequence $3, 5, 7, \ldots, 2n + 1, \ldots$ as the sequence $\{2n + 1\}$.

If the domain of a sequence is a finite set of successive integers, then the sequence is called a **finite sequence.** If the domain is an infinite set of successive integers, then the sequence is called an **infinite sequence.** The sequence $\{2n + 1\}$ discussed above is an infinite sequence.

EXAMPLE 1   **Writing the Terms of a Sequence** Write the first four terms of each sequence:

(A) $a_n = 3n - 2$

(B) $\left\{ \dfrac{(-1)^n}{n} \right\}$

SOLUTION   (A) $1, 4, 7, 10$

(B) $-1, \dfrac{1}{2}, \dfrac{-1}{3}, \dfrac{1}{4}$ ▬

MATCHED PROBLEM 1   Write the first four terms of each sequence:

(A) $a_n = -n + 3$

(B) $\left\{ \dfrac{(-1)^n}{2^n} \right\}$

Explore & Discuss **1**   (A) A multiple-choice test question asked for the next term in the sequence

$$2, 4, 8, \ldots$$

and gave the following choices:

(1) 16   (2) 14   (3) $\frac{25}{2}$

Which is the correct answer?

(B) Compare the first four terms of the following sequences:

(1) $a_n = 2^n$ (2) $b_n = n^2 - n + 2$ (3) $c_n = 5n + \dfrac{6}{n} - 9$

Now, which of the choices in part (A) appears to be correct?

Now that we have seen how to use the general term to find the first few terms in a sequence, we consider the reverse problem. That is, can a sequence be defined just by listing the first three or four terms of the sequence? And can we then use these initial terms to find a formula for the nth term? In general, without other information, the answer to the first question is no. As Explore & Discuss 1 illustrates, many different sequences may start off with the same terms. Simply listing the first three terms (or any other finite number of terms) does not specify a particular sequence. In fact, it can be shown that given any list of m numbers, there are an infinite number of sequences whose first m terms agree with these given numbers.

What about the second question? That is, given a few terms, can we find the general formula for at least one sequence whose first few terms agree with the given terms? The answer to this question is a qualified yes. If we can observe a simple pattern in the given terms, we usually can construct a general term that will produce that pattern. The next example illustrates this approach.

EXAMPLE 2   **Finding the General Term of a Sequence** Find the general term of a sequence whose first four terms are

(A) $3, 4, 5, 6, \ldots$

(B) $5, -25, 125, -625, \ldots$

SOLUTION (A) Since these terms are consecutive integers, one solution is $a_n = n, n \geq 3$. If we want the domain of the sequence to be all natural numbers, another solution is $b_n = n + 2$.

(B) Each of these terms can be written as the product of a power of 5 and a power of -1:

$$5 = (-1)^0 5^1 = a_1$$
$$-25 = (-1)^1 5^2 = a_2$$
$$125 = (-1)^2 5^3 = a_3$$
$$-625 = (-1)^3 5^4 = a_4$$

If we choose the domain to be all natural numbers, a solution is

$$a_n = (-1)^{n-1} 5^n$$

MATCHED PROBLEM 2

Find the general term of a sequence whose first four terms are

(A) $3, 6, 9, 12, \ldots$ 　　　　　　　　 (B) $1, -2, 4, -8, \ldots$

In general, there is usually more than one way of representing the nth term of a given sequence (see the solution of Example 2A). However, unless something is stated to the contrary, we assume that the domain of the sequence is the set of natural numbers N.

■ Series and Summation Notation

If $a_1, a_2, a_3, \ldots, a_n, \ldots$ is a sequence, the expression

$$a_1 + a_2 + a_3 + \cdots + a_n + \cdots$$

is called a **series.** If the sequence is finite, the corresponding series is a **finite series.** If the sequence is infinite, the corresponding series is an **infinite series.** We consider only finite series in this section. For example,

$$1, 3, 5, 7, 9 \qquad \textit{Finite sequence}$$
$$1 + 3 + 5 + 7 + 9 \qquad \textit{Finite series}$$

Notice that we can easily evaluate this series by adding the five terms:

$$1 + 3 + 5 + 7 + 9 = 25$$

Series are often represented in a compact form called **summation notation.** Consider the following examples:

$$\sum_{k=3}^{6} k^2 = 3^2 + 4^2 + 5^2 + 6^2$$
$$= 9 + 16 + 25 + 36 = 86$$

$$\sum_{k=0}^{2} (4k + 1) = (4 \cdot 0 + 1) + (4 \cdot 1 + 1) + (4 \cdot 2 + 1)$$
$$= 1 + 5 + 9 = 15$$

In each case, the terms of the series on the right are obtained from the expression on the left by successively replacing the **summing index k** with integers, starting with the number indicated below the **summation sign Σ** and ending with the number that appears above Σ. The summing index may be represented by letters other than k and may start at any integer and end at any integer greater than or equal to the starting integer. Thus, if we are given the finite sequence

$$\frac{1}{2}, \frac{1}{4}, \frac{1}{8}, \ldots, \frac{1}{2^n}$$

the corresponding series is

$$\frac{1}{2} + \frac{1}{4} + \frac{1}{8} + \cdots + \frac{1}{2^n} = \sum_{j=1}^{n} \frac{1}{2^j}$$

where we have used j for the summing index.

EXAMPLE 3   **Summation Notation**   Write

$$\sum_{k=1}^{5} \frac{k}{k^2 + 1}$$

without summation notation. Do not evaluate the sum.

SOLUTION

$$\sum_{k=1}^{5} \frac{k}{k^2 + 1} = \frac{1}{1^2 + 1} + \frac{2}{2^2 + 1} + \frac{3}{3^2 + 1} + \frac{4}{4^2 + 1} + \frac{5}{5^2 + 1}$$

$$= \frac{1}{2} + \frac{2}{5} + \frac{3}{10} + \frac{4}{17} + \frac{5}{26}$$

MATCHED PROBLEM 3   Write

$$\sum_{k=1}^{5} \frac{k + 1}{k}$$

without summation notation. Do not evaluate the sum.

Explore & Discuss **2**

(A) Find the smallest value of n for which the value of the series

$$\sum_{k=1}^{n} \frac{k}{k^2 + 1}$$

is greater than 3.

(B) Find the smallest value of n for which the value of the series

$$\sum_{j=1}^{n} \frac{1}{2^j}$$

is greater than 0.99. Greater than 0.999.

If the terms of a series are alternately positive and negative, we call the series an **alternating series.** The next example deals with the representation of such a series.

EXAMPLE 4   **Summation Notation**   Write the alternating series

$$\frac{1}{2} - \frac{1}{4} + \frac{1}{6} - \frac{1}{8} + \frac{1}{10} - \frac{1}{12}$$

using summation notation with

(A) The summing index k starting at 1

(B) The summing index j starting at 0

SOLUTION   (A) $(-1)^{k+1}$ provides the alternation of sign, and $1/(2k)$ provides the other part of each term. Thus, we can write

$$\frac{1}{2} - \frac{1}{4} + \frac{1}{6} - \frac{1}{8} + \frac{1}{10} - \frac{1}{12} = \sum_{k=1}^{6} \frac{(-1)^{k+1}}{2k}$$

(B) $(-1)^j$ provides the alternation of sign, and $1/[2(j+1)]$ provides the other part of each term. Thus, we can write

$$\frac{1}{2} - \frac{1}{4} + \frac{1}{6} - \frac{1}{8} + \frac{1}{10} - \frac{1}{12} = \sum_{j=0}^{5} \frac{(-1)^j}{2(j+1)}$$

MATCHED PROBLEM 4 Write the alternating series

$$1 - \frac{1}{3} + \frac{1}{9} - \frac{1}{27} + \frac{1}{81}$$

using summation notation with

(A) The summing index k starting at 1

(B) The summing index j starting at 0

Summation notation provides a compact notation for the sum of any list of numbers, even if the numbers are not generated by a formula. For example, suppose that the results of an examination taken by a class of 10 students are given in the following list:

$$87, 77, 95, 83, 86, 73, 95, 68, 75, 86$$

If we let $a_1, a_2, a_3, \ldots, a_{10}$ represent these 10 scores, the average test score is given by

$$\frac{1}{10} \sum_{k=1}^{10} a_k = \frac{1}{10}(87 + 77 + 95 + 83 + 86 + 73 + 95 + 68 + 75 + 86)$$

$$= \frac{1}{10}(825) = 82.5$$

More generally, in statistics, the **arithmetic mean** $\bar{a}$ of a list of n numbers $a_1, a_2, \ldots, a_n$ is defined as

$$\bar{a} = \frac{1}{n} \sum_{k=1}^{n} a_k$$

EXAMPLE 5 **Arithmetic Mean** Find the arithmetic mean of 3, 5, 4, 7, 4, 2, 3, and 6.

SOLUTION $\bar{a} = \frac{1}{8} \sum_{k=1}^{8} a_k = \frac{1}{8}(3 + 5 + 4 + 7 + 4 + 2 + 3 + 6) = \frac{1}{8}(34) = 4.25$

MATCHED PROBLEM 5 Find the arithmetic mean of 9, 3, 8, 4, 3, and 6.

Answers to Matched Problems
1. (A) 2, 1, 0, −1   (B) $\frac{-1}{2}, \frac{1}{4}, \frac{-1}{8}, \frac{1}{16}$

2. (A) $a_n = 3n$ (B) $a_n = (-2)^{n-1}$ **3.** $2 + \frac{3}{2} + \frac{4}{3} + \frac{5}{4} + \frac{6}{5}$

4. (A) $\sum_{k=1}^{5} \frac{(-1)^{k-1}}{3^{k-1}}$ (B) $\sum_{j=0}^{4} \frac{(-1)^j}{3^j}$ **5.** 5.5

Exercise B-1

A *Write the first four terms for each sequence in Problems 1–6.*

1. $a_n = 2n + 3$

2. $a_n = 4n - 3$

3. $a_n = \dfrac{n + 2}{n + 1}$

4. $a_n = \dfrac{2n + 1}{2n}$

5. $a_n = (-3)^{n+1}$

6. $a_n = \left(-\dfrac{1}{4}\right)^{n-1}$

7. Write the 10th term of the sequence in Problem 1.

8. Write the 15th term of the sequence in Problem 2.

9. Write the 99th term of the sequence in Problem 3.

10. Write the 200th term of the sequence in Problem 4.

In Problems 11–16, write each series in expanded form without summation notation, and evaluate.

11. $\displaystyle\sum_{k=1}^{6} k$

12. $\displaystyle\sum_{k=1}^{5} k^2$

13. $\displaystyle\sum_{k=4}^{7} (2k - 3)$

14. $\displaystyle\sum_{k=0}^{4} (-2)^k$

15. $\displaystyle\sum_{k=0}^{3} \dfrac{1}{10^k}$

16. $\displaystyle\sum_{k=1}^{4} \dfrac{1}{2^k}$

Find the arithmetic mean of each list of numbers in Problems 17–20.

17. 5, 4, 2, 1, and 6

18. 7, 9, 9, 2, and 4

19. 96, 65, 82, 74, 91, 88, 87, 91, 77, and 74

20. 100, 62, 95, 91, 82, 87, 70, 75, 87, and 82

B *Write the first five terms of each sequence in Problems 21–26.*

21. $a_n = \dfrac{(-1)^{n+1}}{2^n}$

22. $a_n = (-1)^n (n - 1)^2$

23. $a_n = n[1 + (-1)^n]$

24. $a_n = \dfrac{1 - (-1)^n}{n}$

25. $a_n = \left(-\dfrac{3}{2}\right)^{n-1}$

26. $a_n = \left(-\dfrac{1}{2}\right)^{n+1}$

In Problems 27–42, find the general term of a sequence whose first four terms agree with the given terms.

27. $-2, -1, 0, 1, \ldots$

28. $4, 5, 6, 7, \ldots$

29. $4, 8, 12, 16, \ldots$

30. $-3, -6, -9, -12, \ldots$

31. $\dfrac{1}{2}, \dfrac{3}{4}, \dfrac{5}{6}, \dfrac{7}{8}, \ldots$

32. $\dfrac{1}{2}, \dfrac{2}{3}, \dfrac{3}{4}, \dfrac{4}{5}, \ldots$

33. $1, -2, 3, -4, \ldots$

34. $-2, 4, -8, 16, \ldots$

35. $1, -3, 5, -7, \ldots$

36. $3, -6, 9, -12, \ldots$

37. $1, \dfrac{2}{5}, \dfrac{4}{25}, \dfrac{8}{125}, \ldots$

38. $\dfrac{4}{3}, \dfrac{16}{9}, \dfrac{64}{27}, \dfrac{256}{81}, \ldots$

39. $x, x^2, x^3, x^4, \ldots$

40. $1, 2x, 3x^2, 4x^3, \ldots$

41. $x, -x^3, x^5, -x^7, \ldots$

42. $x, \dfrac{x^2}{2}, \dfrac{x^3}{3}, \dfrac{x^4}{4}, \ldots$

Write each series in Problems 43–50 in expanded form without summation notation. Do not evaluate.

43. $\displaystyle\sum_{k=1}^{5} (-1)^{k+1} (2k - 1)^2$

44. $\displaystyle\sum_{k=1}^{4} \dfrac{(-2)^{k+1}}{2k + 1}$

45. $\displaystyle\sum_{k=2}^{5} \dfrac{2^k}{2k + 3}$

46. $\displaystyle\sum_{k=3}^{7} \dfrac{(-1)^k}{k^2 - k}$

47. $\displaystyle\sum_{k=1}^{5} x^{k-1}$

48. $\displaystyle\sum_{k=1}^{3} \dfrac{1}{k} x^{k+1}$

49. $\displaystyle\sum_{k=0}^{4} \dfrac{(-1)^k x^{2k+1}}{2k + 1}$

50. $\displaystyle\sum_{k=0}^{4} \dfrac{(-1)^k x^{2k}}{2k + 2}$

Write each series in Problems 51–54 using summation notation with

(A) *The summing index k starting at $k = 1$*

(B) *The summing index j starting at $j = 0$*

51. $2 + 3 + 4 + 5 + 6$

52. $1^2 + 2^2 + 3^2 + 4^2$

53. $1 - \dfrac{1}{2} + \dfrac{1}{3} - \dfrac{1}{4}$

54. $1 - \dfrac{1}{3} + \dfrac{1}{5} - \dfrac{1}{7} + \dfrac{1}{9}$

Write each series in Problems 55–58 using summation notation with the summing index k starting at $k = 1$.

55. $2 + \dfrac{3}{2} + \dfrac{4}{3} + \cdots + \dfrac{n + 1}{n}$

56. $1 + \dfrac{1}{2^2} + \dfrac{1}{3^2} + \cdots + \dfrac{1}{n^2}$

57. $\dfrac{1}{2} - \dfrac{1}{4} + \dfrac{1}{8} - \cdots + \dfrac{(-1)^{n+1}}{2^n}$

58. $1 - 4 + 9 - \cdots + (-1)^{n+1} n^2$

C *In Problems 59–62, discuss the validity of each statement. If the statement is true, explain why. If not, give a counterexample.*

59. For each positive integer n, the sum of the series
$$1 + \dfrac{1}{2} + \dfrac{1}{3} + \cdots + \dfrac{1}{n} \text{ is less than 4.}$$

60. For each positive integer n, the sum of the series
$$\dfrac{1}{2} + \dfrac{1}{4} + \dfrac{1}{8} + \cdots + \dfrac{1}{2^n} \text{ is less than 1.}$$

61. For each positive integer n, the sum of the series
$$\dfrac{1}{2} - \dfrac{1}{4} + \dfrac{1}{8} - \cdots + \dfrac{(-1)^{n+1}}{2^n} \text{ is greater than or equal}$$
to $\dfrac{1}{4}$.

62. For each positive integer n, the sum of the series
$$1 - \dfrac{1}{2} + \dfrac{1}{3} - \dfrac{1}{4} + \cdots + \dfrac{(-1)^{n+1}}{n} \text{ is greater than or}$$
equal to $\dfrac{1}{2}$.

Some sequences are defined by a **recursion formula** *—that is, a formula that defines each term of the sequence in terms of one or more of the preceding terms. For example, if $\{a_n\}$ is defined by*

$$a_1 = 1 \quad and \quad a_n = 2a_{n-1} + 1 \quad for \ n \geq 2$$

then

$$a_2 = 2a_1 + 1 = 2 \cdot 1 + 1 = 3$$
$$a_3 = 2a_2 + 1 = 2 \cdot 3 + 1 = 7$$
$$a_4 = 2a_3 + 1 = 2 \cdot 7 + 1 = 15$$

and so on. In Problems 63–66, write the first five terms of each sequence.

63. $a_1 = 2 \quad$ and $\quad a_n = 3a_{n-1} + 2 \quad$ for $n \geq 2$

64. $a_1 = 3$ and $a_n = 2a_{n-1} - 2$ for $n \geq 2$

65. $a_1 = 1$ and $a_n = 2a_{n-1}$ for $n \geq 2$

66. $a_1 = 1$ and $a_n = -\frac{1}{3}a_{n-1}$ for $n \geq 2$

If A is a positive real number, the terms of the sequence defined by

$$a_1 = \frac{A}{2} \quad \text{and} \quad a_n = \frac{1}{2}\left(a_{n-1} + \frac{A}{a_{n-1}}\right) \quad \text{for } n \geq 2$$

can be used to approximate $\sqrt{A}$ to any decimal place accuracy desired. In Problems 67 and 68, compute the first four terms of this sequence for the indicated value of A, and compare the fourth term with the value of $\sqrt{A}$ obtained from a calculator.

67. $A = 2$

68. $A = 6$

69. The sequence defined recursively by $a_1 = 1$, $a_2 = 1$, $a_n = a_{n-1} + a_{n-2}$ for $n \geq 3$ is called the *Fibonacci sequence.* Find the first ten terms of the Fibonacci sequence.

70. The sequence defined by $b_n = \frac{\sqrt{5}}{5}\left(\frac{1 + \sqrt{5}}{2}\right)^n$ is related to the Fibonacci sequence. Find the first ten terms (to three decimal places) of the sequence $\{b_n\}$ and describe the relationship.

Section B-2 ARITHMETIC AND GEOMETRIC SEQUENCES

- Arithmetic and Geometric Sequences
- nth-Term Formulas
- Sum Formulas for Finite Arithmetic Series
- Sum Formulas for Finite Geometric Series
- Sum Formula for Infinite Geometric Series
- Applications

For most sequences it is difficult to sum an arbitrary number of terms of the sequence without adding term by term. But particular types of sequences—*arithmetic sequences* and *geometric sequences*—have certain properties that lead to convenient and useful formulas for the sums of the corresponding *arithmetic series* and *geometric series*.

■ Arithmetic and Geometric Sequences

The sequence $5, 7, 9, 11, 13, \ldots, 5 + 2(n - 1), \ldots$, where each term after the first is obtained by adding 2 to the preceding term, is an example of an arithmetic sequence. The sequence $5, 10, 20, 40, 80, \ldots, 5(2)^{n-1}, \ldots$, where each term after the first is obtained by multiplying the preceding term by 2, is an example of a geometric sequence.

DEFINITION **Arithmetic Sequence**

A sequence of numbers

$$a_1, a_2, a_3, \ldots, a_n, \ldots$$

is called an **arithmetic sequence** if there is a constant d, called the **common difference,** such that

$$a_n - a_{n-1} = d$$

That is,

$$a_n = a_{n-1} + d \quad \text{for every } n > 1$$

DEFINITION **Geometric Sequence**

A sequence of numbers

$$a_1, a_2, a_3, \ldots, a_n, \ldots$$

is called a **geometric sequence** if there exists a nonzero constant r, called a **common ratio,** such that

$$\frac{a_n}{a_{n-1}} = r$$

That is,

$$a_n = ra_{n-1} \quad \text{for every } n > 1$$

Explore & Discuss **1**   (A) Describe verbally all arithmetic sequences with common difference 2.

(B) Describe verbally all geometric sequences with common ratio 2.

EXAMPLE 1   **Recognizing Arithmetic and Geometric Sequences**   Which of the following can be the first four terms of an arithmetic sequence? Of a geometric sequence?

(A) $1, 2, 3, 5, \ldots$ (B) $-1, 3, -9, 27, \ldots$

(C) $3, 3, 3, 3, \ldots$ (D) $10, 8.5, 7, 5.5, \ldots$

SOLUTION   (A) Since $2 - 1 \neq 5 - 3$, there is no common difference, so the sequence is not an arithmetic sequence. Since $2/1 \neq 3/2$, there is no common ratio, so the sequence is not geometric either.

(B) The sequence is geometric with common ratio -3. It is not arithmetic.

(C) The sequence is arithmetic with common difference 0, and is also geometric with common ratio 1.

(D) The sequence is arithmetic with common difference -1.5. It is not geometric.

MATCHED PROBLEM 1   Which of the following can be the first four terms of an arithmetic sequence? Of a geometric sequence?

(A) $8, 2, 0.5, 0.125, \ldots$ (B) $-7, -2, 3, 8, \ldots$ (C) $1, 5, 25, 100, \ldots$

■ *n*th-Term Formulas

If $\{a_n\}$ is an arithmetic sequence with common difference d, then

$$a_2 = a_1 + d$$
$$a_3 = a_2 + d = a_1 + 2d$$
$$a_4 = a_3 + d = a_1 + 3d$$

This suggests that

THEOREM 1   *n*th TERM OF AN ARITHMETIC SEQUENCE

$$a_n = a_1 + (n - 1)d \qquad \text{for all } n > 1 \tag{1}$$

Similarly, if $\{a_n\}$ is a geometric sequence with common ratio r, then

$$a_2 = a_1 r$$
$$a_3 = a_2 r = a_1 r^2$$
$$a_4 = a_3 r = a_1 r^3$$

This suggests that

THEOREM 2   *n*th TERM OF A GEOMETRIC SEQUENCE

$$a_n = a_1 r^{n-1} \qquad \text{for all } n > 1 \tag{2}$$

EXAMPLE 2   **Finding Terms in Arithmetic and Geometric Sequences**

(A) If the 1st and 10th terms of an arithmetic sequence are 3 and 30, respectively, find the 40th term of the sequence.

(B) If the 1st and 10th terms of a geometric sequence are 3 and 30, find the 40th term to three decimal places.

SOLUTION   (A) First use formula (1) with $a_1 = 3$ and $a_{10} = 30$ to find d:

$$a_n = a_1 + (n - 1)d$$
$$a_{10} = a_1 + (10 - 1)d$$
$$30 = 3 + 9d$$
$$d = 3$$

Now find a_{40}:

$$a_{40} = 3 + 39 \cdot 3 = 120$$

(B) First use formula (2) with $a_1 = 3$ and $a_{10} = 30$ to find r:

$$a_n = a_1 r^{n-1}$$
$$a_{10} = a_1 r^{10-1}$$
$$30 = 3r^9$$
$$r^9 = 10$$
$$r = 10^{1/9}$$

Now find a_{40}:

$$a_{40} = 3(10^{1/9})^{39} = 3(10^{39/9}) = 64{,}633.041$$

MATCHED PROBLEM 2   (A) If the 1st and 15th terms of an arithmetic sequence are -5 and 23, respectively, find the 73rd term of the sequence.

(B) Find the 8th term of the geometric sequence

$$\frac{1}{64}, \frac{-1}{32}, \frac{1}{16}, \dots$$

■ Sum Formulas for Finite Arithmetic Series

If $a_1, a_2, a_3, \dots, a_n$ is a finite arithmetic sequence, then the corresponding series $a_1 + a_2 + a_3 + \cdots + a_n$ is called a *finite arithmetic series*. We will derive two simple and very useful formulas for the sum of a finite arithmetic series. Let d be the common difference of the arithmetic sequence $a_1, a_2, a_3, \dots, a_n$ and let S_n denote the sum of the series $a_1 + a_2 + a_3 + \cdots + a_n$. Then

$$S_n = a_1 + (a_1 + d) + \cdots + [a_1 + (n - 2)d] + [a_1 + (n - 1)d]$$

Reversing the order of the sum, we obtain

$$S_n = [a_1 + (n - 1)d] + [a_1 + (n - 2)d] + \cdots + (a_1 + d) + a_1$$

Something interesting happens if we combine these last two equations by addition (adding corresponding terms on the right sides):

$$2S_n = [2a_1 + (n - 1)d] + [2a_1 + (n - 1)d] + \cdots + [2a_1 + (n - 1)d] + [2a_1 + (n - 1)d]$$

All the terms on the right side are the same, and there are n of them. Thus,

$$2S_n = n[2a_1 + (n - 1)d]$$

and we have the following general formula:

THEOREM 3   SUM OF A FINITE ARITHMETIC SERIES: FIRST FORM

$$S_n = \frac{n}{2}[2a_1 + (n - 1)d] \tag{3}$$

Replacing

$$[a_1 + (n-1)d] \quad \text{in} \quad \frac{n}{2}[a_1 + a_1 + (n-1)d]$$

by a_n from equation (1), we obtain a second useful formula for the sum:

THEOREM 4   SUM OF A FINITE ARITHMETIC SERIES: SECOND FORM

$$S_n = \frac{n}{2}(a_1 + a_n) \tag{4}$$

EXAMPLE 3   **Finding a Sum** Find the sum of the first 30 terms in the arithmetic sequence:

$$3, 8, 13, 18, \ldots$$

SOLUTION   Use formula (3) with $n = 30$, $a_1 = 3$, and $d = 5$:

$$S_{30} = \frac{30}{2}[2 \cdot 3 + (30 - 1)5] = 2{,}265$$

MATCHED PROBLEM 3   Find the sum of the first 40 terms in the arithmetic sequence:

$$15, 13, 11, 9, \ldots$$

EXAMPLE 4   **Finding a Sum** Find the sum of all the even numbers between 31 and 87.

SOLUTION   First, find n using equation (1):

$$a_n = a_1 + (n - 1)d$$
$$86 = 32 + (n - 1)2$$
$$n = 28$$

Now find S_{28} using formula (4):

$$S_n = \frac{n}{2}(a_1 + a_n)$$

$$S_{28} = \frac{28}{2}(32 + 86) = 1{,}652$$

MATCHED PROBLEM 4   Find the sum of all the odd numbers between 24 and 208.

■ Sum Formulas for Finite Geometric Series

If $a_1, a_2, a_3, \ldots, a_n$ is a finite geometric sequence, then the corresponding series $a_1 + a_2 + a_3 + \cdots + a_n$ is called a *finite geometric series*. As with arithmetic series, we can derive two simple and very useful formulas for the sum of a finite geometric series. Let r be the common ratio of the geometric sequence $a_1, a_2, a_3, \ldots, a_n$ and let S_n denote the sum of the series $a_1 + a_2 + a_3 + \cdots + a_n$. Then

$$S_n = a_1 + a_1 r + a_1 r^2 + \cdots + a_1 r^{n-2} + a_1 r^{n-1}$$

If we multiply both sides by r, we obtain

$$rS_n = a_1 r + a_1 r^2 + a_1 r^3 + \cdots + a_1 r^{n-1} + a_1 r^n$$

Now combine these last two equations by subtraction to obtain

$$rS_n - S_n = (a_1r + a_1r^2 + a_1r^3 + \cdots + a_1r^{n-1} + a_1r^n) - (a_1 + a_1r + a_1r^2 + \cdots + a_1r^{n-2} + a_1r^{n-1})$$

$$(r-1)S_n = a_1r^n - a_1$$

Notice how many terms drop out on the right side. Solving for S_n, we have

THEOREM 5 **SUM OF A FINITE GEOMETRIC SERIES: FIRST FORM**

$$S_n = \frac{a_1(r^n - 1)}{r - 1} \qquad r \neq 1 \tag{5}$$

Since $a_n = a_1r^{n-1}$, or $ra_n = a_1r^n$, formula (5) also can be written in the form

THEOREM 6 **SUM OF A FINITE GEOMETRIC SERIES: SECOND FORM**

$$S_n = \frac{ra_n - a_1}{r - 1} \qquad r \neq 1 \tag{6}$$

EXAMPLE 5 **Finding a Sum** Find the sum (to 2 decimal places) of the first ten terms of the geometric sequence:

$$1, 1.05, 1.05^2, \ldots$$

SOLUTION Use formula (5) with $a_1 = 1$, $r = 1.05$, and $n = 10$:

$$S_n = \frac{a_1(r^n - 1)}{r - 1}$$

$$S_{10} = \frac{1(1.05^{10} - 1)}{1.05 - 1}$$

$$\approx \frac{0.6289}{0.05} \approx 12.58$$

MATCHED PROBLEM 5 Find the sum of the first eight terms of the geometric sequence:

$$100, 100(1.08), 100(1.08)^2, \ldots$$

■ Sum Formula for Infinite Geometric Series

Explore & Discuss **2**

(A) For any n, the sum of a finite geometric series with $a_1 = 5$ and $r = \frac{1}{2}$ is given by [see formula (5)]

$$S_n = \frac{5\left[\left(\frac{1}{2}\right)^n - 1\right]}{\frac{1}{2} - 1} = 10 - 10\left(\frac{1}{2}\right)^n$$

Discuss the behavior of S_n as n increases.

(B) Repeat part (A) if $a_1 = 5$ and $r = 2$.

Given a geometric series, what happens to the sum S_n of the first n terms as n increases without stopping? To answer this question, let us write formula (5) in the form

$$S_n = \frac{a_1r^n}{r - 1} - \frac{a_1}{r - 1}$$

It is possible to show that if $-1 < r < 1$, then r^n will approach 0 as n increases. Thus, the first term above will approach 0 and S_n can be made as close as we please to the second term, $-a_1/(r - 1)$ [which can be written as $a_1/(1 - r)$], by taking n sufficiently large. Thus, if the common ratio r is between -1 and 1, we conclude that the sum of an infinite geometric series is

---

THEOREM 7   SUM OF AN INFINITE GEOMETRIC SERIES

$$S_\infty = \frac{a_1}{1 - r} \qquad -1 < r < 1 \qquad\qquad (7)$$

If $r \leq -1$ or $r \geq 1$, then an infinite geometric series has no sum.

APPLICATIONS

EXAMPLE 6   **Loan Repayment**   A person borrows \$3,600 and agrees to repay the loan in monthly installments over a period of 3 years. The agreement is to pay 1% of the unpaid balance each month for using the money and \$100 each month to reduce the loan. What is the total cost of the loan over the 3 years?

SOLUTION   Let us look at the problem relative to a time line:

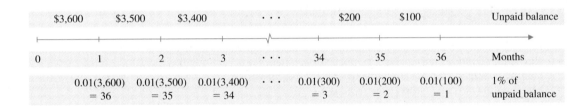

The total cost of the loan is

$$1 + 2 + \cdots + 34 + 35 + 36$$

The terms form a finite arithmetic series with $n = 36$, $a_1 = 1$, and $a_{36} = 36$, so we can use formula (4):

$$S_n = \frac{n}{2}(a_1 + a_n)$$

$$S_{36} = \frac{36}{2}(1 + 36) = \$666$$

We conclude that the total cost of the loan over the period of 3 years is \$666.   ▬

MATCHED PROBLEM 6   Repeat Example 6 with a loan of \$6,000 over a period of 5 years.

EXAMPLE 7   **Economy Stimulation**   The government has decided on a tax rebate program to stimulate the economy. Suppose that you receive \$600 and you spend 80% of this, and each of the people who receive what you spend also spend 80% of what they receive, and this process continues without end. According to the **multiplier principle** in economics, the effect of your \$600 tax rebate on the economy is multiplied many times. What is the total amount spent if the process continues as indicated?

SOLUTION We need to find the sum of an infinite geometric series with the first amount spent being $a_1 = (0.8)(\$600) = \$480$ and $r = 0.8$. Using formula (7), we obtain

$$S_\infty = \frac{a_1}{1 - r}$$

$$= \frac{\$480}{1 - 0.8} = \$2,400$$

Thus, assuming the process continues as indicated, we would expect the $600 tax rebate to result in about $2,400 of spending. ▄

MATCHED PROBLEM 7 | Repeat Example 7 with a tax rebate of $1,000.

Answers to Matched Problems   **1.** (A) The sequence is geometric with $r = \frac{1}{4}$. It is not arithmetic.
(B) The sequence is arithmetic with $d = 5$. It is not geometric.
(C) The sequence is neither arithmetic nor geometric.

2. (A) 139   (B) −2   **3.** −960   **4.** 10,672

5. 1,063.66   **6.** $1,830   **7.** $4,000

Exercise B-2

A In Problems 1 and 2, determine whether the indicated sequence can be the first three terms of an arithmetic or geometric sequence, and, if so, find the common difference or common ratio and the next two terms of the sequence.

1. (A) −11, −16, −21, ...   (B) 2, −4, 8, ...
(C) 1, 4, 9, ...   (D) $\frac{1}{2}, \frac{1}{6}, \frac{1}{18}, ...$

2. (A) 5, 20, 100, ...   (B) −5, −5, −5, ...
(C) 7, 6.5, 6, ...   (D) 512, 256, 128, ...

B In Problem 3–8, determine whether the finite series is arithmetic, geometric, both, or neither. If the series is arithmetic or geometric, find its sum.

3. $\displaystyle\sum_{k=1}^{101} (-1)^{k+1}$

4. $\displaystyle\sum_{k=1}^{200} 3$

5. $1 + \dfrac{1}{2} + \dfrac{1}{3} + \cdots + \dfrac{1}{50}$

6. $3 - 9 + 27 - \cdots - 3^{20}$

7. $5 + 4.9 + 4.8 + \cdots + 0.1$

8. $1 - \dfrac{1}{4} + \dfrac{1}{9} - \cdots - \dfrac{1}{100^2}$

Let $a_1, a_2, a_3, \ldots, a_n, \ldots$ be an arithmetic sequence. In Problems 9–14, find the indicated quantities.

9. $a_1 = 7; d = 4; a_2 = ?; a_3 = ?$

10. $a_1 = -2; d = -3; a_2 = ?; a_3 = ?$

11. $a_1 = 2; d = 4; a_{21} = ?; S_{31} = ?$

12. $a_1 = 8; d = -10; a_{15} = ?; S_{23} = ?$

13. $a_1 = 18; a_{20} = 75; S_{20} = ?$

14. $a_1 = 203; a_{30} = 261; S_{30} = ?$

Let $a_1, a_2, a_3, \ldots, a_n, \ldots$ be a geometric sequence. In Problems 15–24, find the indicated quantities.

15. $a_1 = 3; r = -2; a_2 = ?; a_3 = ?; a_4 = ?$

16. $a_1 = 32; r = -\frac{1}{2}; a_2 = ?; a_3 = ?; a_4 = ?$

17. $a_1 = 1; a_7 = 729; r = -3; S_7 = ?$

18. $a_1 = 3; a_7 = 2,187; r = 3; S_7 = ?$

19. $a_1 = 100; r = 1.08; a_{10} = ?$

20. $a_1 = 240; r = 1.06; a_{12} = ?$

21. $a_1 = 100; a_9 = 200; r = ?$

22. $a_1 = 100; a_{10} = 300; r = ?$

23. $a_1 = 500; r = 0.6; S_{10} = ?; S_\infty = ?$

24. $a_1 = 8,000; r = 0.4; S_{10} = ?; S_\infty = ?$

25. $S_{41} = \displaystyle\sum_{k=1}^{41} (3k + 3) = ?$

26. $S_{50} = \displaystyle\sum_{k=1}^{50} (2k - 3) = ?$

27. $S_8 = \displaystyle\sum_{k=1}^{8} (-2)^{k-1} = ?$

28. $S_8 = \displaystyle\sum_{k=1}^{8} 2^k = ?$

29. Find the sum of all the odd integers between 12 and 68.

30. Find the sum of all the even integers between 23 and 97.

31. Find the sum of each infinite geometric sequence (if it exists).
   (A) $2, 4, 8, \ldots$
   (B) $2, -\frac{1}{2}, \frac{1}{8}, \ldots$

32. Repeat Problem 31 for:
   (A) $16, 4, 1, \ldots$
   (B) $1, -3, 9, \ldots$

C

33. Find $f(1) + f(2) + f(3) + \cdots + f(50)$ if $f(x) = 2x - 3$.

34. Find $g(1) + g(2) + g(3) + \cdots + g(100)$ if $g(t) = 18 - 3t$.

35. Find $f(1) + f(2) + \cdots + f(10)$ if $f(x) = \left(\frac{1}{2}\right)^x$.

36. Find $g(1) + g(2) + \cdots + g(10)$ if $g(x) = 2^x$.

37. Show that the sum of the first n odd positive integers is n^2, using appropriate formulas from this section.

38. Show that the sum of the first n even positive integers is $n + n^2$, using formulas in this section.

39. If $r = 1$, neither the first form nor the second form for the sum of a finite geometric series is valid. Find a formula for the sum of a finite geometric series if $r = 1$.

40. If all of the terms of an infinite geometric series are less than 1, could the sum be greater than 1,000? Explain.

41. Does there exist a finite arithmetic series with $a_1 = 1$ and $a_n = 1.1$ that has sum equal to 100? Explain.

42. Does there exist a finite arithmetic series with $a_1 = 1$ and $a_n = 1.1$ that has sum equal to 105? Explain.

43. Does there exist an infinite geometric series with $a_1 = 10$ that has sum equal to 6? Explain.

44. Does there exist an infinite geometric series with $a_1 = 10$ that has sum equal to 5? Explain.

Applications

45. *Loan repayment.* If you borrow \$4,800 and repay the loan by paying \$200 per month to reduce the loan and 1% of the unpaid balance each month for the use of the money, what is the total cost of the loan over 24 months?

46. *Loan repayment.* If you borrow \$5,400 and repay the loan by paying \$300 per month to reduce the loan and 1.5% of the unpaid balance each month for the use of the money, what is the total cost of the loan over 18 months?

47. *Economy stimulation.* The government, through a subsidy program, distributes \$5,000,000. If we assume that each person or agency spends 70% of what is received, and 70% of this is spent, and so on, how much total increase in spending results from this government action? (Let $a_1 = \$3,500,000$.)

48. *Economy stimulation.* Due to reduced taxes, a person has an extra \$1,200 in spendable income. If we assume that the person spends 65% of this on consumer goods, and the producers of these goods in turn spend 65% on consumer goods, and that this process continues indefinitely, what is the total amount spent (to the nearest dollar) on consumer goods?

49. *Compound interest.* If \$1,000 is invested at 5% compounded annually, the amount A present after n years forms a geometric sequence with common ratio $1 + 0.05 = 1.05$. Use a geometric sequence formula to find the amount A in the account (to the nearest cent) after 10 years. After 20 years. (*Hint:* Use a time line.)

50. *Compound interest.* If \$$P$ is invested at $100r\%$ compounded annually, the amount A present after n years forms a geometric sequence with common ratio $1 + r$. Write a formula for the amount present after n years. (*Hint:* Use a time line.)

Section B-3   BINOMIAL THEOREM

- Factorial
- Development of the Binomial Theorem

The binomial form

$$(a + b)^n$$

where n is a natural number, appears more frequently than you might expect. The coefficients in the expansion play an important role in probability studies. The *binomial formula,* which we will derive informally, enables us to expand $(a + b)^n$ directly for n any natural number. Since the formula involves *factorials,* we digress for a moment here to introduce this important concept.

■ Factorial

For n a natural number, n **factorial,** denoted by $n!$, is the product of the first n natural numbers. **Zero factorial** is defined to be 1. That is,

---

DEFINITION n **Factorial**

$$n! = n \cdot (n - 1) \cdot \cdots \cdot 2 \cdot 1$$
$$1! = 1$$
$$0! = 1$$

It is also useful to note that $n!$ can be defined recursively.

DEFINITION n **Factorial—Recursive Definition**

$$n! = n \cdot (n - 1)! \qquad n \geq 1$$

EXAMPLE 1   **Factorial Forms** Evaluate each:

(A) $5! = 5 \cdot 4 \cdot 3 \cdot 2 \cdot 1 = 120$

(B) $\dfrac{8!}{7!} = \dfrac{8 \cdot 7!}{7!} = 8$

(C) $\dfrac{10!}{7!} = \dfrac{10 \cdot 9 \cdot 8 \cdot 7!}{7!} = 720$

MATCHED PROBLEM 1   Evaluate each:

(A) $4!$ (B) $\dfrac{7!}{6!}$ (C) $\dfrac{8!}{5!}$

The following important formula involving factorials has applications in many areas of mathematics and statistics. We will use this formula to provide a more concise form for the expressions encountered later in this discussion.

THEOREM 1   **FOR n AND r INTEGERS SATISFYING $0 \leq r \leq n$,**

$$C_{n,r} = \frac{n!}{r!(n - r)!}$$

EXAMPLE 2   **Evaluating $C_{n,r}$**

(A) $C_{9,2} = \dfrac{9!}{2!(9 - 2)!} = \dfrac{9!}{2!7!} = \dfrac{9 \cdot 8 \cdot 7!}{2 \cdot 7!} = 36$

(B) $C_{5,5} = \dfrac{5!}{5!(5 - 5)!} = \dfrac{5!}{5!0!} = \dfrac{5!}{5!} = 1$

MATCHED PROBLEM 2   Find

(A) $C_{5,2}$ (B) $C_{6,0}$

■ Development of the Binomial Theorem

Let us expand $(a + b)^n$ for several values of n to see if we can observe a pattern that leads to a general formula for the expansion for any natural number n:

$$(a + b)^1 = a + b$$
$$(a + b)^2 = a^2 + 2ab + b^2$$
$$(a + b)^3 = a^3 + 3a^2b + 3ab^2 + b^3$$
$$(a + b)^4 = a^4 + 4a^3b + 6a^2b^2 + 4ab^3 + b^4$$
$$(a + b)^5 = a^5 + 5a^4b + 10a^3b^2 + 10a^2b^3 + 5ab^4 + b^5$$

INSIGHT

1. The expansion of $(a + b)^n$ has $(n + 1)$ terms.
2. The power of a decreases by 1 for each term as we move from left to right.
3. The power of b increases by 1 for each term as we move from left to right.
4. In each term, the sum of the powers of a and b always equals n.
5. Starting with a given term, we can get the coefficient of the next term by multiplying the coefficient of the given term by the exponent of a and dividing by the number that represents the position of the term in the series of terms. For example, in the expansion of $(a + b)^4$ above, the coefficient of the third term is found from the second term by multiplying 4 and 3; and then dividing by 2 [that is, the coefficient of the third term $= (4 \cdot 3)/2 = 6$].

We now postulate these same properties for the general case:

$$(a + b)^n = a^n + \frac{n}{1}a^{n-1}b + \frac{n(n - 1)}{1 \cdot 2}a^{n-2}b^2 + \frac{n(n - 1)(n - 2)}{1 \cdot 2 \cdot 3}a^{n-3}b^3 + \cdots + b^n$$

$$= \frac{n!}{0!(n - 0)!}a^n + \frac{n!}{1!(n - 1)!}a^{n-1}b + \frac{n!}{2!(n - 2)!}a^{n-2}b^2 + \frac{n!}{3!(n - 3)!}a^{n-3}b^3 + \cdots + \frac{n!}{n!(n - n)!}b^n$$

$$= C_{n,0}a^n + C_{n,1}a^{n-1}b + C_{n,2}a^{n-2}b^2 + C_{n,3}a^{n-3}b^3 + \cdots + C_{n,n}b^n$$

And we are led to the formula in the binomial theorem (a formal proof requires mathematical induction, which is beyond the scope of this book):

THEOREM 2   BINOMIAL THEOREM

For all natural numbers n,

$$(a + b)^n = C_{n,0}a^n + C_{n,1}a^{n-1}b + C_{n,2}a^{n-2}b^2 + C_{n,3}a^{n-3}b^3 + \cdots + C_{n,n}b^n$$

EXAMPLE 3   **Using the Binomial Theorem**   Use the binomial formula to expand $(u + v)^6$.

SOLUTION $(u + v)^6 = C_{6,0}u^6 + C_{6,1}u^5v + C_{6,2}u^4v^2 + C_{6,3}u^3v^3 + C_{6,4}u^2v^4 + C_{6,5}uv^5 + C_{6,6}v^6$
$$= u^6 + 6u^5v + 15u^4v^2 + 20u^3v^3 + 15u^2v^4 + 6uv^5 + v^6$$

MATCHED PROBLEM 3   Use the binomial formula to expand $(x + 2)^5$.

EXAMPLE 4   **Using the Binomial Theorem**   Use the binomial formula to find the sixth term in the expansion of $(x - 1)^{18}$.

SOLUTION

$$\text{Sixth term} = C_{18,5}x^{13}(-1)^5 = \frac{18!}{5!(18-5)!}x^{13}(-1)$$
$$= -8{,}568x^{13}$$

MATCHED PROBLEM 4   Use the binomial formula to find the fourth term in the expansion of $(x-2)^{20}$.

Explore & Discuss **1**   (A) Use the formula for $C_{n,r}$ to find

$$C_{6,0} + C_{6,1} + C_{6,2} + C_{6,3} + C_{6,4} + C_{6,5} + C_{6,6}$$

(B) Write the binomial theorem for $n = 6$, $a = 1$, and $b = 1$, and compare with the results of part (A).

(C) For any natural number n, find

$$C_{n,0} + C_{n,1} + C_{n,2} + \cdots + C_{n,n}$$

Answers to Matched Problems   **1.** (A) 24   (B) 7   (C) 336   **2.** (A) 10   (B) 1
3. $x^5 + 10x^4 + 40x^3 + 80x^2 + 80x + 32$
4. $-9{,}120x^{17}$

Exercise B-3

A   *In Problems 1–20, evaluate each expression.*

1. $6!$ **2.** $7!$ **3.** $\dfrac{10!}{9!}$

4. $\dfrac{20!}{19!}$ **5.** $\dfrac{12!}{9!}$ **6.** $\dfrac{10!}{6!}$

7. $\dfrac{5!}{2!3!}$ **8.** $\dfrac{7!}{3!4!}$ **9.** $\dfrac{6!}{5!(6-5)!}$

10. $\dfrac{7!}{4!(7-4)!}$ **11.** $\dfrac{20!}{3!17!}$ **12.** $\dfrac{52!}{50!2!}$

B

13. $C_{5,3}$ **14.** $C_{7,3}$ **15.** $C_{6,5}$ **16.** $C_{7,4}$
17. $C_{5,0}$ **18.** $C_{5,5}$ **19.** $C_{18,15}$ **20.** $C_{18,3}$

Expand each expression in Problems 21–26 using the binomial formula.

21. $(a + b)^4$ **22.** $(m + n)^5$
23. $(x - 1)^6$ **24.** $(u - 2)^5$
25. $(2a - b)^5$ **26.** $(x - 2y)^5$

Find the indicated term in each expansion in Problems 27–32.

27. $(x - 1)^{18}$; 5th term
28. $(x - 3)^{20}$; 3rd term
29. $(p + q)^{15}$; 7th term
30. $(p + q)^{15}$; 13th term

31. $(2x + y)^{12}$; 11th term
32. $(2x + y)^{12}$; 3rd term

C

33. Show that $C_{n,0} = C_{n,n}$ for $n \geq 0$.

34. Show that $C_{n,r} = C_{n,n-r}$ for $n \geq r \geq 0$.

35. The triangle next is called **Pascal's triangle.** Can you guess what the next two rows at the bottom are? Compare these numbers with the coefficients of binomial expansions.

```
            1
         1     1
       1    2    1
     1    3    3    1
   1    4    6    4    1
```

36. Explain why the sum of the entries in each row of Pascal's triangle is a power of 2. (*Hint:* Let $a = b = 1$ in the binomial theorem.)

37. Explain why the alternating sum of the entries (e.g., $1 - 4 + 6 - 4 + 1$) in each row of Pascal's triangle is equal to 0.

38. Show that $C_{n,r} = \dfrac{n - r + 1}{r} C_{n,r-1}$ for $n \geq r \geq 1$.

39. Show that $C_{n,r-1} + C_{n,r} = C_{n+1,r}$ for $n \geq r \geq 1$.

TABLE I   Basic Geometric Formulas

1. **Similar Triangles**
   (A)   Two triangles are similar if two angles of one triangle have the same measure as two angles of the other.
   (B)   If two triangles are similar, their corresponding sides are proportional:

   $$\frac{a}{a'} = \frac{b}{b'} = \frac{c}{c'}$$

   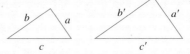

2. **Pythagorean Theorem**

   $$c^2 = a^2 + b^2$$

   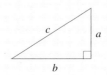

3. **Rectangle**

   $A = ab$ Area
   $P = 2a + 2b$ Perimeter

4. **Parallelogram**

   $h =$ height
   $A = ah = ab \sin \theta$ Area
   $P = 2a + 2b$ Perimeter

   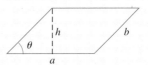

5. **Triangle**

   $h =$ height
   $A = \frac{1}{2}hc$ Area
   $P = a + b + c$ Perimeter
   $s = \frac{1}{2}(a + b + c)$ Semiperimeter
   $A = \sqrt{s(s - a)(s - b)(s - c)}$ Area: Heron's formula

   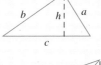
   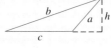

6. **Trapezoid**

   Base a is parallel to base b.

   $h =$ height
   $A = \frac{1}{2}(a + b)h$ Area

   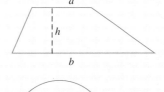

7. **Circle**

   $R =$ radius
   $D =$ diameter
   $D = 2R$
   $A = \pi R^2 = \frac{1}{4}\pi D^2$ Area
   $C = 2\pi R = \pi D$ Circumference
   $\dfrac{C}{D} = \pi$ For all circles
   $\pi \approx 3.141.59$

   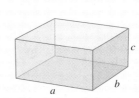

8. **Rectangular Solid**

   $V = abc$ Volume
   $T = 2ab + 2ac + 2bc$ Total surface area

   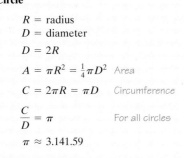

TABLE I *(continued)*

9. Right Circular Cylinder

R = radius of base
h = height
$V = \pi R^2 h$ *Volume*
$S = 2\pi Rh$ *Lateral surface area*
$T = 2\pi R(R + h)$ *Total surface area*

10. Right Circular Cone

R = radius of base
h = height
s = slant height
$V = \frac{1}{3}\pi R^2 h$ *Volume*
$S = \pi Rs = \pi R\sqrt{R^2 + h^2}$ *Lateral surface area*
$T = \pi R(R + s) = \pi R(R + \sqrt{R^2 + h^2})$ *Total surface area*

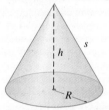

11. Sphere

R = radius
D = diameter
$D = 2R$
$V = \frac{4}{3}\pi R^3 = \frac{1}{6}\pi D^3$ *Volume*
$S = 4\pi R^2 = \pi D^2$ *Surface area*

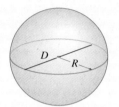

TABLE II   Integration Formulas

[*Note:* The constant of integration is omitted for each integral, but must be included in any particular application of a formula. The variable u is the variable of integration; all other symbols represent constants.]

Integrals Involving u^n

1. $\displaystyle\int u^n \, du = \frac{u^{n+1}}{n+1}, \quad n \neq -1$

2. $\displaystyle\int u^{-1} \, du = \int \frac{1}{u} \, du = \ln|u|$

Integrals Involving $a + bu, a \neq 0$ and $b \neq 0$

3. $\displaystyle\int \frac{1}{a+bu} \, du = \frac{1}{b} \ln|a+bu|$

4. $\displaystyle\int \frac{u}{a+bu} \, du = \frac{u}{b} - \frac{a}{b^2} \ln|a+bu|$

5. $\displaystyle\int \frac{u^2}{a+bu} \, du = \frac{(a+bu)^2}{2b^3} - \frac{2a(a+bu)}{b^3} + \frac{a^2}{b^3} \ln|a+bu|$

6. $\displaystyle\int \frac{u}{(a+bu)^2} \, du = \frac{1}{b^2}\left(\ln|a+bu| + \frac{a}{a+bu} \right)$

7. $\displaystyle\int \frac{u^2}{(a+bu)^2} \, du = \frac{(a+bu)}{b^3} - \frac{a^2}{b^3(a+bu)} - \frac{2a}{b^3} \ln|a+bu|$

8. $\displaystyle\int u(a+bu)^n \, du = \frac{(a+bu)^{n+2}}{(n+2)b^2} - \frac{a(a+bu)^{n+1}}{(n+1)b^2}, \quad n \neq -1, -2$

9. $\displaystyle\int \frac{1}{u(a+bu)} \, du = \frac{1}{a} \ln\left|\frac{u}{a+bu}\right|$

10. $\displaystyle\int \frac{1}{u^2(a+bu)} \, du = -\frac{1}{au} + \frac{b}{a^2} \ln\left|\frac{a+bu}{u}\right|$

11. $\displaystyle\int \frac{1}{u(a+bu)^2} \, du = \frac{1}{a(a+bu)} + \frac{1}{a^2} \ln\left|\frac{u}{a+bu}\right|$

12. $\displaystyle\int \frac{1}{u^2(a+bu)^2} \, du = -\frac{a+2bu}{a^2u(a+bu)} + \frac{2b}{a^3} \ln\left|\frac{a+bu}{u}\right|$

Integrals Involving $a^2 - u^2, a > 0$

13. $\displaystyle\int \frac{1}{u^2 - a^2} \, du = \frac{1}{2a} \ln\left|\frac{u-a}{u+a}\right|$

14. $\displaystyle\int \frac{1}{a^2 - u^2} \, du = \frac{1}{2a} \ln\left|\frac{u+a}{u-a}\right|$

Integrals Involving $(a+bu)$ and $(c+du), b \neq 0, d \neq 0$, and $ad - bc \neq 0$

15. $\displaystyle\int \frac{1}{(a+bu)(c+du)} \, du = \frac{1}{ad-bc} \ln\left|\frac{c+du}{a+bu}\right|$

16. $\displaystyle\int \frac{u}{(a+bu)(c+du)} \, du = \frac{1}{ad-bc}\left(\frac{a}{b} \ln|a+bu| - \frac{c}{d} \ln|c+du| \right)$

17. $\displaystyle\int \frac{u^2}{(a+bu)(c+du)} \, du = \frac{1}{bd}u - \frac{1}{ad-bc}\left(\frac{a^2}{b^2} \ln|a+bu| - \frac{c^2}{d^2} \ln|c+du| \right)$

18. $\displaystyle\int \frac{1}{(a+bu)^2(c+du)} \, du = \frac{1}{ad-bc}\frac{1}{a+bu} + \frac{d}{(ad-bc)^2} \ln\left|\frac{c+du}{a+bu}\right|$

19. $\displaystyle\int \frac{u}{(a+bu)^2(c+du)} \, du = -\frac{a}{b(ad-bc)}\frac{1}{a+bu} - \frac{c}{(ad-bc)^2} \ln\left|\frac{c+du}{a+bu}\right|$

20. $\displaystyle\int \frac{a+bu}{c+du} \, du = \frac{bu}{d} + \frac{ad-bc}{d^2} \ln|c+du|$

(continued)

TABLE II (continued)

Integrals Involving $\sqrt{a + bu}, a \neq 0$ and $b \neq 0$

21. $\int \sqrt{a + bu}\, du = \dfrac{2\sqrt{(a + bu)^3}}{3b}$

22. $\int u\sqrt{a + bu}\, du = \dfrac{2(3bu - 2a)}{15b^2}\sqrt{(a + bu)^3}$

23. $\int u^2\sqrt{a + bu}\, du = \dfrac{2(15b^2u^2 - 12abu + 8a^2)}{105b^3}\sqrt{(a + bu)^3}$

24. $\int \dfrac{1}{\sqrt{a + bu}}\, du = \dfrac{2\sqrt{a + bu}}{b}$

25. $\int \dfrac{u}{\sqrt{a + bu}}\, du = \dfrac{2(bu - 2a)}{3b^2}\sqrt{a + bu}$

26. $\int \dfrac{u^2}{\sqrt{a + bu}}\, du = \dfrac{2(3b^2u^2 - 4abu + 8a^2)}{15b^3}\sqrt{a + bu}$

27. $\int \dfrac{1}{u\sqrt{a + bu}}\, du = \dfrac{1}{\sqrt{a}}\ln\left|\dfrac{\sqrt{a + bu} - \sqrt{a}}{\sqrt{a + bu} + \sqrt{a}}\right|, \quad a > 0$

28. $\int \dfrac{1}{u^2\sqrt{a + bu}}\, du = -\dfrac{\sqrt{a + bu}}{au} - \dfrac{b}{2a\sqrt{a}}\ln\left|\dfrac{\sqrt{a + bu} - \sqrt{a}}{\sqrt{a + bu} + \sqrt{a}}\right|, \quad a > 0$

Integrals Involving $\sqrt{a^2 - u^2}, a > 0$

29. $\int \dfrac{1}{u\sqrt{a^2 - u^2}}\, du = -\dfrac{1}{a}\ln\left|\dfrac{a + \sqrt{a^2 - u^2}}{u}\right|$

30. $\int \dfrac{1}{u^2\sqrt{a^2 - u^2}}\, du = -\dfrac{\sqrt{a^2 - u^2}}{a^2 u}$

31. $\int \dfrac{\sqrt{a^2 - u^2}}{u}\, du = \sqrt{a^2 - u^2} - a\ln\left|\dfrac{a + \sqrt{a^2 - u^2}}{u}\right|$

Integrals Involving $\sqrt{u^2 + a^2}, a > 0$

32. $\int \sqrt{u^2 + a^2}\, du = \dfrac{1}{2}\left(u\sqrt{u^2 + a^2} + a^2\ln|u + \sqrt{u^2 + a^2}|\right)$

33. $\int u^2\sqrt{u^2 + a^2}\, du = \dfrac{1}{8}\left[u(2u^2 + a^2)\sqrt{u^2 + a^2} - a^4\ln|u + \sqrt{u^2 + a^2}|\right]$

34. $\int \dfrac{\sqrt{u^2 + a^2}}{u}\, du = \sqrt{u^2 + a^2} - a\ln\left|\dfrac{a + \sqrt{u^2 + a^2}}{u}\right|$

35. $\int \dfrac{\sqrt{u^2 + a^2}}{u^2}\, du = -\dfrac{\sqrt{u^2 + a^2}}{u} + \ln|u + \sqrt{u^2 + a^2}|$

36. $\int \dfrac{1}{\sqrt{u^2 + a^2}}\, du = \ln|u + \sqrt{u^2 + a^2}|$

37. $\int \dfrac{1}{u\sqrt{u^2 + a^2}}\, du = \dfrac{1}{a}\ln\left|\dfrac{u}{a + \sqrt{u^2 + a^2}}\right|$

38. $\int \dfrac{u^2}{\sqrt{u^2 + a^2}}\, du = \dfrac{1}{2}\left(u\sqrt{u^2 + a^2} - a^2\ln|u + \sqrt{u^2 + a^2}|\right)$

39. $\int \dfrac{1}{u^2\sqrt{u^2 + a^2}}\, du = -\dfrac{\sqrt{u^2 + a^2}}{a^2 u}$

Integrals Involving $\sqrt{u^2 - a^2}, a > 0$

40. $\int \sqrt{u^2 - a^2}\, du = \dfrac{1}{2}\left(u\sqrt{u^2 - a^2} - a^2\ln|u + \sqrt{u^2 - a^2}|\right)$

TABLE II *(concluded)*

41. $\int u^2 \sqrt{u^2 - a^2}\, du = \frac{1}{8}\left[u(2u^2 - a^2)\sqrt{u^2 - a^2} - a^4 \ln|u + \sqrt{u^2 - a^2}|\right]$

42. $\int \frac{\sqrt{u^2 - a^2}}{u^2}\, du = -\frac{\sqrt{u^2 - a^2}}{u} + \ln|u + \sqrt{u^2 - a^2}|$

43. $\int \frac{1}{\sqrt{u^2 - a^2}}\, du = \ln|u + \sqrt{u^2 - a^2}|$

44. $\int \frac{u^2}{\sqrt{u^2 - a^2}}\, du = \frac{1}{2}\left(u\sqrt{u^2 - a^2} + a^2 \ln|u + \sqrt{u^2 - a^2}|\right)$

45. $\int \frac{1}{u^2\sqrt{u^2 - a^2}}\, du = \frac{\sqrt{u^2 - a^2}}{a^2 u}$

Integrals Involving $e^{au}, a \neq 0$

46. $\int e^{au}\, du = \frac{e^{au}}{a}$

47. $\int u^n e^{au}\, du = \frac{u^n e^{au}}{a} - \frac{n}{a}\int u^{n-1} e^{au}\, du$

48. $\int \frac{1}{c + de^{au}}\, du = \frac{u}{c} - \frac{1}{ac}\ln|c + de^{au}|, \quad c \neq 0$

Integrals Involving ln u

49. $\int \ln u\, du = u \ln u - u$

50. $\int \frac{\ln u}{u}\, du = \frac{1}{2}(\ln u)^2$

51. $\int u^n \ln u\, du = \frac{u^{n+1}}{n+1}\ln u - \frac{u^{n+1}}{(n+1)^2}, \quad n \neq -1$

52. $\int (\ln u)^n\, du = u(\ln u)^n - n\int (\ln u)^{n-1}\, du$

Integrals Involving Trigonometric Functions of $au, a \neq 0$

53. $\int \sin au\, du = -\frac{1}{a}\cos au$

54. $\int \cos au\, du = \frac{1}{a}\sin au$

55. $\int \tan au\, du = -\frac{1}{a}\ln|\cos au|$

56. $\int \cot au\, du = \frac{1}{a}\ln|\sin au|$

57. $\int \sec au\, du = \frac{1}{a}\ln|\sec au + \tan au|$

58. $\int \csc au\, du = \frac{1}{a}\ln|\csc au - \cot au|$

59. $\int (\sin au)^2\, du = \frac{u}{2} - \frac{1}{4a}\sin 2au$

60. $\int (\cos au)^2\, du = \frac{u}{2} + \frac{1}{4a}\sin 2au$

61. $\int (\sin au)^n\, du = -\frac{1}{an}(\sin au)^{n-1}\cos au + \frac{n-1}{n}\int (\sin au)^{n-2}\, du, \quad n \neq 0$

62. $\int (\cos au)^n\, du = \frac{1}{an}\sin au(\cos au)^{n-1} + \frac{n-1}{n}\int (\cos au)^{n-2}\, du, \quad n \neq 0$

CHAPTER 1

Exercise 1-1

1. $m = 5$ **3.** $x < -\frac{7}{2}$ **5.** $x \le 4$ **7.** $x < -3$ or $(-\infty, -3)$ **9.** $-1 \le x \le 2$ or $[-1, 2]$

11. $x = \frac{5}{2}$ **13.** $x > -\frac{15}{4}$ **15.** $y = 8$ **17.** $x = 10$ **19.** $y \ge 3$ **21.** $x = 36$ **23.** $m < \frac{36}{7}$ **25.** $x = 10$

27. $3 \le x < 7$ or $[3, 7)$ 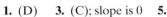 **29.** $-20 \le C \le 20$ or $[-20, 20]$ **31.** $y = \frac{3}{4}x - 3$

33. $y = -(A/B)x + (C/B) = (-Ax + C)/B$ **35.** $C = \frac{5}{9}(F - 32)$ **37.** $B = 3A/2(m - n)$

39. $-2 < x \le 1$ or $(-2, 1]$ **41.** (A) and (C): $a > 0$ and $b > 0$, or $a < 0$ and $b < 0$ (B) and (D): $a > 0$ and $b < 0$, or $a < 0$ and $b > 0$

43. (A) $>$ (B) $<$ **45.** Negative **47.** True **49.** False **51.** True **53.** 3,500 \$15 tickets; 4,500 \$25 tickets

55. \$180,000 in Fund A and \$320,000 in Fund B **57.** \$22,191 **59.** (A) \$420 (B) \$55 **61.** 34 rounds **63.** \$32,000

65. 5,851 books **67.** (B) 6,180 books (C) At least \$11.50 **69.** 5,000 **71.** 12.6 yr

73. Cephalic index is between 66.7 cm and 100 cm

Exercise 1-2

1. (D) **3.** (C); slope is 0 **5.** 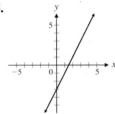 **7.** 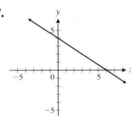 **9.** Slope $= 3$; y intercept $= 1$

11. Slope $= -\frac{3}{7}$; y intercept $= -6$

13. $y = -2x + 3$ **15.** $y = \frac{4}{3}x - 4$

17. x intercept: -1; y intercept: -2; $y = -2x - 2$

19. x intercept: -3; y intercept: 1; $y = x/3 + 1$

21. 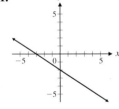 **23.** 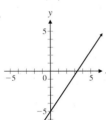 **25.** 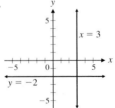 **27.** -4 **29.** $-\frac{3}{5}$ **31.**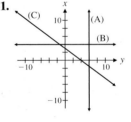

33. 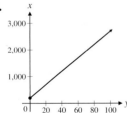 **35.** (A) 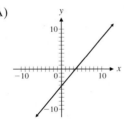 (B) x intercept: 3.5; y intercept: -4.2 (C) 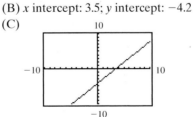 (D) x intercept: 3.5; y intercept: -4.2 (E) $x > 3.5$, or $(3.5, \infty)$

37. $x = 4, y = -3$ **39.** $x = -1.5, y = -3.5$ **41.** $y = -4x + 5$ **43.** $y = \frac{3}{2}x + 1$ **45.** $y = 4.6$

47. (A) $m = \frac{2}{3}$ (B) $-2x + 3y = 11$ (C) Linear function **49.** (A) $m = -\frac{5}{4}$ (B) $5x + 4y = -14$ (C) Linear function

51. (A) Not defined (B) $x = 5$ (C) Neither **53.** (A) $m = 0$ (B) $y = 5$ (C) Constant function

55. The graphs have the same y intercept, $(0, 2)$.

57. (A) $130; $220

(B)
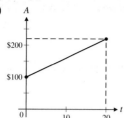

59. $C = 124 + 0.12x$; 1,050 doughnuts

61. (A) $C = 75x + 1,647$

(B)
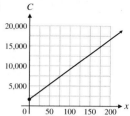

(C) The y intercept, $1,647, is the fixed cost and the slope, $75, is the cost per club.

63. (A) $R = 1.4C - 7$
(B) $137

65. (A) $V = -7,500t + 157,000$
(B) $112,000
(C) During the 12th year
(D)
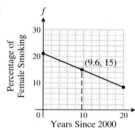

67. (A) $T = -1.84x + 212$
(B) $205.56°F$
(C) 6,522 ft
(D)
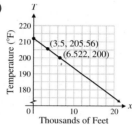

69. (A) $T = 70 - 3.6A$
(B) 10,000 ft

71. $N = -0.008t + 2.8$; 2.4 persons

73. (A) $f = -0.625t + 21$
(B) 2009
(C)

75. (A) $p = 0.000225x + 0.5925$
(B) $p = -0.0009x + 9.39$
(C) $(7,820, 2.352)$
(D)
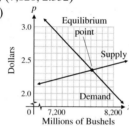

77. (A) $s = \dfrac{2}{5}w$
(B) 8 in
(C) 9 lb

79. (A) $y = 0.5x + 9$
(B) $y = 0.3x + 9$
(C)
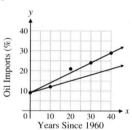

(D) A: 39%; B: 27%

Section 1-3

1. (A) $w = 49 + 1.7h$ (B) The rate of change of weight with respect to height is 1.7 inches per kilogram. (C) 55.8 kg (D) $5'6.5''$

3. (A) $p = 0.44\overline{5}d + 14.7$ (B) The rate of change of pressure with respect to depth is $0.44\overline{5}$ lb/in² per foot. (C) 37 lb/in² (D) 99 ft

5. (A) $a = 2,880 - 24t$ (B) -24 ft/sec (C) 24 ft/sec

7. $s = 0.6t + 331$; the rate of change of the speed of sound with respect to temperature is 0.6 m/s per °C.

9. (A)
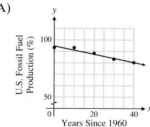

(B) The rate of change of the percentage of oil imports with respect to time is -0.36% per year.
(C) 76.6% of total production
(D) 2029

11. (A)
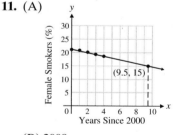

(B) 2009

13. (A)
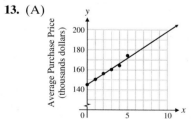

(B) $198,000
(C) The rate of change of average price with respect to time is $5,300 per year.

15. (A)
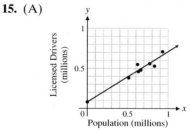

(B) 962,000
(C) 1,048,000

17. (A)
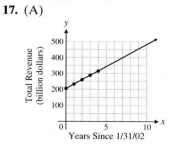

(B) $425.8 billion

19. (A)

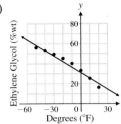

(B) 2°F

(C) 22.75%

21. (A) The rate of change of height with respect to Dbh is 1.37 ft/in.
(B) Height increases by approximately 1.37 ft.
(C) 18 ft (D) 20 in

23. (A) The rate of change average monthly price is $1.56 per year.
(B) $55 per month

25. (A) Male enrollment is increasing at the rate of 40,000 students per year; female enrollment is increasing at the rate of 160,000 students per year.
(B) Male: 7 million; female: 10.5 million   (C) 1981

27. Men: $y = -0.103x + 51.509$; women: $y = -0.145x + 58.286$ The graphs of these linear regression models indicate that women will catch up to men in the year 2129.

29. Supply: $y = 0.2x + 0.87$; demand: $y = -0.15x + 3.5$; equilibrium price = $2.37

Chapter 1 Review Exercise

1. $x = 2.8$ *(1-1)*   **2.** $x = 2$ *(1-1)*   **3.** $y = 1.8 - 0.4x$ *(1-1)*   **4.** $x = \frac{4}{3}y + \frac{7}{3}$ *(1-1)*   **5.** $y < \frac{13}{4}$ or $\left(-\infty, \frac{13}{4}\right)$ *(1-1)*

6. $1 \le x < 3$ or $[1, 3)$ 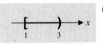 *(1-1)*   **7.** $x \ge \frac{9}{2}$ or $\left[\frac{9}{2}, \infty\right)$ 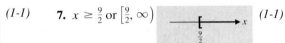 *(1-1)*

8.   *(1-2)*

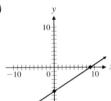

9. $2x + 3y = 12$ *(1-2)*

10. x intercept = 9; y intercept = −6;
slope = $\frac{2}{3}$ *(1-2)*

11. $y = -\frac{2}{3}x + 6$ *(1-2)*

12. Vertical line: $x = -6$; horizontal line: $y = 5$ *(1-2)*

13. (A) $y = -\frac{2}{3}x$ (B) $y = 3$ *(1-2)*

14. (A) $3x + 2y = 1$ (B) $y = 5$
(C) $x = -2$ *(1-2)*

15. $x = \frac{25}{2}$ *(1-1)*   **16.** $u = 36$ *(1-1)*

17. $x = \frac{30}{11}$ *(1-1)*   **18.** $x = 21$ *(1-1)*   **19.** $x = 4$ *(1-1)*

20. $x < 4$ or $(-\infty, 4)$ *(1-1)*

21. $x \ge 1$ or $[1, \infty)$ *(1-1)*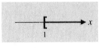

22. $x < -\frac{143}{17}$ or $\left(-\infty, -\frac{143}{17}\right)$ 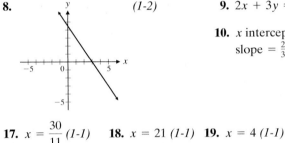 *(1-1)*

23. $1 < x \le 4$ or $(1, 4]$ *(1-1)*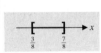

24. $\frac{3}{8} \le x \le \frac{7}{8}$ or $\left[\frac{3}{8}, \frac{7}{8}\right]$ *(1-1)*

25.   *(1-2)*

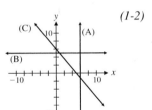

26. The graph of $x = -3$ is a vertical line with x intercept −3, and the graph of $y = 2$ is a horizontal line with y intercept 2. *(1-2)*

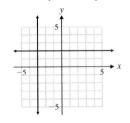

27. (A) An oblique line through the origin with slope −3/4   (B) A vertical line with x intercept −4/3   (C) The x axis   (D) An oblique line with x intercept 12 and y intercept 9 *(1-2)*

28. $\frac{2A - bh}{h}$ *(1-1)*   **29.** $\frac{S - P}{St}$ *(1-1)*   **30.** $a < 0$ and b any real number *(1-1)*   **31.** Less than *(1-1)*

32. The graphs appear to be perpendicular to each other. (It can be shown that if the slopes of two slant lines are the negative reciprocals of each other, then the two lines are perpendicular.) *(1-2)*

33. $75,000 *(1-1)*   **34.** 9,375 CDs *(1-1)*   **35.** (A) $m = 132 - 0.6x$ (B) $M = 187 - 0.85x$
(C) Between 120 and 170 beats per minute (D) Between 102 and 144.5 beats per minute *(1-3)*

36. (A) $V = 224,000 - 15,500t$ (B) $38,000 *(1-3)*   **37.** (A) $R = 0.16C$ (B) $192   (C) $110
(D) The slope is 1.6. This is the rate of change of retail price with respect to cost. *(1-3)*

38. $200; $400 *(1-1)*   **39.** Demand: $p = 5.24 - 0.00125x$; 1,560 bottles *(1-3)*

40. (A)

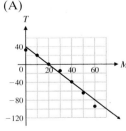

(B) −30°F   (C) 45% *(1-3)*

41. (A) Weekly Internet usage for men is increasing at an annual rate of 1.3 hours per week and for women at an annual rate of 1.5 hours per week.

(C) 2007 *(1-3)*

(B)

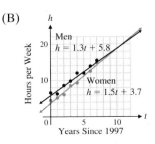

42. (A) The CPI is increasing at the rate of $4.10 per year.
(B) $213.60 *(1-3)*

43. (A) The rate of change of tree height with respect to Dbh is 0.74
(B) Tree height increases by about 0.74 feet.
(C) 21 ft   (D) 16 in *(1-3)*

CHAPTER 2

Exercise 2-1

1. 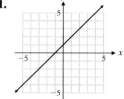   **3.** 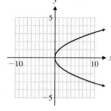   **5.** 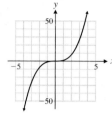   **7.**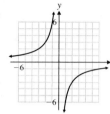

9. Function
11. Not a function
13. Function
15. Function
17. Not a function
19. Function
21. Linear
23. Constant

25. Neither
27. Linear
29. Constant

33. 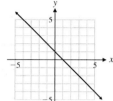   **35.** 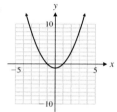   **37.** 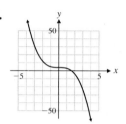   **39.**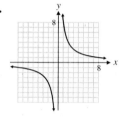

41. 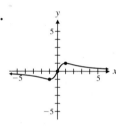   **43.** (A)

x	0	5	10	15	20
f(x)	0	375	500	375	0
g(x)	150	250	350	450	550
f(x) − g(x)	−150	125	150	−75	−550

(B)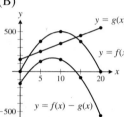

45. $y = 0$
47. $y = 4$
49. $x = -5, 0, 4$
51. $x = -6$
53. 1   **55.** −5
57. 3   **59.** 7
61. −45   **63.** 0
65. All real numbers

67. All real numbers except −4   **69.** $x \leq 7$ **71.** $f(2) = 0$, and 0 is a number; therefore, $f(2)$ exists. On the other hand, $f(3)$ is not defined, since the denominator would be 0; therefore, we say that $f(3)$ does not exist.   **73.** $g(x) = 2x^3 - 5$ **75.** $G(x) = 2\sqrt{x} - x^2$
77. Function f multiplies the domain element by 2 and subtracts 3 from the result.   **79.** Function F multiplies the cube of the domain element by 3 and subtracts twice the square root of the domain element from the result.   **81.** A function with domain R
83. A function with domain R **85.** Not a function; for example, when $x = 1$, $y = \pm 3$ **87.** A function with domain all real numbers except $x = 4$ **89.** Not a function; for example, when $x = 4$, $y = \pm 3$ **91.** 4   **93.** $h - 1$ **95.** 24   **97.** 48   **99.** 27   **101.** −1
103. $4x^2 - 1$ **105.** $x^2 + 2x$ **107.** (A) $4x + 4h - 3$ (B) $4h$ (C) 4   **109.** (A) $4x^2 + 8xh + 4h^2 - 7x - 7h + 6$
(B) $8xh + 4h^2 - 7h$ (C) $8x + 4h - 7$ **111.** (A) $20x + 20h - x^2 - 2xh - h^2$ (B) $20h - 2xh - h^2$ (C) $20 - 2x - h$

113. $P(w) = 2w + \dfrac{50}{w}, w > 0$ **115.** $A(l) = l(50 - l), 0 \leq l \leq 50$

117. $54, $42   **119.** (A) $R(x) = (75 - 3x)x,$
$1 \leq x \leq 20$

(B)
x	R(x)
1	72
4	252
8	408
12	468
16	432
20	300

(C)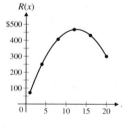

121. (A) $P(x) = 59x - 3x^2 - 125,$
$1 \leq x \leq 20$

(B)
x	P(x)
1	−69
4	63
8	155
12	151
16	51
20	−145

(C)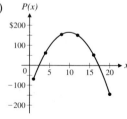

123. (A) $V(x) = x(8 - 2x)(12 - 2x)$
(B) $0 \leq x \leq 4$

(C)
x	V(x)
1	60
2	64
3	36

(D)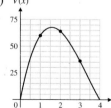

125. (A) The graph indicates that there is a value of x near 2 that will produce a volume of 65. The table shows $x = 1.9$ to one decimal place:

x	1.8	1.9	2
$V(x)$	66.5	65.4	64

(B) $x = 1.93$ to two decimal places

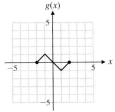

127. $v = \dfrac{75 - w}{15 + w}$; 1.9032 cm/sec

Exercise 2-2

1. Domain: all real numbers; range: all real numbers   **3.** Domain: $[0, \infty)$; range: $(-\infty, 0]$

5. Domain: all real numbers; range: $[0, \infty)$ **7.** Domain: all real numbers; range: all real numbers

9. 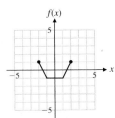   **11.** 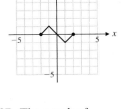   **13.** 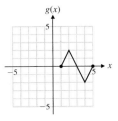   **15.**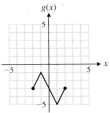

17. 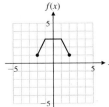   **19.**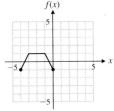

21. The graph of $g(x) = -|x + 3|$ is the graph of $y = |x|$ reflected in the x axis and shifted 3 units to the left.

23. The graph of $f(x) = (x - 4)^2 - 3$ is the graph of $y = x^2$ shifted 4 units to the right and 3 units down.

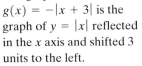

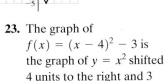

25. The graph of $f(x) = 7 - \sqrt{x}$ is the graph of $y = \sqrt{x}$ reflected in the x axis and shifted 7 units up.

27. The graph of $h(x) = -3|x|$ is the graph of $y = |x|$ reflected in the x axis and vertically expanded by a factor of 3.

29. The graph of the basic function $y = x^2$ is shifted 2 units to the left and 3 units down. Equation: $y = (x + 2)^2 - 3$.

31. The graph of the basic function $y = x^2$ is reflected in the x axis and shifted 3 units to the right and 2 units up. Equation: $y = 2 - (x - 3)^2$.

33. The graph of the basic function $y = \sqrt{x}$ is reflected in the x axis and shifted 4 units up. Equation: $y = 4 - \sqrt{x}$.

35. The graph of the basic function $y = x^3$ is shifted 2 units to the left and 1 unit down. Equation: $y = (x + 2)^3 - 1$.

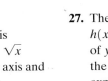

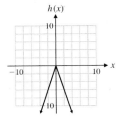

37. $g(x) = \sqrt{x - 2} - 3$ **39.** $g(x) = -|x + 3|$ **41.** $g(x) = -(x - 2)^3 - 1$ **43.**

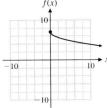   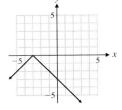   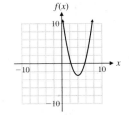   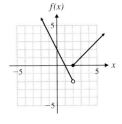

45. 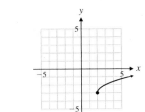   **47.**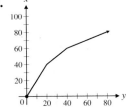

49. The graph of the basic function $y = |x|$ is reflected in the x axis and vertically contracted by a factor of 0.5. Equation: $y = -0.5|x|$.

51. The graph of the basic function $y = x^2$ is reflected in the x axis and vertically expanded by a factor of 2. Equation: $y = -2x^2$.

53. The graph of the basic function $y = \sqrt[3]{x}$ is reflected in the x axis and vertically expanded by a factor of 3. Equation: $y = -3\sqrt[3]{x}$.

55. Reversing the order does not change the result.
57. Reversing the order can change the result.
59. Reversing the order can change the result.

61. (A) The graph of the basic function $y = \sqrt{x}$ is (B)
reflected in the x axis, vertically expanded
by a factor of 4, and shifted up 115 units.

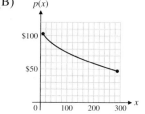

63. (A) The graph of the basic function $y = x^3$ is
vertically contracted by a factor of 0.000 48
and shifted right 500 units and up 60,000 units.

65. (A) $S(x) = \begin{cases} 8.5 + 0.065x & \text{if } 0 \le x \le 700 \\ -9 + 0.09x & \text{if } x > 700 \end{cases}$

(B)

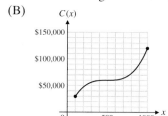

(B)

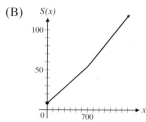

67. (A) $T(x) = \begin{cases} 0.035x & \text{if } 0 \le x \le 30{,}000 \\ 0.0625x - 825 & \text{if } 30{,}000 < x \le 60{,}000 \\ 0.0645x - 945 & \text{if } x > 60{,}000 \end{cases}$

(B)

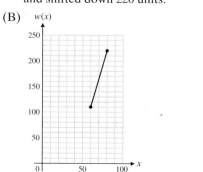

(C) $1,675; $3,570

69. (A) The graph of the basic
function $y = x$ is vertically
expanded by a factor of 5.5
and shifted down 220 units.

(B)

71. (A) The graph of the
basic function
$y = \sqrt{x}$ is vertically
expanded by a
factor of 7.08.

(B)

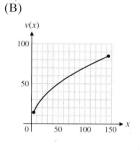

Exercise 2-3

1. $(x - 2)^2 - 1$ **3.** $-(x - 3)^2 + 5$ **5.** The graph of $f(x)$ is the graph of $y = x^2$ shifted right 2 units and down 1 unit.
7. The graph of $m(x)$ is the graph of $y = x^2$ reflected in the x axis, then shifted right 3 units and up 5 units.
9. (A) m (B) g (C) f (D) n
11. (A) x intercepts: 1, 3; y intercept: -3 (B) Vertex: $(2, 1)$ (C) Maximum: 1 (D) Range: $y \le 1$ or $(-\infty, 1]$
  (E) Increasing interval: $x \le 2$ or $(-\infty, 2]$ (F) Decreasing interval: $x \ge 2$ or $[2, \infty)$
13. (A) x intercepts: $-3, -1$; y intercept: 3 (B) Vertex: $(-2, -1)$ (C) Minimum: -1 (D) Range: $y \ge -1$ or $[-1, \infty)$
  (E) Increasing interval: $x \ge -2$ or $[-2, \infty)$ (F) Decreasing interval: $x \le -2$ or $(-\infty, -2]$
15. (A) x intercepts: $3 \pm \sqrt{2}$, y intercept: -7 (B) Vertex: $(3, 2)$ (C) Maximum: 2 (D) Range: $y \le 2$ or $(-\infty, 2]$
17. (A) x intercepts: $-1 \pm \sqrt{2}$, y intercept: -1 (B) Vertex: $(-1, -2)$ (C) Minimum: -2 (D) Range: $y \ge -2$ or $[-2, \infty)$
19. $y = -[x - (-2)]^2 + 5$ or $y = -(x + 2)^2 + 5$ **21.** $y = (x - 1)^2 - 3$
23. Standard form: $(x - 4)^2 - 4$ (A) x intercepts: 2 and 6, y intercept: 12 (B) Vertex: $(4, -4)$ (C) Minimum: -4
  (D) Range: $y \ge -4$ or $[-4, \infty)$
25. Standard form: $-4(x - 2)^2 + 1$ (A) x intercepts: 1.5 and 2.5, y intercept: -15 (B) Vertex: $(2, 1)$ (C) Maximum: 1
  (D) Range: $y \le 1$ or $(-\infty, 1]$
27. Standard form: $0.5(x - 2)^2 + 3$ (A) x intercepts: none, y intercept: 5 (B) Vertex: $(2, 3)$ (C) Minimum: 3
  (D) Range: $y \ge 3$ or $[3, \infty)$
29. (A) $-4.87, 8.21$ (B) $-3.44, 6.78$ (C) No solution
31. 651.0417 **33.** The vertex of the parabola is on the x axis.
35. $g(x) = 0.25(x - 3)^2 - 9.25$ (A) x intercepts: $-3.08, 9.08$; y intercept: -7 (B) Vertex: $(3, -9.25)$ (C) Minimum: -9.25
  (D) Range: $y \ge -9.25$ or $[-9.25, \infty)$
37. $f(x) = -0.12(x - 4)^2 + 3.12$ (A) x intercepts: $-1.1, 9.1$; y intercept: 1.2 (B) Vertex: $(4, 3.12)$ (C) Maximum: 3.12
  (D) Range: $y \le 3.12$ or $(-\infty, 3.12]$
39. $x = -5.37, 0.37$ **41.** $-1.37 < x < 2.16$ **43.** $x \le -0.74$ or $x \ge 4.19$
45. Axis: $x = 2$; vertex: $(2, 4)$; range: $y \ge 4$ or $[4, \infty)$; no x intercepts or $7.61 < x \le 10$

47. (A)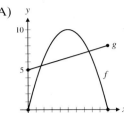

(B) 1.64, 7.61
(C) $1.64 < x < 7.61$
(D) $0 \le x < 1.64$

49. (A)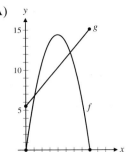

(B) 1.10, 5.57
(C) $1.10 < x < 5.57$
(D) $0 \le x < 1.10$ or
$5.57 < x \le 8$

51. $f(x) = x^2 + 1$ and $g(x) = -(x - 4)^2 - 1$ are two examples. The graphs do not cross the x axis.

53. (A)

x	28	30	32	34	36
Mileage	45	52	55	51	47
$f(x)$	45.3	51.8	54.2	52.4	46.5

(B)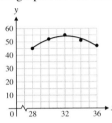

(C) $f(31) = 53.50$ thousand miles; $f(35) = 49.95$ thousand miles

57. (A)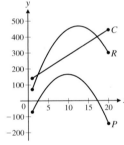

(B) 12,500,000 chips;
$468,750,000
(C) $37.50

59. (A)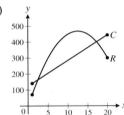

(B) 2,415,000 chips and 17,251,000 chips
(C) Loss: $1 \le x < 2.415$ or $17.251 < x \le 20$; profit: $2.415 < x < 17.251$

61. (A) $P(x) = 59x - 3x^2 - 125$

(C) Intercepts and break-even points: 2,415,000 chips and 17,251,000 chips
(E) Maximum profit is $165,083,000 at a production level of 9,833,000 chips. This is much smaller than the maximum revenue of $468,750,000.

63. $x = 0.14$ cm
65. 10.6 mph

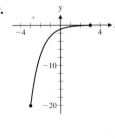

Exercise 2-4

1. (A) k
(B) g
(C) h
(D) f

3.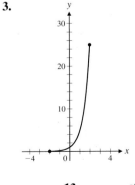

5.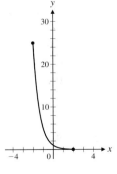

7.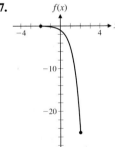

9.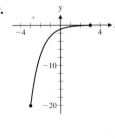

11.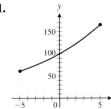

13.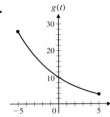

15. 4^{6xy} **17.** e **19.** $8e^{3.6t}$
21. The graph of g is a reflection of the graph f in the x axis.
23. The graph of g is the graph of f shifted 1 unit to the left.
25. The graph of g is the graph of f shifted 1 unit up.
27. The graph of g is the graph of f vertically expanded by a factor of 2 and shifted to the left 2 units.

29. (A)
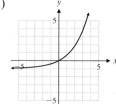
(B)
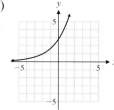
(C)
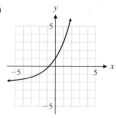
(D)
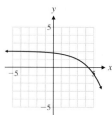

31.
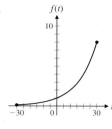

33.
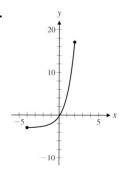

35.
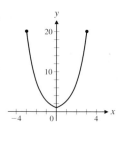

37.
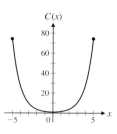

39.
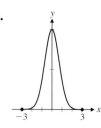

41. $a = 1, -1$
43. $x = 1$
45. $x = -1, 6$
47. $x = 3$
49. $x = 3$
51. $x = -3, 0$

53.

55.
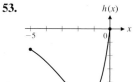

57. $x = 1.40$
59. $x = -0.73$
61. (A) \$2,633.56
  (B) \$7,079.54
63. (A) \$11,871.65
  (B) \$20,427.93
65. (A) \$10,553.57
  (B) \$10,507.42
  (C) \$10,525.03

67. N approaches 2 as t increases without bound.

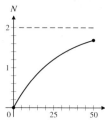

69. (A) \$7,674,000

(B) The model gives an average salary of \$1,773,000 in 2000. Inclusion of the data for 2000 gives an average salary of \$7,997,000 in 2015.

71. (A) 10% (B) 1%
73. (A) $N = 65e^{0.02t}$
  (B) 2002: 60,000,000; 2014: 76,000,000
  (C)

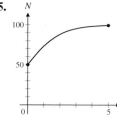

75. (A) $P = 6.5e^{0.0114t}$
  (B) 2015: 7.2 billion; 2030: 8.5 billion
  (C)

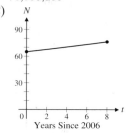

77. (A) 24,238,000,000

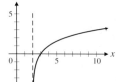

Exercise 2-5

1. $27 = 3^3$ **3.** $10^0 = 1$ **5.** $8 = 4^{3/2}$ **7.** $\log_7 49 = 2$ **9.** $\log_4 8 = \frac{3}{2}$ **11.** $\log_b A = u$ **13.** 0 **15.** 1 **17.** 1 **19.** 3
21. -3 **23.** 3 **25.** $\log_b P - \log_b Q$ **27.** $5 \log_b L$ **29.** q^p **31.** $x = 9$ **33.** $y = 2$ **35.** $b = 10$ **37.** $x = 2$ **39.** $y = -2$
41. $b = 100$ **43.** False **45.** True **47.** True **49.** False **51.** False **53.** $x = 2$ **55.** $x = 8$ **57.** $x = 7$ **59.** No solution
61.

63. The graph of $y = \log_2(x - 2)$ is the graph of $y = \log_2 x$ shifted to the right 2 units.
65. Domain: $(-1, \infty)$; range: all real numbers
67. (A) 3.547 43 (B) $-2.160\,32$ (C) 5.626 29 (D) $-3.197\,04$
69. (A) 13.4431 (B) 0.0089 (C) 16.0595 (D) 0.1514
71. 1.0792 **73.** 1.4595 **75.** 30.6589 **77.** 18.3559

79. Increasing: $(0, \infty)$

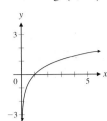

81. Decreasing: $(0, 1]$
Increasing: $[1, \infty)$

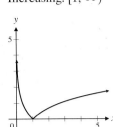

83. Increasing: $(-2, \infty)$

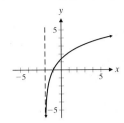

85. Increasing: $(0, \infty)$

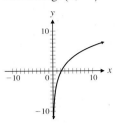

87. Because $b^0 = 1$ for any permissible base $b(b > 0, b \neq 1)$.

89. $y = c \cdot 10^{0.8x}$

91. $x > \sqrt{x} > \ln x$ for $1 < x \leq 16$

93. 4 yr **95.** 9.87 yr; 9.80 yr

97. (A) 5,373

(B) 7,220

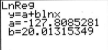

101. 153.9 bushels/acre

103. 891 yr

Chapter 2 Review Exercise

1. *(2-1)*

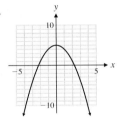

2. *(2-1)*

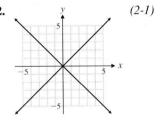

3. *(2-1)*

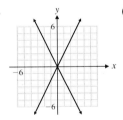

4. (A) Not a function
(B) A function
(C) A function
(D) Not a function *(2-1)*

5. (A) -2 (B) -8
(C) 0 (D) Not defined *(2-1)*

6. $v = \ln u$ *(2-5)*

7. $y = \log x$ *(2-5)*

8. $M = e^N$ *(2-5)* **9.** $u = 10^v$ *(2-5)* **10.** 5^{2x} *(2-4)* **11.** e^{2u^2} *(2-4)* **12.** $x = 9$ *(2-5)* **13.** $x = 6$ *(2-5)* **14.** $x = 4$ *(2-5)*
15. $x = 2.157$ *(2-5)* **16.** $x = 13.128$ *(2-5)* **17.** $x = 1{,}273.503$ *(2-5)* **18.** $x = 0.318$ *(2-5)*
19. (A) $y = 4$ (B) $x = 0$ (C) $y = 1$ (D) $x = -1$ or 1 (E) $y = -2$ (F) $x = -5$ or 5 *(2-1)*

20. (A)

(B)

(C)

(D)

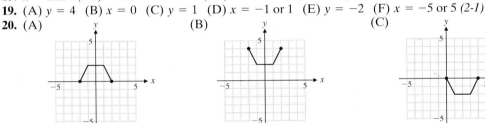

(2-2)

21. $f(x) = -(x - 2)^2 + 4$. The graph of $f(x)$ is the graph of $y = x^2$ reflected in the x axis, then shifted right 2 units and up 4 units. *(2-2)*
22. (A) g (B) m (C) n (D) f *(2-2, 2-3)* **23.** (A) x intercepts: $-4, 0$; y intercept: 0 (B) Vertex: $(-2, -4)$ (C) Minimum: -4
(D) Range: $y \geq -4$ or $[-4, \infty)$ (E) Increasing on $[-2, \infty)$ (F) Decreasing on $(-\infty, -2]$ *(2-3)*

25. *(2-1)*

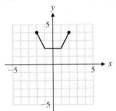

26. Quadratic *(2-3)* **27.** Linear *(2-1)* **28.** None *(2-1, 2-3)* **29.** Constant *(2-1)*
30. $x = 8$ *(2-5)* **31.** $x = 3$ *(2-5)* **32.** $x = 3$ *(2-4)* **33.** $x = -1, 3$ *(2-4)* **34.** $x = 0, \frac{3}{2}$ *(2-4)*
35. $x = -2$ *(2-5)* **36.** $x = \frac{1}{2}$ *(2-5)* **37.** $x = 27$ *(2-5)* **38.** $x = 13.3113$ *(2-5)*
39. $x = 158.7552$ *(2-5)* **40.** $x = 0.0097$ *(2-5)* **41.** $x = 1.4359$ *(2-5)* **42.** $x = 1.4650$ *(2-5)*
43. $x = 92.1034$ *(2-5)* **44.** $x = 9.0065$ *(2-5)* **45.** $x = 2.1081$ *(2-5)* **46.** $x = 2.8074$ *(2-5)*
47. $x = -1.0387$ *(2-5)* **48.** (A) All real numbers except $x = -2$ and 3 (B) $x < 5$ *(2-1)*

49. Function g multiplies a domain element by 2 and then subtracts 3 times the square root of the domain element from the result. *(2-1)*
50. Standard form: $4\left(x + \frac{1}{2}\right)^2 - 4$
(A) x intercepts: $-\frac{3}{2}$ and $\frac{1}{2}$; y intercept: -3
(B) Vertex: $\left(-\frac{1}{2}, -4\right)$ (C) Minimum: -4
(D) Range: $y \geq -4$ or $[-4, \infty)$ *(2-3)*
51. $(-1.54, -0.79)$; $(0.69, 0.99)$ *(2-4, 2-5)*

52. *(2-1)*

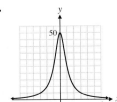

53. *(2-1)*

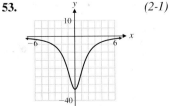

54. 6 *(2-1)*   **55.** −19 *(2-1)*   **56.** 10x − 4 *(2-1)*   **57.** 21 − 5x *(2-1)*   **58.** (A) −1   (B) −1 − 2h   (C) −2h   (D) −2 *(2-1)*
59. (A) $a^2 − 3a + 1$ (B) $a^2 + 2ah + h^2 − 3a − 3h + 1$ (C) $2ah + h^2 − 3h$ (D) $2a + h − 3$ *(2-1)*
60. The graph of function m is the graph of $y = |x|$ reflected in the x axis and shifted to the right 4 units. *(2-2)*
61. The graph of function g is the graph of $y = x^3$ vertically contracted by a factor of 0.3 and shifted up 3 units. *(2-2)*
62. The graph of $y = x^2$ is vertically expanded by a factor of 2, reflected in the x axis, and shifted to the left 3 units. Equation:
$y = -2(x + 3)^2$ *(2-2)*

63. $f(x) = 2\sqrt{x + 3} − 1$ *(2-2)*   **64.** True *(2-3)*   **65.** False *(2-3)*   **66.** False *(2-3)*   **67.** True *(2-4)*   **68.** True *(2-5)*
69. False *(2-5)*   **70.** True *(2-3)*   **71.** False *(2-4)*   **72.** Increasing: $[−2, 4]$ *(2-4)*
73. Decreasing: $[0, ∞)$ *(2-4)*   **74.** Increasing: $(−1, 10]$ *(2-5)*

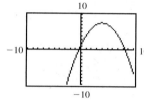

75. 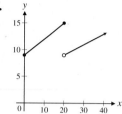 *(2-2)*   **76.** 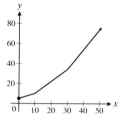 *(2-2)*   **77.** $y = -(x − 4)^2 + 3$
(2-2, 2-3)

78. $f(x) = -0.4(x − 4)^2 + 7.6$
(A) x intercepts: −0.4, 8.4;
y intercept: 1.2
(B) Vertex: (4.0, 7.6)
(C) Maximum: 7.6
(D) Range: $x ≤ 7.6$ or
$(−∞, 7.6]$ *(2-3)*

79.

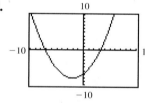

(A) x intercepts: −0.4, 8.4;
y intercept: 1.2
(B) Vertex: (4.0, 7.6)
(C) Maximum: 7.6
(D) Range: $x ≤ 7.6$ or
$(−∞, 7.6]$ *(2-3)*

80. $\log 10^π = π$ and $10^{\log \sqrt{2}} = \sqrt{2}$; $\ln e^π = π$ and $e^{\ln \sqrt{2}} = \sqrt{2}$ *(2-5)*   **81.** $x = 2$ *(2-5)*
82. $x = 2$ *(2-5)*   **83.** $x = 1$ *(2-5)*   **84.** $x = 300$ *(2-5)*   **85.** $y = ce^{-5t}$ *(2-5)*
86. If $\log_1 x = y$, then $1^y = x$; that is, $1 = x$ for all positive real numbers x, which is not
possible. *(2-5)*
87. The graph of $y = \sqrt[3]{x}$ is vertically expanded by a factor of 2, reflected in the x axis, and shifted
1 unit left and 1 unit down. Equation: $y = -2\sqrt[3]{x + 1} − 1$. *(2-2)*
88. $G(x) = 0.3(x + 2)^2 − 8.1$ (A) x intercepts: −7.2, 3.2; y intercept: −6.9
(B) Vertex: $(−2, −8.1)$ (C) Minimum: −8.1   (D) Range: $x ≥ −8.1$ or $[−8.1, ∞)$
(E) Decreasing: $(−∞, −2]$; increasing: $[−2, ∞)$ *(2-3)*

89.

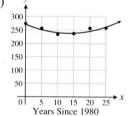

(A) x intercepts: −7.2, 3.2;
y intercept: −6.9
(B) Vertex: $(−2, −8.1)$
(C) Minimum: −8.1
(D) Range: $x ≥ −8.1$ or $[−8.1, ∞)$
(E) Decreasing: $(−∞, −2]$;
increasing: $[−2, ∞)$ *(2-3)*

90. (A)

x	0	5	10	15	20	25
Consumption	271	255	233	236	256	257
$f(x)$	271	251	240	239	246	262

(B)

(C) 300 eggs *(2-3)*

Years Since 1980

92. (A) $S(x) = \begin{cases} 3 & \text{if } 0 ≤ x ≤ 20 \\ 0.057x + 1.86 & \text{if } 20 < x ≤ 200 \\ 0.0346x + 6.34 & \text{if } 200 < x ≤ 1{,}000 \\ 0.0217x + 19.24 & \text{if } x > 1{,}000 \end{cases}$

93. \$6,521.89 *(2-4)*   **94.** \$6,362.50 *(2-4)*
95. 201 months ($≈16.7$ years) *(2-4)*   **96.** 3,424 days ($≈9.4$ years) *(2-4)*

(B) *(2-2)*

97. (A)

(B) $R = C$ for $x = 4.686$ thousand units (4,686 units) and
for $x = 27.314$ thousand units (27,314 units); $R < C$
for $1 ≤ x < 4.686$ or $27.314 < x ≤ 40$; $R > C$ for
$4.686 < x < 27.314$.
(C) Maximum revenue is 500 thousand dollars (\$500,000).
This occurs at an output of 20 thousand units (20,000
units). At this output, the wholesale price is
$p(20) = \$25$. *(2-3)*

98. (A) $P(x) = R(x) - C(x) = x(50 - 1.25x) - (160 + 10x)$

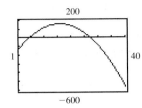

(B) $P = 0$ for $x = 4.686$ thousand units (4,686 units) and for $x = 27.314$ thousand units (27,314 units); $P < 0$ for $1 \le x < 4.686$ or $27.314 < x \le 40$; $P > 0$ for $4.686 < x < 27.314$.

(C) Maximum profit is 160 thousand dollars ($160,000). This occurs at an output of 16 thousand units (16,000 units). At this output, the wholesale price is $p(16) = \$30$. *(2-3)*

99. (A) $A(x) = -\frac{3}{2}x^2 + 420x$

(B) Domain: $0 \le x \le 280$

(C)

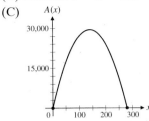

(D) There are two solutions to the equation $A(x) = 25,000$, one near 90 and another near 190.

(E) 86 ft; 194 ft

(F) Maximum combined area is 29,400 ft². This occurs for $x = 140$ ft and $y = 105$ ft. *(2-3)*

100. (A) 2,833 sets   (B) 4,836

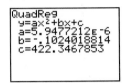 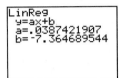

(D) Equilibrium price: $131.59; equilibrium quantity: 3,587 cookware sets *(2-3)*

101. (A) 3,519,000,000

(B) The model gives 7,594,000,000 for the number of wireless subscribers in 2018. *(2-4)*

102. (A) $N = 2^{2t}$ or $N = 4^t$

(B) 15 days *(2-2, 2-3)*

103. $k = 0.009\ 42$; 489 ft *(2-2, 2-3)*

104. (A) 6,139,000 *(2-5)*

105. 23.1 yr *(2-4)*   **106.** (A) $789 billion

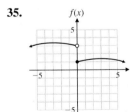 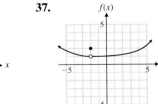

(B) 2018 *(2-4)*

CHAPTER 3

Exercise 3-1

1. (A) 2   (B) 2   (C) 2   (D) 2   **3.** (A) 1   (B) 2   (C) Does not exist   (D) 2   (E) No

5. (A) 1   (B) 2   (C) Does not exist   (D) Does not exist   (E) No

7. (A) 1   (B) 1   (C) 1   (D) 3   (E) Yes, define $g(3) = 1$.

9. (A) -2 (B) -2 (C) -2 (D) 1   (E) Yes   **11.** (A) 2   (B) 2   (C) 2   (D) Does not exist   (E) Yes

13. 12   **15.** 1   **17.** -4 **19.** -1.5 **21.** 3   **23.** 15   **25.** -6 **27.** $\frac{7}{5}$ **29.** 10   **31.** 3   **33.** 16

35.

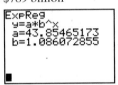

37.

39. (A) 1   (B) 1   (C) 1   (D) 1

41. (A) 2   (B) 1   (C) Does not exist   (D) Does not exist

43. (A) -6 (B) Does not exist   (C) 6

45. (A) 1   (B) -1 (C) Does not exist   (D) Does not exist

47. (A) Does not exist   (B) $\frac{1}{2}$ (C) $\frac{1}{4}$

49. (A) -5 (B) -3 (C) 0

51. (A) 0   (B) -1 (C) Does not exist   **53.** (A) 1   (B) $\frac{1}{3}$ (C) $\frac{3}{4}$ **55.** 3   **57.** 4   **59.** $1/(2\sqrt{2})$

61. Does not exist   **63.** -2

65. (A) $\lim_{x \to 1^-} f(x) = 2$ (B) $\lim_{x \to 1^-} f(x) = 3$ (C) $m = 1.5$ (D) The graph in (A) is broken when it jumps from $(1, 2)$ up to $(1, 3)$. The graph in (B) is also broken when it jumps down from $(1, 3)$ to $(1, 2)$. The graph in (C) is one continuous piece, with no breaks or jumps.

$\lim_{x \to 1^+} f(x) = 3$ $\lim_{x \to 1^+} f(x) = 2$

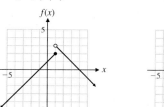
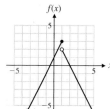
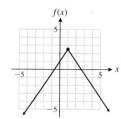

67. $2a$ **69.** $1/(2\sqrt{a})$

71. (A) $F(x) = \begin{cases} 0.99 & \text{if } 0 < x \le 20 \\ 0.07x - 0.41 & \text{if } x \ge 20 \end{cases}$

(B)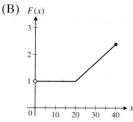

(C) All 3 limits are 0.99.

75. (A) $D(x) = \begin{cases} x & \text{if } 0 \le x < 300 \\ 0.97x & \text{if } 300 \le x < 1,000 \\ 0.95x & \text{if } 1,000 \le x < 3,000 \\ 0.93x & \text{if } 3,000 \le x < 5,000 \\ 0.9x & \text{if } x \ge 5,000 \end{cases}$

(B) $\lim_{x \to 1,000} D(x)$ does not exist because
$\lim_{x \to 1,000^-} D(x) = 970$ and $\lim_{x \to 1,000^+} D(x) = 950$;
$\lim_{x \to 3,000} D(x)$ does not exist because
$\lim_{x \to 3,000^-} D(x) = 2,850$ and $\lim_{x \to 3,000^+} D(x) = 2,790$

77. (A) $F(x) = \begin{cases} 20x & \text{if } 0 < x \le 4,000 \\ 80,000 & \text{if } x \ge 4,000 \end{cases}$

(B) $\lim_{x \to 4,000} F(x) = 80,000$; $\lim_{x \to 8,000} F(x) = 80,000$

79. $\lim_{x \to 5} f(x)$ does not exist; $\lim_{x \to 10} f(x) = 0$;
$\lim_{x \to 5} g(x) = 0$; $\lim_{x \to 10} g(x) = 1$

Exercise 3-2

1. f is continuous at $x = 1$, since $\lim_{x \to 1} f(x) = f(1)$.

3. f is discontinuous at $x = 1$, since $\lim_{x \to 1} f(x) \ne f(1)$.

5. f is discontinuous at $x = 1$, since $\lim_{x \to 1} f(x)$ does not exist.

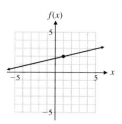
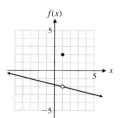
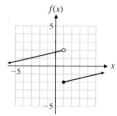

7. (A) 2   (B) 1   (C) Does not exist   (D) 1   (E) No   **9.** (A) 1   (B) 1   (C) 1   (D) 3   (E) No
11. (A) 1   (B) 1   (C) 1   (D) 3   (E) No   **13.** (A) 2   (B) −1   (C) Does not exist   (D) 2   (E) No
15. All x **17.** All x, except $x = -2$ **19.** All x, except $x = -4$ and $x = 1$ **21.** All x

23. All x, except $x = \pm\dfrac{3}{2}$

25. (A) 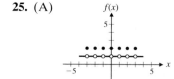   (B) 1   (C) 2   (D) No   (E) All integers
27. $-3 < x < 4$; $(-3, 4)$
29. $x < 3$ or $x > 7$; $(-\infty, 3) \cup (7, \infty)$
31. $x < -2$ or $0 < x < 2$; $(-\infty, -2) \cup (0, 2)$
33. $-5 < x < 0$ or $x > 3$; $(-5, 0) \cup (3, \infty)$

35. (A) $(-4, -2) \cup (0, 2) \cup (4, \infty)$ **37.** (A) $(-\infty, -2.5308) \cup (-0.7198, \infty)$
(B) $(-\infty, -4) \cup (-2, 0) \cup (2, 4)$ (B) $(-2.5308, -0.7198)$
39. (A) $(-\infty, -2.1451) \cup (-1, -0.5240) \cup (1, 2.6691)$
(B) $(-2.1451, -1) \cup (-0.5240, 1) \cup (2.6691, \infty)$
41. $[6, \infty)$ **43.** $(-\infty, \infty)$ **45.** $(-\infty, -3] \cup [3, \infty)$ **47.** $(-\infty, \infty)$

49. Since $\lim_{x\to 1^-} f(x) = 2$ and $\lim_{x\to 1^+} f(x) = 4$, $\lim_{x\to 1} f(x)$ does not exist and f is not continuous at $x = 1$.

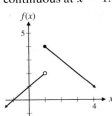

51. This function is continuous for all x.

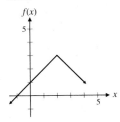

53. Since $\lim_{x\to 0} f(x) = 0$ and $f(0) = 1$, $\lim_{x\to 0} f(x) \neq f(0)$ and f is not continuous at $x = 0$.

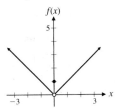

55. Since $\lim_{x\to 2^-} f(x) = 0$ and $\lim_{x\to 2^+} f(x) = 4$, $\lim_{x\to 2} f(x)$ does not exist. Furthermore, $f(2)$ is not defined. Thus, f is not continuous at $x = 2$.

57. Since $f(-1)$ and $f(1)$ are not defined, f is not continuous at $x = -1$ and $x = 1$, even though $\lim_{x\to -1} f(x) = 2$ and $\lim_{x\to 1} f(x) = 2$.

59. (A) Yes (B) Yes (C) Yes (D) Yes   **61.** (A) Yes (B) No (C) Yes (D) No (E) Yes

63. x intercepts: $x = -5, 2$ **65.** x intercepts: $x = -6, -1, 4$ **67.** No, but this does not contradict Theorem 2, since f is discontinuous at $x = 1$.

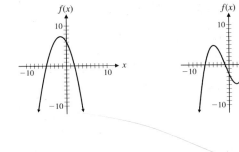

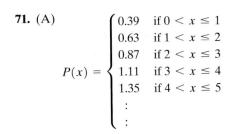

69. The following sketches illustrate that either condition is possible. Theorem 2 implies that one of these two conditions must occur.

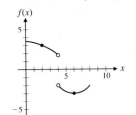

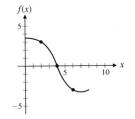

71. (A)
$$P(x) = \begin{cases} 0.39 & \text{if } 0 < x \leq 1 \\ 0.63 & \text{if } 1 < x \leq 2 \\ 0.87 & \text{if } 2 < x \leq 3 \\ 1.11 & \text{if } 3 < x \leq 4 \\ 1.35 & \text{if } 4 < x \leq 5 \\ \vdots \\ \vdots \end{cases}$$

(B)

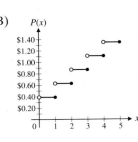

(C) Yes; no

75. (A) $S(x) = \begin{cases} 5 + 0.63x & \text{if } 0 \leq x \leq 50 \\ 14 + 0.45x & \text{if } 50 < x \end{cases}$

(B)

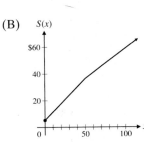

(C) Yes

77. (A)

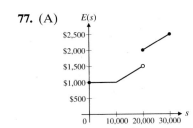

(B) $\lim_{s\to 10,000} E(s) = \$1,000$; $E(10,000) = \$1,000$

(C) $\lim_{s\to 20,000} E(s)$ does not exist; $E(20,000) = \$2,000$

(D) Yes; no

79. (A) t_2, t_3, t_4, t_6, t_7

(B) $\lim_{t\to t_5} N(t) = 7$; $N(t_5) = 7$

(C) $\lim_{t\to t_3} N(t)$ does not exist; $N(t_3) = 4$

Exercise 3-3

1. -2 **3.** $-\infty$ **5.** Does not exist **7.** 0 **9.** (A) $-\infty$ (B) ∞ (C) Does not exist

11. (A) ∞ (B) ∞ (C) ∞ **13.** (A) 3 (B) 3 (C) 3 **15.** (A) $-\infty$ (B) ∞ (C) Does not exist

17. (A) ∞ (B) $-\infty$ **19.** (A) $-\infty$ (B) $-\infty$

21. Discontinuous at $x = -3$; $\lim_{x \to -3^-} f(x) = -\infty$; $\lim_{x \to -3^+} f(x) = \infty$; $x = -3$ is a vertical asymptote

23. Discontinuous at $x = -2$ and $x = 2$; $\lim_{x \to -2^-} h(x) = \infty$; $\lim_{x \to -2^+} h(x) = -\infty$; $\lim_{x \to 2^-} h(x) = -\infty$; $\lim_{x \to 2^+} h(x) = \infty$; $x = -2$ and $x = 2$ are vertical asymptotes

25. No points of discontinuity; no vertical asymptotes

27. Discontinuous at $x = 1$ and $x = 3$; $\lim_{x \to 1^-} H(x) = -\infty$; $\lim_{x \to 1^+} H(x) = \infty$; $\lim_{x \to 3} H(x) = 2$; $x = 1$ is a vertical asymptote

29. Discontinuous at $x = 0$ and $x = 4$; $\lim_{x \to 0} T(x) = -\infty$; $\lim_{x \to 4} T(x) = \infty$; $x = 0$ and $x = 4$ are vertical asymptotes

31. $\dfrac{4}{5}$ **33.** ∞ **35.** 0 **37.** Horizontal asymptote: $y = 2$; vertical asymptote: $x = -2$

39. Horizontal asymptote: $y = 1$; vertical asymptotes: $x = -1$ and $x = 1$

41. No horizontal asymptotes; no vertical asymptotes **43.** Horizontal asymptote: $y = 0$; no vertical asymptotes

45. No horizontal asymptotes; vertical asymptote: $x = 3$ **47.** Horizontal asymptote: $y = 2$; vertical asymptotes: $x = -1$ and $x = 2$

49. Horizontal asymptote: $y = 2$; vertical asymptote: $x = -1$ **51.** $a = 4$ **53.** $a = -2$

55. If $a_n > 0$, the limit is ∞. If $a_n < 0$, the limit is $-\infty$. **57.** $\lim_{x \to \infty} f(x) = \infty$; $\lim_{x \to -\infty} f(x) = \infty$

59. $\lim_{x \to \infty} f(x) = \infty$; $\lim_{x \to -\infty} f(x) = -\infty$ **61.** $\lim_{x \to \infty} f(x) = \infty$; $\lim_{x \to -\infty} f(x) = -\infty$ **63.** $\lim_{x \to \infty} f(x) = -\infty$; $\lim_{x \to -\infty} f(x) = -\infty$

65. (A) $C(x) = 180x + 200$ (B) $\overline{C}(x) = \dfrac{180x + 200}{x}$ (C) (D) \$180 per board

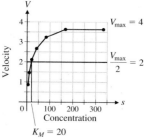

67. (A) $C_e(x) = 950 + 56x$; $\overline{C}_e(x) = \dfrac{950}{x} + 56$ (B) $C_c(x) = 900 + 66x$; $\overline{C}_c(x) = \dfrac{900}{x} + 66$ (C) At $x = 5$ years

(D) At $x = 5$ years (E) $\lim_{x \to \infty} \overline{C}_e(x) = 56$; $\lim_{x \to \infty} \overline{C}_c(x) = 66$

69. The long-term drug concentration is 5 mg/ml. **71.** (A) \$18 million (B) \$38 million (C) $\lim_{x \to 1^-} P(x) = \infty$

73. (C) $V_{\max} = 4$, $K_M = 20$

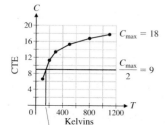

(D) $v(s) = \dfrac{4s}{20 + s}$

(E) $v = \dfrac{12}{7}$ when $s = 15$; $s = 60$ when $v = 3$

75. (A) $C_{\max} = 18$, $M = 150$

(B) $C(T) = \dfrac{18T}{150 + T}$

(C) $C = 14.4$ at $T = 600$ K, $T = 300$ K when $C = 12$

Exercise 3-4

1. (A) -3; slope of the secant line through $(1, f(1))$ and $(2, f(2))$

(B) $-2 - h$; slope of the secant line through $(1, f(1))$ and $(1 + h, f(1 + h))$

(C) -2; slope of the tangent line at $(1, f(1))$

3. (A) 15 (B) 15 (C) $6 + 3h$ (D) 6 (E) 6 (F) 6 (G) $y = 6x - 3$

5. $f'(x) = 0$; $f'(1) = 0$, $f'(2) = 0$, $f'(3) = 0$ **7.** $f'(x) = 3$; $f'(1) = 3$, $f'(2) = 3$, $f'(3) = 3$

9. $f'(x) = -6x; f'(1) = -6, f'(2) = -12, f'(3) = -18$ **11.** $f'(x) = 2x + 6; f'(1) = 8, f'(2) = 10, f'(3) = 12$
13. $f'(x) = 4x - 7; f'(1) = -3, f'(2) = 1, f'(3) = 5$ **15.** $f'(x) = -2x + 4; f'(1) = 2, f'(2) = 0, f'(3) = -2$
17. $f'(x) = 6x^2; f'(1) = 6, f'(2) = 24, f'(3) = 54$ **19.** $f'(x) = -\dfrac{4}{x^2}; f'(1) = -4, f'(2) = -1, f'(3) = -\dfrac{4}{9}$

21. $f'(x) = \dfrac{3}{2\sqrt{x}}; f'(1) = \dfrac{3}{2}, f'(2) = \dfrac{3}{2\sqrt{2}}$ or $\dfrac{3\sqrt{2}}{4}, f'(3) = \dfrac{3}{2\sqrt{3}}$ or $\dfrac{\sqrt{3}}{2}$

23. $f'(x) = \dfrac{5}{\sqrt{x + 5}}; f'(1) = \dfrac{5}{\sqrt{6}}$ or $\dfrac{5\sqrt{6}}{6}, f'(2) = \dfrac{5}{\sqrt{7}}$ or $\dfrac{5\sqrt{7}}{7}, f'(3) = \dfrac{5}{2\sqrt{2}}$ or $\dfrac{5\sqrt{2}}{4}$

25. $f'(x) = \dfrac{6}{(x + 2)^2}; f'(1) = \dfrac{2}{3}, f'(2) = \dfrac{3}{8}, f'(3) = \dfrac{6}{25}$

27. (A) 5   (B) $3 + h$ (C) 3   (D) $y = 3x - 1$ **29.** (A) 5 m/s   (B) $3 + h$ m/s   (C) 3 m/s
31. Yes   **33.** No   **35.** Yes   **37.** Yes
39. (A) $f'(x) = 2x - 4$ (B) $-4, 0, 4$
(C)

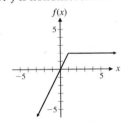

41. $v = f'(x) = 8x - 2; 6$ ft/s, 22 ft/s, 38 ft/s
43. (A) The graphs of g and h are vertical translations of the graph of f. All three functions should have the same derivative.
(B) $2x$
45. (A) The slope of the graph of f is 0 at any point on the graph.

47. f is nondifferentiable at $x = 1$ **49.** f is differentiable for all real numbers   **51.** No   **53.** No

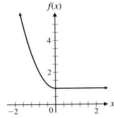

55. $f'(0) = 0$ **57.** 6 s; 192 ft/s
59. (A) $8.75   (B) $R'(x) = 60 - 0.05x$
(C) $R(1,000) = 35,000; R'(1,000) = 10$; At a production level of 1,000 car seats, the revenue is $35,000 and is increasing at the rate of $10 per seat.
61. (A) $S'(t) = 1/\sqrt{t + 10}$
(B) $S(15) = 10; S'(15) = 0.2$. After 15 months, the total sales are $10 million and are increasing at the rate of $0.2 million, or $200,000, per month.
(C) The estimated total sales are $10.2 million after 16 months and $10.4 million after 17 months.
63. (A) $p'(t) = 28t - 6.6$
(B) $p(15) = 3,653.4; p'(15) = 413.4$; In 2010, 3,653.4 thousand tons of zinc are produced and this quantity is increasing at the rate of 413.4 thousand tons per year.
65. (A)
```
QuadReg
y=ax²+bx+c
a=.6666666667
b=24.88484848
c=1071.018182
R²=.9772827077
```

(B) $R(20) = 1,835.4$ billion kilowatts, $R'(20) = 51.6$ billion kilowatts. In 2016, 1,835.4 billion kilowatts will be sold and the amount sold is increasing at the rate of 51.6 billion kilowatts per year.
67. (A) $P'(t) = 12 - 2t$
(B) $P(3) = 107; P'(3) = 6$. After 3 hours, the ozone level is 107 ppb and is increasing at the rate of 6 ppb per hour.
69. (A) $f'(t) = 0.016t - 0.5$
(B) $f(30) = 6.6; f'(30) = -0.02$; In 2010, the number of male infant deaths per 100,000 births is 6.6 and is decreasing at the rate of 0.02 male infant deaths per 100,000 births.

Exercise 3-5

1. 0   **3.** $9x^8$ **5.** $3x^2$ **7.** $-4x^{-5}$ **9.** $\frac{8}{3}x^{5/3}$ **11.** $-\frac{10}{x^{11}}$ **13.** $10x$ **15.** $2.8x^6$

17. $\frac{x^2}{6}$ **19.** 12   **21.** 2   **23.** 9   **25.** 2   **27.** $4t - 3$ **29.** $-10x^{-3} - 9x^{-2}$

31. $1.5u^{-0.7} - 8.8u^{1.2}$ **33.** $0.5 - 3.3t^2$ **35.** $-\frac{8}{5}x^{-5}$ **37.** $3x + \frac{14}{5}x^{-3}$

39. $-\frac{20}{9}w^{-5} + \frac{5}{3}w^{-2/3}$ **41.** $2u^{-1/3} - \frac{5}{3}u^{-2/3}$ **43.** $-\frac{9}{5}t^{-8/5} + 3t^{-3/2}$ **45.** $-\frac{1}{3}x^{-4/3}$ **47.** $-0.6x^{-3/2} + 6.4x^{-3} + 1$

49. (A) $f'(x) = 6 - 2x$ (B) $f'(2) = 2; f'(4) = -2$ (C) $y = 2x + 4; y = -2x + 16$ (D) $x = 3$

51. (A) $f'(x) = 12x^3 - 12x$ (B) $f'(2) = 72; f'(4) = 720$ (C) $y = 72x - 127; y = 720x - 2,215$ (D) $x = -1, 0, 1$

53. (A) $v = f'(x) = 176 - 32x$ (B) $f'(0) = 176$ ft/s; $f'(3) = 80$ ft/s   (C) 5.5 s

55. (A) $v = f'(x) = 3x^2 - 18x + 15$ (B) $f'(0) = 15$ ft/s; $f'(3) = -12$ ft/s   (C) $x = 1$ s, $x = 5$ s

57. $f'(x) = 2x - 3 - 2x^{-1/2} = 2x - 3 - \frac{2}{x^{1/2}}; x = 2.1777$ **59.** $f'(x) = 4\sqrt[3]{x} - 3x - 3; x = -2.9018$

61. $f'(x) = 0.2x^3 + 0.3x^2 - 3x - 1.6; x = -4.4607, -0.5159, 3.4765$

63. $f'(x) = 0.8x^3 - 9.36x^2 + 32.5x - 28.25; x = 1.3050$

69. $8x - 4$ **71.** $-20x^{-2}$ **73.** $-\frac{1}{4}x^{-2} + \frac{2}{3}x^{-3}$ **75.** $2x - 3 - 10x^{-3}$

81. (A) $S'(t) = 0.09t^2 + t + 2$

(B) $S(5) = 29.25, S'(5) = 9.25$. After 5 months, sales are \$29.25 million and are increasing at the rate of \$9.25 million per month.

(C) $S(10) = 103, S'(10) = 21$. After 10 months, sales are \$103 million and are increasing at the rate of \$21 million per month.

83. (A) $N'(x) = 3,780/x^2$

(B) $N'(10) = 37.8$. At the \$10,000 level of advertising, sales are increasing at the rate of 37.8 boats per \$1,000 spent on advertising.

$N'(20) = 9.45$. At the \$20,000 level of advertising, sales are increasing at the rate of 9.45 boats per \$1,000 spent on advertising.

85. (A)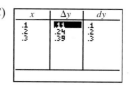
```
CubicReg
 y=ax³+bx²+cx+d
 a=6.266666667
 b=-194.5714286
 c=1544.761905
 d=2571.428571
 ■
```
(B) In 1992, 3,900 limousines were produced and limousine production was decreasing at the rate of 400 limousines per year.

(C) In 1998, 3,900 limousines were produced and limousine production was increasing at the rate of 600 limousines per year.

87. (A) -1.37 beats/min   (B) -0.58 beat/min   **89.** (A) 25 items/h   (B) 8.33 items/h

Exercise 3-6

1. $\Delta x = 3; \Delta y = 45; \Delta y/\Delta x = 15$ **3.** 12   **5.** 12   **7.** $dy = (24x - 3x^2)\,dx$

9. $dy = \left(2x - \frac{x^2}{3}\right)dx$ **11.** $dy = -\frac{295}{x^{3/2}}\,dx$ **13.** (A) $12 + 3\Delta x$ (B) 12   **15.** $dy = 6(2x + 1)^2\,dx$

17. $dy = \frac{9 - x^2}{(x^2 + 9)^2}\,dx$ **19.** $dy = 1.4; \Delta y = 1.44$ **21.** $dy = -3; \Delta y = -3\frac{1}{3}$ **23.** 120 in.³

25. (A) $\Delta y = \Delta x + (\Delta x)^2; dy = \Delta x$ (B)

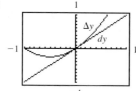

(C)

x	Δy	dy
.1	.11	.1
.2	.24	.2
.3	.39	.3

27. (A) $\Delta y = -\Delta x + (\Delta x)^2 + (\Delta x)^3; dy = -\Delta x$ (B)

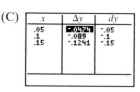

(C)

x	Δy	dy
.05	-.047...	-.05
.1	-.089	-.1
.15	-.1241	-.15

29. True   **31.** False   **33.** $dy = \frac{6x - 2}{3(3x^2 - 2x + 1)^{2/3}}\,dx$ **35.** $dy = 3.9; \Delta y = 3.83$

37. 40-unit increase; 20-unit increase   **39.** $-\$2.50; \$1.25$ **41.** -1.37/min; -0.58/min

43. 1.26 mm²   **45.** 3 wpm   **47.** (A) 2,100 increase   (B) 4,800 increase   (C) 2,100 increase

Exercise 3-7

1. (A) $29.50 (B) $30
3. (A) $420
   (B) $\overline{C}'(500) = -0.24$. At a production level of 500 frames, average cost is decreasing at the rate of 24¢ per frame.
   (C) Approximately $419.76
5. (A) $14.70 (B) $15
7. (A) $P'(450) = 0.5$. At a production level of 450 cassettes, profit is increasing at the rate of 50¢ per cassette.
   (B) $P'(750) = -2.5$. At a production level of 750 cassettes, profit is decreasing at the rate of $2.50 per cassette.
9. (A) $13.50
   (B) $\overline{P}'(50) = \$0.27$. At a production level of 50 mowers, the average profit per mower is increasing at the rate of $0.27 per mower.
   (C) Approximately $13.77
11. (A) $p = 100 - 0.025x$, domain: $0 \le x \le 4,000$ (B) $R(x) = 100x - 0.025x^2$, domain: $0 \le x \le 4,000$
   (C) $R'(1,600) = 20$. At a production level of 1,600 radios, revenue is increasing at the rate of $20 per radio.
   (D) $R'(2,500) = -25$. At a production level of 2,500 radios, revenue is decreasing at the rate of $25 per radio.
13. (A) $p = 200 - \frac{1}{30}x$, domain: $0 \le x \le 6,000$ (B) $C'(x) = 60$
   (C) $R(x) = 200x - (x^2/30)$, domain: $0 \le x \le 6,000$
   (D) $R'(x) = 200 - (x/15)$
   (E) $R'(1,500) = 100$. At a production level of 1,500 saws, revenue is increasing at the rate of $100 per saw.
      $R'(4,500) = -100$. At a production level of 4,500 saws, revenue is decreasing at the rate of $100 per saw.
   (F) Break-even points; (600, 108,000) and (3,600, 288,000)   (G) $P(x) = -(x^2/30) + 140x - 72,000$
      (H) $P'(x) = -(x/15) + 140$
      (I) $P'(1,500) = 40$. At a production level of 1,500 saws, profit is increasing at the rate of $40 per saw. $P'(3,000) = -60$. At a production level of 3,000 saws, profit is decreasing at the rate of $60 per saw.

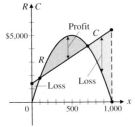

15. (A) $p = 20 - 0.02x$, domain: $0 \le x \le 1,000$ (B) $R(x) = 20x - 0.02x^2$, domain: $0 \le x \le 1,000$
   (C) $C(x) = 4x + 1,400$ (D) Break-even points: (100, 1,800) and (700, 4,200)   (E) $P(x) = 16x - 0.02x^2 - 1,400$
      (F) $P'(250) = 6$. At a production level of 250 toasters, profit is increasing at the rate of $6 per toaster. $P'(475) = -3$. At a production level of 475 toasters, profit is decreasing at the rate of $3 per toaster.

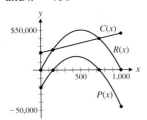

17. (A) $x = 500$ (B) $P(x) = 176x - 0.2x^2 - 21,900$
   (C) $x = 440$
   (D) Break-even points: (150, 25,500) and (730, 39,420); x intercepts for $P(x)$: $x = 150$ and $x = 730$

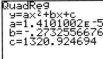

19. (A) $R(x) = 20x - x^{3/2}$
   (B) Break-even points: (44, 588), (258, 1,016)

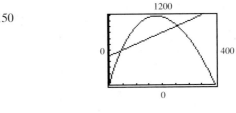

21. (A) 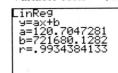   (B) Fixed costs $\approx$ $721,680   (C) (713, 807,703), (5,423, 1,376,227)
      Variable costs $\approx$ $121   (D) $254 \le p \le $1,133

Chapter 3 Review Exercise

1. (A) 16 (B) 8 (C) 8 (D) 4 (E) 4 (F) 4 *(3-3)* **2.** $f'(x) = -3$ *(3-3)*

3. (A) 22 (B) 8 (C) 2 (D) -5 *(3-1)* **4.** (A) 1 (B) 1 (C) 1 (D) 1 *(3-1)*

5. (A) 2 (B) 3 (C) Does not exist (D) 3 *(3-1)* **6.** (A) 4 (B) 4 (C) 4 (D) Does not exist *(3-1)*

7. (A) Does not exist (B) 3 (C) No *(3-2)* **8.** (A) 2 (B) Not defined (C) No *(3-2)*

9. (A) 1 (B) 1 (C) Yes *(3-2)* **10.** 5 *(3-3)* **11.** 5 *(3-3)* **12.** ∞ *(3-3)* **13.** $-\infty$ *(3-3)*

14. 0 *(3-1)* **15.** 0 *(3-1)* **16.** 0 *(3-1)* **17.** Vertical asymptote: $x = 2$ *(3-3)* **18.** Horizontal asymptote: $y = 5$ *(3-3)*

19. $x = 2$ **20.** $f'(x) = 10x$ *(3-4)* **21.** (A) -3 (B) 6 (C) -2 (D) 3 (E) -11 *(3-5)*

22. $x^2 - 10x$ *(3-5)* **23.** $x^{-1/2} - 3 = \dfrac{1}{x^{1/2}} - 3$ *(3-5)* **24.** 0 *(3-5)* **25.** $-\dfrac{3}{2}x^{-2} + \dfrac{15}{4}x^2 = \dfrac{-3}{2x^2} + \dfrac{15x^2}{4}$ *(3-5)*

26. $-2x^{-5} + x^3 = \dfrac{-2}{x^5} + x^3$ *(3-5)* **27.** $f'(x) = 12x^3 + 9x^2 - 2$ *(3-5)* **28.** $\Delta x = 2$, $\Delta y = 10$, $\Delta y/\Delta x = 5$ *(3-6)*

29. 5 *(3-6)* **30.** 6 *(3-6)* **31.** $\Delta y = 0.64$; $dy = 0.6$ *(3-6)* **32.** (A) 4 (B) 6 (C) Does not exist (D) 6 (E) No *(3-2)*

33. (A) 3 (B) 3 (C) 3 (D) 3 (E) Yes *(3-2)* **34.** (A) $(8, \infty)$ (B) $[0, 8]$ *(3-2)* **35.** $(-3, 4)$ *(3-2)*

36. $(-3, 0) \cup (5, \infty)$ *(3-2)* **37.** $(-2.3429, -0.4707) \cup (1.8136, \infty)$ *(3-2)* **38.** (A) 3 (B) $2 + 0.5h$ (C) 2 *(3-4)*

39. $-x^{-4} + 10x^{-3}$ *(3-4)* **40.** $\dfrac{3}{4}x^{-1/2} - \dfrac{5}{6}x^{-3/2} = \dfrac{3}{4\sqrt{x}} - \dfrac{5}{6\sqrt{x^3}}$ *(3-5)* **41.** $0.6x^{-2/3} - 0.3x^{-4/3} = \dfrac{0.6}{x^{2/3}} - \dfrac{0.3}{x^{4/3}}$ *(3-4)*

42. $-\dfrac{3}{5}(-3)x^{-4} = \dfrac{9}{5x^4}$ *(3-5)* **43.** (A) $m = f'(1) = 2$ (B) $y = 2x + 3$ *(3-4, 3-5)* **44.** $x = 5$ *(3-4)* **45.** $x = -5, x = 3$ *(3-5)*

46. $x = -1.3401, 0.5771, 2.2630$ *(3-4)* **47.** ± 2.4824 *(3-5)* **48.** (A) $v = f'(x) = 16x - 4$ (B) 44 ft/sec *(3-5)*

49. (A) $v = f'(x) = -10x + 16$ (B) $x = 1.6$ sec *(3-5)*

50. (A) The graph of g is the graph of f shifted 4 units to the right, and the graph of h is the graph of f shifted 3 units to the left:

(B) The graph of g' is the graph of f' shifted 4 units to the right, and the graph of h' is the graph of f' shifted 3 units to the left:

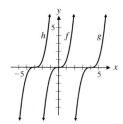

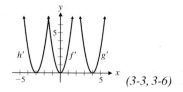

(3-3, 3-6)

51. (A) The graph of g is a vertical stretch of the graph of f.
(B) The graph of g is a vertical translation of the graph of f, and the graph of g' is the same as the graph of f'. *(3-3)*

52. $dy = 0.6928$, $\Delta y = 0.6816$ *(3-6)* **53.** Differential approximation: 2.0083; calculator value: 2.0083 *(3-6)*

54. $(-\infty, \infty)$ *(3-2)* **55.** $(-\infty, 2) \cup (2, \infty)$ *(3-2)* **56.** $(-\infty, -4) \cup (-4, 1) \cup (1, \infty)$ *(3-2)* **57.** $(-\infty, \infty)$ *(3-2)*

58. $[-2, 2]$ *(3-2)* **59.** (A) -1 (B) Does not exist (C) $-\dfrac{2}{3}$ *(3-1)* **60.** (A) $\dfrac{1}{2}$ (B) 0 (C) Does not exist *(3-1)*

61. (A) -1 (B) 1 (C) Does not exist *(3-1)* **62.** (A) $-\dfrac{1}{6}$ (B) Does not exist (C) $-\dfrac{1}{3}$ *(3-1)*

63. (A) 0 (B) -1 (C) Does not exist *(3-1)* **64.** (A) $\dfrac{2}{3}$ (B) $\dfrac{2}{3}$ (C) Does not exist *(3-3)*

65. (A) ∞ (B) $-\infty$ (C) ∞ *(3-3)* **66.** (A) 0 (B) 0 (C) Does not exist *(3-3)* **67.** 4 *(3-1)* **68.** $\dfrac{-1}{(x + 2)^2}$ *(3-1)*

69. (A) $\lim_{x \to -2^-} f(x) = -6$; $\lim_{x \to -2^+} f(x) = 6$; $\lim_{x \to -2} f(x)$ does not exist (B) $\lim_{x \to 0} f(x) = 4$
(C) $\lim_{x \to 2^-} f(x) = 2$; $\lim_{x \to 2^+} f(x) = -2$; $\lim_{x \to 2} f(x)$ does not exist *(3-1)*

70. $2x - 1$ *(3-4)* **71.** $1/(2\sqrt{x})$ *(3-4)* **72.** No *(3-4)* **73.** No *(3-4)* **74.** No *(3-4)* **75.** Yes *(3-4)*

76. Horizontal asymptote: $y = 5$; vertical asymptote: $x = 7$ *(3-3)* **77.** Horizontal asymptote: $y = 0$; vertical asymptote: $x = 4$ *(3-3)* **78.** No horizontal asymptotes; vertical asymptote: $x = 3$ *(3-3)*

79. Horizontal asymptotes: $y = 1$; vertical asymptotes: $x = -2, x = 1$ *(3-3)* **80.** Horizontal asymptote: $y = 1$; vertical asymptotes: $x = -1, x = 1$ *(3-3)*

81. The domain of $f'(x)$ is all real numbers except $x = 0$. At $x = 0$, the graph of $y = f(x)$ is smooth, but it has a vertical tangent. *(3-4)*

82. (A) $\lim_{x \to 1^-} f(x) = 1; \lim_{x \to 1^+} f(x) = -1$ (B) $\lim_{x \to 1^-} f(x) = -1; \lim_{x \to 1^+} f(x) = 1$ (C) $m = 1$

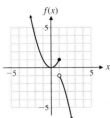   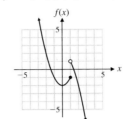   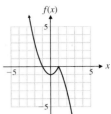

   (D) The graphs in (A) and (B) have discontinuities at $x = 1$; the graph in (C) does not. *(3-2)*

83. (A) 1   (B) -1 (C) Does not exist   (D) No *(3-4)*

84. (A) $S(x) = \begin{cases} 7.47 + 0.4x & \text{if } 0 \le x \le 90 \\ 24.786 + 0.2076x & \text{if } 90 < x \end{cases}$ (B)

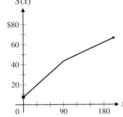

   (C) Yes *(3-2)*

85. (A) \$179.90   (B) \$180 *(3-7)*

86. (A) $C(100) = 9,500; C'(100) = 50$. At a production level of 100 bicycles, the total cost is \$9,500, and cost is increasing at the rate of \$50 per bicycle.

   (B) $\overline{C}(100) = 95; \overline{C}'(100) = -0.45$. At a production level of 100 bicycles, the average cost is \$95, and average cost is decreasing at a rate of \$0.45 per bicycle. *(3-7)*

87. The approximate cost of producing the 201st printer is greater than that of the 601st printer. Since these marginal costs are decreasing, the manufacturing process is becoming more efficient. *(3-7)*

88. (A) $C'(x) = 2; \overline{C}(x) = 2 + \dfrac{9,000}{x}; \overline{C}'(x) = \dfrac{-9,000}{x^2}$

   (B) $R(x) = xp = 25x - 0.01x^2; R'(x) = 25 - 0.02x; \overline{R}(x) = 25 - 0.01x; \overline{R}'(x) = -0.01$

   (C) $P(x) = R(x) - C(x) = 23x - 0.01x^2 - 9,000; P'(x) = 23 - 0.02x;$

   $\overline{P}(x) = 23 - 0.01x - \dfrac{9,000}{x}; \overline{P}'(x) = -0.01 + \dfrac{9,000}{x^2}$

   (D) $(500, 10,000)$ and $(1,800, 12,600)$

   (E) $P'(1,000) = 3$. Profit is increasing at the rate of \$3 per umbrella.
   $P'(1,150) = 0$. Profit is flat.
   $P'(1,400) = -5$. Profit is decreasing at the rate of \$5 per umbrella.

   (F)

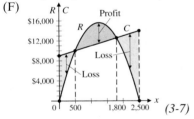

   (3-7)

89. (A) 8   (B) 20 *(3-5)*

90. $N(9) = 27; N'(t) = 3.5$; After 9 months, 27,000 pools have been sold and the total sales are increasing at the rate of 3,500 pools per month. *(3-5)*

91. (A)

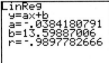

   (B) $N(50) = 38.6; N'(50) = 2.6$. In 2010, natural gas consumption is 38.6 trillion cubic feet and is increasing at the rate of 2.6 trillion cubic feet per year *(3-4)*

92. (A)

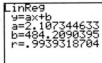

   (B) Fixed costs: \$484.21; variable costs per kringle: \$2.11

   (C) $(51, 591.15), (248, 1,007.62)$

   (D) $\$4.07 < p < \$11.64$ *(3-7)*

93. $C'(10) = -1$; $C'(100) = -0.001$ (3-5)

94. $F(4) = 98.16$; $F'(4) = -0.32$; After 4 hours the patient's temperature is 98.16°F and is decreasing at the rate of 0.32°F per hour. (3-5)

95. (A) 10 items/h   (B) 5 items/h (3-5)

96. (A)

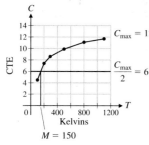

(B) $C(T) = \dfrac{12T}{150 + T}$

(C) $C = 9.6$ at $T = 600$ K, $T = 750$ K when $C = 10$ (3-3)

CHAPTER 4

Exercise 4-1

1. $1,221.40; $1,648.72; $2,225.54

3.

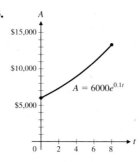

5. 11.55   **7.** 10.99   **9.** 0.14

11.

n	$[1 + (1/n)]^n$
10	2.593 74
100	2.704 81
1,000	2.716 92
10,000	2.718 15
100,000	2.718 27
1,000,000	2.718 28
10,000,000	2.718 28
↓	↓
∞	$e = 2.718\ 281\ 828\ 459\ldots$

13. $\lim_{n\to\infty}(1 + n)^{1/n} = 1$ **15.**

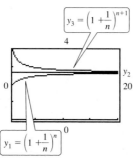

17. (A) $17,349.87   (B) 7.36 yr   **19.** $11,890.41

21. 8.11%   **23.** (A)

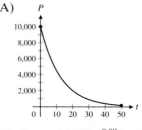

(B) $\lim_{t\to\infty}10,000e^{-0.08t} = 0$

25. 9.90 yr   **27.** 8.66%   **29.** 7.3 yr

31. (A) $A = Pe^{rt}$ (B)

$2P = Pe^{rt}$

$2 = e^{rt}$

$rt = \ln 2$

$t = \dfrac{\ln 2}{r}$

Although r could be any positive number, the restrictions on r are reasonable in the sense that most investments would be expected to earn a return of between 2% and 30%.

(C) The doubling times (in years) are 13.86, 6.93, 4.62, 3.47, 2.77, and 2.31, respectively.

33. $t = -(\ln 0.5)/0.000\ 433\ 2 \approx 1,600$ yr   **35.** $r = (\ln 0.5)/30 \approx -0.0231$ **37.** 53.3 yr   **39.** 1.39%

Exercise 4-2

1. $5e^x + 3$ **3.** $-\dfrac{2}{x} + 2x$ **5.** $3x^2 - 6e^x$ **7.** $e^x + 1 - \dfrac{1}{x}$ **9.** $\dfrac{3}{x}$ **11.** $5 - \dfrac{5}{x}$ **13.** $\dfrac{2}{x} + 4e^x$

15. $f'(x) = \dfrac{1}{x}$; $y = x + 2$ **17.** $f'(x) = 3e^x$; $y = 3x + 3$ **19.** $f'(x) = \dfrac{3}{x}$; $y = \dfrac{3}{e^x}$ **21.** $f'(x) = e^x$; $y = ex + 2$

23. Yes; yes   **25.** No; no   **27.** $f(x) = 10x + \ln 10 + \ln x$; $f'(x) = 10 + \dfrac{1}{x}$ **29.** $f(x) = \ln 4 - 3\ln x$; $f'(x) = -\dfrac{3}{x}$

31. $\dfrac{1}{x \ln 2}$ **33.** $3^x \ln 3$ **35.** $2 - \dfrac{1}{x \ln 10}$ **37.** $1 + 10^x \ln 10$ **39.** $\dfrac{3}{x} + \dfrac{2}{x \ln 3}$ **41.** $2^x \ln 2$

43. $(-0.82, 0.44), (1.43, 4.18), (8.61, 5503.66)$ **45.** $(0.49, 0.49)$ **47.** $(3.65, 1.30), (332, 105.11, 12.71)$

51. $-\$28{,}447/\text{yr}; -\$18{,}664/\text{yr}; -\$11{,}021/\text{yr}$

53. $A'(t) = 5{,}000(\ln 4)4^t;$ $A'(1) = 27{,}726$ bacteria/hr (rate of change at the end of the first hour); $A'(5) = 7{,}097{,}827$ bacteria/hr (rate of change at the end of the fifth hour)

55. At the 40-lb weight level, blood pressure would increase at the rate of 0.44 mm of mercury per pound of weight gain. At the 90-lb weight level, blood pressure would increase at the rate of 0.19 mm of mercury per pound of weight gain.

57. $dR/dS = k/S$

Exercise 4-3

1. $2x^3(2x) + (x^2 - 2)(6x^2) = 10x^4 - 12x^2$ **3.** $(x - 3)(2) + (2x - 1)(1) = 4x - 7$

5. $\dfrac{(x - 3)(1) - x(1)}{(x - 3)^2} = \dfrac{-3}{(x - 3)^2}$ **7.** $\dfrac{(x - 2)(2) - (2x + 3)(1)}{(x - 2)^2} = \dfrac{-7}{(x - 2)^2}$

9. $3xe^x + 3e^x = 3(x + 1)e^x$ **11.** $x^3\left(\dfrac{1}{x}\right) + 3x^2 \ln x = x^2(1 + 3 \ln x)$

13. $(x^2 + 1)(2) + (2x - 3)(2x) = 6x^2 - 6x + 2$ **15.** $(0.4x + 2)(0.5) + (0.5x - 5)(0.4) = 0.4x - 1$

17. $\dfrac{(2x - 3)(2x) - (x^2 + 1)(2)}{(2x - 3)^2} = \dfrac{2x^2 - 6x - 2}{(2x - 3)^2}$ **19.** $(x^2 + 2)2x + (x^2 - 3)2x = 4x^3 - 2x$

21. $\dfrac{(x^2 - 3)2x - (x^2 + 2)2x}{(x^2 - 3)^2} = \dfrac{-10x}{(x^2 - 3)^2}$ **23.** $\dfrac{(x^2 + 1)e^x - e^x(2x)}{(x^2 + 1)^2} = \dfrac{(x - 1)^2 e^x}{(x^2 + 1)^2}$

25. $\dfrac{(1 + x)\left(\dfrac{1}{x}\right) - \ln x}{(1 + x)^2} = \dfrac{1 + x - x \ln x}{x(1 + x)^2}$ **27.** $xf'(x) + f(x)$ **29.** $x^3 f'(x) + 3x^2 f(x)$ **31.** $\dfrac{x^2 f'(x) - 2xf(x)}{x^4}$

33. $\dfrac{f(x) - xf'(x)}{[f(x)]^2}$ **35.** $e^x f'(x) + f(x)e^x = e^x[f'(x) + f(x)]$ **37.** $\dfrac{f(x)\left(\dfrac{1}{x}\right) - (\ln x)f'(x)}{f(x)^2} = \dfrac{f(x) - (x \ln x)f'(x)}{x f(x)^2}$

39. $(2x + 1)(2x - 3) + (x^2 - 3x)(2) = 6x^2 - 10x - 3$ **41.** $(2.5t - t^2)(4) + (4t + 1.4)(2.5 - 2t) = -12t^2 + 17.2t + 3.5$

43. $\dfrac{(x^2 + 2x)(5) - (5x - 3)(2x + 2)}{(x^2 + 2x)^2} = \dfrac{-5x^2 + 6x + 6}{(x^2 + 2x)^2}$ **45.** $\dfrac{(w^2 - 1)(2w - 3) - (w^2 - 3w + 1)(2w)}{(w^2 - 1)^2} = \dfrac{3w^2 - 4w + 3}{(w^2 - 1)^2}$

47. $(1 + x - x^2)e^x + e^x(1 - 2x) = (2 - x - x^2)e^x$ **49.** $f'(x) = (1 + 3x)(-2) + (5 - 2x)(3); y = -11x + 29$

51. $f'(x) = \dfrac{(3x - 4)(1) - (x - 8)(3)}{(3x - 4)^2}; y = 5x - 13$ **53.** $f'(x) = \dfrac{2^x - x(2^x \ln 2)}{2^{2x}}; y = \left(\dfrac{1 - 2 \ln 2}{4}\right)x + \ln 2$

55. $f'(x) = (2x - 15)(2x) + (x^2 + 18)(2) = 6(x - 2)(x - 3); x = 2, x = 3$

57. $f'(x) = \dfrac{(x^2 + 1)(1) - x(2x)}{(x^2 + 1)^2} = \dfrac{1 - x^2}{(x^2 + 1)^2}; x = -1, x = 1$ **59.** $7x^6 - 3x^2$ **61.** $-27x^{-4} = -\dfrac{27}{x^4}$

63. $(w + 1)2^w \ln 2 + 2^w = [(w + 1) \ln 2 + 1]2^w$ **65.** $\dfrac{(4x^2 + 5x - 1)(6x - 2) - (3x^2 - 2x + 3)(8x + 5)}{(4x^2 + 5x - 1)^2} = \dfrac{23x^2 - 30x - 13}{(4x^2 + 5x - 1)^2}$

67. $9x^{1/3}(3x^2) + (x^3 + 5)(3x^{-2/3}) = \dfrac{30x^3 + 15}{x^{2/3}}$ **69.** $\dfrac{(1 + x^2)\dfrac{1}{x \ln 2} - 2x \log_2 x}{(1 + x^2)^2} = \dfrac{1 + x^2 - 2x^2 \ln x}{x(1 + x^2)^2 \ln 2}$

71. $\dfrac{(x^2 - 3)(2x^{-2/3}) - 6x^{1/3}(2x)}{(x^2 - 3)^2} = \dfrac{-10x^2 - 6}{(x^2 - 3)^2 x^{2/3}}$ **73.** $g'(t) = \dfrac{(3t^2 - 1)(0.2) - (0.2t)(6t)}{(3t^2 - 1)^2} = \dfrac{-0.6t^2 - 0.2}{(3t^2 - 1)^2}$

75. $(20x)\dfrac{1}{x \ln 10} + 20 \log x = \dfrac{20(1 + \ln x)}{\ln 10}$ **77.** $x^{-2/3}(3x^2 - 4x) + (x^3 - 2x^2)\left(-\tfrac{2}{3}x^{-5/3}\right) = -\tfrac{8}{3}x^{1/3} + \tfrac{7}{3}x^{4/3}$

79. $\dfrac{(x^2 + 1)[(2x^2 - 1)(2x) + (x^2 + 3)(4x)] - (2x^2 - 1)(x^2 + 3)(2x)}{(x^2 + 1)^2} = \dfrac{4x^5 + 8x^3 + 16x}{(x^2 + 1)^2}$

81. $\dfrac{e^t(1 + \ln t) - (t \ln t)e^t}{e^{2t}} = \dfrac{1 + \ln t - t \ln t}{e^t}$

83. (A) $S'(t) = \dfrac{(t^2 + 50)(180t) - 90t^2(2t)}{(t^2 + 50)^2} = \dfrac{9{,}000t}{(t^2 + 50)^2}$

    (B) $S(10) = 60$; $S'(10) = 4$. After 10 months, the total sales are 60,000 CDs and sales are increasing at the rate of 4,000 CDs per month.

    (C) Approximately 64,000 CDs

85. (A) $\dfrac{dx}{dp} = \dfrac{(0.1p + 1)(0) - 4{,}000(0.1)}{(0.1p + 1)^2} = \dfrac{-400}{(0.1p + 1)^2}$

    (B) $x = 800$; $dx/dp = -16$. At a price level of \$40, the demand is 800 CD players per week and demand is decreasing at the rate of 16 players per dollar.

    (C) Approximately 784 CD players

87. (A) $C'(t) = \dfrac{(t^2 + 1)(0.14) - 0.14t(2t)}{(t^2 + 1)^2} = \dfrac{0.14 - 0.14t^2}{(t^2 + 1)^2}$

    (B) $C'(0.5) = 0.0672$. After 0.5 h, concentration is increasing at the rate of 0.0672 mg/cm^3 per hour.
$C'(3) = -0.0112$. After 3 h, concentration is decreasing at the rate of 0.0112 mg/cm^3 per hour.

89. (A) $N'(x) = \dfrac{(x + 32)(100) - (100x + 200)}{(x + 32)^2} = \dfrac{3{,}000}{(x + 32)^2}$

    (B) $N'(4) = 2.31$; $N'(68) = 0.30$

Exercise 4-4

1. $(3x^2 + 2)^3$ **3.** e^{-x^2} **5.** $y = u^4; u = 3x^2 - x + 5$ **7.** $y = e^u; u = 1 + x + x^2$ **9.** 3   **11.** $(-4x)$ **13.** $2x$

15. $4x^3$ **17.** $6(2x + 5)^2$ **19.** $-8(5 - 2x)^3$ **21.** $5(4 + 0.2x)^4(0.2) = (4 + 0.2x)^4$ **23.** $30x(3x^2 + 5)^4$ **25.** $5e^{5x}$

27. $-18e^{-6x}$ **29.** $(2x - 5)^{-1/2} = \dfrac{1}{(2x - 5)^{1/2}}$ **31.** $-8x^3(x^4 + 1)^{-3} = \dfrac{-8x^3}{(x^4 + 1)^3}$ **33.** $\dfrac{6x}{1 + x^2}$ **35.** $\dfrac{3(1 + \ln x)^2}{x}$

37. $f'(x) = 6(2x - 1)^2; y = 6x - 5; x = \frac{1}{2}$ **39.** $f'(x) = 2(4x - 3)^{-1/2} = \dfrac{2}{(4x - 3)^{1/2}}; y = \dfrac{2}{3}x + 1$; none

41. $f'(x) = 10(x - 2)e^{x^2 - 4x + 1}; y = -20ex + 5e; x = 2$ **43.** $12(x^2 - 2)^3(2x) = 24x(x^2 - 2)^3$

45. $-6(t^2 + 3t)^{-4}(2t + 3) = \dfrac{-6(2t + 3)}{(t^2 + 3t)^4}$ **47.** $\dfrac{1}{2}(w^2 + 8)^{-1/2}(2w) = \dfrac{w}{\sqrt{w^2 + 8}}$ **49.** $12xe^{3x} + 4e^{3x} = 4(3x + 1)e^{3x}$

51. $\dfrac{x^3\left(\dfrac{1}{1 + x}\right) - 3x^2 \ln(1 + x)}{x^6} = \dfrac{x - 3(1 + x)\ln(1 + x)}{x^4(1 + x)}$ **53.** $6te^{3(t^2 + 1)}$ **55.** $\dfrac{3x}{x^2 + 3}$

57. $-5(w^3 + 4)^{-6}(3w^2) = \dfrac{-15w^2}{(w^3 + 4)^6}$ **59.** $5(3x^{1/2} - 1)^4\dfrac{3}{2}x^{-1/2} = \dfrac{15(3\sqrt{x} - 1)^4}{2\sqrt{x}}$ **61.** $4\left(-\dfrac{1}{2}\right)(t^2 - 3t)^{-3/2}(2t - 3) = \dfrac{-4t + 6}{\sqrt{(t^2 - 3t)^3}}$

63. $f'(x) = (4 - x)^3 - 3x(4 - x)^2 = 4(4 - x)^2(1 - x); y = -16x + 48$

65. $f'(x) = \dfrac{(2x - 5)^3 - 6x(2x - 5)^2}{(2x - 5)^6} = \dfrac{-4x - 5}{(2x - 5)^4}; y = -17x + 54$ **67.** $f'(x) = \dfrac{1}{2x\sqrt{\ln x}}; y = \dfrac{1}{2e^x} + \dfrac{1}{2}$

69. $f'(x) = 2x(x - 5)^3 + 3x^2(x - 5)^2 = 5x(x - 5)^2(x - 2); x = 0, 2, 5$

71. $f'(x) = \dfrac{(2x + 5)^2 - 4x(2x + 5)}{(2x + 5)^4} = \dfrac{5 - 2x}{(2x + 5)^3}; x = \frac{5}{2}$ **73.** $f'(x) = (x^2 - 8x + 20)^{-1/2}(x - 4) = \dfrac{x - 4}{(x^2 - 8x + 20)^{1/2}}; x = 4$

75. $f'(x) = g'(x) = \dfrac{8x}{x^2 + 3}$ **77.** $18x^2(x^2 + 1)^2 + 3(x^2 + 1)^3 = 3(x^2 + 1)^2(7x^2 + 1)$

79. $\dfrac{24x^5(x^3 - 7)^3 - (x^3 - 7)^4 6x^2}{4x^6} = \dfrac{3(x^3 - 7)^3(3x^3 + 7)}{2x^4}$ **81.** $\dfrac{1}{\ln 2}\left(\dfrac{6x}{3x^2 - 1}\right)$ **83.** $(2x + 1)(10^{x^2 + x})(\ln 10)$

85. $\dfrac{12x^2 + 5}{(4x^3 + 5x + 7)\ln 3}$ **87.** $2^{x^3 - x^2 + 4x + 1}(3x^2 - 2x + 4)\ln 2$ **89.** $\dfrac{(x - 3)^{1/2}(2) - x(x - 3)^{-1/2}}{x - 3} = \dfrac{x - 6}{(x - 3)^{3/2}}$

91. $\frac{1}{2}[(2x - 1)^3(x^2 + 3)^4]^{-1/2}[8x(2x - 1)^3(x^2 + 3)^3 + 6(x^2 + 3)^4(2x - 1)^2] = (2x - 1)^{1/2}(x^2 + 3)(11x^2 - 4x + 9)$

93. (A) $C'(x) = (2x + 16)^{-1/2} = \dfrac{1}{(2x + 16)^{1/2}}$

(B) $C'(24) = \frac{1}{8}$, or \$12.50. At a production level of 24 calculators, total cost is increasing at the rate of \$12.50 per calculator and the cost of producing the 25th calculator is approximately \$12.50.
$C'(42) = \frac{1}{10}$, or \$10.00. At a production level of 42 calculators, total cost is increasing at the rate of \$10.00 per calculator and the cost of producing the 43rd calculator is approximately \$10.00.

95. (A) $\dfrac{dx}{dp} = 40(p + 25)^{-1/2} = \dfrac{40}{(p + 25)^{1/2}}$

(B) $x = 400$ and $dx/dp = 4$. At a price of \$75, the supply is 400 speakers per week and supply is increasing at the rate of 4 speakers per dollar.

97. (A) After 1 hr, the concentration is decreasing at the rate of 1.60 mg/mL per hour; after 4 hr, the concentration is decreasing at the rate of 0.08 mg/mL per hour.

(B)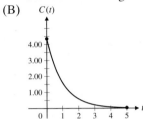

99. 2.27 mm of mercury/yr; 0.81 mm of mercury/yr; 0.41 mm of mercury/yr

101. (A) $f'(n) = n(n - 2)^{-1/2} + 2(n - 2)^{1/2}$

(B) $f'(11) = \frac{29}{3} = 9.67$. When the list contains 11 items, the learning time is increasing at the rate of 9.67 min per item.
$f'(27) = \frac{77}{5} = 15.4$. When the list contains 27 items, the learning time is increasing at the rate of 15.4 min per item.

Exercise 4-5

1. $y' = -\dfrac{3}{5}$ **3.** $y' = \dfrac{3x}{2}$ **5.** $y' = 10x$; 10   **7.** $y' = \dfrac{2x}{3y^2}$; $\dfrac{4}{3}$ **9.** $y' = -\dfrac{3}{2y + 2}$; $-\dfrac{3}{4}$ **11.** $y' = -\dfrac{y}{x}$; $-\dfrac{3}{2}$

13. $y' = -\dfrac{2y}{2x + 1}$; 4   **15.** $y' = \dfrac{6 - 2y}{x}$; -1 **17.** $y' = \dfrac{2x}{e^y - 2y}$; 2   **19.** $y' = \dfrac{3x^2 y}{y + 1}$; $\dfrac{3}{2}$ **21.** $y' = \dfrac{6x^2 y - y \ln y}{x + 2y}$; 2

23. $x' = \dfrac{2tx - 3t^2}{2x - t^2}$; 8   **25.** $y'|_{(1.6,\,1.8)} = -\dfrac{3}{4}$; $y'|_{(1.6,\,0.2)} = \dfrac{3}{4}$ **27.** $y = -x + 5$ **29.** $y = \frac{2}{5}x - \frac{12}{5}$; $y = \frac{3}{5}x + \frac{12}{5}$

31. $y' = -\dfrac{1}{x}$ **33.** $y' = \dfrac{1}{3(1 + y)^2 + 1}$; $\dfrac{1}{13}$ **35.** $y' = \dfrac{3(x - 2y)^2}{6(x - 2y)^2 + 4y}$; $\dfrac{3}{10}$ **37.** $y' = \dfrac{3x^2(7 + y^2)^{1/2}}{y}$; 16

39. $y' = \dfrac{y}{2xy^2 - x}$; 1   **41.** $y = 0.63x + 1.04$ **43.** $p' = \dfrac{1}{2p - 2}$ **45.** $p' = -\dfrac{\sqrt{10{,}000 - p^2}}{p}$ **47.** $\dfrac{dL}{dV} = \dfrac{-(L + m)}{V + n}$

Exercise 4-6

1. 30   **3.** $-\frac{16}{3}$ **5.** $-\frac{16}{7}$ **7.** Decreasing at 9 units/sec   **9.** Approx. -3.03 ft/sec   **11.** $dA/dt \approx 126$ ft²/sec

13. 3,768 cm³/min   **15.** 6 lb/in.²/hr   **17.** $-\frac{9}{4}$ ft/sec   **19.** $\frac{20}{3}$ ft/sec   **21.** 0.0214 ft/sec; 0.0135 ft/sec; yes, at $t = 0.000\,19$ sec

23. 3.835 units/sec   **25.** (A) $dC/dt = \$15{,}000$/wk   (B) $dR/dt = -\$50{,}000$/wk   (C) $dP/dt = -\$65{,}000$/wk

27. $ds/dt = \$2{,}207$/wk   **29.** (A) $dx/dt = -12.73$ units/month   (B) $dp/dt = \$1.53$/month   **31.** Approximately 100 ft³/min

Exercise 4-7

1. 0   **3.** $\dfrac{1}{x}$ **5.** $\dfrac{1}{x + 50}$ **7.** $\dfrac{100 - x}{100x - 0.5x^2}$ **9.** $-\dfrac{2}{1 + 2e^{2x}}$ **11.** $\dfrac{28 + 3 \ln x}{25x + 3x \ln x}$

13. (A) Inelastic (B) Unit elasticity (C) Elastic   **15.** (A) Inelastic (B) Unit elasticity (C) Elastic

17. (A) $x = 6{,}000 - 200p$ $0 \le p \le 30$ (B) $E(p) = \dfrac{p}{30 - p}$ (C) $E(10) = 0.5$; 5% decrease

(D) $E(25) = 5$; 50% decrease   (E) $E(15) = 1$; 10% decrease

19. (A) $x = 3{,}000 - 50p$ $0 \le p \le 60$ (B) $R(p) = 3{,}000p - 50p^2$ (C) $E(p) = \dfrac{p}{60 - p}$

(D) Elastic on (30, 60); inelastic on (0, 30)   (E) Increasing on (0, 30); decreasing on (30, 60)
(F) Decrease   (G) Increase

21. Elastic on $(10, 30)$; inelastic on $(0, 10)$ **23.** Elastic on $(48, 72)$; inelastic on $(0, 48)$
25. Elastic on $(25, 25\sqrt{2})$; inelastic on $(0, 25)$ **27.** $R(p) = 20p(10 - p)$ **29.** $R(p) = 40p(p - 15)^2$
31. $R(p) = 30p - 10p\sqrt{p}$

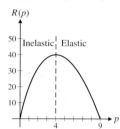

33. $\dfrac{3}{2}$ **35.** $\dfrac{1}{2}$ **37.** k **39.** \$25 per day   **41.** Increase   **43.** Decrease   **45.** \$2.50
47.   **49.** -0.13 robbery annually per 1,000 population age 12 and older.

Chapter 4 Review Exercise

1. \$3,136.62; \$4,919.21; \$12,099.29 *(4-1)*   **2.** $\dfrac{2}{x} + 3e^x$ *(4-2)*   **3.** $2e^{2x-3}$ *(4-4)*   **4.** $\dfrac{2}{2x + 7}$ *(4-4)*

5. (A) $y = \ln(3 + e^x)$ (B) $\dfrac{dy}{dx} = \dfrac{e^x}{3 + e^x}$ *(4-4)*   **6.** $y' = \dfrac{9x^2}{4y}; \dfrac{9}{8}$ *(4-5)*   **7.** $dy/dt = 216$ *(4-6)*

8. (A) $x = 1{,}000 - 25p$ (B) $\dfrac{p}{40 - p}$ (C) 0.6; demand is inelastic and insensitive to small changes in price.
    (D) $1{,}000p - 25p^2$ (E) Revenue increases *(4-7)*   **9.** -10 *(4-2)*

10. $\displaystyle\lim_{n \to \infty}\left(1 + \dfrac{2}{n}\right)^n = e^2 \approx 7.389\,06$ *(4-1)*   **11.** $\dfrac{7[(\ln z)^6 + 1]}{z}$ *(4-4)*   **12.** $x^5(1 + 6 \ln x)$ *(4-3)*   **13.** $\dfrac{e^x(x - 6)}{x^7}$ *(4-3)*

14. $\dfrac{6x^2 - 3}{2x^3 - 3x}$ *(4-4)*   **15.** $(3x^2 - 2x)e^{x^3-x^2}$ *(4-4)*   **16.** $\dfrac{1 - 2x \ln 5x}{xe^{2x}}$ *(4-4)*   **17.** $y = -x + 2$; $y = -ex + 1$ *(4-4)*

18. $y' = \dfrac{3y - 2x}{8y - 3x}; \dfrac{8}{19}$ *(4-5)*   **19.** $x' = \dfrac{4tx}{3x^2 - 2t^2}; -4$ *(4-5)*   **20.** $y' = \dfrac{1}{e^y + 2y}; 1$ *(4-5)*   **21.** $y' = \dfrac{2xy}{1 + 2y^2}; \dfrac{2}{3}$ *(4-5)*

22. $dy/dt = -2$ units/sec *(4-6)*   **23.** 0.27 ft/sec *(4-6)*   **24.** $dR/dt = 1/\pi \approx 0.318$ in./min *(4-6)*
25. Elastic for $5 < p < 15$; inelastic for $0 < p < 5$ *(4-7)*
26.   **27.** (A) $y = [\ln(4 - e^x)]^3$
    (B) $\dfrac{dy}{dx} = \dfrac{-3e^x[\ln(4 - e^x)]^2}{4 - e^x}$ *(4-4)*

(4-7)

28. $2x(5^{x^2-1})(\ln 5)$ *(4-4)*   **29.** $\left(\dfrac{1}{\ln 5}\right)\dfrac{2x - 1}{x^2 - x}$ *(4-4)*   **30.** $\dfrac{2x + 1}{2(x^2 + x)\sqrt{\ln(x^2 + x)}}$ *(4-4)*   **31.** $y' = \dfrac{2x - e^{xy}y}{xe^{xy} - 1}; 0$ *(4-5)*

32. The rate of increase of area is proportional to the radius R, so the rate is smallest when $R = 0$, and has no largest value. *(4-6)*
33. Yes, for $-\sqrt{3}/3 < x < \sqrt{3}/3$ *(4-6)*   **34.** (A) 15 yr   (B) 13.9 yr *(4-1)*
35. $A'(t) = 10e^{0.1t}$; $A'(1) = \$11.05$/yr; $A'(10) = \$27.18$/yr *(4-1)*   **36.** $R'(x) = (1{,}000 - 20x)e^{-0.02x}$ *(4-4)*

37. $p' = \dfrac{-(5{,}000 - 2p^3)^{1/2}}{3p^2}$ *(4-5)*   **38.** $dR/dt = \$110$/day *(4-6)*   **39.** Increase price *(4-7)*   **40.** 0.02378 *(4-7)*

41. -1.111 mg/mL per hour; -0.335 mg/mL per hour *(4-4)*   **42.** $dR/dt = -3/(2\pi)$; approx. 0.477 mm/day *(4-6)*
43. (A) Increasing at the rate of 2.68 units/day at the end of 1 day of training; increasing at the rate of 0.54 unit/day after 5 days of training
    (B) 7 days *(4-4)*
44. $dT/dt = -1/27 \approx -0.037$ min/operation hour *(4-6)*

CHAPTER 5

Exercise 5-1

1. $(a, b); (d, f); (g, h)$　　**3.** $(b, c); (c, d); (f, g)$　　**5.** c, d, f　　**7.** b, f

9. Local maximum at $x = a$; local minimum at $x = c$; no local extrema at $x = b$ and $x = d$　　**11.** e　　**13.** d　　**15.** f

17. c　　**19.** Decreasing on $(-\infty, 1)$; increasing on $(1, \infty)$; $f(1) = -2$ is a local minimum

21. Increasing on $(-\infty, -4)$; decreasing on $(-4, \infty)$; $f(-4) = 7$ is a local maximum　　**23.** Increasing for all x; no local extrema

25. Increasing on $(-\infty, -2)$ and $(3, \infty)$; decreasing on $(-2, 3)$; local maximum at $x = -2$, local minimum at $x = 3$

27. Decreasing on $(-\infty, 1)$; increasing on $(1, \infty)$; $f(1) = 4$ is a local minimum

29. Increasing on $(-\infty, 2)$; decreasing on $(2, \infty)$; local maximum at $x = 2$

31. Increasing on $(-\infty, 8)$; decreasing on $(8, \infty)$; local maximum at $x = 8$

33. Increasing on $(0, e)$; decreasing on (e, ∞); local maximum at $x = e$

35. Decreasing on $(-\infty, -1)$ and $(1.5, 4)$; increasing on $(-1, 1.5)$ and $(4, \infty)$; local maximum at $x = 1.5$; local minima at $x = -1$ and $x = 4$

37. Critical value: $x = -0.39$; decreasing on $(-\infty, -0.39)$; increasing on $(-0.39, \infty)$; local minimum at $x = -0.39$

39. Critical values: $x = -0.77, 1.08, 2.69$; decreasing on $(-\infty, -0.77)$ and $(1.08, 2.69)$; increasing on $(-0.77, 1.08)$ and $(2.69, \infty)$; local minima at $x = -0.77$ and $x = 2.69$; local maximum at $x = 1.08$

41. Critical values: $x = 1.34, 2.82$; decreasing on $(0, 1.34)$ and $(2.82, \infty)$; increasing on $(1.34, 2.82)$; local minimum at $x = 1.34$; and local maximum at $x = 2.82$

43. Critical values: $0.36, 2.15$; increasing on $(-\infty, 0.36)$ and $(2.15, \infty)$; decreasing on $(0.36, 2.15)$; local maximum at $x = 0.36$; local minimum at $x = 2.15$

45. Critical values: $x = -0.05, 0, 0.13, 2.55$; decreasing on $(-\infty, -0.05)$ and $(0.13, 2.55)$; increasing on $(-0.05, 0.13)$ and $(2.55, \infty)$; local minima at $x = -0.05$ and 2.55; local maximum at $x = 0.13$

47. Increasing on $(-\infty, 4)$
Decreasing on $(4, \infty)$
Horizontal tangent at $x = 4$

49. Increasing on $(-\infty, -1), (1, \infty)$
Decreasing on $(-1, 1)$
Horizontal tangents at $x = -1, 1$

51. Decreasing for all x
Horizontal tangent at $x = 2$

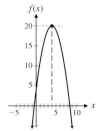
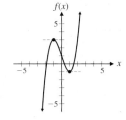
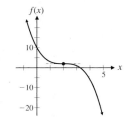

53. Decreasing on $(-\infty, -3)$ and $(0, 3)$; increasing on $(-3, 0)$ and $(3, \infty)$

55.

57.

59.

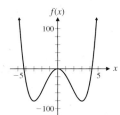
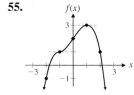
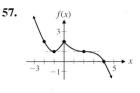
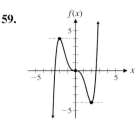

61.　　**63.** g_4　　**65.** g_6　　**67.** g_2　　**69.** Increasing on $(-1, 2)$; decreasing on $(-\infty, -1)$ and $(2, \infty)$; local minimum at $x = -1$; local maximum at $x = 2$

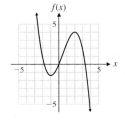

71. Increasing on $(-1, 2)$ and $(2, \infty)$; decreasing on $(-\infty, -1)$; local minimum at $x = -1$

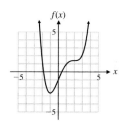

73. Increasing on $(-2, 0)$ and $(3, \infty)$; decreasing on $(-\infty, -2)$ and $(0, 3)$; local minima at $x = -2$ and $x = 3$; local maximum at $x = 0$

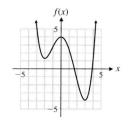

75. $f'(x) > 0$ on $(-\infty, -1)$ and $(3, \infty)$; $f'(x) < 0$ on $(-1, 3)$; $f'(x) = 0$ at $x = -1$ and $x = 3$

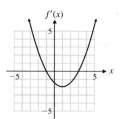

77. $f'(x) > 0$ on $(-2, 1)$ and $(3, \infty)$; $f'(x) < 0$ on $(-\infty, -2)$ and $(1, 3)$; $f'(x) = 0$ at $x = -2$, $x = 1$, and $x = 3$

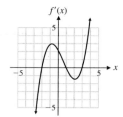

79. Critical values: $x = -2$, $x = 2$; increasing on $(-\infty, -2)$ and $(2, \infty)$; decreasing on $(-2, 0)$ and $(0, 2)$; local maximum at $x = -2$; local minimum at $x = 2$

81. Critical value: $x = -2$; increasing on $(-2, 0)$; decreasing on $(-\infty, -2)$ and $(0, \infty)$; local minimum at $x = -2$

83. Critical values: $x = 0$, $x = 4$; increasing on $(-\infty, 0)$ and $(4, \infty)$; decreasing on $(0, 2)$ and $(2, 4)$; local maximum at $x = 0$; local minimum at $x = 4$

85. Critical values: $x = 0$, $x = 4$, $x = 6$; increasing on $(0, 4)$ and $(6, \infty)$; decreasing on $(-\infty, 0)$ and $(4, 6)$; local maximum at $x = 4$; local minima at $x = 0$ and $x = 6$

87. Critical value: $x = 2$; increasing on $(2, \infty)$; decreasing on $(-\infty, 2)$; local minimum at $x = 2$

89. Critical value: $x = 0$; decreasing on $(-\infty, 0)$; increasing on $(0, \infty)$; $f(0) = 0$ is a local minimum

91. (A) There are no critical values and no local extrema. The function is increasing for all x.

(B) There are two critical values, $x = \pm\sqrt{-k/3}$. The function increases on $(-\infty, -\sqrt{-k/3})$ to a local maximum at $x = -\sqrt{-k/3}$, decreases on $(-\sqrt{-k/3}, \sqrt{-k/3})$ to a local minimum at $x = \sqrt{-k/3}$, and increases on $(\sqrt{-k/3}, \infty)$.

(C) The only critical value is $x = 0$. There are no local extrema. The function is increasing for all x.

93. (A) The marginal profit is positive on $(0, 600)$, 0 at $x = 600$, and negative on $(600, 1,000)$.

(B)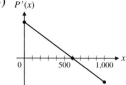

95. (A) The price decreases for the first 15 months to a local minimum, increases for the next 40 months to a local maximum, and then decreases for the remaining 15 months.

(B)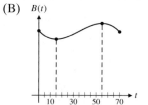

97. (A) $\overline{C}(x) = 0.05x + 20 + \dfrac{320}{x}$

(B) Critical value: $x = 80$; decreasing for $0 < x < 80$; increasing for $80 < x < 150$; local minimum at $x = 80$

99. $P(x)$ is increasing over (a, b) if $P'(x) = R'(x) - C'(x) > 0$ over (a, b); that is, if $R'(x) > C'(x)$ over (a, b).

101. Critical value: $t = 2$; increasing on $(0, 2)$; decreasing on $(2, 24)$; $C(2) = 0.07$ is a local maximum.

103. Critical value: $t = 7$; increasing for $0 < t < 7$; decreasing for $7 < t < 24$; local maximum at $t = 7$.

Exercise 5-2

1. (A) $(a, c), (c, d), (e, g)$ (B) $(d, e), (g, h)$ (C) $(d, e), (g, h)$ (D) $(a, c), (c, d), (e, g)$ (E) $(a, c), (c, d), (e, g)$
(F) $(d, e), (g, h)$ (G) d, e, g (H) d, e, g

3. (C)   **5.** (D)   **7.** $12x - 8$ **9.** $4x^{-3} - 18x^{-4}$ **11.** $2 + \dfrac{9}{2}x^{-3/2}$

13. $8(x^2 + 9)^3 + 48x^2(x^2 + 9)^2 = 8(x^2 + 9)^2(7x^2 + 9)$

15. $-2e^{-x^2} + 4x^2e^{-x^2} = 2e^{-x^2}(2x^2 - 1)$

17. $\dfrac{x^3\left(-\dfrac{2}{x}\right) - (1 - 2\ln x)3x^2}{x^6} = \dfrac{6\ln x - 5}{x^4}$

19. Concave upward for all x; no inflection points

21. Concave downward on $(-\infty, \frac{4}{3})$; concave upward on $(\frac{4}{3}, \infty)$; inflection point at $x = \frac{4}{3}$

23. Concave downward on $(-\infty, 0)$ and $(6, \infty)$; concave upward on $(0, 6)$; inflection points at $x = 0$ and $x = 6$

25. Concave upward on $(-2, 4)$; concave downward on $(-\infty, -2)$ and $(4, \infty)$; inflection points at $x = -2$ and $x = 4$

27. Concave upward on $(-\infty, \ln 2)$; concave downward on $(\ln 2, \infty)$; inflection point at $x = \ln 2$

29. 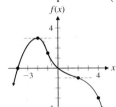   31. 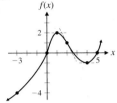   33. 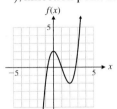   35.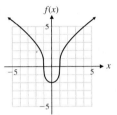

37. Domain: All real numbers
 y intercept: 16; x intercepts: $2 - 2\sqrt{3}, 2, 2 + 2\sqrt{3}$
 Increasing on $(-\infty, 0)$ and $(4, \infty)$
 Decreasing on $(0, 4)$
 Local maximum at $x = 0$, local minimum at $x = 4$
 Concave downward on $(-\infty, 2)$
 Concave upward on $(2, \infty)$
 Inflection point at $x = 2$

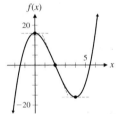

39. Domain: All real numbers
 y intercept: 2; x intercept: -1
 Increasing on $(-\infty, \infty)$
 Concave downward on $(-\infty, 0)$
 Concave upward on $(0, \infty)$
 Inflection point at $x = 0$

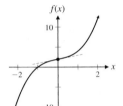

41. Domain: All real numbers
 y intercept: 0; x intercepts: 0, 4
 Increasing on $(-\infty, 3)$
 Decreasing on $(3, \infty)$
 Local maximum at $x = 3$
 Concave upward on $(0, 2)$
 Concave downward on $(-\infty, 0)$ and $(2, \infty)$
 Inflection points at $x = 0$ and $x = 2$

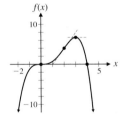

43. Domain: All real numbers
 y intercept: 0; x intercepts: 0, 1
 Increasing on $(0.25, \infty)$
 Decreasing on $(-\infty, 0.25)$
 Local minimum at $x = 0.25$
 Concave upward on $(-\infty, 0.5)$ and $(1, \infty)$
 Concave downward on $(0.5, 1)$
 Inflection points at $x = 0.5$ and $x = 1$

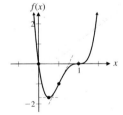

45. Domain: All real numbers
y intercept: 27; x intercepts: $-3, 3$
Increasing on $(-\infty, -\sqrt{3})$ and $(0, \sqrt{3})$
Decreasing on $(-\sqrt{3}, 0)$ and $(\sqrt{3}, \infty)$
Local maxima at $x = -\sqrt{3}$ and $x = \sqrt{3}$
Local minimum at $x = 0$
Concave upward on $(-1, 1)$
Concave downward on $(-\infty, -1)$ and $(1, \infty)$
Inflection points at $x = -1$ and $x = 1$

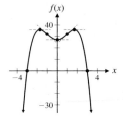

47. Domain: All real numbers
y intercept: 16; x intercepts: $-2, 2$
Decreasing on $(-\infty, -2)$ and $(0, 2)$
Increasing on $(-2, 0)$ and $(2, \infty)$
Local minima at $x = -2$ and $x = 2$
Local maximum at $x = 0$
Concave upward on $(-\infty, -2\sqrt{3}/3)$ and $(2\sqrt{3}/3, \infty)$
Concave downward on $(-2\sqrt{3}/3, 2\sqrt{3}/3)$
Inflection points at $x = -2\sqrt{3}/3$ and $x = 2\sqrt{3}/3$

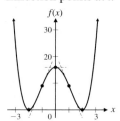

49. Domain: All real numbers
y intercept: 0; x intercepts: 0, 1.5
Decreasing on $(-\infty, 0)$ and $(0, 1.25)$
Increasing on $(1.25, \infty)$
Local minimum at $x = 1.25$
Concave upward on $(-\infty, 0)$ and $(1, \infty)$
Concave downward on $(0, 1)$
Inflection points at $x = 0$ and $x = 1$

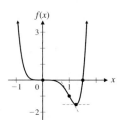

51. Domain: All real numbers
y intercept: 0; x intercept: 0
Increasing on $(-\infty, \infty)$
Concave downward on $(-\infty, \infty)$

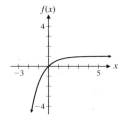

53. Domain: All real numbers
y intercept: 5
Decreasing on $(-\infty, \ln 4)$
Increasing on $(\ln 4, \infty)$
Local minimum at $x = \ln 4$
Concave upward on $(-\infty, \infty)$

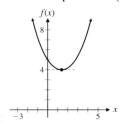

55. Domain: $(0, \infty)$
x intercept: e^2
Increasing on $(-\infty, \infty)$
Concave downward on $(-\infty, \infty)$

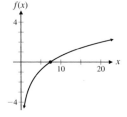

57. Domain: $(-4, \infty)$
x intercept: $e^2 - 4$
Increasing on $(-4, \infty)$
Concave downward on $(-4, \infty)$

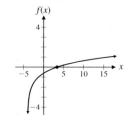

59.

x	f'(x)	f(x)
$-\infty < x < -1$	Positive and decreasing	Increasing and concave downward
$x = -1$	x intercept	Local maximum
$-1 < x < 0$	Negative and decreasing	Decreasing and concave downward
$x = 0$	Local minimum	Inflection point
$0 < x < 2$	Negative and increasing	Decreasing and concave upward
$x = 2$	Local maximum	Inflection point
$2 < x < \infty$	Negative and decreasing	Decreasing and concave downward

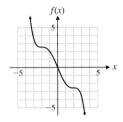

61.

x	f'(x)	f(x)
$-\infty < x < -2$	Negative and increasing	Decreasing and concave upward
$x = -2$	Local maximum	Inflection point
$-2 < x < 0$	Negative and decreasing	Decreasing and concave downward
$x = 0$	Local minimum	Inflection point
$0 < x < 2$	Negative and increasing	Decreasing and concave upward
$x = 2$	Local maximum	Inflection point
$2 < x < \infty$	Negative and decreasing	Decreasing and concave downward

63. Domain: All real numbers
x intercepts: $-1.18, 0.61, 1.87, 3.71$
y intercept: -5
Decreasing on $(-\infty, -0.53)$ and $(1.24, 3.04)$
Increasing on $(-0.53, 1.24)$ and $(3.04, \infty)$
Local minima at $x = -0.53$ and $x = 3.04$
Local maximum at $x = 1.24$
Concave upward on $(-\infty, 0.22)$ and $(2.28, \infty)$
Concave downward on $(0.22, 2.28)$
Inflection points at $x = 0.22$ and $x = 2.28$

65. Domain: All real numbers
y intercept: 100; x intercepts: $8.01, 13.36$
Increasing on $(-0.10, 4.57)$ and $(11.28, \infty)$
Decreasing on $(-\infty, -0.10)$ and $(4.57, 11.28)$
Local maximum at $x = 4.57$
Local minima at $x = -0.10$ and $x = 11.28$
Concave upward on $(-\infty, 1.95)$ and $(8.55, \infty)$
Concave downward on $(1.95, 8.55)$
Inflection points at $x = 1.95$ and $x = 8.55$

67. Domain: All real numbers
x intercepts: $-2.40, 1.16$; y intercept: 3
Increasing on $(-\infty, -1.58)$
Decreasing on $(-1.58, \infty)$
Local maximum at $x = -1.58$
Concave downward on $(-\infty, -0.88)$ and $(0.38, \infty)$
Concave upward on $(-0.88, 0.38)$
Inflection points at $x = -0.88$ and $x = 0.38$

69. Domain: All real numbers
x intercepts: $-6.68, -3.64, -0.72$; y intercept: 30
Decreasing on $(-5.59, -2.27)$ and $(1.65, 3.82)$
Increasing on $(-\infty, -5.59)$, $(-2.27, 1.65)$, and $(3.82, \infty)$
Local minima at $x = -2.27$ and $x = 3.82$
Local maxima at $x = -5.59$ and $x = 1.65$
Concave upward on $(-4.31, -0.40)$ and $(2.91, \infty)$
Concave downward on $(-\infty, -4.31)$ and $(-0.40, 2.91)$
Inflection points at $x = -4.31$, $x = -0.40$, and $x = 2.91$

71. If $f'(x)$ has a local extremum at $x = c$, then $f'(x)$ must change from increasing to decreasing or from decreasing to increasing at $x = c$. Thus, the graph of $y = f(x)$ must change concavity at $x = c$, and there must be an inflection point at $x = c$.

73. If there is an inflection point on the graph of $y = f(x)$ at $x = c$, then $f(x)$ must change concavity at $x = c$. Consequently, $f'(x)$ must change from increasing to decreasing or from decreasing to increasing at $x = c$, and $x = c$ is a local extremum for $f'(x)$.

75. The graph of the CPI is concave upward.

77. The graph of $y = C'(x)$ is positive and decreasing. Since marginal costs are decreasing, the production process is becoming more efficient as production increases.

79. (A) Local maximum at $x = 60$ (B) Concave downward on the whole interval $(0, 80)$

81. (A) Local maximum at $x = 1$
(B) Concave downward on $(-\infty, 2)$;
concave upward on $(2, \infty)$

83. Increasing on $(0, 10)$; decreasing on $(10, 15)$; point of diminishing returns is $x = 10$, max $T'(x) = T'(10) = 500$

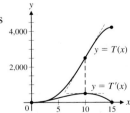

85. Increasing on $(24, 36)$; decreasing on $(36, 45)$; point of diminishing returns is $x = 36$, max $N'(x) = N'(36) = 3,888$

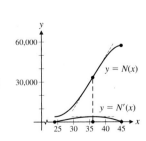

87. (A)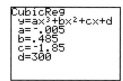

(B) 32 ads to sell 574 cars per month

89. (A) Increasing on $(0, 10)$; decreasing on $(10, 20)$
(B) Inflection point at $t = 10$
(C)
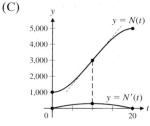

(D) $N'(10) = 300$

91. (A) Increasing on $(5, \infty)$; decreasing on $(0, 5)$
(B) Inflection point at $n = 5$

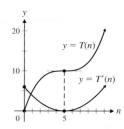

(C) $T'(5) = 0$

Exercise 5-3

1. $\frac{8}{3}$ **3.** $\frac{1}{2}$ **5.** 1   **7.** 4   **9.** 0   **11.** ∞ **13.** ∞ **15.** 5   **17.** ∞ **19.** 8   **21.** 0

23. ∞ **25.** ∞ **27.** $\frac{1}{3}$ **29.** -2 **31.** $-\infty$ **33.** 0   **35.** 0   **37.** 0   **39.** $\frac{1}{4}$ **41.** $\frac{1}{3}$ **43.** 0

45. 0   **47.** ∞ **49.** 1   **51.** 1

Exercise 5-4

1. (A) $(-\infty, b), (0, e), (e, g)$ (B) $(b, d), (d, 0), (g, \infty)$ (C) $(b, d), (d, 0), (g, \infty)$ (D) $(-\infty, b), (0, e), (e, g)$ (E) $x = 0$
(F) $x = b, x = g$ (G) $(-\infty, a), (d, e), (h, \infty)$ (H) $(a, d), (e, h)$ (I) $(a, d), (e, h)$ (J) $(-\infty, a), (d, e), (h, \infty)$
(K) $x = a, x = h$ (L) $y = L$ (M) $x = d, x = e$

3.
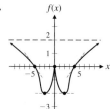

5.
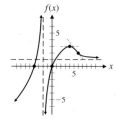

7.
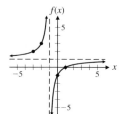

9.
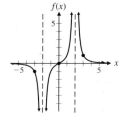

11. Domain: All real numbers, except 3
y intercept: -1; x intercept: -3
Horizontal asymptote: $y = 1$
Vertical asymptote: $x = 3$
Decreasing on $(-\infty, 3)$ and $(3, \infty)$
Concave upward on $(3, \infty)$
Concave downward on $(-\infty, 3)$

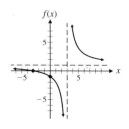

13. Domain: All real numbers, except 2
y intercept: 0; x intercept: 0
Horizontal asymptote: $y = 1$
Vertical asymptote: $x = 2$
Decreasing on $(-\infty, 2)$ and $(2, \infty)$
Concave downward on $(-\infty, 2)$
Concave upward on $(2, \infty)$

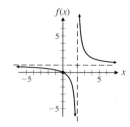

15. Domain: $(-\infty, \infty)$
y intercept: 10
Horizontal asymptote: $y = 5$
Decreasing on $(-\infty, \infty)$
Concave upward on $(-\infty, \infty)$

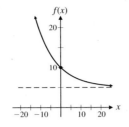

17. Domain: $(-\infty, \infty)$
y intercept: 0; x intercept: 0
Horizontal asymptote: $y = 0$
Increasing on $(-\infty, 5)$
Decreasing on $(5, \infty)$
Local maximum at $x = 5$
Concave upward on $(10, \infty)$
Concave downward on $(-\infty, 10)$
Inflection point at $x = 10$

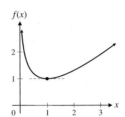

19. Domain: $(-\infty, 1)$
y intercept: 0; x intercept: 0
Vertical asymptote: $x = 1$
Decreasing on $(-\infty, 1)$
Concave downward on $(-\infty, 1)$

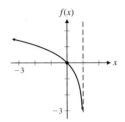

21. Domain: $(0, \infty)$
Vertical asymptote: $x = 0$
Increasing on $(1, \infty)$
Decreasing on $(0, 1)$
Local minimum at $x = 1$
Concave upward on $(0, \infty)$

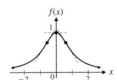

23. Domain: All real numbers, except ± 2
y intercept: 0; x intercept: 0
Horizontal asymptote: $y = 0$
Vertical asymptotes: $x = -2, x = 2$
Decreasing on $(-\infty, -2)$, $(-2, 2)$, and $(2, \infty)$
Concave upward on $(-2, 0)$ and $(2, \infty)$
Concave downward on $(-\infty, -2)$ and $(0, 2)$
Inflection point at $x = 0$

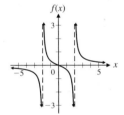

25. Domain: All real numbers
y intercept: 1
Horizontal asymptote: $y = 0$
Increasing on $(-\infty, 0)$
Decreasing on $(0, \infty)$
Local maximum at $x = 0$
Concave upward on $(-\infty, -\sqrt{3}/3)$ and $(\sqrt{3}/3, \infty)$
Concave downward on $(-\sqrt{3}/3, \sqrt{3}/3)$
Inflection points at $x = -\sqrt{3}/3$ and $x = \sqrt{3}/3$

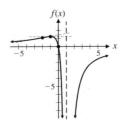

27. Domain: All real numbers except -1 and 1
y intercept: 0; x intercept: 0
Horizontal asymptote: $y = 0$
Vertical asymptote: $x = -1$ and $x = 1$
Increasing on $(-\infty, -1)$, $(-1, 1)$, and $(1, \infty)$
Concave upward on $(-\infty, -1)$ and $(0, 1)$
Concave downward on $(-1, 0)$ and $(1, \infty)$
Inflection point at $x = 0$

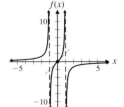

29. Domain: All real numbers except 1
y intercept: 0; x intercept: 0
Horizontal asymptote: $y = 0$
Vertical asymptote: $x = 1$
Increasing on $(-\infty, -1)$ and $(1, \infty)$
Decreasing on $(-1, 1)$
Local maximum at $x = -1$
Concave upward on $(-\infty, -2)$
Concave downward on $(-2, 1)$ and $(1, \infty)$
Inflection point at $x = -2$

31. Domain: All real numbers except 0
Vertical asymptote: $x = 0$
Increasing on $(-\infty, -\sqrt{2})$ and $(\sqrt{2}, \infty)$
Decreasing on $(-\sqrt{2}, 0)$ and $(0, \sqrt{2})$
Local minimum at $x = \sqrt{2}$
Local maximum at $x = -\sqrt{2}$
Concave upward on $(0, \infty)$
Concave downward on $(-\infty, 0)$

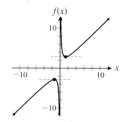

33. Domain: All real numbers except 0
Horizontal asymptote: $y = 1$
Vertical asymptote: $x = 0$
Increasing on $(0, 4)$
Decreasing on $(-\infty, 0)$ and $(4, \infty)$
Local maximum at $x = 4$
Concave upward on $(6, \infty)$
Concave downward on $(-\infty, 0)$ and $(0, 6)$
Inflection point at $x = 6$

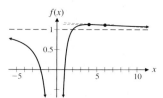

35. Domain: All real numbers except 1
y intercept: 0; x intercept: 0
Vertical asymptote: $x = 1$
Increasing on $(-\infty, 0)$ and $(2, \infty)$
Decreasing on $(0, 1)$ and $(1, 2)$
Local maximum at $x = 0$
Local minimum at $x = 2$
Concave upward on $(1, \infty)$
Concave downward on $(-\infty, 1)$

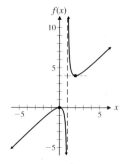

37. Domain: All real numbers except $-3, 3$
y intercept: $-\dfrac{2}{9}$
Horizontal asymptote: $y = 3$
Vertical asymptotes: $x = -3, x = 3$
Increasing on $(-\infty, -3)$ and $(-3, 0)$
Decreasing on $(0, 3)$ and $(3, \infty)$
Local maximum at $x = 0$
Concave upward on $(-\infty, -3)$ and $(3, \infty)$
Concave downward on $(-3, 3)$

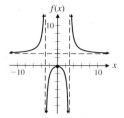

39. Domain: All real numbers except 2
y intercept: 0; x intercept: 0
Vertical asymptote: $x = 2$
Increasing on $(3, \infty)$
Decreasing on $(-\infty, 2)$ and $(2, 3)$
Local minimum at $x = 3$
Concave upward on $(-\infty, 0)$ and $(2, \infty)$
Concave downward on $(0, 2)$
Inflection point at $x = 0$

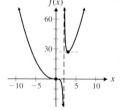

41. Domain: All real numbers
y intercept: 3; x intercept: 3
Horizontal asymptote: $y = 0$
Increasing on $(-\infty, 2)$
Decreasing on $(2, \infty)$
Local maximum at $x = 2$
Concave upward on $(-\infty, 1)$
Concave downward on $(1, \infty)$
Inflection point at $x = 1$

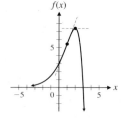

43. Domain: $(-\infty, \infty)$
y intercept: 1
Horizontal asymptote: $y = 0$
Increasing on $(-\infty, 0)$
Decreasing on $(0, \infty)$
Local maximum at $x = 0$
Concave upward on $(-\infty, -1)$ and $(1, \infty)$
Concave downward on $(-1, 1)$
Inflection points at $x = -1$ and $x = 1$

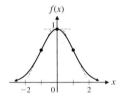

45. Domain: $(0, \infty)$
x intercept: 1
Increasing on $(e^{-1/2}, \infty)$
Decreasing on $(0, e^{-1/2})$
Local minimum at $x = e^{-1/2}$
Concave upward on $(e^{-3/2}, \infty)$
Concave downward on $(0, e^{-3/2})$
Inflection point at $x = e^{-3/2}$

47. Domain: $(0, \infty)$
x intercept: 1
Vertical asymptote: $x = 0$
Increasing on $(1, \infty)$
Decreasing on $(0, 1)$
Local minimum at $x = 1$
Concave upward on $(0, e)$
Concave downward on (e, ∞)
Inflection point at $x = e$

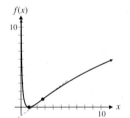

49. Domain: All real numbers except $-4, 2$
y intercept: $-1/8$
Horizontal asymptote: $y = 0$
Vertical asymptote: $x = -4, x = 2$
Increasing on $(-\infty, -4)$ and $(-4, -1)$
Decreasing on $(-1, 2)$ and $(2, \infty)$
Local maximum at $x = -1$
Concave upward on $(-\infty, -4)$ and $(2, \infty)$
Concave downward on $(-4, 2)$

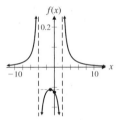

51. Domain: All real numbers except $-2, 2$
y intercept: 0; x intercept: 0
Horizontal asymptote: $y = 0$
Vertical asymptote: $x = -2, x = 2$
Decreasing on $(-\infty, -2)$, $(-2, 2)$, and $(2, \infty)$
Concave upward on $(-2, 0)$ and $(2, \infty)$
Concave downward on $(-\infty, -2)$ and $(0, 2)$
Inflection point at $x = 0$

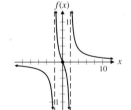

53. Domain: All real numbers except 4
y intercept: 0; x intercept: 0
Vertical asymptote: $x = 4$
Increasing on $(-\infty, 4)$ and $(12, \infty)$
Decreasing on $(4, 12)$
Local minimum at $x = 12$
Concave upward on $(0, 4)$ and $(4, \infty)$
Concave downward on $(-\infty, 0)$
Inflection point at $x = 0$

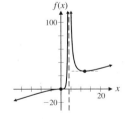

55. Domain: All real numbers except $-\sqrt{3}, \sqrt{3}$
y intercept: 0; x intercept: 0
Vertical asymptote: $x = -\sqrt{3}, x = \sqrt{3}$
Increasing on $(-3, -\sqrt{3}), (-\sqrt{3}, \sqrt{3})$, and $(\sqrt{3}, 3)$
Decreasing on $(-\infty, -3)$ and $(3, \infty)$
Local maximum at $x = 3$
Local minimum at $x = -3$
Concave upward on $(-\infty, -\sqrt{3})$ and $(0, \sqrt{3})$
Concave downward on $(-\sqrt{3}, 0)$ and $(\sqrt{3}, \infty)$
Inflection point at $x = 0$

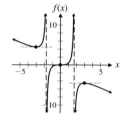

57. Domain: All real numbers except 0
Vertical asymptote: $x = 0$
Oblique asymptote: $y = x$
Increasing on $(-\infty, -2)$ and $(2, \infty)$
Decreasing on $(-2, 0)$ and $(0, 2)$
Local maximum at $x = -2$
Local minimum at $x = 2$
Concave upward on $(0, \infty)$
Concave downward on $(-\infty, 0)$

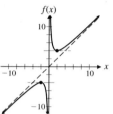

59. Domain: All real numbers except 0
x intercept: $\sqrt[3]{4}$
Vertical asymptote: $x = 0$
Oblique asymptote: $y = x$
Increasing on $(-\infty, -2)$ and $(0, \infty)$
Local maximum at $x = -2$
Decreasing on $(-2, 0)$
Concave downward on $(-\infty, 0)$ and $(0, \infty)$

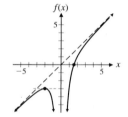

61. Domain: All real numbers except 0
x intercepts: $-\sqrt{3}, \sqrt{3}$
Vertical asymptote: $x = 0$
Oblique asymptote: $y = x$
Increasing on $(-\infty, 0)$ and $(0, \infty)$
Concave upward on $(-\infty, 0)$
Concave downward on $(0, \infty)$

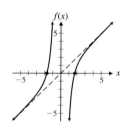

63. Domain: All real numbers except 0
Vertical asymptote: $x = 0$
Oblique asymptote: $y = x$
Increasing on $(-\infty, -2)$ and $(2, \infty)$
Decreasing on $(-2, 0)$ and $(0, 2)$
Local maximum at $x = -2$
Local minimum at $x = 2$
Concave upward on $(0, \infty)$
Concave downward on $(-\infty, 0)$

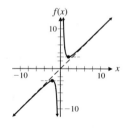

65. Domain: All real numbers except 2, 4
y intercept: $-3/4$; x intercept: -3
Vertical asymptote: $x = 4$
Horizontal asymptote: $y = 1$
Decreasing on $(-\infty, 2), (2, 4)$, and $(4, \infty)$
Concave upward on $(4, \infty)$
Concave downward on $(-\infty, 2)$ and $(2, 4)$

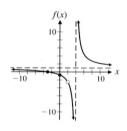

67. Domain: All real numbers except $-3, 3$
y intercept: $5/3$; x intercept: 2.5
Vertical asymptote: $x = 3$
Horizontal asymptote: $y = 2$
Decreasing on $(-\infty, -3)$, $(-3, 3)$, and $(3, \infty)$
Concave upward on $(3, \infty)$
Concave downward on $(-\infty, -3)$ and $(-3, 3)$

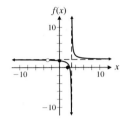

69. Domain: All real numbers except $-1, 2$
y intercept: 0; x intercepts: $0, 3$
Vertical asymptote: $x = -1$
Increasing on $(-\infty, -3)$, $(1, 2)$, and $(2, \infty)$
Decreasing on $(-3, -1)$ and $(-1, 1)$
Local maximum at $x = -3$
Local minimum at $x = 1$
Concave upward on $(-1, 2)$ and $(2, \infty)$
Concave downward on $(-\infty, -1)$

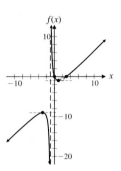

71. Domain: All real numbers except 1
y intercept: -2; x intercept: -2
Vertical asymptote: $x = 1$
Horizontal asymptote: $y = 1$
Decreasing on $(-\infty, 1)$ and $(1, \infty)$
Concave upward on $(1, \infty)$
Concave downward on $(-\infty, 1)$

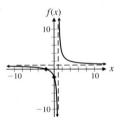

73.

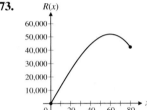

75. (A) Increasing on $(0, 1)$
(B) Concave upward on $(0, 1)$
(C) $x = 1$ is a vertical asymptote
(D) The origin is both an x and a y intercept
(E)

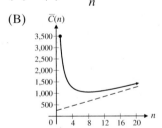

77. (A) $\overline{C}(n) = \dfrac{3,200}{n} + 250 + 50n$

(B)

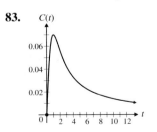

(C) 8 yr

79. (A)

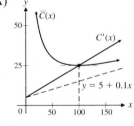

(B) 25 at $x = 100$

81. (A)

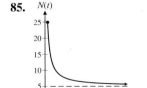

(B) Minimum average cost is \$4.35 when 177 pizzas are produced daily.

83.

(graph of $C(t)$)

85.

(graph of $N(t)$)

Exercise 5-5

1. Min $f(x) = f(0) = 0$; Max $f(x) = f(10) = 14$ **3.** Min $f(x) = f(0) = 0$; Max $f(x) = f(3) = 9$
5. Min $f(x) = f(1) = f(7) = 5$; Max $f(x) = f(10) = 14$ **7.** Min $f(x) = f(1) = f(7) = 5$; Max $f(x) = f(3) = f(9) = 9$
9. Min $f(x) = f(5) = 7$; Max $f(x) = f(3) = 9$ **11.** Min $f(x) = f(1) = 2$; no maximum
13. Max $f(x) = f(-3) = 18$; no minimum   **15.** No absolute extrema   **17.** Max $f(x) = f(3) = 54$; no minimum
19. No absolute extrema   **21.** Min $f(x) = f(0) = 0$; no maximum   **23.** Max $f(x) = f(1) = 1$; Min $f(x) = f(-1) = -1$
25. Min $f(x) = f(0) = -1$, no maximum   **27.** Min $f(x) = f(2) = -2$ **29.** Max $f(x) = f(2) = 4$
31. Min $f(x) = f(2) = 0$ **33.** No maximum   **35.** Max $f(x) = f(2) = 8$ **37.** Min $f(x) = f(4) = 22$

39. Min $f(x) = f(\sqrt{10}) = 14/\sqrt{10}$ **41.** Min $f(x) = f(2) = \dfrac{e^2}{4} \approx 1.847$ **43.** Min $f(x) = f(3) = \dfrac{27}{e^3} \approx 1.344$

45. Max $f(x) = f(e^{1.5}) = 2e^{1.5} \approx 8.963$ **47.** Min $f(x) = f(e^{2.5}) = \dfrac{e^5}{2} \approx 74.207$ **49.** Max $f(x) = f(1) = -1$

51. (A) Max $f(x) = f(5) = 14$; Min $f(x) = f(-1) = -22$ (B) Max $f(x) = f(1) = -2$; Min $f(x) = f(-1) = -22$
  (C) Max $f(x) = f(5) = 14$; Min $f(x) = f(3) = -6$
53. (A) Max $f(x) = f(0) = 126$; Min $f(x) = f(2) = -26$ (B) Max $f(x) = f(7) = 49$; Min $f(x) = f(2) = -26$
  (C) Max $f(x) = f(6) = 6$; Min $f(x) = f(3) = -15$
55. (A) Max $f(x) = f(-1) = 10$; Min $f(x) = f(2) = -11$ (B) Max $f(x) = f(0) = f(4) = 5$; Min $f(x) = f(3) = -22$
  (C) Max $f(x) = f(-1) = 10$; Min $f(x) = f(1) = 2$
57. Local minimum   **59.** Unable to determine   **61.** Neither   **63.** Local maximum

Exercise 5-6

1. Exactly in half   **3.** 15 and -15 **5.** A square of side 25 cm; maximum area $= 625$ cm^2
7. If x and y are the dimensions of the rectangle and A is the fixed area, the model is: Minimize $C = 2Bx + 2AB/x, x > 0$.
This mathematical problem always has a solution. This agrees with our economic intuition that there should be a cheapest way to build the fence.
9. If x and y are the dimensions of the rectangle and C is the fixed amount to be spent, the model is:
Maximize $A = x(C - 2Bx)/(2B), 0 \le x \le C/(2B)$. This mathematical problem always has a solution. This agrees with our economic intuition that there should be a largest area that can be enclosed with a fixed amount of fencing.
11. (A) Maximum revenue is \$125,000 when 500 phones are produced and sold for \$250 each.
  (B) Maximum profit is \$46,612.50 when 365 phones are produced and sold for \$317.50 each.
13. (A) Max $R(x) = R(3,000) = \$300,000$
  (B) Maximum profit is \$75,000 when 2,100 sets are manufactured and sold for \$130 each.
  (C) Maximum profit is \$64,687.50 when 2,025 sets are manufactured and sold for \$132.50 each.
15. (A)

```
QuadReg
y=ax²+bx+c
a=-2.352941ᴇ-5
b=-.0325964781
c=288.9535407
```

(B)

```
LinReg
y=ax+b
a=53.50318471
b=82245.22293
```

(C) The maximum profit is \$118,996 when the price per saw is \$195.

17. (A) \$4.80 (B) \$8   **19.** \$35; \$6,125   **21.** 40 trees; 1,600 lb   **23.** $(10 - 2\sqrt{7})/3 = 1.57$ in. squares
25. 20 ft by 40 ft (with the expensive side being one of the short sides)   **27.** 8 production runs per year
29. 10,000 books in 5 printings   **31.** (A) $x = 5.1$ mi (B) $x = 10$ mi   **33.** 4 days; 20 bacteria/cm^3
35. 50 mice per order   **37.** 1 month; 2 ft   **39.** 4 yr from now

Chapter 5 Review Exercise

1. $(a, c_1), (c_3, c_6)$ (5-1, 5-2)   **2.** $(c_1, c_3), (c_6, b)$ (5-1, 5-2)   **3.** $(a, c_2), (c_4, c_5), (c_7, b)$ (5-1, 5-2)   **4.** c_3 (5-1)
5. c_1, c_6 (5-1)   **6.** c_1, c_3, c_5 (5-1)   **7.** c_4, c_6 (5-1)   **8.** c_2, c_4, c_5, c_7 (5-2)
9. (5-2)

10. (5-2)

11. $f''(x) = 12x^2 + 30x$ (5-2)   **12.** $y'' = 8/x^3$ (5-2)

13. Domain: All real numbers
y intercept: 0; x intercepts: 0, 9
Increasing on $(-\infty, 3)$ and $(9, \infty)$
Decreasing on $(3, 9)$
Local maximum at $x = 3$
Local minimum at $x = 9$
Concave upward on $(6, \infty)$
Concave downward on $(-\infty, 6)$
Inflection point at $x = 6$ *(5-4)*

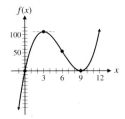

14. Domain: All real numbers
y intercept: 16; x intercepts: $-4, 2$
Increasing on $(-\infty, -2)$ and $(2, \infty)$
Decreasing on $(-2, 2)$
Local maximum at $x = -2$
Local minimum at $x = 2$
Concave upward on $(0, \infty)$
Concave downward on $(-\infty, 0)$
Inflection point at $x = 0$ *(5-4)*

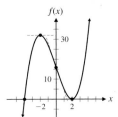

15. Domain: All real numbers
y intercept: 0; x intercepts: 0, 4
Increasing on $(-\infty, 3)$
Decreasing on $(3, \infty)$
Local maximum at $x = 3$
Concave upward on $(0, 2)$
Concave downward on $(-\infty, 0)$ and $(2, \infty)$
Inflection points at $x = 0$ and $x = 2$ *(5-4)*

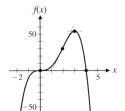

16. Domain: all real numbers
y intercept: -3; x intercepts: $-3, 1$
No vertical or horizontal asymptotes
Increasing on $(-2, \infty)$
Decreasing on $(-\infty, -2)$
Local minimum at $x = -2$
Concave upward on $(-\infty, -1)$ and $(1, \infty)$
Concave downward on $(-1, 1)$
Inflection points at $x = -1$ and $x = 1$ *(5-4)*

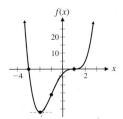

17. Domain: All real numbers, except -2
y intercept: 0; x intercept: 0
Horizontal asymptote: $y = 3$
Vertical asymptote: $x = -2$
Increasing on $(-\infty, -2)$ and $(-2, \infty)$
Concave upward on $(-\infty, -2)$
Concave downward on $(-2, \infty)$ *(5-4)*

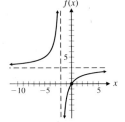

18. Domain: All real numbers
y intercept: 0; x intercept: 0
Horizontal asymptote: $y = 1$
Increasing on $(0, \infty)$
Decreasing on $(-\infty, 0)$
Local minimum at $x = 0$
Concave upward on $(-3, 3)$
Concave downward on $(-\infty, -3)$ and $(3, \infty)$
Inflection points at $x = -3, 3$ *(5-4)*

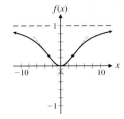

19. Domain: All real numbers except $x = -2$
y intercept: 0; x intercept: 0
Horizontal asymptote: $y = 0$
Vertical asymptote: $x = -2$
Increasing on $(-2, 2)$
Decreasing on $(-\infty, -2)$ and $(2, \infty)$
Local maximum at $x = 2$
Concave upward on $(4, \infty)$
Concave downward on $(-\infty, -2)$ and $(-2, 4)$
Inflection point at $x = 4$ *(5-4)*

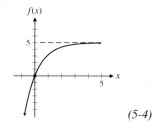

20. Domain: All real numbers
y intercept: 0; x intercept: 0
Increasing on $(-\infty, \infty)$
Concave upward on $(-\infty, -3)$ and $(0, 3)$
Concave downward on $(-3, 0)$ and $(3, \infty)$
Inflection points at $x = -3, 0, 3$ *(5-4)*

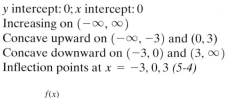

21. Domain: All real numbers
y intercept: 0; x intercept: 0
Horizontal asymptote: $y = 5$
Increasing on $(-\infty, \infty)$
Concave downward on $(-\infty, \infty)$

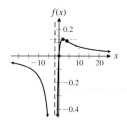

(5-4)

22. Domain: $(0, \infty)$
x intercept: 1
Increasing on $(e^{-1/3}, \infty)$
Decreasing on $(0, e^{-1/3})$
Local minimum at $x = e^{-1/3}$
Concave upward on $(e^{-5/6}, \infty)$
Concave downward on $(0, e^{-5/6})$
Inflection point at $x = e^{-5/6}$

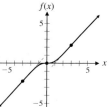
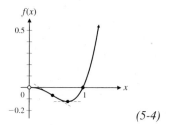

(5-4)

23. 3 *(5-3)*   **24.** $-\frac{1}{5}$ *(5-3)*   **25.** $-\infty$ *(5-3)*
26. 0 *(5-3)*   **27.** 0 *(5-3)*   **28.** 1 *(5-3)*
29. 0 *(5-3)*   **30.** 0 *(5-3)*   **31.** 1 *(5-3)*
32. 2 *(5-3)*

33.

x	$f'(x)$	$f(x)$
$-\infty < x < -2$	Negative and increasing	Decreasing and concave upward
$x = -2$	x intercept	Local minimum
$-2 < x < -1$	Positive and increasing	Increasing and concave upward
$x = -1$	Local maximum	Inflection point
$-1 < x < 1$	Positive and decreasing	Increasing and concave downward
$x = 1$	Local minimum	Inflection point
$1 < x < \infty$	Positive and increasing	Increasing and concave upward

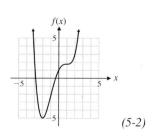

(5-2)

34. (C) *(5-2)*   **35.** Local maximum at $x = -1$; local minimum at $x = 5$ *(5-5)*
36. Min $f(x) = f(2) = -4$; Max $f(x) = f(5) = 77$ *(5-5)*   **37.** Min $f(x) = f(2) = 8$ *(5-5)*
38. Max $f(x) = f(e^{4.5}) = 2e^{4.5} \approx 180.03$ *(5-5)*
39. Max $f(x) = f(0.5) = 5e^{-1} \approx 1.84$ *(5-5)*
40. Yes. Since f is continuous on $[a, b]$, f has an absolute maximum on $[a, b]$. But each endpoint is a local minimum; hence, the absolute maximum must occur between a and b. *(5-5)*
41. No, increasing/decreasing properties apply to intervals in the domain of f. It is correct to say that $f(x)$ is decreasing on $(-\infty, 0)$ and $(0, \infty)$. *(5-1)*
42. A critical value for $f(x)$ is a partition number for $f'(x)$ that is also in the domain of f. For example, if $f(x) = x^{-1}$, then 0 is a partition number for $f'(x) = -x^{-2}$, but 0 is not a critical value for $f(x)$ since 0 is not in the domain of f. *(5-1)*

43. Max $f'(x) = f'(2) = 12$ *(5-2, 5-5)*

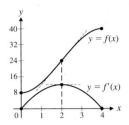

44. Each number is 20; minimum sum is 40 *(5-6)*

45. Domain: All real numbers
x intercepts: 0.79, 1.64; y intercept: 4
Increasing on $(-1.68, -0.35)$ and $(1.28, \infty)$
Decreasing on $(-\infty, -1.68)$ and $(-0.35, 1.28)$
Local minima at $x = -1.68$ and $x = 1.28$
Local maximum at $x = -0.35$
Concave downward on $(-1.10, 0.60)$
Concave upward on $(-\infty, -1.10)$ and $(0.60, \infty)$
Inflection points at $x = -1.10$ and $x = 0.60$ *(5-4)*

46. Domain: All real numbers
x intercepts; 0, 11.10; y intercept: 0
Increasing on $(1.87, 4.19)$ and $(8.94, \infty)$
Decreasing on $(-\infty, 1.87)$ and $(4.19, 8.94)$
Local maximum at $x = 4.19$
Local minima at $x = 1.87$ and $x = 8.94$
Concave upward on $(-\infty, 2.92)$ and $(7.08, \infty)$
Concave downward on $(2.92, 7.08)$
Inflection points at $x = 2.92$ and $x = 7.08$ *(5-4)*

47. Max $f(x) = f(1.373) = 2.487$ *(5-5)*
48. Max $f(x) = f(1.763) = 0.097$ *(5-5)*
49. (A) For the first 15 months, the graph of the price is increasing and concave downward, with a local maximum at $t = 15$. For the next 15 months, the graph of the price is decreasing and concave downward, with an inflection point at $t = 30$. For the next 15 months, the graph of the price is decreasing and concave upward, with a local minimum at $t = 45$. For the remaining 15 months, the graph of the price is increasing and concave upward.

(B) $p(t)$

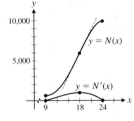

(5-2)

50. (A) Max $R(x) = R(10,000) = \$2,500,000$
(B) Maximum profit is \$175,000 when 3,000 stoves are manufactured and sold for \$425 each.
(C) Maximum profit is \$119,000 when 2,600 stoves are manufactured and sold for \$435 each. *(5-6)*
51. (A) The expensive side is 50 ft; the other side is 100 ft.   (B) The expensive side is 75 ft; the other side is 150 ft. *(5-6)*
52. \$49; \$6,724 *(5-6)*   **53.** 12 orders/yr *(5-6)*   **54.** Min $\overline{C}(x) = \overline{C}(200) = 50$

55. Min $\overline{C}(x) = \overline{C}(e^5) \approx \$49.66$ *(5-4)*
56. A maximum revenue of \$18,394 is realized at a production level of 50 units at \$367.88 each. *(5-6)*

57. $R(x)$

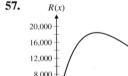

(5-3)

58. \$549.15; \$9,864 *(5-6)*   **59.** \$1.52 *(5-6)*

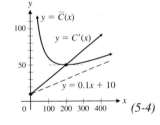

(5-4)

60. 20.39 feet *(5-6)*

61. (A)
```
QuadReg
y=ax²+bx+c
a=.0061285714
b=.1224285714
c=102.2
```

(B) Min $\overline{C}(x) = \overline{C}(129) = \$1.71$ *(5-4)*

62. Increasing on $(0, 18)$; decreasing on $(18, 24)$; point of diminishing returns is $x = 18$,
max $N'(x) = N'(18) = 972$ *(5-2)*

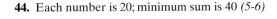

63. (A)
```
CubicReg
y=ax³+bx²+cx+d
a=-.01
b=.83
c=-2.3
d=221
```

(B) 28 ads to sell 588 refrigerators per month *(5-2)*
64. 3 days *(5-1)*   **65.** 2 yr from now *(5-1)*

CHAPTER 6

Exercise 6-1

1. $(x^3/3) + C$ **3.** $(x^8/8) + C$ **5.** $2x + C$ **7.** $-5u^{-1} + C$ **9.** $e^2t + C$ **11.** $(z^{50}/50) + C$
13. $\pi x^2 + C$ **15.** $8 \ln |u| + C$ **17.** $10x^{3/2} + C$ **19.** $-21t^{-1/3} + C$ **21.** $\frac{1}{2}x^2 - \frac{2}{3}x^{3/2} + C$ **23.** $y = 40x^5 + C$
25. $P = 24x - 3x^2 + C$ **27.** $y = \frac{1}{3}u^6 - u^3 - u + C$ **29.** $y = e^x + 3x + C$ **31.** $x = 5 \ln|t| + t + C$
33. True   **35.** False   **37.** True   **39.** False   **41.** False
43. No, since one graph cannot be obtained from another by a vertical translation.
45. Yes, since one graph can be obtained from another by a vertical translation.   **47.** $(5x^2/2) - (5x^3/3) + C$
49. $6x + (5x^3/3) + (x^5/5) + C$ **51.** $2\sqrt{u} + C$ **53.** $-(x^{-2}/8) + C$ **55.** $4 \ln |u| + u + C$ **57.** $5e^z + 4z + C$
59. $x^3 + 2x^{-1} + C$ **61.** $2x^5 + 2x^{-4} - 2x + C$ **63.** $2x^{3/2} + 4x^{1/2} + C$ **65.** $\frac{3}{5}x^{5/3} + 2x^{-2} + C$ **67.** $(e^x/4) - (3x^2/8) + C$
69. $-4z^{-3} - \frac{5}{2}z^{-2} - 3 \ln|z| + C$ **71.** $\frac{2}{5}x^3 - \frac{2}{3}\ln|x| + C$ **73.** $y = x^2 - 3x + 5$ **75.** $C(x) = 2x^3 - 2x^2 + 3,000$
77. $x = 40\sqrt{t}$ **79.** $y = -2x^{-1} + 3 \ln|x| - x + 3$ **81.** $x = 4e^t - 2t - 3$ **83.** $y = 2x^2 - 3x + 1$ **85.** $x^2 + x^{-1} + C$
87. $\frac{1}{2}x^2 + x^{-2} + C$ **89.** $e^x - 2 \ln|x| + C$ **91.** $M = t + t^{-1} + \frac{3}{4}$ **93.** $y = 3x^{5/3} + 3x^{2/3} - 6$ **95.** $p(x) = 10x^{-1} + 10$
97. x^3 **99.** $x^4 + 3x^2 + C$ **107.** $\overline{C}(x) = 15 + \dfrac{1,000}{x}; C(x) = 15x + 1,000; C(0) = \$1,000$

109. (A) The cost function increases from 0 to 8, is concave downward from 0 to 4, and is concave upward from 4 to 8. There is an inflection point at $x = 4$.
(B) $C(x) = x^3 - 12x^2 + 53x + 30; C(4) = \$114,000; C(8) = \$198,000$ (C)
(D) Manufacturing plants are often inefficient at low and high levels of production.

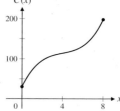

111. $S(t) = 2,000 - 15t^{5/3}; 80^{3/5} \approx 14$ mo
113. $S(t) = 2,000 - 15t^{5/3} - 70t; t \approx 8.92$ mo
115. $L(x) = 4,800x^{1/2}; L(25) = 24,000$ labor-hours
117. $W(h) = 0.0005h^3; W(70) = 171.5$ lb
119. 19,400

Exercise 6-2

1. $\frac{1}{3}(3x + 5)^3 + C$ **3.** $\frac{1}{6}(x^2 - 1)^6 + C$ **5.** $-\frac{1}{2}(5x^3 + 1)^{-2} + C$ **7.** $e^{5x} + C$ **9.** $\ln|1 + x^2| + C$
11. $\frac{2}{3}(1 + x^4)^{3/2} + C$ **13.** $\frac{1}{11}(x + 3)^{11} + C$ **15.** $-\frac{1}{6}(6t - 7)^{-1} + C$ **17.** $\frac{1}{12}(t^2 + 1)^6 + C$ **19.** $\frac{1}{2}e^{x^2} + C$
21. $\frac{1}{5}\ln|5x + 4| + C$ **23.** $-e^{1-t} + C$ **25.** $-\frac{1}{18}(3t^2 + 1)^{-3} + C$ **27.** $\frac{1}{3}(4 - x^3)^{-1} + C$
29. $\frac{2}{5}(x + 4)^{5/2} - \frac{8}{3}(x + 4)^{3/2} + C$ **31.** $\frac{2}{3}(x - 3)^{3/2} + 6(x - 3)^{1/2} + C$ **33.** $\frac{1}{11}(x - 4)^{11} + \frac{2}{5}(x - 4)^{10} + C$
35. $\frac{1}{8}(1 + e^{2x})^4 + C$ **37.** $\frac{1}{2}\ln|4 + 2x + x^2| + C$ **39.** $-\frac{1}{12}(x^4 + 2x^2 + 1)^{-3} + C$
41. (A) Differentiate the right side to get the integrand on the left side.
(B) Wrong, since $\dfrac{d}{dx}[\ln|2x - 3| + C] = \dfrac{2}{2x - 3} \neq \dfrac{1}{2x - 3}$. If $u = 2x - 3$, then $du = 2\,dx$. The integrand was not adjusted for the missing constant factor 2.
(C) $\displaystyle\int \frac{1}{2x - 3}dx = \frac{1}{2}\int \frac{2}{2x - 3}dx = \frac{1}{2}\ln|2x - 3| + C$ Check: $\dfrac{d}{dx}\left[\dfrac{1}{2}\ln|2x - 3| + C\right] = \dfrac{1}{2x - 3}$
43. (A) Differentiate the right side to get the integrand on the left side.
(B) Wrong, since $\dfrac{d}{dx}[e^{x^4} + C] = 4x^3e^{x^4} \neq x^3e^{x^4}$. If $u = x^4$, then $du = 4x^3\,dx$. The integrand was not adjusted for the missing constant factor 4.
(C) $\displaystyle\int x^3e^{x^4}\,dx = \frac{1}{4}\int 4x^3e^{x^4}\,dx = \frac{1}{4}e^{x^4} + C$ Check: $\dfrac{d}{dx}\left[\dfrac{1}{4}e^{x^4} + C\right] = x^3e^{x^4}$
45. (A) Differentiate the right side to get the integrand on the left side.
(B) Wrong, since $\dfrac{d}{dx}\left[\dfrac{(x^2 - 2)^2}{3x} + C\right] = \dfrac{3x^4 - 4x^2 - 4}{3x^2} \neq 2(x^2 - 2)^2$. If $u = x^2 - 2$, then $du = 2x\,du$. It appears that the student invalidly moved a variable factor across the integral sign as follows:
$$\int 2(x^2 - 2)^2\,dx = \frac{1}{x}\int 2x(x^2 - 2)^2\,dx.$$
(C) $\displaystyle\int 2(x^2 - 2)^2\,dx = \int (2x^4 - 8x^2 + 8)\,dx = \frac{2}{5}x^5 - \frac{8}{3}x^3 + 8x + C$
Check: $\dfrac{d}{dx}\left[\dfrac{2}{5}x^5 - \dfrac{8}{3}x^3 + 8x + C\right] = 2x^4 - 8x^2 + 8 = 2(x^2 - 2)^2$

47. $\frac{1}{9}(3x^2 + 7)^{3/2} + C$ **49.** $\frac{1}{8}x^8 + \frac{4}{5}x^5 + 2x^2 + C$ **51.** $\frac{1}{9}(x^3 + 2)^3 + C$ **53.** $\frac{1}{4}(2x^4 + 3)^{1/2} + C$ **55.** $\frac{1}{4}(\ln x)^4 + C$
57. $e^{-1/x} + C$ **59.** $x = \frac{1}{3}(t^3 + 5)^7 + C$ **61.** $y = 3(t^2 - 4)^{1/2} + C$ **63.** $p = -(e^x - e^{-x})^{-1} + C$
67. $p(x) = 2,000/(3x + 50)$; 250 bottles **69.** $C(x) = 12x + 500 \ln(x + 1) + 2,000$; $\overline{C}(1,000) = \$17.45$
71. (A) $S(t) = 10t + 100e^{-0.1t} - 100, 0 \le t \le 24$ (B) $S(12) \approx \$50$ million (C) 18.41 mo
73. $Q(t) = 100 \ln(t + 1) + 5t, 0 \le t \le 20$; $Q(9) \approx 275$ thousand barrels **75.** $W(t) = 2e^{0.1t}$; $W(8) \approx 4.45$ g
77. (A) $-1,000$ bacteria/mL per day (B) $N(t) = 5,000 - 1,000 \ln(1 + t^2)$; 385 bacteria/mL (C) 7.32 days
79. $N(t) = 100 - 60e^{-0.1t}, 0 \le t \le 15$; $N(15) \approx 87$ words/min **81.** $E(t) = 12,000 - 10,000(t + 1)^{-1/2}$; $E(15) = 9,500$ students

Exercise 6-3

1. $y = 3x^2 + C$ **3.** $y = 7 \ln|x| + C$ **5.** $y = 50e^{0.02x} + C$ **7.** $y = \dfrac{x^3}{3} - \dfrac{x^2}{2}$ **9.** $y = e^{-x^2} + 2$
11. $y = 2 \ln|1 + x| + 5$
13. Figure B. When $x = 1$, the slope $dy/dx = 1 - 1 = 0$ for any y. When $x = 0$, the slope $dy/dx = 0 - 1 = -1$ for any y. Both are
consistent with the slope field shown in Figure B.
15. $y = \dfrac{x^2}{2} - x + C$; $y = \dfrac{x^2}{2} - x - 2$

17.
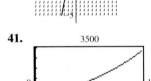
19. $y = Ce^{2t}$ **21.** $y = 100e^{-0.5x}$ **23.** $x = Ce^{-5t}$ **25.** $x = -(5t^2/2) + C$
27. Figure A. When $y = 1$, the slope $dy/dx = 1 - 1 = 0$ for any x. When $y = 2$, the slope
$dy/dx = 1 - 2 = -1$ for any x. Both are consistent with the slope field shown in Figure A.
29. $y = 1 - e^{-x}$

31.
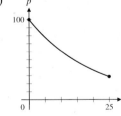
33.
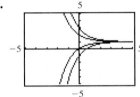
35. $y = \sqrt{25 - x^2}$ **37.** $y = -3x$ **39.** $y = 1/(1 - 2e^{-t})$

41.
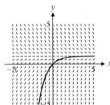
43.
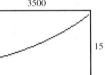
45.
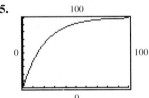
47.
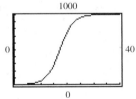

49. Apply the second-derivative test to $f(y) = ky(M - y)$. **51.** 1999 **53.** $A = 1,000e^{0.08t}$ **57.** $A = 8,000e^{0.06t}$
59. (A) $p(x) = 100e^{-0.05x}$ **61.** (A) $N = L(1 - e^{-0.051t})$ **63.** $I = I_0e^{-0.00942x}$; $x \approx 74$ ft
(B) \$60.65 per unit (B) 22.5% (C) 32 days **65.** (A) $Q = 3e^{-0.04t}$ (B) $Q(10) = 2.01$ mL
(C) (D) (C) 27.47 hr (D)

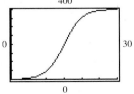

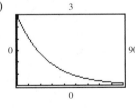

67. 0.023 117 **69.** Approx. 24,200 yr **71.** 104 times; 67 times
73. (A) 7 people; 353 people (B) 400
(C)

Exercise 6-4

1.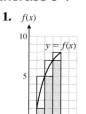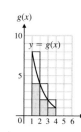

3. Figure A: $L_3 = 13$, $R_3 = 20$; Figure B: $L_3 = 14$, $R_3 = 7$
5. $L_3 \leq \int_1^4 f(x)\,dx \leq R_3$; $R_3 \leq \int_1^4 g(x)\,dx \leq L_3$; since $f(x)$ is increasing, L_3 underestimates the area and R_3 overestimates the area; since $g(x)$ is decreasing, the reverse is true.
7. In both figures, the error bound for L_3 and R_3 is 7.
9. $S_5 = -260$ **11.** $S_4 = -1{,}194$ **13.** $S_3 = -33.01$ **15.** $S_6 = -38$
17. -2.475 **19.** 4.266 **21.** 2.474 **23.** -5.333 **25.** 1.067 **27.** -1.066
29. 15 **31.** 58.5 **33.** -54 **35.** 248 **37.** 0 **39.** -183
41. False   **43.** False   **45.** False

47. $L_{10} = 286{,}100 \text{ ft}^2$; error bound is $50{,}000 \text{ ft}^2$; $n \geq 200$
49. $L_6 = -3.53$, $R_6 = -0.91$; error bound for L_6 and R_6 is 2.63. Geometrically, the definite integral over the interval $[2, 5]$ is the sum of the areas between the curve and the x axis from $x = 2$ to $x = 5$, with the areas below the x axis counted negatively and those above the x axis counted positively.
51. Increasing on $(-\infty, 0]$; decreasing on $[0, \infty)$ **53.** Increasing on $[-1, 0]$ and $[1, \infty)$; decreasing on $(-\infty, -1]$ and $[0, 1]$
55. $n \geq 22$ **57.** $n \geq 104$ **59.** $L_3 = 2{,}580$, $R_3 = 3{,}900$; error bound for L_3 and R_3 is 1,320
61. (A) $L_5 = 3.72$; $R_5 = 3.37$ (B) $R_5 = 3.37 \leq \int_0^5 A'(t)\,dt \leq 3.72 = L_5$
63. $L_3 = 114$, $R_3 = 102$; error bound for L_3 and R_3 is 12

Exercise 6-5

1. (A) $F(15) - F(10) = 375$
(B)

3. (A) $F(15) - F(10) = 85$
(B)

5. 5 **7.** 5 **9.** 2 **11.** $-\frac{7}{3} \approx -2.333$
13. 2 **15.** $\frac{1}{2}(e^2 - 1) \approx 3.195$
17. $2 \ln 3.5 \approx 2.506$ **19.** $\frac{3}{4}$ **21.** 12 **23.** -2
25. 14 **27.** $5^6 = 15{,}625$ **29.** $\ln 4 \approx 1.386$
31. $20(e^{0.25} - e^{-0.5}) \approx 13.550$ **33.** $\frac{1}{2}$
35. $\frac{8}{3} \approx 2.667$ **37.** $\frac{1}{2}(1 - e^{-1}) \approx 0.316$
39. $-\frac{3}{2} - \ln 2 \approx -2.193$

41. (A) Average $f(x) = 250$
(B)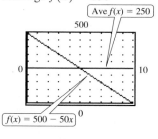

43. (A) Average $f(t) = 2$
(B)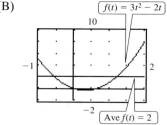

45. (A) Average $f(x) = \frac{45}{28} \approx 1.61$
(B)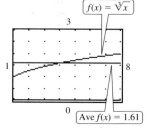

47. (A) Average $f(x) = 2(1 - e^{-2}) \approx 1.73$
(B)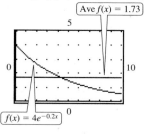

49. $\frac{1}{6}(15^{3/2} - 5^{3/2}) \approx 7.819$ **51.** $\frac{1}{2}(\ln 2 - \ln 3) \approx -0.203$ **53.** 0
55. 4.566 **57.** 2.214 **61.** $\displaystyle\int_{300}^{900} \left(500 - \frac{x}{3}\right) dx = \$180{,}000$
63. $\int_0^5 500(t - 12)\,dt = -\$23{,}750$; $\int_5^{10} 500(t - 12)\,dt = -\$11{,}250$

65. (A)

(B) $6{,}505$ **67.** Useful life $= \sqrt{\ln 55} \approx 2$ yr; total
profit $= \frac{51}{22} - \frac{5}{2}e^{-4} \approx 2.272$ or $\$2{,}272$
69. (A) $\$420$ (B) $\$135{,}000$

71. (A)

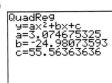

```
QuadReg
y=ax²+bx+c
a=3.074675325
b=-24.98073593
c=55.56363636
```

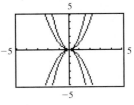

(B) $100,505

73. $50e^{0.6} - 50e^{0.4} - 10 \approx \$6.51$
75. 4,800 labor-hours
77. (A) $I = -200t + 600$
   (B) $\frac{1}{3}\int_0^3 (-200t + 600)\,dt = 300$
79. $100 \ln 11 + 50 \approx 290$ thousand barrels;
   $100 \ln 21 - 100 \ln 11 + 50 \approx 115$ thousand barrels
81. $2e^{0.8} - 2 \approx 2.45$ g; $2e^{1.6} - 2e^{0.8} \approx 5.45$ g

83. $10°C$ **85.** $0.6 \ln 2 + 0.1 \approx 0.516$; $(4.2 \ln 625 + 2.4 - 4.2 \ln 49)/24 \approx 0.546$

Chapter 6 Review Exercise

1. $3x^2 + 3x + C$ *(6-1)*   **2.** 50 *(6-5)*   **3.** -207 *(6-5)*   **4.** $-\frac{1}{8}(1 - t^2)^4 + C$ *(6-2)*   **5.** $\ln|u| + \frac{1}{4}u^4 + C$ *(6-1)*

6. 0.216 *(6-5)*   **7.** e^{-x^2} *(6-1)*   **8.** $\sqrt{4 + 5x} + C$ *(6-1)*   **9.** $y = f(x) = x^3 - 2x + 4$ *(6-3)*

10. (A) $2x^4 - 2x^2 - x + C$ (B) $e^t - 4\ln|t| + C$ *(6-1)*   **11.** $R_2 = 72$; error bound for R_2 is 48 *(6-4)*

12. $\int_1^5 (x^2 + 1)\,dx = \frac{136}{3} \approx 45.33$; actual error is $\frac{80}{3} \approx 26.67$ *(6-5)*   **13.** $L_4 = 30.8$ *(6-4)*   **14.** 7 *(6-5)*

15. Width $= 2 - (-1) = 3$; height $=$ average $f(x) = 7$ *(6-5)*   **16.** $S_4 = 368$ *(6-4)*   **17.** $S_5 = 906$ *(6-4)*   **18.** -10 *(6-4)*

19. 0.4 *(6-4)*   **20.** 1.4 *(6-4)*   **21.** 0 *(6-4)*   **22.** 0.4 *(6-4)*   **23.** 2 *(6-4)*   **24.** -2 *(6-4)*   **25.** -0.4 *(6-4)*

26. (A) $1;1$ (B) $4;4$ *(6-3)*

27. $dy/dx = (2y)/x$; the slopes computed in Problem 26A are compatible with the slope field shown. *(6-3)*

29. $y = \frac{1}{4}x^2$; $y = -\frac{1}{4}x^2$ *(6-3)*   **30.** *(6-3)*   **31.** *(6-3)*

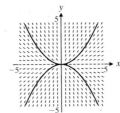

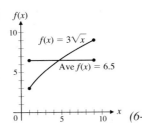

32. $\frac{2}{3}(2)^{3/2} \approx 1.886$ *(6-5)*   **33.** $\frac{1}{6} \approx 0.167$ *(6-5)*   **34.** $-5e^{-t} + C$ *(6-1)*   **35.** $\frac{1}{2}(1 + e^2)$ *(6-1)*   **36.** $\frac{1}{6}e^{3x^2} + C$ *(6-2)*

37. $2(\sqrt{5} - 1) \approx 2.472$ *(6-5)*   **38.** $\frac{1}{2}\ln 10 \approx 1.151$ *(6-5)*   **39.** 0.45 *(6-5)*   **40.** $\frac{1}{48}(2x^4 + 5)^6 + C$ *(6-2)*

41. $-\ln(e^{-x} + 3) + C$ *(6-2)*   **42.** $-(e^x + 2)^{-1} + C$ *(6-2)*   **43.** $y = f(x) = 3\ln|x| + x^{-1} + 4$ *(6-2, 6-3)*

44. $y = 3x^2 + x - 4$ *(6-3)*

45. (A) Average $f(x) = 6.5$ (B)

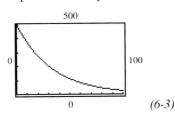

(6-5)

46. $\frac{1}{3}(\ln x)^3 + C$ *(6-2)*
47. $\frac{1}{8}x^8 - \frac{2}{5}x^5 + \frac{1}{2}x^2 + C$ *(6-2)*
48. $\frac{2}{3}(6 - x)^{3/2} - 12(6 - x)^{1/2} + C$ *(6-2)*
49. $\frac{1,234}{15} \approx 82.267$ *(6-5)*
50. $\frac{64}{15} \approx 4.267$ *(6-5)*
51. $y = 3e^{x^3} - 1$ *(6-3)*
52. $N = 800e^{0.06t}$ *(6-3)*

53. Limited growth

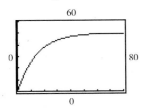

(6-3)

54. Exponential decay

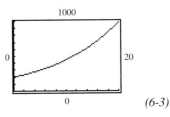

(6-3)

55. Unlimited growth

(6-3)

56. Logistic growth

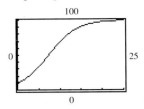

(6-3)

57. 1.167 *(6-5)*   **58.** 99.074 *(6-5)*   **59.** -0.153 *(6-5)*

60. $L_2 = \$180,000$; $R_2 = \$140,000$; $\$140,000 \leq \int_{200}^{600} C'(x)\,dx \leq \$180,000$ *(6-4)*

61. $\int_{200}^{600} \left(600 - \frac{x}{2}\right)dx = \$160,000$ *(6-5)*   **62.** $\int_{10}^{40} \left(150 - \frac{x}{10}\right)dx = \$4,425$ *(6-5)*

63. $P(x) = 100x - 0.01x^2$; $P(10) = \$999$ *(6-3)*

64. $\int_0^{15} (60 - 4t)\,dt = 450$ thousand barrels *(6-5)*

65. 109 items *(6-5)*   **66.** $16e^{2.5} - 16e^2 - 8 \approx \$68.70$ *(6-5)*

67. Useful life $= 10 \ln \frac{20}{3} \approx 19$ yr; total profit $= 143 - 200e^{-1.9} \approx 113.086$ or $\$113,086$ *(6-5)*

68. $S(t) = 50 - 50e^{-0.08t}$; $50 - 50e^{-0.96} \approx \$31$ million; $-(\ln 0.2)/0.08 \approx 20$ mo *(6-3)*   **69.** 1 cm^2 *(6-3)*   **70.** 800 gal *(6-5)*

71. (A) 145 million   (B) About 46 years *(6-3)*

72. $\dfrac{-\ln 0.04}{0.000\ 123\ 8} \approx 26,000$ yr *(6-3)*   **73.** $N(t) = 95 - 70e^{-0.1t}$; $N(15) \approx 79$ words/min *(6-3)*

CHAPTER 7

Exercise 7-1

1. $\int_a^b g(x)\,dx$ **3.** $\int_a^b [-h(x)]\,dx$

5. Since the shaded region in Figure C is below the x axis, $h(x) \le 0$; thus, $\int_a^b h(x)\,dx$ represents the negative of the area.

7. 24   **9.** 18   **11.** 51   **13.** 42   **15.** 6   **17.** 0.833   **19.** 2.350   **21.** 1   **23.** 0.432   **25.** $\int_a^b [-f(x)]\,dx$

27. $\int_b^c f(x)\,dx + \int_c^d [-f(x)]\,dx$ **29.** $\int_c^d [f(x) - g(x)]\,dx$ **31.** $\int_a^b [f(x) - g(x)]\,dx + \int_b^c [g(x) - f(x)]\,dx$

33. Find the intersection points by solving $f(x) = g(x)$ on the interval $[a, d]$ to determine b and c. Then observe that $f(x) \ge g(x)$ over $[a, b]$, $g(x) \ge f(x)$ over $[b, c]$, and $f(x) \ge g(x)$ over $[c, d]$. Thus,

area $= \int_a^b [f(x) - g(x)]\,dx + \int_b^c [g(x) - f(x)]\,dx + \int_c^d [f(x) - g(x)]\,dx$.

35. 2.5   **37.** 7.667   **39.** 12   **41.** 15   **43.** 32   **45.** 36   **47.** 9   **49.** 2.832   **51.** $\int_{-3}^3 \sqrt{9 - x^2}\,dx$; 14.137

53. $\int_0^4 \sqrt{16 - x^2}\,dx$; 12.566   **55.** $\int_{-2}^2 2\sqrt{9 - x^2}\,dx$; 12.566   **57.** 18   **59.** 1.858   **61.** 52.616   **63.** 8   **65.** 101.75

67. 8   **69.** 17.979   **71.** 5.113   **73.** 8.290   **75.** 3.166   **77.** 1.385

79. Total production from the end of the fifth year to the end of the 10th year is $50 + 100 \ln 20 - 100 \ln 15 \approx 79$ thousand barrels.

81. Total profit over the 5-yr useful life of the game is $20 - 30e^{-1.5} \approx 13.306$, or $\$13,306$.

83. 1935: 0.412; 1947: 0.231; income was more equally distributed in 1947.

85. 1963: 0.818; 1983: 0.846; total assets were less equally distributed in 1983.

87. (A) $f(x) = 0.3125x^2 + 0.7175x - 0.015$ (B) 0.104   **89.** Total weight gain during the first 10 hr is $3e - 3 \approx 5.15$ g.

91. Average number of words learned during the second 2 hr is $15 \ln 4 - 15 \ln 2 \approx 10$.

Exercise 7-2

1. 0.43   **3.** 744.99   **5.** 10.27   **7.** 151.75   **9.** 93,268.66   **11.** (A) 10.72 (B) 3.28 (C) 10.72

13. (A) .75 (B) .11   **15.** 8 yr   **17.** (A) .11 (B) .10   **19.** $P(t \ge 12) = 1 - P(0 \le t \le 12) = .89$

(C)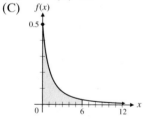

21. $\$12,500$

23. If $f(t)$ is the rate of flow of a continuous income stream, then the total income produced from 0 to 5 yr is the area under the graph of $y = f(t)$ from $t = 0$ to $t = 5$.

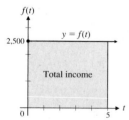

25. $8,000(e^{0.15} - 1) \approx \$1,295$

27. If $f(t)$ is the rate of flow of a continuous income stream, then the total income produced from 0 to 3 yr is the area under the graph of $y = f(t)$ from $t = 0$ to $t = 3$.

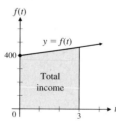

29. $\$255,562$; $\$175,562$ **31.** $20,000(e^{0.25} - e^{-0.08}) \approx \$7,218$ **33.** $\$875$

35. Clothing store: FV $= 120,000(e^{0.5} - 1) \approx \$77,847$; computer store: FV $= 200,000(e^{0.5} - e^{0.25}) \approx \$72,939$; the clothing store is the better investment.

37. Bond: FV $= 10,000e^{0.4} \approx \$14,918$; business: FV $= 25,000(e^{0.4} - 1) \approx \$12,296$; the bond is the better investment.

39. $\$55,230$ **41.** $\dfrac{k}{r}(e^{rT} - 1)$ **43.** $\$625,000$

45. The shaded area is the consumers' surplus and represents the total savings to consumers who are willing to pay more than $150 for a product but are still able to buy the product for $150.

47. $9,900

49. The area of the region PS is the producers' surplus and represents the total gain to producers who are willing to supply units at a lower price than $67 but are still able to supply the product at $67.

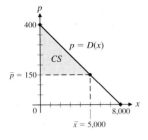

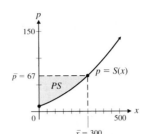

51. $CS = \$3,380$; $PS = \$1,690$ **53.** $CS = \$6,980$; $PS = \$5,041$ **55.** $CS = \$7,810$; $PS = \$8,336$

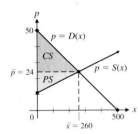

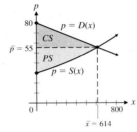

57. $CS = \$8,544$; $PS = \$11,507$ **59.** (A) $\bar{x} = 21.457$; $\bar{p} = \$6.51$ (B) $CS = 1.774$ or $\$1,774$; $PS = 1.087$ or $\$1,087$

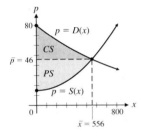

Exercise 7-3

1. $\frac{1}{3}xe^{3x} - \frac{1}{9}e^{3x} + C$ **3.** $\frac{x^3}{3}\ln x - \frac{x^3}{9} + C$ **5.** $u = x + 2;\ \dfrac{(x+2)(x+1)^6}{6} - \dfrac{(x+1)^7}{42} + C$ **7.** $-xe^{-x} - e^{-x} + C$

9. $\frac{1}{2}e^{x^2} + C$ **11.** $(xe^x - 4e^x)\big|_0^1 = -3e + 4 \approx -4.1548$ **13.** $(x\ln 2x - x)\big|_1^3 = (3\ln 6 - 3) - (\ln 2 - 1) \approx 2.6821$

15. $\ln(x^2 + 1) + C$ **17.** $(\ln x)^2/2 + C$ **19.** $\frac{2}{3}x^{3/2}\ln x - \frac{4}{9}x^{3/2} + C$

21. $(x + 3)\dfrac{(x-1)^7}{7} - \dfrac{(x-1)^8}{56} + C$ **23.** $\dfrac{x^5}{5} - \dfrac{2x^3}{3} + x + C$

25. The integral represents the negative of the area between the graph of $y = (x - 3)e^x$ and the x axis from $x = 0$ to $x = 1$.

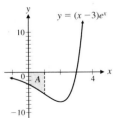

27. The integral represents the area between the graph of $y = \ln 2x$ and the x axis from $x = 1$ to $x = 3$.

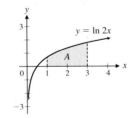

29. $(x^2 - 2x + 2)e^x + C$

31. $\dfrac{xe^{ax}}{a} - \dfrac{e^{ax}}{a^2} + C$

33. $\left(-\dfrac{\ln x}{x} - \dfrac{1}{x}\right)\Big|_1^e = -\dfrac{2}{e} + 1 \approx 0.2642$

35. $6\ln 6 - 4\ln 4 - 2 \approx 3.205$

37. $xe^{x-2} - e^{x-2} + C$

39. $\frac{1}{2}(1 + x^2)\ln(1 + x^2) - \frac{1}{2}(1 + x^2) + C$

41. $(1 + e^x)\ln(1 + e^x) - (1 + e^x) + C$

43. $x(\ln x)^2 - 2x\ln x + 2x + C$

45. $x(\ln x)^3 - 3x(\ln x)^2 + 6x\ln x - 6x + C$

47. 2   **49.** $\dfrac{1}{3}$ **51.** 1.56   **53.** 34.98   **55.** $\int_0^5 (2t - te^{-t})\, dt = \$24$ million

57. The total profit for the first 5 yr (in millions of dollars) is the same as the area under the marginal profit function, $P'(t) = 2t - te^{-t}$, from $t = 0$ to $t = 5$.

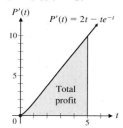

59. $3,278   **61.** 0.264

63. The area bounded by $y = x$ and the Lorenz curve $y = xe^{x-1}$, divided by the area under the graph of $y = x$ from $x = 0$ to $x = 1$, is the Gini index of income concentration. The closer this index is to 0, the more equally distributed income is; the closer the index is to 1, the more concentrated income is in a few hands.

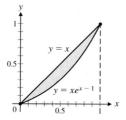

65. $S(t) = 1,600 + 400e^{0.1t} - 40te^{0.1t}$; 15 mo

67. $977

69. The area bounded by the price–demand equation, $p = 9 - \ln(x + 4)$, and the price equation, $y = \bar{p} = 2.089$, from $x = 0$ to $x = \bar{x} = 1,000$, represents the consumers' surplus. This is the amount consumers who are willing to pay more than $2,089 save.

71. 2.1388 ppm

73. $N(t) = -4te^{-0.25t} - 40e^{-0.25t} + 80$; 8 wk; 78 words/min

75. 20,980

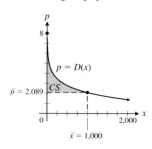

Exercise 7-4

1. $\ln\left|\dfrac{x}{1+x}\right| + C$ **3.** $\dfrac{1}{3+x} + 2\ln\left|\dfrac{5+2x}{3+x}\right| + C$ **5.** $\dfrac{2(x-32)}{3}\sqrt{16+x} + C$ **7.** $-\ln\left|\dfrac{1+\sqrt{1-x^2}}{x}\right| + C$

9. $\dfrac{1}{2}\ln\left|\dfrac{x}{2+\sqrt{x^2+4}}\right| + C$ **11.** $\dfrac{1}{3}x^3 \ln x - \dfrac{1}{9}x^3 + C$ **13.** $x - \ln|1 + e^x| + C$ **15.** $9\ln\frac{3}{2} - 2 \approx 1.6492$

17. $\dfrac{1}{2}\ln\frac{12}{5} \approx 0.4377$ **19.** $\ln 3 \approx 1.0986$ **21.** $-\dfrac{\sqrt{4x^2+1}}{x} + 2\ln|2x + \sqrt{4x^2+1}| + C$ **23.** $\dfrac{1}{2}\ln|x^2 + \sqrt{x^4-16}| + C$

25. $\dfrac{1}{6}(x^3\sqrt{x^6+4} + 4\ln|x^3 + \sqrt{x^6+4}|) + C$ **27.** $-\dfrac{\sqrt{4-x^4}}{8x^2} + C$ **29.** $\dfrac{1}{5}\ln\left|\dfrac{3+4e^x}{2+e^x}\right| + C$

31. $\dfrac{2}{3}(\ln x - 8)\sqrt{4 + \ln x} + C$ **33.** $\dfrac{1}{5}x^2e^{5x} - \dfrac{2}{25}xe^{5x} + \dfrac{2}{125}e^{5x} + C$ **35.** $-x^3e^{-x} - 3x^2e^{-x} - 6xe^{-x} - 6e^{-x} + C$

37. $x(\ln x)^3 - 3x(\ln x)^2 + 6x \ln x - 6x + C$ **39.** $\dfrac{64}{3}$ **41.** $\dfrac{1}{2}\ln\frac{9}{5} \approx 0.2939$ **43.** $\dfrac{-1 - \ln x}{x} + C$

45. $\sqrt{x^2 - 1} + C$ **47.** 31.38   **49.** 5.48   **51.** $3,000 + 1,500\ln\frac{1}{3} \approx \$1,352$

53.

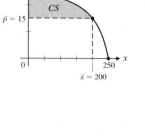

55. $C(x) = 200x + 1,000\ln(1 + 0.05x) + 25,000$; 608; $198,773

57. $100,000e - 250,000 \approx \$21,828$

59. 0.1407

61. As the area bounded by the two curves gets smaller, the Lorenz curve approaches $y = x$ and the distribution of income approaches perfect equality—all persons share equally in the income available.

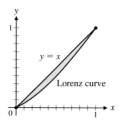

63. $S(t) = 1 + t - \dfrac{1}{1+t} - 2\ln|1+t|$; $24.96 - 2\ln 25 \approx \$18.5$ million

65. The total sales (in millions of dollars) over the first 2 yr (24 mo) is the area under the graph of $y = S'(t)$ from $t = 0$ to $t = 24$.

67. $P(x) = \dfrac{2(9x-4)}{135}(2+3x)^{3/2} - 2{,}000.83$; 54; \$37,932

69. $100\ln 3 \approx 110$ ft **71.** $60\ln 5 \approx 97$ items

73. The area under the graph of $y = N'(t)$ from $t = 0$ to $t = 12$ represents the total number of items learned in that time interval.

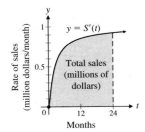

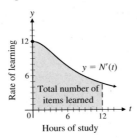

Chapter 7 Review Exercise

1. $\int_a^b f(x)\,dx$ (7-1)   **2.** $\int_b^c [-f(x)]\,dx$ (7-1)   **3.** $\int_a^b f(x)\,dx + \int_b^c [-f(x)]\,dx$ (7-1)

4. Area $= 1.153$

5. $\frac{1}{4}xe^{4x} - \frac{1}{16}e^{4x} + C$ (7-3, 7-4)   **6.** $\frac{1}{2}x^2 \ln x - \frac{1}{4}x^2 + C$ (7-3, 7-4)

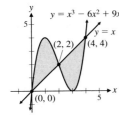

(7-1)

7. $\dfrac{(\ln x)^2}{2} + C$ (6-2)   **8.** $\dfrac{\ln(1+x^2)}{2} + C$ (7-2)   **9.** $\dfrac{1}{1+x} + \ln\left|\dfrac{x}{1+x}\right| + C$ (7-4)

10. $-\dfrac{\sqrt{1+x}}{x} - \dfrac{1}{2}\ln\left|\dfrac{\sqrt{1+x}-1}{\sqrt{1+x}+1}\right| + C$ (7-4)   **11.** 12 (7-1)   **12.** 40 (7-1)   **13.** 34.167 (7-1)

14. 0.926 (7-1)   **15.** 12 (7-1)   **16.** 18.133 (7-1)   **17.** $\int_a^b [f(x) - g(x)]\,dx$ (7-1)

18. $\int_b^c [g(x) - f(x)]\,dx$ (7-1)   **19.** $\int_b^c [g(x) - f(x)]\,dx + \int_c^d [f(x) - g(x)]\,dx$ (7-1)

20. $\int_a^b [f(x) - g(x)]\,dx + \int_b^c [g(x) - f(x)]\,dx + \int_c^d [f(x) - g(x)]\,dx$ (7-1)

21. Area $= 20.833$

22. 1 (7-3, 7-4)   **23.** $\frac{15}{2} - 8\ln 8 + 8\ln 4 \approx 1.955$ (7-4)

24. $\frac{1}{6}(3x\sqrt{9x^2 - 49} - 49\ln|3x + \sqrt{9x^2 - 49}|) + C$ (7-4)

25. $-2te^{-0.5t} - 4e^{-0.5t} + C$ (7-3, 7-4)   **26.** $\frac{1}{3}x^3 \ln x - \frac{1}{9}x^3 + C$ (7-3, 7-4)

27. $x - \ln|1 + 2e^x| + C$ (7-4)

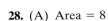 (7-1)

28. (A) Area $= 8$ (B) Area $= 8.38$

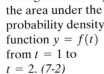   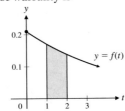

(7-1)

29. $\frac{1}{3}(\ln x)^3 + C$ (6-2)

30. $\frac{1}{2}x^2(\ln x)^2 - \frac{1}{2}x^2 \ln x + \frac{1}{4}x^2 + C$ (7-3, 7-4)

31. $\sqrt{x^2 - 36} + C$ (6-2)

32. $\frac{1}{2}\ln|x^2 + \sqrt{x^4 - 36}| + C$ (7-4)

33. $50\ln 10 - 42\ln 6 - 24 \approx 15.875$ (7-3, 7-4)

34. $x(\ln x)^2 - 2x\ln x + 2x + C$ (7-3, 7-4)

35. $-\frac{1}{4}e^{-2x^2} + C$ (6-2)

36. $-\frac{1}{2}x^2 e^{-2x} - \frac{1}{2}xe^{-2x} - \frac{1}{4}e^{-2x} + C$ (7-3, 7-4)

37. 1.703 (7-1)   **38.** (A) .189 (B) .154 (7-2)

39. The probability that the product will fail during the second year of warranty is the area under the probability density function $y = f(t)$ from $t = 1$ to $t = 2$. (7-2)

40. $R(x) = 65x - 6[(x+1)\ln(x+1) - x]$; 618/wk; \$29,506 (7-3)

41. (A)   (B) \$8,507 (7-2)

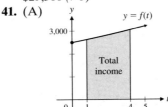

42. (A) \$20,824 (B) \$6,623 *(7-2)*

43. (A)

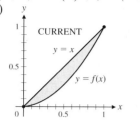

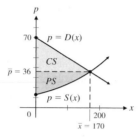

(B) More equitably distributed, since the area bounded by the two curves will have decreased.
(C) Current = 0.3; projected = 0.2; income will be more equitably distributed 10 years from now. *(7-1)*

44. (A) $CS = \$2,250$; $PS = \$2,700$ (B) $CS = \$2,890$; $PS = \$2,278$

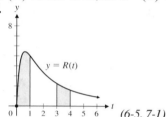

(7-2)

45. (A) 25.403 or 25,403 lb (B) $PS = 121.6$ or \$1,216 *(7-2)*   **46.** 4.522 mL; 1.899 mL *(6-5, 7-4)*

47.

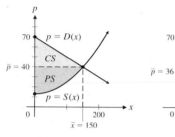

48. .667; .333 *(7-2)*

49. The probability that the doctor will spend more than an hour with a randomly selected patient is the area under the probability density function $y = f(t)$ from $t = 1$ to $t = 3$. *(7-2)*

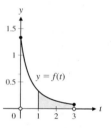

(6-5, 7-1)

50. 45 thousand *(6-5, 7-1)*   **51.** .368 *(7-2)*

CHAPTER 8

Exercise 8-1

1. -4 **3.** 11   **5.** 4   **7.** Not defined   **9.** 59   **11.** -1 **13.** -64 **15.** 16π **17.** 791   **19.** 0.192

21. 118   **23.** 36,095.07   **25.** $f(x) = x^2 - 7$ **27.** $f(y) = 38y + 20$ **29.** $f(y) = -2y^3 + 5$ **31.** $y = 2$ **33.** $x = -\dfrac{1}{2}, 1$

35. $-1.926, 0.599$ **37.** $2x + h$ **39.** $2y^2$ **41.** $E(0, 0, 3); F(2, 0, 3)$

43. (A) In the plane $y = c$, c any constant, $z = x^2$.
(B) The y axis; the horizontal line parallel to the y axis and passing through the point $(1, 0, 1)$; the horizontal line parallel to the y axis and passing through the point $(2, 0, 4)$ (C) A parabolic "trough" lying on top of the y axis

45. (A) Upper semicircles whose centers lie on the y axis   (B) Upper semicircles whose centers lie on the x axis
(C) The upper hemisphere of radius 6 with center at the origin

47. (A) $a^2 + b^2$ and $c^2 + d^2$ both equal the square of the radius of the circle.
(B) Bell-shaped curves with maximum values of 1 at the origin
(C) A bell, with maximum value 1 at the origin, extending infinitely far in all directions.

49. \$4,400; \$6,000; \$7,100   **51.** $R(p, q) = -5p^2 + 6pq - 4q^2 + 200p + 300q; R(2, 3) = \$1,280; R(3, 2) = \$1,175$

53. 30,065 units   **55.** (A) \$272,615.08   (B) 12.2%

57. $T(70, 47) \approx 29$ min; $T(60, 27) = 33$ min   **59.** $C(6, 8) = 75; C(8.1, 9) = 90$ **61.** $Q(12, 10) = 120; Q(10, 12) \approx 83$

Exercise 8-2

1. $f_x(x, y) = 4$ **3.** $f_y(x, y) = 6$ **5.** $f_y(x, y) = -3x + 4y$ **7.** $f_x(x, y) = -8x + 5y$ **9.** $\dfrac{\partial z}{\partial x} = 3x^2 + 8xy$

11. $\dfrac{\partial z}{\partial y} = -3y^2 + 1$ **13.** $\dfrac{\partial z}{\partial y} = 20(5x + 2y)^9$ **15.** $\dfrac{\partial z}{\partial x} = 10(x^2 - 4xy + y^2)^4(x - 2y)$ **17.** 9

19. 3   **21.** -4 **23.** 0   **25.** $f_{xx}(x, y) = 0$ **27.** $f_{xy}(x, y) = 0$ **29.** $f_{yy}(x, y) = 6y$

31. $f_{yx}(x, y) = 48x^5y^3 + 63x^2y^6 = 3x^2y^3(16x^3 + 21y^3)$ **33.** $f_{xy}(x, y) = y^2e^{xy^2}(2xy) + e^{xy^2}(2y) = 2y(1 + xy^2)e^{xy^2}$

35. $f_{yy}(x, y) = \dfrac{2 \ln x}{y^3}$ **37.** $f_{xx}(x, y) = 80(2x + y)^3$ **39.** $f_{xy}(x, y) = 720xy^3(x^2 + y^4)^8$ **41.** $C_x(x, y) = 6x + 10y + 4$

Answers A-49

43. 2 **45.** $C_{xx}(x, y) = 6$ **47.** $C_{xy}(x, y) = 10$ **49.** 6 **51.** $S_y(x, y) = \frac{x^3}{y} + 8ye^x$ **53.** $-1 + 8e^{-1} \approx 1.943$

55. $S_{yx}(x, y) = \frac{3x^2}{y} + 8ye^x$ **57.** $S_{yy}(x, y) = -\frac{x^3}{y^2} + 8e^x$ **59.** $3 + 8e^{-1} \approx 5.943$ **61.** $1 + 8e^{-1} \approx 3.943$

63. $f_x(x, y) = 6x(x^2 - y^3)^2; f_y(x, y) = -9y^2(x^2 - y^3)^2$ **65.** $f_x(x, y) = 24xy(3x^2y - 1)^3; f_y(x, y) = 12x^2(3x^2y - 1)^3$

67. $f_x(x, y) = 2x/(x^2 + y^2); f_y(x, y) = 2y/(x^2 + y^2)$ **69.** $f_x(x, y) = y^4e^{xy^2}; f_y(x, y) = 2xy^3e^{xy^2} + 2ye^{xy^2}$

71. $f_x(x, y) = 4xy^2/(x^2 + y^2)^2; f_y(x, y) = -4x^2y/(x^2 + y^2)^2$

75. $f_{xx}(x, y) = 2y^2 + 6x; f_{xy}(x, y) = 4xy = f_{yx}(x, y); f_{yy}(x, y) = 2x^2$

77. $f_{xx}(x, y) = -2y/x^3; f_{xy}(x, y) = (-1/y^2) + (1/x^2) = f_{yx}(x, y); f_{yy}(x, y) = 2x/y^3$

79. $f_{xx}(x, y) = (2y + xy^2)e^{xy}; f_{xy}(x, y) = (2x + x^2y)e^{xy} = f_{yx}(x, y); f_{yy}(x, y) = x^3e^{xy}$

81. $x = 2$ and $y = 4$ **83.** $x = 1.200$ and $y = -0.695$

85. (A) $-\frac{13}{3}$ (B) The function $f(0, y)$, for example, has values less than $-\frac{3}{13}$.

87. (A) $c = 1.145$ (B) $f_x(c, 2) = 0; f_y(c, 2) = 92.021$

89. $f_{xx}(x, y) + f_{yy}(x, y) = (2y^2 - 2x^2)/(x^2 + y^2)^2 + (2x^2 - 2y^2)/(x^2 + y^2)^2 = 0$ **91.** (A) $2x$ (B) $4y$

93. $P_x(1,200, 1,800) = 24$; profit will increase approx. \$24 per unit increase in production of type A calculators at the (1,200, 1,800) output level; $P_y(1,200, 1,800) = -48$; profit will decrease approx. \$48 per unit increase in production of type B calculators at the (1,200, 1,800) output level

95. $\partial x/\partial p = -5$: a \$1 increase in the price of brand A will decrease the demand for brand A by 5 lb at any price level (p, q); $\partial y/\partial p = 2$: a \$1 increase in the price of brand A will increase the demand for brand B by 2 lb at any price level (p, q)

97. (A) $f_x(x, y) = 7.5x^{-0.25}y^{0.25}; f_y(x, y) = 2.5x^{0.75}y^{-0.75}$

(B) Marginal productivity of labor $= f_x(600, 100) \approx 4.79$; Marginal productivity of capital $= f_y(600, 100) \approx 9.58$

(C) Capital

99. Competitive **101.** Complementary

103. (A) $f_w(w, h) = 6.65w^{-0.575}h^{0.725}; f_h(w, h) = 11.34w^{0.425}h^{-0.275}$

(B) $f_w(65, 57) = 11.31$: for a 65-lb child 57 in. tall, the rate of change in surface area is 11.31 in.^{2} for each pound gained in weight (height is held fixed); $f_h(65, 57) = 21.99$: for a child 57 in. tall, the rate of change in surface area is 21.99 in.^{2} for each inch gained in height (weight is held fixed)

105. $C_W(6, 8) = 12.5$: index increases approx. 12.5 units for a 1-in. increase in width of head (length held fixed) when $W = 6$ and $L = 8$; $C_L(6, 8) = -9.38$: index decreases approx. 9.38 units for a 1-in. increase in length (width held fixed) when $W = 6$ and $L = 8$.

Exercise 8-3

1. $f_x(x, y) = 4; f_y(x, y) = 5$; the functions $f_x(x, y)$ and $f_y(x, y)$ never have the value 0.

3. $f_x(x, y) = -1.2 + 4x^3; f_y(x, y) = 6.8 + 0.6y^2$; the function $f_y(x, y)$ never has the value 0.

5. $f(-2, 0) = 10$ is a local maximum. **7.** $f(-1, 3) = 4$ is a local minimum. **9.** f has a saddle point at $(3, -2)$.

11. $f(3, 2) = 33$ is a local maximum. **13.** $f(2, 2) = 8$ is a local minimum. **15.** f has a saddle point at $(0, 0)$.

17. f has a saddle point at $(0, 0); f(1, 1) = -1$ is a local minimum.

19. f has a saddle point at $(0, 0); f(3, 18) = -162$ and $f(-3, -18) = -162$ are local minima.

21. The test fails at $(0, 0); f$ has saddle points at $(2, 2)$ and $(2, -2)$. **23.** f has a saddle point at $(0.614, -1.105)$.

25. $f(x, y)$ is nonnegative and equals 0 when $x = 0$, so f has a local minimum at each point of the y axis.

27. (B) Local minimum **29.** 2,000 type A and 4,000 type B; Max $P = P(2, 4) = \$15$ million

31. (A) When $p = \$10$ and $q = \$12$, $x = 56$ and $y = 16$; when $p = \$11$ and $q = \$11$, $x = 6$ and $y = 56$.

(B) A maximum weekly profit of \$288 is realized for $p = \$10$ and $q = \$12$.

33. $P(x, y) = P(4, 2)$ **35.** 8 in. by 4 in. by 2 in. **37.** 20 in. by 20 in. by 40 in.

Exercise 8-4

1. Max $f(x, y) = f(3, 3) = 18$ **3.** Min $f(x, y) = f(3, 4) = 25$

5. $F_x = -3 + 2\lambda = 0$ and $F_y = 4 + 5\lambda = 0$ have no simultaneous solution.

7. Max $f(x, y) = f(3, 3) = f(-3, -3) = 18$; Min $f(x, y) = f(3, -3) = f(-3, 3) = -18$

9. Maximum product is 25 when each number is 5. **11.** Min $f(x, y, z) = f(-4, 2, -6) = 56$

13. Max $f(x, y, z) = f(2, 2, 2) = 6$; Min $f(x, y, z) = f(-2, -2, -2) = -6$ **15.** Max $f(x, y) = f(0.217, 0.885) = 1.055$

17. $F_x = e^x + \lambda = 0$ and $F_y = 3e^y - 2\lambda = 0$ have no simultaneous solution.

19. Maximize $f(x, 5)$, a function of just one independent variable.

21. (A) Max $f(x, y) = f(0.707, 0.5) = f(-0.707, 0.5) = 0.47$

23. 60 of model A and 30 of model B will yield a minimum cost of \$32,400 per week.

25. (A) 8,000 units of labor and 1,000 units of capital; Max $N(x, y) = N(8,000, 1,000) \approx 263,902$ units

(B) Marginal productivity of money ≈ 0.6598; increase in production $\approx 32,990$ units

27. 8 in. by 8 in. by $\frac{8}{3}$ in. **29.** $x = 50$ ft and $y = 200$ ft; maximum area is 10,000 ft^2

Exercise 8-5

1.

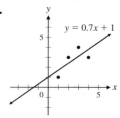

$y = 0.7x + 1$

3.

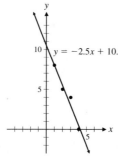

$y = -2.5x + 10.5$

5.

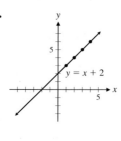

$y = x + 2$

7. $y = -1.5x + 4.5$; $y = 0.75$ when $x = 2.5$ **9.** $y = 2.12x + 10.8$; $y = 63.8$ when $x = 25$
11. $y = -1.2x + 12.6$; $y = 10.2$ when $x = 2$ **13.** $y = -1.53x + 26.67$; $y = 14.4$ when $x = 8$
15.

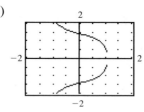

$y = 0.75x^2 - 3.45x + 4.75$

21. (A) $y = 1.52x - 0.16$; $y = 0.73x^2 - 1.39x + 1.30$ (B) The quadratic function
23. The normal equations form a system of 4 linear equations in the 4 variables a, b, c, and d, which can be solved by Gauss–Jordan elimination.

25. (A) $y = 514.2x - 2,634$ (B) 7,650,000 vehicles
27. (A) $y = -0.48x + 4.38$ (B) $6.56 per bottle
29. (A) $P = -0.66T + 48.8$ (B) 11.18 beats/min
31. (A) $y = 0.0121x + 56.35$ (B) $58.77°F$
33. (A) $D = -3.1A + 54.6$ (B) 45%   **35.** (A) $y = 0.0858x + 10.84$ (B) 21.48 ft

Exercise 8-6

1. (A) $3x^2y^4 + C(x)$ (B) $3x^2$ **3.** (A) $2x^2 + 6xy + 5x + E(y)$ (B) $35 + 30y$

5. (A) $\sqrt{y + x^2} + E(y)$ (B) $\sqrt{y + 4} - \sqrt{y}$ **7.** (A) $\dfrac{\ln x \ln y}{x} + C(x)$ (B) $\dfrac{2\ln x}{x}$

9. 9   **11.** 330   **13.** $(56 - 20\sqrt{5})/3$ **15.** 1   **17.** 16   **19.** 49   **21.** $\frac{1}{8}\int_1^5 \int_{-1}^1 (x + y)^2 \, dy \, dx = \frac{32}{3}$

23. $\frac{1}{15}\int_1^4 \int_2^7 (x/y) \, dy \, dx = \frac{1}{2}\ln\frac{7}{2} \approx 0.6264$ **25.** $\frac{4}{3}$cubic units   **27.** $\frac{32}{3}$cubic units   **29.** $\int_0^1 \int_1^2 xe^{xy} \, dy \, dx = \frac{1}{2} + \frac{1}{2}e^2 - e$

31. $\int_0^1 \int_{-1}^1 \dfrac{2y + 3xy^2}{1 + x^2} \, dy \, dx = \ln 2$

35. (A) $\frac{1}{3} + \frac{1}{4}e^{-2} - \frac{1}{4}e^2$ (B)

(C) Points to the right of the graph in part (B) are greater than 0; points to the left of the graph are less than 0.

37. $\dfrac{1}{0.4}\int_{0.6}^{0.8} \int_5^7 \dfrac{y}{1 - x} \, dy \, dx = 30 \ln 2 \approx \$20.8$ billion

39. $\frac{1}{10}\int_{10}^{20} \int_1^2 x^{0.75}y^{0.25} \, dy \, dx = \frac{8}{175}(2^{1.25} - 1)(20^{1.75} - 10^{1.75}) \approx 8.375$ or 8,375 units

41. $\frac{1}{192}\int_{-8}^8 \int_{-6}^6 [10 - \frac{1}{10}(x^2 + y^2)] \, dy \, dx = \frac{20}{3}$insects/ft²   **43.** $\frac{1}{8}\int_{-2}^2 \int_{-1}^1 [100 - 15(x^2 + y^2)] \, dy \, dx = 75$ ppm

45. $\frac{1}{10,000}\int_{2,000}^{3,000} \int_{50}^{60} 0.000\,013\,3xy^2 \, dy \, dx \approx 100.86$ ft   **47.** $\frac{1}{16}\int_8^{16} \int_{10}^{12} 100\frac{x}{y} \, dy \, dx = 600 \ln 1.2 \approx 109.4$

Exercise 8-7

1. $R = \{(x, y)|0 \le y \le 4 - x^2, 0 \le x \le 2\}$
$R = \{(x, y)|0 \le x \le \sqrt{4 - y}, 0 \le y \le 4\}$

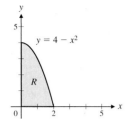

3. R is a regular x region:
$R = \{(x, y)|x^3 \le y \le 12 - 2x, 0 \le x \le 2\}$

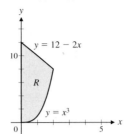

5. R is a regular y region: $R = \{(x, y)|\frac{1}{2} y^2 \le x \le y + 4, -2 \le y \le 4\}$

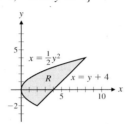

7. $\frac{1}{2}$ **9.** $\frac{39}{70}$ **11.** R consists of the points on or inside the rectangle with corners $(\pm 2, \pm 3)$; both
13. R is the arch-shaped region consisting of the points on or inside the rectangle with corners $(\pm 2, 0)$ and $(\pm 2, 2)$ that are not inside the circle of radius 1 centered at the origin; regular x region
15. $\frac{56}{3}$ **17.** $-\frac{3}{4}$ **19.** $\frac{1}{2}e^4 - \frac{5}{2}$

21. $R = \{(x, y)|0 \le y \le x + 1, \quad 0 \le x \le 1\}$
$\int_0^1 \int_0^{x+1} \sqrt{1 + x + y}\, dy\, dx = (68 - 24\sqrt{2})/15$

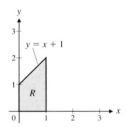

23. $R = \{(x, y)|0 \le y \le 4x - x^2, 0 \le x \le 4\}$
$\int_0^4 \int_0^{4x-x^2} \sqrt{y + x^2}\, dy\, dx = \frac{128}{5}$

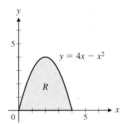

25. $R = \{(x, y)|1 - \sqrt{x} \le y \le 1 + \sqrt{x}, 0 \le x \le 4\}$
$\int_0^4 \int_{1-\sqrt{x}}^{1+\sqrt{x}} x(y - 1)^2\, dy\, dx = \frac{512}{21}$

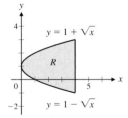

27. $\int_0^3 \int_0^{3-y} (x + 2y)\, dx\, dy = \frac{27}{2}$

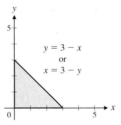

29. $\int_0^1 \int_0^{\sqrt{1-y}} x\sqrt{y}\, dx\, dy = \frac{2}{15}$

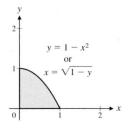

31. $\int_0^1 \int_{4y^2}^{4y} x\, dx\, dy = \frac{16}{15}$

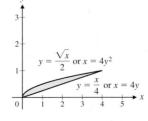

33. $\int_0^4 \int_0^{4-x} (4 - x - y)\, dy\, dx = \frac{32}{3}$

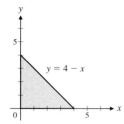

35. $\int_0^1 \int_0^{1-x^2} 4\, dy\, dx = \frac{8}{3}$

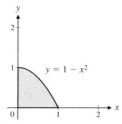

37. $\int_0^4 \int_0^{\sqrt{y}} \frac{4x}{1 + y^2}\, dx\, dy = \ln 17$

39. $\int_0^1 \int_0^{\sqrt{x}} 4ye^{x^2}\, dy\, dx = e - 1$

41. $R = \{(x, y) | x^2 \le y \le 1 + \sqrt{x}, 0 \le x \le 1.49\}$

$\int_0^{1.49} \int_{x^2}^{1+\sqrt{x}} x\, dy\, dx \approx 0.96$

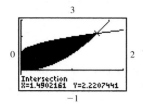

43. $R = \{(x, y) | y^3 \le x \le 1 - y, \ 0 \le y \le 0.68\}$

$\int_0^{0.68} \int_{y^3}^{1-y} 24xy\, dx\, dy \approx 0.83$

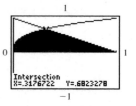

45. $R = \{(x, y) | e^{-x} \le y \le 3 - x, \ -1.51 \le x \le 2.95\}$ Regular x region

$R = \{(x, y) | -\ln y \le x \le 3 - y, \ 0.05 \le y \le 4.51\}$ Regular y region

$\int_{-1.51}^{2.95} \int_{e^{-x}}^{3-x} 4y\, dy\, dx = \int_{0.05}^{4.51} \int_{-\ln y}^{3-y} 4y\, dx\, dy \approx 40.67$

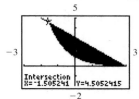

Chapter 8 Review Exercise

1. $f(5, 10) = 2{,}900$; $f_x(x, y) = 40$; $f_y(x, y) = 70$ (8-1, 8-2)   **2.** $\partial^2 z/\partial x^2 = 6xy^2$; $\partial^2 z/\partial x\, \partial y = 6x^2 y$ (8-2)
3. $2xy^3 + 2y^2 + C(x)$ (8-6)   **4.** $3x^2 y^2 + 4xy + E(y)$ (8-6)   **5.** 1 (8-6)
6. $f_x(x, y) = 5 + 6x + 3x^2$; $f_y(x, y) = -2$; the function $f_y(x, y)$ never has the value 0. (8-3)
7. $f(2, 3) = 7$; $f_y(x, y) = -2x + 2y + 3$; $f_y(2, 3) = 5$ (8-1, 8-2)   **8.** $(-8)(-6) - (4)^2 = 32$ (8-2)
9. $(1, 3, -\frac{1}{2}), (-1, -3, \frac{1}{2})$ (8-4)   **10.** $y = -1.5x + 15.5$; $y = 0.5$ when $x = 10$ (8-5)   **11.** 18 (8-6)   **12.** $\frac{8}{5}$ (8-7)
13. $f_x(x, y) = 2xe^{x^2+2y}$; $f_y(x, y) = 2e^{x^2+2y}$; $f_{xy}(x, y) = 4xe^{x^2+2y}$ (8-2)
14. $f_x(x, y) = 10x(x^2 + y^2)^4$; $f_{xy}(x, y) = 80xy(x^2 + y^2)^3$ (8-2)
15. $f(2, 3) = -25$ is a local minimum; f has a saddle point at $(-2, 3)$. (8-3)   **16.** Max $f(x, y) = f(6, 4) = 24$ (8-4)
17. Min $f(x, y, z) = f(2, 1, 2) = 9$ (8-4)   **18.** $y = \frac{116}{165}x + \frac{100}{3}$ (8-5)   **19.** $\frac{27}{5}$ (8-6)   **20.** 4 cubic units (8-6)
21. 0 (8-6)   **22.** (A) 12.56   (B) No (8-6)
23. $F_x = 12x^2 + 3\lambda = 0$, $F_y = -15y^2 + 2\lambda = 0$, and $F_\lambda = 3x + 2y - 7 = 0$ have no simultaneous solution. (8-4)   **24.** 1 (8-7)
25. (A) $P_x(1, 3) = 8$; profit will increase \$8,000 for a 100-unit increase in product A if the production of product B is held fixed at an output level of (1, 3).
(B) For 200 units of A and 300 units of B, $P(2, 3) = $100 thousand is a local maximum. (8-2, 8-3)
26. 8 in. by 6 in. by 2 in. (8-3)   **27.** $y = 0.63x + 1.33$; profit in sixth year is \$5.11 million (8-4)
28. (A) Marginal productivity of labor ≈ 8.37; marginal productivity of capital ≈ 1.67; management should encourage increased use of labor.
(B) 80 units of labor and 40 units of capital; Max $N(x, y) = N(80, 40) \approx 696$ units; marginal productivity of money ≈ 0.0696; increase in production ≈ 139 units

(C) $\dfrac{1}{1{,}000} \int_{50}^{100} \int_{20}^{40} 10x^{0.8}y^{0.2}\, dy\, dx = \dfrac{(40^{1.2} - 20^{1.2})(100^{1.8} - 50^{1.8})}{216} = 621$ items (8-4)

29. $T_x(70, 17) = -0.924$ min/ft increase in depth when $V = 70$ ft^3 and $x = 17$ ft *(8-2)*

30. $\frac{1}{16}\int_{-2}^{2}\int_{-2}^{2}[100 - 24(x^2 + y^2)]\, dy\, dx = 36$ ppm *(8-6)*   **31.** 50,000 *(8-1)*

32. $y = \frac{1}{2}x + 48$; $y = 68$ when $x = 40$ *(8-5)*

33. (A) $y = 0.4933x + 25.20$ (B) 84.40 people/mi^2 (C) 89.30 people/mi^2; 97.70 people/mi^2 *(8-5)*

34. (A) $y = 1.069x + 0.522$ (B) 64.68 yr   (C) 64.78 yr; 64.80 yr *(8-5)*

CHAPTER 9

Exercise 9-1

1. $\pi/10$ rad   **3.** $\pi/2$ rad   **5.** 3π rad   **7.** I   **9.** II   **11.** III   **13.** -1 **15.** 0   **17.** 0   **19.** 120°   **21.** 360°

23. 135°   **25.** $1/\sqrt{2}$, or $\sqrt{2}/2$ **27.** 1/2   **29.** 1/2   **31.** 0   **33.** -1 **35.** $\sqrt{3}/2$ **37.** $-1/\sqrt{2}$, or $-\sqrt{2}/2$

39. 1/2   **41.** 0.1411   **43.** -0.6840 **45.** 0.7970   **47.** $3\pi/20$ rad   **49.** 15°   **51.** $1/\sqrt{3}$, or $\sqrt{3}/3$ **53.** 1   **55.** -2

57. $-\sqrt{2}$ **59.** Not defined   **61.** Not defined   **63.** $\sqrt{2}$ **65.** $-\sqrt{3}$

69. 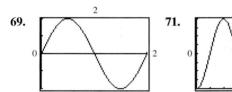   **71.** 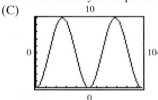   **73.** $\left\{ x \mid x \neq \pm\dfrac{\pi}{2}, \pm\dfrac{3\pi}{2}, \pm\dfrac{5\pi}{2}, \ldots \right\}$

75. $\left\{ x \mid x \neq \pm\dfrac{\pi}{2}, \pm\dfrac{3\pi}{2}, \pm\dfrac{5\pi}{2}, \ldots \right\}$

81. (A) $P(13) = 5$, $P(26) = 10$, $P(39) = 5$, $P(52) = 0$

(B) $P(30) \approx 9.43$, $P(100) \approx 0.57$; thus, 30 weeks after January 1 the profit on a week's sales of bathing suits is $943, and 100 weeks after January 1 the profit on a week's sales of bathing suits is $57.

(C)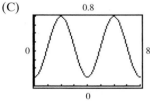

83. (A) $V(0) = 0.10$, $V(1) = 0.45$, $V(2) = 0.80$, $V(3) = 0.45$, $V(7) = 0.45$

(B) $V(3.5) \approx 0.20$, $V(5.7) \approx 0.76$; thus, the volume of air in the lungs of a normal seated adult 3.5 sec after exhaling is approximately 0.20 L and 5.7 sec after exhaling is approx. 0.76 L.

(C)

85. (A) $-5.6°$ (B) $-4.7°$

Exercise 9-2

1. $-\sin t$ **3.** $3x^2 \cos x^3$ **5.** $t \cos t + \sin t$ **7.** $(\cos x)^2 - (\sin x)^2$ **9.** $5(\sin x)^4 \cos x$ **11.** $\dfrac{\cos x}{2\sqrt{\sin x}}$

13. $-\dfrac{x^{-1/2}}{2}\sin\sqrt{x} = \dfrac{-\sin\sqrt{x}}{2\sqrt{x}}$ **15.** $f'\left(\dfrac{\pi}{6}\right) = \cos\dfrac{\pi}{6} = \dfrac{\sqrt{3}}{2}$

17. Increasing on $[-\pi, 0]$; decreasing on $[0, \pi]$; concave upward on $[-\pi, -\pi/2]$ and $[\pi/2, \pi]$; concave downward on $[-\pi/2, \pi/2]$; local maximum at $x = 0$; $f'(x) = -\sin x$; $f(x) = \cos x$

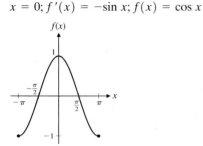

19. $\dfrac{(\cos x)^2 + (\sin x)^2}{(\cos x)^2} = \dfrac{1}{(\cos x)^2} = (\sec x)^2$ **21.** $\dfrac{x \cos\sqrt{x^2 - 1}}{\sqrt{x^2 - 1}}$

23. $2e^x \cos x$ **25.**

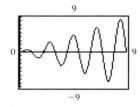

27.

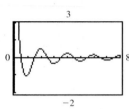

29.

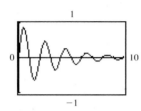

31. (A) $P'(t) = \dfrac{5\pi}{26}\sin\dfrac{\pi t}{26}, 0 < t < 104$

(B) $P'(8) = \$0.50$ hundred, or \$50 per week; $P'(26) = \$0$ per week; $P'(50) = -\$0.14$ hundred, or $-\$14$ per week

(C)

t	$P(t)$	
26	\$1,000	Local maximum
52	\$0	Local minimum
78	\$1,000	Local maximum

(D)

t	$P(t)$	
0	\$0	Absolute minimum
26	\$1,000	Absolute maximum
52	\$0	Absolute minimum
78	\$1,000	Absolute maximum
104	\$0	Absolute minimum

(E) Same answer as for part (C)

33. (A) $V'(t) = \dfrac{0.35\pi}{2}\sin\dfrac{\pi t}{2}, 0 \le t \le 8$ (B) $V'(3) = -0.55$ L/sec; $V'(4) = 0.00$ L/sec; $V'(5) = 0.55$ L/sec

(C)

t	$V(t)$	
2	0.80	Local maximum
4	0.10	Local minimum
6	0.80	Local maximum

(D)

t	$V(t)$	
0	0.10	Absolute minimum
2	0.80	Absolute maximum
4	0.10	Absolute minimum
6	0.80	Absolute maximum
8	0.10	Absolute minimum

(E) Same answer as for part (C)

Exercise 9-3

1. $-\cos t + C$ **3.** $\frac{1}{3}\sin 3x + C$ **5.** $\frac{1}{13}(\sin x)^{13} + C$ **7.** $-\frac{3}{4}(\cos x)^{4/3} + C$ **9.** $\frac{1}{3}\sin x^3 + C$ **11.** 1   **13.** 1

15. $\sqrt{3}/2 - \frac{1}{2} \approx 0.366$ **17.** 1.4161   **19.** 0.0678   **21.** $e^{\sin x} + C$ **23.** $\ln|\sin x| + C$ **25.** $-\ln|\cos x| + C$

27. (A)

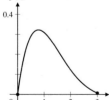

(B) $L_6 \approx 0.498$ **29.** (A) \$520 hundred, or \$52,000

(B) \$106.38 hundred, or \$10,638

(C)

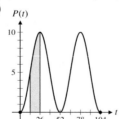

31. (A) 104 tons

(B) 31 tons

(C)

Chapter 9 Review Exercise

1. (A) $\pi/6$ (B) $\pi/4$ (C) $\pi/3$ (D) $\pi/2$ *(9-1)*   **2.** (A) -1 (B) 0 (C) 1 *(9-1)*   **3.** $-\sin m$ *(9-2)*

4. $\cos u$ *(9-2)*   **5.** $(2x - 2)\cos(x^2 - 2x + 1)$ *(9-2)*   **6.** $-\frac{1}{3}\cos 3t + C$ *(9-3)*

7. (A) $30°$ (B) $45°$ (C) $60°$ (D) $90°$ *(9-1)*   **8.** (A) $\frac{1}{2}$ (B) $\sqrt{2}/2$ (C) $\sqrt{3}/2$ *(9-1)*

9. (A) -0.6543 (B) 0.8308 *(9-1)*   **10.** $(x^2 - 1)\cos x + 2x \sin x$ *(9-2)*   **11.** $6(\sin x)^5 \cos x$ *(9-2)*

12. $(\cos x)/[3(\sin x)^{2/3}]$ *(9-2)*   **13.** $\frac{1}{2}\sin(t^2 - 1) + C$ *(9-3)*   **14.** 2 *(9-3)*   **15.** $\sqrt{3}/2$ *(9-3)*   **16.** -0.243 *(9-3)*

17. $-\sqrt{2}/2$ *(9-2)*   **18.** $\sqrt{2}$ *(9-3)*

19. (A)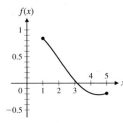

(B) $R_4 \approx 0.121$ *(6-4, 9-3)*

20. $\pi/12$ *(9-1)*   **21.** (A) -1 (B) $-\sqrt{3}/2$ (C) $-\frac{1}{2}$ *(9-1)*
22. $1/(\cos u)^2 = (\sec u)^2$ *(9-2)*   **23.** $-2x(\sin x^2)e^{\cos x^2}$ *(9-2)*
24. $e^{\sin x} + C$ *(9-3)*   **25.** $-\ln|\cos x| + C$ *(9-3)*   **26.** 15.2128 *(9-3)*
27. *(9-2, 9-3)*   **28.** *(9-2, 9-3)*

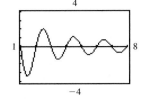

29. *(9-2, 9-3)*

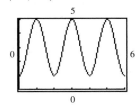

30. (A) $R(0) = \$5$ thousand; $R(2) = \$4$ thousand; $R(3) = \$3$ thousand; $R(6) = \$1$ thousand
   (B) $R(1) = \$4.732$ thousand, $R(22) = \$4$ thousand; thus, the revenue is \$4,732 for a month of sweater sales 1 month after January 1, and \$4,000 for a month of sweater sales 22 months after January 1. *(9-1)*

31. (A) $R'(t) = -\dfrac{\pi}{3}\sin\dfrac{\pi t}{6}, 0 \le t \le 24$

   (B) $R'(3) = -\$1.047$ thousand, or $-\$1,047$/mo; $R'(10) = \$0.907$ thousand, or \$907/mo; $R'(18) = \$0.000$ thousand
   (C)   (D)   (E) Same answer as for part (C) *(9-2)*

t	$P(t)$	
6	\$1,000	Local minimum
12	\$5,000	Local maximum
18	\$1,000	Local minimum

t	$P(t)$	
0	\$5,000	Absolute maximum
6	\$1,000	Absolute minimum
12	\$5,000	Absolute maximum
18	\$1,000	Absolute minimum
24	\$5,000	Absolute maximum

32. (A) \$72 thousand, or \$72,000   (B) \$6.270 thousand, or \$6,270   (C) *(9-3)*

APPENDIX A

Self-Test on Basic Algebra

1. (A) $(y + z)x$ (B) $(2 + x) + y$ (C) $2x + 3x$ *(A-1)*   **2.** $x^3 - 3x^2 + 4x + 8$ *(A-2)*   **3.** $x^3 + 3x^2 - 2x + 12$ *(A-2)*
4. $-3x^5 + 2x^3 - 24x^2 + 16$ *(A-2)*   **5.** (A) 1 (B) 1 (C) 2 (D) 3 *(A-2)*   **6.** (A) 3 (B) 1 (C) -3 (D) 1 *(A-2)*
7. $14x^2 - 30x$ *(A-2)*   **8.** $6x^2 - 5xy - 4y^2$ *(A-2)*   **9.** $4a^2 - 12ab + 9b^2$ *(A-2)*   **10.** $4xy - 2y^2$ *(A-2)*
11. $9x^6 - 12x^3y + 4y^2$ *(A-2)*   **12.** $x^3 - 6x^2y + 12xy^2 - 8y^3$ *(A-2)*   **13.** (A) 4.065×10^{12} (B) 7.3×10^{-3} *(A-5)*
14. (A) 255,000,000 (B) 0.000 406 *(A-5)*   **15.** (A) T (B) F *(A-1)*   **16.** 0 and -3 are two examples of infinitely many. *(A-1)*
17. $6x^5y^{15}$ *(A-5)*   **18.** $3u^4/v^2$ *(A-5)*   **19.** 6×10^2 *(A-5)*   **20.** x^6/y^4 *(A-5)*   **21.** $u^{7/3}$ *(A-6)*   **22.** $3a^2/b$ *(A-6)*   **23.** $\frac{5}{9}$ *(A-5)*
24. $x + 2x^{1/2}y^{1/2} + y$ *(A-6)*   **25.** $6x + 7x^{1/2}y^{1/2} - 3y$ *(A-6)*   **26.** $(3x - 1)(4x + 3)$ *(A-3)*   **27.** $(4x - 3y)(2x - 3y)$ *(A-3)*
28. Not factorable relative to the integers *(A-3)*   **29.** $3n(2n - 5)(n + 1)$ *(A-3)*   **30.** $(x - y)(7x - y)$ *(A-3)*

31. $3x(2x + 1)(2 - x)$ *(A-3)*   **32.** $\dfrac{12a^3b - 40b^2 - 5a}{30a^3b^2}$ *(A-4)*   **33.** $\dfrac{7x - 4}{6x(x - 4)}$ *(A-4)*   **34.** $\dfrac{-8(x + 2)}{x(x - 4)(x + 4)}$ *(A-4)*

35. $2x + y$ *(A-4)*   **36.** $\dfrac{-1}{7(7 + h)}$ *(A-4)*   **37.** $\dfrac{xy}{y - x}$ *(A-6)*

38. (A) Subtraction (B) Commutative $(+)$ (C) Distributive (D) Associative $(\cdot)$ (E) Negatives (F) Identity $(+)$ *(A-1)*

39. $6x^{2/5} - 7(x - 1)^{3/4}$ *(A-6)*   **40.** $2\sqrt{x} - 3\sqrt[3]{x^2}$ *(A-6)*   **41.** $2 - \frac{3}{2}x^{-1/2}$ *(A-6)*   **42.** $\sqrt{3x}$ *(A-6)*   **43.** $\sqrt{x} + \sqrt{5}$ *(A-6)*

44. $\dfrac{1}{\sqrt{x - 5}}$ *(A-6)*   **45.** $\dfrac{1}{\sqrt{u + h} + \sqrt{u}}$ *(A-6)*   **46.** $x = 0, 5$ *(A-7)*   **47.** $x = \pm\sqrt{7}$ *(A-7)*   **48.** $x = -4, 5$ *(A-7)*   **49.** $x = 1, \frac{1}{6}$ *(A-7)*

Exercise A-1

1. vu **3.** $(3 + 7) + y$ **5.** $u + v$ **7.** T   **9.** T   **11.** F   **13.** T   **15.** T   **17.** T   **19.** T   **21.** F   **23.** T   **25.** T   **27.** No
29. (A) F (B) T (C) T   **31.** $\sqrt{2}$ and π are two examples of infinitely many.   **33.** (A) N, Z, Q, R (B) R (C) Q, R (D) Q, R
35. (A) F, since, for example, $2(3 - 1) \neq 2 \cdot 3 - 1$ (B) F, since, for example, $(8 - 4) - 2 \neq 8 - (4 - 2)$
  (C) T   (D) F, since, for example, $(8 \div 4) \div 2 \neq 8 \div (4 \div 2)$.
37. $\frac{1}{11}$ **39.** (A) $2.166\,666\,666\ldots$ (B) $4.582\,575\,69\ldots$ (C) $0.437\,500\,000\ldots$ (D) $0.261\,261\,261\ldots$

Exercise A-2

1. 3   **3.** $x^3 + 4x^2 - 2x + 5$ **5.** $x^3 + 1$ **7.** $2x^5 + 3x^4 - 2x^3 + 11x^2 - 5x + 6$ **9.** $-5u + 2$ **11.** $6a^2 + 6a$ **13.** $a^2 - b^2$
15. $6x^2 - 7x - 5$ **17.** $2x^2 + xy - 6y^2$ **19.** $9y^2 - 4$ **21.** $-4x^2 + 12x - 9$ **23.** $16m^2 - 9n^2$ **25.** $9u^2 + 24uv + 16v^2$
27. $a^3 - b^3$ **29.** $x^2 - 2xy + y^2 - 9z^2$ **31.** 1   **33.** $x^4 - 2x^2y^2 + y^4$ **35.** $-40ab$ **37.** $-4m + 8$ **39.** $-6xy$
41. $u^3 + 3u^2v + 3uv^2 + v^3$ **43.** $x^3 - 6x^2y + 12xy^2 - 8y^3$ **45.** $2x^2 - 2xy + 3y^2$ **47.** $x^4 - 10x^3 + 27x^2 - 10x + 1$
49. $4x^3 - 14x^2 + 8x - 6$ **51.** $m + n$ **53.** No change   **55.** $(1 + 1)^2 \neq 1^2 + 1^2$; either a or b must be 0
57. $0.09x + 0.12(10,000 - x) = 1,200 - 0.03x$ **59.** $10x + 30(3x) + 50(4,000 - x - 3x) = 200,000 - 100x$
61. $0.02x + 0.06(10 - x) = 0.6 - 0.04x$

Exercise A-3

1. $3m^2(2m^2 - 3m - 1)$ **3.** $2uv(4u^2 - 3uv + 2v^2)$ **5.** $(7m + 5)(2m - 3)$ **7.** $(4ab - 1)(2c + d)$ **9.** $(2x - 1)(x + 2)$
11. $(y - 1)(3y + 2)$ **13.** $(x + 4)(2x - 1)$ **15.** $(w + x)(y - z)$ **17.** $(a - 3b)(m + 2n)$ **19.** $(3y + 2)(y - 1)$
21. $(u - 5v)(u + 3v)$ **23.** Not factorable   **25.** $(wx - y)(wx + y)$ **27.** $(3m - n)^2$ **29.** Not factorable
31. $4(z - 3)(z - 4)$ **33.** $2x^2(x - 2)(x - 10)$ **35.** $x(2y - 3)^2$ **37.** $(2m - 3n)(3m + 4n)$ **39.** $uv(2u - v)(2u + v)$
41. $2x(x^2 - x + 4)$ **43.** $(2x - 3y)(4x^2 + 6xy + 9y^2)$ **45.** $xy(x + 2)(x^2 - 2x + 4)$ **47.** $[(x + 2) - 3y][(x + 2) + 3y]$
49. Not factorable   **51.** $(6x - 6y - 1)(x - y + 4)$ **53.** $(y - 2)(y + 2)(y^2 + 1)$ **55.** $3(x - y)^2(5xy - 5y^2 + 4x)$
57. True   **59.** False

Exercise A-4

1. $8d^6$ **3.** $\dfrac{15x^2 + 10x - 6}{180}$ **5.** $\dfrac{15m^2 + 14m - 6}{36m^3}$ **7.** $\dfrac{1}{x(x - 4)}$ **9.** $\dfrac{x - 6}{x(x - 3)}$ **11.** $\dfrac{-3x - 9}{(x - 2)(x + 1)^2}$ **13.** $\dfrac{2}{x - 1}$

15. $\dfrac{5}{a - 1}$ **17.** $\dfrac{x^2 + 8x - 16}{x(x - 4)(x + 4)}$ **19.** $\dfrac{7x^2 - 2x - 3}{6(x + 1)^2}$ **21.** $\dfrac{x(y - x)}{y(2x - y)}$ **23.** $\dfrac{-17c + 16}{15(c - 1)}$ **25.** $\dfrac{1}{x - 3}$ **27.** $\dfrac{-1}{2x(x + h)}$

29. $\dfrac{x - y}{x + y}$ **31.** (A) Incorrect (B) $x + 1$ **33.** (A) Incorrect (B) $2x + h$ **35.** (A) Incorrect (B) $\dfrac{x^2 - x - 3}{x + 1}$

37. (A) Correct **39.** $\dfrac{-2x - h}{3(x + h)^2 x^2}$ **41.** $\dfrac{x(x - 3)}{x - 1}$

Exercise A-5

1. $2/x^9$ **3.** $3w^7/2$ **5.** $2/x^3$ **7.** $1/w^5$ **9.** $4/a^6$ **11.** $1/a^6$ **13.** $1/8x^{12}$ **15.** 8.23×10^{10} **17.** 7.83×10^{-1} **19.** 3.4×10^{-5}
21. 40,000 **23.** 0.007 **25.** 61,710,000 **27.** 0.000 808 **29.** 1 **31.** 10^{14} **33.** $y^6/25x^4$ **35.** $4x^6/25$ **37.** $4y^3/3x^5$
39. $\frac{7}{4} - \frac{1}{4}x^{-3}$ **41.** $\frac{5}{2}x^2 - \frac{3}{2} + 4x^{-2}$ **43.** $\dfrac{x^2(x - 3)}{(x - 1)^3}$ **45.** $\dfrac{2(x - 1)}{x^3}$ **47.** 2.4×10^{10}; 24,000,000,000 **49.** 3.125×10^4; 31,250

51. 64 **55.** uv **57.** $\dfrac{bc(c + b)}{c^2 + bc + b^2}$ **59.** (A) 1.105×10^{12} (B) 1.4747 (C) 0.6781 **61.** (A) \$28,479 (B) \$1,205 (C) 4.23%

63. (A) 9×10^{-6} (B) 0.000 009 (C) 0.0009% **65.** 1,478,437

Exercise A-6

1. $6\sqrt[5]{x^3}$ **3.** $8xy\sqrt[5]{xy^4}$ **5.** $\sqrt{x^2 + y^2}$ (not $x + y$) **7.** $5x^{3/4}$ **9.** $(2x^2y)^{3/5}$ **11.** $x^{1/3} + y^{1/3}$ **13.** 5 **15.** 64 **17.** -7
19. -16 **21.** $\frac{8}{125}$ **23.** $\frac{1}{27}$ **25.** $x^{2/5}$ **27.** m **29.** $2x/y^2$ **31.** $xy^2/2$ **33.** $1/24x^{7/12}$ **35.** $2x + 3$ **37.** $30x^5\sqrt{3x}$ **39.** 2
41. $12x - 6x^{35/4}$ **43.** $3u - 13u^{1/2}v^{1/2} + 4v$ **45.** $36m^{3/2} - \dfrac{6m^{1/2}}{n^{1/2}} + \dfrac{6m}{n^{1/2}} - \dfrac{1}{n}$ **47.** $9x - 6x^{1/2}y^{1/2} + y$ **49.** $\frac{1}{2}x^{1/3} + x^{-1/3}$

51. $\frac{2}{3}x^{-1/4} + \frac{1}{3}x^{-2/3}$ **53.** $\frac{1}{2}x^{-1/6} - \frac{1}{4}$ **55.** $4n\sqrt{3mn}$ **57.** $\dfrac{2(x + 3)\sqrt{x - 2}}{x - 2}$ **59.** $7(x - y)(\sqrt{x} + \sqrt{y})$ **61.** $\dfrac{1}{xy\sqrt{5xy}}$
63. $\dfrac{1}{\sqrt{x + h} + \sqrt{x}}$ **65.** $\dfrac{1}{(t + x)(\sqrt{t} + \sqrt{x})}$ **67.** $x = y = 1$ is one of many choices.
69. $x = y = 1$ is one of many choices. **71.** False **73.** False **75.** False **77.** True **79.** True **81.** False **83.** $\dfrac{x + 8}{2(x + 3)^{3/2}}$

85. $\dfrac{x - 2}{2(x - 1)^{3/2}}$ **87.** $\dfrac{x + 6}{3(x + 2)^{5/3}}$ **89.** 103.2 **91.** 0.0805 **93.** 4,588 **95.** (A) and (E); (B) and (F); (C) and (D)

Exercise A-7

1. $\pm\sqrt{11}$ **3.** $-\frac{4}{3}, 2$ **5.** $-2, 6$ **7.** 0, 2 **9.** $3 \pm 2\sqrt{3}$ **11.** $-2 \pm \sqrt{2}$ **13.** $0, \frac{15}{2}$ **15.** $\pm\frac{3}{2}$ **17.** $\frac{1}{2}, -3$ **19.** $\left(-1 \pm \sqrt{5}\right)/2$
21. $\left(3 \pm \sqrt{3}\right)/2$ **23.** No real solution **25.** $\left(-3 \pm \sqrt{11}\right)/2$ **27.** $\pm\sqrt{3}$ **29.** $-\frac{1}{2}, 2$ **31.** $(x - 2)(x + 42)$
33. Not factorable in the integers **35.** $(2x - 9)(x + 12)$ **37.** $(4x - 7)(x + 62)$ **39.** $r = \sqrt{A/P} - 1$
41. If $c < 4$, there are two distinct real roots; if $c = 4$, there is one real double root; and if $c > 4$, there are no real roots.
43. 1,575 bottles at \$4 each **45.** 13.64% **47.** 8 ft/sec; $4\sqrt{2}$ or 5.66 ft/sec

APPENDIX B

Exercise B-1

1. 5, 7, 9, 11 **3.** $\frac{3}{2}, \frac{4}{3}, \frac{5}{4}, \frac{6}{5}$ **5.** 9, -27, 81, -243 **7.** 23 **9.** $\frac{101}{100}$ **11.** $1 + 2 + 3 + 4 + 5 + 6 = 21$ **13.** $5 + 7 + 9 + 11 = 32$
15. $1 + \frac{1}{10} + \frac{1}{100} + \frac{1}{1,000} = \frac{1,111}{1,000}$ **17.** 3.6 **19.** 82.5 **21.** $\frac{1}{2}, -\frac{1}{4}, \frac{1}{8}, -\frac{1}{16}, \frac{1}{32}$ **23.** 0, 4, 0, 8, 0 **25.** 1, $-\frac{3}{2}, \frac{9}{4}, -\frac{27}{8}, \frac{81}{16}$ **27.** $a_n = n - 3$
29. $a_n = 4n$ **31.** $a_n = (2n - 1)/2n$ **33.** $a_n = (-1)^{n+1}n$ **35.** $a_n = (-1)^{n+1}(2n - 1)$ **37.** $a_n = \left(\frac{2}{5}\right)^{n-1}$ **39.** $a_n = x^n$
41. $a_n = (-1)^{n+1}x^{2n-1}$ **43.** $1 - 9 + 25 - 49 + 81$ **45.** $\frac{4}{7} + \frac{8}{9} + \frac{16}{11} + \frac{32}{13}$ **47.** $1 + x + x^2 + x^3 + x^4$

49. $x - \dfrac{x^3}{3} + \dfrac{x^5}{5} - \dfrac{x^7}{7} + \dfrac{x^9}{9}$ **51.** (A) $\displaystyle\sum_{k=1}^{5}(k + 1)$ (B) $\displaystyle\sum_{j=0}^{4}(j + 2)$ **53.** (A) $\displaystyle\sum_{k=1}^{4}\dfrac{(-1)^{k+1}}{k}$ (B) $\displaystyle\sum_{j=0}^{3}\dfrac{(-1)^j}{j + 1}$ **55.** $\displaystyle\sum_{k=1}^{n}\dfrac{k + 1}{k}$

57. $\displaystyle\sum_{k=1}^{n}\dfrac{(-1)^{k+1}}{2^k}$ **59.** False **61.** True **63.** 2, 8, 26, 80, 242 **65.** 1, 2, 4, 8, 16

67. $1, \frac{3}{2}, \frac{17}{12}, \frac{577}{408}$; $a_4 = \frac{577}{408} \approx 1.414\,216$, $\sqrt{2} \approx 1.414\,214$ **69.** 1, 1, 2, 3, 5, 8, 13, 21, 34, 55

Exercise B-2

1. (A) Arithmetic, with $d = -5$; $-26, -31$ (B) Geometric, with $r = 2$; $-16, 32$ (C) Neither (D) Geometric, with $r = \frac{1}{3}$; $\frac{1}{54}, \frac{1}{162}$

3. Geometric; 1   **5.** Neither   **7.** Arithmetic; 127.5   **9.** $a_2 = 11, a_3 = 15$ **11.** $a_{21} = 82, S_{31} = 1{,}922$ **13.** $S_{20} = 930$

15. $a_2 = -6, a_3 = 12, a_4 = -24$ **17.** $S_7 = 547$ **19.** $a_{10} = 199.90$ **21.** $r = 1.09$ **23.** $S_{10} = 1{,}242, S_\infty = 1{,}250$

25. 2,706   **27.** -85 **29.** 1,120   **31.** (A) Does not exist (B) $S_\infty = \frac{8}{5} = 1.6$ **33.** 2,400   **35.** 0.999

37. Use $a_1 = 1$ and $d = 2$ in $S_n = (n/2)[2a_1 + (n-1)d]$.   **39.** $S_n = na_1$ **41.** No   **43.** Yes

45. $\$48 + \$46 + \cdots + \$4 + \$2 = \$600$ **47.** About \$11,670,000   **49.** \$1,628.89; \$2,653.30

Exercise B-3

1. 720   **3.** 10   **5.** 1,320   **7.** 10   **9.** 6   **11.** 1,140   **13.** 10   **15.** 6   **17.** 1   **19.** 816

21. $C_{4,0}a^4 + C_{4,1}a^3b + C_{4,2}a^2b^2 + C_{4,3}ab^3 + C_{4,4}b^4 = a^4 + 4a^3b + 6a^2b^2 + 4ab^3 + b^4$

23. $x^6 - 6x^5 + 15x^4 - 20x^3 + 15x^2 - 6x + 1$ **25.** $32a^5 - 80a^4b + 80a^3b^2 - 40a^2b^3 + 10ab^4 - b^5$ **27.** $3{,}060x^{14}$

29. $5{,}005p^9q^6$ **31.** $264x^2y^{10}$ **33.** $C_{n,0} = \dfrac{n!}{0!\,n!} = 1; C_{n,n} = \dfrac{n!}{n!\,0!} = 1$ **35.** 1 5 10 10 5 1; 1 6 15 20 15 6 1

A

Abscissa, 14, 43
Absolute equality, 429, 459
Absolute inequality, 429, 459
Absolute maxima and minima, 334–342, 355
 defined, 335
 examples, 335
 extreme value theorem, 336
 finding an absolute extremum on an open interval
  (example), 340–341
 local extrema, testing (example), 339–340
 locating absolute extrema, 336–338
  on a closed interval, 336
 second derivative and, 338–341, 355
 second-derivative test for absolute extremum, 340
Absolute value, defined, 65
Absolute value function, 65, 131
Amount (future value), 101
Analytic geometry, fundamental theorem of, 14
Angles, 534–535
 initial side, 534, 552
 special, finding exact values for, 537
 in standard position, 535, 552
 terminal side, 534, 552
Antiderivatives, 360–361, 417
Antidifferentiation, partial, 510
Area between two curves, 424–428, 459
 income distribution (example), 428–431
Average cost, modeling, 327–330
Average profit (profit per unit), 204
Average rate of change, 30, 209
 defined, 170
Average revenue (revenue per unit), 204
Average value, 411–413, 530
 average price (example), 413
 of a continuous function f over $[a, b]$, 412
 of a function, 412
 over rectangular regions, 514–515

B

Base e exponential functions, 96–97
Base, exponential functions, 94
Basic differentiation properties, 183–191, 209
 constant function rule, 183
 constant functions, differentiating, 184
 constant multiple property, 185–186
 power rule, 184–185
 sum and difference properties, 186–187
Basic elementary functions, 64–65, 119
 domains of, 65
 evaluating, 64–65
 graph, range, and domain of, 65
Best fit, 33
Bioclimatic rule for temperate climates, 508
Body surface area (BSA):
 defined, 30
 estimating, 30
Boiling point, 27
Break-even analysis, 9, 55
 example, 9–10
Break-even points, 84, 119
 defined, 202
Briggsian logarithms, 111
British thermal unit (Btu), defined, 154fn

C

Calculator evaluation of logarithms, 111–113
Calculus:
 fundamental theorem of, 359, 394,
  404–417, 419

change in cost vs. area under marginal cost
  (example), 406
 change in profit (example), 409
 definite integrals, evaluating, 407–411
 theorem, 407
 useful life (example), 410
multivariable, 463–532
original term for, 128
Cartesian coordinate system, 14, 43
Cavanaugh, Joseph, 532
Cephalic index, 472
Chain rule, 241–250, 271
 composite functions, 241–242
 defined, 241, 244–245
 general derivative rules, 247–248, 271
 general power rule, 241, 242–244, 271
 partial derivatives using, 474–475
 reversing, 372–374
 using, 245–247
Change-of-base formula for logarithms, 228fn, 271
Change-of-base formulas, 111
Change-of-variable method, 376, 417
Closed interval, 6
Cobb–Douglas production function for a
  product, 532
Combined transformations, 69–70
Common logarithms, 111, 121
Completing the square, 79, 119
Compound interest, 100–102, 120
Compound interest formula, 102, 270
Concavity, 293–297
 concave downward, 294–295
 concave upward, 294–295
 example, 296
 summary, 295
Constant e, 217–218, 270
Constant function rule, 183
Constant functions, 52, 119
 continuity properties, 148
 differentiating, 184
Constant multiple property, 185–186
Constant of integration, 362
Constant of integration, 417
Constraint, 490
Consumers' surplus, 440–443, 460
 defined, 440
 and equilibrium price, 442–443
 example, 440–441
Continuity, 144–155, 209
 continuous at the point $x = c$, 145
 continuous on the open interval (a, b), 145
 defined, 144–145
 discontinuous function, 144–145
 function continuous:
  on the closed interval $[a, b]$, 147
  on the half-closed interval $[a, b]$, 147–148
  on the left, 147
  on the right, 147
 of functions defined by graphs, 146
 of functions defined by equations, 147
 general continuity properties, 148
 one-sided, 147
 properties, 148–151
  solving inequalities by using, 149
  using, 148–149
 sign properties on an interval (a, b), 149
Continuity properties, constant functions, 148
Continuous compound interest, 218–221, 270,
  386–387
 computing, 219
 doubling time, computing, 221
 formula, 219

growth of an investment, graphing, 219–220
 growth time, computing, 220–221
Continuous compound interest formula, 219
Continuous income stream, 436–437, 459–460
 future value of, 437–440, 460
Continuous random variable, 434–435
Coordinate axes, 14
Coordinates, 14
Cosecant functions, 538, 552
Cosine functions, 535–536, 552
 evaluating, 536–537
 graphs of, 538
Cosine of θ, 535
Cost function, defined, 201
Costs, 9, 55, 119
Cotangent functions, 538, 552
Coterminal angles, 534
Critical points, 483
 multiple, 484–486
Critical values, 277, 355
 and partition numbers, 278–280
Cube function, 65
Cube root function, 65
Curve fitting, 31
Curve sketching, 301–304
Curves, 31
Curve-sketching techniques, 321–334, 355
 average cost, modeling, 327–330
 graphing strategy:
  modifying, 321, 355
  using, 321–327
 oblique asymptote, 328, 355

D

Dashed boxes, use of, in text, 4fn, 53fn
Definite integral, 359, 394–404
 defined, 400
 example, 400
 left and right sums, approximating areas
  by, 395–398
 as a limit of sums, 398–400
 properties of, 400–401
Definite integrals, 418
 average value, 411–413
 defined, 418
 evaluating, 407–411
 recognizing, 411–413
 and substitution techniques, 407–409
Degree measure 1, defined, 534, 552
Degree of a polynomial, 87
Degree–radian conversion, 535, 552
Degree–radian measure, angles, 534–535
Dependent variables, 50, 119, 528
Derivatives, 169–182, 209
 average rate of change, defined, 170, 209
 defined, 174
 of exponential and logarithmic functions,
  224–232, 271
  cable TV subscribers, 230
  derivative of e^x, 224–225
  derivative of ln x, 225–226
  exponential and logarithmic models, 229–230
  logarithmic and exponential functions, 227–228
  price–demand model (example), 229
 finding, 175–177, 177
  four-step process, 175–176
 instantaneous rate of change, 172
 interpretations of, 175
 involving sine and cosine (example), 542
 nonexistence of, 178–179
 notation, 183

Derivatives (*continued*)
  of products, 232–235
    differentiating a product, 233
    finding, 234–235
    product rule, 233
    tangent lines, 234
  of products and quotients, 271
  of quotients, 235–238
    differentiating quotients, 236
    finding, 237
    quotient rule, 235
    sales analysis (example), 237–238
  rate of change, 169–172
  revenue analysis (example), 169–170
  sales analysis (example), 177–178
  of secant, 543
  slope of the tangent line, 172–174
    finding, 176–177
  tangent line slopes, finding, 176
  of trigonometric functions, 542–547
  velocity (example), 170–172
    instantaneous rate of change, 172, 209
    instantaneous velocity, 171
Diameter of a tree at breast height (dhb), 34
Difference quotient, 54
  limit of, 139–141
Differentiability, 174–175
Differential equations, 384–394, 418
  continuous compound interest, 386–387
  defined, 384
  exponential growth law, 387
  exponential growth phenomena,
      comparison of, 390–391
  first-order equations, 384–385
  population growth, 387–388
  radioactive decay, 388–389
    archaeology (example), 389–390
  second-order equations, 384
  slope fields and, 384–386
Differentials, 192–198, 209, 375
  approximations using, 195–197
  cost–revenue (example), 196–197
  defined, 194, 375
  increment notation, 192–193
  increments, comparing differentials
      and, 195–196
Discrete random variable, 435
Domain of a function, finding, 52, 54
Domains:
  of basic elementary functions, 65
  of exponential functions, 94, 94–95, 120
  of logarithmic functions, 108
  of polynomial functions, 87, 120
  of rational functions, 88
Double inequality, 6
Double integrals:
  average value over rectangular regions, 514–515
  defined, 512–514
  evaluating, 513, 523–525
  of exponential functions, 514
  integrand, 513, 530
  over general regions, 530
  over rectangular regions, 509–519, 529
  over regular regions, 522–526, 529–530
  region of integration, 513, 530
  volume and, 515–517
Doubling times, 113

E

Elastic demand, 266, 271
Elasticity of demand, 263–270, 271
  defined, 265
  example, 266
  relative rate of change, 263–264
  and revenue, 267–268
  theorem, 265

Elementary functions, 63–76
  basic, 64–65
    evaluating, 64–65
  beginning library of, 64–65
  reflection, 68–69, 119
  vertical and horizontal shifts, 65–67
  vertical shrink of a graph, 68–69, 119
  vertical stretch of a graph, 68–69, 119
End behavior of a polynomial, 161–162
Endpoint solution, 350
Endpoints, 6
Equations:
  functions specified by, 50–52
  in two variables, 47–48
Equilibrium point, 24, 43
Equilibrium price, 24, 43, 442, 460
Equilibrium quantity, 24, 43, 442, 460
Equivalent equations, 3
Error bound, 396, 418
Error in an approximation, 396, 418
Euler, Leonhard, 218
Even positive integer, continuity properties, 148
Explicit rule, 251
Exponential and logarithmic functions:
  derivatives of, 224–232, 271
    cable TV subscribers, 230
    derivative of e^x, 224–225
    derivative of ln x, 225–226
    exponential and logarithmic models,
        229–230
    logarithmic and exponential functions, 227–228
    price–demand model example, 229
Exponential equations, solving, 113
Exponential functions, 93–105, 120
  base, 94
  with base e, 96–97
  compound interest application, 100–102
  defined, 94, 120
  domains of, 94–95, 120
  exponential growth, 98
  exponential regression, 99
  graphing techniques, 95–96
  graphs of, 94–95
    basic properties, 95
  growth and decay applications, 97–100
  logarithmic–exponential conversions, 108
  properties of, 96
  range, 94, 120
Exponential growth law, 387, 418
Exponential growth phenomena, comparison of,
    390–391
Exponential decay, 387
Extrapolation, 34
Extreme value theorem, 336

F

First-degree (linear) equations, defined, 3, 42
First-degree (linear) inequality, defined, 3, 42
First-derivative test, 281–284, 355
First-order differential equations, 384–385, 418
First-order partial derivatives, 477, 528
Fixed costs, 9, 55
Formula, for a particular variable, solving, 5
Function notation, 52–53
Function of two independent variables, 15, 528
Functions, 47–63
  absolute value function, 65
  constant, 52, 119
  cost function, 55
  cube function, 65
  cube root function, 65
  defined, 49–50, 119
  domain, finding, 52, 54
  elementary, 63–76
  and equations, 50–51
  equations in two variables, 47–48

evaluation, 53–54
exponential, 93–105
finding values of, from its graph, 129–130
function notation, 52–53
  using, 55
identity function, 65
linear, 52
logarithmic, 105–126
notation, 52–53
one-to-one, 106–107
piecewise-defined, 70, 119
polynomial functions, 86–87
price-demand function, 55
profit function, 55
quadratic, 76–93
rational functions, 86–87
revenue function, 55
specified by equations, 50–52
  defined, 50
  graph of, 50
square function, 65
square root function, 65
vertical-line test for, 51–52
Fundamental theorem of analytic geometry, 14
Fundamental theorem of calculus, 359, 394,
    404–417, 419
  change in cost vs. area under marginal cost
      (example), 406
  change in profit (example), 409
  definite integrals, evaluating, 407–411
  theorem, 407
  useful life (example), 410
Future value of continuous income stream,
    437–440, 460

G

General continuity properties:, 148
General derivative rules, 247–248
  chain rule, 247–248, 271
General indefinite integral formulas, 376
General power rule, 241, 242–244, 271
  chain rule, 241, 242–244, 271
General regions, double integrals over, 530
Gini, Corrado, 429
Gini index of income concentration, 429, 459
Graphing calculators, 16–17
  linear regression feature, 502
  manuals for, 33fn
  numerical integration on, 410–411
  numerical integration on, 419
  TRACE routine, 16–17
  zero routine, 16–17
Graphing strategy:
  modifying, 321, 355
  using, 321–327
Graphs, 13–21, 129–130, 274–342
  absolute maxima and minima, 334–342
  analyzing, 300
  of $Ax + By = C$, 14–15
  concavity, 293–297
  critical values, 277, 355
    and partition numbers, 278–280
  curve sketching, 301–304
  economics applications, 284–291
    agricultural exports and imports (example),
        284–285
    revenue analysis (example), 285–287
  of an equation, 15
  of exponential functions, 94–95
    basic properties, 95
  finding values of a function from, 129–130
  first-derivative test, 281–284, 355
  of functions specified by equations, 50
  horizontal lines, 15, 17, 21, 43
  increasing and decreasing functions,
      275–280

inflection points, 297–299, 355
  defined, 297, 355
  locating, 298–299
L'Hôpital's rule, 311–320
local extrema, 280–281
parabolas, 76
point of diminishing returns, 304–307
of polynomial functions, 87
of quadratic functions, 76–77
of rational functions, 88–90
second derivative, 293–311
  notation, 295
transformations of, 65
  combining, 69–70
using intercepts to graph a line, 16
vertical lines, 15, 17, 21, 43
Growth time, computing, 220–221

H

Horizontal asymptotes, 209
  limits at infinity:
    finding, 162–164
    of rational functions, 88, 120, 163
Horizontal axis, 14
Horizontal lines, 15, 17, 21, 43
  graphs, 15, 17, 21, 43
Horizontal shifts, 65–67
Horizontally translating (shifting) a graph, 67, 69, 119

I

Identity function, 65
Implicit differentiation, 251–255, 271
  defined, 251–252
  examples, 253–255
Increasing and decreasing properties of a function,
    275–280, 355
Increment notation, 192–193
Increments, 192–193
  comparing differentials and, 195–196
Indefinite integrals, 359, 394, 417
  applications, 366–368
    advertising, 367–368
    cost function, 367
    curves, 366–367
  of basic functions, 362
  defined, 362
  formulas, 361–366
    general, 376
    using, 364–366
  properties, 363–364
    using, 364–366
Independent variables, 50, 119, 528
Indeterminate form, limits, 138
Inelastic demand, 266, 271
Inequality notation, 7
  and line graphs, 7
Infinite limits, 155–159, 209
  defined, 156
  vertical asymptotes:
    defined, 156, 209
    locating, 157–159
    of rational functions, 157
Inflection points, 297–299, 355
  defined, 297, 355
  locating, 298–299
Initial side, angles, 534, 552
Instantaneous rate of change, 172, 209
Instantaneous velocity, 171
Integral calculus, 359
Integral sign, 362
Integrand, 362, 400, 418, 530
  double integrals, 513, 530
Integration, 359–422
  area between two curves, 424–428, 459
    income distribution (example), 428–431

constant of, 362, 417
consumers' surplus, 440–443, 460
  defined, 440
  and equilibrium price, 442–443
  example, 440–441
integration-by-parts formula, 460
lower limit of, 400, 418
by parts, 446–452, 460
  defined, 446
  examples, 446–448
  integration-by-parts formula, 446
  repeated use of (example), 448–449
  selection of u and dv, 448
  using (example), 449–450
producers' surplus, 440–443
  defined, 440–443
  and equilibrium price, 442–443
  example, 441–442, 455–456
reduction formulas, 454–455
region of, 530
reversing the order of, 525–526
by substitution, 372–383, 417
  example, 376–377
  method of substitution, 375
  procedure, 376
  reversing the chain rule, 372–374
of trigonometric functions, 547–551
upper limit of, 400, 418
using tables, 452–459, 460
  substitution and integral tables, 453–454
Intercepts, of polynomial and rational
    functions, 284
Interest, 100
Interest rates, 100
Interpolation, 34, 43
Interval notation, 6–7
  and line graphs, 7
Inverse functions, 106
Inverse of a function, 107
Iterated integrals, 513, 530
  evaluating, 511–512

L

Lagrange multipliers, 491
  maxima and minima using, 490–497, 529
LCD (least common denominator), 4–5
Leading term, 161
Least squares approximation, 499–509
  educational testing (example), 503–505
  energy consumption (example), 505
  normal equations, 501–502
  theorem, 501
Least squares line, 500
Left sum, 398, 418
  approximating areas by, 395–398
  error bounds, 397
  limits of, 397
Left-hand limit, 132
Leibniz, Gottfried Wilhelm von, 128
L'Hôpital, Marquis de, 312
L'Hôpital's rule, 311–320, 321, 355
  for 0/0 indeterminate forms, 313–314,
    318–320
  applicability of (example), 315–316
  defined, 312
  examples, 314–315
  and limit properties, 312
    limits involving exponential and
      logarithmic forms, 313
    limits involving powers of $x - c$ (example),
      312–313
  limits at infinity, 316–318
  one-sided limits, 316–317
  repeated application of, 316
  theorem, 314
Libby, Willard, 388

Limits, 129–144, 209
  absolute value function, 131
  algebraic approach, 134–139
  analyzing, 130, 132
    graphically, 133–134
  defined, 129, 131
  of difference quotients, 139–141
  evaluating, 136–138
  existence of, 133
  functions and graphs, 129–130
  graphical approach, 130–134
  indeterminate form, 138
  infinite, 155–159, 209
  at infinity, 159–164, 209
  left-hand limit (limit from the left), 132
  one-sided, 132
  of polynomial and rational functions, 136
  properties of, 134–135
    using, 135–136
  of a quotient, 138–139
  right-hand limit (limit from the right), 132
  two-sided, 132
  unrestricted, 132
  0/0 indeterminate form, 138
Limits at infinity, 159–164, 209
  horizontal asymptotes:
    finding, 162–164
    of rational functions, 163
  and l'Hôpital's rule, 316–318
  polynomial functions, 160–162
    end behavior of a polynomial, 161–162
  power functions, 160
  rational functions, 163
  using the symbol ∞, 159
Line graphs, and interval notation/inequality
    notation, 7
Linear equations, 3–5, 42
  CD break-even analysis (example), 9–10
  Consumer Price Index (CPI) (example), 10–11
  defined, 3
  purchase price (example), 9
  solving, 4–5
  standard form, 3, 42, 43
  in two variables, standard form, 15
    graph of, 15
    solution, 15
    solution set, 15
Linear functions, 52
Linear inequality, 5–11
  defined, 3
  double inequality, 6
    solving, 8
  interval notation, 6–7
  properties of, 6
  sense of, reversal of, 6
  solving, 8
Linear regression, 29–37, 499, 529
  carbon monoxide emissions (example), 34
  defined, 33
  forestry (example), 35–36
  slope as a rate of change, 29–31
Linearly related variables, 29
  rate of change of, 30
Lines, 13–21
  Cartesian coordinate system, 14, 43
  cost equation (example), 21–22
  equations of, 19–20, 21
  horizontal, 15, 17, 21
  horizontal axis, 14
  point-slope form, 20–21, 43
  slope of, 17–19
  slope-intercept form of a line, 19–20, 21
  supply and demand (example), 22–24
  use of term, 13
  using intercepts to graph, 16
  vertical, 15, 17, 21
  vertical axis, 14

Local extrema, 280–282, 355
  defined, 280
  existence of, 281
  finding, 484
    multiple critical points, 484–486
  first-derivative test for, 282
  locating (example), 283–284
  package design (example), 487–488
  and partial derivatives, 483
  of polynomial and rational functions, 284
  profit (example), 486
  second–derivative test for, 483
  turning point, 280
Local maximum, 280, 482, 529
Local minimum, 280, 482, 529
Logarithmic functions, 105–126, 120–121
  with base 2, 107
  calculator evaluation of logarithms, 111–113
  defined, 106, 108, 225–226
  domain of, 108, 120
  inverse functions, 106, 120
  log to the base b of x, 108
  logarithmic equations, solving, 110–111
  logarithmic–exponential conversions, 108
  properties of, 109–110, 121
    using, 110
  range of, 108, 120
  uses of, 106
Logarithmic regression, 115, 121
Logarithmic–exponential conversions, 108
Logarithms, calculator evaluation of, 111–113
Lorenz chart, 428
Lorenz curve, 428–429, 459
  distribution of income (example), 429–430
Lower limit of integration, 400, 418

M
Marginal analysis, 210
  average profit (profit per unit), 204
  average revenue (revenue per unit), 204
  in business and economics, 199–208, 210
  cost analysis (example), 199–200, 204–205
  marginal average cost, 204, 210
  marginal cost, 199, 210
    and exact cost, 200
  marginal profit, 199
  marginal revenue, 199
  production strategy (example), 201
Marginal average cost, 204
Marginal productivity:
  of capital, 475
  of labor, 475
  of money, 494
Marginal profit, defined, 203
Marginal revenue, defined, 202
Mathematical model, 29, 56
Maxima and minima, 482–490, 529
  least squares approximation, 499–509
  method of least squares, 499–509, 529
  using Lagrange multipliers, 490–497, 529
Maximum value, 79
Method of least squares, 499–509, 529
Method of substitution, 375, 417
Metric tonne, 34fn
Midpoint sums, 399
Multiplier principle, 518
Multivariable calculus, 463–532
  functions of several variables, 528
    examples of, 465–467
  functions of two or more independent variables, 463–465
  maxima and minima, 482–490, 529
  partial derivatives, 473–482
    defined, 473
    example, 474
    of f with respect to x, 473

of f with respect to y, 474
  first-order, 477
  productivity (example), 475–476
  profit (example), 475
  second-order, 476–478
    using the chain rule (example), 474–475
  three-dimensional coordinate systems, 467–470

N
Napierian logarithms, 111
Natural logarithms, 111, 113, 121
Negative angle, 534, 552
Negative of the principal square root, 51fn
Newton, Isaac, 128
Nondifferentiability, 178–179
Normal equations, 501–502
  for quadratic regression, set of, 503
Normal probability density function, 436
Numerical integration routine, computing areas with, 428

O
Oblique asymptote, 328, 355
Odd positive integer, continuity properties, 148
One-sided continuity, 147–148
One-sided limits, 132
  and l'Hôpital's rule, 316–317
One-to-one functions, 106–107
Open interval, 6
Optimization, 342–354, 355
  area and perimeter, 343–345
  inventory control, 350–351
  inventory control problem, example, 350–351
  problems, 342
  profit, maximizing, 346–348
  revenue, maximizing, 346, 348–349
Ordered pair of real numbers, 14fn
Ordinate, 14, 43
Origin, 14

P
Parabolas, 76, 119
  axis, 80
  vertex, 80
Partial antiderivative, 510–511
  evaluating, 511
Partial derivatives, 473–482
  defined, 473
  example, 474
  of f with respect to x, 473, 528
  of f with respect to y, 474, 528
  first-order, 477, 528
  and local extrema, 483
  productivity (example), 475–476
  profit (example), 475
  second-order, 476–478, 529
  using the chain rule (example), 474–475
Partition numbers, 149–150
  and critical values, 278–280
Percentage rate of change, 271
Period of a function, 538, 552
Periodic functions, 538, 552
Piecewise-defined functions, 70, 119
Point of diminishing returns, 304–307
  maximum rate of change (example), 304–307
Point-by-point plotting, 47–48, 119
Point-slope form, 20–21, 43
  defined, 20
  using, 21
Polynomial functions, 86–87, 120
  and asymptotes, 321
  continuity properties, 148
  continuous graph of, 87

defined, 86
  domain of, 87, 120
  domains of, 87, 120
  graphs of, 87
Polynomials, limits of, 136
Population growth, 387–388
Positive angle, 534, 552
Power functions:
  defined, 184
  differentiating, 185
Power rule, 184–185
Predictions, 34
Price–demand equation, defined, 201
Price-demand functions, 55, 119
Principal (present value), 101
Principal square root, 51fn
Probability density functions, 434–436, 459
  defined, 434
  duration of telephone calls (example), 435
Producers' surplus, 440–443, 460
  defined, 441
  and equilibrium price, 442–443
  example, 441–442
Product rule, 233
Products, derivatives of, 232–235
  differentiating a product, 233
  finding, 234–235
  product rule, 233
  tangent lines, 234
Products and quotients, derivatives of, 271
Profit, 119
Profit function, defined, 203
Profit-loss analysis, 55, 119
Properties:
  continuity, 148–151
  limits, 134–135

Q
Quadrants, 14
Quadratic functions, 76–93, 119
  analyzing, 81–82
  defined, 76
  graphs of, 76–77
  intercepts, equations, and inequalities, 77–79
  parabola, 76, 119
    axis, 80
    vertex, 80
  properties of, and their graphs, 79–82
  vertex form, 79–82
Quadratic regression, 120
Quotients:
  derivatives of, 235–238
    differentiating quotients, 236
    finding, 237
    quotient rule, 235
    sales analysis (example), 237–238
  limits of, 138–139

R
Radian measure 1, defined, 534–535, 552
Radioactive decay, 388–389
  archaeology (example), 389–390
Range, 536
  of basic elementary functions, 65
  defined, 119
  of exponential functions, 94, 120
  of logarithmic functions, 108
Rate, 101
Rate of change, 30, 43, 169–172
  average, 30
Rate of descent, 30
  finding, 30–31
Rational functions, 86–87, 120
  continuity properties, 148
  defined, 88

domain of, 88
find the asymptotes of, 164
graphs of, 88–90
horizontal asymptote, 88, 120
limits at infinity, 163
limits of, 136
vertical asymptote, 88
Real-number domains, sine and cosine
   functions with, 536
Rectangular coordinate system, 14
Rectangular regions:
   average value over, 514–515
   double integrals over, 509–519, 529
   regular x region, 530
   regular y region, 530
Reduction formulas, 454–455
Reflection, 68–69, 119
Region of integration, 530
   double integrals, 513
Regression analysis, 31, 43, 428, 459, 499
Regression line, 33, 500
Regular regions, 519–522
   defined, 520
   describing (example), 521–522
   double integrals over, 519–526, 529–530
   regular x region, 519–520
      describing (example), 521
      test for, 522
   regular y region, 520
      test for, 522
Related rates, 257, 271
   and business, 260–261
   and motion, 257–258
      example, 258–260
   related-rates problem, 257, 271
      suggestions for solving, 258
Relative growth rate, 97–98, 387, 418
Relative rate of change, 271
Residuals, 499, 529
Revenue function, defined, 202
Revenues, 9, 55, 119
   defined, 201
Richter scale, 106
Riemann, Georg, 398
Riemann sums, 398–399, 418
   defined, 398
   limit of, 399
Right sum, 398, 418
   approximating areas by, 395–398
   error bounds, 397
   limits of, 397
Right-hand limit, 132

S

Saddle point, 482
Scatter plot, 32, 43
Secant, derivatives of, 543
Secant functions, 538, 552
Secant line:
   defined, 172
   slope of, 173
Second derivative:
   absolute maxima and minima and, 338–341, 355
   graphs, 293–311
      notation, 295
Second-order equations, 384
Second-order partial derivatives,
   476–478, 529

Set of normal equations for quadratic
   regression, 503
Short ton, defined, 34fn
Sign chart, constructing, 150
Sine functions, 535–536, 552
   evaluating, 536–537
   graphs of, 538
Sine of θ, 535
Slope:
   defined, 17
   finding, 18–19
   geometric interpretation of, 18
   of a graph, 174
   of a line, 17–19
   as a rate of change, 29–31
   run, 17
   of secant line, 173
   of a tangent line, 172–174
Slope fields:
   defined, 384–385, 418
   differential equations and, 384–386
Slope of a graph, 174
Slope of secant line, 173
Slope of the tangent line, 172–174
   finding, 176–177
Slope-intercept form of a line, 19–20, 21
   defined, 20
   using, 20
Solution:
   of an equation, defined, 3
   of an equation in two variables,
      15, 42
   of an inequality, 6
Solution set:
   defined, 3, 6
   of an equation in two variables, 15, 42
Solving an equation, use of term, 3
Solving an inequality, 6
Square function, 65
Square root function, 65
Standard form, 79
Standard position, angles in, 535, 552
Substitution:
   and integral tables, 453–454
   integration by, 372–383, 417
      example, 376–377
      method of substitution, 375, 417
      procedure, 376
      reversing the chain rule, 372–374
         using substitution (example), 375–376
      price-demand (example), 380–381
      techniques, 377–380
Sum and difference properties, 186–187
Sum of the squares of the residuals, 500–501,
   503, 507, 529
Summation notation, 398
Surface, 468

T

Table of integrals, 460
   defined, 452
   using, 452–453
Tangent functions, 538, 552
Tangents, 188–189
Terminal side, angles, 534, 552
Test number, 150
Texas Instruments graphing calculator
   (TI-83 family), 17fn

Therm, defined, 154fn
Three-dimensional coordinate systems, 528
Transformations, of graphs, 65
Trigonometric functions:
   angles, 534–535
   cosine functions, 535–536
   defined, 533
   definite integrals and (example), 549
   derivative formulas, 542–544
   derivatives of, 542–547, 552
   domain, 536
   indefinite integrals and (example), 548
   integral formulas, 547–550
      area under a sine curve, 547–548
      total revenue (example), 549–550
   integration of, 547–551, 552
   revenue (example), 544–545
   review, 534–541
   sine functions, 535–536
   slope (example), 543
Turning point, 280

U

Union, 147
Unit elasticity, 266, 271
Unrestricted limits, 132
Upper limit of integration, 400, 418

V

Variable costs, 9
Vertex, angles, 534
Vertex form, quadratic functions, 79–82
Vertical and horizontal shifts, 65–67
Vertical asymptotes:
   defined, 156, 209
   locating, 157–159
   of rational functions, 88, 157
Vertical axis, 14
Vertical lines, 15, 17, 21, 43
   graphs of, 15, 17, 21, 43
Vertical shifts, 65–67
Vertical shrink of a graph, 68–69, 119
Vertical stretch of a graph, 68–69, 119
Vertical-line test, for functions, 51–52, 119
Vertically translating (shifting) a graph,
   66, 69, 119
Volume:
   double integrals and, 515–517
   example, 516
   under a surface, 515

W

Weierstrass, Karl, 131fn
Word problems, procedure for
   solving, 8–9

X

x axis, 14
x coordinate, 14, 43

Y

y axis, 14
y coordinate, 14, 43
y intercept, 15

APPLICATIONS INDEX

Business & Economics

Advertising, 104, 190–191, 262, 311, 358, 367–368, 393
  point of diminishing returns, 310, 358
  and sales, 472, 480
Agricultural exports and imports, 284–285
Automation–labor mix for minimum cost, 489
Automobile production, 92
Average and marginal costs, 333
Average cost, 166, 198, 292, 357, 358, 415
Average income, 273
Average price, 458

Beef production, 507
Break-even analysis, 9–10, 12–13, 84–85, 92–93, 208, 214
Break-even point, 84–85, 92, 93
Budgeting for least cost, 498
Budgeting for maximum production, 498

Cable television, 40–41
Car rental, 353
Certificate of deposit (CD), 104
Cobb–Douglas production function, 518
Communications, 125
Compound interest, 101, 104, 118
Computer purchase price, 9
Construction, 125, 357
Construction costs, 333, 353, 354, 358
Consumer Price Index (CPI), 10–11, 45
Consumers' surplus, 451, 457, 462
Continuous compound interest, 222, 273, 392
Continuous income stream, 451, 458, 462
Cost, 415–416, 421, 458
Cost analysis, 27, 206, 214, 309–310
Cost function, 55, 57, 62, 84, 92, 93, 124, 250, 371, 382–383, 471–472
Cost, revenue, and profit rates, 262

Demand equations, 273, 480
Demand function, 416
Depreciation, 27, 100
Distribution of wealth, 433
Doubling rate, 223
Doubling time, 114–115, 117, 223, 273

Electricity consumption, 182
Electricity rates, 75, 124
Employee training, 167, 214, 333, 404, 415
Energy consumption, 28–29, 38
Energy costs, 166–167
Equilibrium point, 23, 24, 28, 118, 125
Equilibrium price, 24, 42, 125
Equipment rental, 12, 154–155

Fixed costs, 9–10, 12, 13, 27
Flight conditions, 27
Flight navigation, 27–28
Future value, 472
  of a continuous income stream, 462

Growth time, 223

Home ownership rates, 115–116
Hospital costs, 74

Income, 154
Income distribution, 433, 451, 458, 462
Individual retirement account (IRA), 12
Inflation, 12, 309
Internet growth, 105
Internet usage, 45
Inventory, 416, 421
Inventory control, 350–351, 353, 357
Investments, 44

Labor costs and learning, 371–372, 416
Laboratory management, 354
Linear depreciation, 44

Maintenance costs, 415
Manufacturing, 27, 353
Manufacturing costs, 354, 371
Marginal analysis, 273, 293, 357–358
Marketing, 383, 422, 458
Markup, 27
Maximizing profit, 489, 507–508, 531
Maximum profit, 352, 358
Maximum revenue, 270, 353, 358
Maximum revenue and profit, 352, 357
Maximum shipping volume, 490
Maximum volume, 490, 498, 499
Mineral consumption, 182
Mineral production, 182
Minimizing average costs, 334
Minimizing cost, 489
Minimizing material, 531
Minimum material, 489
Money growth, 124
Motor vehicle production, 191, 507
Multiplier principle, 518

Natural-gas consumption, 214
Natural-gas rates, 154, 213

Oil production, 383, 416, 432–433
Operational costs, 353–354

Package design, 472, 496–497
Packaging, 62–63
Performance evaluation, 76
Point of diminishing returns, 310
Postal rates, 154
Present value, 101, 222–223
Price analysis, 292, 357
Price–demand, 198, 262, 393, 458
Price-demand analysis, 44, 61, 74, 74–75, 124, 125
Price-demand equation, 191, 382
Price–demand function, 83–84
Price–supply, 393, 421
Price–supply equation, 240, 250
Pricing, 9, 12, 31–32, 44, 56–58
Producers' surplus, 451, 457, 462
Product demand, 481
Product mix for maximum profit, 489
Product warranty, 461
Production, 451
  point of diminishing returns, 310
  scheduling, 92
Productivity, 472, 480–481, 498, 532

Profit, 39–40, 62, 181, 310, 451, 458, 531–532, 546
  maximizing, 346–348
Profit analysis, 206, 292
Profit and production, 421
Profit function, 55, 57–58, 62, 84, 85, 93, 124, 421, 480
Profit-loss analysis, 55, 93, 124–125

Rate of change:
  of cost, 269
  of revenue, 273
Rental income, 353, 357
Replacement time, 333
Resale value, 232
Resource depletion, 421
Revenue, 39, 62, 92, 181, 310, 333, 371, 458, 544–545, 546–547
  and elasticity, 269–270, 273
  maximizing, 346, 348–350
  and profit, 198
  and profit functions, 480
Revenue analysis, 206–207, 285–287
Revenue, cost, and profit, 207–208, 472
Revenue function, 55–56, 62, 83–84, 92–93, 124, 383, 461, 472
Rule of 72, 92

Sales analysis, 182, 190, 214, 237–238, 240, 371, 451
Sales commissions, 12
Sales salary plus commission, 44
Salvage value, 232
Savings, 104
Seasonal business cycle, 540–541, 551
Simple interest, 27
Sports salaries, 104
State income tax, 75
Supply and demand, 22–24, 42, 118
  for soybeans, 42
Supply function, 416

Telecommunications, 125
Telephone rates, 143–144
Tire mileage, 91–92
Total revenue, 549–550

Useful life, 415, 421, 433

Volume discount, 144

Wholesale pricing, 56–58

Life Sciences

Agriculture, 28, 118, 126, 353, 499
Air pollution, 182, 508–509
Animal supply, 155

Bacteria control, 354, 358
Bacterial growth, 98, 232
Biochemistry, 167–168
Biology, 250, 383, 393, 416, 507–508
Biophysics, 257
Bird flights, 354
Blood flow, 472, 481
Blood flow rate, 93